# BIOCHEMICAL ENGINEERING

# BIOCHEMICAL ENGINEERING

JOHN S. BAILEY
S.C. BHATIA

CBS PUBLISHERS & DISTRIBUTORS PVT. LTD.
NEW DELHI • BENGALURU • CHENNAI • KOCHI • MUMBAI • PUNE

ISBN: 978-81-239-1677-4 (PB)
ISBN: 978-81-239-1678-1 (HB)

**First Edition: 2009**
Reprint: 2015



*Published by:*
Satish Kumar Jain for CBS Publishers & Distributors Pvt. Ltd.,
4819/XI Prahlad Street, 24 Ansari Road, Daryaganj, New Delhi - 110002
delhi@cbspd.com, cbspubs@airtelmail.in • www.cbspd.com
Ph.: 23289259, 23266861, 23266867 • Fax: 011-23243014

*Corporate Office:* 204 FIE, Industrial Area, Patparganj, Delhi - 110 092
Ph: 49344934 • Fax: 011-49344935
E-mail: publishing@cbspd.com • publicity@cbspd.com

*Branches:*
- ***Bengaluru:*** 2975, 17th Cross, K.R. Road, Bansankari 2nd Stage, Bengaluru - 70
  Ph: +91-80-26771678/79 • Fax: +91-80-26771680
  E-mail: cbsbng@gmail.com, bangalore@cbspd.com
- ***Chennai:*** No. 7, Subbaraya Street, Shenoy Nagar, Chennai - 600030
  Ph: +91-44-26681266, 26680620 • Fax: +91-44-42032115 • E-mail: chennai@cbspd.com
- ***Kochi:*** 36/14, Kalluvilakam, Lissie Hospital Road, Kochi - 682018
  Ph: +91-484-4059061-65 • Fax: +91-484-4059065 • E-mail: cochin@cbspd.com
- ***Mumbai:*** 83-C, Dr. E. Moses Road, Worli, Mumbai - 400018
  Ph: +91-9833017933, 022-24902340/41 • E-mail: mumbai@cbspd.com
- ***Pune:*** Bhuruk Prestige, Sr. No. 52/12/2+1+3/2,
  Narhe, Haveli (Near Katraj-Dehu Road Bypass), Pune - 411041
  Ph: +91-20-64704058/59, 32342277 • E-mail: pune@cbspd.com

*Representatives:*

- Hyderabad: 0-9885175004 • Kolkata: 0-9831437309, 0-9051152362
- Nagpur: 0-9021734563 • Patna: 0-9334159340 • Vijayawada: 0-9000660880

*Printed at:*
India Binding House, Noida (UP)

# Preface

Biochemical engineering is a branch of chemical engineering which deals with the design and construction of unit processes and unit operations that involve biological organisms or molecules. In other words biochemical engineering is the application of chemical engineering principles to understand model, design and development processes for environmental remediation—to work in other areas that combine biochemistry, microbiology and chemical engineering. Thus, a biochemical process industry being multidisciplinary one is incomplete without a chemical engineer, a microbiologist and biochemist. All the three need to interact together in a biochemical industry and biochemical engineer plays a major role in developing processes.

The understanding of biochemical engineering is strengthen with brief discussions on fundamentals of biochemistry and microbiology discussed in chapters 1 and 2. The subject matter is well organised and covers the concepts right from the beginning. Chapter 3 is devoted to enzymes, their properties, types and their kinetics based on different models and theories whereas chapter 4 discusses the techniques of immobilisation of enzymes along with their advantages and applications. Chapter 5 acquaints the readers with industrial applications of enzymes. Chapter 6 introduces various fermentation processes such as aerobic, anaerobic along with design and construction aspects. Chapter 7 focuses on biotechnology of industrial micro-organisms. Various types of industrial micro-organisms along with cell growth, metabolism and its regulations are discussed. Chapter 8 concentrates on isolation, preservation and improvement of industrially important micro-organisms. Chapter 9 deals with media for industrial fermentation, the media selection plays an important role and is the backbone for the design of fermentor and discusses the requirements and sources of carbon, nitrogen and energy required for the medium. Chapter 10 introduces students with sterilisation and highlights sterilisation processes and its engineering aspects.

Chapter 11 is devoted to development of inocula for industrial fermentation. Development of inocula for yeast, bacterial and mycelial processes is discussed. Chapter 12 makes in-depth study of design of a fermentor along with design and analysis of bioreactors. The scale-up criteria for bioreactors along with choice of bioreactors is also highlighted based on the physico-chemical properties of fermentation broths. Chapter 13 deals with aeration and agitation in bioprocess. Most of industrial fermentation follow aerobic biological process. So it is essential to have basic understanding of aeration and transfer of air bubbles in the liquid media. Agitation is also important to properly mix the reactor contents, to obtain uniform concentration throughout the reactor. Considering this aspects, stress is laid on oxygen transfer in liquid. The other important topics covered in this chapter include power requirements in aeration, factors effecting oxygen transfer coefficient, types of agitators and spargers. Chapter 14 focuses on optimisation of fermentation processes by quantitative analysis. Chapter 15 is devoted to biosensors in bioprocess monitoring and control. This chapter discusses basic components of on-line process monitoring and control along with transducers and biosensors based on thermal effects. Chapter 16 deals with fermentation process validation which is an integral component of current good manufacturing practices. The topic covered in this chapter are fermentation process validation, its main parameters for validation like raw materials, pH, temperature, air flow rate, etc. Chapter 17 concentrates on instrumentation and control and discusses the methods of measuring process variables along with on-line analysis of other chemical factors and control systems. Chapter 18 focuses on computer applications in fermentation technology. Chapter 19 discusses the principles of effluent treatment. Effluent treatment systems are becoming mandatory in bioprocess industries. This chapter discusses physical, chemical biological method of treatment along with recovery of by-products. Also brief mention is made about industrial waste-water treatment. Chapter 20 focuses on recovery and purification of fermentation products. The chapter discusses the removal of microbial cells and solid matter, foam separation, centrifugation, liquid-liquid extraction and recovery of solvents. If a fermentation process is to yield a product at a competitive price, the chosen micro-organism or animal cell culture should give the desired end-product in predictable and economically adequate quantities. Considering this chapter 21 is devoted to fermentation economics.

This reference textbook is essential reading for all students, teachers, professionals, researchers and industrialists involved with chemical engineering, biochemical engineering, environmental science, microbiology, biotechnology and life sciences. It is also a valuable source of information for those preparing for or already associated with pollution prevention, food processing, agricultural engineering, etc. The reference textbook also caters to the requirement of the syllabus prescribed by various Indian universities for undergraduate and postgraduate students pursuing these courses. Constructive suggestions are always welcome from users of this book.

Diagrams, figures, tables and index supplement the text. All the topics have been covered in a cogent and lucid style to help the reader grasp the information quickly and easily.

**John S. Bailey**
**S. C. Bhatia**

# Contents at a Glance

# Contents

Chapter 1

# Fundamentals of Biochemistry

## INTRODUCTION

Biochemistry is the study of life cyclic processes in terms of chemicals. It is not the study of chemistry of living organisms. It obtains information from the basic micro-organisms viz., bacteria, viruses, algae and fungi. It draws information on energetics from basic thermodynamics. It couples them to show a path how life cycle proceeds with the mutual cooperation of various activities of the living beings. Energy is required for various metabolic activities in the living organisms.

This energy is released by breaking of some of the high energy storing molecules, usually phosphate containing molecules. Oxidation of NADH (nicotinamide adenine dinucleotide, reduced form) in the mitochondria is one of the main reactions.

Some of the chemical/biochemical reactions in the living organisms are facilitated by another type of compounds known as enzymes. The facilitation of the reaction is known as catalysis and hence enzymes are known as biocatalysts or biological catalysts.

The cells themselves contain some of the enzymes. The cell regulates the type of enzyme and its number by which it regulates both the type of biochemical reaction and its rate. Genetic engineering deals with the information-flow of the living organisms. Biochemistry, coupled with genetic engineering principles, plays a vital role in devising techniques for manipulating most of the biochemical transformations. Thus, of late, the theoretical knowledge of biochemistry is being diverted to applied aspects in regulating the cell metabolism.

Living organisms contain various biomolecules which are the building blocks of the cell and also help in storing and releasing energy for biotransformations and at the same time carrying forward genetic information. They include carbohydrates, lipids, proteins, nucleic acids, vitamins, etc.

Biomolecules: Living organisms contain a large number of biomolecules which are essentially composed of carbon and nitrogen. The bacteria which are the smallest cells contain a large number of (5000) different kinds of compounds including 3000 kinds of proteins, 1000 kinds of nucleic acids. These biomolecules have high molecular weights and are complex in structure. They form the basic building blocks for the cells and serve various purposes.

For example, the amino acids serve not only as building blocks of protein molecules but also as precursors of hormones, alkaloids, porphyrins, pigments, etc.

Similarly, nucleotides serve not only as building blocks of nucleic acids but also as co-enzymes and as energy carrying molecules.

## LIPIDS

Lipids are substances that can be extracted from plant or animal matter by organic solvents, such as chloroform, diethyl ether or toluene. The most common lipids are fats and oils. In addition to fats and oils, numerous other biological materials, including waxes, cholesterol and some vitamins and hormones, are classified as lipids. The general structure of fats and oils is shown in Fig. 1.1. These compounds, called triglycerides, are esters formed from the alcohol glycerol, $CH_2(OH)CH(OH)CH_2(OH)$ and a long-chain fatty acid such as stearic acid, $CH_3(CH_2)_{16}C(O)OH$.

```
     H       O
     |       ‖
  H—C—O—C—R
     |       O
     |       ‖
  H—C—O—C—R
     |       O
     |       ‖
  H—C—O—C—R
     |
     H
```

**Fig. 1.1.** General formula of fats and oils. The R group is from a fatty acid and is a hydrocarbon chain, such as $-(CH_2)_{16}CH_3$.

Lipids are a class of compounds which are fatty/oily in nature and present in cells and tissues. They have the following distinct characteristic features:

1. They are insoluble in water.
2. They are soluble in non-polar solvents like *n*-hexane, alcohol, chloroform, benzene, etc. and hence can be extracted from the organisms using these solvents.
3. They release a lot of energy on breakdown and hence are considered as the energy storage media.
4. They contain a large proportion of C-H bonds.
5. Upon saponification, they release fatty acids and glycerol and hence glycerol is considered as a binding agent for fatty acids.
6. They are synthesised by the cells from sugars.

Some lipid compounds, such as vitamins and hormones, have intense biological activity. Lipids also include a heterogeneous group of structural component. Some lipids are combined with other classes of compounds and they are known as:

1. Lipoproteins.
2. Proteolipids.
3. Phosphatidopeptides.
4. Lipoaminoacids.
5. Lipopolysaccharides

Lipids are toxicologically important for several reasons. Some toxic substances interfere with lipid metabolism, leading to detrimental accumulation of lipids. Some triglyceride lipids, for example, those formed from the unsaturated fatty acid, oleic acid, $CH_3(CH_2)_7(CH{=}CH)(CH_2)_7C(O)OH$, contain alkenyl carbon-carbon bonds, which are prone to oxidation through the action of toxic substances, such as carbon tetrachloride, $CCl_4$. Many toxic organic compounds are poorly soluble in water, but are lipid-soluble, so that bodies of lipids in organisms serve to dissolve and store toxicants.

As biomolecules, they are a constituent of cell walls and form a protective coating to the cell and encourage some species-specific reactions. They are also energy earners and release energy as and when the cell requires it.

### Fatty Acids

The fatty acids are generally available in the esterified form in cells and tissues and are rarely available in free form. They are named on the basis of the hydrocarbon from which they are derived. These hydrocarbons may be saturated or unsaturated. Accordingly, the fatty acids are also classified as saturated and unsaturated fatty acids. In saturated fatty acids, the carbon atoms hold as many hydrogen atoms as possible in addition to the carboxylic grouping, whereas in the unsaturated form, the carbon atoms hold double and triple bonds also. Fatty acids differ from each other primarily in chain length and in the number and position of their unsaturated bonds. They are generally represented by a shorthand notation based on the length of the carbon chain and the number, position and configuration of the double bonds. For example, palmitic acid is symbolised as 16:0 and oleic acid as $18:1^{\Delta 9}$. Tables 1.1 and 1.2 highlights the nomenclature, melting points and neutralisation values of saturated and unsaturated fatty acids.

**Table 1.1.** Nomenclature, melting points (mp), and neutralisation values[a] (NV) of saturated fatty acids.

| *Abbreviation* | *Systematic name* | *Trivial name* | *mp (°C)* | *NV* |
|---|---|---|---|---|
| 4:0 | Butanoic | Butyric | –7.9 | 637 |
| 5:0 | Pentanoic | Valeric | –33.5 | 549 |
| 6:0 | Hexanoic | Caproic | –3.4 | 483 |
| 7:0 | Heptanoic | Enanthic | –7.1 | 430 |
| 8:0 | Octanoic | Caprylic | 16.7 | 389 |
| 9:0 | Nonanoic | Pelargonic | 12.5 | 355 |
| 10 | Decanoic | Capric | 31.6 | 326 |
| 11 | Undecanoic | Undecylic | 28.7 | 301 |
| 12 | Dodecanoic | Lauric | 44.2 | 280 |
| 13 | Tridecanoic | Tridecylic | 41.1 | 262 |
| 14 | Tetradecanoic | Myristic | 54.4 | 246 |
| 15 | Pentadecanoic | Pentadecylic | 52.1 | 231 |
| 16 | Hexadecanoic | Palmitic | 62.9 | 219 |
| 17 | Heptadecanoic | Margaric | 61.3 | 207 |
| 18 | Octadecanoic | Stearic | 69.6 | 197 |
| 19 | Nonadecanoic | Nonadecylic | 68.6 | 188 |
| 20 | Eicosanoic | Arachidic | 75.4 | 180 |
| 21 | Heneicosanoic | – | 74.3 | 172 |
| 22 | Docosanoic | Behenic | 80.0 | 165 |
| 23 | Tricosanoic | – | 79.1 | 158 |
| 24 | Tetracosanoic | Lignoceric | 84.2 | 152 |

[a] Neutralisation value is equal to milligrams of KOH required to neutralise 1 gram of the acid.

**Table 1.2.** Nomenclature, melting points (mp), and neutralisation values[a] (NV) of some common unsaturated fatty acids.

| *Abbreviation* | *Systematic name* | *Common name* | *mp (°C)* | *NV* |
|---|---|---|---|---|
| 14:1 *n*–5 c | *cis*-9-Tetradecenoic | Myristoleic | – | 247.5 |
| 16:1 *n*–7 c | *cis*-9-Hexadecenoic | Palmitoleic | –0.5 | 220.5 |
| 16:1 *n*–7 tr | *trans*-9-Hexadecenoic | Palmitelaidic | 31 | 220.5 |
| 18:1 *n*–12 c | *cis*-6-Octadecenoic | Petroselinic | 30 | 198.6 |
| 18:1 *n*–9 c | *cis*-9-Octadecenoic | Oleic | 13–16 | 198.6 |
| 18:1 *n*–9 tr | *trans*-9-Octadecenoic | Elaidic | 43.7 | 198.6 |
| 18:1 *n*–7 tt | *trans*-11-Octadecenoic | Vaccenic | 44 | 198.6 |
| 18:2 *n*–6 c | *cis*-9,12-Octadecenoic | Linoleic | –5.0 | 200.06 |
| 18:3 *n*–6 c | *cis*-6,9,12-Octadecatrienoic | Gamma linolenic | 201.51 | – |
| 18:4 *n*–3 c | *cis*-9,12,15-Octadecatetraenoic | Alpha linolenic | –11.0 | 201.51 |
| 18:4 *n*–3 c | *cis*-6,9,12,15-Octadecatetraenoic | Steridonic | – | 202.9 |
| 20:3 *n*–6 c | *cis*-8,11,14-Eicosatrienoic | Dihomo-gamma-linolenic | – | 167.7 |
| 20:4 *n*–6 c | *cis*-8,11,14-Eicosatetraenoic | Arachidonic | –49.5 | 184.28 |
| 20:5 *n*–3 c | *cis*-5,8,11,14,17-Eicosapentaenoic | EPA | – | – |
| 22:1 *n*–9 c | *cis*-13-Docosenoic | Erucic | 33.5 | 165.7 |
| 22:5 *n*–3 c | *cis*-7,10,13,16,19-Docosapentaenoic | Clupanodonic | – | – |
| 22:6 *n*–3 c | *cis*-4,7,10,13,16,19-Docosahexaenoic | DHA | – | – |

[a] Neutralisation value is equal to milligrams of KOH required to neutralise 1 gram of the acid.

### Lipids Containing Glycerol

Fatty acid esters of glycerol are called acyl-glycerols or glycerides. Depending upon the number of hydroxyl groups of the glycerol esterified, it is known as monoacyl-, diacyl- or triacyl-glycerol.

### Phospholipids

Phospholipids consist of a glycerol connected to a phosphate group. They consist of two fatty acids linked to glycerol molecules. They are generally crystal white solids. They turn brown when exposed to air due to polymerisation of unsaturated fatty acids. They are characterised also by the presence of nitrogen group. They generally possess a negative charge on the phosphate group. Most of the phospholipids are derivatives of L-α-phosphoric acid

$$
\begin{array}{l}
RCOOCH_2 \\
\quad\;\; | \\
R'COOCH \quad O \\
\quad\;\; | \qquad\;\; \| \\
\quad\;\; CH_2O\,POH \\
\qquad\qquad\;\; | \\
\qquad\qquad OH
\end{array}
$$

L-α-Phosphatidic acid

L-α-phosphatidic acid on hydrolysis yields one mole each of glycerol, phosphoric acid and two moles of fatty acid.

## Lipids Not Containing Glycerols

### *Sphingolipids*

Plant and animal cells contain sphingolipids as important components. The sphingolipids are present in the brain and nerve tissues in large quantities. They are lipids containing sphingosine in place of glycerol as their backbone. *Sphingomyelins* are the sphingolipids in higher animals which contain phosphoryl ethanolamine or phosphoryl choline.

### Terpenes

Terpenes contain multiples of the five-carbon hydrocarbon isoprene (2-methyl-1,3 butadiene) and they are classified based on the number of isoprene units.

$$CH_2{=}\overset{\displaystyle CH_3}{\overset{|}{C}}{-}CH{=}CH_2$$

Isoprene

Terpenes are known for their characteristic odour and flavour and are mostly present in the essential oils of plants.

### Steroids

Steroids are a group of compounds derived from the saturated tetracyclic hydrocarbon perhydrocyclopentanophenanthrene. Many steroids have different functions and distinct metabolic activity. Mostly they are derived from natural sources. All steroids originate from the linear triterpenesqualene which cyclises rcadily.

### Lipoproteins

Lipoproteins are a class of compounds formed by the association of certain lipids with proteins. Hence they exhibit properties characteristic of both lipids and proteins. They are classified into the following two major groups:

1. Transport lipoproteins.
2. Membrane systems.

The transport lipoproteins are so named because they carry water-insoluble lipids to various organs of the body through blood. They are also known as plasma lipoproteins and are further classified into the following three groups depending upon the type of lipid:

1. Very low density lipoproteins (VLDL).
2. Low density lipoproteins (LDL).
3. High density lipoproteins (HDL).

Membrane systems are the lipoproteins present in the membrane of the cell structure. They contain about 40 per cent lipid and 60 per cent protein. They are mainly meant for protecting the cell from external environment. They also facilitate the transport of certain enzymes and other transport systems.

## Lipids Combined with Other Compounds

Lipids combined with other classes of compounds, in addition to lipoproteins, are as follows:

1. Proteolipids: Proteolipids contain proteins associated with lipids in which lipid content is more.
2. Phosphatidopeptides: Lipid category of phosphatides associated with peptides are phosphatido-peptides.
3. Lipoamino acids: Lipids associated with amino acids are lipoamino acids.

4. Lipopolysaccharides: Lipids found in nature combined with polysaccharides are lipopolysaccharides.

## PROTEINS

The word protein is derived from a Greek word meaning 'of the first rank'. They are considered to be the beginners and builders of biochemical reaction. Thus, they occupy a central position in the construction and functioning of living matter. All enzymes are proteins and they catalyse biochemical reactions as biocatalysts. Collagen is an elongated protein which is responsible for the tensile strength of skin and bones. In addition, proteins perform various other functions as well; viz., they help in transport of oxygen into blood and transmission of nerve impulses, etc. Proteins constitute a major part of the animal tissues.

Proteins are a large molecules consisting of as many as twenty amino acids. Their molecular weights 'range' from 50000 to few millions. The amino acids are linked through

$$-NH-\underset{\displaystyle R_1}{\underset{|}{CH}}-\overset{\displaystyle O}{\overset{\|}{C}}-NH-\underset{\displaystyle R_2}{\underset{|}{CH}}-\overset{\displaystyle O}{\overset{\|}{C}}-$$

Peptide linkage

Thus, proteins are the basic building blocks of biological material. Proteins have essential functions in living systems; they constitute enzymes and most of the cytoplasm inside the cell. Proteins consist of huge molecules with molecular masses of several tens of thousands that are composed primarily of biopolymers of amino acids. There are 20 naturally-occurring amino acids having the general formula shown by the first structure in Fig. 1.2 where R represents a group ranging from H atom (in the amino acid glycine) to moderately complex structures. Five examples of the 20 amino acids found in most proteins are shown in Fig. 1.2.

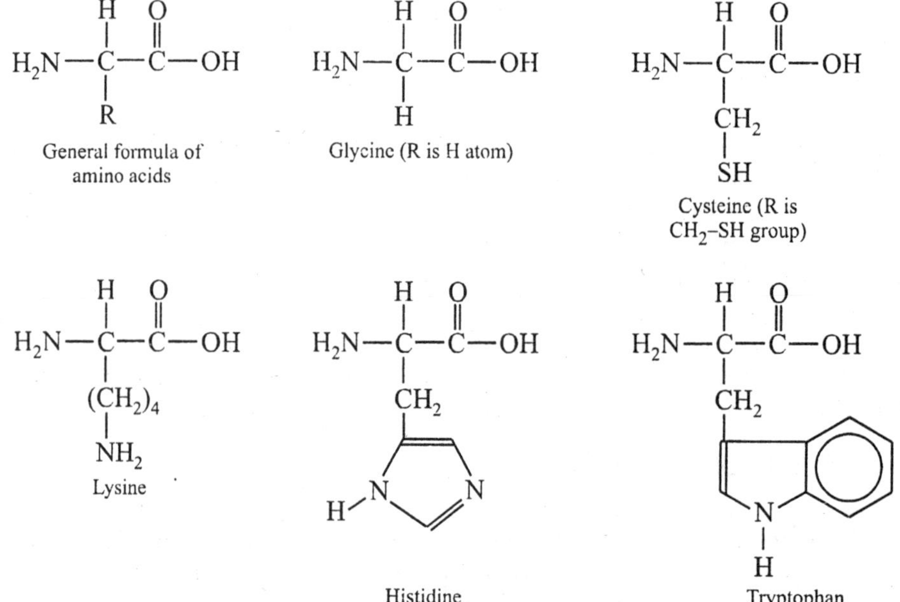


**Fig. 1.2.** General formula and five specific examples of amino acids.

Figure 1.3 which illustrates a tripeptide consisting of only three amino acids, shows how amino acids form biopolymers; protein molecules contain polypeptides formed from many amino acid molecules. The amino acids in proteins are joined at peptide linkages outlined by dashed lines in Fig. 1.3.

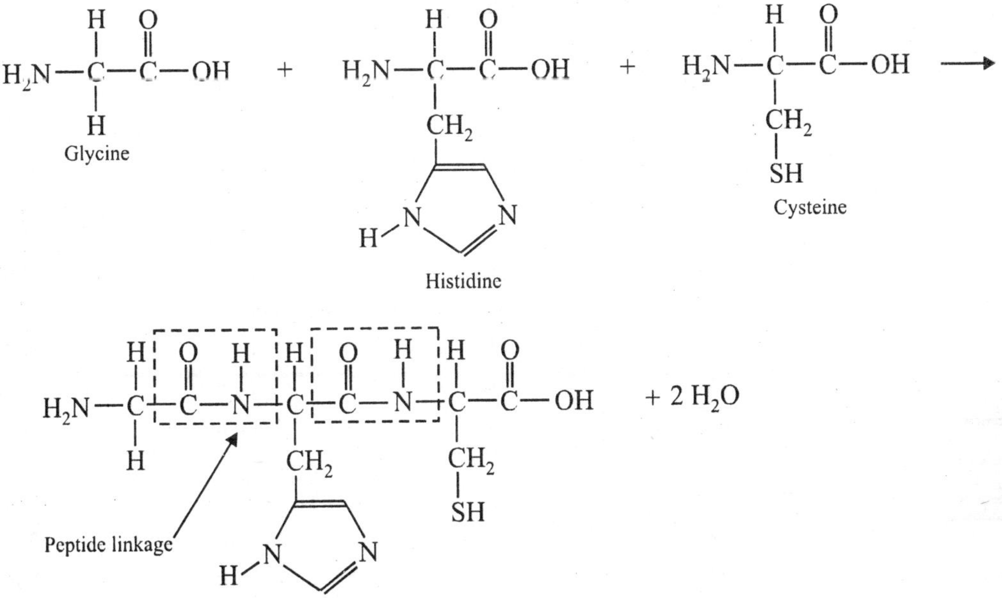

Glycine + Histidine + Cysteine →

Peptide linkage

$+ 2\,H_2O$

**Fig. 1.3.** A tripeptide formed by the linking of three amino acids. The peptide linkages with which the amino acids are joined are outlined by dashed lines.

The structures of protein molecules determine the behaviour of proteins in crucial areas such as the processes by which the body's immune system recognises substances that are foreign to the body. Proteinaceous enzymes depend upon their structures for the very specific functions of the enzymes. The order of amino acids in the protein molecule determines its primary structure. Secondary and tertiary protein structures depend upon the ways in which the polypeptide molecules are bent and folded by hydrogen bonding. The loss of a protein's secondary and tertiary structure is called denaturation and may be caused by heat or the action of foreign chemicals. Some corrosive poisons act by denaturing proteins.

## Amino Acids

Amino acids are the building blocks of proteins. There are twenty amino acids present in all living organisms, which is a clear manifestation of the biological unity in all living organisms. These twenty amino acids have different structural configurations and are represented by three lettered symbols (Table 1.3). They are also represented by single alphabets which help in writing the amino acid sequence in the proteins. Some proteins contain a large combination of several amino acids.

**Table 1.3.** Amino acid symbols, chemical group and charge on the molecule.

| *Amino acid* | *Three-letter symbol* | *One-letter symbol* | *Chemical group* | *Remarks* |
|---|---|---|---|---|
| Alanine | Ala | A | Aliphatic | Non polar |
| Arginine | Arg | R | Aliphatic | Basic |

*(Cont'd ...)*

| *Amino acid* | *Three-letter symbol* | *One-letter symbol* | *Chemical group* | *Remarks* |
|---|---|---|---|---|
| Asparagine | Asn | N | Aliphatic | Polar |
| Aspartic acid | Asp | D | Aliphatic | Acidic |
| Asparagine/Aspartic acid | Asx | B | Aliphatic | – |
| Cysteine | Cys | C | Aliphatic | Polar |
| Glutamine | Gln | Q | – | – |
| Glutamic acid | Glu | E | Aliphatic | Acidic |
| Glutamine/glutamic acid | Glx | Z | – | – |
| Glycine | Gly | G | Aliphatic | Polar |
| Histidine | His | H | Aliphatic | Basic |
| Isoleucine | Ile | I | Aliphatic | Non polar |
| Leucine | Leu | L | Aliphatic | Non polar |
| Lysine | Lys | K | Aliphatic | Basic |
| Methionine | Met | M | Aliphatic | Non polar |
| Phenylalanine | Phe | F | Aromatic | Non polar |
| Proline | Pro | P | Hectrocyclic | Non polar |
| Serine | Ser | S | Aliphatic | Polar |
| Threonine | Thr | T | Aliphatic | Polar |
| Tryptophan | Trp | W | Hectrocyclic | Non polar |
| Tyrosine | Tyr | Y | Aromatic | Polar |
| Valine | Val | V | Aliphatic | Non polar |

The amino acids have the structure:

$$
\begin{array}{c}
NH_2 \\
| \\
R-C-COOH \\
| \\
H
\end{array}
$$

These are known as α-amino acids in which the amino group and the carboxyl group are attached to the same carbon atom. Different amino acids possess different groups in place of R. Amino acids are also classified as acidic, basic and neutral amino acids.

## Analysis of Amino Acid Mixtures

Chromatographic separation methods are useful for the separation, identification and quantitative analysis of amino acid mixtures. A few chromatographic separation methods are described as follows.

### *Partition chromatography*

The relative distribution of the solute between two immiscible phases is taken as an advantage for separation of the mixture of amino acids. It is similar to counter-current extraction principle. When a solute is brought into contact with a two solvent system (in which the two solvents are immiscible) the solute gets distributed between the two solvents depending upon the partition coefficient of the solute.

The immiscible solvent systems could be phenol-water or *n*-butanol-water, etc. Subsequently, the solute is separated from the solvents.

### *Amino acid analyser*

In this technique of amino acid analysis, the entire procedure of cation exchange chromatography is automated so that elution, collection of fractions, analysis of fractions and recording data are all performed automatically.

Other chromatographic methods include:

1. Column chromatography.
2. Paper chromatography.
3. TLC.
4. Ion-exchange chromatography.

## Purification of Proteins

The characteristic properties of proteins which are used to separate mixtures of proteins are:

1. Molecular size.
2. Solubility,
3. Electric charge,
4. Differences in absorption characteristics.
5. Biological affinity for other molecules.

The following are some methods employed to separate proteins.

### *Dialysis and ultra-filtration*

Proteins can be separated from low molecular weight solutes by dialysis. This method employs a semipermeable membrane on which the bigger protein molecules are retained. Small solute molecules and water pass through the membrane. Thus separation takes place in ultra-filtration pressure or centrifugal force is used to filter the aqueous medium and small solute molecules through a semipermeable membrane. Cellophane and other synthetic materials are employed as membranes.

### *Gel filtration*

Separation is carried out based on the relative size of the proteins. Gel filtration or molecular-sieve chromatography consists of a column packed with a polysaccharide called *Sephadex*. The hydrated sephadex gel will have molecular size of 0.1 mm in diameter. Thus, it holds smaller protein molecules. The bigger protein molecules pass through the column and get separated out.

### *Ion-exchange chromatography*

In this method, the separation is based on the charge of the proteins. The ion-exchange column is packed with negatively charged carboxylate ions. They retain the positively charged proteins at pH below. The negatively charged proteins will pass through the column and get separated. Subsequently, the adsorbed proteins are separated from the column by pouring a solution of sodium chloride.

### *Affinity chromatography*

The method is based on the ligand-ligand interaction. The chromatographic column consists of a polysaccharide molecule (like agarose). The protein mixture is passed through the column. Those proteins, which have the property of binding non-covalently to ligand molecules of agarose, will get attached to the agarose particles. Other proteins, which do not have the affinity for such ligand interaction, will pass through the column and get separated. This method is extensively used for separation of enzyme proteins.

### *Electrophoresis*

The isoelectric point of a protein is that value of pH at which the net electric charge on the protein molecule is zero (i.e., the positive charge equals the negative charge). Hence, at isoelectric point, the mobility of the protein under an electric field will be zero. This property of a protein is used to separate a protein mixture by the electrophoresis method. The protein mixture is subjected to electric field in the retarding gel support. Proteins with different isoelectric points will have different mobilities. Thus, the separation takes place in the electrophoresis method. This method can also be effectively used as an analytical tool for proteins.

## CARBOHYDRATE

In addition to proteins, two other major kinds of substances that are synthesised and broken down by metabolic processes are carbohydrates and lipids. The nature and importance of these materials are discussed briefly in this section. (Lipids have already been discussed in the beginning of this chapter.)

Carbohydrates have the approximate simple formula $CH_2O$ and include a diverse range of substances composed of simple sugars such as glucose:

```
           CH2OH
             |
             C————————O
     H     / |          \     H
     |   /   H            \   |
     C                        C
     | \    OH      H      /  |
    HO   \   |      |    /    OH
             C——————C
             |      |
             H      OH
```

Glucose molecule

High-molecular-mass polysaccharides, such as starch and glycogen ('animal starch'), are biopolymers of simple sugars. The major functions of carbohydrates are to store and transfer energy, processes with which toxic substances may interfere.

They contain carbon, hydrogen and oxygen, with the ratio of hydrogen and oxygen being 2:1. They have a general chemical formula: $C_x(H_2O)_y$ and may be classified as:

1. Monosaccharides.
2. Oligosaccharides (di- and tri-saccharides).
3. Poly-saccharides.
4. High molecular weight poly-saccharides.

### Monosaccharides

Formaldehyde (HCHO) and hydroxy-acetaldehyde (OH—$CH_2$—CHO) conform to the empirical formula of carbohydrates, $(CH_2O)_n$, n being equal to 1 and 2 respectively.

The smallest three-carbon sugar molecules are glyceraldehyde and dihydroxyacetone and they are termed as trioses.

$$\begin{array}{c} HC{=}O \\ | \\ CHOH \\ | \\ CH_2OH \end{array} \qquad \begin{array}{c} CH_2OH \\ | \\ C{=}O \\ | \\ CH_2OH \end{array}$$

Glyceraldehyde (Aldotriose) Dihydroxyacetone (Ketotriose)

All the simple monosaccharides are white crystalline solids, sweet in taste and are freely soluble in water, but are insoluble in non-polar solvents. The carbon skeleton of the common monosaccharides is unbranched and each carbon atom except one, contains a hydroxyl group. The carbon atom has a carbonyl oxygen which is combined in an aldehyde or ketal linkage. If the carbonyl group is at the end of the chain, the monosaccharide is an aldehyde derivative and such monosaccharides are called aldoses. If it is at any other position, the monosaccharide is a ketone derivative and is called as ketoses. Among monosaccharides, the hexoses are abundant. Aldopentoses are important components of nucleic acids and various polysaccharides.

### Disaccharides

Disaccharides are anhydrides of two monosaccharides, which include sucrose, maltose, lactose and cellobiose.

#### *Sucrose*

Sucrose occurs in sugarcane or sugar beets and it is disaccharide of glucose and fructose. Sucrose yields an equimolar mixture of glucose and fructose on hydrolysis. The hydrolysis of sucrose is called the inversion of sucrose because during the reactions, the sign of rotation changes from dextro to laevo. It is an enzymatic-catalysed reaction by invertase. The equimolar mixture of glucose and fructose is called 'invert sugar'.

It is sweeter than sucrose in taste. Hence hydrolysis of sucrose is carried out in confectionary industry. This is present naturally in honey.

### Oligosaccharides

Oligosaccharides are anhydrides of several monosaccharide residues. Compounds containing ten or less monosaccharide units are called as oligosaccharides, while those containing more than ten are called polysaccharides.

### Polysaccharides

Polysaccharides are also known as glycans. They are anhydrides of one or more monosaccharides in which a large number of units are combined. Many polysaccharides are composed of only one repeating monosaccharide and even where a large number exists, they appear to be arranged in a systematic fashion. Homopolysaccharides possess one repeating unit, while heteropolysaccharides contain more than one monosaccharide unit. D-Glucose is the most prevalent monosaccharide unit though polysaccharides of D-mannose, D-fructose, D- and L-galactose, D-xylose and D-arabinose are common. D-glucosamine, D-galactosamine, D-glucuronic acid, N-acetylmuramic acid and N-acetylneuraminic acid are the monosaccharide derivaties which are commonly found in natural polysaccharides.

### *Cellulose*

Cellulose is the compound which gives strength to plant tissues. On acid hydrolysis of cellulose, glucose is formed in yields of 95–96 per cent. The cellulose is abundantly available in nature with the purest source being cotton containing at least 90 per cent pure cellulose. It is insoluble in water. Its molecular weight is of the order of 50000 and 25,00,000 corresponding to 300–15000 glucose residues per molecule.

On complete hydrolysis, cellulose yields D-glucose. Partial hydrolysis yields cellobiose. Cellulase enzyme can hydrolyse cellulose to D-glucose. But cellulase enzyme is not secreted in the digestive tract of mammals except cows.

## Nucleic Acids

Nucleic acids are the polymers of nucleotides, which are the units that store genetic information in each cell. Thus, nucleic acids contain all the hereditary information. Hence they are also known as informational biopolymers. When the cells divide, a copy of the parent cell's nucleic acids is passed on to the daughter cell.

## Nucleotides

Nucleotides are the building blocks of nucleic acids and they are also energy carriers. They contain essentially the following three components:

1. Phosphoric acid.
2. Ribose or deoxyribose (five-carbon sugars).
3. A nitrogenous base.

The nitrogenous base is derived either from purine or pyrimidine.

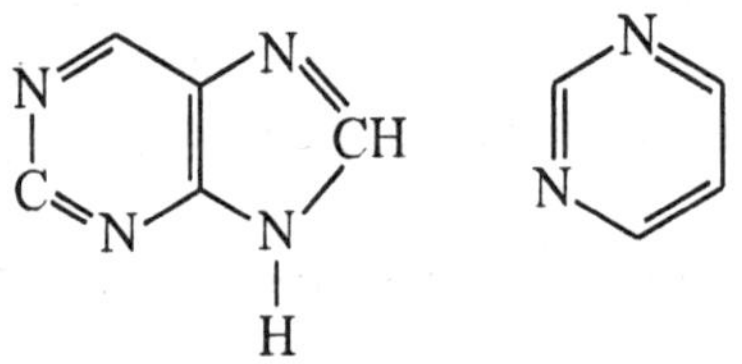

N N CH C N N H

Purine

N N

Pyrimidine

O HO—P—O—CH$_2$ O Base OH H H H H

OH H (or OH)-Ribonucleotide

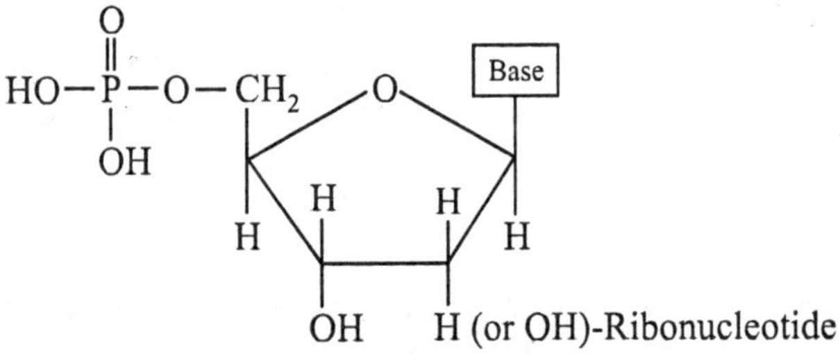

Deoxyribonucleotide

The five nitrogen bases found in DNA and RNA are adenine, guanine (purines) and cytosine, thymine and uracil (pyrimidines). Ribonucleic acid (RNA) is a polymer of ribose containing nucleotides while deoxyribonucleic acid (DNA) is a polymer of deoxyribose containing nucleotides.

DNA contains 2-deoxy 2-D-ribose whereas RNA contains D-ribose sugars which are in their furanose form in nucleotides. The base in the form of pyridine or pyrimidine is joined to the pentose sugar by a β-N-glycosyl bond between carbon 1 of pentose and nitrogen atom of the bases. This is known as nucleoside.

The phosphate group is joined to the pentose of the nucleoside in the form of the ester at the carbon 5 position. When phosphate group is added to the nucleoside, it is called nucleotide. Thus nucleotides are the 5′-phosphates of the corresponding nucleosides.

Ribonucleotides and deoxyribonucleotides occur in free form in cells and most of them are 5′-monophosphates. Some of them also occur as 5′-diphosphates and 5′-triphosphates. Adenosine may occur in all the three forms as adenosine 5′-monophosphate (AMP), adenosine 5′ diphosphate (ADP) and adenosine 5′-triphosphate (ATP). ATP is a carrier of phosphate and pyrophosphate and is involved in the transfer of chemical energy. ATP and ADP are interconverted by dephosphorylation and rephosphorylation during respiration. ATP-ADP system is responsible for transferring phosphate groups in the cell. Nucleoside triphosphates (NTPs) function as energy-rich precursors of mononucleotide units in the biosynthesis of DNA and RNA.

## DNA

Deoxyribonucleic acid (DNA) consists of deoxyribonucleotides linked with covalent bonding. The backbone of DNA consists of alternating phosphate diesters and pentose groups with the covalent bonding in which the continuity is maintained by the phosphodiester bridges. The purines and pyrimidines constitute the side chains. Different sources of DNA molecules differ in terms of the ratio of the four major types of nucleotide monomers. Hence, their molecular weight will also differ.

DNA is present in procaryotic cells, eucaryotic cells and bacteria in different forms. In eucaryotic cells it is present in single chromosome while in procaryotic cells, it is present in different chromosomes as different molecules. Hence its molecular weight will also change considerably. In bacteria, it is generally present in the nucleus. In some cases, it may be present in the cytoplasm in the form of extra-chromosomal DNA. In eucaryotic cells also, in some cases it may be present in the mitochondria also; but in such cases its molecular weight and base composition are different from that of the DNA present in the nucleus of the cell.

## RNA

Ribonucleic acid (RNA) was discovered by Frankel-Conrat and Williams and was assigned a single helical structure. In this respect, it differs from DNA, which has a double helical structure. Unlike DNA, it contains ribose sugars and the nitrogen group is provided by uracil.

There are three major types of RNAs, namely:

1. Messenger RNA (mRNA).
2. Ribosomal RNA (rRNA).
3. Transfer RNA (tRNA).

Each type of RNA has a characteristic range of molecular weights in between 25,000–11,00,000. Each RNA occurs in multiple molecular forms. RNA is present in the cytoplasm in bacteria. Mitochondrial RNAs and tRNAs are different from the extramitochondrial froms.

## VITAMINS

Vitamins are a group of organic compounds which are required in trace amounts to stimulate body growth and also to keep away certain types of diseases. They are also present in certain enzymes to assist them in their catalytic activity. Hence they are also known as co-enzymes. Vitamins are present in living cells in addition to lipids, proteins, carbohydrates, nucleic acids, etc. Since most of the vitamins contain amines, the name vitamin was derived from *Vitamine* which indicated amines required for health. Subsequently some vitamins were found not to contain amines, but the name vitamin was retained.

Vitamins are classified into two groups, viz., water-soluble and fat-soluble. Table 1.4 gives the list of vitamins that are co-enzyme precarious and Table 1.5 of the water and fat soluble vitamins.

**Table 1.4.** Vitamins that are co-enzyme precursors.

| *Vitamin* | *Co-enzyme* | *Human deficiency disease* |
|---|---|---|
| Biotin | Biocytin | *a* |
| Cobalamin ($B_{12}$) | Cobalamin ($B_{12}$) co-enzymes | Pernicious anemia |
| Folic acid | Tetrahydrofolate | Megaloblastic anemia |
| Nicotinamide | Nicotinamide co-enzymes | Pellagra |
| Pantothenate | Co-enzyme A | *a* |
| Pyridoxine ($B_6$) | Pyridoxal phosphate | *a* |
| Riboflavin ($B_2$) | Flavin co-enzymes | *a* |
| Thiamine ($B_1$) | Thiamine pyrophosphate | Beriberi |

[a]No specific name, deficiency in humans is rare or unobserved.

**Table 1.5.** Vitamins and their co-enzyme forms.

| *Type* | *Co-enzyme or active form* | *Function promoted* |
|---|---|---|
| Water-soluble | | |
| Thiamin | Thiamin pyrophosphate (TPP) | Aldehyde-group transfer |
| Riboflavin | Flavin mononucleotide (FMN) | Hydrogen-atom (electron) transfer |
| | Flavin adenine dinucleotide (FAD) | Hydrogen-atom (electron) transfer |
| Nicotinic acid | Nicotinamide adenine dinucleotide (NAD) | Hydrogen-atom (electron) transfer |
| | Nicotinamide adenine dinucleotide phosphate (NADP) | Hydrogen-atom (electron) transfer |
| Pantothenic acid | Coenzyme A (CoA) | Acyl-group transfer |
| Pyridoxine | Pyridoxal phosphate | Amino-group transfer |
| Biotin | Biocytin | Carboxyl transfer |
| Folic acid | Tetrahydropholic acid | One-carbon-group transfer |
| Vitamin $B_{12}$ | Co-enzyme $B_{12}$ | 1,2 shift of hydrogen atoms |
| Lipoic acid | Lipoyllysine | Hydrogen-atom and acyl-group transfer |
| Ascorbic acid | – | Co-factor in hydroxylation |
| Fat soluble | | |
| Vitamin A | 11-*cis*-Retinal | Visual cycle |
| Vitamin D | 1,25-Dihydroxycholecalciferol | Calcium and phosphate metabolism |
| Vitamin E | – | Antioxidant |
| Vitamin K | – | Prothrombin biosynthesis |

## ENZYMES

Enzymes are proteinaceous substances with highly specific structures that enable biochemical reactions to occur. After participating in a chemical reaction, an enzyme is regenerated intact and can take part in additional reactions. A substance that behaves in such a manner to enable chemical reactions to occur is called a catalyst. Enzymes are very specific in their ability to interact with particular substances or class of substances called substrates. This specificity is based upon the unique shapes of enzymes is illustrated in Fig. 1.4. As shown in the figure, an enzyme recognises a particular substrate by its molecular structure and binds to it to produce an enzyme-substrate complex. This complex then breaks apart to form one or more products different from the original enzyme, regenerating the unchanged enzyme, which is then available to catalyse additional reactions.

Enzymes are named for what they do. For example, lipase enzymes cause lipid triglycerides to dissociate and form glycerol and fatty acids. Since these enzymes catalyse the breakdown of biopolymers by the addition of water, they belong to a major system of enzymes called hydrolysing enzymes. Other major systems of enzymes are oxidoreductase enzymes that catalyse biological oxidation-reduction reactions; transferase enzymes that move chemical groups from one molecule to another; lyase enzymes that remove chemical groups without hydrolysis and participate in the formation of C=C bonds or addition of species to such bonds; isomerase enzymes, which enable conjunction of isomers without loss or gain of atoms; and ligase enzymes that work in conjunction with ATP (adenosine triphosphate, a high-energy molecule that plays a crucial role in energy-yielding, glucose-oxidising metabolic processes) to link molecules together with the formation of bonds such as carbon-carbon or carbon-sulphur bonds.

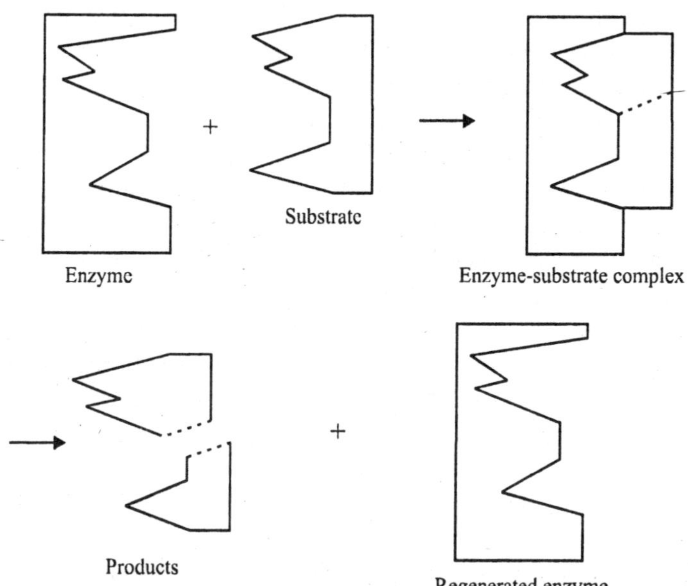


**Fig. 1.4.** Enzyme structure in enzyme-catalysed biochemical reactions.

A major mechanism of toxicity is the alteration or destruction of enzymes by agents such as cyanide, heavy metals or insecticidal parathion. An enzyme that has been destroyed obviously cannot perform its designated function, whereas one that has been altered may either not function at all or may act improperly. Toxicants can affect enzymes in several ways. Parathion, for example, bonds covalently to the nerve enzyme acetylcholinesterase, which can then no longer serve to stop nerve impulses. Heavy metals tend to bind to sulphur atoms in enzymes (such as sulphur from the amino acid cysteine shown in Fig. 1.2) thereby altering the shape and function of the enzyme. Enzymes are denatured by some poisons so that the enzyme no longer has its crucial specific shape.

Chapter 2

# Fundamentals of Microbiology

## INTRODUCTION

Small living creatures called micro-organisms interact in numerous ways with human activities. On the large scale of the biosphere, which consists of all regions of the earth containing life, micro-organisms play a primary role in the capture of energy from the sun. Their biological activities also complete critical segments of the cycles of carbon, oxygen, nitrogen and other elements essential for life. Microbes are also responsible for many human, animal and plant diseases.

In this text we concentrate primarily on mankind's use of microbes. These versatile biological catalysts have served mankind for millennia. Fermented foods such as cheese, bread, yoghurt and soya have long contributed to mankind's nutrition. Pasteur and Tyndall identified micro-organisms as the critical, active agents in prior fermentation practice and initiated the emergence of microbiology as a science. From these beginnings, further work by Buchner, Neuberg and Weizmann led to processes for production of ethanol, glycerol and other chemicals in the early 20th century.

In the 1940s complementary developments in biochemistry, microbial genetics and engineering ushered in the era of antibiotics with tremendous relief to mankind's suffering and mortality.

This period marks the birth of biochemical engineering, the engineering of processes using catalysts, feedstocks and/or sorbents of biological origin. Biotechnology began to change from empirical art to predictive, optimised design.

A later generation of fermentation processes produced steroids for birth control and for treatment of arthritis and inflammation. Methods for cultivation of plant and animal cells made possible mass production of vaccines and other useful biological agents. Clearly, mankind's successful harnessing and direction of cellular activities has had many health, social, environmental and economic impacts on past and contemporary human civilisation.

An interwoven fabric of research in molecular biology and microbial genetics has led to fundamental understanding of many of the controls and catalysts involved in complex biochemical syntheses conducted by living cells. On this foundation of basic knowledge, the methods of recombinant DNA technology have been erected. It is difficult to imagine the scope and magnitude of the eventual benefits of these marvelous tools. New vaccines and drugs have already been produced, but these are only the beginnings of revolutionary developments to come.

Our challenge in learning biochemical engineering is to understand and analyse the processes of biotechnology so that we can design and operate them in a rational way. To reach this goal, however, a basic working knowledge of cell growth and function is required. These factors and others peculiar to biological systems usually dominate biochemical process engineering. Consider for a moment that a living micro-organism may be viewed in an approximate conceptual sense as an expanding chemical reactor which takes in chemical species called nutrients from its environment, grows, reproduces and releases products into its surroundings. In instances such as sewage treatment, consumption of nutrients (here the organic sewage material) is the engineering objective. When microbes are grown for food sources or supplements, it is the mass of microbial matter produced which is desired. For a sewage-treatment process, on the other hand, this microbial matter produced by nutrient consumption constitutes an undesirable solid waste and its amount should be minimised.

Finally, the products formed and released during cellular activity are of major concern in many industrial and natural contexts, including penicillin and ethanol manufacture. The relative rates of nutrient utilisation, growth and release of products depend strongly on the type of cells involved and on the temperature, composition and motion of their environment. Understanding these interactions requires a foundation built upon biochemistry, biophysics and cell biology.

Whenever possible we shall extend our study of biological processes beyond qualitative understanding to determine quantitative mathematical representations. These mathematical models will often be extremely oversimplified and idealised, since even a single micro-organism is a very complicated system. Nevertheless, basic concepts in microbiology will serve as a guide in formulating models and checking their validity, just as basic knowledge in fluid mechanics is useful when correlating the friction factor with the Reynolds number.

Microbiology is the study of living organisms too small to be seen clearly by the naked eye. As a rough rule of thumb, most micro-organisms have a diameter of 0.1 mm or less. Present knowledge indicates that even the simplest micro-organism houses chemical reactors, information and control systems, and mass-transfer operations of amazing sophistication, efficiency and organisation. These conclusions have been reached in numerous experimental studies involving methods adapted from the physical sciences. Since this approach has proved so fruitful, the applicability of the principles of chemistry and physics to biological systems is now a widespread working hypothesis within the life sciences. The term biophysics is sometimes used to indicate explicitly the union of the biological and physical sciences.

A development critical to the understanding of living systems started in 1838, when Schleiden and Schwann first proposed the cell theory. This theory stated that all living systcms are composed of cells and their products. Thus, the concept of a basic module, or building block, for life emerged. This notion of a common denominator permits an important decomposition in the analysis of living systems: first the component parts, the cells, can be studied and then this knowledge is used to try to understand the complete organism.

The value of this decomposition rests on the fact that cells from a wide variety of organisms share many common features in their structure and function. In many instances this permits successful extrapolation of knowledge gained from experiments on cells from one organism to cells of other types. This existence of common cellular characteristics also simplifies our task of learning how micro-organisms behave. By concentrating on the apparently universal features of cellular function, a basic framework for understanding all living systems can be established.

## CELL

The cell is the structural and functional unit of all known living organisms. It is the smallest unit of an organism that is classified as living, and is sometimes called the building block of life. Some organisms, such as most bacteria, are unicellular (consist of a single cell). Other organisms, such as humans, are multicellular. (Humans have an estimated 100 trillion or 1014 cells; a typical cell size is 10 μm; a typical cell mass is 1 nanogram.) The largest known cell is an ostrich egg. In 1837 before the final cell theory was developed, a Czech Jan Evangelista Purkyne observed small 'granules' while looking at the plant tissue through a microscope. The cell theory, first developed in 1839 by Matthias Jakob Schleiden and Theodor Schwann, states that all organisms are composed of one or more cells. All cells come from preexisting cells. Vital functions of an organism occur within cells, and all cells contain the hereditary information necessary for regulating cell functions and for transmitting information to the next generation of cells.

There are two distinct types of cells: prokaryotic cells (e.g., bacterial cell) and eukaryotic cells (e.g., plant or animal cell). Some major features of the eukaryotic cell are shown in Fig. 2.1. The main difference between the two is a well-defined nucleus surrounded by a membranous nuclear envelope present only in eukaryotic cells.

Despite this difference they share a number of common features: the genetic information is stored in genes, proteins serve as their main structural material, ribosomes are used to synthesise proteins, adenosine triphosphate is the main source of metabolic energy to sustain various cellular processes and a cell membrane that controls the flow of substances into and out of the cell.

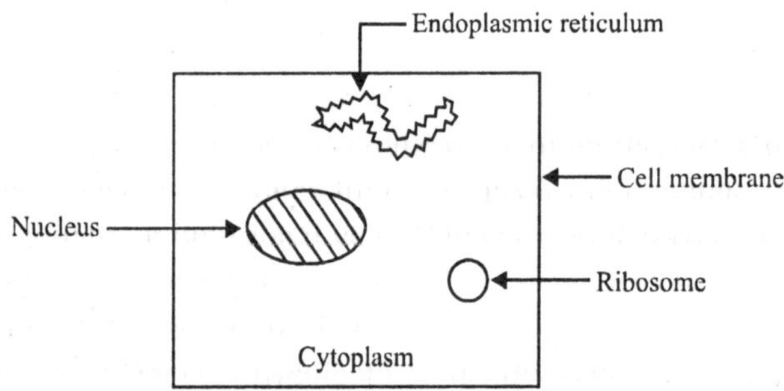


**Fig. 2.1.** Major features of the eukaryotic cell.

Eukaryotic cells are filled with cytoplasm and contain a number of different structures. Energy conversion and utilisation are mediated by cell mitochondria. Ribosomes participate in protein synthesis. Some toxicants are metabolised by enzymatic processes that involve the endoplasmic reticulum. The cell membrane and nucleus are of particular importance in respect to toxic substances and are discussed below.

### Cell Membrane

The cell membrane encloses the cell and regulates the passage of ions, nutrients, metabolic products, and lipid-soluble ('fat-soluble') substances into and out of it. Cell membranes are composed in part of phospholipids that are arranged with their hydrophilic ('water-seeking') heads on the cell membrane surfaces and their hydrophobic ('water-repelling') tails inside the membrane. Cell membranes contain bodies of proteins that are involved in the transport of some substances through the membrane. One reason the cell membrane is very important in toxicology is because it regulates the passage of toxicants and their products into and out of the cell interior.

Furthermore, when its membrane is damaged by toxic substances, a cell may not function properly and the organism may be harmed.

### Cell Nucleus

The cell nucleus contains deoxyribonucleic acid, DNA, a basic unit of which is shown in Fig. 2.2. Through its DNA the nucleus is the 'control center' of the cell that regulates cell division and protein synthesis. DNA is a polymeric substance with a molecular mass in the range of 6 to 16 million, the basic repeating unit of which is a nucleotide. As shown in Fig. 2.2, a nucleotide is composed of phosphoric acid, deoxyribose (a sugar) and one of four organic bases (adenine, guanine, cytosine and thymine). DNA determines heredity, cell reproduction and protein synthesis in cells. When DNA is damaged by foreign substances, various toxic effects, including mutations, cancer, birth defects and defective immune system function may occur.

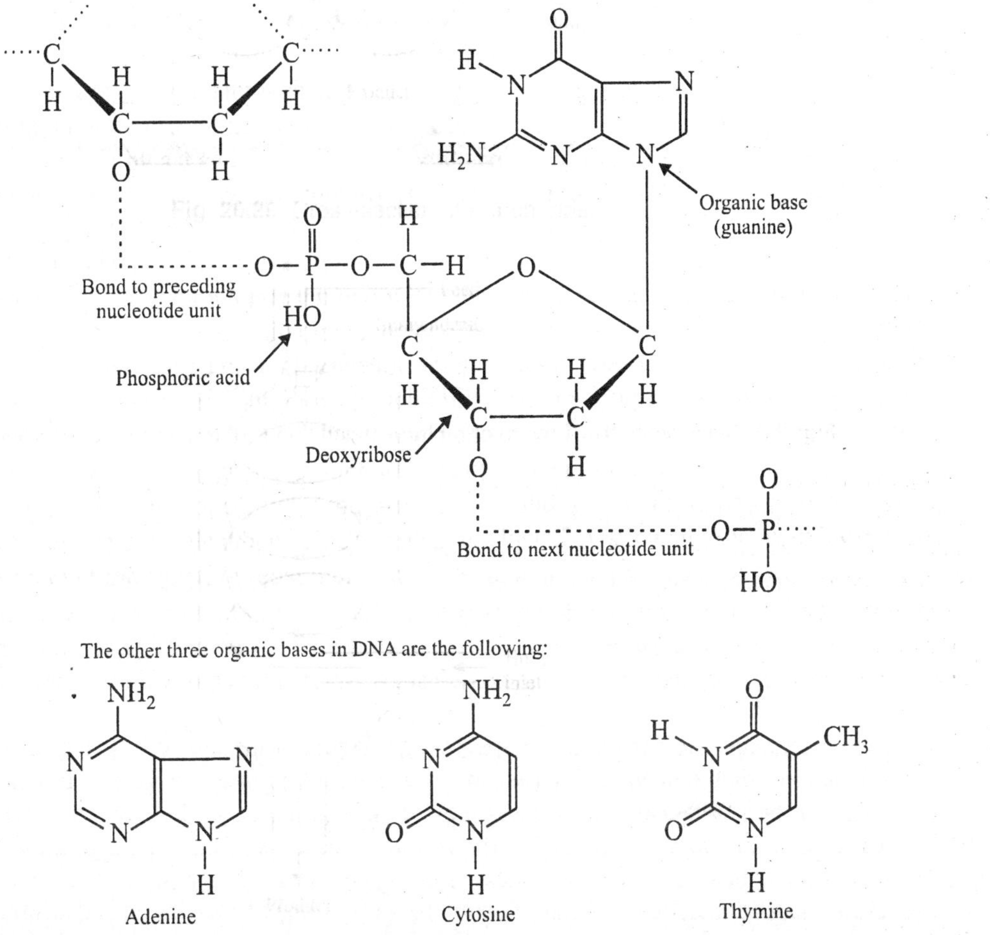

Bond to preceding nucleotide unit

Organic base (guanine)

Phosphoric acid

Deoxyribose

Bond to next nucleotide unit

The other three organic bases in DNA are the following:

Adenine

Cytosine

Thymine

**Fig. 2.2.** Unit of nucleotide in DNA consisting of phosphoric acid, one of four organic bases, and deoxyribose.

## IMPORTANT TYPES OF CELL

In this section we shall briefly review the classification of the kingdom of *protists*, which consists of all living things with a very simple biological organisation compared to plants and animals. All unicellular (single-celled) organisms belong to this kingdom and organisms containing multiple cells which are all of the same type are also classed as protists. Plants and animals, on the other hand, are distinguished by a diversity of cell types. A classification of plant and animal cells that can be grown on solid or in liquid nutrient media is included at the end of this section.

Figure 2.3 shows a breakdown of the protist kingdom into groupings convenient for our purposes. These classifications show differences in several characteristics including the following: energy and nutritional requirements, growth and product release rates, method of reproduction and capability and means of motion. All these factors are of great practical importance in applications. Also significant in classification are differences in the morphology or the physical form and structure, between these various types of organisms. The morphology of a micro-organism has an influence on the rate of nutrient mass transfer to it and also can profoundly affect the fluid mechanics of a suspension containing the organism. Obviously then we must examine each group in Fig. 2.3 individually.

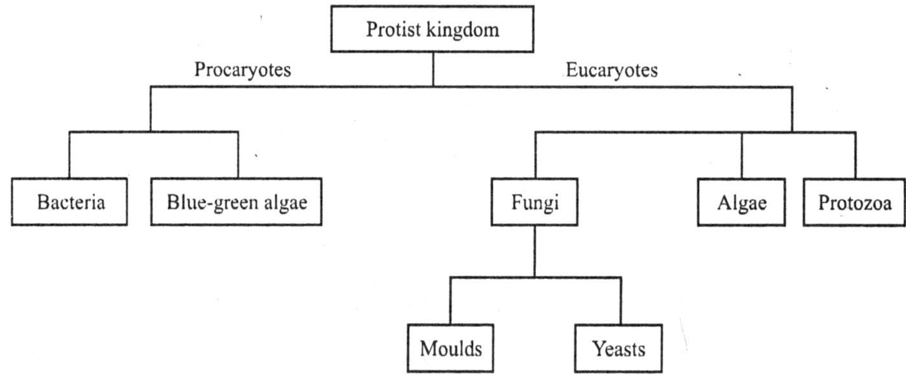


**Fig. 2.3.** Classifications of micro-organisms belonging to the kingdom of protists.

Taxonomy is the art of biological classification. The basic unit in this classification scheme is the species which is characterised by a high degree of similarity in physical and biochemical properties and significant differences from the properties of related organisms. Biological species are assigned a two-word latinised name in which the capitalised first word is the *genus* or generic name and the second word is the specific name, often a descriptive term. For example, an extensively studied bacterium found in the human intestine has the name *Escherichia* (generic name) *coli* (specific, descriptive name). The species name is italicised, and, if the generic name is clear from context, it is usually abbreviated to its first letter: *E. coli.*

In order to organise the cataloging of species and genera, a hierarchical system of taxonomy has been developed in which related genera first are grouped into families, then related families collected in orders, orders in turn organised into classes, next classes gathered in divisions or phyla, and finally phyla grouped into kingdoms. In Table 2.3, for example, protist designates a kingdom, fungi constitute a division, and yeasts belong to a common class. Often, gradations in properties between micro-organisms are sufficiently smooth that detailed classification becomes somewhat artificial and arbitrary, especially for bacteria and yeasts.

## Bacteria

The bacteria are relatively small organisms usually enclosed by rigid walls. In many species the outer surface of the cell wall is covered with a slimy, gummy coating called a capsule or slime layer. Bacteria are typically unicellular, but they may exist in three basic morphological forms (Fig. 2.4). Most cannot utilise light energy, are capable of motion (motile) and reproduce by division into two daughter cells (binary fission). Still, many exceptions to each of these rules are known.

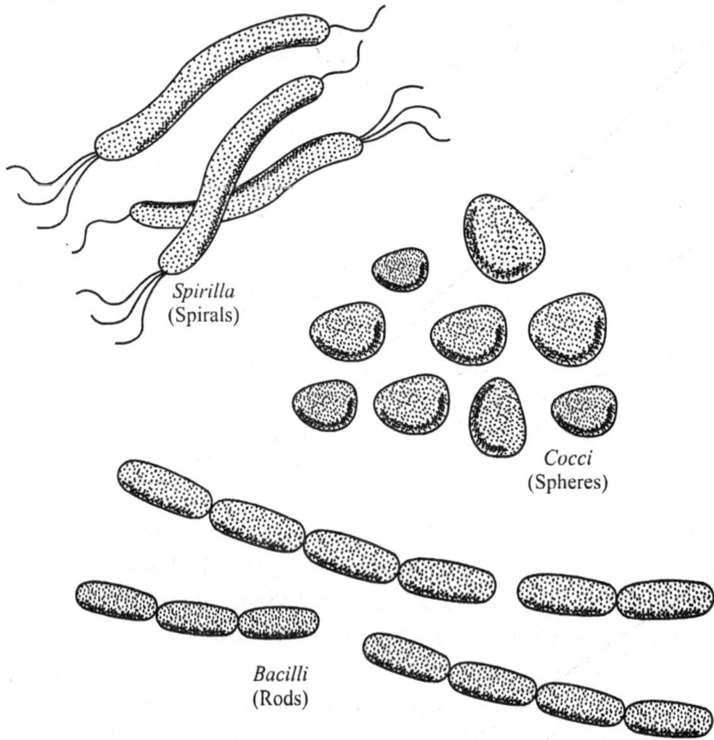


**Fig. 2.4.** The three forms of bacteria.

Whether or not oxygen is supplied to the cells is especially important in commercial exploitation of micro-organisms. In an aerobic process, oxygen is provided, usually as air, for use by the micro-organisms. Manufacture of vinegar, some antibiotics and animal-feed supplements are among the important microbial applications which employ aeration. The sparing solubility of oxygen in the aqueous media typical of these systems has major implications in process design. The protists function without oxygen in an anaerobic process such as production of some alcohols or digestion of organic wastes.

Especially important in commercial utilisation and control of bacteria is their ability to form endospores under adverse conditions. Spores are dormant forms of the cell, capable of resisting heat, radiation and poisonous chemicals. When the spores are returned to surroundings suitable for cell function, they can germinate to give normal, functioning cells. This normal, biologically active cell state is often called the vegetative form in order to distinguish it from the spore form. There are two major groups of sporeforming bacteria.

The aerobic *Bacillus* species are extremely widespread and adaptable. Several *Clostridium* species, which normally function under anaerobic conditions, die in the presence of oxygen in the vegetative state but form spores unaffected by oxygen. Some bacteria whose vegetative forms are rapidly killed at 45°C can form spores which survive boiling in water for several hours. Therefore, when we are attempting to kill micro-organisms by heating (heat sterilisation) the sporeforming capability of bacteria demands use of higher temperature, typically achieved by boiling under pressure in an autoclave to give $T > 120°C$.

The blue-green algae will not be discussed here since they are not of great commerical significance. They are important, however, in the overall operation of natural aquatic systems since they participate in the nitrogen cycle.

### Fungi

The kingdom fungi includes some of the most important organisms, both in terms of their ecological and economic roles. By breaking down dead organic material, they continue the cycle of nutrients through ecosystems. In addition, most vascular plants could not grow without the symbiotic fungi or mycorrhizae, that inhabit their roots and supply essential nutrients. Other fungi provide numerous drugs (such as penicillin and other antibiotics), foods like mushrooms, truffles and morels, and the bubbles in bread, champagne and beer.

Fungi also cause a number of plant and animal diseases: in humans, ringworm, athlete's foot, and several more serious diseases are caused by fungi. Because fungi are more chemically and genetically similar to animals than other organisms, this makes fungal diseases very difficult to treat. Plant diseases caused by fungi include rusts, smuts and leaf, root and stem rots and may cause severe damage to crops. However, a number of fungi, in particular the yeasts, are important 'model organisms' for studying problems in genetics and molecular biology.

### Yeasts

Yeasts form one of the important subgroups of fungi. Fungi, like bacteria, are widespread in nature although they usually live in the soil and in regions of lower relative humidity than bacteria. They are unable to extract energy from sunlight and usually are free-living. Although most fungi have a relatively complex morphology, yeasts are distinguished by their usual existence as single, small cells from 5 to 30 μm long and from 1 to 5 μm wide.

The various paths of reproduction of yeasts are asexual (budding and fission) as shown in Fig. 2.5. In budding, a small offspring cell begins to grow on the side of the original cell; physical separation of mature offspring from the parent may not be immediate and formation of clumps of yeast cells involving several generations is possible.

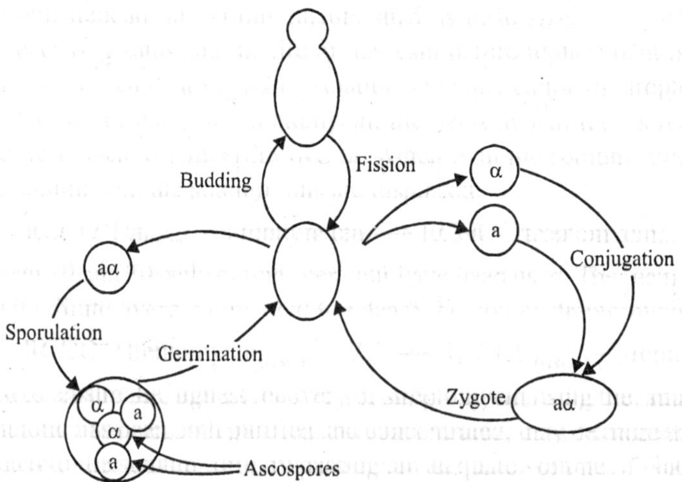


**Fig. 2.5.** Various paths of reproduction of yeasts.

Fission occurs by division of the cell into two new cells. Sexual reproduction occurs by conjugation of two haploid cells (each having a single set of chromosomes) with dissolution of the adjoining wall to form a diploid cell (two sets of chromosomes per cell) zygote. The nucleus in the diploid cell may undergo one or several divisions and form ascospores; each of these eventually becomes an individual new haploid cell, which may then undergo subsequent reproduction by budding, fission or sexual fusion again. The ascospores, which here are a normal stage in the reproductive cycle of these organisms, should not be confused with the endospores, discussed above, which are a defense mechanism against hostile environments.

In the production of alcoholic beverages, yeasts are the only important industrial microbes. In addition to supplying the consumer market for beer and wine, anaerobic yeast activities produce industrial alcohol and glycerol. The yeasts themselves are also grown for baking purposes and as protein supplements to animal feed.

## Moulds

Moulds are higher fungi with a vegetative structure called a mycelium. As illustrated in Fig. 2.6, the mycelium is a highly branched system of tubes. Within these enclosing tubes is a mobile mass of cytoplasm containing many nuclei. The mycelium may consist of more than one cell of related types. The long, thin filaments of cells within the mycelium are called hyphae. In some cases the mycelium may be very dense. This property, coupled with moulds, oxygen-supply requirements for normal function, can cause complexities in their cultivation since the mycelium can represent a substantial mass-transfer resistance.

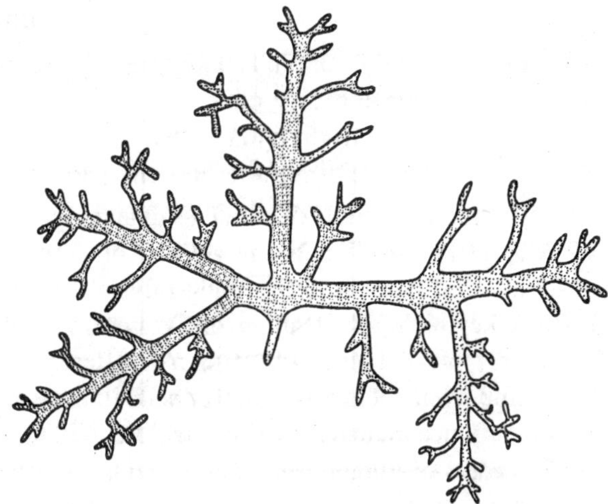

**Fig. 2.6.** The mycelial structure of moulds. A dense mycelium can cause conditions near its centre to differ considerably from those at the outer extremities.

Moulds, like yeasts, do not contain chlorophyll, and they are generally nonmotile. Reproduction, which may be sexual or asexual, is typically accomplished by means of spores. Spore properties form an important component in fungi classification.

The most important classes of moulds industrially are *Aspergillus* and *Penicillium* (Fig. 2.7). Major useful products of these organisms include antibiotics (biochemical compounds which kill certain micro-organisms; or inhibit their growth), organic acids and biological catalysts.

The strain *Aspergillus niger* normally produces oxalic acid ($HO_2CCO_2H$). Limitation of both phosphate nutrient and certain metals such as copper, iron and manganese results in a predominant yield of citric acid [$HOOCCH_2COH(COOH)CH_2COOH$]. This limitation method is the basis for the commercial biochemical citric acid process. Thus *A. niger* is an interesting example differentiating approaches to biochemical-reactor design and optimisation from those of non-biological reactors: a much greater selectivity can sometimes be achieved in the biological system by minor alteration of feed composition to the reactor.

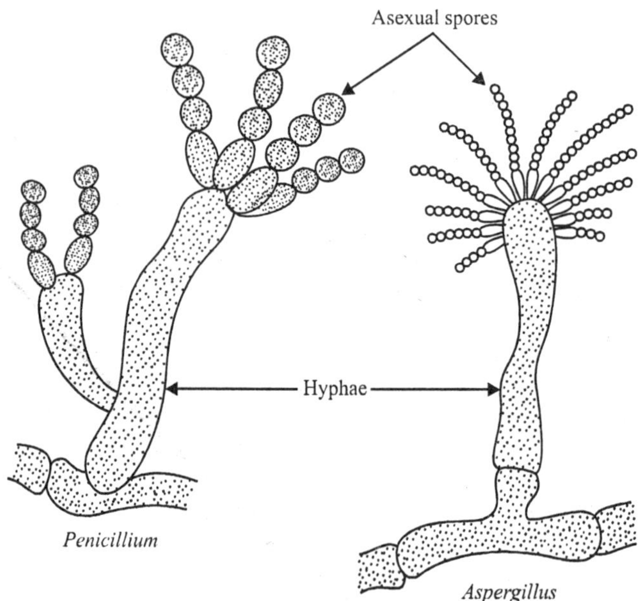


**Fig. 2.7.** The hyphae of *aspergillus* and *penicillium*, two industrially important varieties of moulds.

This example (as well as that of penicillin) should motivate us to learn the essentials of cell structure, metabolism, and its function. Without this background in cellular processes and characteristics, our skills as process engineers, which are well suited to design and analysis of many aspects of biochemical processes, could be wasted because key biological features of the system would be ignored.

Penicillin production offers an example of a second fundamental difference between microbiological and non-biological reactors. Major improvement in production of penicillin has arisen by use of ultraviolet irradiation of Penicillium spores to produce mutants of the original Penicillium strain. Cell mutation by various techniques may result in orders-of-magnitude improvements in desired yield. Recombinant DNA technology provides carefully controlled and novel genetic alterations in certain organisms. This approach has made possible manufacture of valuable animal products (proteins) in simple bacteria. These examples indicate the central importance of genetics to the biochemical engineer. This subject occupies a large fraction of the present efforts in both university and industrial biological research. The practical importance of designing genetic modifications (or in other situations, of avoiding such changes) emphasises the need for close cooperation between engineers, biologists and biochemists in biochemical process design and evaluation. The history of penicillin production is in itself a story of joint development of new techniques, including deep-submerged production, solvent extraction of a delicate product on a large scale, air-sterilisation procedure for high volumetric flow rates and isolation of mutants with high penicillin yields.

Before leaving bacteria and fungi, we should mention the *actinomycetes*, a group of micro-organisms with some properties of both fungi and bacteria. These organisms are extremely important for antibiotic manufacture. Although formally classified as bacteria, actinomycetes resemble fungi in their formation of long, highly branched hyphae. Also, design of antibiotic production processes utilising actinomycetes is very similar to those involving moulds. One difference, however, is the susceptibility of actinomycetes to infection and disease by viruses which also can attack bacteria.

### Algae and Protozoa

These relatively large eucaryotes have sophisticated and highly organised structures. For example, *Euglena* has flagella for locomotion, lacks a rigid wall and has an eyespot sensitive to light. The cell, guided by the eyespot, moves in response to stimulus by illumination—clearly a valuable asset since most algae require energy in the form of light. Many diatoms (another kind of algae) have exterior skeletons of complex architecture which are impregnated with silica. These skeletons are widely employed as filter aids in industry.

Considerable commercial interest in algae is concentrated on their possible exploitation as foodstuffs and food supplements. In Japan, several processes for algae food cultivation are in operation today. Also important in Asia is use of seaweed in the human diet. While not micro-organisms, many seaweeds are actually multicellular algae. Like the simpler blue-green algae, eucaryotic algae serve a vital function in the cycles or matter on earth.

Just as algae may be viewed as primitive plants, protozoa, which cannot exploit sunlight's energy, are in a sense primitive animals. The habitats, morphology and activities of protozoa span a broad spectrum. For example, some trypanosomes carry serious disease, including African sleeping sickness. The *Trichonympha* inhabit the intestines of termites and assist them in digesting wood. While the amoeba has a changing, amorphous shape, the heliozoa have an internal skeleton and definite form.

Although protozoa are not now employed for industrial manufacture of either cells or products, their activities are significant among the micro-organisms which participate in biological waste treatment. These processes, widely employed in urban communities and large industrial plants throughout the world, are suprisingly complicated from a microbiological viewpoint. Since a complex mixture of different nutrients and microbes are present in sewage or industrial wastes, a correspondingly large collection of different protists are present and indeed necessary in treatment operations. These diverse organisms compete for nutrients, devour each other and interact in numerous ways characteristic of a small-scale ecological system.

### Animal and Plant Cells

Many vaccines and other useful biochemicals are produced by growth of animal cells in process reactors; i.e., by cell propagation outside of the whole animal. Improvements in cultivation techniques for these tissue-derived cells and emerging methods for genetic manipulation of animal and plant cells offer great potential for expanded commercial utilisation of these higher cells. The reactors in which 'tissue' cultures may be propagated may be quite similar to microbial reactors, admitting a unified treatment of cell kinetics and biochemical reactors.

When a piece of animal tissue, perhaps after disruption to break the cells apart, is placed in appropriate nutrient liquid, many cell types, such as blood cells, die within a few days, weeks or months. Other cells multiply and are called primary cell lines.

Often these cells can be 'passaged' by transfer to fresh medium after which further cell multiplication occurs giving a secondary cell line. Some secondary cells can be passaged apparently indefinitely; these cells are then dubbed an established, permanent or stable cell line. Many cell lines have been developed from the epithelial tissues (skin and tissues which cover organs and line body cavities), connective tissues and blood and lymph of several animals including man, hamster, monkey and mouse.

Some cultured tissue cells can be grown suspended in liquid, but growth of most cell lines requires attachment to a solid surface. This anchorage dependence poses stringent restrictions on scale-up for production of vaccines and other biological products from animal cell culture. Microcarrier culture techniques have greatly increased the volumetric productivity of reactors for anchorage-dependent culture cells.

It is also possible to grow certain plant cells in culture, either as a callus (a lump of undifferentiated plant tissue growing on solid nutrient medium) or as aggregated cells in suspension. Since plants produce many commercially important compounds including perfumes, dyes, medicinals and opiates, there is significant potential for future applications of plant cell culture. Cultured plant cells can also catalyse highly specific useful transformations. Cultured plant cells have several potential agricultural applications including whole plant regeneration. Greater knowledge of plant biology and of requirements and constraints in plant cell culture are needed before significant commercialisation of this approach occurs.

Chapter 3

# Enzymes and Enzyme Kinetics

## INTRODUCTION

Enzymes are biomolecules that catalyse (i.e., increase the rates of) chemical reactions. Almost all enzymes are proteins. In enzymatic reactions, the molecules at the beginning of the process are called substrates, and the enzyme converts them into different molecules, the products. Almost all processes in a biological cell need enzymes in order to occur at significant rates. Since enzymes are extremely selective for their substrates and speed up only a few reactions from among many possibilities, the set of enzymes made in a cell determines which metabolic pathways occur in that cell.

Like all catalysts, enzymes work by lowering the activation energy ($E_a$ or $\Delta G$) for a reaction, thus dramatically increasing the rate of the reaction. Most enzyme reaction rates are millions of times faster than those of comparable uncatalysed reactions. As with all catalysts, enzymes are not consumed by the reactions they catalyse, nor do they alter the equilibrium of these reactions. However, enzymes do differ from most other catalysts by being much more specific. Enzymes are known to catalyse about 4000 biochemical reactions. A few RNA molecules called ribozymes catalyse reactions, with an important example being some parts of the ribosome. Synthetic molecules called artificial enzymes also display enzyme-like catalysis.

Enzyme activity can be affected by other molecules. Inhibitors are molecules that decrease enzyme activity; activators are molecules that increase activity. Many drugs and poisons are enzyme inhibitors. Activity is also affected by temperature, chemical environment (e.g., pH) and the concentration of substrate.

Some enzymes are used commercially, for example, in the synthesis of antibiotics. In addition, some household products use enzymes to speed up biochemical reactions (e.g., enzymes in biological washing powders break down protein or fat stains on clothes; enzymes in meat tenderisers break down proteins, making the meat easier to chew).

The kinetic behaviour of enzymes has been studied in detail for a century, beginning with the classic work of Henri and Michaelis and Menten. The objectives have been threefold: to gain an understanding of the mechanisms of enzyme action; to illuminate the physiological roles of enzyme-catalysed reactions; and, mainly in recent years, to manipulate enzyme properties for biotechnological ends. Experimental practice has been overwhelmingly dominated by the first of these aims; most experiments have been designed as if shedding light on the mechanism was the principal or even the only, objective. However, although much valuable information has been obtained in this way, there are some important aspects of enzyme function that are obscured when working mainly with isolated enzymes in conditions far removed from those that exist in the cell. For this reason the latter part of this chapter will be devoted to a discussion of enzymes as they behave in complex mixtures.

## MICHAELIS-MENTEN KINETICS

The starting point for any discussion of enzyme kinetics is the Michaelis-Menten equation, which expresses the initial rate $v$ of a reaction at a concentration $a$ of the substrate transformed in a reaction catalysed by an enzyme at total concentration $e_0$:

$$v = \frac{k_0 e_0 a}{K_m + a} = \frac{Va}{K_m + a} \qquad \text{... (3.1)}$$

The parameters are $k_0$, the catalytic constant, and $K_m$, the Michaelis constant. The form shown in the middle is more fundamental than that on the right, but the second form, in which $k_0 e_0$ is written as the limiting rate $V$, is often used because the enzyme concentration in meaningful units is often not known. $V$ is the limit that v approaches as the enzyme becomes saturated (that is, when a becomes very large) and $K_m$ is the value of a at which $v = 0.5V$ (that is, at which the rate is half-maximal). The ratio $k_0/K_m$ is called the specificity constant and given the symbol $k_A$: It is a more fundamental constant than $K_m$ in the analysis of enzyme mechanisms (i.e., it has a simpler mechanistic meaning), but the equation is usually written in terms of $V$ (or $k_0$) and $K_m$ nonetheless.

The principal assumption implied by equation 3.1 is that the rate of the reverse reaction is negligible: This may be because the reaction is irreversible for practical purposes, but even with reversible reactions the reverse reaction can be made negligible by measuring the rate in the absence of products and by extrapolating the rate back to zero time, that is by estimating the initial rate at a time when no products have accumulated. Even if there is no significant reverse reaction, products can still affect the rate of the forward reaction because product inhibition (discussed later) is a common phenomenon.

Another assumption is that the reaction is allowed sufficient time to reach a steady state because all reactions pass through an initial acceleration phase known as the transient state. This phase is normally very brief (a few milliseconds), and in practice enzymes are usually studied, under steady-state conditions. Note, however, that this is made experimentally possible by working with extremely small enzyme concentrations compared with those that may exist in the cell. The use of very low enzyme concentrations has two important consequences: First, it normally means that the enzyme concentration can be neglected in comparison with the substrate concentration and second it makes the steady-state rate sufficiently slow to be easily measured and the steady-state phase long enough to be meaningful. If high enzyme concentrations were used, the transient state would be as brief as before, but the steady state would also be very brief, so that there would be no period in which one could adequately treat the rate as constant.

Equation 3.1 may be derived from the following model:

$$A + E \underset{k_{-1}}{\overset{k_1}{\rightleftharpoons}} EA \underset{k_{-2}}{\overset{k_2}{\rightleftharpoons}} EP \underset{k_{-3}}{\overset{k_3}{\rightleftharpoons}} E + P \qquad \text{... (3.2)}$$

which assumes that the reaction passes through an enzyme—substrate complex EA, which undergoes catalytic transformation to an enzyme-product complex EP, which then breaks down to form products. Although real enzyme mechanisms may be more complicated than this, every reaction passes through steps of substrate binding, chemical transformation, and product release. In simple introductory treatments the second and third steps are often treated as a single step, but conceptually they are clearly distinct. In reactions of more than one substrate, the steps do not necessarily occur in the order one might guess;

that is, some products may be released before all substrates have bound, but the general principle that any reaction involves the same three kinds of step remains valid.

The Michaelis-Menten parameters can be defined as follows in terms of the rate constants shown in equation 3.2:

$$k_0 + \frac{k_2 k_3}{k\ _2 + k_2 + k_3};\ k_A = \frac{k_1 k_2 k_3}{k\ _1 k\ _2 + k\ _1 k_3 + k_2 k_3};\ K_m = \frac{k\ _1 k\ _2 + k\ _1 k_3 + k_2 k_3}{k_1(k\ _2 + k_2 + k_3)} \quad \text{... (3.3)}$$

Note that none of the three parameters has a simple transparent meaning. The interpretations commonly attributed to them depend on additional simplifying assumptions that are not always correct. For example, $K_m$ is often said to be equal to the equilibrium constant $k_{-1}/k_1$ for dissociation of A from EA, but the expression in equation 3.3 does not take this form unless $k_2$ is very small. As there is no good reason for $k_2$ to be small and indeed ideas of evolutionary optimisation of enzyme function lead one to expect the opposite when the enzyme is acting on its natural physiological substrate, it follows that $K_m$ should not, in general, be regarded as a measure of the equilibrium dissociation constant.

Despite these difficulties in providing a detailed mechanistic meaning to $K_m$ it does provide a measure of the tightness of substrate binding in the steady state, as it is quite correct to take $a/K_m$ as equal to the ratio of the sum of concentrations of all enzyme complexes (i.e., both EA and EP) over the concentration of free enzyme. Similarly $k_0$ provides a valid measure of the capacity of the enzyme substrate complex to react to give products, even if it cannot be interpreted as the rate constant for a unique step in the mechanism. The reason for the term specificity constant for $k_A$, that is, its relationship to enzyme specificity, will become clear once inhibition has been considered.

## Graphical Analysis

Equation 3.1 defines a hyperbolic dependence of rate on substrate concentration, as illustrated in Fig. 3.1. The initial steep rise in $v$ as $a$ increases from zero is rapidly transformed into the phenomenon of saturation, whereby further increases in $a$ produce smaller and smaller increases in $v$, which approaches, but does not reach or exceed, the limiting rate $V$.

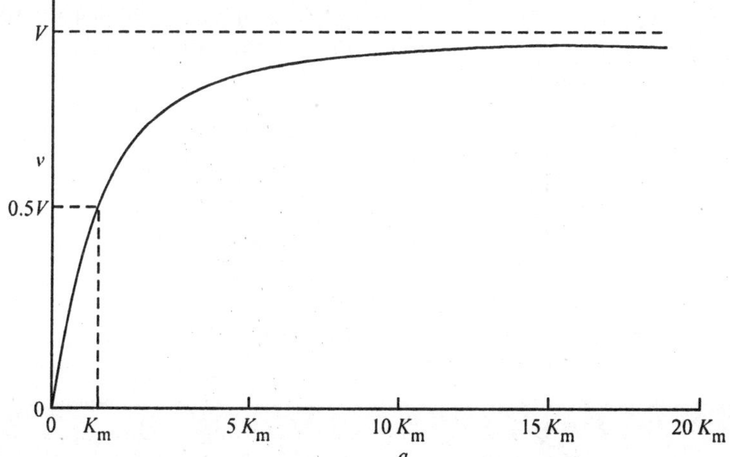


**Fig. 3.1.** Michaelis-Menten dependence of rate on substrate concentration. The curve is a rectangular hyperbola through the origin, approaching a limit of $v = V$ at saturation. The rate is $0.5V$ at 'a' substrate concentration equal to the Michaelis constant, $K_m$, but note that the approach to the limit is slow, so that, for example, even at '$a$' = 10 $K_m$ the rate is still nearly 10 per cent less than $V$.

The rectangular hyperbola makes this type of plot inconvenient for estimating the values of the kinetic parameters (because the line does not approach the saturation limit closely enough at reasonable values of $a$ to allow direct measurement of $V$). For this purpose, therefore, it is usual to transform equation 3.1 into one of the following three forms, which underlie the three straight-line plots illustrated in Figs. 3.2 through 3.4:

$$\frac{1}{v} = \frac{1}{V} + \frac{K_m}{V}\frac{1}{a} \quad \text{... (3.4)}$$

$$\frac{a}{v} = \frac{K_m}{V} + \frac{1}{V}a \quad \text{... (3.5)}$$

$$v = V \quad K_m \frac{v}{a} \quad \text{... (3.6)}$$

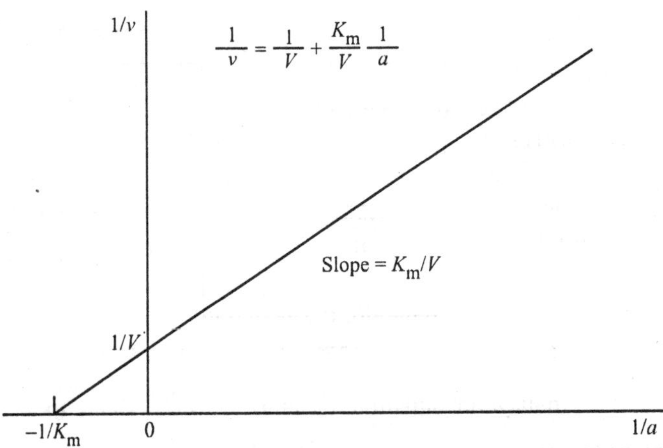


**Fig. 3.2.** The double-reciprocal plot. This is the most widely used method of plotting the Michaelis-Menten equation as a straight line. However, the severe distortion of any experimental errors in the original data causes it to give a misleading impression.

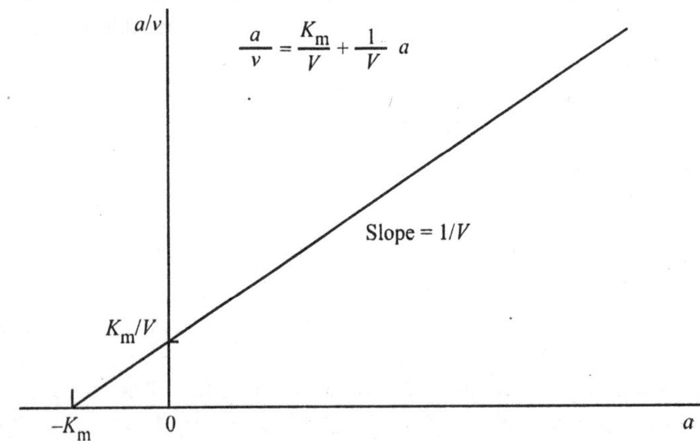


**Fig. 3.3.** The plot of $a/v$ against $a$. This alternative to the plot shown in Fig. 3.2 produces much less distortion of the experimental error.

The double-reciprocal plot, illustrated in Fig. 3.2 and based on equation 3.4, is the most widely used, but it is also the least satisfactory because it distorts the effect of experimental error to such an extent that it is difficult to form any visual judgement of where the best line should be drawn. The other two plots are better, and the plot of '$v$' against v/a (Fig. 3.4, equation 3.6) has the particular advantage that the entire observable range of '$v$' values, from 0 to $V$, is mapped onto a finite range of graph; this makes it easy to judge by eye if an experiment has been well-designed. On the other hand, it has the disadvantage that $v$, normally the less reliable measurement, contributes to both coordinates, and errors in '$v$' cause deviations along lines through the origin, rather than parallel with one or the other axis.

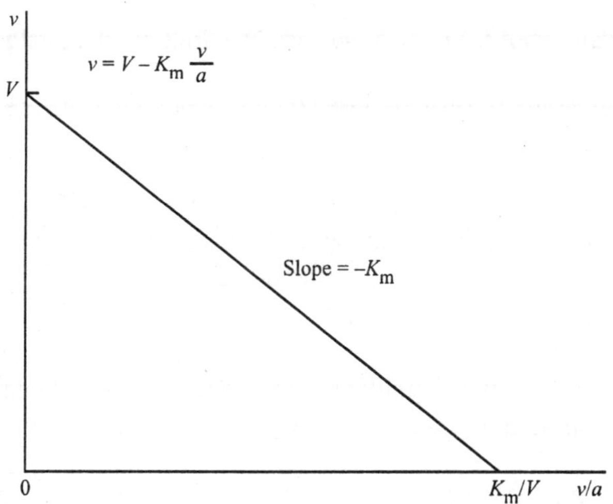


**Fig. 3.4.** The plot of '$v$' against $v/a$. The third way of plotting the Michaelis-Menten equation as a straight line also avoids the error-distorting property of the plot shown in Fig. 3.2 and maps the entire range of observable rates (from 0 to $V$) onto a finite range of paper. This is a desirable property because it makes it impossible to disguise deficiencies in the experimental design.

In modern practice it is usually best to regard these plots as for illustration purposes only, and to use suitable computer programmes for the actual parameter estimation. For this purpose it is not sufficient just to apply unweighted linear regression to the straight-line plots, as this suffers from the same statistical distortions as the plots themselves. The following two equations for calculating best-fit values of $K_m$ and $V$ give satisfactory results if the '$v$' values have uniform coefficient of variation (uniform standard deviation expressed as a percentage), as is usually at least approximately correct:

$$K_m = \frac{\Sigma v^2 \Sigma(v/a) \quad \Sigma(v^2/a)\Sigma v}{\Sigma(v^2/a^2)\Sigma v \quad \Sigma(v^2/a)\Sigma(v/a)} \quad \text{... (3.7)}$$

$$V = \frac{\Sigma(v^2/a^2)\Sigma v^2 \quad [\Sigma(v^2/a)]^2}{\Sigma(v^2/a^2)\Sigma v \quad \Sigma(v^2/a)\Sigma(v/a)} \quad \text{... (3.8)}$$

Each summation is made over all observations.

## Two-Substrate Reactions

Enzymes that catalyse reactions of a single substrate are only a small minority of all the enzymes known, but the Michaelis-Menten equation remains useful for examining the kinetics of the more common case of a reaction with two substrates and (often but not necessarily) two products, because such a reaction normally obeys Michaelis-Menten kinetics when only one substrate concentration is varied at a time. This is illustrated by the following typical equation for such a reaction:

$$v = \frac{k_0 e_0 ab}{K_{iA}K_{mB} + K_{mB}a + K_{mA}b + ab} \qquad \text{... (3.9)}$$

Although at first sight this appears quite different from equation 3.1, it can be arranged in the same form if one of the two substrate concentrations, for example b, is treated as a constant:

$$v = \frac{\left(\dfrac{k_0 b}{K_{mB} + b}\right) e_0 a}{\dfrac{K_{iA}K_{mB} + K_{mA}b}{K_{mB} + b} + a} \qquad \text{... (3.10)}$$

The two fractions in this equation can be regarded as 'apparent' values of the Michaelis–Menten parameters for A, that is, the equation can be written as:

$$v = \frac{k_0^{app} e_0 a}{K_{mA}^{app} + a} \qquad \text{... (3.11)}$$

with

$$k_0^{app} = \frac{k_0 b}{K_{mB} + b};\ k_A^{app} = \frac{(k_0/K_{mA})b}{(K_{iA}K_{mB}/K_{mA}) + b};\ k_{mA}^{app} = \frac{K_{iA}K_{mB} + K_{mA}b}{K_{mB} + b} \qquad \text{... (3.12)}$$

Notice that the expressions for the apparent values of $k_0$ and $k_A$ are both individually of Michaelis-Menten form with respect to '$b$', whereas that for the apparent value of $K_m$ is more complicated:

> 'This behaviour is quite typical and is one of the reasons why $k_0$ and $k_A$ should regarded as more fundamental parameters than $K_m$. More generally, the concept of apparent parameters pervades the analysis of simple cases in steady-state enzyme kinetics, being important for the study of reactions with multiple substrates and inhibition.'

## INHIBITION AND ACTIVATION

### Inhibition

For most enzyme-catalysed reactions, molecules exist that resemble the substrate closely enough to bind to the enzyme, but not closely enough to undergo a chemical reaction. Such a molecule is known as a competitive inhibitor and causes competitive inhibition, characterised by a rate equation of the following form:

$$v = \frac{k_0 e_0 a}{K_m(1 + i/K_{ic}) + a} \qquad \text{... (3.13)}$$

in which '$i$' is the concentration of the inhibitor and $K_{ic}$ is the competitive inhibition constant. (The qualification 'competitive' and the second subscript c are usually omitted if only this simplest kind of inhibition is being considered.

Inhibitors can interfere with catalysis as well as with substrate binding. In the simplest case, an inhibitory term affects the variable term in the denominator of the Michaelis-Menten equation, instead of the constant term:

$$v = \frac{k_0 e_0 a}{K_m + a(1 + i/K_{iu})} \qquad \text{... (3.14)}$$

This is called uncompetitive inhibition and the inhibition constant $K_{iu}$ is the uncompetitive inhibition constant. This is important as a limiting case of inhibition, but in its pure form it is not at all common. Much more often one has mixed inhibition, when both competitive and uncompetitive effects occur simultaneously:

$$v = \frac{k_0 e_0 a}{K_m(1 + i/K_{ic}) + a(1 + i/K_{iu})} \qquad \text{... (3.15)}$$

There is no particular reason for the two inhibition constants $K_{ic}$ and $K_{iu}$ to be equal, and most of the mechanisms one might propose to account for mixed inhibition lead one to expect them to be different, yet the case where $K_{ic} = K_{iu}$ is often given an undeserved prominence in discussions of inhibition, largely because experiments done many years ago suggested that it was a more common phenomenon than it is. This is called noncompetitive inhibition and its rate equation is the same as equation 3.15, but with both $K_{ic}$ and $K_{iu}$ written simply as $K_i$.

All of these kinds of inhibition are conveniently discussed in terms of apparent Michaelis-Menten parameters. In the general case (equation 3.15), these are as follows:

$$k_0^{app} = \frac{k_0}{1 + i/K_{iu}}; \; k_A^{app} = \frac{kA}{1 + i/K_{ic}}; \; k_m^{app} = \frac{K_m(1 + i/K_{ic})}{1 + i/K_{iu}} \qquad \text{... (3.16)}$$

Note that the first two expressions have the same form and both simplify to independence of i in the event that one or other inhibition term is negligible. The expression for the apparent value of $K_m$ is more complicated, especially when one considers how it varies with the different types of inhibition: It increases with the concentration of a competitive inhibitor, it decreases as the concentration of an uncompetitive inhibitor increases, it may change in either direction as the concentration of a mixed inhibitor increases, or it is independent of inhibitor concentration if the inhibition is noncompetitive. In general it is simplest to regard $k_A$ as the parameter affected by competitive inhibition, negligibly so when the competitive component is negligible, $k_0$ as the parameter affected by uncompetitive inhibition, negligibly so when the uncompetitive component is negligible, and $K_m$ just as the ratio of the two, so $K_m = k_0/k_A$.

The effects of the different kinds of inhibition on the common plots as illustrated in Figs. 3.2 through 3.4 follows naturally from equation 3.16. Any competitive effect affects the apparent value of $k_A$, hence, it increases the slope of the plot of $1/v$ against $1/a$ (Fig. 3.2), it increases the ordinate intercept of the plot of $a/v$ against a (Fig. 3.3) and it decreases the abscissa intercept of the plot of $v$ against $v/a$ (Fig. 3.4). Conversely, any uncompetitive effect increases the ordinate intercept of the plot of $1/v$ against $1/a$, increases the slope of the plot of $a/v$ against $a$, and decreases the ordinate intercept of the plot of '$v$' against $v/a$. When both components of the inhibition are present, both kinds of effects occur. As an illustration we may consider just one example, the effect of competitive inhibition on the plot of $1/v$ against $1/a$: Plots made at various different inhibitor concentrations produce a family of straight lines intersecting on the ordinate axis, as shown in Fig. 3.5, the lack of effect on the ordinate intercept being a direct consequence of the lack of effect on the apparent value of V.

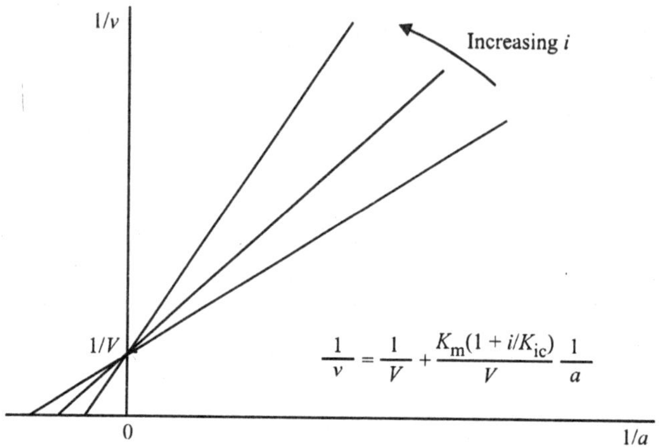


**Fig. 3.5.** Effect of competitive inhibition on the double-reciprocal plot.

## Specificity

Specificity is the most fundamentally important property of enzymes. Although one is often impressed by the catalytic effectiveness of enzymes, accelerating a reaction is not, in reality, difficult: Heating the reaction mixture in a sealed tube is an efficient way of accelerating virtually any reaction, essentially without limit.

What is difficult is to accelerate one selected reaction without at the same time accelerating a great mass of unwanted reactions. What is important about an enzyme, therefore, is not that it is an excellent catalyst for a small set of reactions, but that it is an extremely bad catalyst—virtually without any activity—for all other reactions. In other words, the essential properties of enzymes are that they act under very mild conditions and are highly specific.

In the past, specificity was often assessed by comparing the kinetic parameters for different reactions measured in isolation from one another, which led to sterile arguments as to whether specificity was best measured in terms of $k_0$, $K_m$, or some combination of the two. This type of argument was resolved once it was realised that the only meaningful way of defining specificity is as a property of an enzyme that allows it to discriminate between substrates that are mixed together.

The simplest way to consider this is with a model similar to that for competitive inhibition, except that one assumes that both molecules are capable of reacting. The equation for reaction of one substrate A in the presence of a competing substrate A′ follows an equation similar to that for competitive inhibition (equation 3.13):

$$v = \frac{k_0 e_0 a}{K_m(1 + a'/K'_m) + a} \quad \text{... (3.17)}$$

with the inhibitor concentration replaced by a′, the concentration of A′, and the inhibition constant by $K'_m$, the Michaelis constant for the reaction of A′ considered in isolation. The rate of reaction of A′ is given by the same equation with an obvious transposition of symbols:

$$v' = \frac{k'_0 e_0 a'}{K'_m(1 + a/K_m) + a'} \quad \text{... (3.18)}$$

It can then be seen that the ratio of rates is the ratio of substrate concentrations multiplied by the ratio of specificity constants:

$$\frac{v}{v'} = \frac{k_0 / K_m}{k'_0/K_m} \frac{a}{a'} = \frac{k_A a}{k'_A a'} \qquad \text{... (3.19)}$$

This result, which is still less well-known than its importance merits, is the reason for the term specificity constant. Note that although inspection of equation (3.1) suggests that $k_A$ is no more than the parameter that defines the rate at very low substrate concentrations, no assumption about the magnitudes of the concentrations was made in arriving at equation (3.19). It is thus valid at all concentrations and the specificity constant measures specificity at all concentrations, not just low ones.

## Activation

Activation is the opposite from inhibition, in which a reaction proceeds more rapidly in the presence of a particular molecule than in its absence. It is less common than inhibition, and discussion is complicated by the fact that a variety of quite different phenomena have been termed 'activation'. The most important of these is a confusion between true activation and the case where the activator is really a component of the substrate. Numerous ATP-handling enzymes are said to be activated by magnesium ions, when in reality the complex MgATP is the true substrate, that is, the species that reacts with the enzyme. Other metal ions, such as the zinc in a number of enzymes, may be true activators as they bind to the enzyme itself and confer catalytic properties on it.

Another misuse of the term activation relates to coenzymes such as NAD in many dehydrogenases: Alcohol dehydrogenase, for example, may be said to be activated by NAD, but although consideration of the metabolic pathway in which the reaction occurs may make it convenient to regard ethanol as the substrate and NAD as the coenzyme, this is meaningless when the reaction is considered in isolation. So far as alcohol dehydrogenase is concerned, it catalyses a reaction that requires two substrates, ethanol and oxidised NAD; the reaction will not proceed unless both are present, and neither has any more reason to be called the substrate than the other.

When such improper uses of the term are excluded, there remain a number of enzymes for which the true inverse of inhibition occurs. In the simplest cases the equations are just the inverse of inhibition equations, with terms of the form $i/K_i$ replaced by ones of the form $K_x/x$ (for an activator X with activation constant $K_x$). However, the simplest cases constitute a smaller proportion of the whole than they do for inhibition. This is because whereas most inhibitors inhibit completely, in the sense that enzyme species with inhibitor bound retain no activity as long as the inhibitor remains bound, many enzymes subject to activation retain some activity in the absence of the activator. As a result, full analysis of activation is often more complicated than it is for inhibition, but this will not be discussed further here.

## Irreversible Inhibition

The types of inhibition considered so far are examples of reversible inhibition; the inhibitor binds reversibly and catalytic activity returns when the inhibitor is released. Irreversible inhibition also occurs, in which the inhibitor either binds so tightly that for practical purposes it cannot be removed, or reacts with the enzyme and converts it irreversibly to a form that has no catalytic activity. These two cases are conceptually different, and the former is more correctly called tight-binding inhibition, rather than irreversible inhibition. However, they are not easy to distinguish in practice, and have similar practical effects and, hence, similar practical uses.

Although irreversible inhibition has played a smaller part than reversible inhibition in the academic study of enzyme mechanisms, it has far greater industrial and pharmacological importance. This is because competitive inhibitors, the most common kind of reversible inhibitors, are almost completely ineffective in complete physiological systems, for reasons to be considered shortly. By contrast, whenever irreversible (or tight-binding) inhibition occurs in a physiological system, it can be expected to have profound effects. Many toxic and pharmacologically active substances owe their effects to irreversible inhibition.

Both tight-binding and irreversible inhibition manifest themselves in ways that allow them to be confused with noncompetitive inhibition, as in equation (3.15) with the two inhibition constants equal. This is because the practical effect of irreversible inhibition is not on any of the kinetic parameters in the Michaelis-Menten equation, but on $e_0$, the total enzyme concentration. However, a decrease in $e_0$ can be confused with a decrease in the apparent value of $k_0$, as they occur as a product in equation (3.1). Although uncompetitive inhibition affects $k_0$, it does so without affecting $k_A$ and thus also changes $K_m$. Decreasing $k_0$ without affecting $K_m$, similar to what one would observe if $e_0$ decreases is a definition of noncompetitive inhibition.

## Inhibitory Effects in Metabolic Systems

Competitive and uncompetitive inhibition are sufficiently similar in their effects in artificial experiments on isolated enzymes that they are often not distinguished, and an uncompetitive component in mixed inhibition often passes unnoticed. Many inhibitors are described in the literature as competitive inhibitors in the absence of any real evidence of the type of inhibition. This sort of confusion can easily lead to the entirely false idea that they are similar in their effects in systems where the inhibited enzyme is mixed with other enzymes and catalyses a step in the middle of a pathway.

In a typical experiment *in vitro*, one decides the concentrations of the various components in advance and measures the rate that results; however, this is very different from what happens in the cell. To a first approximation, an enzyme catalysing a step in the middle of a pathway must transform its substrate at the rate at which it arrives, that is, within certain limits it has little or no effect on the rate of its reaction, but instead determines the concentrations of the metabolites around it. (This is an oversimplification, but is useful for discussion.)

It is useful, therefore, to transform equations (3.13) and (3.14) into expressions for $a$ in terms of $i$:

$$a = \frac{vK_m(1 + i/K_{ic})}{k_0e_0 \quad v} \qquad \text{... (3.20)}$$

$$a = \frac{vK_m}{k_0e_0 \quad v(1 + i/K_{iu})} \qquad \text{... (3.21)}$$

However, similar equations (3.13) and (3.14) may seem, their transformed versions are drastically different. Equation (3.20) shows a linear dependence of $a$ on $i$, which means that increasing i can never result in uncontrolled increases in $a$. This is illustrated in Fig. 3.6(a). Even at an inhibition concentration equal to the inhibition constant, the substrate concentration is only doubled. By contrast, the curve defined by equation (3.21) is a rectangular hyperbola [Fig. 3.6(b)] that produces a steep and uncontrolled rise in substrate concentration at quite moderate inhibitor concentrations: The point is that in competitive inhibition, rises in substrate and inhibitor concentrations oppose one another—not only does the inhibitor compete with the substrate, but equally, the substrate competes with the inhibitor. In uncompetitive inhibition, however, these effects potentiate one another.

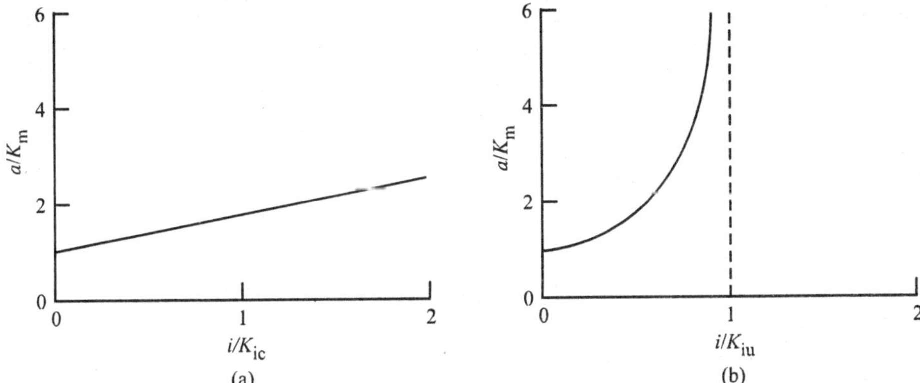


**Fig. 3.6.** Effects of (a) competitive and (b) uncompetitive inhibition on the concentration of substrate in a constant-rate system. Both curves are drawn for the case of $a = K_m$ in the absence of inhibitor. Note that both kinds of inhibitor have quantitatively equal effects at very low concentrations, but the initial slope is maintained indefinitely if the inhibition is competitive, whereas it rapidly becomes infinite if the inhibition is uncompetitive.

It follows that although it is relatively easy to find molecules that will act as competitive inhibitors, it is also largely useless as a strategy for designing pesticides or drugs because it is correspondingly easy for the organism to counteract the effect of the inhibition. To produce major metabolic effects one needs uncompetitive inhibitors, irreversible inhibitors, or tight-binding inhibitors: None of these are as easy to produce as weakly binding competitive inhibitors, but they are far more effective.

## NON-MICHAELIS-MENTEN BEHAVIOUR

All of the cases considered so far can be regarded as generalisations of the Michaelis-Menten equation (3.1). However, although many enzymes do behave in this way, at least as a first approximation, there are some important exceptions. It is simple to calculate from equation 3.1 that if $a = K_m/9$ then $v = 0.1V$ and if $a = 9K_m$ then $v = 0.9V$; in other words, spanning the 10–90 per cent range of available rates requires an 81-fold range of substrate concentrations, almost two orders of magnitude. Similar calculations may be done with any of the equations of the Michaelis-Menten type for additional substrates, inhibitors, or activators. Their implication is that as long as enzymes follow Michaelis-Menten kinetics, their rates cannot be adequately varied by manipulating concentrations of substrates, for example, because effective regulation will often require sensitivity to small changes in signals—certainly changes much smaller than two orders of magnitude.

A second difficulty arises from the fact that inhibition of the types considered commonly derives from structural similarities between inhibitors and substrates or products, whereas there is no reason to expect the molecules needed for metabolic signals to resemble the substrates or products of the enzymes that need to respond to the signals. In reality, the concentration of the end product of a pathway often serves as such a signal: too low and the pathway needs to be activated; too high and it needs to be inhibited. It is often found, therefore, that the enzyme that catalyses the first committed step of a pathway, that is, the first step after a branch point, in the branch that leads to the end product in question, is inhibited by that end product.

For inhibition of this kind to be possible, the enzyme must have a specific binding site for the end product, independent of the binding sites for substrates and products.

Such a site is called an allosteric site, and the phenomenon is called allosteric inhibition. Because the need for it often coincides with the need for higher sensitivity than is provided by Michaelis-Menten kinetics and the common kinds of inhibition, allosteric inhibition is often cooperative: This means that the equations that define it are more complicated than those considered above, allowing, for example, a change from 10 per cent to 90 per cent inhibited over a concentration range much smaller than 81-fold, and typically less than 10-fold (though rarely, if ever, less than 3-fold) (Fig. 3.7).

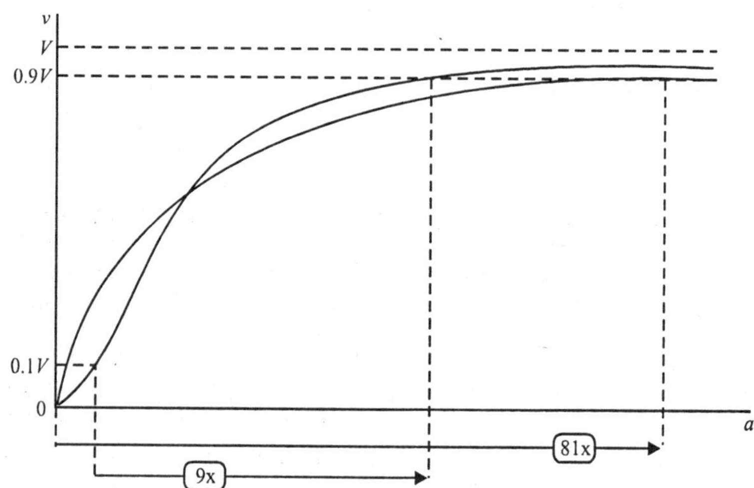


**Fig. 3.7.** Non-Michaelis-Menten kinetics. For an enzyme obeying the Michaelis-Menten equation ( Fig. 3.1), an 81-fold increase in substrate concentration is needed to bring the rate from 10 per cent to 90 per cent of *V*. If the enzyme shows positive cooperativity the curve typically becomes sigmoid (S-shaped), and this range of substrate concentrations is decreased (to nine-fold in the example, but in strongly cooperative cases it can be as small as three-fold).

The fact that allosteric and cooperative behaviour are often found together has led many authors to treat the two terms as synonymous, a tendency encouraged by the fact that in one of the most widely accepted models of non-Michaelis-Menten behaviour, that of Monod, Wyman and Changeux, both result from the same structural properties attributed to the enzyme. Nonetheless, most careful authors consider the two properties to be conceptually distinct and not necessarily occurring together, so the two terms should be considered distinct as well.

## KINETICS OF MULTIENZYME SYSTEMS

As noted in the introduction, nearly all kinetic studies of enzymes have been carried out using isolated enzymes, and although this has been very valuable for arriving at a good understanding of the nature of enzyme catalysis, it is quite inadequate as a guide to how systems of enzymes will behave. One cannot assume that the flux through a metabolic pathway is a property of a unique enzyme catalysing the rate-limiting step, and that the properties of the pathway as a whole can be deduced from studies of the kinetics of this one enzyme in isolation.

Space does not permit a full analysis of this subject, which has developed from a classic paper by Kacser and Burns, but it should suffice to examine an example of a pathway in which biotechnologists have attempted to increase the flux by identifying the rate-limiting enzyme and using genetic manipulation to increase its activity. Tryptophan biosynthesis in yeast provides such an example, tryptophan being a commercially valuable metabolite for which increased production would be very desirable. The tryptophan pathway consists of five enzymes, and in the classical model any one of these could be the 'key' enzyme catalysing the rate-limiting step. However, when the activity of each of these enzymes was increased in turn, either singly or at the same time as others in the pathway, the results were trivial: Increases of enzyme activity of 20-fold or greater produced flux increases of perhaps 30 per cent. Only when all five enzyme activities were increased (or all but one, apparently unimportant, activity) was there a substantial increase in flux, which even then was much smaller than the increase in enzyme activity.

The object here is not to analyse in detail why manipulation of tryptophan biosynthesis did not produce the desired results, but to use it to illustrate the point that the whole approach is misconceived. One cannot treat the kinetic behaviour of systems of enzymes as if it were determined by the properties of a single component. Moreover, abundant evidence exists to show that all organisms have evolved regulatory mechanisms to control metabolic fluxes and concentrations to satisfy their own requirements. Artificially trying to force more activity in a pathway by increasing the activities of certain enzymes simply stimulates the regulatory mechanisms to resist. Tryptophan biosynthesis is just one example of a general result and similar efforts in other pathways—for example, increased alcohol production in yeast and increased starch production in potatoes—have produced similar results.

The essential point is that flux control is not a property of a single enzyme in a pathway, but is shared among all the enzymes in the pathway. Strictly speaking, it is shared among all the enzymes in the cell, but it is usually safe to assume that enzymes catalysing reactions remote from the pathway of interest have very little effect on the flux through it, so to a first approximation one can consider flux control to be shared among the enzymes of the pathway.

To express this idea in quantitative terms, one can define a flux control coefficient for any enzyme by the following equation:

$$C_i^J = \frac{\partial \ln J}{\partial p} \bigg/ \frac{\partial \ln v_i}{\partial p} \qquad \text{... (3.22)}$$

This definition compares the effect on the flux J through a pathway of some perturbation of the activity of the *i*th enzyme; represented by a change in the parameter *p*, with the effect the same perturbation would have on the rate $v_i$ of the same reaction if it were considered in isolation. The identity of the parameter *p* does not have to be specified, because as long as it affects only one enzyme, the control coefficient defined by equation (3.22) is independent of the manner in which the flux and isolated rate are perturbed. However, to make the definition more concrete, consider the case where *p* is the logarithm of the enzyme concentration, $e_i$.

As most reactions are considered under conditions where the rate is proportional to the enzyme concentration, it will often be true to write $d\ln v_i = d\ln e_i$, so that the denominator in equation (3.22) has a value of unity and the whole equation simplifies to the following:

$$C_i^J = \frac{\partial \ln J}{\partial \ln e_i} \qquad \text{... (3.23)}$$

This equation is less abstract and simpler to understand than equation (3.22) and in the past was often used as a primary definition of a control coefficient. However, this is not recommended, first because it is not always true that the isolated rate is proportional to the enzyme concentration and second because equation (3.23) can give the false impression that control coefficients are concerned only with effects brought about by changes in enzyme concentration.

Flux control coefficients are measures of how much the flux through a pathway is dependent on the activities of the individual enzymes.

Mathematical analysis shows that they satisfy a property called the summation relationship:

$$\Sigma C_i^J = 1 \qquad \text{... (3.24)}$$

The interpretation of this relationship would be straightforward if all flux control coefficients were always positive, but in reality negative flux control coefficients also occur; for example, if a pathway contains a branch point, the enzymes in one branch normally have negative flux control coefficients for flux through the other. Nonetheless, in practice, negative flux control coefficients are nearly always small in magnitude, so although the sum of positive flux control coefficients may be greater than 1, it is rarely much greater than 1, so that the idea of sharing flux control among all the enzymes of a system is reasonably accurate. It follows that we should not expect any enzyme to have complete control over flux, and the closer to complete control any enzyme approaches, the less control all of the others have, taken together.

Moreover, a flux control coefficient is not a constant, but tends to decrease when the activity of an enzyme is increased. In other words, even if one can identify one enzyme that has a large proportion of the total flux control, increasing its activity will tend to decrease that proportion, so that the amount the flux can be increased by increasing the activity of one enzyme will always be small. Once this is understood, the failure to increase the flux to tryptophan (in the example mentioned) or to achieve significant flux increases (in other similar examples), ceases to be a mysterious result, but is simply what was to be expected.

Chapter 4

# Immobilised Enzyme Technology

## INTRODUCTION

Nelson and Griffin found that an enzyme in water-insoluble form showed catalytic activity. They reported that invertase extracted from yeast was adsorbed on charcoal and the adsorbed enzyme showed the same activity as native enzyme. Sumner found that urease from jack bean became water insoluble on standing in 30 per cent alcohol and sodium chloride for 1 to 2 days at room temperature and that the water-insoluble urease was active.

Thus, it has been known for a long time that enzymes in water-insoluble form show catalytic activity. However, the first attempt to immobilise an enzyme to improve its properties for a particular application was not made until 1953, when Grubhofer and Schleith immobilised enzymes such as carboxypeptidase, diastase, pepsin and ribonuclease by using diazotised polyaminopolystyrene resin.

On the other hand, before this, immobilisation of physiologically active protein was carried out by Micheel and Ewers. Further, Campbell prepared immobilised antigen by binding albumin to diazotised *p*-aminobenzylcellulose.

Subsequently, a number of reports on the preparation and application of immobilised antigens and antibodies appeared. Such studies may also be applicable to immobilised enzymes.

Immobilised enzymes are defined as 'enzymes physically confined or localised in a certain defined region of space with retention of their catalytic activities and which can be used repeatedly and continuously'. Accordingly, enzymes modified to a water-insoluble form by suitable techniques satisfy this definition of immobilised enzymes. In addition, when an enzyme reaction using a substrate of high molecular weight is carried out in a reactor equipped with a semipermeable ultrafiltration membrane without leakage of the enzyme from the reactor, this can also be considered as a kind of immobilised enzyme system.

As functional groups involved in enzyme immobilisation, free, amino and carboxyl groups, the sulfhydryl group of cystein, the imidazole group of histidine, phenolic groups and hydroxyl groups of serine and threonine may be considered. For immobilisation of an enzyme, it is necessary that functional groups in the active center should not be involved in the reaction leading to immobilisation of the enzyme.

Further, because the tertiary structure of enzyme protein is maintained by relatively weak binding forces, such as hydrogen, hydrophobic and ionic bonds, it is necessary to carry out the immobilisation reaction under mild conditions.

Methods for enzyme immobilisation can be classified into three basic categories as follows:

1. Carrier-binding method: the binding of enzymes to water-insoluble carriers such as polysaccharide derivatives, synthetic polymers, and porous glass, etc.
2. Cross-linking method: intermolecular cross-linking of enzymes by means of bifunctional or multifunctional reagents such as glutaraldehyde, bisdiazobenzidine and hexamethylene diisocyanate, etc.
3. Entrapping method: incorporating enzymes into the lattice of a semipermeable gel on enclosing the enzymes in a semipermeable polymer membrane, such as collagen, gelatine cellulose triacetate polyacrylamide, and κ-carrageenan, etc.

These methods are shown schematically in Fig. 4.1. The carrier-binding method is the oldest immobilisation method for enzymes, and many papers have been published on this method. When enzymes are immobilised in this way, care is required regarding the selection of carriers as well as in binding techniques. The carrier-binding method can be further divided into three categories according to the binding mode of the enzyme, that is, physical adsorption, ionic binding, and covalent binding.

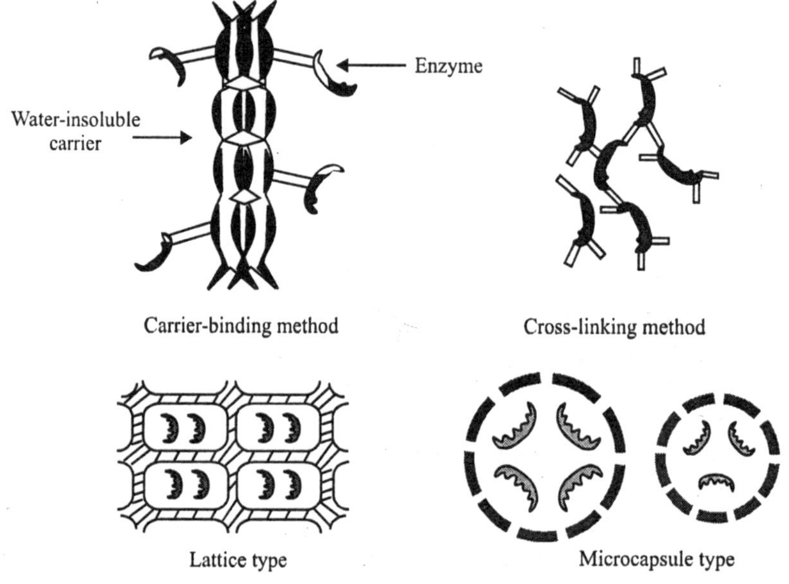


**Fig. 4.1.** Schematic diagrams of immobilised enzymes.

The cross-linking method is based on the formation of chemical bonds, but water-insoluble carriers are not used in this method. The immobilisation of enzymes is performed by the formation of intermolecular cross-linkages between the enzyme molecules by means of bifunctional or multifunctional reagents.

The entrapping method is based on confining enzymes in the lattice of a polymer matrix or enclosing enzymes in semipermeable membranes, and it can be classified into the lattice and microcapsule types. This method differs from the covalent binding and cross-linking method in that the enzyme itself does not bind to the gel matrix or membrane. Thus, this method may have wide applicability. However, if a chemical polymerisation reaction is used for entrapping, relatively severe conditions are required and lose of enzyme activity occurs in some cases. Therefore, it is necessary to select the most suitable conditions for the immobilisation of various enzymes.

Preparation methods and characteristics of immobilised enzymes are summarised broadly in Table 4.1, though there are many exceptions.

**Table 4.1.** Preparation and characteristics of immobilised enzyme.

| *Characteristic* | *Carrier-binding method* | | | *Cross-linking method* | *Entrapping method* |
|---|---|---|---|---|---|
| | *Physical adsorption* | *Ionic binding* | *Covalent binding* | | |
| Preparation | Easy | Easy | Difficult | Difficult | Difficult |
| Enzyme activity | Low | High | High | Moderate | High |
| Substrate specificity | Unchangeable | Unchangeable | Changeable | Changeable | Unchangeable |
| Binding force | Weak | Moderate | Strong | Strong | Strong |
| Regeneration | Possible | Possible | Impossible | Impossible | Impossible |
| General applicability | Low | Moderate | Moderate | Low | High |
| Cost of immobilisation | Low | Low | High | Moderate | Low |

Immobilisation of enzymes by covalent binding or cross-linking is carried out under relatively severe conditions in comparison with those of physical adsorption or ionic binding. Thus, in the former cases, conformational change of the enzyme structure and partial destruction of the active center may occur. Accordingly, unless immobilisation of an enzyme by covalent binding is carried out under well-controlled conditions, immobilised enzyme having high activity cannot be obtained. However, the binding forces between the enzyme and carrier are strong, and the enzyme cannot easily be lost from carriers even in the presence of substrates or salts at high concentrations. However, when the activity of enzymes immobilised by covalent binding decreases on long-term operation, regeneration is impossible.

On the other hand, immobilisation of enzymes by the ionic-binding method can be achieved simply under mild conditions. Accordingly, in this case, preparations having relatively high activity are obtained. However, the binding forces between enzyme and carrier are weak in comparison with those in the covalent-binding methods. Therefore, leakage of the enzyme from the carrier may occur after changes in the ionic strength, pH of the substrate, or product solution. This type of immobilised enzyme can be regenerated when the enzyme activity decreases after prolonged operation. Thus, the ionic-binding method is advantageous in comparison with the covalent-binding method, particularly when expensive carriers or enzymes are used.

Unlike the ionic-binding, covalent-binding and cross-linking methods in the entrapping method, no binding between enzyme and carrier should occur in theory. Therefore, in many cases, preparations having high activity are obtained. However, in this method, regeneration of activity losses is impossible, as in the case of the covalent-binding method. A major disadvantage of the entrapping method is that its application is limited to small molecular substrate and product molecules; the entrapped enzyme shows little or no activity towards macromolecular substrates.

Although a number of enzyme immobilisation methods have been studied, no ideal general methods applicable for the immobilisation of many enzymes have yet been developed. Each method has specific disadvantages. Therefore, in practice, it is necessary to find a suitable method and conditions for the immobilisation of a particular enzyme in the light of the intended application.

Because of the special characteristics of immobilised enzymes, a considerable amount of work has been carried out on application of immobilised enzymes as solid catalysts having high substrate specificity. Immobilised enzymes are used in various fields, such as chemical processes, analysis, medical treatment, food processing, affinity chromatography and so forth. For the efficient utilisation of such immobilised systems, chemical engineering studies are necessary.

The use of enzymes in food as well as in pharmaceutical and chemical industries has increased steadily since the early 1970s. Applications range from synthesis in the milligram scale for natural or non-natural oligosaccharides up to industrial synthesis for the production of bulk chemicals such as acrylamide. Some examples are listed in Table 4.2. This section highlights some of the most important aspects for the application of immobilised enzymes and presents selected examples.

**Table 4.2.** Examples of the industrial use of immobilised enzymes.

| *Enzyme* | *Substrate* | *Product* | *Scale (tonne $a^{-1}$)* |
|---|---|---|---|
| Penicillin amidase | Penicillin G | 6-Amino penicillanic acid | 6000 |
| Cephalosporin amidase | Glutaryl-7-ACA | 7-ACA | – |
| Amino acylase | Acyl-D-L-amino acid | L-amino acid | Approximately 300 |
| Aspartase | Fumaric acid | L-aspartic acid | 1200 |
| Fumarase | Fumaric acid | L-malic acid | 360 |
| Lactase (galactosidase) | Lactose GalGlc | Low lactose milk Gal + Glc | – |
| Aspartase β-decarboxylase | Aspartic acid | L-alanine | 120 |
| Nitrile hydratase | Acrylonitrile | Acrylamide | 30000 |
| Glucose isomerase | Glucose | High fructose corn syrup | 80,00,000 |
| Lipase (expected) | Rac-1-phenylethyl-amine | (S )-1-phenylethyl-amine | 200 |

The most common definition for immobilised enzymes that proposed by Katchalski-Katzir is: 'Enzymes physically confined or localised in a certain defined region of space with retention of their catalytic activities, which can be used repeatedly and continuously'. According to this definition, three types of immobilised enzymes can be distinguished:

1. Heterogenisation of the soluble enzyme by coupling to an insoluble support by adsorption or covalent binding, by cross-linking of the enzyme or entrapment in a lattice or in microcapsules such as alginate beads.
2. Retention of the enzyme by means of ultrafiltration membrane.
3. Use of whole cells for biotransformations using their enzyme apparatus.

Here because of the importance the first type is considered. Membrane reactors have been reviewed quite recently, as have applications of whole cell processes. Retention of the cells within the reactor may be achieved by membrane separation or by the same immobilisation methods that are used for isolated enzymes. In principle, the cell itself can be regarded as a form of native immobilisation of enzymes.

The major goal behind immobilisation (in its most common definition) is the recovery of the biocatalysts, separation from products and reactants and subsequent reuse in batch or continuous processes. The latter is especially important for a reduction of the catalyst costs. Enzymes immobilised on a support often show enhanced stability when compared with the soluble form. When considering the use of soluble or carrier-fixed enzymes, the following topics have to be addressed:

1. Additional costs for support and chemicals performing the immobilisation have to be balanced against the increase of stability.
2. Loss of activity during the immobilisation step.
3. When the catalyst is immobilised only by adsorption or entrapment without covalent attachment, its leakage from the carrier support has to be examined and compared with the overall deactivation rate.
4. Mass transfer limitations for enzymes on a support may cause problems when adjusting of the pH is necessary during the reaction.

5. With soluble enzymes, higher volumetric activities at high catalyst concentrations are possible, enabling conversion of poor substrates at reasonable rates.
6. Whereas membrane reactors can be easily sterilised before use, this is not possible for reactors with carrier-fixed enzymes. To prevent microbial contamination, these processes are quite often operated at higher temperatures.
7. When enzymes are to be used together with organic solvents to increase reactant or product solubility or to alter their kinetics, it may become necessary to immobilise them on a support. The support will at the same time act as a water pool to maintain the enzymatic activity. In such systems, water-insoluble organic solvents have less effect on the enzyme stability than water-soluble solvents.

Table 4.3 summarises some arguments in favour or against immobilisation of biocatalysts, namely enzymes. The final decision for a certain reactor design should be based on an optimisation process covering all relevant factors contributing to the overall costs, including investment, catalyst consumption, or productivity.

**Table 4.3.** Comparison of processes using soluble or carrier-fixed enzymes.

| *Enzyme fixed on a carrier or membrane* | *Soluble enzyme retained by a membrane* |
|---|---|
| Continuous or batch process | Continuous or batch process |
| Improved stability | Lower enzyme stability |
| Loss of activity during immobilisation | No loss of activity during immobilisation |
| Immobilisation costs | No costs for immobilisation |
| Mass transfer limitations (heterogeneous reaction) | No mass transfer limitation (homogeneous reaction) |
| Established technology[a] | New technology[b] |
| | Sterile process possible |
| Higher investment costs | Easy supply of new enzyme |

[a]Immobilisation protocol may be different for each enzyme.
[b]Can be used for different enzymes without modification.

## REACTORS FOR IMMOBILISED ENZYMES

The types of reactors used for immobilised enzymes are summarised in Fig. 4.2. The principles developed for general heterogeneous catalysis in synthetic chemistry are valid, resulting in well-known reactor configurations. Differences between enzyme catalysis and other systems result from the nature of the biocatalyst and reaction medium. For example, soft particles containing the biocatalyst, such as alginate beads, may limit the pressure drop in fixed-bed reactors.

The decision as to specific reactor design will be based on a careful analysis of the kinetic properties of the reaction system. For example, if the enzyme shows a strong substrate-surplus inhibition, a continuously operated reactor with complete backmixing working at high conversion is advantageous. A reaction with strong product inhibition may utilise a batch reactor or a plug flow reactor to achieve higher volume and catalyst specific productivities. An extractive bioreactor may be used if substrates and products show different solubilities [Fig. 4.2(i) and Fig. 4.2(j)]. By using this reactor configuration, the destabilising effect of organic solvents may also be overcome, because the enzyme is separated from the organic phase, which is used to extract the insoluble product. The aqueous phase containing the enzyme will be saturated until the maximum solubility of the substrate is reached.

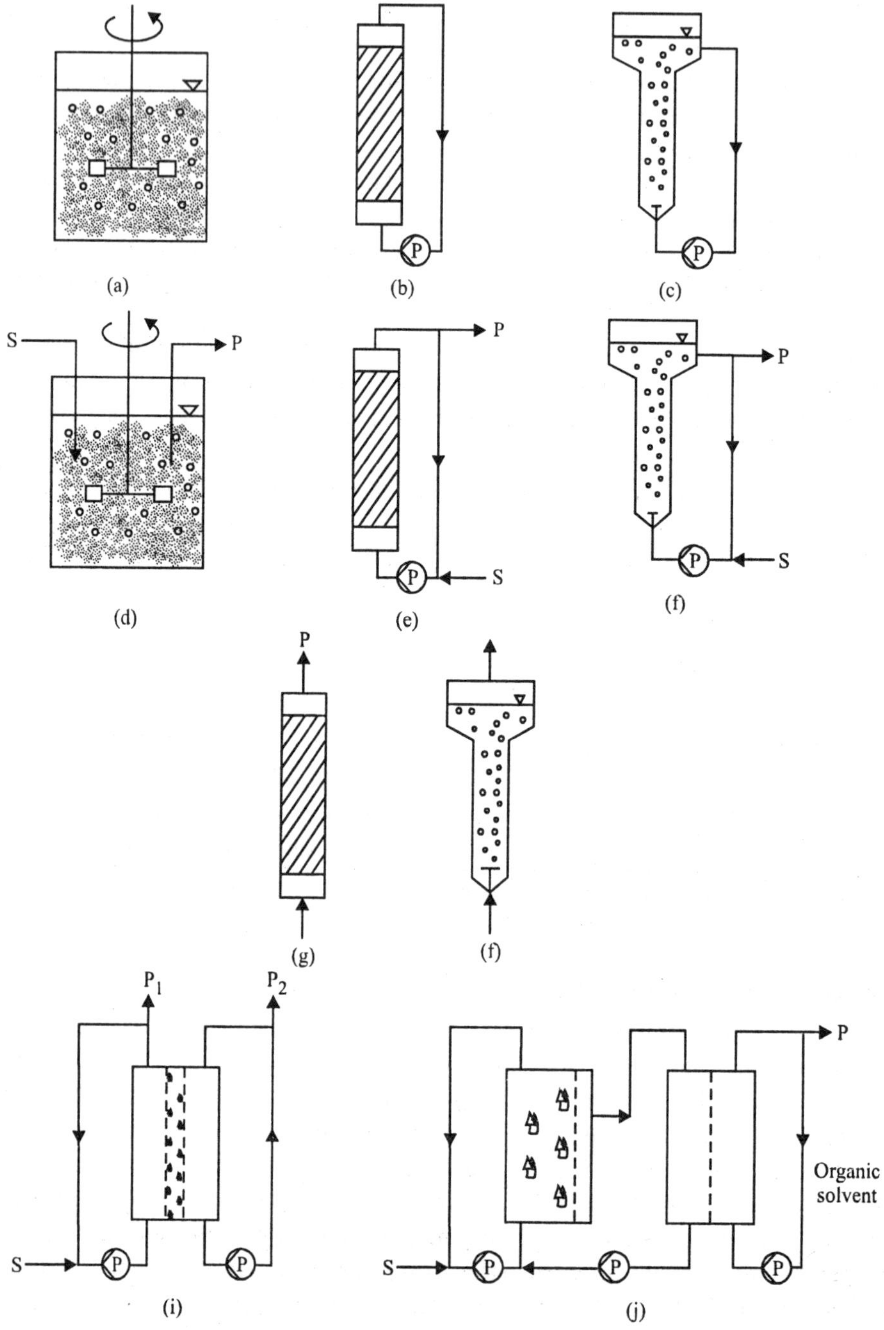


**Fig. 4.2.** Reactors for immobilised enzymes. (a-c) Batch reactors with complete backmixing; (a) stirred-tank reactor; (b) fixed-bed reactor; (c) fluidised-bed reactor; (d-f) continuously operated reactors with complete backmixing; (g-h) continuously operated reactors with plug-flow behaviour; (i) reactor with the enzyme immobilised in or on a membrane that may at the same time separate two phases such as water and organic solvent; and (j) reactor with physically separated enzyme and organic solvent in order to prevent denaturation of the protein (the membrane reactor shown may be replaced by one of the types for immobilised enzymes).

Reactions using biocatalysts are normally performed in aqueous solution at temperatures between 10°C and 80°C and at ambient pressure. Due to the inhibition of some enzymes by heavy metals, the materials of construction must not release these elements. Reactors are operated under conditions that prevent microbial contamination. The reactor itself as well as the substrate may be sterilised prior to reaction by using chemical agents (ethanol, formaldehyde, ethyleneoxide, Velcorin) or steam. Ultraviolet rays or γ-rays may be used to sterilise the immobilised enzyme on its support. Alternatively, the immobilisation may be performed under sterile conditions. Antibacterial agents may be added to the reaction mixture to prevent microbial growth while the reactor is running. In some cases, the reactants may act as sterilants or inhibitors of microbial growth, such as ketones or alcohols. At higher concentrations (more than 500 mmol/l), solutions may become autosterile because of osmotic pressure effects. Industrial processes are often performed at elevated temperatures, above 55°C, reducing the danger of microbial contamination.

For a constant product quality and reproducibility of downstream processing, the reactor should be operated at constant conversion. To overcome the deactivation per unit of time that shows all biocatalysts as a result of denaturation processes, either the residence time has to be increased or fresh enzyme has to be supplied. The latter is especially easy for soluble enzymes. For carrier-fixed enzymes, a combination of both methods is used, as discussed later.

## SELECTED EXAMPLES OF REACTIONS USING IMMOBILISED ENZYMES

One of the first enzymes shown to be active after immobilisation by adsorption on charcoal was invertase. This enzyme is still used for the large-scale production of invert sugar by hydrolysis of sucrose. One kilogram of immobilised enzyme supports production of approximately 10000 kg of invert sugar.

### Production of L-Amino Acids Using Amino Acylase

Enantiomerically pure L-amino acids are required for use in parenteral nutrition, as food additives or as chiral synthons. In addition to isolation from natural sources and direct fermentation, racemic mixtures of L- and D-amino acid can be resolved by enzymatic reaction with one of the two enantiomers. Figure 4.3 illustrates this for racemic amino acid using amino acylase I from *Aspergillus*. The unreacted D-enantiomer can be racemised easily and recycled to increase the overall yield.

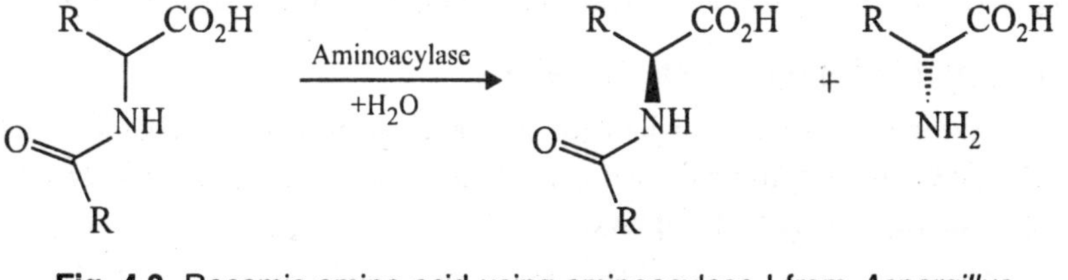

R, $CO_2H$, NH, O, R —Aminoacylase, $+H_2O$→ R, $CO_2H$, NH, O, R + R, $CO_2H$, $NH_2$

**Fig. 4.3.** Racemic amino acid using aminoacylase I from *Aspergillus*.

### Production of 6-Amino Penicillanic Acid and 7-Amino Cephalosporanic Acid Using Amidases

6-Amino penicillanic acid (6-APA) and 7-amino cephalosporanic acid (7-ACA) are important intermediates for semisynthetic β-lactam antibiotics. Both compounds can be obtained readily by amidase-catalysed cleavage of the corresponding amides that are accessible by fermentation.

Whereas penicillin amidase has been used for production of 6-APA since the early 1970s, cephalosporin C amidase has been used just recently in a chemoenzymatic synthesis of 7-APA. Amidases have been prepared from several different sources, such as *Pseudomonas*, *Bacillus* and (recombinant)

*Escherichia coli*. In general, the enzyme is immobilised on and in various supports such as polyacrylonitrile fibers, polyacrylamid beads. For 6-APA batch production, typically a 10 per cent solution of penicillin G is circulated through the bioreactor or processed in a stirred-tank reactor. One cycle lasts about 3 hours. The lifetime for the enzyme is more than 360 cycles. The space-time yield for this process is about 800 g $L^{-1}$ $d^{-1}$. One of the main problems for both processes is the liberation of a strong acid. To avoid large pH shift, an effective method to control and maintain the desired pH is needed. High recirculation rates are used for the enzyme immobilised on hollow fibers. For the enzyme immobilised in beads, both a small bead diameter (typically around 100 to 600 μm) and good control of film and pore diffusion by high fluid velocities are needed.

### Production of High-Fructose Corn Syrup Using Glucose Isomerase

The production of high-fructose corn syrup, a mixture of glucose and fructose, is the largest enzymatic process in commercial practice so far utilising immobilised biocatalysts as shown in Fig. 4.4. The enzyme is available from different bacterial sources. By site-directed mutagenesis, more thermostable mutants have been obtained. Most of the enzyme preparations commercially available are cross-linked with glutaraldehyde. The enzyme is used in fixed-bed reactors and the feed flow is adjusted to overcome enzyme deactivation and ensure a constant product quality (conversion). Different columns of different enzyme age are operated in parallel, resulting in an average conversion of 45 per cent in a process of uniform throughput. The equilibrium conversion is about 50 per cent. If the residual activity has dropped to about 10 per cent of the initial activity, the enzyme is replaced. For the most-stable commercially available enzyme preparations, the half-life is more than 100 days. Typical temperatures range from 50°C to 60°C with 50 per cent substrate solution.

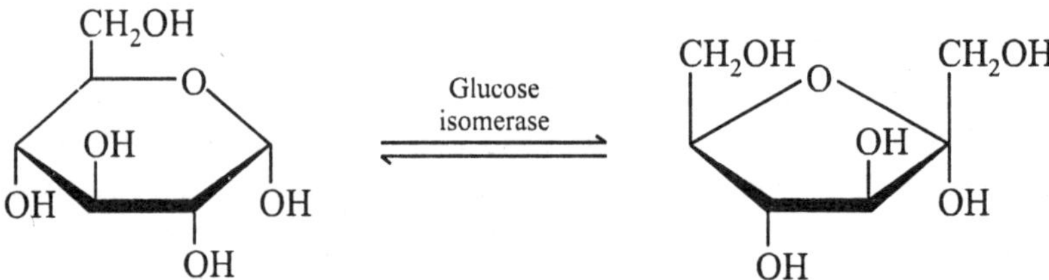


**Fig. 4.4.** Production of high-fructose corn syrup using glucose is isomerase.

### Production of Acrylamide Using Nitrile Hydratase

Biotransformations have been also established for the production of bulk chemicals. The enzymatic production of acrylamide from acrylonitrile using nitrile hydratase from *Rhodococcus rhodochrous* has replaced the copper-catalysed process in Japan (Fig. 4.5).

CN —Nitrile hydratase, $+H_2O$→ $NH_2$, O

**Fig. 4.5.** Enzymatic production of acrylamide from acrylonitrile using nitrile hydratase from *Rhodococcus rhodochrous.*

For this process, whole cells were immobilised in a cationic acrylamide-based polymer gel. Typical reaction conditions range from 0°C to 15°C and pH 6.5 to 8.0. Acrylonitrile is added gradually in order to achieve a final concentration of acrylamide of 400 g $L^{-1}$. The conversion of acrylonitrile is more than 99 per cent, with a selectivity for acrylamide of more than 99 per cent. Because of its broad substrate specificity, the nitrile hydratase may be also used for production of nicotinamide, isonicotinamide or pyrazinamide.

### Production of Glycerol-3-Phosphate Using Glycerol Kinase

Glycerol-3-phosphate (G3P) is a useful starting material for the synthesis of phospholipids. These compounds are of importance for numerous medicinal, diagnostic and research applications. Their potential applications include microemulsion and liposomal drug delivery systems. The enzymic synthesis of G3P using glycerol kinase has been commercialised by Genzyme. Adenosine triphosphate (ATP) is used as the phosphorylating agent and regeneration of ATP is achieved by pyruvate kinase (Fig. 4.6) using phosphoenol pyruvate (PEP).

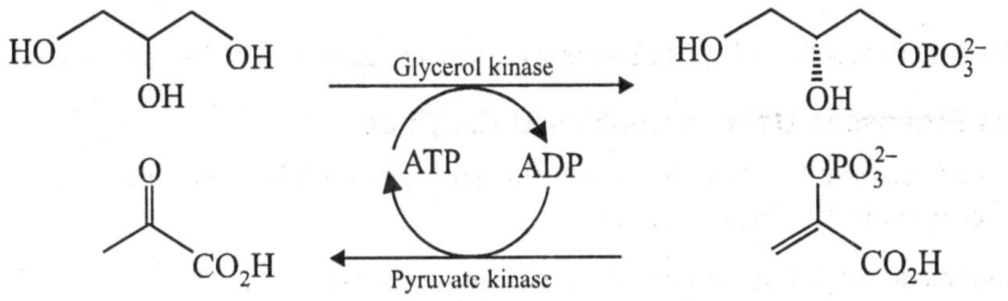

**Fig. 4.6.** Production of glycerol-3-phosphate using glycerol kinase.

The enzymes used are immobilised on a solid support. Especially for the glycerol kinase the stability was increased dramatically when compared with the native form. Its half-life in aqueous solution is about 15 minutes, because of dissociation of the subunits. The reaction is typically performed in a batch reactor containing the immobilised enzyme in a fixed bed. Multikilogram-scale production typically uses 1 mol $L^{-1}$ glycerol and 1 mol $L^{-1}$ phosphoenol pyruvate. For this process, space-time yield is about 81 g $L^{-1}$ $d^{-1}$. Catalytic amounts of ATP (1 to 10 mol%) are used, resulting in total turnover numbers up to 100. These and even higher values are attainable without requiring coimmobilisation of the cofactors. NADH values of about 1000 may be reached easily. These processes are economically feasible because of the low cost of the cofactors.

### Production of (2R,3S)-Methyl-*p*-Methoxyphenylglycidate Using a Lipase

Optically pure (2R,3S)-methyl-*p*-methoxyphenylglycidate is used as chiral building block for the synthesis of diltiazem. Sepracor has developed a process for enantio-selective hydrolysis of the racemic ester using a lipase, which has reached a 50 t $a^{-i}$ scale as shown in Fig. 4.7. The reaction is performed in a multiphase membrane reactor with the enzyme reversibly entraped within the porous matrix of an asymmetric hollow fiber ultrafiltration membrane. This reactor is shown in Fig. 4.1(i). An aqueous buffer is circulated on one side of the hydrophilic membrane, providing an aqueous environment for the enzyme. A 17 per cent solution of the racemic ester dissolved in toluene is circulated on the other side of the membrane. Despite its lower enantioselectivity, *Candida cylindracea* lipase is being used; lipases from other sources were not retained sufficiently within the membrane. The reactor is used in a batch mode. The space-time yield, based on the membrane area, is about 125 g $m^{-2}$ $d^{-1}$.

This type of reactor has also been used for the hydrolysis of butter oil or the production of peroxycarboxylic acids.

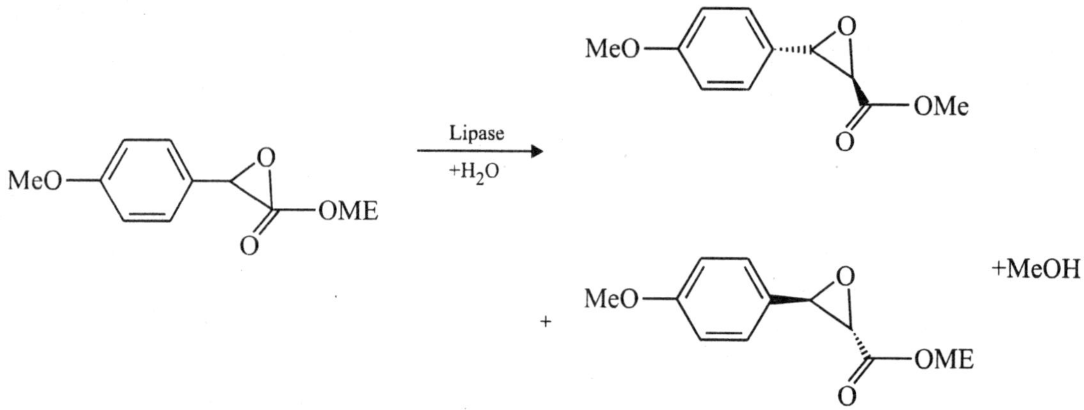

**Fig. 4.7.** Production of (2R,3S)-methyl-*p*-methoxyphenylglycidate using a lipase.

## Miscellaneous Processes Using Immobilised Enzymes

Table 4.4 lists some reactions that have been performed using immobilised enzymes to give an impression of the range of enzymatic biotransformation.

**Table 4.4.** Further examples for application of immobilised enzymes.

| *Enzyme* | *Substrate* | *Product* |
|---|---|---|
| Glucose kinase | Glc, ATP | Glucose-6-phosphate |
| Oxynitrilase | Aldehydes, HCN | Cyanohydrines |
| Transaminase | Ketone | Amine |
| Pyranose oxidase | Sugar | Keto-sugars |
| Hydantoinase | Hydantoin | Amino acid |
| Oxidoreductase | Benzoylformate | Mandelate |
| Sialidase | Neu5Ac, Gal | Neu5Ac-Gal |
| Glucosidase | Glc, octanol | Octylglycoside |
| Chymotrypsin | Amino acids | Oligopeptides |
| Glycosyltransferase | Carbohydrate | Oligosaccharide |

Chapter 5

# Industrial Applications of Enzymes

## INTRODUCTION

Enzymes are natural catalysts that have made a substantial impact on the food processing and food ingredients industries over the past 25 years. This has occurred because of the high catalytic specificity of enzymes and because they are considered safe and can be contained within finished food products. The specificity of the reactions catalysed by enzymes enhances further the purity of the finished products. Early commercial enzymes were simple hydrolases, such as amylase for the starch industry. Now, many much cleverer enzymes are becoming available such as glucose isomerase, glucose oxidase, alcohol oxidases etc.

Enzymes are being utilised in diverse products like confectionary, beer, dairy products, food proteins, fat oils, marine products etc.

Enzymes chosen for a specific process must be selected to assure their suitability for the given technical and commercial application.

To summarise the various reaction parameters:

1. The pH must be adjusted to meet enzyme characteristics; or when the process pH cannot be changed, the appropriate enzyme must be selected.
2. Reaction velocity can be influenced by the reaction temperature. Operating at temperatures just below inactivation reactions can be significantly accelerated. Enzymatic activity, however, can still occur under refrigeration.
3. If, for technical reasons, the reaction time is fixed, the reaction can be governed by the temperature or by the amount of enzyme employed.

Figure 5.1 illustrates the interaction of reaction parameters in practice. Table 5.1 lists some of the major technical enzyme applications.

## FOOD INDUSTRY

### Enzymes in the Starch and Sugar Industries

Starch itself, as a natural product, is relatively insignificant in everyday use. At one time, starching shirts and shirt collars (i.e., the use of starch in laundering) had widespread application. With time, this practice has lost importance and now when starch is mentioned, it is usually listed as an additive in puddings and sauces and more recently, as a basic commodity for the chemical industry.

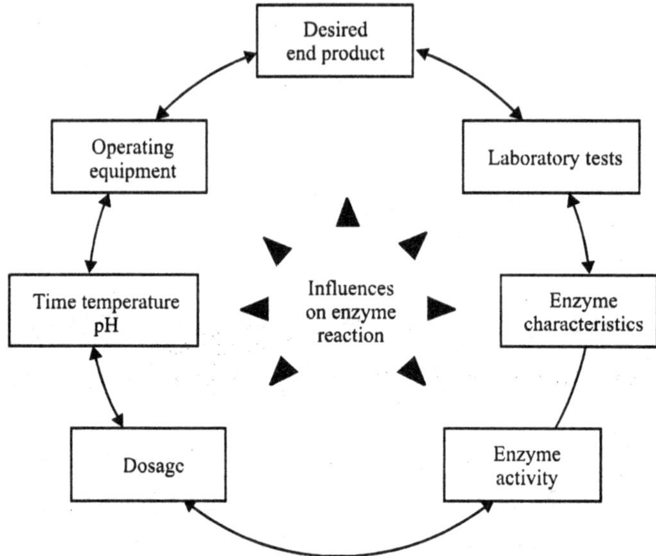


**Fig. 5.1.** Interaction between reaction parameters.

**Table 5.1.** Survey of technical enzyme applications.

| *Process or product* | *Enzyme preparations* |
|---|---|
| Alcohol | α-Amylases, β-amylases, amyloglucosidase, β-glucanase, proteases, cellulases |
| Amino acids | Amino acid acylase (immobilised), proteases with peptidases |
| Animal feed | Proteases, amylases, hemicellulases, β-glucanases |
| Beer | α-Amylases, β-glucanases, amyloglucosidase, proteases |
| Berries | Pectinases |
| Bread | α-Amylases, β-amylase, amyloglucosidase, fungal proteases pentosanases |
| Bread preservation | Bacterial α-amylase, branching enzyme, β-glucanases, pentosanases, β-amylase |
| Cheese ripening | Lipases, proteases, peptidases |
| Cheesemaking | Animal and microbial rennet, lysozyme |
| Chicken feed | β-Glucanases |
| Chitin breakdown | Chitinase |
| Cider | Clarification: pectin esterase |
| Citrus juices | Lowering viscosity or clarification: pectinases |
| | Eliminating bitterness: naringinase |
| Citrus slicing | Cellulases |
| Cleaning lipid traps | Lipases, proteases |
| Coffee | Extracts: β-glucanases |
| Confections | Invertase |
| Cookies, crackers | α-Amylases, proteases |
| Dential hygiene | Dextranase, mutase, glucanases, peroxidases |
| Detergents | Proteases, amylases |

*(Cont'd ...)*

| *Process or product* | *Enzyme preparations* |
|---|---|
| Dextran breakdown | Sugar industry: dextranase |
| Dextrins | From starch: α-amylases |
| Egg products | Preservation: glucose oxidase, catalase |
| | Whippability: lipases |
| Fish curing | Pancreatin |
| Fish protein | Lowering viscosity of water extracts: proteases |
| Flavours | From fruits: pectinases, hemicellulases, cellulases |
| | From milk products: lipases, proteases |
| | In baked goods: lipases |
| Flax processing | Pectinases, macerases |
| Flour processing | At the mill: fungal α-amylase |
| | Malt flour: proteases, lipoxygenase |
| Fructose production | From inulin: inulinase |
| | From glucose: glucose isomerase |
| Fruit juices | Pectinase, hemicellulases, cellulases, amylases, amyloglucosidase, arabanase |
| Fruit, vegetable pulps | Macerases |
| Gelatine hydrolysis | Proteases |
| Glucose production | From starch: α-amylases, amyloglucosidase |
| Glucuronic acid production | Glucose oxidase, catalase |
| Invert-sugar production | Invertase |
| Lactose | Milk hydrolysis: yeast lactase |
| Laundry detergents | Proteases, cellulases, lipases, amylases |
| Leather manufacture | Proteases, carbohydrases, lipases |
| Lipid hydrolysis | Lipases |
| Lipid transesterification | Lipases |
| Lipid separation | Proteases |
| Maceration | Plants: pectinases, hemicellulases, amylases |
| Maize gluten | Isolation: cellulases, β-glucanases |
| Maltose syrup | α- and β-amylases |
| Meat tenderising | Proteases, especially papain |
| Meat trimmings | Proteases with peptidases |
| Milk | Cold sterilisation: $H_2O_2$/catalase |
| Oils | Olive and citrus oil production: pectinases |
| | Hydrolysis, transesterification: lipases |
| Phytic acid, phytate | Hydrolytic liberation of phosphate and calcium: phytase |
| Plant pigments | Anthocyanin production—pectinases |
| Removing bitterness | From citrus juices |
| | From protein hydrolysates: peptidases |
| Removing sizing | Textiles: α-amylases, cellulases |
| | Films: bacterial proteases |
| Rubber | Proteases |

*(Cont'd ...)*

| *Process or product* | *Enzyme preparations* |
|---|---|
| Silage | Grasses; cellulases, β-glucanases |
| Soft drinks | Stabilisation: glucose oxidase |
| Soyabean hydrolysis | Proteases, α-glucosidase |
| Tea | Extraction: tannase |
| Wafers | Bacterial proteases |
| Waste-water | Proteases, cellulases, complex enzyme preparations |
| Whey processing | Lactose hydrolysis: lactases |
| | Whey protein modification: proteases |
| Winemaking | Maceration: pectinases |
| | Clarification: pectinases, proteases, β-glucanases |
| Yeast hydrolysis | β-glucanases for lysing cell walls |

Generally, most people are unaware of the incredibly broad spectrum of industrial processes involving the utilisation of starch and starch products. A few examples of processing and use can be cited:

'The isolation, modification and digestion of purified starch for the food industry (usually for desserts, sauces, dressings, fillings, soups, ready-to-serve dishes, gum and jelly products); for the production of flours, baked goods, breakfast cereals, beer and other alcoholic beverages; and in the manufacture of paper, textiles, adhesives and for other technical applications. Recently, starch has become an important raw material for the chemical industry.'

Starch, second only to cellulose, is the most abundant polysaccharide in nature and the most important in industrial utilisation. It represents 0.05 per cent of the total biomass. Only about 2 per cent of the total available starch, present primarily as starch granules isolated from the plant biomass, is purified and processed further. Several of the most interesting features of starch are its insolubility at room temperature, its ability to solubilise when heated and the forming of a gel, depending on the concentration, when its 'pasting point' is reached. The temperature varies with the type and source of the starch.

To understand the following, it is important to note that starch modification or conversion, whether performed chemically or enzymatically, can generally be accomplished only in the gelatinised state and that the desired product properties can be achieved only under these conditions.

It should also be pointed out that the starch granules generated during the technical extraction processes will contain, in spite of costly separation and washing steps, not only amylose and amylopectin but also a series of companion compounds that are not carbohydrates yet are constitutive parts of the granules. These include lipids and phospholipids, fatty acids, proteins, celluloses and hemicelluloses (pentosans, β-glucans), phosphoric acid (especially in potatoes), minerals and of course, water. The amount of such starch 'impurities' are normally quite small and will vary with the starch source as shown in Table 5.2. These substances can, nevertheless, play a significant role in determining the success or failure of a large-scale technical operation employing enzymatic starch conversion.

Until recently maize was the most important cereal for starch production because the isolated raw material was best suited for enzymatic or other conversion processes. This, however, meant that European industries have had to import and stockpile expensive dried maize of precisely defined quality. In addition, the grain had to be softened in soaking stations before separation into the technically important fractions was possible. The machinery used for separating maize embryos, pericarps and silk as well as the maize protein (gluten) includes mills, cyclone separators, jet and other washers, decanters, centrifuges, presses and dryers. Valuable by-products of corn processing are corn oil, maize gluten and feed mixes.

**Table 5.2.** Composition of various starches.

| *Component* | *Maize* | *Waxy Maize* | *Wheat* | *Rice* | *Potato* | *Tapioca* |
|---|---|---|---|---|---|---|
| Amylose (% of starch) | 21–30 | 0–1 | 19–25 | 17–19 | 18–23 | 17–19 |
| Amylopectin (% of starch) | 70–79 | 99–100 | 75–81 | 81–83 | 77–82 | 81–83 |
| Lipids (% of dry solids) | 0.4–0.8 | 0.1–0.2 | 0.8 | 0.8 | 0.1 | 0.1 |
| Protein (% of dry solids) | 0.35 | 0.2–0.3 | 0.4–0.8 | 0.45 | 0.1 | 0.1 |
| Ash (% of dry solids) | 0.1 | 0.07 | 0.15–0.3 | 0.5 | 0.2 | 0.2 |

One commercial process for producing wheat starch is shown in Fig. 5.2. First, the wheat is dry-milled to flour; then, after adding water and intensive kneading, relatively expensive but effective fractionation operations follow.

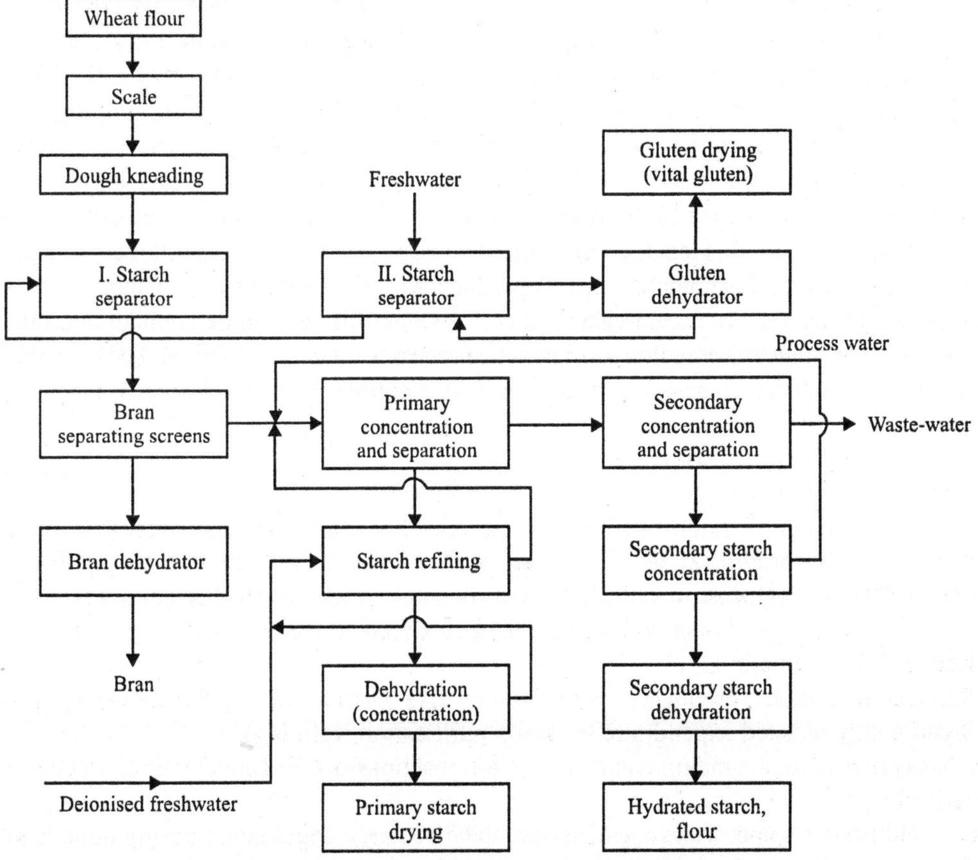


**Fig. 5.2.** Isolation of wheat starch from wheat flour.

Products include a *prima*- and a *secunda*-grade starch (distinguished by degree of purity) as well as wheat gluten (primarily as 'vital gluten'). The latter is much in demand as a constituent of baked goods and because of its relatively good market price (despite the usual price fluctuations), assures the profitability of processing wheat.

Obtaining starch from potato or other tubers is quite a different process that involves grinding stations; however, this procedure will not be discussed here because potato starch is unimportant in enzymatic processing. In Third World countries, extraction of tapioca root and broken rice plays a significant role; these starches have a purity satisfactory for enzymatic conversion. The hydrolysis of the starch molecule—that is, its more or less complete digestion into maltodextrins, maltotriose, maltose and particularly into glucose with in organic acids, such as hydrochloric and sulphuric acids—has been known for a long time. It was first attempted by the pharmacist Kirchoff who, during Napoleon's continental blockade, sought to produce gum arabica. Thus, a process formed under duress and once converted to an industrial scale had become, for many middle European countries, the alternative to sugarcane imports from their overseas colonial possessions. From that time on, starch conversion products (primarily syrups) were used as sweeteners for food. However, the acid catalysed conversion is a crude and poorly controlled process and produces, because of the by-products formed, a very contaminated product by today's standards. Even with costly purification steps, it still does not qualify for many potential applications.

Therefore, considerable interest was created when plant and microbial starch hydrolysing enzymes became available. The utilisation of malt and malt extracts was attempted quite early. But only at the onset of this century, Takamine was able to produce a fungal amylase on an industrial scale that allowed significant improvements in starch hydrolysis. The saccharification enzyme, amyloglucosidase (glucoamylase), entered the market in the 1960s and in the mid-1970s, a bacterial α-amylase with adequate thermostability and a glucose isomerase arrived; de-branching enzymes such as pullulanase finally became available during the 1980s. All the research and development costs were certainly justified because an extensive spectrum of technical enzymes with very different characteristics are now available that have a strong impact on the manufacture of products as well as on process technology.

Although syrups derived from acid-catalysed conversions still have some commercial importance for reasons of economy, the introduction of enzymatic processes undoubtedly triggered the rapid growth in the starch industry during the past few decades. Significant improvements in process technology and product quality were due to:

1. Lower energy costs for conversion.
2. Lower requirements for corrosion-resistant facilities.
3. Increased amount of starch conversion [determined as DE value or dextrose (Dextrose is the common or commercial term for D-glucose) equivalent]; increased product yields (e.g., 1080 kg of D-glucose monohydrate from 1000 kg of dried starch); this represents a 20 per cent increase over the yield obtained with the acid-catalysed conversion method.
4. Reduced discolouration.
5. Reduced retrograded products (arising from reverse reactions during the conversion process).
6. Significantly reduced expenditure for syrup purification (refining).

Three basic enzymatic degradative and conversion reactions occur: Liquefaction, saccharification and isomerisation.

Figure 5.3 illustrates some of the chemistry involved in starch degradation during liquefaction and saccharification employing the technically most significant enzymes, thermostable α-amylase and amyloglucosidase.

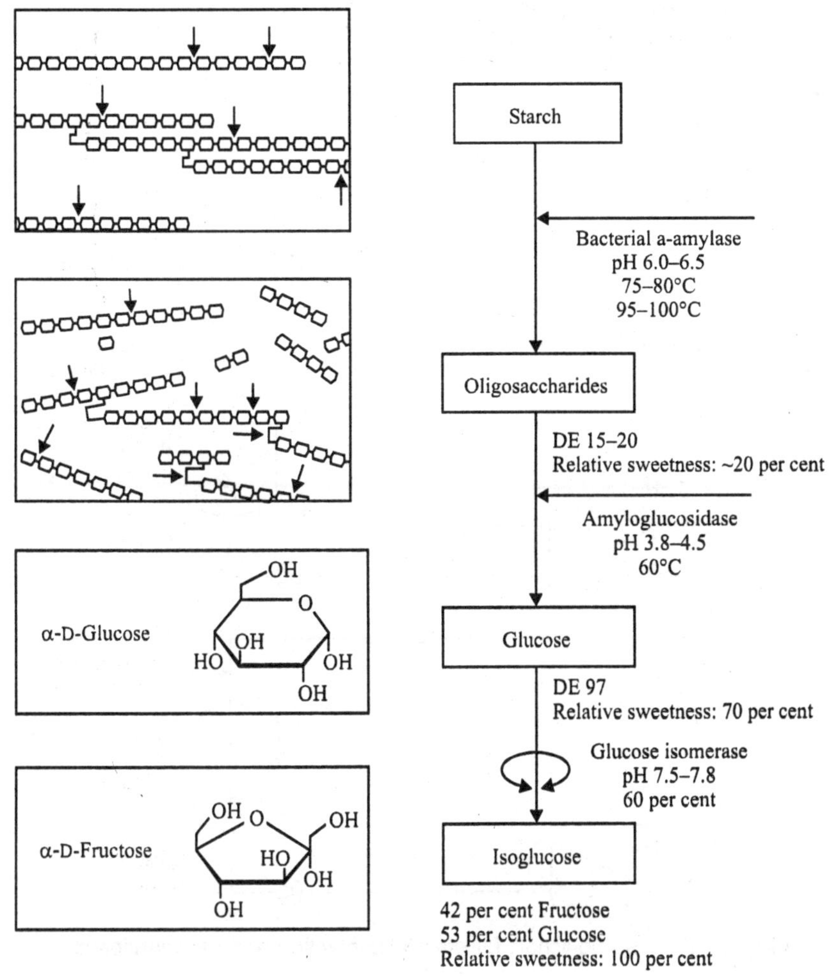


**Fig. 5.3.** Enzyme-catalysed conversion of starch to glucose and fructose.

A flowchart for a large-scale technical process is shown in Fig. 5.4. This best illustrates how the respective enzyme characteristics are utilised during the process and how they permit a very modern process design with continuous operation or one requiring special process staging.

The liquefaction phase processes the starch slurry through the hydration and pasting temperatures, reduces its high viscosity and splits the starch polymer into relatively uniform medium-sized fragments.

In practice, this is done by first funnelling the end product of starch production, the starch slurry (30–35 per cent solids), into a mixing tank to which cofactors such as $Ca^{2+}$ (as $CaCl_2$) are added to increase the thermal tolerance of bacterial α-amylases (Optiamyl [SOEG] or Tenase [SOE]: maximum 400 ppm $Ca^{2+}$; Optitherm [SOEG] or Taka-Therm [SOE]: maximum 100 ppm $Ca^{2+}$), provided the hardness of the water used in the process had not already furnished these optimal conditions.

Further, the pH is adjusted to 6.0; in addition to setting the optimal activity of the individual enzyme, it is important to note that at pH > 6.2, the terminal reducing groups of the starch fragments easily isomerise and a sizable quantity of maltulose (4-α-glucopyranosyl-D-fructose) can be detected at the conclusion of the starch to glucose conversion process.

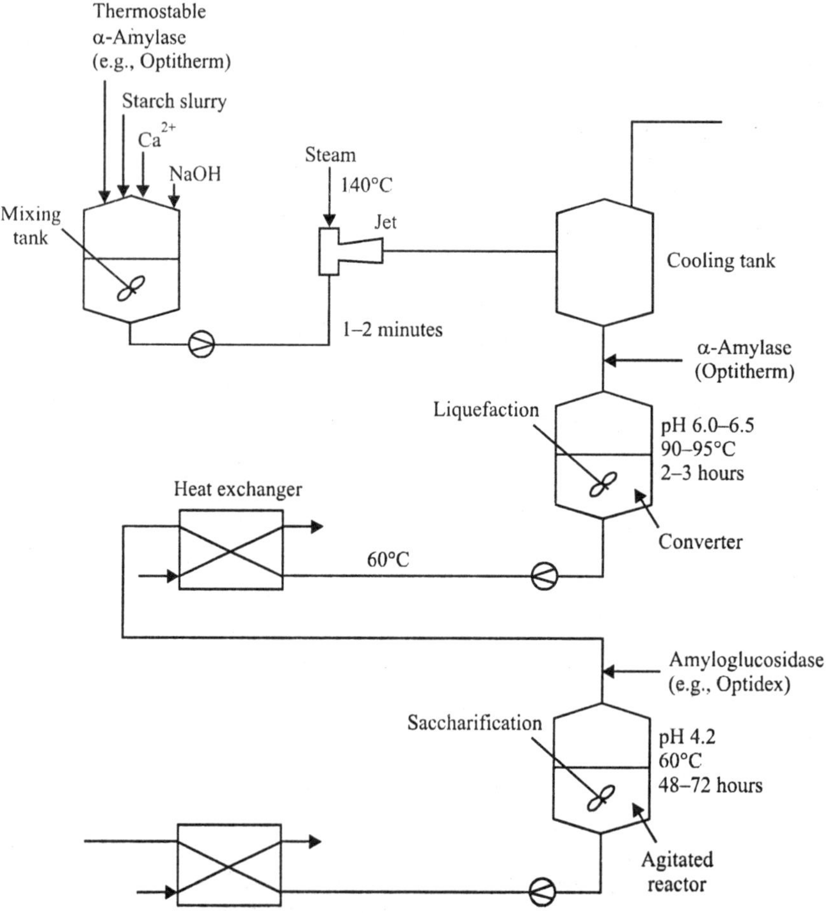


**Fig. 5.4.** Process flowchart for starch liquefaction and saccharification.

Subsequently, all or part of the liquefying enzyme, such as Optitherm-L 420 [SOEG] or Taka-Therm [SOE], is added to the mixing tank or is continuously fed into the succeeding pipeline system. On passing the slurry through one or more jet cookers, which are steam-injection systems, most of the starch granules immediately disintegrate and dissolve on continued heating and mechanical shearing. The temperatures and retention times used in the process rests are inversely proportional; that is there are processes employing relatively low temperatures (105–107°C) for 5–10 minutes duration and others using high temperatures of 120–130°C for just a few seconds. Because the latter conditions will inactivate even a thermally stable α-amylase, a second addition of Optitherm-L 420 or Taka-Therm into the expansion cooler or at a later stage is required. However, the total amount of liquefaction enzyme required in either case lies between 0.5 and 0.8 liter of Optitherm-L 420 or Taka-Therm per metric tonne of dry starch.

Conversion to dextrin is performed in well-insulated pipelines, continuously stirred tank cascades or in simple holding tanks over a 2–3 hours period. Customarily, the system is slowly cooled from 100°C to 95 or 90°C. The endpoint of the conversion is determined by either a DE (dextrose equivalent) measurement and/or the absence of an iodine reaction or by the filterability of the preparation. Opinions as to which DE value indicates complete conversion differ widely; some industries consider a DE of 8 or 10 as adequate, whereas others feel that a DE of 15–16 is mandatory. In any case, the DE value as well as jet cooker conditions determine if a syrup after liquefaction and especially after saccharification can be easily and economically filtered, that is, if it can be filtered rapidly without product loss.

Disintegration of the starch granules and the degree of the starch molecule's hydrolysis is, in view of what is known today, very critical because the native composition of starch granules includes, among others, amylose-lipid complexes. When amylose has not been adequately hydrolysed, these amylose-lipid associations that dissolve only at temperatures above 100°C can reassociate under unfavourable conditions at temperatures below 80°C. In such cases, especially at relatively low saccharification temperatures, turbidity arises from very fine particles of retrograded starch; these quickly clog the diatomaceous-earth-coated rotary vacuum filters during syrup clarification.

In addition to the process just described, there are historical methods that employ either total liquefaction or acid hydrolysis with temperatures usually about 150°C prior to the jet cooker stage. For some raw materials such as wheat and potato starches, separating the insoluble components becomes easier; however, it has been repeatedly demonstrated that this does not compensate for the disadvantages of acid hydrolysis described earlier.

When producing maltodextrin products (DE 16-20; roller- or spray-dried), the use of enzymes such as Optiamyl-L [SOEG] or Tenase [SOE] are preferred because of their lower optimal temperature and their concomitantly lower inactivating temperature.

The viscosity can be precisely controlled by the enzyme dose; usually 1.0–3.0 liters of Optiamyl-L or Tenase is added to 840 tonne of dry starch. The enzyme must be rapidly inactivated, usually at 110–120°C, so that the product characteristics such as the ability to absorb water and thicken are retained.

The product of liquefaction or dextrinisation is a mixture of oligosaccharides containing small amounts of mono-, di- and trisaccharides and is, for that reason, a superior substrate for the sweetening process. This process can be carried out with one or several different enzymes, depending on the type of syrup composition and properties desired.

The production of glucose syrups and D-glucose is by and large the most important saccharification process. An amyloglucosidase from *Aspergillus niger* is usually employed with a maximum activity at pH 4.0–4.5 and 60°C. The extent of starch hydrolysis (expressed as DE value) and time required for saccharification are strongly dependent on the substrate concentration and the amount of enzyme. With more enzyme, oligosaccharide hydrolysis will be completed more rapidly; however, this may trigger a potent reverse reaction catalysed by amyloglucosidase leading to maltose and maltotriose. The proper amyloglucosidase isozyme I : II ratio was long thought to be critical in achieving optimal starch hydrolysis; it has since been shown, however, that a greater percentage of isozyme I will provide limited improvements only with high-molecular-weight substrates (low DE value after hydrolysis). Splitting the α-1,6-linkages in amylopectin fragments is also largely catalysed by isozyme I; this reaction, however, is not limiting; that is, it will not significantly hamper complete starch hydrolysis down to the glucose unit.

In addition to the main enzyme fractions, other components of amyloglucosidase preparations can influence the course of saccharification. Primarily transglucosidase should be mentioned in that this enzyme can generate panose from two molecules of maltase by splitting off glucose. This trisaccharide with an a-1,6 linkage can, in turn, again be hydrolysed by amyloglucosidase to isomaltose and glucose. Isomaltose cannot be enzymatically hydrolysed any further and therefore, systems with high levels of contaminating transglucosidase activity can considerably lower the yield of D-glucose. Today, however, one can assume that commercially available amyloglucosidase preparations do not contain transglucosidase in amounts that will affect saccharification yields. This also applies to proteolytic activities of the preparations that, because of the low pH and temperature conditions during saccharification, are generally not thought to be too significant in discolouring syrups.

As seen in Fig. 5.4, saccharification is started by cooling the syrup to 60°C with heat exchangers and adjusting the pH to 4.0–4.5 with hydrochloric acid. The saccharification enzyme, such as 0.6-0.7 liter Optidex-L 300 [SOEG] or Diazyme-L [SOE) per metric tonne of dry matter, is added either continuously to the pipelines or directly to the saccharification tanks. The latter are well insulated and the contents may be at rest or may be periodically or continuously stirred. The converters are filled and emptied in batches or are run in continuous operation. To avoid recycling of syrups, which would lower glucose yields, 6–12 conversion tanks are required for continuous operation. The time required to reach optimal DE values (97–98 DE) is approximately 48–72 hours. Syrups with 96 per cent glucose or more can be obtained with 2–3 per cent disaccharides (maltose and isomaltose) and with 1–2 per cent higher sugars. Frequently, the pH value drops a little because of microbial contamination and acid formation. In situations where the temperature in the converters cannot be monitored precisely and held at a constant 60°C, a massive contamination with yeasts and moulds leading to layers of film cannot be excluded.

In the past few years debranching enzymes such as pullulanases and isoamylases have attained some importance. These enzymes are capable of breaking the α-1,6 bonds within the starch polymer and thus support the action of amyloglucosidase. This means that the saccharification rate will increase slightly with an equal amount of amyloglucosidase or that with less amyloglucosidase and therefore fewer back reactions, these enzymes produce very good yields and as experience has shown, the longer reaction times will effect a very uniform saccharification. Debranching enzymes are synthesised by many bacterial species with optimal activities that are usually within pH 5.0–6.0 and 50–55°C. These characteristics make these enzymes relatively unsuitable for use with the saccharification conditions described previously. One exception is the pullulanase isolated from *Bacillus acidopullulyticus*, which is very compatible with amyloglucosidase and therefore is used by itself or in enzyme mixtures by several starch processing industries.

High expectations were held for an immobilised amyloglucosidase in the late 1980s. However, all products offered very quickly showed that the required DE values of 96.0 or higher could not be obtained because of diffusion resistance. The relatively high level of reversion products formed was a consequence of increased enzyme density in the carrier-bound preparations. On the other hand, amyloglucosidase preparations, offered in liquid form, are rather inexpensive, so the technological expenditure necessary for a total continuous operation with its concomitant risks (substrate purity and downtimes) cannot be justified. However, implementation of immobilised amyloglucosidase in the production of other syrup varieties, to be described subsequently, has proved worthwhile in individual cases.

After completing the sweetening process in the manner described, inactivation of the enzyme is not necessary; the same holds for the liquefaction phase of starch conversion. The subsequent syrup purification steps can cause enzyme inactivation and prevent, except with long holding times, the loss of glucose yield due to amyloglucosidase-catalysed back reactions. Sugar refining starts by precipitating insoluble components such as proteins and lipids on increasing the pH (with sodium carbonate solution) and by removing the particulates with separators or diatomaceous-earth-coated vacuum rotary filters—provided the formation of retrograded starch described earlier does not prevent this.

In addition, the removal of minerals and colour is essential and accomplished in most operations by passing the syrup through a combination of strongly acidic and weak alkaline anion exchange resins which are coupled in series. Treatment with activated charcoal is conducted when there is a high demand for light syrups. The absorbent is added either prior to the first separatory step or at the final ultra-purification step.

Glucose syrups with a high DE value can be stored and transported only at 60°C because they tend to readily crystallise after having been concentrated in the evaporator to 75–78 per cent solids.

To obtain ultrapure α-D-glucose monohydrate, crystallising vats with cooling mantels and cooling coils are used. Up to 60 per cent of the glucose can be crystallised by gradual cooling to 30°C. The crystallised glucose and mother liquor (hydrol) are separated from each other with sieve centrifuges. The crystallisation process itself and subsequent additional purification and drying procedures require a great deal of intuitive skills and are part of the best-kept secrets of this industry.

A whole series of production possibilities exist between the two extremes of glucose and maltodextrin manufacture; these products include glucose syrups with low, normal, intermediate and high degrees of saccharification tailored by the dose of amyloglucosidase and the duration of saccharification; however, a rapid inactivation of amyloglucosidase is required, by either lowering the pH (pH < 2.5–3), heat treatment (5–10 minutes at 95–100°C) or a combination of both.

These liquid products seldom become turbid; however, they do crystallise easier than maltodextrins and are more hygroscopic in their dried state.

Maltose-forming enzymes such as β-amylases from barley or barley malt have long been used to provide additional saccharified products for specific applications. However, fungal α-amylase from *Aspergillus niger* has become an interesting alternative for producing high-maltose syrups. Both enzymes are employed under the following conditions. The initial liquefaction (a low DE value will increase the final maltose content) is reduced to 48–52 per cent solids and adjusted to pH 5.0–5.5. Saccharification commences on addition of 160–240 ml of enzyme such as fungal amylase-L 40000 [SOEG] to 1 tonne of starch and proceeds in either batch or continuous operation at 55°C (60°C) for 40–48 hours. The liquor with a DE value of 48–52 is composed of:

1. Glucose 9–12 per cent.
2. Maltose 48–52 per cent.
3. Maltotriose 10–15 per cent.
4. Higher oligomers 21–33 per cent.

This composition is stabilised by the rapid inactivation of the saccharification enzyme (acidification, thermal treatment or a combination of both).

The advantage of β-amylase derived from higher plants over the fungal α-amylases is that they are less expensive and accomplish their tasks in a significantly shorter time (12–24 hours). However, they, too, are unable to hydrolyse the α-1,6 linkage of amylopectin and hydrolysis stops at the β-limit dextrins. A complete hydrolysis of the branched starch molecules with an increased maltose yield of 70 per cent and above can be achieved only by adding microbial pullulanases; these should be adapted as much as possible to the pH and temperature optima of the maltogenic enzymes.

In the manufacture of high-conversion syrups, both amyloglucosidase and fungal α-amylase are used; accordingly, the combined characteristics of glucose and maltose syrups are obtained.

Here, as in the cases described earlier, a broad range of possibilities exist for the individual commercial user. Normally, the process is started with a substrate of 38–42 DE (derived from acid hydrolysis), which, after concentration to 48–52 per cent solids, is saccharified at pH 5.0–5.2 and 42–60°C with 60 ml fungal amylase-L 40000 [SOEG] and 25 ml Optidex-L 300 [SOEG] or Diazyme-L [SOE]. After 40–48 hours, the following approximate syrup composition at DE 63–70 is obtained:

1. Glucose 30–35 per cent.
2. Maltose 40–45 per cent.
3. Maltotriose 8–10 per cent.
4. Higher oligomers 20–22 per cent.

To stabilise this composition, rapid enzyme inactivation is required, for example 2–3 minutes above 100°C or 20–30 minutes at 85–90°C.

Refining methods for high-maltose and high-conversion syrups are essentially those used for glucose syrups. In continuous well-controlled operations, the thermal inactivation step can be eliminated and the sweetening enzymes completely removed with activated charcoal.

The methods of starch liquefaction and saccharification discussed so far clearly indicate that syrups can be produced with any combination of oligo-, di- and monosaccharides and thus can be tailor-made for a multitude of applications:

1. Maltodextrins: Maltodextrins are marketed as white-brown, spray-dried powders; they are slightly sweet, serve to provide bulk and can be viewed as easily digestible component of food. They are used primarily in baby food and special diets, in dried soups, as a coffee whitener and as a flavour carrier.
2. Glucose syrups: Glucose syrups with low or intermediate DE values are utilised not only for their DE-dependent sweetness but also for their nutritional value and special functional properties such as fermentability, viscosity, absorbing and releasing moisture, bulk, foam stabilisation and ability to prevent sugar crystallisation which again depends on the DE value. The syrups are of prime importance in the manufacture of sweets of all kinds (especially caramels, jelly confections, chocolates), marmalades and jams, fruit syrups, refreshment beverages, baked goods, fruit juices, ice cream, canned foods and margarine.
3. Maltose syrups: Maltose syrups have properties similar to those of glucose syrups, but with a higher viscosity and lower hygroscopicity. They have less tendency to brown and crystallise and are preferentially used in ice cream, hard and soft caramel candies, baked goods and brewery mashes.
4. High-conversion syrups: High-conversion syrups are significantly sweeter than maltose syrups and promote browning, moisture uptake and crystallisation. These features make them particularly suitable for marmalades, fruit preserves, sweets in general, sauces and dressings. Furthermore, they have some importance in making beer and malt liquors.
5. Glucose syrups with high DE and glucose content: Glucose crystallises from high-glucose syrups as its anhydride or monohydrate. Crystalline glucose exists in the $\alpha$ form but when in solution it can mutarotate to its $\beta$ form with both isomers present at equilibrium. Glucose is by far the most important sugar in nature and therefore many applications for glucose exist.

Glucose and glucose syrups with high DE values are frequently utilised for their physical and chemical properties. Because of their high nutritional value, they are often substituted for sucrose in manufacturing baked goods (shortbread, biscuits and yeast pastries, breads for toasting, waffles, rusk and cookies), for beverages, jams and fruit preserves, fish, meat and sausages and serve as an important ingredient in sweets.

Crystalline glucose is particularly important in the medical and dietetic area. As human blood sugar, it quickly enters the bloodstream and provides energy during heavy physical exertion by athletes. Because glucose can be well tolerated, it is readily administered to patients and babies and can, when fed intravenously, acquire therapeutic functions, including detoxification, adjusting electrolyte and water imbalance, preventing or controlling heart disease and managing diabetic comas.

Because α-D-glucose can quite easily be converted chemically or by fermentation, this monosaccharide is also used as the raw material for producing sorbitol and mannitol (via reduction), glucuronic acid (via oxidation), vitamin C (microbially) and isoglucose (via enzyme isomerisation). Glucose is also involved in other biotechnological production processes and most recently as a reaction participant in the manufacture of plastics.

Glucose isomerisation is the latest step in enzymatic starch conversion. The enzyme catalysing this reaction, glucose isomerase (actually xylose isomerase), is capable of converting about 50 per cent of the D-glucose in a given syrup into D-fructose.

Isomerisation is economically feasible when the fructose yield has reached about 42 per cent (degree of isomerisation about 46 per cent); but reaching the complete reaction equilibrium would require an excessive amount of time. The reaction product is called isoglucose or high-fructose syrup (HFS) and its sweetening power and composition are similar to those of invert sugar.

This success, however, was realised only when it became possible to immobilise the enzyme and reduce enzyme costs to an acceptable level (based on the soluble products from first-generation isomerases). Second-generation glucose isomerases, always intracellular enzymes, were developed by using producer organisms as enzyme delivery systems, cross-linking or fixing the intracellular enzymes with glutaraldehyde. Third-generation glucose isomerases are characterised by highly purified enzyme fractions tightly bound to porous, highly inert organic (DEAE cellulose) or inorganic ($SiO_2$, $Al_2O_3$) carriers.

Isomerisation with immobilised enzyme products is usually accomplished with a fixed-bed reactor, usually designed as a well-insulated column assembly furnished with a perforated bottom support, inlet, outlet and observation windows. The substrate is pumped downward through the column. Filling the columns with a second-generation glucose isomerase is somewhat more labour-intensive than when using the third-generation enzyme preparations because it involves swelling the cells containing the fixed enzyme and carefully washing out all colour forming agents.

At a specific substrate flow rate that permits the formation of 42 per cent fructose, a filled column assembly is started at full efficiency (usually measured as the rate of voided volumes). In the course of days and weeks the efficiency drops and the substrate flow has to be reduced—or the temperature increased—to assure constant fructose formation. In most of the enzyme preparations marketed, the activity drops to 50 per cent after 50–55 days (half-life). Continued utilisation of the system is economical only for the second half-life (residual activity 20–25 per cent). The total amount of product formed over this period determines productivity and is expressed as kilograms of isoglucose (42 per cent fructose) per kilogram of enzyme or its standard activity. Second-generation glucose isomerases have productivities of 1.5–5 tonnes of isoglucose per kilogram of enzyme and therefore, require significantly larger columns (up to 150 cm in diameter and 10 meter in height) than do third-generation immobilised enzymes, which permit maximum yields of 22 tonne of isoglucose per kilogram of enzyme, such as Optisweet 22 [SOEG]. The height of the columns is limited primarily by the pressure gradient between substrate input and output.

To guarantee a sufficiently uniform commercial isoglucose production, several column assemblies (up to eight) are coupled either in parallel or in series and are periodically filled and emptied. Therefore, product formation will follow a sawtoothed pattern and the unavoidable fluctuations must be compensated with buffer tanks. Figure 5.5 shows the course of isomerisation.

Isomerisation

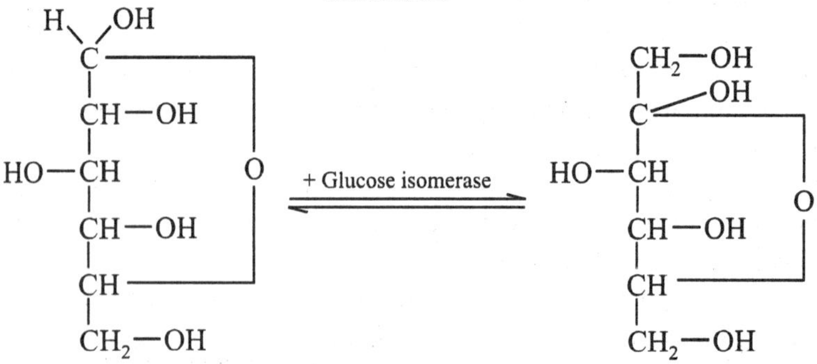

```
   H  OH                                              CH2—OH
    \/                                                |    OH
    C ──────────┐                                     C ────────┐
    |           |                                     |         |
    CH—OH       |                                     |         |
    |           |                                HO—CH          |
HO—CH           O      + Glucose isomerase            |         O
    |           |     ⇌                               CH—OH     |
    CH—OH       |                                     |         |
    |           |                                     CH ───────┘
    CH ─────────┘                                     |
    |                                                 CH2—OH
    CH2—OH
```

D-Glucose D-Fructose

**Fig. 5.5.** Flowchart of the isomerisation step with immobilised glucose isomerases.

Glucose syrup of sufficiently high purity (e.g., <5 ppm $Ca^{2+}$) with 45–50 per cent solids is adjusted to a pH of 7.5 (the effect of pH on productivity and the cofactors $Mg^{2+}$ and $SO_2$ are added to provide optimal protection of the enzyme from denaturation. After sterilisation, the substrate is heated to about 60°C with heat exchangers and discharged into the isomerisation columns. The temperatures of the charge can be slightly higher because the endothermic reaction in the columns will cause a drop of about 2.5°C by the time it is discharged from the column. Similar to the syrups described earlier, isoglucose, after lowering the pH to 4.0–4.5 to suppress colour formation, is purified with activated charcoal, filtered, demineralised and reduced in an evaporator to 70–72 per cent solids. In contrast to the glucose syrups, the low-viscosity isoglucose can be stored and transported at 35°C without recrystallisation.

Many countries have started to manufacture isoglucose products with more than 42 per cent fructose content, such as HFS 55 (containing 55 per cent fructose). This percentage, however, cannot be met by the technology described here because the reaction equilibrium for glucose isomerases is approximately 50 per cent as mentioned earlier. Therefore, a procedure for the large-scale enrichment of fructose by partition column chromatography and continuously returning the unwanted glucose to the isomerisation process has been applied. The higher sweetening power of HFS 55 is particularly advantageous in refrigerated soft drinks and can, if necessary, reduce the caloric content of the sugar added. Isoglucose is further used as an additive in marmalades, jams, canned fruits, fruit juices and milk products.

Isolation, purification and subsequent enzymatic liquefaction and saccharification of starch is the current state of the art in enzymatic processing and is practiced worldwide. The question of whether a high degree of starch purity is always necessary was asked rather early. Professor Tegge, from the Germany coinine the term 'direct enzymatic saccharification' and for utilising readily available and inexpensive raw materials such as maize flour, potato peels, broken rice and millet flour in the enzymatic conversion to glucose syrups. It is important, however, not to impose unrealistic demands on the products so obtained. Further, the enzyme preparations used cannot have contaminating proteolytic activities that will have an unfavourable impact on syrup colour.

In an age of conscientious nutrition, natural sweetening of foodstuffs is also considered. Addition of all kinds of enzymatically processed flour to baby food, breakfast cereals and baked goods has become quite customary. Bacterial α-amylases are also used in potato juice production. Moreover, there is a sizable effort to recover the starch from bakery waste by enzymatic degradation, making recycling old bread and bakery leftovers possible. As shown in Figure 5.6, the process technology needed for such recycling is quite elaborate; however, it need not be because often simultaneous hydrolysis and saccharification without subsequent refining suffice to produce suitable additives for doughmaking.

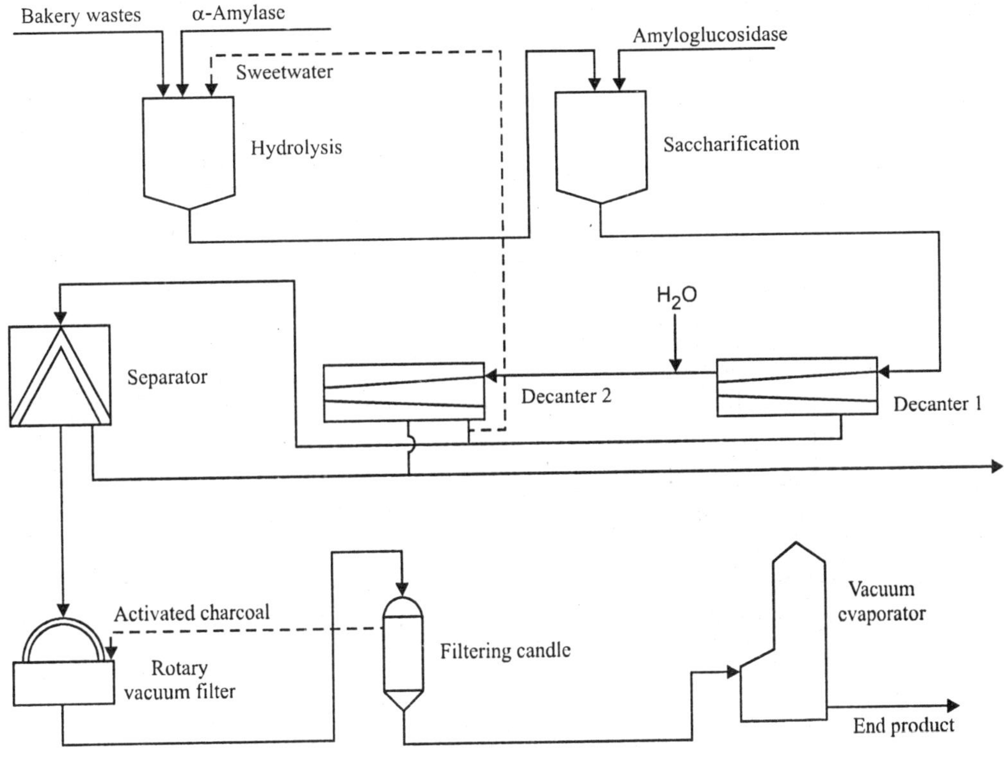


**Fig. 5.6.** Saccharification of bakery wastes.

Wastes from potato processing can be converted into animal feed as shown in Fig. 5.7 or alternatively, can serve as browning agents or be fermented into alcohol.

Cacao powder also contains starch; the basic properties of many chocolate products can be improved with enzymatic treatment.

Some Third World countries have begun to obtain high-grade sweeteners from broken rice (i.e., inferior grains accumulated during rice harvesting).

Furthermore, direct enzymatic saccharification has long been a well-known process in the fermentation industry. Inexpensive starch-containing raw materials are optimally degraded by enzymes for a specific microbial production strain prior to delivery to the fermenter. Preliminary and prudent steps have been initiated for utilising amylases and other enzymes in treating waste-water and wastes present in urban clarification plants.

When discussing the enzymatic conversion of starch, the processes used in the textile and paper industries must be mentioned as well. Starch is an important raw material in the paper industry. It is used in bonding paper pulp, in sizing of paper surfaces and as a thickening agent in paints. Modified starches are almost always used. Modification can be done either chemically, thermochemically or with bacterial α-amylases such as the α-amylase-LP [SOEG]. The latter process has since become the most important and many paper manufacturers have small plants with converters for the enzyme-catalysed hydrolysis of starch producing the desired viscosity and hygroscopicity.

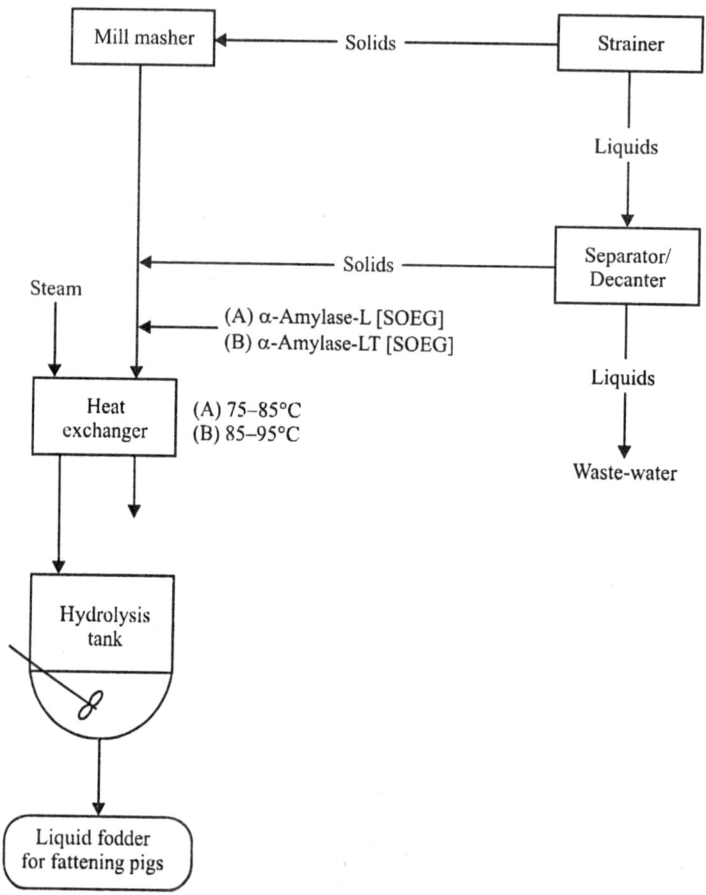


**Fig. 5.7.** Processing wastes from the production of potato chips, french fries and similar products.

Future of enzymes in the starch industry: Only a few high-performance starch processing enzymes have been added in recent years. The latest really important process improvement was the application of the pullulanases and other debranching activities during saccharification. Recently, serious efforts were undertaken to improve the efficacy of the amyloglucosidase-pullulanase mixture on saccharification time or yield by optimising the natural composition of the amyloglucosidase preparations. The ratio of amyloglucosidase isozymes I and II can be adjusted to favour the prevailing desired level of saccharification either during or after fermentation (DE after hydrolysis, either high or low; Dx after saccharification within a given range; content of di- and trisaccharides). In addition, an acid-stable α-amylase is formed during this process that is particularly well suited to attack the small amounts of high-molecular-weight polysaccharides that are usually present as a consequence of starch degradation. As a rule, almost all amyloglucosidase preparations on the market invariably contain α-amylase activity as well; however, depending on the biological strain and fermentation characteristics, this enzyme is usually present in a relatively low concentration. An increase in acid α-amylase content during *Aspergillus niger* fermentation (in this case a multiple of the standard amount is required) causes the sweetening enzyme to increase glucose yields by an added 0.5–1.0 per cent or to significantly reduce the time required to attain the desired degree of degradation.

Factors dependent on the raw materials play an important role in other interesting new developments. For saccharified wheat starch, the formation of gel-like precipitates make it quite difficult to separate the insoluble components (sediment) in an economically feasible manner. Comparisons of product data from various enzyme producers provided the first clue to the fact that secondary activities in sweetening enzyme preparations can significantly improve the flocculation properties and subsequent filtration behaviour of sweetened syrups and prevent foaming and sensory changes in the syrups. It has since been learned that the degradation of companion substances in wheat starch is eminently important in this context. Examples are the enzymatic cleaving of wheat lysophospholecithin, which has excellent emulsification properties and is present in high concentrations; and the cleaving of pentosans, which, when present even at very low concentrations, will increase the syrup's viscosity to such extent that particulates remain suspended. The more recent amyloglucosidase preparations such as Optidex-LW (SOEG] or Diaxyme-L [SOE] contain the required activities of both lysophospholipase and pentosanase in sufficient quantity. Naturally, they can also be added individually in other sweetening processes.

Switching to native maize and wheat varieties led to the development of enzyme cocktails for improving the separation of starch from gluten or from fibrous material or for the reduction of waste-water generated during wheat processing. These multienzyme mixtures exhibit predominantly cellulase, pentosanase, β-glucanase and xylanase activities, such as Pentosanase-L [SO; SOE]; functionally, they are considered to be technical aids only. Surely is it the dream of every process engineer or operations manager in the starch industry and also of every enzyme technologist, to be able to run hydrolysis, saccharification and isomerisation in a continuous operation under completely standardised conditions. Modern methods of biotechnology suggest that it may be possible to modify the properties of specific enzymes, either chemically or by genetic engineering, to such an extent that they can be tailor-made for specific applications. It would be particularly attractive to be able to balance the pH characteristics of the amylases and isomerases, thus eliminating both the addition of acids and bases and consequently the salts formed during the refining process. A first step in this direction is the effort to employ heat-stable α-amylase at a pH below 6.0, thereby simultaneously and significantly reducing the formation of maltulose (4-α-D-glucopyranosyl-D-fructose).

Currently, there is no method available to shift the pH for isomerisation into the neutral range. The upper process temperatures are determined by the pasting point of the starches (60–80°C) and the lower, by the range for microbial contamination (55–60°C). Further, it would be most desirable to remove the rather paradoxical situation of the essential addition of $Ca^{2+}$ to activate α-amylase and then remove it almost completely prior to isomerisation because of its inhibitory effect. For a complete continuous process operation, utilisation of an immobilised amyloglucosidase should also be considered. This enzyme, as well as glucose isomerase, should—according to the most recent studies—be combined in a single-column reactor; loss of activity during prolonged operating times can be compensated for by the stepwise or continuous addition of soluble or carrier-bound enzyme.

### *Enzymes for wet milling of maize and cereal grains*

When maize and cereal grains are used for starch processing, the starch granules must be separated from the other kernel components. This is accomplished by the wet-milling process. The kernels are soaked prior to the milling step so that the components are evenly removed. During soaking (30–50 hours at 45–50°C) the granules imbibe about 45 per cent water. About 6 per cent of the maize solids are solubilised in the sulphur dioxide-containing steepwater. This steepwater, the first by-product of the process, is used by the industry as an essential fermentation nutrient in other processes. After drawing off the steepwater, the softened kernels enter the milling process started with the addition of cellulase.

The advantages of using enzymes are as follows:

1. Increasing starch yield.
2. Increasing milling capacity.
3. Reducing steeping time.
4. Reducing drying costs for gluten by increasing the solids in the separated wet gluten.

The enzymes currently employed in this process are cellulases isolated from *Trichoderma* sp. The preparations used in the process discussed here must be free of amylase activity so that no losses arise during starch refining. Further, the preparations must not show any or only an extremely low, proteolytic activity, as proteolytic activity would lower the quality of the gluten isolated. Gluten quality is particularly important for wheat gluten. Prior to use, the enzyme preparations should be tested for proteolytic activity with a sufficiently sensitive method. The method measuring the drop in viscosity of a gelatin solution is well suited for such an assay.

Tests to establish the effectiveness of technical cellulose preparations and their economic advantage must be conducted on an industrial scale. The criteria can include the increase in starch yield and the increase of the wet-gluten dry matter. In maize starch processing, particularly the bonds between gluten and fibrous components must be broken. The success of this separation depends on the amount of water absorbed by the kernel. It must be borne in mind that maize kernel quality and steeping time will affect yield. An important improvement accomplished with enzymes is the reduction in water content of the gluten protein when obtained by either centrifugation or extraction.

### *Enzyme dosing*

Generally, liquid enzyme preparations are used at the onset of wet-milling processes. To detect the effects clearly, pilot production runs are initiated with relatively high doses of enzyme; for example, 500 ml of *Trichoderma* cellulase with a specific activity of 2000 C units per milligram of protein (e.g., 500 mg Rohament 7069 [RM]) is added to one tonne of starchy raw material.

Depending on the results obtained, the enzyme dose is lowered. Typically, good results are obtained with 40–50 per cent of the enzyme quantity cited above.

An alternative to this application is the addition of the enzyme preparation to a gluten suspension prior to vacuum filter centrifugation. A reaction time of 1 hour at 40°C is required in this case to digest the polysaccharides, which significantly hamper filtration.

### *Wheat starch*

For the isolation of wheat starch, the use of enzymes has the following advantages:

1. Increasing the yield of a-starch.
2. Improving extraction of water from bran.
3. Saving energy during drying of the gluten.

## Enzymes in the Brewing Industry

Brewing beer is almost as old as baking bread. The Sumerians produced an alcoholic beverage from fermenting bread grain. The Assyrians are known to have brewed beer about 4000 B.C.

The processes involved in brewing beer include malting in the malt house, wort preparation and fermentation of the wort in the brewery.

### *Malt house*

The objective of the malting process is the enzymatic digestion of barley or wheat grains during germination. First the barley is cleaned and then steeped. This increases the kernel's water content of 44–48 per cent. The steeped barley kernels are spread approximately 30 cm deep over the malt-house floor and kept at a temperature of about 15°C. Adequate aeration is mandatory for germination so that the low temperatures can be maintained. Germination is rapid and is completed after about 6–7 days. The 'green malt' generated is kiln-dried as carefully as possible to retain high enzymatic activity. The temperatures used in kiln drying determine the malt quality, which, in turn, imparts the individual characteristic to a beer. Approximately 70–80 kg of dry malt is obtained from 100 kg of barley. The use of enzymes in the malting process has been described many times, but they are rarely employed, although soaking in a solution of *Trichoderma* cellulases is believed to help solubilise the malt during the mashing process. Gibberellic acid does, however, significantly increase the formation of amylolytic activity during malting.

### Brewery

Enzymes play a central role in brewing. In the traditional brewing process, the enzymes are formed during the malting operation. They are necessary to convert the high-molecular-weight polymer constituents of grains into low-molecular-weight compounds. These endogenous enzymes are the malt α- and β-amylases, which convert starch into fermentable sugars and non-fermentable dextrins, as well as proteases and peptidases that digest grain proteins into low-molecular-weight peptides and free amino acids. The latter are needed as nutrients for the yeast during fermentation. The protein digests are also involved in the beer's aroma and stability. The β-glucanases formed during the malting process are of special significance in successful brewing.

Further, there are a number of other important endogenous enzymes, namely, various phosphatases and the pentosanases that hydrolyse gums during malting.

Modern brewing technology employs, in addition to the malted barley described above, raw barley and other raw grains such as rice, corn or sorghum. The necessity of partially or completely replacing malt as a brewing addendum becomes important when malt is unavailable or excessively expensive. In such cases, the native malt enzymes are replaced with technical enzyme preparations (Fig. 5.8).

Ground malt or malt together with a given amount of raw grain and water are mashed at 37°C. The mash is then held at different temperatures for various time periods (protease rests) to extract and solubilise the contents of malt and raw grain.

At the end of the mashing process, on clarification, the spent grain is removed from the wort. The wort is delivered to the wort kettle and boiled together with hops or hop extract. During this process, the endogenous and exogenous enzymes are inactivated and certain protein fractions are coagulated and precipitated.

At this stage, solubilisation and conversion of the hop constituents occur. Subsequently, the wort is filtered and cooled to the fermentation temperature. On the addition of yeast, fermentation is initiated; when completed, the yeast is removed and the green beer is transferred to the lager tank. The beer must then again be filtered prior to barreling or bottling.

### Mashing process

Various mashing methods are used for solubilising the constituents of malted or raw grains; in part, they have a traditional origin; examples are the Bavarian three-step mashing process (decoction or boiler mashing) and the British infusion mashing.

Alternating thermal and enzymatic steps are employed in all the procedures in order to solubilise the high-molecular-weight polymers of the grains.

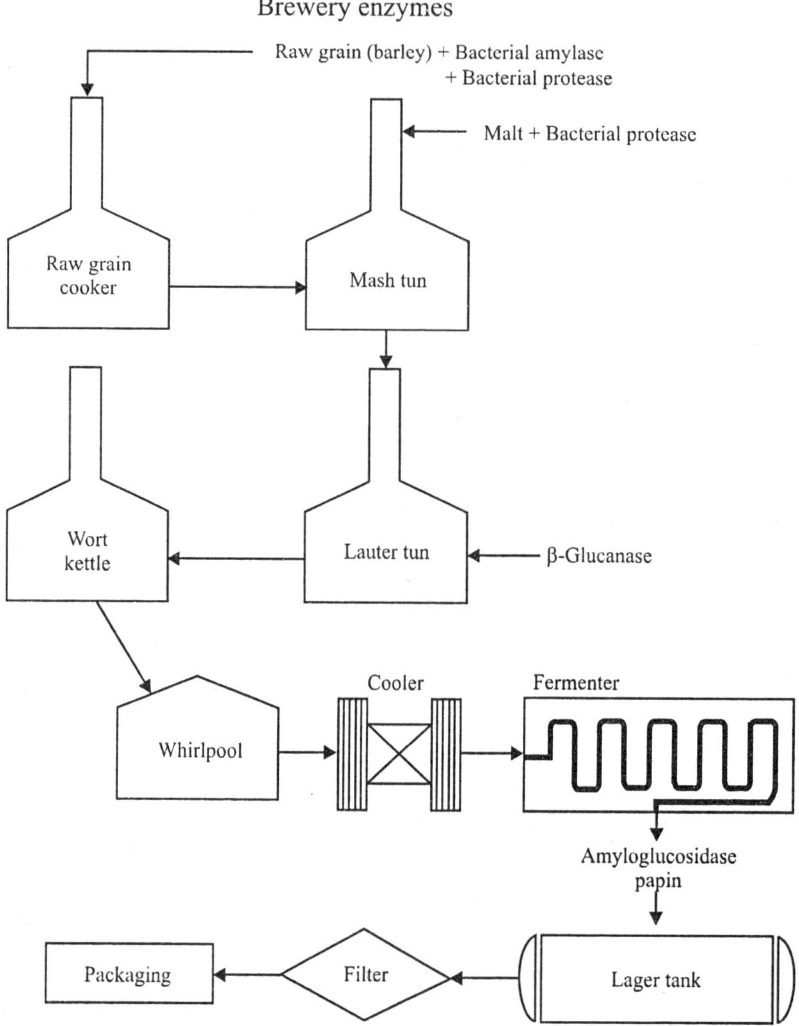


**Fig. 5.8.** General brewing schematic.

### *Decoction mashing*

Decoction mashing includes a physical as well as an enzymatic malt extraction. The three-step mashing procedure (Fig. 5.9) is the best known and oldest method; the current decoction methods have been derived from it.

In decoction procedures, depending on the mashing process employed, batches of mash are withdrawn and heated to 100°C; these boiled batches are then recombined with the main mash lot. This accomplishes pasting of a portion of the starch that then can be converted while the enzymes in the non-heated mash remain active.

This process is still in use today in brewing dark beers. In contrast to malt, the raw grains must first undergo pre-treatment (raw-grain milling) to increase the availability of their constituents (Figs. 5.10 and 5.11).

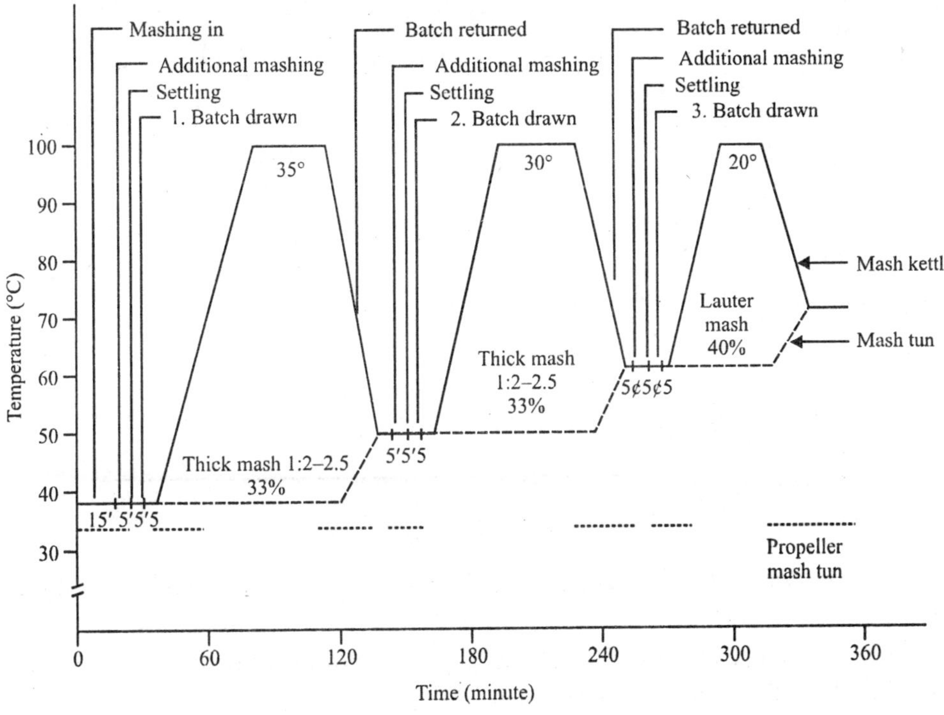


**Fig. 5.9.** Three-step mash process (340 minutes duration).

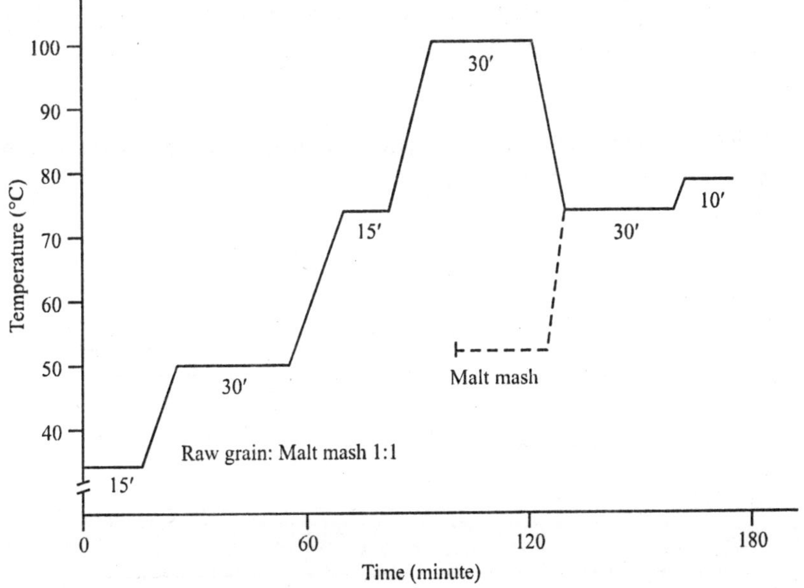


**Fig. 5.10.** Method for processing 20 per cent raw grain.

Depending on the type and amount of raw grain, various mashing processes can be implemented. First, the raw grain is mashed with malt at a ratio of 1:1 and temperature of 35°C; the mash is then

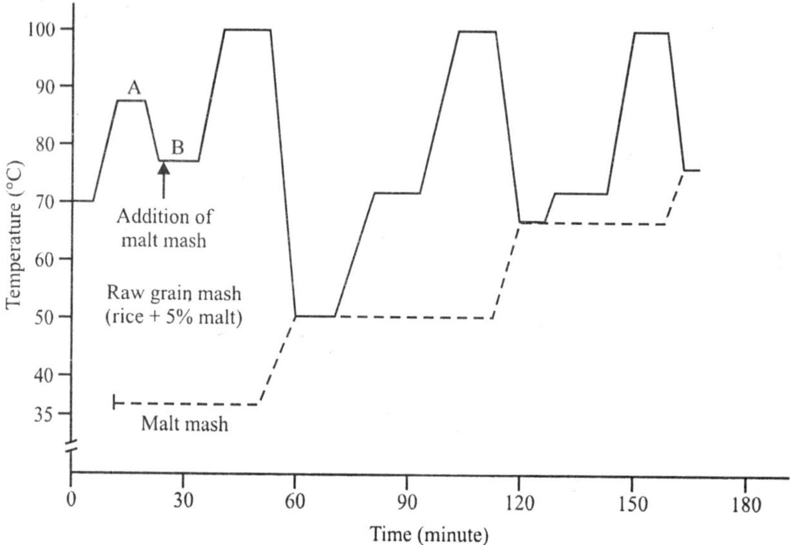


**Fig. 5.11.** Method of processing raw grain with prior pasting (~20 per cent rice): (a) pasting; and (b) liquefaction of raw grain.

successively heated to temperatures of 50, 62 and 72°C and held at each temperature for a determined period of time (rests) and then ultimately boiled at 100°C.

The remaining malt is subsequently mashed and held at 50°C in the first rest. The boiled mash is cooled to 65–70°C and mixed with the pure malt mash.

Thus, an additional saccharification step can be conducted at 70–75°C. Subsequently, another rest can follow at 78°C, after which the lautering process is initiated.

### *Infusion mashing*

The raw grain or malt constituents are converted and solubilised only by enzymatic reactions. The boiler step is not used in this process (Fig. 5.12).

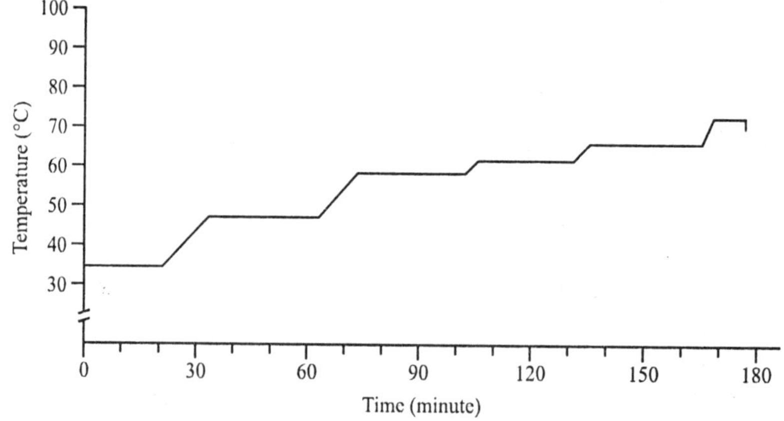


**Fig. 5.12.** Infusion method (180 minutes duration).

The infusion method includes the following steps: first, a protein rest at 50°C for 30 minutes. Holding this step for a longer period is not required because the proteolytic activity and the activity of the endo-β-glucanases decrease. The maltose rest can be carried out in several steps in order to fully exploit the β-amylase activity. The rest of 70–72°C is usually held only until the iodine reaction is negative. However, because the boiling used in the decoction method is not used in this process, the rest time is extended to permit better saccharification. The mash is subsequently heated to 78°C and pumped into the lauter tun.

### *Brewing with enzymes*

In traditional brewing (Fig. 5.13), malt quality depends on several factors. These include the properties of the particular barley cultivar and where it is grown, climatic conditions and the malting process itself. All these factors strongly influence the enzymatic activities of the malt. High-grade malts endowed with an optimal complement of enzymes are very expensive.

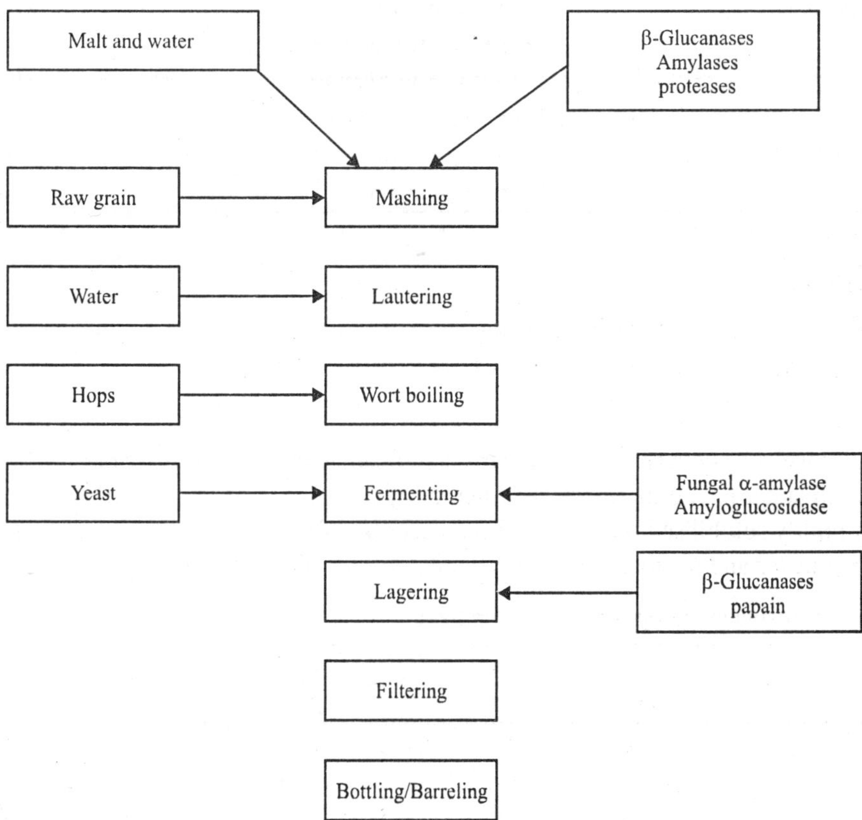


**Fig. 5.13.** Use of enzyme preparations in traditional beer brewing.

Variations in malt quality can give rise to difficulties during processing, such as during lautering of the mash, filtration of the wort or in the green beer. However, endogenous malt enzyme deficiencies can be compensated for by the targeted use of enzyme preparations with specific activities.

If barley or other cereals are brewed, the addition of exogenous enzyme preparations is absolutely essential for a reliable and economical process. Table 5.3 lists the types and functions of enzyme preparations and these preparations are employed in brewing.

**Table 5.3.** Purpose and site of enzyme use.

| | *Application* | |
|---|---|---|
| *Enzyme* | *Purpose* | *Site* |
| Bacterial proteases | Protein hydrolysis for increasing FAN[a] | Mash tun |
| Papain | Cold stabilisation | Lager tank |
| Bacterial α-amylase | Starch hydrolysis | Raw grain cooker |
| | Starch hydrolysis | Mash tun |
| Fungal α-amylase | Improving fermentation | Mash tun |
| | | Fermenter |
| Amyloglucosidase | Improving fermentation | Mash tun |
| | Light beer | Fermenter |
| Bacterial β-glucanases | Improving extract | Mash tun |
| | Shorter purification times | Mash tun |
| | Improving wort filtration | Mash tun |
| | Improving beer filtration | Lager tank |
| Fungal β-glucanases | Improving extract | Mash tun |
| | Improving filtration | Mash tun |

[a]Free amino nitrogen.

### Brewing with malt

If only poor-quality malt is available for brewing, the malt constituents can be solubilised with a modified mashing method or by the addition of enzymes.

#### *β-Glucan digestion*

High-molecular-weight β-glucans are particularly significant in that they increase mash viscosity, thereby hampering mash lautering. Furthermore, the filterability of wort and beer is poor. Also, hazes, even gels formed from poorly solubilised malts, have been described for finished beer. Steiner has reported the amounts of β-glucans and pentosans in barley, malt, wort and beer (Table 5.4).

**Table 5.4.** Gum contents in barley, malt, hops and beer.

| | *Contents (g/100 g)* | | |
|---|---|---|---|
| | *Total gums* | *β-Glucan* | *Pentosan* |
| Barley | 6.4 | 4.8 | 1.6 |
| Malt | 2.3 | 1.1 | 1.2 |
| Wort | 4.7 | 4.1 | 0.6 |
| Beer | 4.9 | 4.0 | 0.9 |

Usually, some β-glucans are digested during malting; however, a large amount is solubilised at increased temperatures during mashing.

β-glucan digestion is limited because the β-glucanases in malt are heat-labile and the poorly solubilised malt itself contains an inadequate amount of these enzymes.

However, the addition of β-glucanase preparations reported by Enari and Markkanen can compensate for this deficiency (e.g., 150 gram Rohamalt MG Plus [RM] per tonne of malt) (Table 5.5).

**Table 5.5.** Analysis of a tank of beer brewed with malt.

| | *Tank bear* | |
|---|---|---|
| *Characteristics* | *Without Rohamalt MG plus* | *With Rohamalt MG plus* |
| Turbidity (FU[a]) | 16 | 13 |
| Viscosity (mPA · s) | 1.68 | 1.63 |
| β-Glucan (mg/liter) | 373 | 45 |
| Filterability (g) | 81 | 104 |
| Filtering efficiency (ml/h) | 5.3 | 6.2 |

[a]Formazan units (turbidity standard).

### *Proteolysis*

Proteins are inadequately digested in poor-quality malt. This results in a deficiency of the amino acids that must serve as nutrients for the yeast. In addition, undigested proteins will cause hazing in beer; consequently, problems of stability will arise. Sorensen and Fullbrook have studied the effects of various microbial proteases on proteolysis in mash. They found that alkaline bacterial protease is strongly inhibited by barley extracts. However, a neutral bacterial protease proved to be considerably more effective in this system.

A dose of 0.2–0.5 kg of enzyme preparation with an activity of 1.0 Anson unit per metric ton of malt is sufficient to generate adequate proteolysis during the protein conversion step. The soluble nitrogen content increases significantly and beer stability is improved.

### *Fermentability*

The addition of amylolytic enzymes to wort improves its fermentability. Heat-stable bacterial α-amylase can be used for the maltose rest during the mashing process (e.g., 0.3–0.5 kg Rohalase HT [RM]/T malt). Continuing the starch conversion in the wort, 100–200 grams of fungal α-amylase with an amylolytic activity of 45000 SKB/g (e.g., Rohalase M3 [RM] or Fungamyl 1600 S [NO]) is added to each metric ton of wort.

## ***Use of raw grain in brewing***

To reduce brewing costs, an increasing number of brewers employ raw grain, ostensibly obtaining the same results (extracts, fermentable substances, brew stability, etc.) as with malt. Raw grains employed are sorghum, rice, wheat, maize, wheat-B starch and sugar syrup. Only limited amounts of raw grain are permitted as brewing addenda in individual countries; in Europe, it varies from country to country, allowing 20–30 per cent raw grain.

Raw grain must be milled finer than malt but not so fine as to cause clumping. When using both malt and raw grain, each must be milled separately. If the raw grain portion is greater than 15 per cent, then a coarser malt grist must be used to avoid problems in lautering. Mash filters are recommended when the raw grain portion reaches 30–40 per cent. The type of raw grain processing depends on the raw grain selected. Thus, maize and rice grains must be mashed with lots of water because both imbibe considerable amounts of water. Using enzymes causes a rapid decrease in viscosity of the gelatinised starch. The iodine-reduction test measures the extent of starch conversion. Iodine colour indication occurs only when the solubilised starch is pasted (gelatinised). The temperature at which the raw grain conversion must be conducted depends on the pasting temperature. Iodine colour indication for sorghum or rice is observed only at temperatures above 80°C (Table 5.6).

**Table 5.6.** Pasting temperatures.

| *Raw grain* | *Pasting temperature (°C)* |
|---|---|
| Barley | 62–74 |
| Barley malt | 63–64 |
| Malt | 64 |
| Maize | 68–74 |
| Rice | 67–75 |
| Wheat | 56–66 |
| Potatoes | 52–58 |
| Sorghum | 69–75 |

If a beer is brewed from malt and raw grain, parallel mashing of raw grain and malt is required. Subsequently, both are combined and saccharified at 68–72°C.

This temperature must be held until the iodine test standard is reached. However, a 45 minutes rest is recommended. An α-amylase at 0.5 kg per metric tonne of grain with an activity of 6000 SKB/g (Rohalase AF [RM]) or depending on the pasting temperature, a heat-stable bacterial α-amylase (e.g., Rohalase AT [RM]), at 0.05 kg/T grain with an activity of 4000 SKB/g, should be used for raw grain conversion. The raw-grain conversion is conducted either under 1–2 bar pressure or at normal atmospheric pressure. If the malt has been extracted under pressure, problems in beer flavour can arise when the malt; grain ratio is greater than 1:1. Other variations in flavour arise from the use of oil-rich semolinas that, on lipid hydrolysis, generate fatty acids. These can also impair formation of the beer's head. Reduced amylolytic activity results in inadequate starch hydrolysis. The availability of assimilable nitrogen for the yeast must be assured because of the very low solubility of raw grain protein.

### *Brewing with unmalted barley*

To reduce the cost of energy associated with malting, raw barley is used. However, using more than 15 per cent unmalted barley causes problems during the brewing process because of limited enzymatic activity, primarily because the β-glucanase and proteases are lacking. A two-step mash process employing both malt and barley produces poorer results than when mashing with malt alone.

However, when enzymes are utilised, brewing with 30–40 per cent unmalted barley is entirely possible. When compared to a standard mash (100 per cent malt), the β-glucan is digested just as well with added β-glucanases. This way, a smooth filtration of wort and green beer is guaranteed. A bacterial α-amylase with an activity of 6000 SKB/g (e.g., Rohalase AF [RM]) should be added to the raw grain cooker at 0.5 kg/T grain. To assure that an adequate amount of fermentable sugar remains available in the total mash, an amyloglucosidase, such as Rohalase HT [RM] or Diase [GB], should be added.

### *Beer Production with 30 per cent barley and 70 per cent malt*

Barley is mashed in the presence of a bacterial α-amylase, such as Rohalase AF [RM] or BAN [NO] at 50°C; malt is mashed in parallel. After boiling the raw grain, both mashes are combined and the standard one-step mash process in the mash turn begins. To improve protein hydrolysis, a bacterial protease with an activity of 200 PU/mg (e.g., Rohalase P [RM]) can be added at 0.3–0.5 kg/T to the proteolytic step, which will yield about a 20 per cent increase of formalin nitrogen.

***Brewing with corn***

The use of technical enzymes in brewing beer from malt and corn has been successful and is outlined in the flowchart in Fig. 5.14. The problems that normally develop with the addition of corn can be avoided by dosing the given enzymes appropriately.

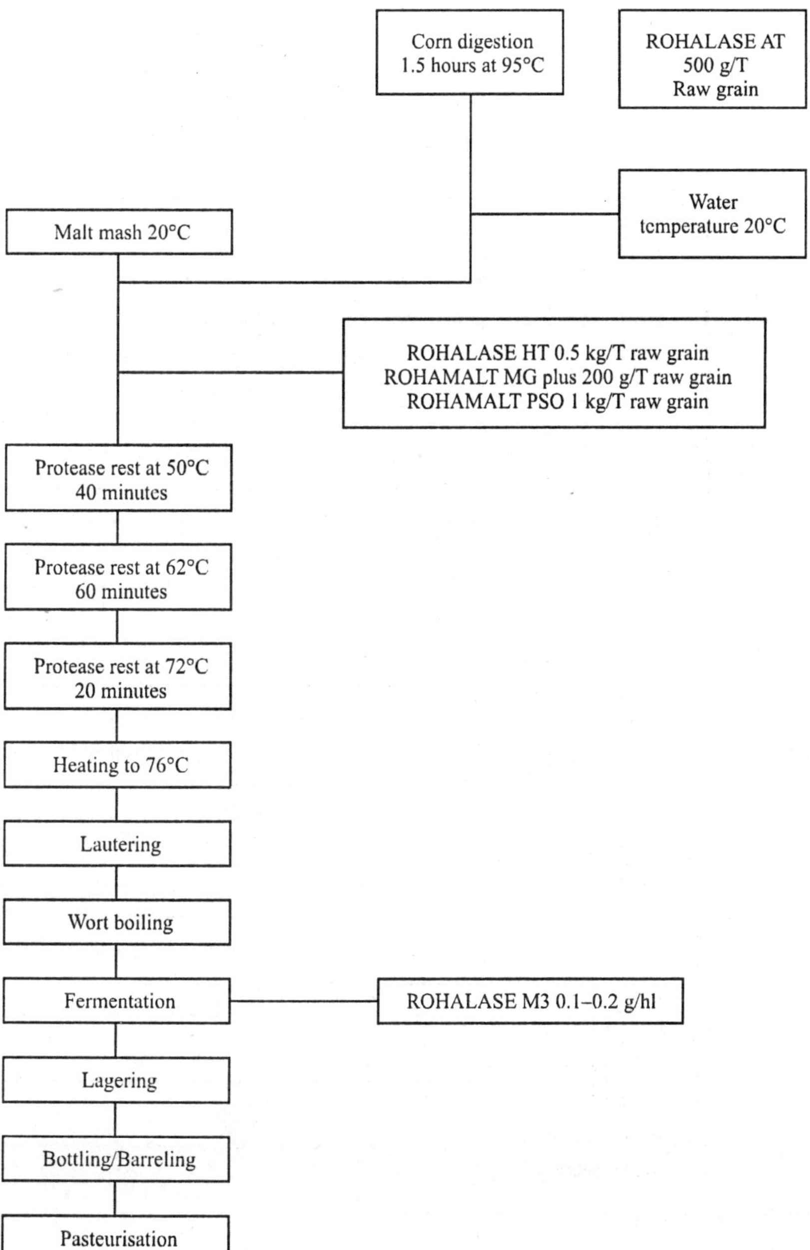


**Fig. 5.14.** Flowchart for brewing beer with malt corn.

Difficulties again include poor fermentability because of a very limited starch hydrolysis, the lack of free amino nitrogen due to inadequate proteolysis and poor filterability because of excessive amounts of highly viscous β-glucans.

### Brewing with 100 per cent sorghum

Brewing beer exclusively with sorghum (Fig. 5.15) has only recently become important because importation of malt has been forbidden in some non-European countries. The key problem in this case is the poor fermentability of the wort because the amino nitrogen for adequate nutritional support of the yeast is limited.

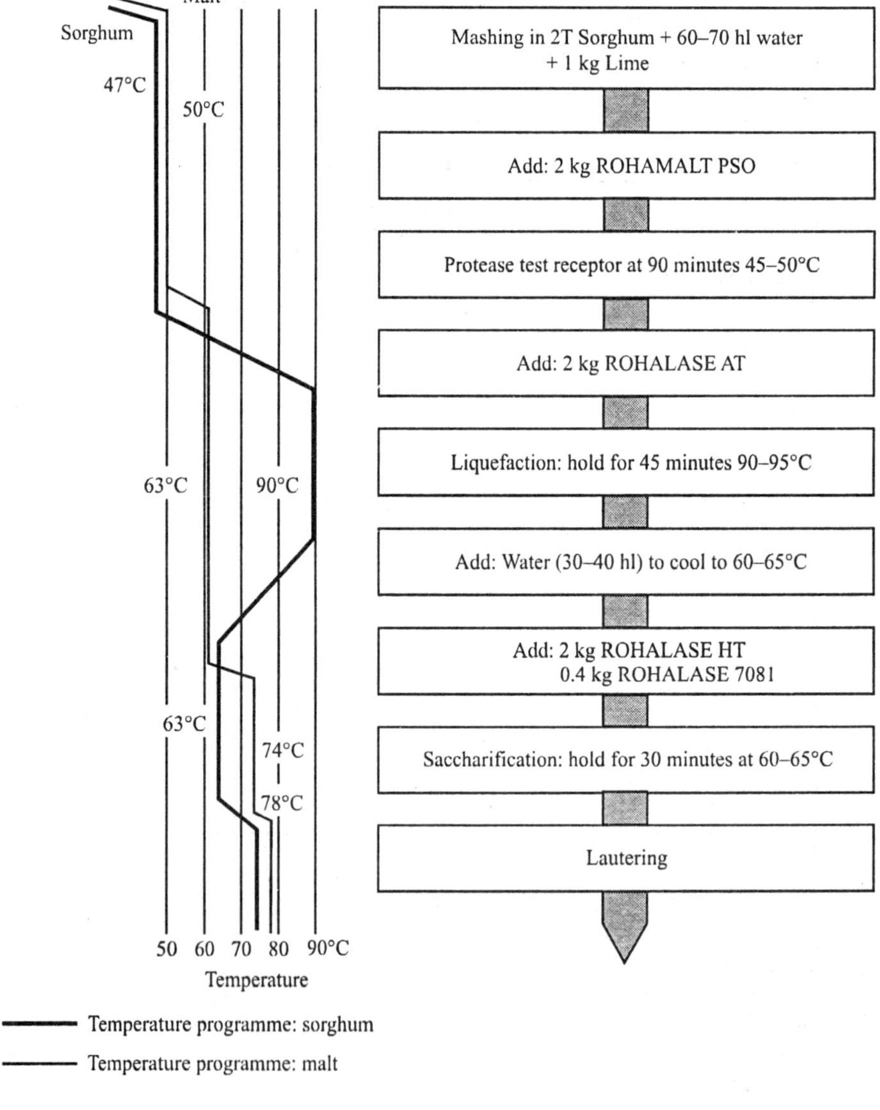


**Fig. 5.15.** Flowchart and mashing steps for brewing beer with sorghum.

Addition of a bacterial protease at 1 kg/T grain with an activity of 250 PU/mg (e.g., Rohamalt PSO [RM]) helps to liberate sufficient amino nitrogen. Because of the high pasting temperature of sorghum starch, a heat-stable α-amylase at 1 kg/T grain with an activity of 3000 SKB/g should be used for raw grain extraction. For a better saccharification of mash, a combination of amyloglucosidase and fungal α-amylase is advantageous.

### *Analytic monitoring of the mashing process*

#### *Proteolysis*

Proteolysis is monitored by determining free amino nitrogen (FAN). With a FAN content of 150–180 kg per tonne based on a 12 per cent wort, an adequate nutrition of yeast is assured. In addition, sufficient high-molecular-weight protein is still present to permit good stability of the head and body of the beer.

#### *Starch hydrolysis*

The simplest way to determine the degree of starch hydrolysis is the iodine test. Inadequate hydrolysis is indicated by the development of a gray film formed by the various amounts of dextrins, which react with iodine to form a reddish to reddish-purple colour. A poorly saccharified wort shows slow settling and poor clarification. Such conditions cause a drop in pH during fermentation, an increase in alcohol content and increased haze.

This brings secondary fermentation to a halt and damages the beer's flavour by producing, for instance, a yeast diacetyl flavour.

### *β-Glucan digestion*

Because direct determination of the β-glucan content is complicated, filtration measurements are usually conducted with wort and green beer. These are then compared with the filtration data obtained with standard brews. Varying malt and raw grain qualities will typically give rise to long lautering times, caused by the high β-glucan content. This, however, is not attributable to β-glucan alone but also to the coagulates formed from proteins, polyphenols and β-glucan.

## Difficulties in Green-Beer Filtration

After completion of storage, the beer is again filtered. This can occasionally give rise to unexpected delays in the brewery's production run. Gumlike substances and hemicelluloses such as pentosans and β-glucans are responsible for poor filtration performance. Filterability is often unsatisfactory when poorly solubilised malt with little hemicellulase activity or when mixtures of raw grain are used. Improved filtration is achieved by adding hemicellulase preparations such as Rohamalt F [RM], Ceremix [NO] or Filtrase [GB] to the lager tank shortly before terminating lagering.

## Cold Stabilisation of Beer

Beer stored at very low temperatures will develop a haze. The haze will disappear when the beer is warmed. Aged beer and especially pasteurised beers tend to develop a haze when chilled. This turbidity is the consequence of reactions between oppositely charged lipophilic colloids that precipitate in the same way as would an excessively dehydrated saturated solution. During warming, the precipitates again dissolve as a result of rehydration.

The most important reactants are the polyphenols, which can form complex associations with carbohydrates and polypeptides.

Pepsin, ficin, bromelain, fungal and bacterial proteases have been used, in addition to papain, to prevent cold haze. However, papain by itself or in combination with other proteases is the most frequently employed enzyme in cold stabilisation. Papain activity is determined by the BAPA method (BAPA, *Na*-benzoyl-L-arginine-*p*-nitranilide), in which the activity of one BAPA unit corresponds to the amount of enzyme that converts 1 μmol of substrate per minute under defined reaction conditions.

Papain is an 'intelligent' enzyme in that it acts only when haze is present or it may act to prevent the development of the cold turbidity that previously hydrolysed the protein. At a dose of 200–300 BAPA units of papain per hectoliter (26.4 gals) beer, e.g., Rohamalt Stab [RM] or Collupulin [GB], the amount of foam and its stability is not reduced. The Forcier test [*MEBAK Analysenmethoden*] is used for determining the haze stability of a beer. The beer is subjected to a cold-heat treatment and the haze that develops is measured.

Other future options for cold stabilisation may be the use of immobilised proteases or the use of tannase and polyphenol oxidase.

### Light-Beer Production

Light beer, as distinguished from standard beer, is characterised by less wort and lower alcohol and residual sugar content and thus fewer calories.

Light beer is brewed with malt or malt with a maximum of 50 per cent raw barley grain. The production process is the same as that for standard beer; however, after boiling the wort and at the start of fermentation, a heat-stable amyloglucosidase is added, such as Rohalase AS [RM], which converts dextrins during fermentation into glucose that is subsequently fermented by yeast.

To avoid secondary saccharification in beer, the use of a heat-labile amyloglucosidase (e.g., amyloglucosidase from *Rhizopus*) is recommended; this enzyme is then completely inactivated on pasteurisation.

### Diacetyl in Beer

Diacetyl formation can be prevented by addition of diacetyl reductase (Maturex [NO]) at the start of primary fermentation, when most of the α-acetolactate is being formed. The reaction is strongly pH-dependent. The activity at pH 5.0 is about three times higher than that at pH 4.0. In primary fermentation, the pH drops after the first day from 5.2 to 4.5 ([NO], technical information).

### Enzymatic Alcohol Production

Alcoholic beverages as well as alcohol for medicine and industry are produced in many countries from different starchy raw materials. Historically, the development started with producing brandy from sugar and starch. For many centuries, juices containing sugar or honey were fermented and the alcohol produced was subsequently concentrated by distillation.

Fortified spirits are directly distilled from wine without further processing. As such, depending on their origin, they enter the market as brandy or cognac.

Sugarcane and sugarbeet molasses often serve as the raw materials for alcoholic beverages and technical alcohol. Other raw materials include fruits such as cherries, pears, apples (for Calvados), plums (for Slivovic) and agave juice. The latter is the source for Mexican pulque.

Many types of alcoholic beverages are produced from starchy materials, including whiskey from malted barley and bourbon whiskey from corn and rye. Aquavit is obtained from potatoes or barley and vodka from rye, wheat or potatoes. In East Asia sake is produced from rice.

Alcohol for industrial use is synthesised from ethylene by hydrogenation (OXO process). It is used as a solvent for paints and cosmetics and as a raw material for chemical compounds. It is employed in trade and domestically as fuel.

However, renewable natural energy sources such as sugar, starch, cellulose and hemicellulose can serve as raw materials for producing alcohol.

In addition, whey, a waste product from cheesemaking and an environmental burden, can be converted to ethanol by fermentation.

Naturally, cellulose, like starch, can also targeted as a raw material. Many research projects dealt with the enzymatic degradation of cellulose to fermentable sugars. However, the enzymatic conversion of cellulose still depends on a costly pre-treatment of the fiber and thus the cellulose-based technology is considerably more expensive than that of starch.

## Starch Degradation

The starch cannot be fermented directly. Starch must first be liquefied before it can be saccharified. This is accomplished with green malt, roasted barley malt, fungal malt and microbial amylases.

Prior to 1945, distilleries in Germany used only malt for saccharification; since then fungal malt has also been used. Even earlier, mould cultures served as a source of saccharifying enzymes in the Orient. In the amylo process, steeped rice kernels are inoculated with spores of *Rhizopus delemar* and the mash, saccharified during an incubation of 24–30 hours at about 37°C, is then fermented with yeast.

Fungal malt is the end product of a top-fermenting *Aspergillus oryzae* on a heat bran nutrient medium. This procedure was later replaced by fungal amylase preparations obtained from bottom-fermenting species.

In addition to malt and the β-amylase preparations from barley or sweet potatoes the following microbially derived enzymes are employed in distilleries:

1. For liquefaction:
   (a) Bacterial α-amylase.
   (b) Heat-stable bacterial α-amylase.
   (c) Fungal α-amylase.
2. For saccharification:
   (a) Amyloglucosidase from *A. niger* or from *Rhizopus* sp.
   (b) Fungal α-amylase by itself or in combination with amyloglucosidase.
   (c) Pullulanase for complete conversion in combination with other amylases.
3. For reducing viscosity of the mash: β-glucanases.

## Alcohol Production

The following basic steps can be distinguished:

1. Processing the raw material by milling and heating or treating with acid or alkali.
2. Liquefying starch with α-amylases or with acid to dextrins.
3. Saccharifying dextrins with β-amylases or amyloglucosidase to fermentable sugars.
4. Fermenting saccharified mash.
5. Distilling fermented mash. This yields unrefined alcohol and the wash as a by-product.
6. Rectifying crude spirits (first run, distilled alcohol, rectified alcohol and tails).

## Starch Conversion

During the past 30 years, green malt and dry malt have been replaced in the distilleries by microbial enzymes for liquefying and saccharifying starch.

Short liquefaction periods for starchy raw materials are desired to ensure that saccharification can proceed during fermentation in 2–3 days. Until recently, the Henze steam process (Fig. 5.16) which involved pressure-cooking potatoes at 140–180°C, was the process of choice.

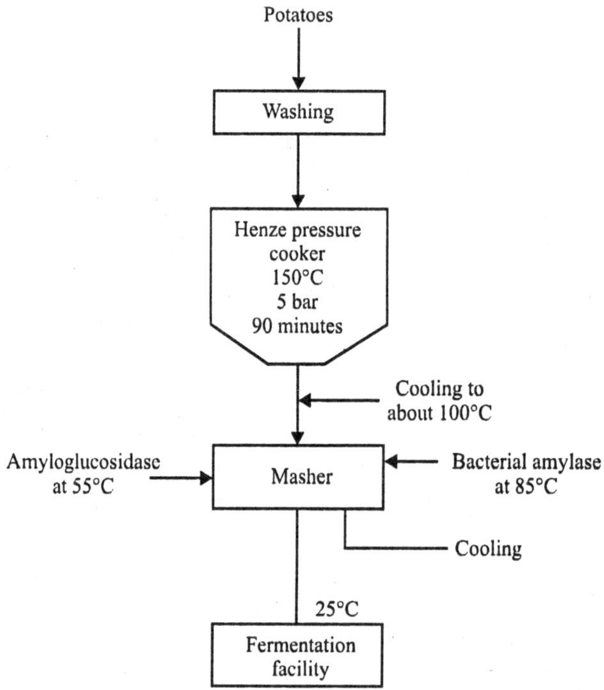


**Fig. 5.16.** Flowchart of the Henze process.

Now, potato processing with a 'cold-mash process' has gained importance. After chopping of the raw material in a blender, liquefaction is accomplished on addition of bacterial α-amylase from *B. subtilis*, at 65–75°C with a temperature maximum of 85°C (e.g., Optimash-L 840 [SO]). The energy requirement is thereby considerably reduced, particularly when energy recovery is possible. Milling raw grain into a fine meal is a prerequisite for the 'mashing at saccharification temperatures' method employed in grain distilleries.

Ultrafine meals can be obtained using carborundum milling disks. In processing of rye and wheat, the grist is slurried in cold water (about 15°C) and the pH, if too low, is adjusted to 5.6–5.8. When malt is used, 4–5 per cent malt is required (at 57°C) or a corresponding amount of technical enzyme preparation may be added. This, process requires the addition of disinfectants.

### *Boiling process*

After reduction of the raw material into a finely ground meal, the liquefying enzymes are added and the starch mass is steam-heated to 80°C. Frequently, a protein rest at 50°C is staged to ensure that this temperature range is passed through slowly.

Experts in the field have different opinions about the yields and energy savings in milling or processing the raw materials. A few examples without evaluation are described below:

### *High-pressure steam process*

Pressure cooking serves to totally free the starch grains from their cellular associations. This occurs only on heating the mash in a pressurised vessel at temperatures >100°C. It also allows the introduction of thermostable bacterial α-amylases that can tolerate temperatures in excess of 100°C.

In the potato distillery—when exhausting the heated mash at constant temperature—water and lime are added to adjust the pH in the mash tub. After starting the discharge process and adjusting the temperature to 70°C, one-third of the allotted enzyme is added. After an additional 10 minutes of exhausting, the remaining two-thirds of the enzyme are added. Alternatively, the rapid exhaust procedure is efficiently conducted at 90–100°C and on cooling to 80–85°C, one-third of the enzyme is added. The remaining enzyme is added after the mash has cooled to 70°C. In the grain distillery, white lime and water are added to the mash tub. After exhausting and cooling to 70–75°C, the entire enzyme lot (e.g., Optimash L 840 [SO] at 0.275 liter/T starch) is added. Given a high starch content, an aliquot of the enzymes is added at 80°C to rapidly lower the viscosity. The rest of the enzyme is then added at 70°C.

In the American batch process, soaked ground corn (about 25 per cent dry substance) is treated with thermostable α-amylase (e.g., Termamyl 60 L [NO] at 0.2–0.3 kg per tonne starch), the mash is then heated with steam to about 150°C. Although the enzyme will be inactivated within a short time, it helps maintain a lower viscosity in the boiler. After this extraction, additional enzyme at 0.35–0.7 kg/T starch is added on cooling. When the liquefaction is completed, amyloglucosidase is added at 60°C to initiate saccharification. In a process yielding ethanol for fuel, 100 kg of ground corn in 400 liters of water plus a thermostable bacterial α-amylase are initially heated to 65–71°C; then the temperature is raised to 160°C. On cooling to about 65°C, a fungal amylase from *A. niger* is added and the mash is saccharified with amyloglucosidase from *R. niveus* at 32°C.

### *Continuous processes*

A variety of these processes exist and were developed with the aid of technical enzymes. The advantages of continuous operation include better process control and better utilisation of facility capacity. Figure 5.17 illustrates the continuous alcohol production process.

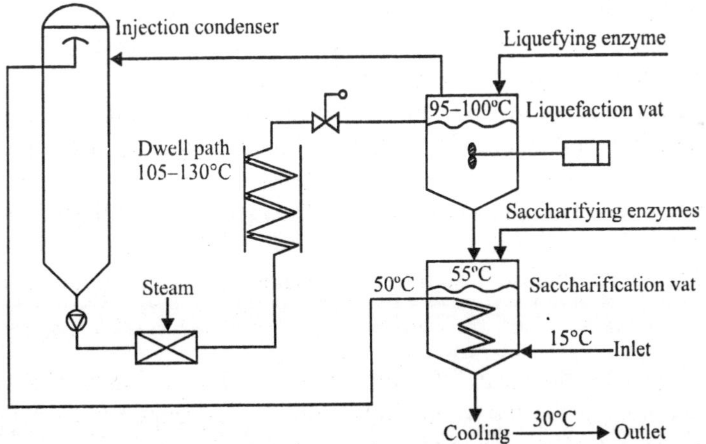


**Fig. 5.17.** Continuous alcohol production process.

### Baking Industry

Wheat and other cereal grains which are used as sources of flour for baking purposes contain a number of naturally occurring enzymes that are essential for germination and growth. The most common source of flour for baking purposes comes from wheat and the major constituent of flour is starch. The most important enzymes contained in flour for the baking process are the amylases and proteases. The amylases act on the damaged starch from the milling process to produce dextrins and low molecular weight sugars. These sugars are then made available to the yeast during the baking process. The protease enzyme acts on the protein material breaking down the protein into peptides and amino acids which are the nutrients for yeast to grow and speed up the fermentation process. Fungal $\alpha$-amylase, cereal $\beta$-amylases along with fungal proteases are ideally suited for the use in baking process.

Some of the enzyme modifications of baking flour are:

1. Increased bread volume, improved crust colour, improved crumb structure and anti-staling effects due to the action of amylase on starch.
2. Gluten modification, dough viscosity reduction and energy saving, improved machinability in biscuit production due to the action of protease on protein.

Recent advances in the application of enzymes in the baking industry are:

1. To improve quality of fiber-enriched wheat breads by enzymic treatment of the fiber fraction.
2. To improve bread making quality of Dorum wheat flour by utilising amylases and proteases.
3. Application of enzymes in baking technology for maintaining the quality and freshness of bread during free storage.

### Application of Glucose Oxidase

Glucose oxidase is an oxidoreductase, oxidising D-glucose to gluconic acid. It is used in the food industry primarily to prevent changes in the colour and flavour of food products—both during processing and in storage. It is well-known that the oxidation which occurs when oxygen is present causes significant deterioration of foods. Hence there is a need for addition of chemical antioxidants such as ascorbic acid and glucose oxidase. Because of its mild action, glucose oxidase is widely used for this purpose:

1. It is utilised as an antioxidant in citrus drinks, canned soft drinks, beer, wine, mayonnaise, salad dressings and in dried foods like instant coffee, cake mixes, milk powders, dried soups, active dried yeast, concentrated aroma components and other dried foods which are very susceptible to oxidative deterioration.
2. The second most important market for glucose oxidase and catalase in the food industry is the removal of glucose from egg whites and whole eggs.
3. Glucose oxidase with ascorbic acid can be used for oxidising flour.
4. Glucose oxidase in conjunction with cellulase should protect against infection in fermented fruit mashes.
5. Glucose oxidase is widely used as a diagnostic agent in medicine in the measurement of glucose levels in patients.
6. It is also used to detect glucose in starch hydrolysates, fermentation liquids and several foods.

### Animal Feed

Enzymes are utilised as a feed supplement for poultry, pigs, etc. They aid in breakdown of organic feed substrates, thereby encouraging the bioavailability of otherwise trapped nutrients, while improving liveweight and feed conversion efficiency in diets based on corn, soyabean meal, wheat meal, rapeseed meal, etc. Depending on the substrate used in the feed the enzyme composition also varies. Usually a predetermined mixture of phytases, xylanases, cellulases, amylases, pectinases and proteases are used based on the kind of feed composition and the animal for which it is used. Recent advances are:

1. Feed value improvement of chicken diets by enzyme supplementation.
2. Productivity of broiler chickens and efficiency of food consumption due to addition of vitamins and enzyme preparations.
3. The improvement of the nutritive value of barley for chicks by enzyme supplementation.
4. The effect of non-starch polysaccharide hydrolysing enzymes in linseed diets on chicken and pullets.
5. Use of non-starch polysaccharide hydrolysing enzymes as feed supplements for laying hens.

## Food and Drink Industries

The traditional use of yeasts (e.g., *Saccharomyces carlsbergensis*) in the baking and brewing industries arose because they contain the enzymes for alcoholic fermentation: In common with other organisms, they metabolise hexose sugars to produce pyruvate, but, whereas animals convert this to lactate under anaerobic conditions, the anaerobic end product in yeasts is ethanol, with carbon dioxide being evolved:

$$\underset{\text{Pyruvic acid}}{CH_3.CO.CO_2H} \xrightarrow[Mg^{2+},\ TPP]{\text{Pyruvate decarboxylase}} \underset{\text{Acetaldehyde}}{CH_3.COH} \xrightarrow[NADH \;\rightarrow\; NAD^+]{\text{Alcohol dehydrogenase}} \underset{\text{Ethanol}}{CH_3.CH_2OH} + CO_2$$

In the baking of bread, the preliminary process involves the mixing of wheat flour (mainly starch and protein) with yeast and water. Starch consists of D-glucose units linked by α-1,4 glycosidic bonds, with α-1,6 bonds at branching points; the enzymes α-amylase and β-amylase present in the flour cleave some of the α-1,4 bonds, the eventual products being glucose, maltose (a disaccharide) and some oligosaccharides which cannot be broken down further because of the presence of α-1,6 bonds. Glucose and maltose can then be metabolised by the yeast and carbon dioxide is formed which distends the protein framework of the dough, ready for baking. However, wheat flour often has a low α-amylase content, so it may be supplemented with malt flour or a more recent development, by fungal α-amylase (from *Aspergillus oryzae*). Wheat α-amylase is very heat stable, so it continues to act for a time during the baking process; this may lead to too much starch breakdown taking place and cause the bread to be somewhat soggy. For this reason, wheat α-amylase is sometimes inactivated by a brief treatment with superheated steam prior to supplementation of the flour with the more heat-labile fungal α-amylase. Proteases from *Aspergillus oryzae* may be introduced to cause strictly limited breakdown of the wheat protein (gluten), thus shortening mixing time and enabling a smoother, more uniform dough to be obtained. Malt flour also contains proteases, but in variable amounts, so its use might lead to excessive gluten breakdown.

In brewing, the main starting material is malt, produced by allowing barley seed to germinate in moist conditions. The reserve starch is broken down by the amylase present to give, among other products, glucose and maltose. The grains may then be roasted to prevent further growth and to add flavour, after which the soluble material present is extracted by water to produce the wort. This is further flavoured with hops and then yeast is added to produce ethanol by alcoholic fermentation of the glucose and maltose.

Bacterial α-amylase (from Bacillus subtilis), which is even more heat stable than wheat α-amylase, is of increasing importance in the brewing industry. It may be used to produce a wort from unmalted barley (and maize), although some malt (a minimum of 20 per cent) is usually added. Fungal *exo*-1,4-α-glucosidase, better known as amyloglucosidase or glucamylase (e.g., from *Aspergillus niger*) may also be added, since this can cleave α-1,6 as well as α-1,4 glycosidic bonds and so increase the yield of glucose and maltose from starch. Malt contains some protease activity, so bacterial proteases are usually added when a wort is being prepared from unmalted barley.

In the industrial production of glucose from starch, the latter is first solubilised and partly degraded by bacterial α-amylase (or, but less commonly than hitherto, by HCl) and then treated with fungal amyloglucosidase (free or immobilised). This can cleave both types of glycosidic bond found in starch, so gives a good yield of glucose. Glucose may also be obtained from cellulose-containing waste products by treatment with cellulase (e.g., from *Aspergillus niger*); as a further possibility, it may be produced, together with galactose, by the action of β-galactosidase (lactase) on lactose, which is present in whey and so is a major by-product of cheese manufacturing.

Invert sugar, a mixture of glucose and fructose, is produced from sucrose by the action of yeast β-fructofuranosidase, better known as invertase (an enzyme which can only be extracted by disruption of the yeast cell wall). Sucrose crystallises from solution more readily than invert sugar at the high concentrations required for jam making, so the latter is widely used for this purpose. Also, in confectionery, crystalline sucrose is coated with chocolate and invertase, causing the subsequent conversion of sucrose to invert sugar and hence liquefication to form a 'soft center'.

Invert sugar may also be produced from glucose by the action of glucose isomerase (bacterial or fungal), an enzyme now thought to be identical to xylose isomerase. At the present time, conversion of glucose to fructose is the biggest industrial application of immobilised enzymes. Fructose is much sweeter than glucose, so high-fructose syrup can be used to replace sucrose as a sweetener, where it is economical to do so. It may also find increased use in this respect because of fears about the adverse effects of sucrose on health, regardless of whether fructose can be proved to be any safer.

Certain amino acids essential for animal growth (e.g., methionine) are found in low amounts in the proteins of some vegetable-based foodstuffs, so may have to be added to the diet. However, animals can only utilise L-amino acids, whereas D- and L-amino acids, in equal amounts, are produced by chemical processes. Dietary supplements usually consist of such mixtures, so half is wasted, even if no other harm is done. The procedure introduced by Chibata and colleagues for the production of L-amino acids from racemic mixtures is of great significance: Acetyl D- and L-amino acids are passed through a system containing aminoacylase immobilised on a DEAE-Sephadex (or other) carrier, causing the deacylation of the L-form alone, which can then be separated; the unchanged D-form is then subjected to racemisation and the process repeated.

The clarification of cider, wines and fruit juices (e.g., apple) is usually achieved by treatment with fungal pectinases. The pectins of fruit and vegetables play an important role in jam making and other processes by bringing about gel formation. However, they cause fruit drinks to be cloudy by preventing the flocculation of suspended particles.

Pectinases are a group of enzymes including polygalacturonases, which break the main chains of pectins and pectinesterases, which hydrolyse methyl esters. Their action releases the trapped particles and allows them to flocculate.

Cheese production involves the conversion of the milk protein κ-casein, to paracasein by a defined, limited hydrolysis catalysed by chymosin (rennin). In the presence of $Ca^{2+}$, para-casein clots may be separated from the whey, after which the clot is allowed to mature under controlled conditions, still in the presence of chymosin, to form cheese. Since chymosin can only be extracted from calves killed before they are weaned (pepsin is produced instead of chymosin after weaning), the enzyme is in short supply, so there has been a large-scale search for an acceptable substitute. Proteases from animals (pepsin), plants (ficain and papain) and over a thousand micro-organisms have been tried, either on their own or mixed with calf chymosin; several have been largely, if not completely, successful (e.g., a protease from *Mucor rouxii*).

Papain is sometimes used as a meat tenderiser; some South American natives have traditionally wrapped their meat in leaves of papaya, the fruit from which papain is extracted. Papain (and other proteases) may also be used in the brewing industry to prevent chill hazes, caused by precipitation of complexes of protein and tannin at low temperatures.

## Flavouring and Colouring

Flavouring substances represent between 10 and 15 per cent of the weight of world use of food additives, representing up to 25 per cent of the value of the total food additives market. Industrial enzymes find a variety of applications, ranging from the general processes for the production of the raw flavour ingredients such as sugar and protein hydrolysates, to highly specific enzyme reactions targeted at identified flavour compounds. It is possible that the intense sweeteners which are derived from extracts of exotic plants are already being processed with the aid of enzymes degrading the tissues of the source materials. Sucrose production utilises several enzymes which facilitate the degradation of mucilages produced by microbial contamination of juice prior to processing. Fructose is becoming increasingly available by either the processing of sucrose or the isomerisation of glucose from starch processing. Recently the enzymatic hydrolysis of insulin has been proposed as a potentially useful industrial process and enzymes to carry out the conversion are now commercially available. Dextrose and maltose/dextrose syrups are key sources of flavouring for many applications covering baking, confectioners, processing fruits, jams, soups, snack and convenience foods, dietary products and breakfast cereals. Recent advances include production of maltose syrup by bioconversion of cassava starch.

### *Enzymic processing for colour*

Colour materials represent 0.5–0.8 per cent of the total weight of food additives and some 5 per cent of the value of food additives. The use of enzymes in the extraction of natural colours is not very diverse at present, but it is receiving considerable attention at the research and development level following the rapid increase in the availability of complex enzyme preparations for the degradation of plant tissues under mild conditions.

#### *Future developments*

1. Use of esterases, lipases and proteases with specific action in cheese flavour development.
2. Peptidases will contribute to the subtlety of control of bitterness and flavour components in the hydrolysis of low cost legume and cereal proteins.

## Fruit Juice

The first enzymes to be used by the fruit juice industry were the pectolytic enzymes for the clarification of apple juice. Since then fruit juice processing has developed into a major, highly technological industry, covering not only core fruits but also stone fruits, citrus and tropical fruits, berries, grapes and even vegetables. The functions of pectolytic enzymes have become more specialised and other enzymes such as amylases and cellulases also now form an integral part of today's fruit juice technology. Cellulases either as a single preparation or as part of the pectolytic enzyme preparation are useful tools for speeding up the extraction of colour from the fruit or for the total liquefaction of the plant tissue.

Some of the recent advances are:

1. Food processing of tropical fruits like mango, papaya, pineapple, banana, etc., by enzyme preparations.
2. Application of cellulases in strawberry juice production.

### Proteins

The concept of enzymatic modification of protein is not new, as illustrated by the dairy and leather industries. However, the high cost of animal protein has led to development of the enzymatic treatment of proteins to increase their value and availability and for their recovery from scrap or waste materials. A wide range of enzymes and substrates are currently available for examination and the number of modified proteins resulting from these processes grows daily. Enzymatic treatment of protein is carried out for a number of different reasons. The protein may be modified so as to increase its value; alternatively, the aim may be extraction of the protein to turn it into more usable form. Application of plant and animal proteases is shown in Table 5.7 and microbial proteases in Table 5.8.

**Table 5.7.** Application of plant and animal proteases.

| *Enzyme* | *Distribution within industry* |
|---|---|
| Papain | Beer chillproofing, meal tenderisation, protein hydrolysates, baking |
| Ficin | Meat tenderisation, protein hydrolysates |
| Bromelain | Meat tenderisation, digestive aid, clinical applications |
| Trypsin | Leather manufacture, protein hydrolysates, clinical application, membrane cleaning |
| Pepsin | Functional protein hydrolysates, cheese production |
| Rennin | Cheese production, casein production |

**Table 5.8.** Application of microbial proteases.

| *Enzyme* | *Distribution within industry* |
|---|---|
| Fungal acid | Baking, photographic, brewing and fermentation industries, protein hydrolysates |
| Fungal neutral | As for fungal acid |
| Fungal alkaline | Little application within industry |
| Bacterial neutral | Brewing, gelatine, photographic and meat industry, protein hydrolysates, cheese ripening |
| Bacterial alkaline | Gelatine, leather and meat industry, protein hydrolysates, enzyme detergents |

Recent advances in this area include enzyme coagulation of recombined mixture of milk proteins.

### Vegetable Protein Modification

Proteins which are available to mankind from animals and plants are found in various forms. Vegetable proteins in bulk, exhibit a number of desirable functional properties. Enzymes are generally used to obtain useful derivatives from proteins with low or no functionality enzymic hydrolysis. Highly flavoured savoury extracts are obtained from the residual cakes of soya/peanuts. These are further processed for nutritious and delicious food products by enzymes. Ingredients for soups, dry and canned vegetable products, vegetable extracts, pet foods, improved animal feeds, etc. use enzymes in their preparations.

Proteins such as leaf concentrates, peas, lupins and similar crops are also improved by the use of enzymes in extraction and processing. A range of enzyme formulation offer a wide choice for vegetable protein modification. These formulations are custom tailored to suite the customer needs. Conversion of complex vegetable protein to highly modified hydrolysates of many small peptides is possible.

Recent advances in extraction include:

1. CGTase and α-amylase for potato waste treatment.
2. Enzyme modification as tool for alteration of safety and quality of soya proteins.
3. Enzyme hydrolysis of proteins in corn gluten meal.
4. Production of feed from cellulose containing material by enzymic treatment.
5. Enzymic degradation of carbohydrates in vegetable protein isolation.
6. An enzymic process for the physical refining of seed oils by degumming of seed oils.

### Citrus Processing

Citrus fruits such as oranges, lemons, limes, grape fruits find wide applications in healthcare, cosmetics, pharmaceuticals, health foods, beverages and brewery. Enzymes enhance the efficiency of the mechanical process of citrus fruit processing. The numerous important products from citrus fruits include juice, concentrates, pulp wash, peel juice, clarified juice, essential oils, aroma, essences and pectin. The two types of enzymes used in citrus processing are:

1. Enzymes for cytolysis which help citrus processors improve quality, yield and processing efficiency.
2. Glucose oxidase enzymes which helps preserve the colour and flavour of citrus products.

Pectolytic enzymes which contains pectinases such as pectin lyase and polygalacturonase acts to depolymerise pectin which is predominantly present in cell wall and juice of the citrus fruit. It also has significant side activities such as hemicellulase, which improves viscosity reduction and naringinase, which debitters through the hydrolysis of naringin. Macerating enzymes which is a mixture of pectinase, cellulase and hemicellulase is suited for many applications in citrus processing.

The cellulase and hemicellulase increase the degradation of complex carbohydrates, resulting in improved extraction of juice, flavour, colour, cloud and essential oils. It also has significant amounts of naringinase activity. Glucose oxidase preparation contains glucose oxidase and catalase for the efficient removal of oxygen. Glucose oxidase is used for maintaining colour and flavour in citrus beverages. The different applications of enzymes in citrus processing include—juice concentration, pulpwashing and pulpwash concentration, peel washing and peelwash concentration, breaking of citrus oil emulsions, juice clarification, peel removal, production of cloudy peel products, comminuted products and peel juice, treatment of by-products, equipment cleaning, deoxygenation (to prevent degradation of flavour and inhibit microbial growth), debittering (with naringinase and hesperidinase).

### Enzymes for the Dairy Industry

Milk is a multi component secretion of the mammary gland containing important sources of nutrition: protein, fat, lactose, together with small amounts of minerals and vitamins. Milk is also a source of large number of enzymes (about 40) albeit in small quantities some of which have important implication with regard to milk processing. The application of enzymes in milk processing is well established through the use of rennet which probably represents the earliest use of enzyme technology in the food industry.

More recently β-galactosidase has found commercial use in the hydrolysis of lactose in milk products. These two applications represent the major use of enzyme technology in the dairy industry. The active, milk coagulating enzyme present in the calf rennet extract is an acid protease, designated as renin or chymosin.

Renin causes the casein in milk to coagulate in cheese making. Renin causes a slow decomposition during the ripening process, when desirable and characteristic flavours are obtained. Enzymes used for cheese maturation and flavouring preparations other than renin are: proteases from lactobacillus, trypsin and pepsin.

Other enzymes used in dairy technology are:

1. Proteases and lipases in cheese ripening process for consistent flavour character and balance in the final products.
2. Lipases in the manufacture of certain Italian cheese varieties (characteristics flavour).
3. Lipases modified butter fat products as a source of major flavour components of butter.
4. Catalase for degrading excess hydrogen peroxide remaining in the milk after peroxide treatment (used as chemical sterilant).
5. Lactose degradation by lactase used in the production of baby food, milk powder for ice-cream and bread, feed stuffs and articles of food which have a whey basis, whole milk concentrate, etc.

The use of enzymes in dairy industry favours:

1. Increase in number of ingredients to be used in food processes.
2. Enzyme assisted product diversification in place of rigid chemical and physical processing methods.
3. Reduction in energy costs by value addition in the process methods used in the dairy industry.
4. Reduction in size of installations, especially when working with immobilised systems.

Some of the enzymes used in dairy products are shown in Table 5.9.

**Table 5.9.** Enzymes used in dairy products.

| *Enzyme* | *Effect* |
|---|---|
| Rennin | Cleavage of proteins and clotting of milk |
| Catalase | Milk sterilisation |
| Glucose oxidase | Removal of oxygen by oxidation of glucose to gluconic acid |
| Lactase | Cleavage of lactose to glucose and galactose |

## PLANT TISSUES

There are many examples of the enzymic modification of plant tissues relating to well-defined applications. The majority of the enzymes used are standard industrial preparations and are slotted into well-defined applications, so that degradation of structural cell wall materials or carbohydrates or proteins, can be achieved in the appropriate sequence, or individually and exclusively, according to the overall process objective. Some proven applications of enzymes for industrial processing of plant tissues are shown in Table 5.10.

### Membrane Cleaning

Industrial filtration systems incur high operating costs when they require frequent downtime for cleaning of the filter surfaces. Enzymatic cleaning solutions can be prepared which are of modest ionic strength and pH values of between 5 and 9 and which will remove many filter soils without hazard, while producing a very modest effluent load on discharge from the system. The choice of enzyme washing system will depend upon several factors. These include the nature of the soiling materials, the acceptable pH and temperature range of the system, the available washing cycle time and the critical chemical standard applied to the process being cleaned (e.g., food products: pharmaceuticals or other high purity products). Bacterial alkaline protease with detergent, trypsin, papain, α-amylases, β-glucanases, cellulases, pectinases, etc. have been conceptualised for different applications.

**Table 5.10.** Some proven applications of enzymes for industrial processing of plant tissues.

| *Application* | *Action* |
|---|---|
| Alginates | Extraction and/or rehydration (cellulase, pentonase, pectinase) |
| Analytical samples | Solubilisation of complex materials, unbinding of protein or carbohydrate complexes (subtilisin, proteases, pectinases, lipase, amylases) |
| Baking | Dough and gluten modification (amylases, proteases, cellulases) |
| Citrus cloud | Partial degradation of pectin (modified pectinases) |
| Citrus juice | Reduction of viscosity of gums for concentration (cellulase, pectinase) |
| Fiber | Deactivation of protease inhibitors, processing and analyses (cellulase, pectinase, hemicellulase) |
| Gums | Isolation and purification (amylase, protease) |

## DETERGENTS

Enzyme based detergents have a significant marketshare ranging from 30 to 60 per cent in most industrialised countries. Most commercial detergent enzymes are alkaline bacterial proteinases of the same type. For heavy duty formulations and liquid detergents, enzymes with a greater alkali tolerance are preferable. In enzymatic machine dishwashing the detergents require an alkali-resistant and moderately heat-resistant α-amylase. Recent advances in the application of enzymes for detergents are:

1. Application of cellulases and lipases in laundry detergents for soil removal, softening of textiles and colour brightening.
2. Application of enzymes in detergent for removing fish odour.
3. Enzyme mixture of proteases, cellulases; amylases, lipases and peroxidases for both laundry and dishwashing applications.
4. Detergent compositions comprising pectin lyase for overall cleaning performance and stain and soil removal of fruit and vegetable juice.
5. Utilising antibodies specific to oxidoreductase, lipase, amylase, protease, cellulase to have controlled removal of stains.
6. Enzymes as detergent materials with bleaching agent resistance and adaptable to low temperature cleaning application of the oxidoreductases of detergents.
7. Isopeptidase enzyme for the removal of body soils from fabrics.

### Paper Industry

The paper industry is growing at a phenomenal rate of 7–8 per cent annually since globalisation started in India. Mills are looking for modern technologies to produce world-class products to cater to domestic and international markets. Thus, the Indian pulp and paper industry is undergoing a sea change to keep pace with consuming industries to meet the ever-increasing demand.

### Traditional Virgin Paper Making

Wood consists of three major natural polymers—cellulose, hemicellulose and lignin. In the pulp mill, these polymers are separated and washed.

The first step in converting wood fibre to paper is the formation of a pulp suspension, produced by mechanical or chemical process.

Mechanical pulping is an attrition process in which fibres are separated mechanically with the input of large amounts of energy. Mechanical pulps contain all the wood components, including lignin. Although they are cheap to produce, they have the drawback of becoming darker when exposed to sunlight. Therefore, they are used mainly in the manufacture of newsprint and magazine paper.

In chemical pulping, wood chips are cooked in chemicals until the lignin dissolves and releases the wood fibers. The dominant chemical pulping process is the Kraft process, which gives a dark brown pulp caused by lignin residues. Before it can be used for the manufacture of fine paper grades, this dark pulp must undergo a bleaching process. Chlorine or chlorine compounds are used as bleaching agent, which result in effluents rich in chlorates and chlorinated organic compounds that are harmful to the environment.

### Pulp Bleaching

Viikari reported that endoxylanases decrease chemicals needed for bleaching kraft pulp. Since then many researchers have confirmed their results and the technology is being utilised on commercial scale today. Hemicellulase treatments are effective on both hardwoods and softwoods. Although promising results were obtained initially with mannanases, they are not as effective as xylanases. Rossari offers *Rossapase*, a blend of xylanases, cellulases and hemicellulases for effective bleaching of pulp.

There are at least two explanations for how *Rossapase* enhance pulp bleaching. The first model suggests that it increase access of bleaching chemicals to pulp fibers by removing precipitated xylan. The uncoated fibers are then more susceptible to bleaching chemicals and lignin extraction. Essentially, this model proposes that xylan physically entraps lignin and chromophores in the pulp matrix.

The second model suggests that *Rossapase* release chromophores and lignin from the cellulosic pulp matrix by breaking covalent linkages between the hemicellulose and lignin moieties. Rather than postulating a physical entrapment, it holds that residual lignin and chromophores are chemically bound in the pulp.

Recent evidence supports the role of xylanase in breaking lignin-carbohydrate bonds. Hemicellulose hydrolysis release the bound chromophores and lignin, but xylan removal, *per se*, is not good because it decreases the pulp yield and if carried to an extreme, xylan removal can decrease pulp strength. So the objective is not to take out the xylan, but rather to remove the chromophores and lignin. The reason for using enzymes is to do that in a more specific, economical and environmentally benign manner.

### Desirable Enzyme Characteristics

Several features determine which xylanases are effective for bleaching:

1. It is desirable to have a low molecular weight xylanase that is able to penetrate the fiber. An enzyme must have access to the substrate if it is going to release chromophoric groups and lignin.
2. It must have an appropriate pH optimum. Kraft pulps are alkaline and the enzyme must be able to function at the ambient pH.
3. The enzyme should have an alkaline isoelectric point (pl) in order to bind to negatively-charged fiber surfaces.
4. The enzyme must be thermal stable at the temperature prevailing in the pulp stock.
5. It should possess appropriate substrate specificity. The objective is to release chromophores and extract residual lignin—not remove the bulk xylan.

### Enzyme penetration

Enzyme penetration throughout the cellulose fiber is essential in order to fully access and extract residual lignin. Complete penetration might not be achieved in a single enzyme application. Enzyme penetration is determined in part by molecular weight, but other factors, such as substrate binding, may be more critical.

High substrate affinity may be a positive factor when enzymes are applied to pulps in a dilute suspension (because they would keep the enzyme in the pulp rather than in free solution), but at the relatively high consistencies used in bleaching plants (> 10 per cent total solids), they could work against enzyme diffusion to the interior of the fiber.

### Alkaline pH optimum

The second important characteristic is that the enzymes should have an alkaline pH optimum.

### Alkaline isoelectric point

A third feature that's useful is an alkaline isoelectric point. With an alkaline pH, the enzyme binds more readily to the fibers under the pH prevailing in the pulp and this is important for determining its ability to attack the substrate. Pulp fibers are negatively charged due to the presence of sugar acids and when an enzyme has an alkaline pH and hence a positive charge, it binds more effectively.

### Thermal stability

Most microbial xylanases are stable at 50–60°C and this is within the range of prevailing temperatures in pulp mills. The combination of alkaline and thermal stability, however, is more difficult to come.

## Classification of Xylanases

Xylanases can be classified in three ways:

### Based on molecular weight and pH

Information for this sort of classification can readily be obtained during purification and initial characterisation.

### Based on crystal structure

This can be derived by a determination of DNA sequence. Xylanases can be structurally classified into family F (now known as glycosidase family 10) and family G (now known as family 11).

### Based on kinetic properties, substrate specificity or product profiles

Virtually all xylanases are 'endo' acting, as readily determined by chromatography, but the more detailed determination of kinetic properties—measuring the relative reaction rates on various substrates and determining the kinetics of intermediate product formation—is much less common. Family 10 xylanases occasionally exhibit endocellulase activity; they generally have a higher molecular weight and they occasionally possess a cellulose binding domain. Members of family 10 act on both PNP-xylobioside and PNP-cellobiocide, however, the overall catalytic efficiency on PNP-xylobioside is about 50 times higher.

This suggests that family 10 enzymes act mainly on xylan. Even though all xylanases are endo acting, they can show variations in their product profiles. Some enzymes form predominantly xylose and xylobiose and others predominantly (or exclusively) form xylotriose and other higher oligosaccharide products.

This difference appears to result from the number of substrate-binding subsites on the enzyme surface. The number of pyranose rings that the enzyme will bind effectively determines the nature of the oligoproducts. Family 11 xylanases are true xylanases. They don't have cellulase activity; they consistently exhibit a low molecular weight and they can have either a high or low pl.

### *Chromophore release*

Some xylanases release chromophores from pulps in rather significant amounts, e.g., that from Streptomyces. The crude enzyme mixture releases material that absorbs very strongly in the UV and also into the visible region. The latter absorptivity shows up as a colour that one sees in the pulp. Chromophore release correlates with enzyme dose over the effective range of enzyme addition. In fact we can also see that the kappa number following enzyme treatment and alkali extraction decreases with the enzyme dose. Absorptivity is much greater in the UV region than in the viable region, so monitoring UV absorptivity is a sensitive means to determine the extent of enzyme action on pulp. In short, treatment with *Rossapase* has two indirect effects:

1. It is possible to wash out more lignin from the pulp.
2. The pulp becomes more susceptible to bleaching chemicals. This is called 'bleach boosting' and significantly reduces the need for chemicals in the subsequent bleaching stage.

Thus *Rossapase* helps to achieve the desired level of brightness of the finished pulp using-less-chlorine.

## De-inking of Recycled Paper

Apart from producing virgin paper, recycling of used paper, particularly Mixed Office Waste (MOW), possesses the greatest potential in meeting the demand, in terms of quality and volume, under the Indian situation.

Despite greater availability of MOW, its conversion to higher grade pulp is much restricted owing to difficulties in removing toner inks, which result into bigger visible specks in the final product. De-inking of non-impact printed papers (xerographic/laser) is very difficult, even after several cleaning/ screening cycles in chemical de-inking processes. Current de-inking methods used in recycling MOW mostly involve extraction under alkaline conditions (Fig. 5.18).

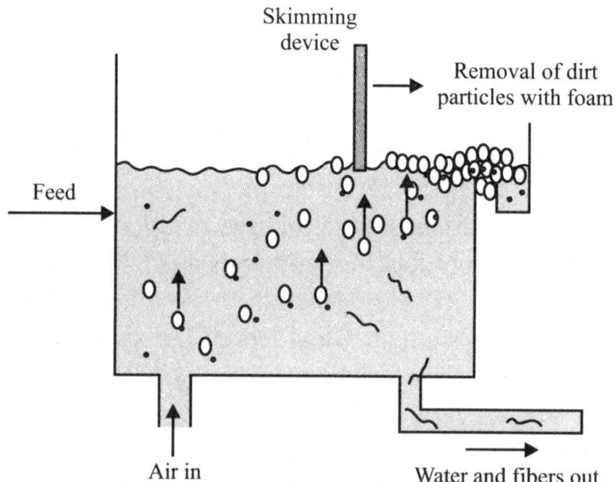


**Fig. 5.18.** De-inking of recycled paper.

### *Conventional de-inking*

Conventional de-inking processes use alkali and detergents. The recovered paper is first dissolved in water and separated from the non-fiber impurities. The fiber is then progressively cleaned by flotation process, where air is blown into the solution. The ink adheres to air bubbles and rises to the surface from where it is separated. After the ink is removed, the fiber is bleached with hydrogen peroxide. Flotation de-inking is the most important method of recycling paper worldwide.

### *Enzymatic de-inking*

Enzyme enhanced de-inking processes show promise for avoiding excessive use of alkaline chemicals and surfactants and yield better quality paper in terms of brightness or non-yellowing. Several research studies have shown that regardless of ink types, enzyme treatment tends to reduce ink particle size when coupled with mechanical pulping, compared to conventional de-inking.

In addition to removal of ink, it also helps improve the strength of paper sheets and freeness. It reduces fine content of the recycled fiber.

A variety of enzymes are being used for de-inking, viz., cellulases, hemicellulases, lipases, esterases, pectinases, amylases and lignolytic enzymes. *Rossapase* is a blend of all the useful enzymes.

Hemicellulases, cellulases and lignolytic enzymes in *Rossapase* alter the fiber surface or bond in the vicinity of ink particles, thereby freeing the ink for removal by washing or flotation. The toner particles adhering to the paper surface released by the enzymatic treatment are further separated from the pulp by flotation.

De-inking with enzymes at acidic to neutral pH reduces the overall chemical requirements and minimises the yellowing of reclaimed paper normally associated using alkaline de-inking. It can reduce the pulping time, thereby saving energy and potentially increasing production. Besides decrease in the ink particle size, milder alkaline conditions prevalent in enzymatic de-inking bears a positive impact on control of stickies at mill scale. Last, but not the least, reduction in quantity of non-ionic surfactants, which biodegrade slowly in effluent treatment plants (ETP) is an additional positive environmental effect.

### Enzymatic De-inking: FAQ

With acid-sized (sulphite) paper, good results are obtained using a *Trichoderma reesei* acid cellulose. With alkaline sized papers, one has to adjust the pH to within the activity range of the enzyme. It is more economical to use a neutral cellulase enzyme (*Humicola insolens*) in order to save alkali. Recycled fibers contain large amounts of calcium carbonate and its alkaline pKa generally necessitates the use of a neutral to alkaline enzyme.

We get high CMCase [carboxymethylcellulase] enzyme from bacteria and FPase (filter paper activity) from fungi. This mixture of bacterial and fungal cellulase prove a best choice for de-inking. Synergism between endo- and exo-acting cellulases generally gives the highest overall activity.

Similarity between the mechanisms in enzyme de-inking and enzyme detergency: Principal enzymes used in detergents are proteinases and lipases. Cellulases are also used to reduce 'pill' formation in cotton fabrics. However, cellulases may also help to release stains from amorphous cellulose or cellulose microfibrils on the fabric surface.

The mechanism for enzymatic de-inking of paper fibers is similar, but it depends, in part, on the types of contaminants being removed. The inks do not usually stain the fibers themselves, but rather form particles or micro particles that are either embedded in the paper fibre or intertwined and adherent with the fibres.

In the case of small particles that are ground into the fibres, the cellulases could release them by removing the microfibrilar matrix in which they are embedded. This can be done fairly easily. Once the microfibrils are gone, the fibers would have a lower specific surface area and would release water (and particles) more readily. In the case of larger particles (such as toner particles), which are trapped in the plastic matrix (rather than the particles trapped in the fibre), either the fibers break off from the toner particle or the presence of the enzyme allows the fibre to slip free from the adherent toner. The mechanical action of the pulper is synergistic and important to remove toner particles by cellulases.

### Recent Advances in the Application of Enzymes for Paper and Pulp

Some of the recent advances in the application of enzymes for paper and pulp are:

1. Effect of peroxidase and phenol oxidases on bleaching of kraft pulp.
2. Use of catalase in hydrogen peroxide bleaching of kraft pulp.
3. Effect of enzymic pre-treatment on the diffusion of sodium hydroxide in wood.
4. Application of hemicellulolytic enzymes from extremely thermophilic bacteria for pulp bleaching.
5. Enzyme pre-treatment on conventional kraft pulping.
6. Enhancing the freeness of pulp made from secondary fiber by using a mixture of cellulase and pectinase enzymes to the pulp.
7. Bleaching of kraft pulp with lignin degrading enzymes.
8. Bleachability of recycled fibers de-inked with enzyme preparations.
9. Use of cellulase to improve surface smoothness in the production of single face glossy paper.
10. De-inking of mixed office paper, old newspaper and vegetable oil based ink printed paper using cellulases, xylanases and lipases.
11. Enzyme de-inking of non-impact printed white office paper with α-amylase.
12. De-inking of laser printed paper using enzyme mixtures of cellulases and xylanases.

### Application in Other Industires

To give an example, immobilised enzymes may be components of artificial kidney machines, which are used to remove urea and other waste products from the body, where kidney disease prevents this being done by natural processes: Urea enters the machine from the blood, by dialysis (termed haemodialysis) and is converted to $CO_2$ and $NH_4^+$ by immobilised urease; toxic $NH_4^+$ is then either trapped on ion exchange resins or incorporated into glutamate by the action of immobilised glutamate dehydrogenase linked to alcohol dehydrogenase to ensure coenzyme recycling, before the fluid is returned to the blood stream.

Enzymes may also be of great value in the pharmaceutical industry, e.g., for the conversion of naturally occurring penicillin G (blenzyl penicillin) to 6-amino penicillanic acid (6-APA). This reaction, which is catalysed by bacterial penicillin amidase, cleaves the side chain of the substrate at mildly alkaline pH values:

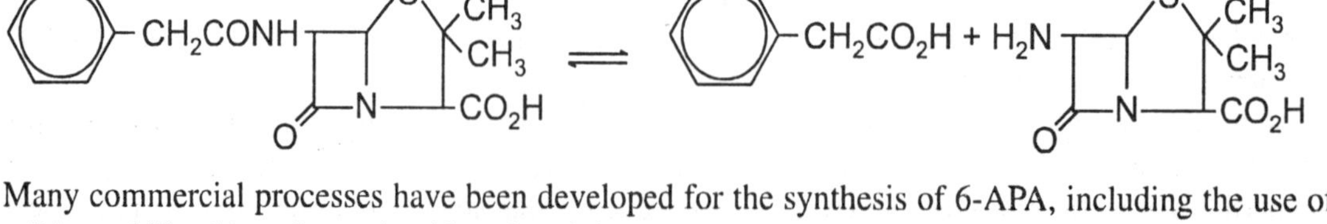

$$C_6H_5{-}CH_2CONH{-}(\text{S, } CH_3, CH_3, \text{N, } CO_2H, \text{O}) \rightleftharpoons C_6H_5{-}CH_2CO_2H + H_2N{-}(\text{S, } CH_3, CH_3, \text{N, } CO_2H, \text{O})$$

Many commercial processes have been developed for the synthesis of 6-APA, including the use of *E. coli* immobilised in polyacrylamide gel and the enzyme entrapped in fibers of cellulose acetate. The importance of 6-APA is that new side chains can be attached to give a variety of semi-synthetic penicillins, e.g., ampicillin (α-methyl benzyl penicillin), a broad-spectrum antibiotic. This may be done by chemical processes or by the use of the enzyme under slightly acidic conditions, to favour the back reaction. Enzymes from different bacteria may be applied to catalyse the forward and back reactions, to make use of their slightly different characteristics and specificities: For example, a mutant of *Kluyvera citrophila* has been used for 6-APA production and one of *Pseudomonas melanogenum* for ampicillin synthesis.

Washing powders incorporating bacterial proteases have been available for several years. Their commercial importance waned because of fears about the effect of enzyme dust on the respiratory system, but this problem was overcome by containment in granules which rupture only on contact with water. The enzymes in question, subtilisins from *Bacillus subtilis* mutants, are stable to alkali, high temperature (e.g., 65°C), detergents and bleaches. They will attack blood and other protein stains.

Bacterial proteases are also used in the leather and textile industries to loosen hair (or wool) and enable it to be separated from hide.

Thus, biotechnology will affect pulp and paper processing in increasing ways. We have already seen commercialisation of enzyme enhanced bleaching of chemical pulps. Enzymatic pitch removal is emerging as a commercial process in Japan. Enzymatic fiberisation during beating has been long recognised as a practical process and the use of enzymes to increase drainage rates for recycled fibers likewise is a useful and economic means to increase paper machine throughput.

In conclusion, one could obtain a higher pulp yield and recover much more in terms of higher value products than is possible with contemporary technologies.

Chapter 6

# Introduction to Fermentation Biotechnology

## INTRODUCTION

The term 'fermentation' is derived from the Latin verb *fervere*, to boil, thus describing the appearance of the action of yeast on extracts of fruit or malted grain. The boiling appearance is due to the production of carbon dioxide bubbles caused by the anaerobic catabolism of the sugars present in the extract. However, fermentation has come to have different meanings to biochemists and to industrial microbiologists. Its biochemical meaning relates to the generation of energy by the catabolism of organic compounds, whereas its meaning in industrial microbiology tends to be much broader.

The catabolism of sugars in an oxidative process which results in the production of reduced pyridine nucleotides which must be re-oxidised for the process to continue. Under aerobic conditions, re-oxidation of reduced pyridine nucleotide occurs by electron transfer, via the cytochrome system, with oxygen acting as the terminal electron acceptor. However, under anaerobic conditions, reduced pyridine nucleotide oxidation is coupled with the reduction of an organic compound, which is often a subsequent product of the catabolic pathway. In the case of the action of yeast on fruit or grain extracts, NADH is regenerated by reduction of pyruvic acid to ethanol. Different microbial taxa are capable of reducing pyruvate to a wide range of end products, as illustrated in Fig. 6.1. Thus, the term fermentation has been used in a strict biochemical sense to mean an energy-generation process in which organic compounds act as both electron donors and terminal electron acceptors.

The production of alcohol by the action of yeast on malt or fruit extracts has been carried out on a large scale for very many years and was the first 'industrial' process for the production of a microbial metabolite. Thus, industrial microbiologists have extended the term fermentation to describe any process for the production of product by the mass culture of a micro-organism. Brewing and the production of organic solvents may be described as fermentations in both senses of the word but the description of an aerobic process as a fermentation is obviously using the term in the broader, microbiological context and it is in this sense that the term is used in this chapter.

## RANGE OF FERMENTATION PROCESSES

There are five major groups of commercially important fermentations:

1. Those that produce microbial cells (or biomass) as the product.
2. Those that produce microbial enzymes.
3. Those that produce microbial metabolites.

4. Those that produce recombinant products.
5. Those that modify a compound which is added to the fermentation—the transformation process.

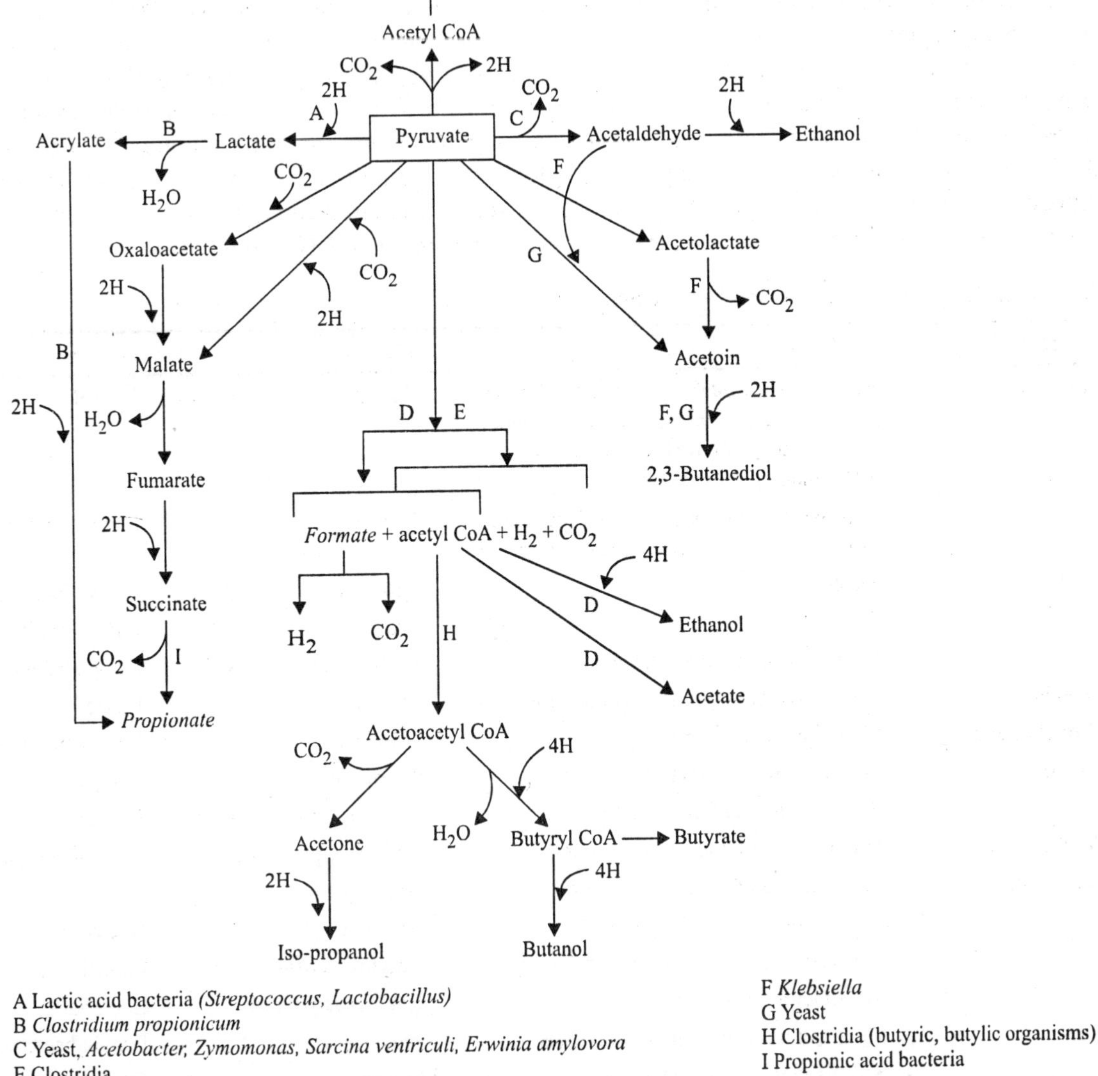


A Lactic acid bacteria *(Streptococcus, Lactobacillus)*
B *Clostridium propionicum*
C Yeast, *Acetobacter, Zymomonas, Sarcina ventriculi, Erwinia amylovora*
E Clostridia
F *Klebsiella*
G Yeast
H Clostridia (butyric, butylic organisms)
I Propionic acid bacteria

**Fig. 6.1.** Bacterial fermentation products of pyruvate. Pyruvate formed by the catabolism of glucose in further metabolised by pathways which are characteristic of particular organisms and which serve as a biochemical aid to identification. End products of fermentations are italicised.

## Microbial Biomass

The commercial production of microbial biomass may be divided into two major processes: the production of yeast to be used in the baking industry and the production of microbial cells to be used as human or animal food (single-cell protein).

## Microbial Enzymes

Enzymes have been produced commercially from plant, animal and microbial sources. However, microbial enzymes have the enormous advantage of being able to be produced in large quantities by established fermentation techniques. Also, it is infinitely easier to improve the productivity of a microbial system compared with a plant or animal one. Furthermore, the advent of recombinant DNA technology has enabled enzymes of animal origin to be synthesised by micro-organisms. The uses to which microbial enzymes have been put are summarised in Table 6.1, from which it may be seen that the majority of applications are in the food and related industries. Enzyme production is closely controlled in micro-organisms and in order to improve productivity these controls may have to be exploited or modified. Such control systems as induction may be exploited by including inducers in the medium, whereas repression control may be removed by mutation and recombination techniques. Also, the number of gene copies coding for the enzyme may be increased by recombinant DNA techniques.

**Table 6.1.** Commercial applications of enzymes.

| *Industry* | *Application* | *Enzyme* | *Source* |
|---|---|---|---|
| Baking and milling | Reduction of dough viscosity, acceleration of fermentation, increase in loaf volume, improvement of crumb softness and maintenance of freshness | Amylase | Fungal |
| | Improvement of dough texture, reduction of mixing time, increase in loaf volume | Protease | Fungal/bacterial |
| Brewing | Mashing | Amylase | Fungal/bacterial |
| | Chillproofing | Protease | Fungal/bacterial |
| | Improvement of fine filtration | β-Glucanase | Fungal/bacterial |
| Cereals | Pre-cooked baby foods, breakfast foods | Amylase | Fungal |
| Chocolate and cocoa | Manufacture of syrups | Amylase | Fungal/bacterial |
| Coffee | Coffee bean fermentation | Pectinase | Fungal |
| | Preparation of coffee concentrates | Pectinase, hemicellulase | Fungal |
| Confectionery | Manufacture of soft center candies | Invertase, pectinase | Fungal/bacterial |
| Corn syrup | Manufacture of high-maltose syrups | Amylase | Fungal |
| | Production of low DE syrups | Amylase | Bacterial |
| | Production of glucose from corn syrup | Amyloglycosidase | Fungal |
| | Manufacture of fructose syrups | Glucose isomerase | Bacterial |
| Dairy | Manufacture of protein hydrolysates | Protease | Fungal/bacterial |
| | Stabilisation of evaporated milk | Protease | Fungal |
| | Production of whole milk concentrates, ice cream and frozen desserts | Lactase | Yeast |
| | Curdling milk | Protease | Fungal/bacterial |
| Eggs, dried | Glucose removal | Glucose oxidase | Fungal |
| Fruit juices | Clarification | Pectinases | Fungal |
| | Oxygen removal | Glucose oxidase | Fungal |

*(Cont'd ...)*

| *Industry* | *Application* | *Enzyme* | *Source* |
|---|---|---|---|
| Laundry | Detergents | Protease, lipase | Bacterial |
| Leather | Dehairing, baiting | Protease | Fungal/bacterial |
| Meat | Tenderisation | Protease | Fungal |
| Pharmaceutical | Digestive aids | Amylase, protease | Fungal |
| | Anti-blood clotting | Streptokinase | Bacterial |
| | Various clinical tests | Numerous | Fungal/bacterial |
| Photography | Recovery of silver from spent film | Protease | Bacterial |
| Protein hydrolysates | Manufacture | Proteases | Fungal/bacterial |
| Soft drinks | Stabilisation | Glucose oxidase, catalase | Fungal |
| Textiles | Desizing of fabrics | Amylase | Bacterial |
| Vegetables | Preparation of purees and soups | Pectinase, amylase, cellulase | Fungal |

## Microbial Metabolites

The growth of a microbial culture can be divided into a number of stages. After the inoculation of a culture into a nutrient medium there is a period during which growth does not appear to occur; this period is referred to as the lag phase and may be considered as a time of adaptation. Following a period during which the growth rate of the cells gradually increases the cells grow at a constant, maximum rate and this period is known as the log or exponential phase. Eventually, growth ceases and the cells enter the so-called stationary phase. After a further period of time the viable cell number declines as the culture enters the death phase. As well as this kinetic description of growth, the behaviour of a culture may also be described according to the products which it produces during the various stages of the growth curve. During the log phase of growth the products produced are essential to the growth of the cells and include amino acids, nucleotides, proteins, nucleic acids, lipids, carbohydrates, etc. These products are referred to as the primary products of metabolism and the phase in which they are produced (equivalent to the log or exponential phase) as the trophophase.

Many products of primary metabolism are of considerable economic importance and are being produced by fermentation, as illustrated in Table 6.2. The synthesis of primary metabolites by wild-type micro-organisms is such that their production is sufficient to meet the requirements of the organism. Thus, it is the task of the industrial microbiologist to modify the wild-type organism and to provide cultural conditions to improve the productivity of these compounds.

During the deceleration and stationary phases some microbial cultures synthesise compounds which are not produced during the trophophase and which do not appear to have any obvious function in cell metabolism. These compounds are referred to as the secondary compounds of metabolism and the phase in which they are produced (equivalent to the stationary phase) as the idiophase. It is important to realise that secondary metabolism may occur in continuous cultures at low growth rates and is a property of slow-growing, as well as non-growing cells. When it is appreciated that micro-organisms grow at relatively low growth rates in their natural environments, it is tempting to suggest that it is the idiophase state that prevails in nature rather than the trophophase, which may be more of a property of micro-organisms in culture. The interrelationships between primary and secondary metabolism are illustrated in Fig. 6.2, from which it may be seen that secondary metabolites tend to be elaborated from the intermediates and products of primary metabolism.

**Table 6.2.** Some primary products of microbial metabolism and their commercial significance.

| *Primary metabolite* | *Commercial significance* |
|---|---|
| Ethanol | 'Active ingredient' in alcoholic beverages |
| | Used as a motor-car fuel when blended with petroleum |
| Citric acid | Various uses in the food industry |
| Glutamic acid | Flavour enhancer |
| Lysine | Feed supplement |
| Nucleotides | Flavour enhancers |
| Phenylalanine | Precursor of aspartame, sweetener |
| Polysaccharides | Applications in the food industry |
| | Enhanced oil recovery |
| Vitamins | Feed supplements |

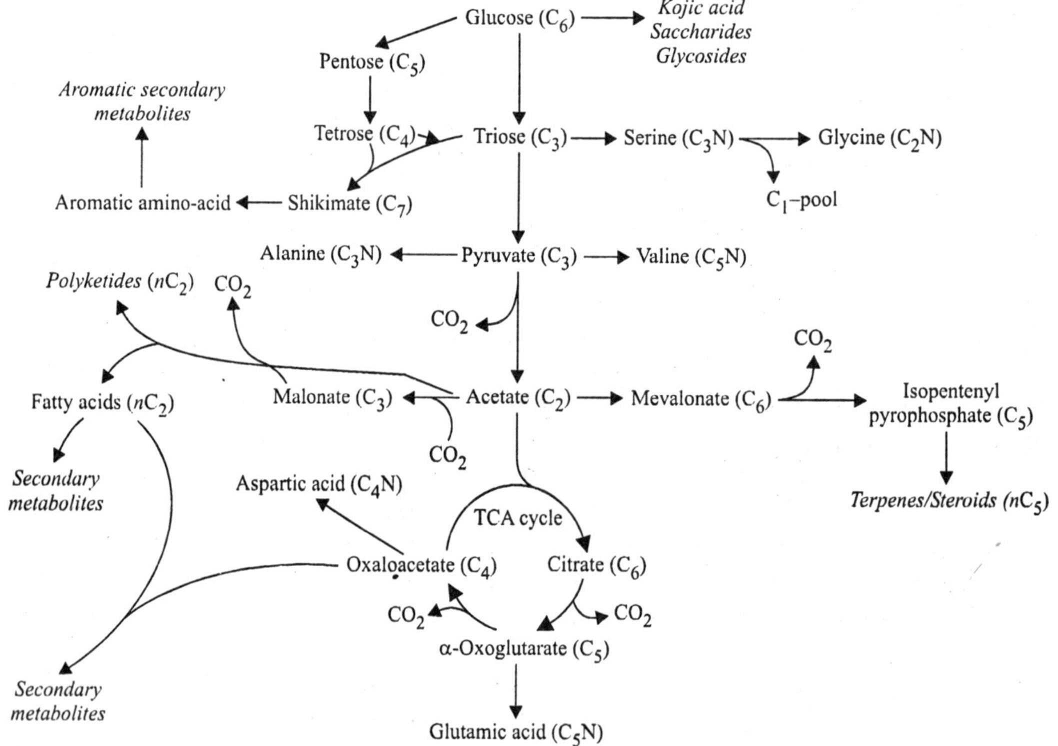


**Fig. 6.2.** The interrelationships between primary and secondary metabolism. Primary catabolic routes are shown in heavy lines and secondary products are italicised.

Although the primary biosynthetic routes illustrated in Fig. 6.2 are common to the vast majority of micro-organisms, each secondary product would be synthesised by only a very few different microbial species. Thus, Fig. 6.2 is a representation of the secondary metabolism exhibited by a very wide range of different micro-organisms. Also, not all micro-organisms undergo secondary metabolism—it is common amongst the filamentous bacteria and fungi and the sporing bacteria but it is not found, for example, in the Enterobacteriaceae. Thus, the taxonomic distribution of secondary metabolism is quite different from that of primary metabolism. It is important to appreciate that the classification of microbial products into primary and secondary metabolites is a convenient, but in some cases, artificial system. To quote, the classification 'should not be allowed to act as a conceptual straitjacket, forcing the reader to consider all products as either primary or secondary metabolites'. It is sometimes difficult to categorise a product as primary or secondary and the kinetics of synthesis of certain compounds may change depending on the cultural conditions.

The physiological role of secondary metabolism in the producer cells has been the subject of considerable debate, but the importance of these metabolites to the fermentation industry is the effects they have on organisms other than those that produce them. Many secondary metabolites have antimicrobial activity, others are specific enzyme inhibitors, some are growth promoters and many have pharmacological properties.

Thus, the products of secondary metabolism have formed the basis of a number of fermentation processes. As is the case for primary metabolites, wild-type micro-organisms tend to produce only low concentrations of secondary metabolites, their synthesis being controlled by induction, catabolite repression and feedback systems.

### Recombinant Products

The advent of recombinant DNA technology has extended the range of potential fermentation products. Genes from higher organisms may be introduced into microbial cells such that the recipients are capable of synthesising 'foreign' (or heterologous) proteins. A wide range of microbial cells have been used as hosts for such systems including *Escherichia coli, Saccharomyces cerevisiae* and filamentous fungi. Products produced by such genetically engineered organisms include interferon, insulin, human serum albumin, factors (VIII and IX), epidermal growth factor, calf chymosin and bovine somatostatin. Important factors in the design of these processes include the secretion of the product, minimisation of the degradation of the product and control of the onset of synthesis during the fermentation, as well as maximising the expression of the foreign gene.

### Transformation Processes

Microbial cells may be used to convert a compound into a structurally related, financially more valuable, compound. Because micro-organisms can behave as chiral catalysts with high positional specificity and stereospecificity, microbial processes are more specific than purely chemical ones and enable the addition, removal or modification of functional groups at specific sites on a complex molecule without the use of chemical protection. The reaction which may be catalysed include dehydrogenation, oxidation, hydroxylation, dehydration and condensation, decarboxylation, amination, deamination and isomerisation. Microbial processes have the additional advantage over chemical reagents of operating at relatively low temperatures and pressures without the requirement of potentially polluting heavy-metal catalysts. Although the production of vinegar is the most well-established microbial transformation process (conversion of ethanol to acetic acid) the majority of these processes involve the production of high-value compounds including steroids, antibiotics and prostaglandins.

The anomaly of the transformation fermentation process is that a large biomass has to be produced to catalyse a single reaction. Thus, many processes have been streamlined by immobilising either the whole cells or the isolated enzymes which catalyse the reactions, on an inert support. The immobilised cells or enzymes may then be considered as catalysts which may be re-used many times.

## CHRONOLOGICAL DEVELOPMENT OF THE FERMENTATION INDUSTRY

The chronological development of the fermentation industry may be represented as five overlapping stages as illustrated in Table 6.3. The development of the industry prior to 1900 is represented by stage 1, where the products were confined to potable alcohol and vinegar. Although beer was first brewed by the ancient Egyptians, the first true large-scale breweries date from the early 1700s when wooden vats of 1500 barrels capacity were introduced. Even some process control was attempted in these early breweries, as indicated by the recorded use of thermometers in 1757 and the development of primitive heat exchangers in 1806. By the mid-1800s the role of yeasts in alcoholic fermentation had been demonstrated independently by Schwann and Kutzing but it was Pasteur who eventually convinced the scientific world of the obligatory role of these micro-organisms in the process. During the late 1800s Hansen started his pioneering work at the Carlsberg brewery and developed methods for isolating and propagating single yeast cells to produce pure cultures and established sophisticated techniques for the production of starter cultures. However, use of pure cultures did not spread to the British ale breweries and it is true to say that many of the small, traditional, ale-producing breweries still use mixed yeast cultures at the present time but, nevertheless, succeed in producing high quality products.

Vinegar was originally produced by leaving wine in shallow bowls or partially filled barrels where it was slowly oxidised to vinegar by the development of a natural flora. The appreciation of the importance of air in the process eventually led to the development of the 'generator' which consisted of a vessel packed with an inert material (such as coke, charcoal and various types of wood shavings) over which the wine or beer was allowed to trickle. The vinegar generator may be considered as the first 'aerobic' fermenter to be developed. By the late 1800s to early 1900s the initial medium was being pasteurised and inoculated with 10 per cent good vinegar to make it acidic and therefore resistant to contamination, as well as providing a good inoculum. Thus, by the beginning of the twentieth century the concepts of process control were well established in both the brewing and vinegar industries.

Between the years 1900 and 1940 the main new products were yeast biomass, glycerol, citric acid, lactic acid, acetone and butanol. Probably the most important advances during this period were the developments in the bakers' yeast and solvent fermentations. The production of bakers' yeast is an aerobic process and it was soon recognised that the rapid growth of yeast cells in a rich wort led to oxygen depletion in the medium which, in turn, resulted in ethanol production at the expense of biomass formation. The problem was minimised by restricting the initial wort concentration such that the growth of the cells was limited by the availability of the carbon source rather than oxygen. Subsequent growth of the culture was then controlled by adding further wort in small amounts. This technique is now called fed-batch culture and is widely used in the fermentation industry to avoid conditions of oxygen limitation. The aeration of these early yeast cultures was also improved by the introduction of air through sparging tubes which could be steam cleaned.

The development of the acetone-butanol fermentation during the First World War by the pioneering efforts of Weizmann led to the establishment of the first truly aseptic fermentation. All the processes discussed so far could be conducted with relatively little contamination provided that a good inoculum was used and reasonable standards of hygiene employed. However, the anaerobic butanol fermentation was susceptible to contamination in the early stages by aerobic bacteria and by acid-producing anaerobic ones, once anaerobic conditions had been established in the later stages of the process. The fermenters employed were vertical cylinders with hemispherical tops and bottoms constructed from mild steel. They could be steam sterilised under pressure and were constructed to minimise the possibility of contamination. Two-thousand-hecta-litre fermenters were commissioned which presented the problems

**Table 6.3.** The stages in the chronological development of the fermentation industry.

| *Stage* | *Main products* | *Vessels* | *Process control* | *Culture method* | *Quality control* | *Pilot plant* | *Strain selection* |
|---|---|---|---|---|---|---|---|
| 1. Pre-1900 | Alcohol | Wooden, up to 1500 barrels capacity | Use of thermometer, hydrometer and heat exchangers | Batch | Virtually nil | Nil | Pure yeast cultures used at the Carlsberg brewery (1886) |
| | | Copper used in later breweries | | | | | |
| | Vinegar | Barrels, shallow trays, trickle filters | | Batch | Virtually nil | Nil | Fermentations inoculated with 'good' vinegar |
| 2. 1900–1940 | Bakers' yeast glycerol, citric acid, lactic acid and acetone/ butanol | Steel vessels of up to 200 $m^3$ for acetone/butanol | pH electrodes with off-line control | Batch and fed-batch system | Virtually nil | Virtually nil | Pure cultures used |
| | | Air spargers used for bakers' yeast | Temperature control | | | | |
| | | Mechanical stirring used in small vessels | | | | | |
| 3. 1940-date | Penicilling streptomycin, other antibiotics, gibberelin, amino acids, nucleotides, transformations, enzymes | Mechanically aerated vessels, operated aseptically—true fermenters | Sterilisable pH and oxygen electrodes. Use of control loops which were later computerised | Batch and fed-batch common. Continuous culture introduced for brewing and some primary metabolites | Very important | Becomes common | Mutation and selection programmes essential |
| 4. 1964-date | Single-cell protein using hydrocarbon and other feedstocks | Pressure cycle and pressure jet vessels developed to overcome gas and heat exchange problems | Use of computer linked control loops | Continuous culture with medium recycle | Very important | Very important | Genertic engineering of producer strains attempted |
| 5. 1979-date | Production of heterologous proteins by microbial and animal cells. Monoclonal antibodies produced by animal cells | Fermenters developed in stages 3 and 4. Animal cells reactors developed | Control and sensors developed in stages 3 and 4 | Batch, fed-batch or continuous. Continuous perfusion developed for animal cell processes | Very important | Very important | Introduction of foreing genes into microbial and animal cell hosts. *In vitro* recombinant DNA techniques used in the improvement of stage 3 products |

of inoculum development and the maintenance of aseptic conditions during the inoculation procedure. The techniques developed for the production of these organic solvents were a major advance in fermentation technology and paved the way for the successful introduction of aseptic aerobic processes in the 1940s.

The third stage of the development of the fermentation industry arose as a result of the wartime need to produce penicillin in submerged culture under aseptic conditions. The production of penicillin is an aerobic process which is very vulnerable to contamination. Thus, although the knowledge gained from the solvent fermentations was exceptionally valuable, the problems of sparging a culture with large volumes of sterile air and mixing a highly viscous broth had to be overcome. Also, unlike the solvent fermentations, penicillin was synthesised in very small quantities by the initial isolates and this resulted in the establishment of strain-improvement programmes which became a dominant feature of the industry in subsequent years. Process development was also aided by the introduction of pilot-plant facilities which enabled the testing of new techniques on a semi-production scale. The development of a large-scale extraction process for the recovery of penicillin was another major advance at this time. The technology established for penicillin fermentation provided the basis for the development of a wide range of new processes. This was probably the stage when the most significant changes in fermentation technology took place resulting in the establishment of many new processes over the period, including other antibiotics, vitamins, gibberellin, amino acids, enzymes and steroid transformations. From 1960s onwards microbial products were screened for activities other than simply antimicrobial properties and screens became more and more sophisticated. These screens have evolved into those operating today utilising miniaturised culture systems, robotic automation and elegant assays.

In the early 1960s the decisions of several multi-national companies to investigate the production of microbial biomass as a source of feed protein led to a number of developments which may be regarded as the fourth stage in the progress of the industry. The largest mechanically stirred fermentation vessels developed during stage 3 were in the range 80,000 to 1,50,000 $dm^3$. However, the relatively low selling price of microbial biomass necessitated its production in much larger quantities than other fermentation products in order for the process to be profitable. Also, hydrocarbons were considered as potential carbon sources which would result in increased oxygen demands and high heat outputs by these fermentations. These requirements led to the development of the pressure jet and pressure cycle fermenters which eliminated the need for mechanical stirring. Another feature of these potential process was that they would have to be operated continuously if they were to be economic. At this time batch and fed-batch processes were common in the industry but the technique of growing an organism continuously by adding fresh medium to the vessel and removing culture fluid had been applied only to a very limited extent on a large scale. The brewers were also investigating the potential of continuous culture at this time, but its application in that industry was short-lived. Several companies persevered in the biomass field and a few processes came to fruition, of which the most long-lived was the ICI Pruteen animal feed process which utilised a continuous 30,00,000 $dm^3$ pressure cycle fermenter for the culture of *Methylophilus methylotrophus* with methanol as carbon source. The operation of an extremely large continuous fermenter for time periods in excess of 100 days presented a considerable aseptic operation problem, far greater than that faced by the antibiotic industry in the 1940s. The aseptic operation of fermenters of this type was achieved as a result of the high standards of fermenter construction, the continuous sterilisation of feed streams and the utilisation of computer systems to control the sterilisation and operation cycles, thus minimising the possibility of human error. However, although the Pruteen process was a technological triumph it became an economic failure because the product was out-priced by soyabean and fishmeal. Eventually, in 1989, the plant was demolished, marking the end of a short, but very exciting, era in the fermentation industry.

Whilst biomass is a very low-value, high-volume product, the fifth stage in the progress of the industry resulted in the establishment of very high-value, low-volume products. The development in *in vitro* genetic manipulation, commonly known as genetic engineering, enabled the expression of human and mammalian genes in micro-organisms, thereby enabling the large scale production of human proteins which could then be used therapeutically.

According to Dykes, it was the small, venture-capital biotechnology companies that pioneered the development of heterologous proteins for therapeutic use. The established pharmaceutical companies used the new genetic engineering techniques to help in the discovery of natural products and in the rational design of drugs; for example, mammalian receptor proteins have been cloned and used in *in vitro* detection systems.

Insulin and human growth hormone have been the two most successful products but other products have far greater potential. Erythropoietin and the myeloid colony stimulating factors (CSFs) control the product of blood cells by stimulating the proliferation, differentiation and activation of specific cell types. Erythropoietin has been used to treat renal-failure anaemia and may have application in the treatment of the platelet deficiency associated with cancer chemotherapy; it is expected to become the top-selling therapeutic protein. Granulocyte-colony stimulating factor (G-CSF) is used during cancer chemotherapy. A number of different growth factors are involved in wound healing and recombinant forms of these proteins would be expected to yield significant returns in the coming years.

The commercial exploitation of recombinant proteins has necessitated the design of contained production facilities. Thus, these processes are drawing on the experience of vaccine fermentations where pathogenic organisms have been grown on relatively large scales. Also, recombinant proteins have been classified as biologicals, not as drugs and thus come under the same regulatory authorities as do vaccines. The major difference between the approval of drugs and biologicals is that the process for the production of a biological must be precisely specified and carried out in a facility that has been inspected and licensed by the regulatory authority, which is not the case for the production of drugs (antibiotics, for example). Thus, any changes which a manufacturer wishes to incorporate into a licensed process must receive regulatory approval. For drugs, only major changes require approval prior to implementation. The result of these containment and regulatory requirements is that the cost of developing a recombinant protein process is extremely high. Buckland illustrated this point in his claim that 'It now costs as much to build a 3000 $dm^3$ scale facility for Biologics as for a 200,000 $dm^3$ scale facility for an antibiotic. Also, even though titres are now reasonable for a recombinant protein (1 g $dm^{-3}$), the cost of manufacture $kg^{-1}$ of bulk drug is about two orders of magnitude higher than that of an antibiotic at 10 g $dm^{-3}$ titre'. Also, the development time for a recombinant protein is considerably longer than that for an antibiotic. For example, Bader claimed that it is feasible for an antibiotic plant to begin production four years after the initiation of the plant design whereas seven years would be required before a recombinant protein could be produced.

The exploitation of genetic engineering approximately coincided with another major development in biotechnology which influenced the progress of the fermentation industry—the production of monoclonal antibiodies. The availability of monoclonal antibodies opened the door to sophisticated analytical techniques and raised hopes for their use as therapeutic agents. Although the promise of therapeutic agents has yet to be realised (only one monoclonal antibody has been licensed for clinical use, OKT3, used in the treatment of acute renal allograft rejection, their use as tools in biological research has increased exponentially. Thus, animal cell culture processes were established to produce monoclonals on a commercial scale.

Subsequently, animal cells were also used as hosts for the production of some human proteins, especially where post-translational modification was essential for protein activity. Although these animal cell processes were based on microbial fermentation technology a number of novel problems had to be solved—animal cells are extremely fragile compared with microbial cells, the achievable cell density is very much less than in a microbial process and the media are very complex.

The outstanding developments in recombinant fermentations (stage 5) have tended to overshadow the progress which has been made in recent years in establishing new fermentations based on conventional microbial products (the continuing development of stage 4). However, the appreciation by the pharmaceutical industry that the activity of microbial metabolites extended well beyond antibacterials has resulted in a number of new microbial products reaching the marketplace in early 1990s. Buckland listed four secondary metabolites such as, cyclosporin, an immunoregulant used to control rejection of transplated organs; imipenem, a modified carbapenem which has the widest antimicrobial spectrum of any antibiotic; lovastatin, a drug used for reducing cholesterol levels and ivermectin, an anti-parasitic drug which has been used to prevent 'African River Blindness' as well as in veterinary practice. Buckland summarised these developments succinctly, 'One of the best kept secrets (unintentionally kept as a secret) in biochemical engineering was that working on secondary metabolites was a fascinating, important and rewarding experience.

Furthermore the four products listed added together have higher sales than all of the recombinant products added together'. Thus, it is still relevant to heed Foster's warning 'never underestimate the power of the microbe'.

## COMPONENT PARTS OF A FERMENTATION PROCESS

Regardless of the type of fermentation (with the possible exception of some transformation processes) an established process may be divided into six basic component parts:

1. The formulation of media to be used in culturing the process organism during the development of the inoculum and in the production fermenter.
2. The sterilisation of the medium, fermenters and ancillary equipment.
3. The production of an active, pure culture in sufficient quantity to inoculate the production vessel.
4. The growth of the organism in the production fermenter under optimum conditions for product formation.
5. The extraction of the product and its purification.
6. The disposal of effluents produced by the process.

The interrelationships between the six component parts are illustrated in Fig. 6.3.

However, one must also visualise the research and development programme which is designed to gradually improve the overall efficiency of the fermentation. Before a fermentation process is established a producer organism has to be isolated, modified such that it produces the desired product in commercial quantities, its cultural requirements determined and the plant designed accordingly. Also, the extraction process has to be established.

The development programme would involve the continual improvement of the process organism, the culture medium and the extraction process.

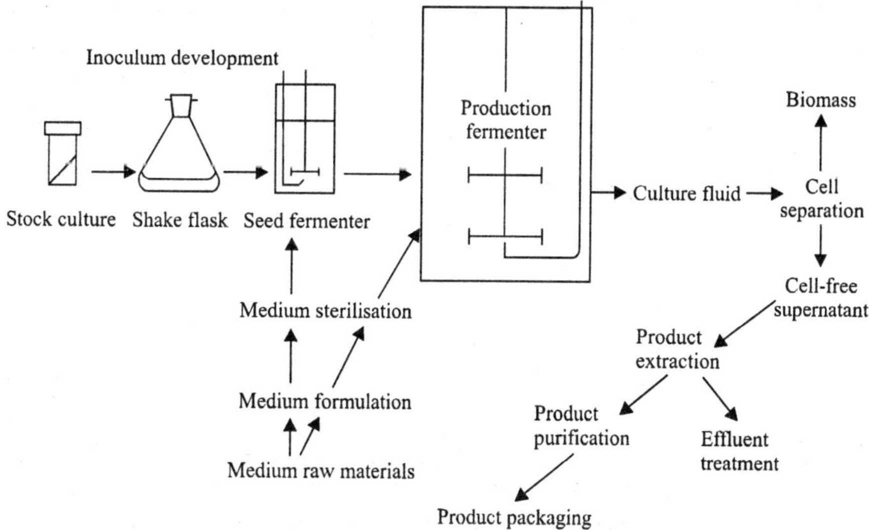


**Fig. 6.3.** A generalised schematic representation of a typical fermentation process.

## ROLE OF TURBULENCE IN FERMENTATION

This section raises the need for a systematic evaluation of non-linear interaction between various turbulent-driven physical processes in fermentors. Various turbulent/mixing studies are formalised based on a multidimensional framework. This formalisation leads us to the limitations of many existing approaches and scale-up rules to mixing problems. The results on the effects of micromixing and fluid shear on yeast cells will be presented as an example to show the possibility of a non-linear interaction. A micro-environmental approach is proposed whereby the effects of multiple turbulent-driven physical processes in the micro-environments of cells can be evaluated.

Fermentation systems are required to be scaled-up from laboratory-scale fermentors (few liters) to production-scale fermentors (5000 liters to 300,000 liters) with reproducibility of the performance. To fulfil this requirement, the local conditions around each cell in the production-scale should be similar to that found optimum in the laboratory scale fermentor. These conditions are temperature, pH, dissolved oxygen, nutrient concentration, osmotic pressure and shearing forces. This requirement is rarely fulfilled and is the main cause of irreproducibility during the traditional step-wise scale-up. Other causes, which will not be discussed here, include strain degeneration during inoculum train, media degradation during sterilisation and constraints placed by the downstream processing. The irreproducibility of local conditions on scale-up is due to the pronounced inhomogeneous distribution of mixing power or turbulence in large fermentors and is termed here as 'mixing effect/turbulence problem'. The variation in local conditions on scale-up affects various types of fermentation processes to different degrees depending on the type of species, the medium composition and the nature of turbulence. The biological effects on cells typically measured are yield, productivity and lysis. Thus the simplest view of mixing problem can be represented in Fig. 6.4. Most processes use a stirred tank bioreactor which is taken as the model system in this section. The approach and results presented here, however, should be applicable to any other reactor format which employs turbulences-driven physical processes.

Traditional scale-up rules have been available for at least three decades which are based on maintaining a constant value of one of the following variables: power per unit volume, tip speed, mixing time and mass transfer rate. Scale-up of fermentation, however, still could be frequently disappointing. For example, scale-up based on a constant mass transfer coefficient for vitamin $B_{12}$ fermentation shows an optimum value of $k_l a$ for operation. Results from the scale-up of penicillin fermentations based on constant (P/V) suggest that product yields at different scales will be significantly different, both higher and lower, depending on the value of (P/V). The main cause of limitation of these scale-up rules is that they consider bulk average values for physical processes and cellular properties.

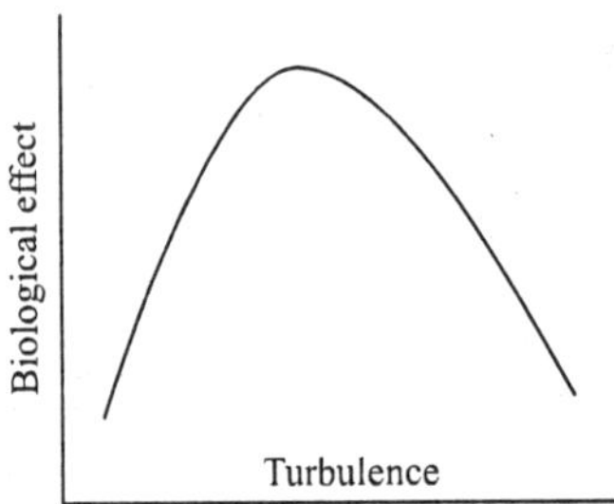


**Fig. 6.4.** Turbulence in fermentation.

This limitation of the scale-up rules has been the motivation for a number of mixing/turbulence studies published over the last two decades to evaluate the effects of local conditions on fermentations. The premise for these studies and traditional scale-up rules has been that only one of the local conditions is most critical to one property of cells. This approach, therefore, is limited. The main reason is that the physical processes driven by turbulence cannot be segmented into 'independent' physical processes such as micro and macro-mixing, gas-liquid transfer, fluid forces and heat transfer. In addition, the physiological effects on cells cannot be simply measured as an overall yield. This overall effect could be the cumulative effect of multiple biological responses, each affected by various physical processes to different degrees. Traditional scale-up rules and mixing studies fail to address such non-linear interaction which may lead to either amplification or cancellation of effects by various physical processes.

## A MULTIDIMENSIONAL PROBLEM

The turbulence-driven physical processes are caused by five intrinsic operating variables. These are agitation (or power input), aeration (or flow rate), scale of equipment, internals and cells/medium used. The physical processes will map the operating region based on their impact on the physiology of cells. Figure 6.5 shows how each of these individual physical processes can restrict the operating range on scale-up. For example as the scale of operation increases, the upper limits on the agitation power and aeration will decrease due to constraints placed by the shear forces and the cost. The lower limits on agitation and aeration required for minimum suspension and gas-liquid mass transfer will increase as the scale of operation increases.

The contours of the operating map in Fig. 6.5 makes the turbulence problem to be a three-dimensional with the scale of equipment as the third axis. In addition to the scale of operation, other intrinsic variables such as the internals and type of cells/medium used will make the turbulence problem to be multi-dimensional. For example, internals such as impeller, sparger and cooling coils would affect the physical processes. Also, if cells/medium show non-Newtonian rheology then depending on the local shear rate, the rheology of the suspension will vary. Such a multi-dimensional framework for the turbulence problem, of which Fig. 6.5 is a simplified view will encompass all the mixing studies performed in the litereture. These mixing studies can be broadly divided into three types (Fig. 6.6).

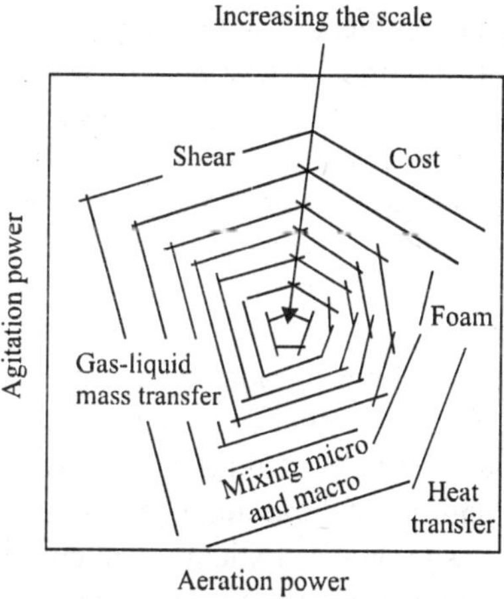


**Fig. 6.5.** Multi-dimensional role of turbulence in fermentation.

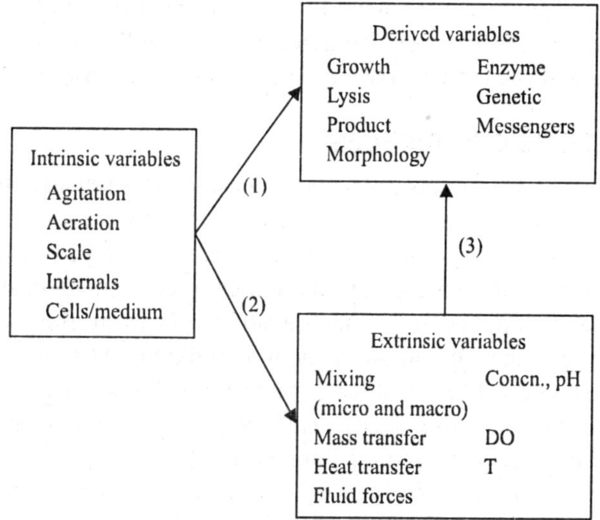


**Fig. 6.6.** Relationships between various studies.

## Biological Characterisation of the Fermentor

In these studies, derived variables or biological effects of primary interest including growth, product yields and lysis are measured as a function of intrinsic variables. Such studies, where one or multiple intrinsic variables are varied simultaneously, are numerous in the literature. These studies could satisfy scale-up requirement for low volume fermentations but they would never elucidate the mechanisms at the levels of physical processes and its physiological consequences.

## Physical Characterisation of the Fermentor

In these studies, extrinsic variables quantifying the physical processes are measured as a function of intrinsic variables using cell suspension broth or simulated broth. Extensive studies, for example, have been performed to obtain correlation to predict $k_la$ and mixing times as a function of agitation, aeration, type of impeller and nature of cells/medium. These studies represent a cross-section of the multi-dimensional framework represented in Fig. 6.5, for example $k_la$ as a function of agitation and aeration. Very limited studies have been performed to determine distributions of circulation times, micro-mixing levels, $k_la$, fluid shear and heat transfer in production fermentors. These type #2 studies provide important data needed to perform mechanistic studies described below.

### *Mechanistic studies*

These studies, performed over the last three decades by many groups, determine the physiological effect(s) of a single extrinsic variable while others are maintained constant. For example, one of the earliest study attempted to evaluate the role of mixing time with respect to the feeding point or micromixing, in the fermentor. The majority of these studies are based on a scale-down approach whereby small-scale strategies are designed to simulate a 'critical' local condition or extrinsic variable found in production fermentor. Examples include simulation of the spatial gradients in concentrations of the carbon source, oxygen and pH, micromixing at the feeding point and fluid stresses. Many of these studies did not consider the data obtained from type #2 studies for physical characterisation of fermentors to simulate realistic local conditions. Other major limitations of all these studies were that the possibility of the interaction between various extrinsic variables exist. We will illustrate this possibility showing results from micromixing and shear studies using *Saccharomyces cerevisiae*.

## INTERACTION OF EXTRINSIC VARIABLES

*Saccharomyces cerevisiae* or baker's yeast, was chosen because it is sensitive to glucose and dissolved oxygen levels. The biomass yield of yeast culture is reduced by more than half if the glucose level exceeds 0.05–0.2 g/l ('glucose effect') or the dissolved oxygen levels dips below 0.1 mg/l ('anaerobic'). To maintain sufficient mixing and mass transfer in production vessels, adequate agitation and aeration power (1–4 w/l) is supplied without regards to the sensitivity of cells to fluid shear effects. Two sets of experiments were performed; one each to study the micromixing and shear effects with no previous indication of possible interaction. These studies were performed in a stirred and aerated bioreactor of volume 3 liter operated in chemostat mode. Dissolved oxygen was maintained above 20 per cent in all experiments to avoid any interaction with mass transfer process.

1. Micromixing effects: The goal of this experiment was to determine whether the micromixing of the incoming concentrated glucose (30 g/l) stream with the yeast cells will affect the biomass yield. Chemostat experiments were performed at dilution rates ranging from 0.16 $hr.^{-1}$ to 0.32 $hr.^{-1}$ at two agitation speeds of 800 rpm (1 w/l) and 1100 rpm (3 w/l) using one Rushton turbine. The feed was added at the injection point away from the impeller (IP #2).
2. Shear effects: The goal of this experiment was to determine whether excess turbulent shear affected physiology of yeast cells. A chemostat was performed at a dilution rate of 0.27 $hr.^{-1}$ and steady states were achieved at two agitation speeds, 650 rpm (3 w/l) and 1200 rpm (10.5 w/l) using two Rushton turbines on the shaft. The glucose concentration in the feed was 5 g/l and was added away from the impeller.

### Micromixing Effects

Figure 6.7 shows biomass yield ($Y_{x/s}$) as a function of dilution rate for two agitation speeds, 800 rpm and 1100 rpm. Data shows the onset of the glucose effect at a critical dilution rate and an hysteresis effect. Changing the agitation speeds showed these two metabolic effects to be sensitive to the mixing state. The onset of glucose effect occurs at a lower dilution rate, from 0.26 to 0.28 hr.$^{-1}$ 800 rpm, compared to that at a dilution rate from 0.30 to 0.33 hr.$^{-1}$ for 1100 rpm during the forward direction of the experiment. Similar effects are observed in the reverse direction of the experiment. Consistently, the cytochrome $aa_3$ enzyme which is repressed by excess glucose showed higher activity at 1100 rpm compared to 800 rpm beyond the dilution rate of 0.24 hr.$^{-1}$.

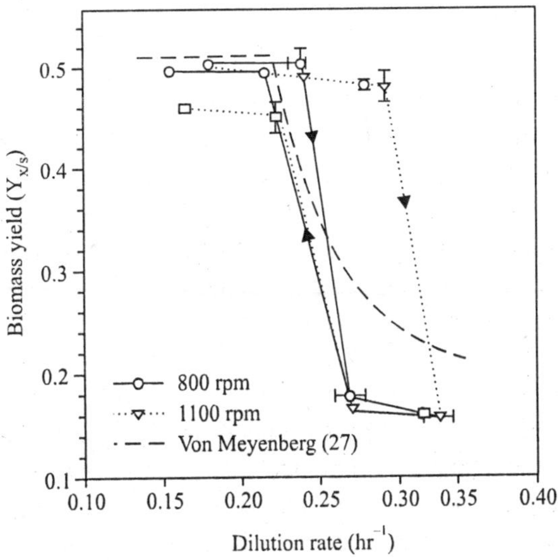


**Fig. 6.7.** Micromixing effects in fermentation.

Changing the agitation speed results in changes in both the macro- and micro-aspects of mixing, mass transfer and shear forces. The small volume of the fermentor and dissolved oxygen control ensures that any changes in the macromixing and mass transfer rates would be insignificant. The questions then remains is that whether the micromixing at the injection point is the only relevant turbulence-driven physical process or whether other physical processes such as shear forces also play a role. Measurements of micromixing levels using a dye system showed that the Kolmogoroff's microscale at the injection point can be changed more effectively by changing its location (from 15 μm near the impeller IP#1 to 45 μm away from the impeller IP#2 at 800 rpm) than by changing the stirrer speed at the injection point away from the impeller (from 45 μm at 1100 rpm to 55 μm at 800 rpm). Additional chemostat experiments at a constant agitation speed of 800 rpm ware performed where the injection point was changed from a location away from the impeller (IP#2) to a location in the impeller region (IP#1). This shift in the location of the injection point produced effects on biomass yield, cytochrome $aa_3$ enzyme activity and hysteresis consistent with the effects produced by changing agitation speed. Thus additional turbulence provided only at the feed point (by better injection point in the impeller region) or throughout the reactor (by increasing the stirrer speed) seems to have similar effects on respiratory enzyme and biomass yield. These results indicate that micromixing at the injection point may not be the only turbulent-driven process controlling the overall biological effect.

## Shear Effects

Chemostat experiments were performed at one dilution rate of 0.27 hr.$^{-1}$ and steady states were achieved at two agitation speed, 650 rpm and 1200 rpm. Although the biomass yield remained unchanged at these two conditions, a number of physiological responses representing possible shear effects were observed. A 20 per cent increase in the incorporation of glucan in the cell size was observed at higher agitation speed. Correspondingly the average cell wall was increased and surface area per unit volume was reduced. Other biological assays showed increased compression modulus of walls, increased resistance to enzymatic degradation of walls and reduced elastic coefficients at the higher agitation speed. The details on method of assays and numerical results are described by Villeneuve. All these effects on the cell wall indicate that yeast cells sense and counter the increased exposure to shear forces at 1200 rpm by building mechanically stronger walls and reducing effective surface area per unit volume. To compensate for the increased cell wall strength related metabolic activity, other metabolic activities are reduced at higher agitation speed. These reduced metabolic activities include reduction in synthesis of proteins, trehalose, glycogen and heat shock proteins. These data clearly demonstrates that yeast cells, hitherto considered insensitive to shear forces, are responding to higher agitation speed. This response, however, could be result of either a complex chain of events caused by induction/repression of certain genes or a simple case of selection of shear resistant cell lines.

These multiple physiological responses did not have measurable impact on the biomass yield at the higher agitation speed of 1200 rpm (10.5 w/l). In contrast, the first set of micromixing experiments showed a higher biomass yield of 45 per cent for 1100 rpm (3 w/l) compared to 15 per cent yield for 800 rpm (1 w/l). The potential causes for the differences in biological effects produced by shear and micromixing experiments are the total power input and inlet glucose concentration. The shear experiment had a very low inlet glucose level of 5 g/l compare to 30 g/l used in micromixing experiment. This could avoid the glucose effect and hence reduction in biomass yield as seen in the 'shear effects' experiments. Whether increased agitation in micromixing experiments caused any physiological effects from shear forces, it is not known yet. Further experiments will be performed to confirm this interaction of micromixing and shear.

## MICRO-ENVIRONMENTAL APPROACH: NEW DIRECTIONS

The experimental data discussed above showed that an interaction between two or more turbulent-driven physical processes is probably non-linear. No systematic approach is yet available to impose defined and relevant values for more than one extrinsic variable simultaneously on cells. Here micro-environmental approach can be applied. The sizes of microscales (50–300 μm) in a typical industrial fermentors are much larger than the size of microbial cells which range from 1 to 5 μm. As a microbial cell, embedded in its microenvironment, circulates in the fermentor, it would be exposed to a range of extrinsic variables in its microenvironment. The overall performance of the fermentor would depend on the response of individual cells to their respective microenvironment. The traditional rules of thumb for scale-up neglect the interaction of the cells with their microenvironment by averaging the hydrodynamics through the use of the engineering variables. A rational strategy to aid scale-up studies, therefore, should be based on the evaluation of interactions of physiological consequences of various physical processes (extrinsic variables) present in the micro-environments of cells. Based on the list of extrinsic variables in Fig. 6.5 and data from our laboratory we propose the following studies for interaction effects.

### Micromixing and Shear

A stirred bioreactor with a recycle loop will be employed where feeding point will be located in the loop region. Such a system is already being used in laboratory. The micromixing in the feeding region will be varied at defined levels by creating turbulence behind a wire mesh in the tube. The cell broth recirculates between a recycle loop and the stirred bioreactor. The turbulent shear can now be varied in the stirred tank by varying agitation and introducing additional turbines.

### Micromixing and Mass Transfer

An annulus type tubular bioreactor will be employed with liquid flow inside and air/oxygen flow on the outside. The micromixing levels in the feeding region will be varied as described earlier. The mass transfer rates will be varied across the membrane by changing partial pressure of oxygen in the gas phase, membrane type and the liquid flow rate.

### Micromixing and Macro-mixing

A stirred bioreactor will be employed with multiple injection point which will produce a range of micromixing levels. Macro-mixing will be imposed using a pulse feeding based on a Monte Carlo method to simulate the circulation time distribution. Similar strategies can be developed for other types of interaction, for example, mass transfer and shear for cell culture system, heat transfer and mass transfer for highly viscous fermentation system, mass transfer and heat transfer for anaerobic fermentation. In addition to these experimental strategies, computer models are also being developed to evaluate the interaction, for example, interaction between mixing and mass transfer in fermentors.

## MATHEMATICAL MODEL FOR A FERMENTATION PROCESS CARRIED OUT IN A FLUIDISED BED REACTOR

A mathematical model describing the dynamics of a fermentation process carried out in a fluidised bed bioreactor with immobilised yeast is proposed. The model is a set of coupled partial differential equations, derived from the unsteady state mass balances of the species involved in the fermentation: substrate, biomass and product. The orthogonal collocation method is used to solve the resulting model. A series of experimental runs for fermentations using glucose was performed to validate the model. The model predicted satisfactorily the dynamic behaviour of the fermentation species under various operating conditions.

Cell immobilisation and fluidised bed reactors are often selected to increase fermentation yields. This selection is mainly due to the advantages that these two alternatives offer over traditional free cell systems and batch and continuous reactors. On one hand, cell immobilisation leads to high reaction rates by virtue of high cell loadings in the support material, reduces contamination risks, keeps high cellular densities and dilution rates and eases product extraction. Fluidised bed reactors, on the other hand, offer additional advantages: reduction of diffusional barriers, re-uses of biocatalyst, reduction of product inhibition effects and that can be operated in continuous or semi-continuous manners. These advantages suggest that any fermentation carried out in a fluidised bed using immobilised cells will produce results which will be very difficult to improve with any other cell system and/or reactor configuration. One way to demonstrate this is by using a theoretical model of the system which will allow us to understand the behaviour of the system and to identify the limiting steps and hence focus on process improvement in the proper areas.

In this section, we propose a mathematical model for a glucose fermentation carried out in a fluidised bed bioreactor with immobilised yeast. The model is used to analyse the effect of process variables and reactor configuration on the yield. Reactor performance is evaluated on a basis of effluent ethanol concentration. The model is a set of coupled non-linear partial differential equations which are solved by means of the orthogonal collocation method and an integration routine. A series of experimental runs for various operating conditions are also performed to validate the model.

### Model Development

Several considerations are taken into account to simplify the mathematical model and facilitate its numerical solution. The species considered in the model are: substrate, product and biomass. The model is based on the Levenspiel's dispersion model that includes convective and diffusive transport through a column. With these considerations, the unsteady state mass balances of substrate and product are given by:

$$\varepsilon\frac{\partial S_L}{\partial t} = D_{S_L}\frac{\partial^2 S_L}{\partial z^2} - u_0\frac{\partial S_L}{\partial z} - R_S$$
$$\varepsilon\frac{\partial P_L}{\partial t} = D_{P_L}\frac{\partial^2 P_L}{\partial z^2} - u_0\frac{\partial P_L}{\partial z} + R_P \quad \text{... (6.1)}$$

where $u_0$ is the superficial velocity; $\varepsilon$ is the bed porosity; $z$ represents the axial direction and $t$ is time, $S_L$, $P_L$ are the substrate and product concentrations in the bulk liquid and $D_{S_L}$, $D_{P_L}$, are the effective diffusivities in the bulk liquid. $R_S$ and $R_P$ denote substrate and product rate expressions, respectively. For spherical particles, these expressions are given by:

$$R_S = \frac{3(1-\varepsilon)}{R_0} D_{SP} \left.\frac{\partial S_I}{\partial r}\right|_{r=R_0}$$
$$R_P = \frac{3(1-\varepsilon)}{R_0} D_{PP} \left.\frac{\partial P_I}{\partial r}\right|_{r=R_0} \quad \text{... (6.2)}$$

where, $S_I$ and $P_I$, are substrate and product concentrations inside the bead, $D_{S_I}$, $D_{P_I}$, are, respectively, the substrate and product diffusivities inside the bead, r represents the radial direction and $R_0$ is the bead radius. To evaluate $R_S$ and $R_P$ we assume: fermentation species in the bulk liquid and on the biocatalyst surface have the same concentration, homogeneous metabolite and substrate diffusion in the bead radial direction, uniform yeast distribution at the start of the fermentation, the amount of cells leaving the bead is negligible and constant concentration profile in the bead center. With these considerations the transient behaviour of substrate and product inside the beads is described by:

$$\frac{\partial S_I}{\partial t} = D_{SP}\left(\frac{\partial^2 S_I}{\partial r^2} + \frac{2}{r}\frac{\partial S_I}{\partial r}\right) - r_S$$
$$\frac{\partial P_I}{\partial t} = D_{PP}\left(\frac{\partial^2 P_I}{\partial r^2} + \frac{2}{r}\frac{\partial P_I}{\partial r}\right) + r_P \quad \text{... (6.3)}$$

here, $r_S$ and $r_P$ denote substrate and product rates, respectively.

If the beads are very small, the quasi-steady state approximation can be applied to equations (6.3). This approximation allows us to incorporate $R_S$ and $R_P$ in (6.1) as an efficiency factor that can be expressed as:

$$\eta_S = \frac{3\, D_{SP}\, dS_I/dr|_{r=R_0}}{R_0\, r_S(S_L)} \quad \text{... (6.4)}$$

Thus, the equations (6.1), (6.2) and (6.4) are combined to give:

$$\varepsilon\frac{\partial S_L}{\partial t} = D_{S_L}\frac{\partial^2 S_L}{\partial z^2} - u_0\frac{\partial S_L}{\partial z} - (1-\varepsilon)\eta_S r_S(S_L)$$
$$\varepsilon\frac{\partial P_L}{\partial t} = D_{P_L}\frac{\partial^2 P_L}{\partial z^2} - u_0\frac{\partial P_L}{\partial z} + (1-\varepsilon)\eta_P r_P(P_L) \quad \text{... (6.5)}$$

here, $r_S$ and $r_P$ are, respectively, related to product formation and/or cell mass concentration. We use Monod kinetic expressions for $r_S$ and $r_P$, since it is well-known that fermentation with *Saccharomyces cerevisiae* follow that kind of kinetics. Furthermore, if we consider that the endogenous metabolism is negligible, the substrate concentration, product formation and cellular growth due to the fermentation can be described by:

$$r_S = \frac{dS_L}{dt} = \frac{\mu_{max} S_L}{(S_L + K_S)Y_{XS}} X \quad \text{... (6.6)}$$

$$r_P = \frac{dP_L}{dt} = \frac{\nu_{max} S_L}{(S_L + K_S)} X$$

$$\frac{\partial X}{\partial t} = \frac{\mu_{max} S_L}{(S_L + K_S)} X \quad \text{... (6.7)}$$

where $X$ represents the cellular concentration per bead volume, $\mu_{max}$ is the maximum growth rate; $K_S$ is the Monod constant and $V_{max} = \mu_{max} Y_{PX}$. Roels and Kossen related $Y_{XS}$ and $Y_{PX}$ to stoichiometry. For ethanol production during anaerobic growth, they found that:

$$Y_{PX} = \frac{1-1.22Y_{XS}}{1.96Y_{XS}} \quad \text{... (6.8)}$$

The boundary conditions are given by:

$$S_L = S_0;\ P_L = 0 \text{ at } (0, z)$$
$$\frac{\partial S_L}{\partial z} = 0;\ \frac{\partial P_L}{\partial z} = 0 \text{ at } (t, L_e)$$
$$S_L \quad \frac{D_{S_L}}{u_0}\frac{dS_L}{dz} = S_L^i \quad \text{... (6.9)}$$
$$P_L \quad \frac{D_{P_L}}{u_0}\frac{dP_L}{dz} = P_L^i \text{ at } (t, 0)$$

here $L_e$ is the expanded bed length of the bioreactor and $S_L^i$ and $P_L^i$ are the input substrate and product concentrations, respectively. For cellular mass the initial condition is:

$$X = X_0 \text{ at } t = 0 \quad \text{... (6.10)}$$

Our experimental set up also includes a recirculation stream and a tank for monitoring purposes (see Fig. 6.8). The dynamic behaviour for glucose and ethanol in the tank can be modelled as a unsteady state continuous stirred tank. Here we assume that no reaction is taking place in the tank. Hence, the species mass balances are:

$$V\frac{dS_L^i}{dt} = F\left(S_L^S - S_L^i\right)$$
$$V\frac{dP_L^i}{dt} = F\left(P_L^S - P_L^i\right) \qquad \text{... (6.11)}$$

where $S_L^S$ and $P_L^S$ are the substrate and product outlet concentrations for the bioreactor, while $S_L^i$ and $P_L^i$ are the substrate and product outlet concentrations for the recirculation tank. The initial conditions associated to equation (6.11) are:

$$S_L^i = S_0 \quad \text{at} \quad t = 0$$
$$P_L^i = 0 \quad \text{at} \quad t = 0 \qquad \text{... (6.12)}$$

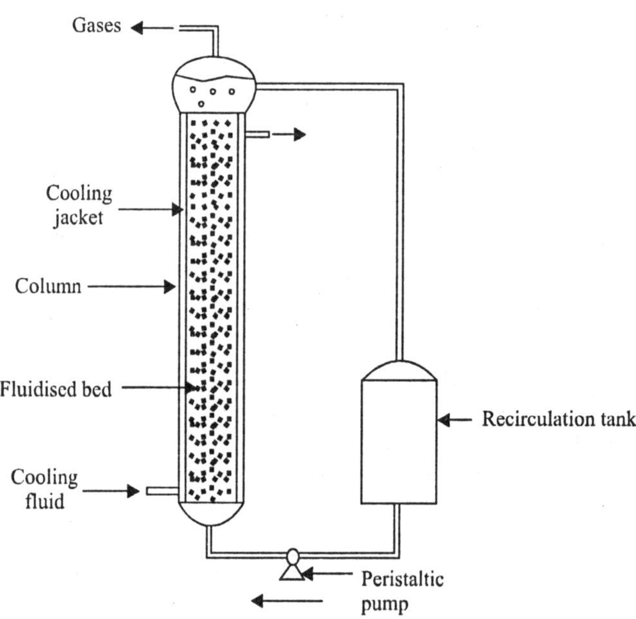


**Fig. 6.8.** Fluidised bed bioreactor.

## Experimental Section

The yeast used in our experiments was *Saccharomyces cerevisiae* ATCC-4126. The fermentation medium was solution of glucose of 30 g/l and other nutrients. The yeast was immobilised in calcium alginate. The yeast was mixed with a two mass per cent sodium alginate solution, which was dropped into a two molar calcium chloride solution. The beads formed were cured in the solution for two hours. The fluidised bed reactor was one meter glass tube with a 20 cm$^2$ cross-sectional area. Ethanol samples were analysed by gas chromatography.

Experiments were carried out for various flow rates (eight to twenty ml/s) with a two litres of fermentation volume in the column. The recirculating tank had a volume of seven litres. The modelling equations were solved by a numerical procedure that combined the spatial discretisation by the orthogonal collocation method and a routine to integrate the resulting set of ordinary differential equations (ODE). The application of the orthogonal collocation method in many chemical engineering problems is very well documented in the literature (Villadsen and Michelsen; Finlayson). In this work, we follow Finlayson's procedure to transform the modelling equations in a set of ODE's which are then solved by the MATLAB integration routine ode45 (Math Works). Chemical and physical properties used in the simulations were taken from other studies with the same type of immobilised yeast and they are reported in Table 6.4. The yield factor $Y_{XS}$ was taken from the work of Aiba, while $Y_{PS}$ was calculated by stoichiometry (equation 6.8).

**Table 6.4.** Operating conditions and typical parameters.

| *Parameter* | *Value* |
|---|---|
| $\mu_{max}$ | 0.113 hr.$^{-1}$ |
| $K_S$ | 4.21 kg/m$^3$ |
| $Y_{XS}$ | 0.121 kg/kg |
| $D_{S_L}$ | 2.4 × 10$^{-6}$ m$^2$/hr. |
| $D_{P_L}$ | 3.6 × 10$^{-6}$ m$^2$/hr. |
| $D_{S_i}$ | 2.4 × 10$^{-6}$ m$^2$/hr. |
| $D_{P_i}$ | 3.96 × 10$^{-6}$ m$^2$/hr. |
| $L_s$ | 0.8 m |
| $R_0$ | 1.5 × 10$^{-3}$ m |
| $\varepsilon$ | 0.85 |
| Residence time | 5.45 × 10$^{-2}$ hr. |
| $S_0$ | 30 kg/m$^3$ |

## Results

The effect of the number of internal collocation points on the solution of the mathematical model is shown in Fig. 6.9.

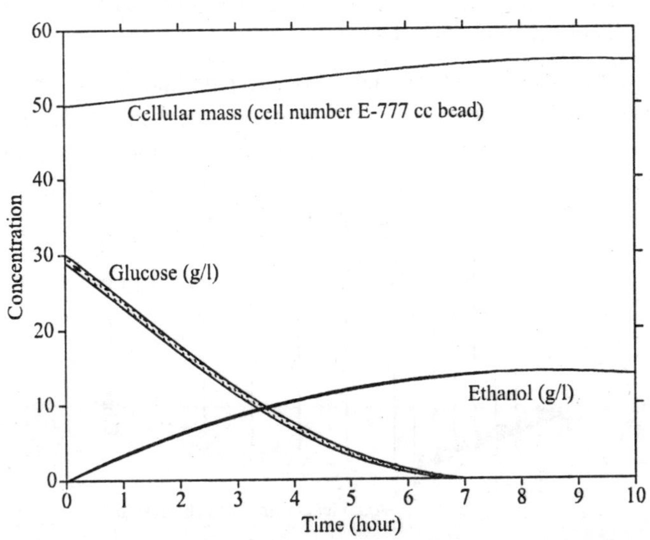


**Fig. 6.9.** Simulation results using one, two and three internal collocation points.

No appreciable differences were found when two to six internal collocation points were used. No attempts were made to solve the modelling equations with more than 6 collocation points. Figure 6.10 shows the simulation results of substrate and product concentrations for three collocation points and in the recirculation tank. We can see that the solutions at the collocation points generate practically the same curve. The maximum error in all cases was less than one per cent. These results can be used to confirm that no reaction takes place in the tank.

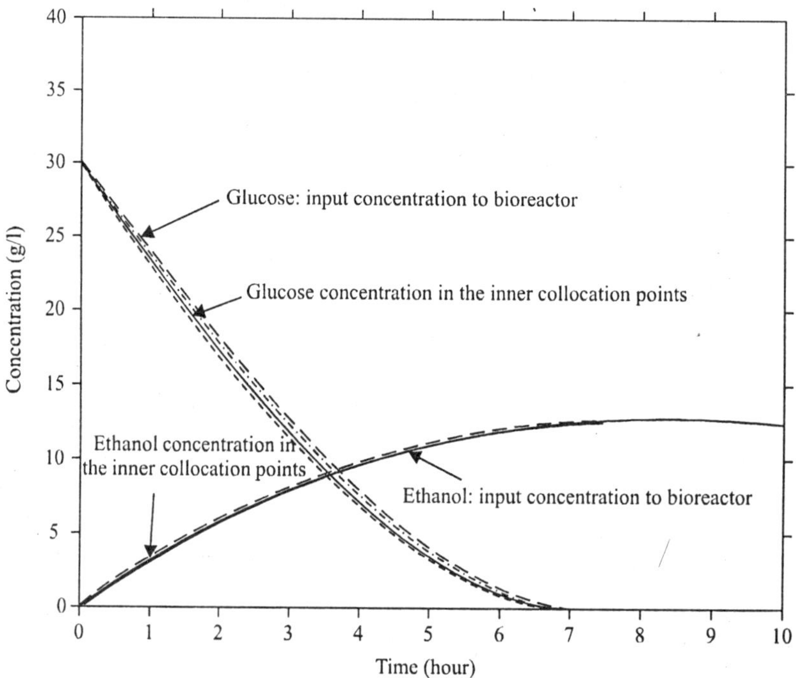


**Fig. 6.10.** Simulation results using one, two and three internal collocation points. The concentration differences in the bulk liquid in the axial direction is negligible.

We use Fig. 6.11 to confirm that the orthogonal collocation method did not exhibit instability problems during the solution procedure. This was probably due to the operating conditions selected in this work which allowed to avoid the effects of nonlinear kinetics and diffusional limitations.

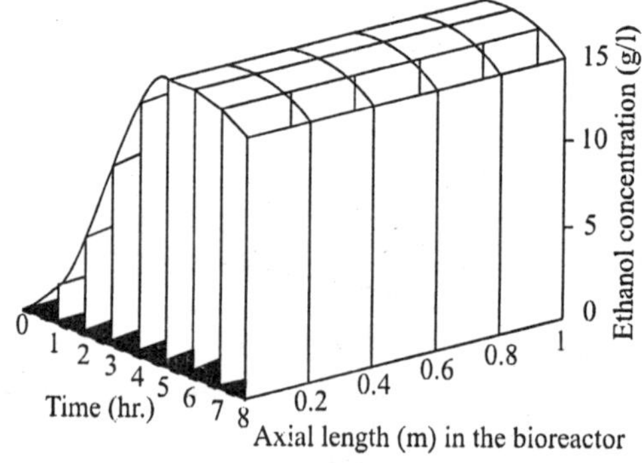


**Fig. 6.11.** Simulation results for ethanol concentration using 5 internal collocation points: a tridimensional perspective.

In order to validate the model a series of experimental runs were carried out by varying the flow rate. Figure 6.12 shows the comparison of both experimental and predicted ethanol concentrations as a function of time for several flow rates.

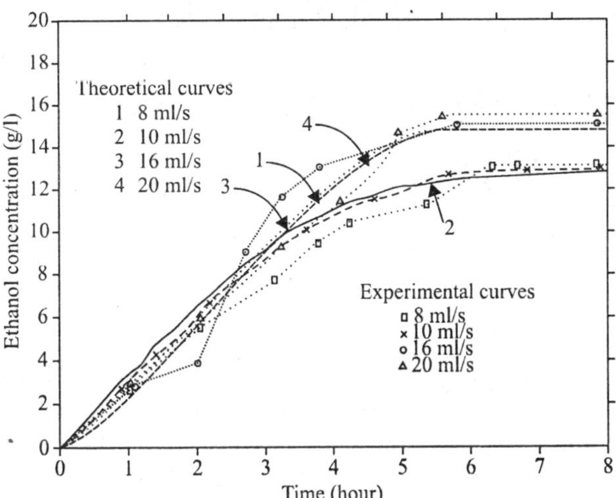


**Fig. 6.12.** Comparison between simulation and experimental results for the ethanol fermentation from glucose using *Saccharomyces cerevisiae* ATCC-4126 immobilised in calcium alginate.

This comparison showed that the model yielded satisfactory results for the flow rates studied. We found a ± two per cent error for the 20 ml/s flow rate which is acceptable within the experimental error (± five per cent). However, the error was higher (± seven per cent) for the eight ml/s flow rate. This result may indicate that diffusional limitations have to be considered in model for low flow rates (less than eight ml/s).

Thus, this mathematical model has described the dynamic behaviour of the fermentation process carried out in a fluidised bed reactor with immobilised yeast. The model incorporated axial dispersion, convective transport and Monod kinetics to predict the dynamic behaviour of substrate, ethanol and biomass. The model was solved by means of the numerical procedure that combined the orthogonal collocation method and an integration routine. Experimental data and theoretical results were compared to validate the model and test the solution procedure. It was found that the model satisfactorily predicted the experimental data for a number of internal collocation points. The solution procedure was found to be fast and accurate and that can be extended to solve more complex cases (e.g., system with diffusional limitations, product and substrate inhibitions, etc.). The model can also provide a useful tool in deriving an optimal set of operating conditions after some key performance criteria have been decided upon.

## Fermentation Biotechnology: Future Prospects

Today the pace of progress in fermentation biotechnology is fast and furious, particularly since the advent of genetic engineering and the recent advances in computer sciences and process control. The subject currently encompasses a host of different scientific disciplines, including microbiology, biochemistry, molecular genetics, bioinformatics and, last but not least, engineering. As we enter a new era of scientific progress, in which the use of renewable resources for the production of desirable end products is recognised as an urgent need, fermentation biotechnology has a major role.

Chapter 7

# Biotechnology of Industrial Micro-organisms

## INTRODUCTION

Commercially important products of industrial fermentations fall into four major categories—microbial cells, large molecules such as enzymes and polysaccharides, primary products and secondary metabolites which are not needed for growth. Cells used to produce these products exhibit a variety of physiological and biochemical properties. Commercial production of fermentation products has mainly involved use of a range of species of bacteria, yeasts and fungi. Recent advances in the development of techniques for cultivation of animal and plant cells has enabled these more complex cells to be used also in fermentation processes. In this chapter the properties of the major species involved are discussed and general aspects of microbial growth and metabolism relevant to industrial cell-culture processes are also considered.

## INDUSTRIAL MICRO-ORGANISMS

Two major classes of cells, prokaryotes and eukaryotes, exist in nature and both types are used in industrial fermentation processes. Eukaryotic cells have a distinct nucleus surrounded by a membrane. Nuclear DNA is associated with protein and exists as definite structures called chromosomes. The cells also contain other structures or organelles having specific physiological or biochemical functions. In contrast, prokaryotes lack a well-defined nucleus so that the genetic material in the form of double-stranded DNA is not separated from other cell constituents by its own membrane. These cells also lack other specialised organelles present in eukaryotes, such as mitochondria and the enzymes associated with these organelles in eukaryotes are found in the protoplasm and plasma membrane of prokaryotes. The cellular properties of prokaryotes and eukaryotes are compared in Table 7.1. Bacterial cells, which are prokaryotic and fungal, yeast, animal and plant cells, which are eukaryotic are used in fermentation processes.

**Table 7.1.** Selected characteristics which differentiate between prokaryotes and eukaryotes.

| *Characteristic* | *Prokaryotes* | *Eukaryotes* |
|---|---|---|
| *Organelles* | | |
| Nucleus | – | + |
| Mitochondria | – | + |
| Endoplasmic reticulum | – | + |

*(Cont'd...)*

| *Characteristic* | *Prokaryotes* | *Eukaryotes* |
|---|---|---|
| *Genome* | | |
| No. of DNA molecules | 1[a] | >1 |
| DNA in organelles | – | + |
| DNA as chromosome structures | – | + |

[a]Bacteria may contain small DNA fragments (plasmids) in addition to the single major genome.

Micro-organisms are also distinguished based on their oxygen requirements. Strict aerobes, such as *Streptomyces* and most filamentous fungi, can metabolise and grow only in the presence of atmospheric oxygen. Strict anaerobes, such as Clostridia, can only grow in the absence of oxygen. Facultative organisms, including industrial yeasts, can grow aerobically or anaerobically.

Bacteria involved in fermentation processes are mainly chemo-organotrophs, that is, they obtain their energy and carbon by oxidation of organic compounds. Some properties of the most important genera involved and some example species are listed in Table 7.2. Bacteria are distinguished as Gram-positive or Gram-negative based on their response to Gram-staining, which binds to peptidoglycan. The cell envelope of Gram-positive bacteria consists largely of peptidoglycan, 15–80 nm thick, covering a single cytoplasmic membrane. In Gram-negative cells this peptidoglycan layer is only 2–3 nm thick. Gram-negative cells have two membranes, an outer membrane and cytoplasmic membrane, separated by a periplasmic space. These different structures are illustrated in Fig. 7.1. Some prokaryotes, including the important antibiotic-producing *Streptomyces*, have filaments or mycelia.

**Table 7.2.** Properties of chemo-organotrophic bacterial genera of importance in industrial fermentations.

| | *Properties* | | | | |
|---|---|---|---|---|---|
| *Genus* | *Gram reaction* | *Endospores* | *Shape* | *Aerobic*[a] | *Example species (products)* |
| *Enterobacteria* | – | – | Rod | ± | *E. coli* (recombinant proteins) |
| *Pseudomonas* | – | – | Rod | + | *P. ovalis* (gluconic acid), *P. fluorescens* (2-oxogluconate) |
| *Xanthomonas* | – | – | Rod | + | *X. compestris* (xanthan gum) |
| *Acetobacter* | – | – | Rod | + | *A. aceti* (acetic acid) |
| *Alcaligenes* | – | – | Rod/coccus | + | *A. faecalis* (curdlan gum) |
| *Bacillus* | + | + | Rod | + | *B. subtilis, B. licheniformis, B. amyloliquefaciens* etc. (extracellular enzymes), *B. coagulans* (glucose isomerase), *B. licheniformis* (bacitracin), *B. brevis* (gramicidin S), *B. thuringiensis* (insecticide) |
| *Clostridium* | + | + | Rod | – | *C. acetobutylicum* (acetone, butanol), *C. thermocellum* (acetate) |
| *Leuconostoc* | + | – | Spherical | ± | *L. mesenteroides* (dextran) |
| *Lactobacillus* | + | – | Rod | – or ± | *L. delbruckii* (lactic acid) |
| *Corynebacterium* | + | – | Irregular | + | *C. glutamicum* (amino acids), *C. simplex* (steroid conversion) |
| *Mycobacterium* | + | – | Branching filaments No mycelia | + | *Mycobacterium* spp. (steroid transformation) |

*(Cont'd...)*

| Genus | Gram reaction | Endospores | Shape | Aerobic[a] | Example species (products) |
|---|---|---|---|---|---|
| | *Properties* | | | | |
| *Nocardia* | + | – | Fragmenting mycelia | + | *N. mediterranei* (rifamycins), *Nocardia* spp. (steroid transformation) |
| *Streptomyces* | + | – | Intact mycelia | + | *S. erythreus* (erythromycin), *S.aureofaciens, S. rimosus* (tetracyclines), *S. fradiae* (glucose isomerase), *S. roseochromogens* (steroid hydroxylations) |

[a]Aerobic: + = aerobic; – = anaerobic; ± = both.

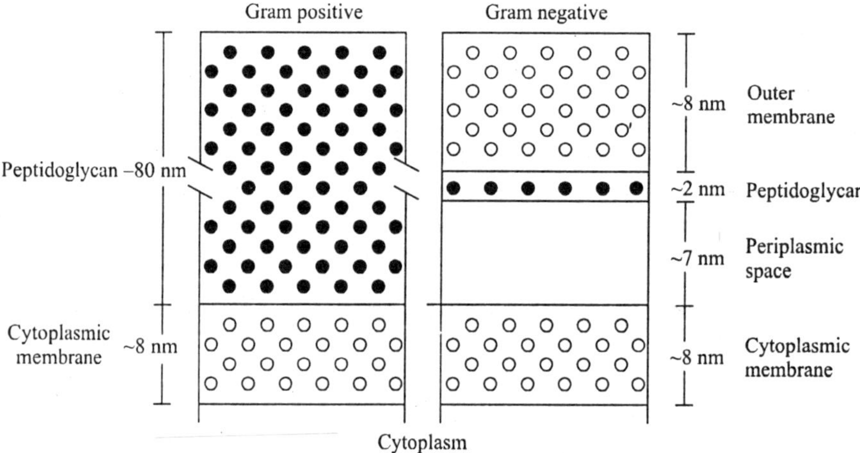


**Fig. 7.1.** Schematic comparison of the cell envelope composition of gram-positive and gram-negative bacteria.

Bacterial reproduction occurs by asexual cell division as shown diagrammatically for *Eschericia coli* in Fig. 7.2. The single circular DNA chromosome is replicated and a septum or dividing membrane develops in the middle of the cell, separating the new and old chromosome. The septum develops into a wall and the cells separate. Spores are produced by spore-forming bacteria usually in response to unfavourable environmental conditions. Spores are more resistant to heat and toxic chemicals than vegetative cells. When bacterial spores are transferred to appropriate media they germinate to reform vegetative cells.

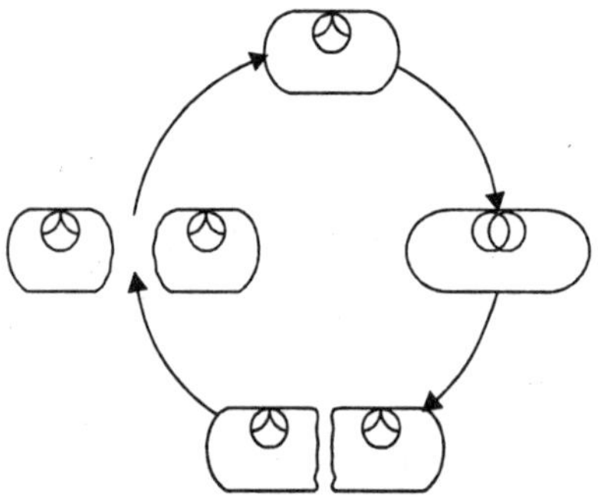

**Fig. 7.2.** Asexual division in *Escherichia coli.*

The fungi are also chemo-organotrophs which have branched filamentous rigid hyphae usually of diameter 2–18 μm. Higher fungi typically have cross-walls or septa at intervals along the hyphae, but these are characteristically absent from the lower fungi. The most important fungi involved in industrial fermentations are from two of the main classification groups, the aseptate Zygomycotina, which includes the *Mucor* and *Rhizopus* genera and the septate Deuteromycotina or Fungi Imperfecti, which include the genera *Trichoderma, Aspergillus, Penicillum, Aureobasidium* and *Fusarium* (Table 7.3).

**Table 7.3.** Characteristics of fungi of importance in industrial fermentations.

| *Classification group* | *Mycelium* | *Spores* | *Genus* | *Drawing* | *Colony appearance* |
|---|---|---|---|---|---|
| Zygomycotina (order: mucorales) | Aseptate thin cottony or felted covers of mycelium | Asexual spores formed in sporangium | *Mucor* |  | *M. miehei* and *M. pusillus* produce grey colonies |
| | | | *Rhizopus* |  | Grows with extreme rapidity filling petri dish with cotton-like mass of mycelium |
| Deuteromycotina (order: Moniales) | Septate filamentous fungi forming loose mycelium on solid or liquid substrate. *Trichoderma, Aspergillus* and *Penicillium* have light coloured mycelium. *Aureobasidium* has dark mycelium. *Fusarium* has light or dark mycelium | Asexual spores, sexual reproduction rare or unknown | *Trichoderma* |  | Most species have colonies that spread rapidly forming a somewhat thin layer of mycelium with green conidial patches |
| | | | *Aspergillus* |  | Rapidly spreading colonies with mycelium, white at first, frequently developing bright yellow areas and producing dark brown or black conidial heads. *A. oryzae* colonies are yellowish to brownish-green |
| | | | *Penicillium* | 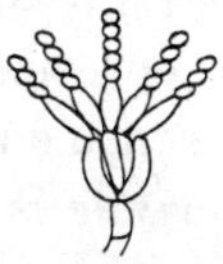 | *P. chrysogenum:* Broadly spreading colonies, bluegreen to bright green with |

*(Cont'd...)*

| *Classification group* | *Mycelium* | *Spores* | *Genus* | *Drawing* | *Colony appearance* |
|---|---|---|---|---|---|
| | | | | | broad white margin during growing period. Usually becomes greyish or purplish brown in age with overgrowth of white or rosy hyphae. Reverse yellow |
| | | | *Aureobasidium* |  | *A. pullulans:* Light to pink and brown hyphae becoming darker with age |
| | | | *Fusarium* |  | *F. graminearum:* Grey-pink to red colonies becoming brown-red with white or grey-yellow sectors |

The Zygomycotina produce haploid vegetative mycelia. Asexual spores are formed in a sporangium. They also reproduce sexually. Deuteromycotina produce haploid vegetative mycelia. Asexual spores are born on conidia which vary according to species. Asexual spores are usually induced by certain environmental conditions. For example, spore-bearing conidiophores are initiated in *Aspergillus niger* when nitrogen becomes limiting.

The major fungal cell-wall constituents in the Zygomycotina are chitosan and chitin while glucan and chitin predominate in Deuteromycotina. Vegetative growth of filamentous fungi involves hyphal extension and occurs at the hyphal tip. Branching also occurs with the development of new extension zones. The length that hyphae grow and the amount of branching depends on the growth environment. Shear forces in fermenters can cause hyphal fragmentation and may result in production of more highly-branched shorter mycelia.

In fungi, the hyphal tip is the main area of protoplasm activity (Fig. 7.3). The cell wall, overlying the plasmalemma, is only 50 nm thick in the apical region. Behind the tip, the cell wall may thicken to about 125 nm and the protoplasmic compartments gradually become more vacuolated as they become more distant from the apex.

Yeasts are micro-fungi which generally exist in the form of single cells and which reproduce by budding (Fig. 7.4). Some yeasts only form individual cells and sometimes short chains while others have a range of cell forms including various filament types. A feature of a growing population of yeast cells is the presence of buds which are produced when the cell divides. The daughter cell starts as a small bud, increases until it is similar in size to the mother cell and then separates. In yeast, the cell envelope includes the cytoplasmic membrane, consisting of lipids, proteins and mannans, a periplasmic space and the cell wall, containing some protein and a large amount of glucan and mannan.

*Saccharomyces cerevisiae* is the most widely used yeast in industrial fermentation through its applications in alcohol production and baking. *Saccharomyces cerevisiae* cannot degrade lactose and another yeast, *Kluyveromyces lactis*, which contains the necessary lactose-transporting and degrading

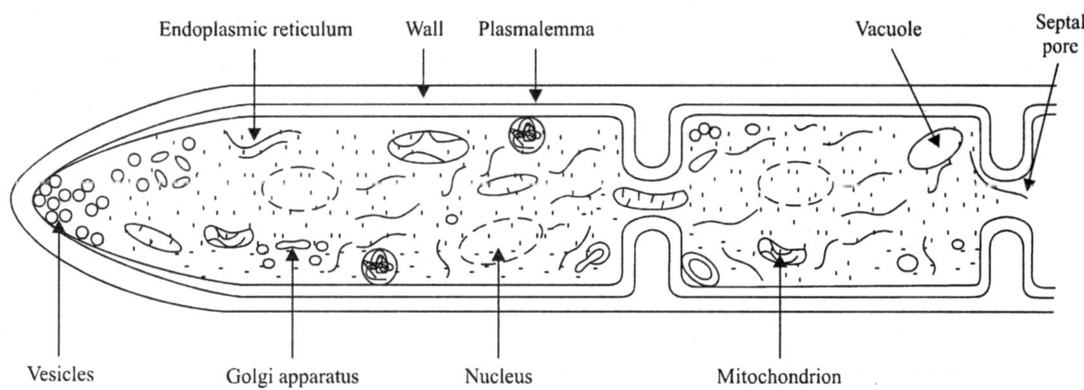


**Fig. 7.3.** Diagrammatic representation of a fungal hypha.

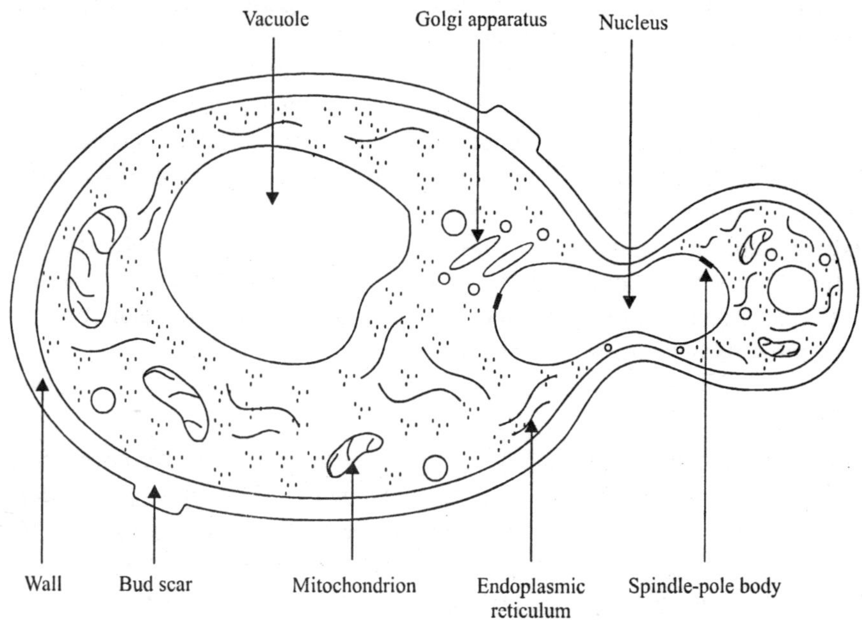


**Fig. 7.4.** Diagrammatic representation of a budding yeast.

enzymes, is used to produce alcohol or biomass from whey. Other important industrial yeasts include *Candida utilis* and *Endomycopsis fibuliger.*

Mammalian cells are generally more complex than fungal or yeast cells. Because they normally exist in controlled isotonic environments, they do not possess tough outer cell walls. The external surface of the plasma membrane contains a coat or glycocalyx consisting of short oligosaccharide chains linked to intrinsic membrane lipids and proteins and to some absorbed proteins. In tissues other than reproductive tissues, cells are diploid and cell division takes of the order of 24 hours. Depending on cell type, animal cells can grow in cultures as monolayers attached to surfaces (anchorage-dependent) or as suspended cells (anchorage-independent) in slowly-rotated, gently-stirred media. Mammalian cell nutritional requirements are very complex and cells are also very sensitive to fluctuations in parameters such as temperature, pH, dissolved oxygen and $CO_2$.

Cell lines prepared directly from animal tissues generally do not survive, although this varies. They also usually exhibit contact inhibition, that is, they stop growing when they touch. However, after repeated passaging, some primary cell lines become transformed, that is, they multiply faster and grow to higher densities than primary cells, can be cultured indefinitely and lose the property of contact inhibition. Tumour cell lines are already transformed.

Hybridoma cells are constructed by fusing a myeloma cell with an antibody producing cell, such that the hybridised cell has the characteristics of transformed cells and can synthesise a single antibody or monoclonal antibody. These cells are used both in cell culture or implanted in the peritoneum of animals to produce monoclonal antibodies.

Many plant cell types can grow on a solid medium or in suspension using techniques similar to those used for animal cell culture. In addition to the complex media requirements which include plant extracts and auxins, carbohydrate must be added to media, as photosynthesis by cell in culture is not as efficient as in whole plants. Plant tissue grows on solid media as callus, which develops as a mixture of parenchymatous cells. On repeated subculture, the cells tend to grow faster, have less complex media requirements, may change their chromosome number and be less able to differentiate. Suspension culture is the most widely used method of plant cell cultivation. The major advantage of suspension culture is that a homogenous environment can be established throughout the reactor.

Viruses are submicroscopic agents which infect plants, animals and bacteria. They consist of genomes made up of DNA or RNA enclosed within a protein coat, containing identical submits, which are sometimes covered by an outer membrane of lipoprotein. Viruses have no cytoplasm or cytoplasmic membrane and they replicate inside specific host cells by using the biosynthetic systems of the host under the direction of the viral genome. Viral vaccines are produced by infection of cultured host cells. Bacterial viruses (bacteriophages) which infect fermentation process microbial strains have caused serious problems particularly in cheese starter cultures.

## CELL GROWTH

Microbial growth is an integral part of nearly all fermentation processes. We have already seen that bacteria reproduce by binary fission to produce two daughter cells of equal size. A doubling of the number of yeast cells normally results from yeast budding mechanisms. Fungi grow by hyphal extension and the morphology of the fungal mycelium can vary depending on the environment. Fungi grow on the surface of solid media producing a mycelial mat, the nature of which will reflect media composition. In submerged culture, fungal cells may grow attached to suspended nutrient particles, as diffuse filamentous mycelia or as dense pellets.

The morphological form of growth invariably influences rate of growth and product formation. Because of limitations in nutrient and product diffusion inside fungal pellets, growth and metabolism occur predominantly at the periphery.

When an organism is inoculated into a fixed volume of medium, the culture passes through a number of growth phase as indicated in Fig. 7.5. After inoculation there is a lag phase, where the organism takes time to adapt to the environmental conditions. Following a short period of time during which the growth rate of cells gradually increases, the cells grow at a constant maximum rate. This growth phase is called the exponential or log phase.

As growth continues, nutrients are consumed and products are excreted by the organism. The growth rate decreases and finally growth ceases often due to depletion of an essential nutrient or build up of a toxic product. This phase is described as the stationary phase.

Growth measured by increase in cell mass may be described by:

$$\frac{dx}{dt} = x \quad \alpha x \qquad \text{... (7.1)}$$

where,

$x$ = the cell concentration (mg $cm^{-3}$)
$t$ = the incubation time (hour)
$\mu$ = the specific growth rate ($hr.^{-1}$)
$\alpha$ = the specific rate of lysis or endogenous metabolism ($hr.^{-1}$)

When nutrient and other cell conditions are favourable $\mu \gg \alpha$ and equation (7.1) becomes:

$$\frac{dx}{dt} = \mu x \qquad \text{... (7.2)}$$

Integration of equation (7.2) gives:

$$x_t = x_0 e^{\mu t} \qquad \text{... (7.3)}$$

where,

$x$ = cell concentration at time 0
$x_t$ = cell concentration after a time interval of t (hour)

On taking the natural logarithms of equation (7.2) we obtain

$$\ln x_t = \ln x_0 + \mu t \qquad \text{... (7.4)}$$

If $t = t_d$, the doubling time, then $x_t = 2x$

$$\ln 2x_0 = \ln x_0 + \mu t_d \qquad \text{... (7.5)}$$

The time taken for the biomass to double, $t_d$, is given by:

$$t_d = \frac{\ln 2}{\mu} \qquad \text{... (7.6)}$$

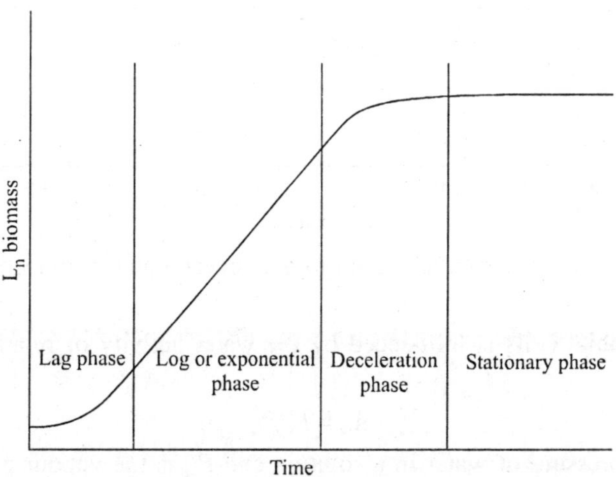


**Fig. 7.5.** Growth of micro-organisms in batch culture.

### Factors Affecting the Growth Rate

Growth rate varies with microbial cell type and also in response to physical and chemical environmental conditions. In general, the cell doubling time increases with increasing organism complexity. Average doubling times for animal and plant cells are substantially longer than for yeasts and moulds which in turn are longer than for bacteria.

As with chemical and enzymatic reactions, cell growth varies as a function of temperature. Most micro-organisms will grow over a range of 25–30°C although the actual temperature at which a particular organism grows depends on its psychrophilic, mesophilic, moderate thermophilic or extreme thermophilic nature as indicated in Fig. 7.6. Animal and plant cells have much narrower temperature ranges for growth. Within the growth range, growth rate increases with increasing temperature until a maximum growth rate is reached above which growth rate falls off rapidly due to increased rate of microbial death. The effect of temperature on growth rate may be described by the Arrhenius equation

$$\mu = Ae^{E_a/RT} \qquad \text{... (7.7)}$$

where,

$A$ = the arrhenius constant
$E$ = the activation energy (kcal mol$^{-1}$)
$R$ = the universal gas constant
$T$ = the absolute temperature

pH influences microbial growth rate in a manner similar to its influence on enzyme activity. Most micro-organisms grow over a pH range of 3 to 4 pH units.

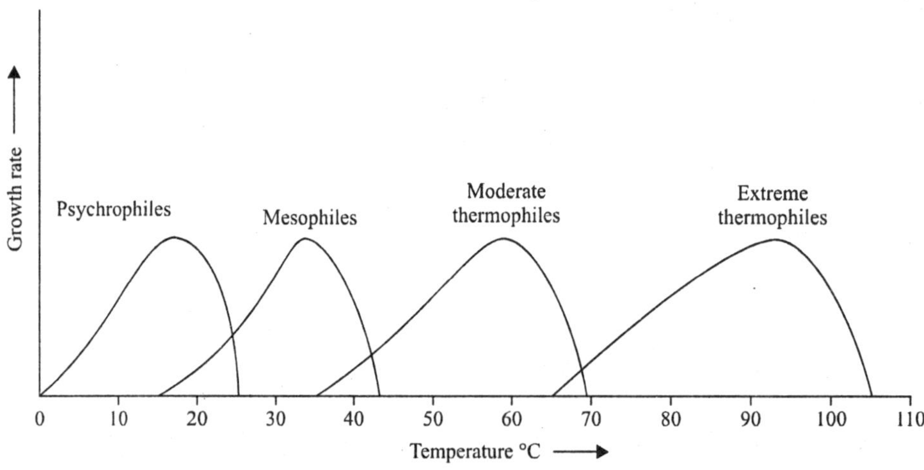


**Fig. 7.6.** Effect of temperature on specific growth rate of psychrophiles, mesophiles and thermophilic micro-organisms.

Growth rate of microbial cells is influenced by the water activity or relative humidity which is defined as:

$$A_w = P_s/P_w \qquad \text{... (7.8)}$$

where $P_s$ is the vapour pressure of water in a solution and $P_w$ is the vapour pressure of pure water. Bacteria require an $A_w$ of $\geqslant 0.95$ whereas moulds can grow at a lower water activity of $\geqslant 0.7$.

As with any chemical reaction, growth rate will depend on the concentration of chemical nutrients and can be described by the Monod equation (Fig. 7.7):

$$\mu = \mu_{max} \frac{s}{k_s + s} \qquad \text{... (7.9)}$$

where,

$\mu_{max}$ = the maximum specific growth rate

$s$ = the residual concentration of the limiting substrate

$K_s$ = the saturation constant and is equivalent to the substrate concentration at $\mu = 0.5\ \mu_{max}$

Typical $K_s$ values for various carbon sources are in the range 1–10 mg $dm^{-3}$. Cells will grow at rates close to $\mu_{max}$ if the limiting carbon source is greater than 10 $K_s$ or 10–100 mg $dm^{-3}$. High carbohydrate concentrations may inhibit growth due to osmotic effects which, as was seen earlier with water activity, have a more severe effect on bacteria than fungi.

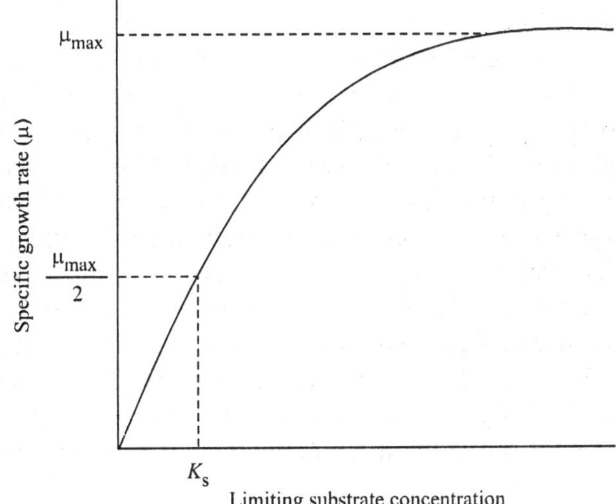


**Fig. 7.7.** Effect of limiting substrate concentration on specific growth rate.

It has been observed that the total amount of cell mass formed during cell growth is often proportional to the amount of substrate utilised

$$\gamma_{x/s} = \frac{\text{Biomass produced (g)}}{\text{Substrate consumed (g)}} \qquad \text{... (7.10)}$$

where $\gamma_{x/s}$ = the biomass yield co-efficient.

## METABOLISM

### Primary Metabolism

Microbial growth depends on the cell's ability to use the nutrients in its surroundings to synthesise the macromolecular components of cellular structures and also the many low-molecular-weight compounds required for cellular activity. Intermediary metabolism is concerned with the reactions which transform carbon and nitrogen compounds entering the cell either into new cell material or into products which are excreted. The synthesis of these compounds requires energy and most of the cells involved in industrial fermentations are hetetrotophs which obtain this energy from breakdown of organic compounds. In aerobic or respiratory processes, organisms are able to completely oxidise some of the substrates to $CO_2$ and $H_2O$, resulting in the provision of maximum energy for conversion of the remaining substrates into new cell mass. In anaerobic or fermentative metabolism, cells are less efficient in converting organic substrates into cellular materials and usually excrete partially degraded intermediates.

Energy-producing or catabolic pathways generate ATP and reduced co-enzymes needed for various biosynthetic reactions and chemical intermediates used as starting points of biosynthesis.

Sugars are broken down by one of three pathways, the Embden-Meyerhof-Parnas (EMP) pathway, the hexose monophosphate (HMP) pathway and the Entner-Doudoroff (ED) pathway. The EMP route, which converts glucose to two molecules of pyruvate, via triose phosphate intermediates, occurs most widely in animal, plant, fungal, yeast and bacterial cells. As it does not produce precursors for biosynthesis of aromatic amino acids, DNA and RNA micro-organisms which solely use this pathway for glucose utilisation must be supplied with growth factors. The ED pathway has a relatively restricted occurrence. A few bacteria, including *Pseudomonas* species, which do not metabolise glucose via the EMP pathway, utilise this pathway. One cycle of the HMP pathway results in the net conversion of glucose to $CO_2$ with production of 12 molecules of NADPA + $H^+$. Micro-organisms can use this pathway to produce pyruvate. The importance of the HMP pathway lies especially in provision of precursors of aromatic amino acids and vitamins and the supply of NADPH and $H^+$ needed for many biosynthetic reactions.

A significant end-product of all three pathways above is pyruvic acid, which is channelled into the tricarboxylic acid (TCA) cycle in aerobic metabolism and is likewise the precursor of the acid, alcohol and other end-products of anaerobic metabolism. A large number of micro-organisms meet their energy needs via the TCA cycle. Acetyl CoA (2C) formed from pyruvate condenses with oxaloacetate (4C) to form citrate (6C) which is converted back to oxaloacetate over the course of the cycle with production of two molecules of $CO_2$ (Fig. 7.8a). Pyruvate is thus oxidised to $CO_2$ with formation of reduced co-enzymes which can be reoxidised by passing their electrons along the election-transport chain to electron acceptors such as oxygen. During this respiration process, some of the energy produced is used to form ATP needed for biosynthesis.

TCA cycle intermediates also serve as precursors for biosynthesis of many amino acids, some organic acids and other fermentation products. When cycle intermediates are diverted into biosynthesis in this way, it is necessary to restore the cycle intermediates so as to maintain the level of oxaloacetate required to condense with incoming acetyl groups from pyruvate. A number of anaplerotic (replenishing) reactions exist which serve this function. The main ones involve carboxylation of pyruvate or phosphoenol pyruvate to form oxaloacetate or malate. Micro-organisms grown on carbon sources such as fatty acids, acetate or their precursors, use the TCA cycle for production of both energy and biosynthetic precursors. Acetyl CoA is formed which condenses with oxaloacetate to enter the cycle. Oxaloacetate and malate, by being converted to phosphoenol pyruvate and pyruvate can be used as precursors of hexoses and pentoses using reversible reactions of the EMP and HMP pathways. Again in the absence of anaplerotic reactions TCA cycle intermediates would be depleted. In this case the introduction of two enzymatic reactions leads to another cyclic mechanism, known as the glyoxylate cycle, which replenishes $C_4$ acids using incoming acetyl CoA. These key anaplerotic reactions are summarised in Fig. 7.8. The glyoxylate cycle may also utilise acetyl CoA produced from hexose via pyruvate to replenish TCA cycle intermediates.

Hexoses are converted to single-cell protein (SCP) by yeasts and fungi generally by use of a combination of the EMP and HMP pathways followed by the TCA cycle and respiration. Production of

SCP from alkanes involves alkane oxidation to acetate units and its assimilation by the glyoxylate and TCA cycles, with enzymes of the EMP and HMP pathway used for biosynthesis of hexoses, pentoses and other cell constituents. In production of SCP from methanol by *Methylophilus* species, the substrate is converted to formaldehyde which is in turn oxidised by the ribulose monophosphate pathway.

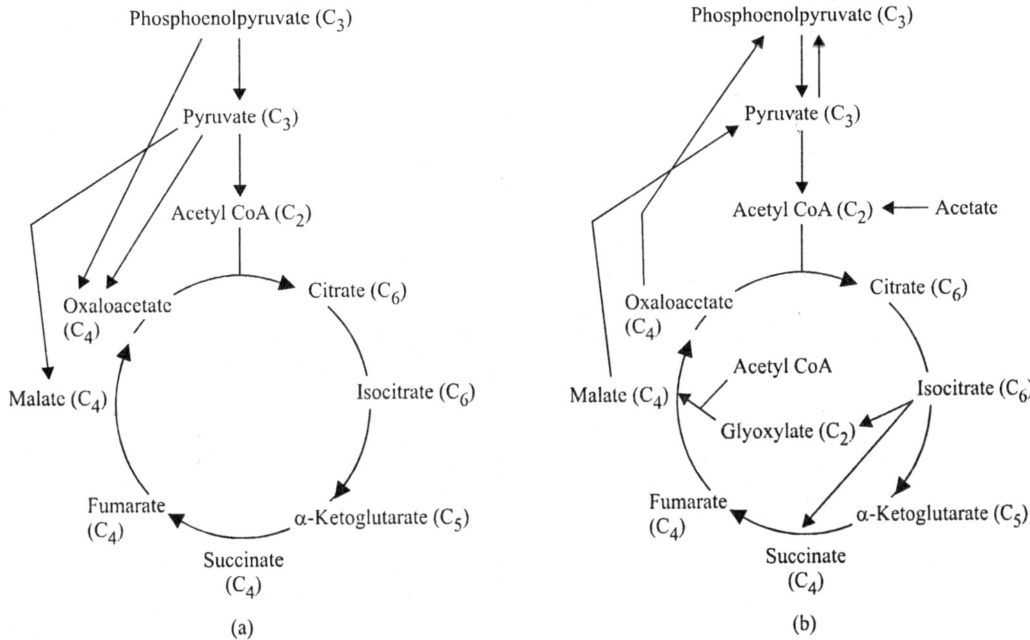


**Fig. 7.8.** (a) Anaplerotic reactions replenishing TCA intermediates when an organism is grown on carbohydrates. (b) Anaplerotic and other key reactions for synthesis of hexose and pentose precursors when an organism is cultured on substrates which generate acetate as a major carbon source.

In the citric acid fermentation process involving *Aspergillus niger*, hexoses are converted, via the EMP pathway, to pyruvate and acetyl CoA which condenses with oxaloacetate to form citrate in the first step of the TCA cycle. Further citrate metabolism is prevented by inhibition of a TCA enzyme downstream from citrate and oxaloacteate needed for the condensation reaction is replenished by carboxylation of pyruvate. The EMP pathway and pyruvate carboxylation also feature in the process for production of itaconic acid from hexose by *A. terreus*. Production of glutamic acid and lysine by *Corynebacterium* and *Brevibacterium* species involves the EMP pathway, the TCA cycle and several anaplerotic mechanisms.

Some of the major anaerobic metabolic pathways and the end-products formed are summarised in Fig. 7.9. Pyruvate is produced via the EMP pathway in important anaerobic fermentations involving strains of *Saccharomyces, Lactobacillus* and *Clostridium* species. In yeast alcoholic fermentations, the pyruvate is converted to acetaldehyde and then to ethanol. NADH + $H^+$ formed during glycolysis is reoxidised by the enzyme alcohol dehydrogenase during conversion of acetaldehyde to ethanol. Addition of sodium bisulphite to the yeast alcoholic fermentation results in the formation of an acetaldehyde-sulphite complex and the reduction of the glycolytic intermediate dihydroxyacetone phosphate to glycerol phosphate, leading to glycerol production. In the production of L-(+)-lactic acid by industrial fermentation using *Lactobacillus delbruckii*, the pyruvate formed from hexose is converted to lactate by the enzyme L-lactate dehydrogenase. While acetic acid can be produced biologically by aerobic conversion of ethanol to acetate the overall stoichiometric yield from glucose is 0.66. *Clostridium thermoaceticum* ferments 1 mole of glucose stoichiometrically to 3 moles of acetate via pyruvate. *Clostridium acetobutylicum* converts pyruvate to mixtures of acetone and butanol. Anaerobic digestion in waste treatment processes involves the interaction of three microbial groups in a complex series of reactions. Organic acids, aldehydes and esters produced by degradation of polymeric material and monomers are converted to acetate, $CO_2$ and hydrogen. Methane and $CO_2$ are then produced from the acetate and hydrogen.

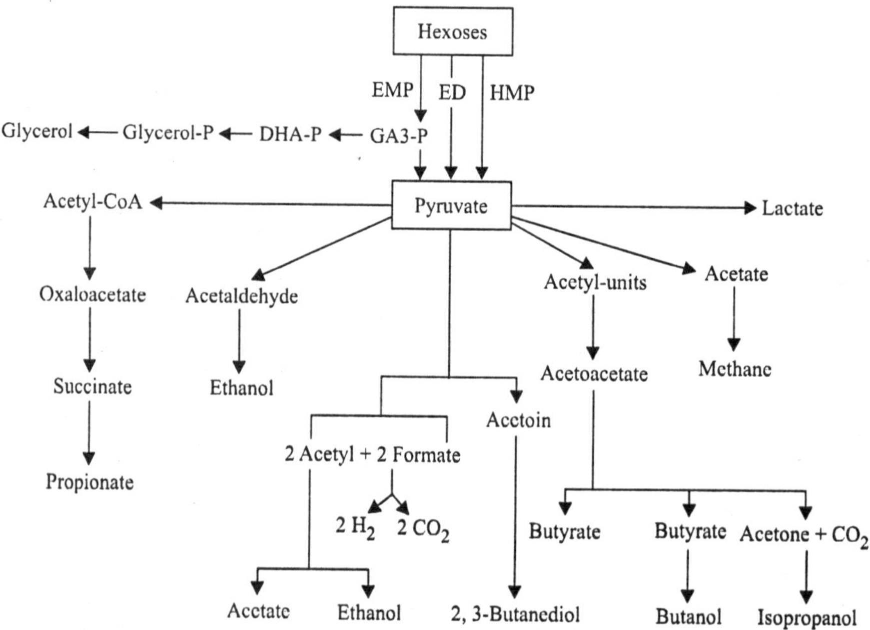


**Fig. 7.9.** Major anaerobic metabolic pathways and end-products formed.

## Secondary Metabolism

Secondary metabolites are compounds which are not required for cellular biosynthesis and which do not play a direct role in energy metabolism. Consequently, the reason that secondary metabolites exist is unclear. Some secondary metabolites, namely antibiotics, are advantageous to the organism in that they inhibit other organisms in their surroundings from competing for available nutrients. Others may act as effectors for differentiation, agents of symbiosis or factors in sexual cycles. They tend to be synthesised as families of related compounds.

Secondary metabolite biosynthesis is closely interrelated to the primary metabolism of the producing cells. Precursors of secondary metabolites are usually normal or modified primary metabolites such as organic acids, isoprene units, aliphatic and aromatic amino acids, sugar derivatives, cyclitol derivatives and purine or pyrimidine bases. The most important group of microbial secondary metabolites produced by industrial fermentation are the antibiotics. Other examples of commercial microbial secondary metabolites include gibberellic acid, ergot alkaloids, antitumour substances, immunomodulators and insecticides. A number of high-value secondary metabolites are extracted from plant material. Examples include the pharmaceuticals vinblastine, vincristine, ajmalicine, digitalis and codeine, the flavour or fragrance compounds quinine, jasmine and spearmint and the pyrethrin insecticides. It is possible that some of these will be produced by suspension culture in the future. A certain degree of cell aggregation and cell differentiation seems to be necessary to obtain higher yields of plant secondary metabolites. For large-scale plant-cell culture this is not always desirable because generally the culture period for differentiation is rather longer than that of undifferentiated cells which adds to costs and increases risks of microbial contamination.

## REGULATION OF METABOLISM

A particular cell has the genetic potential to form over 1000 enzymes. Clearly with such a complex network of metabolic pathways operating in the cell, these enzymes must be synthesised in the correct proportions and must operate in a coordinated manner to enable the cell to operate as an efficient unit. Micro-organisms are able to alter their composition and metabolism in response to environmental changes. The major regulatory mechanisms which confer this flexibility operate at both the level of enzyme synthesis and at the level of enzyme action.

In a given fermentation the flux of substrate through the complex network of catabolic and anabolic pathways will be influenced by a variety of mechanisms. Selected rate-limiting enzymes are strategically placed in the various pathways and these key enzymes are usually sensitive to induction repression and activation/inhibition mechanisms. Over-production of primary metabolites results from high differential flux rates on the upstream side leading to accumulation of the metabolite and reduced or negligible flux rates downstream which would cause metabolite degradation or utilisation. The processes regulating secondary metabolite production are more complex and less well understood but there is evidence that many of the mechanisms are similar to those controlling primary metabolism.

Clearly, in the development of fermentation process for over-production of a particular metabolite, it is essential that account is taken of the various regulatory processes, in the design of optimised environmental conditions and in the isolation or manipulation of microbial production strains.

### Enzyme Synthesis and Degradation

The total concentration of a particular intracellular enzyme present in cells is regulated by the relative rates of synthesis and degradation or inactivation of that enzyme. Proteins in cells are continually being degraded by proteolytic enzymes, such that each protein has an average limited half-life. Proteins also vary in their susceptibility to proteolytic attack. Net synthesis of an enzyme occurs when the rate of synthesis is greater than the rate of degradation.

The general mechanism of protein synthesis in prokaryotes is illustrated in Fig. 7.10a. A single RNA polymerase recognises and binds to specific DNA sites called promoters, initiating mRNA synthesis and transcription of an operon. Beyond the end of the structural gene a termination region causes RNA polymerase to cease transcription and dissociate from the DNA.

In prokaryotes many mRNAs are polycistronic, that is, they encode multiple polypeptide chains (almost all eukaryotic mRNAs are moncistronic). Synthesis of each protein is initiated by the binding of ribosomes to a specific site on the mRNA and translation starts even before the RNA polymerase has reached any signal for termination.

In some cases bacterial transcription is initiated at a constant (usually slow) rate without regulation. In other cases, the rate at which RNA synthesis is initiated is governed by proteins that interact in a site specific manner with DNA regions termed operators, which are located near promoter sites. The regulatory proteins that bind to operators are called repressors or activators depending on whether they decrease or increase the rate of transcription initiation. Some operations are controlled by both activator and repressor proteins and in prokaryotes the promoter-operator sites often control the transcription of DNA encoding for more than one protein. Most regulation of protein synthesis takes place through control of the rate of initiation of transcription. However, in some cases regulation takes place at the level of transcription termination or at translation initiation sites.

The mechanism of protein synthesis in eukaryotes is illustrated in Fig 7.10b. Transcription occurs in the nucleus and the primary RNA transcript is extensively modified in the nucleus before it emerges into the cytoplasm to associate with ribosomes. Control of protein synthesis in eukaryotes may be exerted at the level of RNA processing as well as at transcription or translation. There is very little translational regulation in prokaryotes.

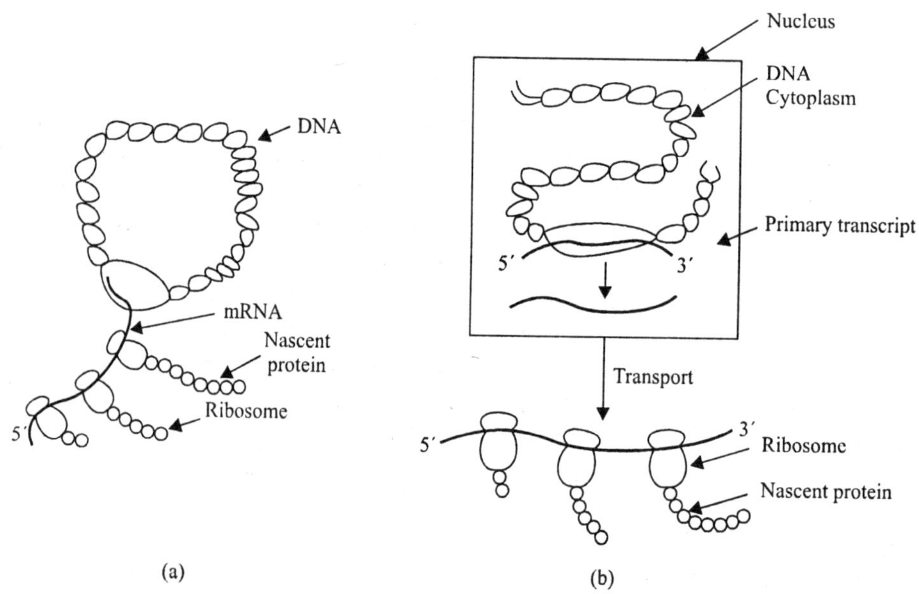


**Fig. 7.10.** Protein synthesis in prokaryotes (a) and eukaryotes (b).

In both prokaryotes and eukaryotes the stability or half-life of mRNA will also influence the rate of protein synthesis. Different mRNAs have different stabilities so that the effect on overall protein synthesis is complex.

### *Induction*

Some enzymes are produced constitutively, that is, they are always made in substantial amounts when cells are growing. Other enzymes are inducible, that is, they are formed only in the presence of key substrates in the medium. The structural genes of inducible enzymes are normally inactive or operate at a very low 'basal' rate in the absence of substrate. When the substrate is present, the structural gene is switched on and transcription and translation occur. This process is illustrated using the Jacob and Monod model for induction of β-galactosidase by *E. coli* (Fig. 7.11).

A regulator gene synthesises a repressor which binds to an operator gene in the absence of inducer, preventing RNA polymerase from moving along the DNA to transcribe the structural gene. The inducer binds to the repressor, removing it from the operator gene and allowing transcription to take place.

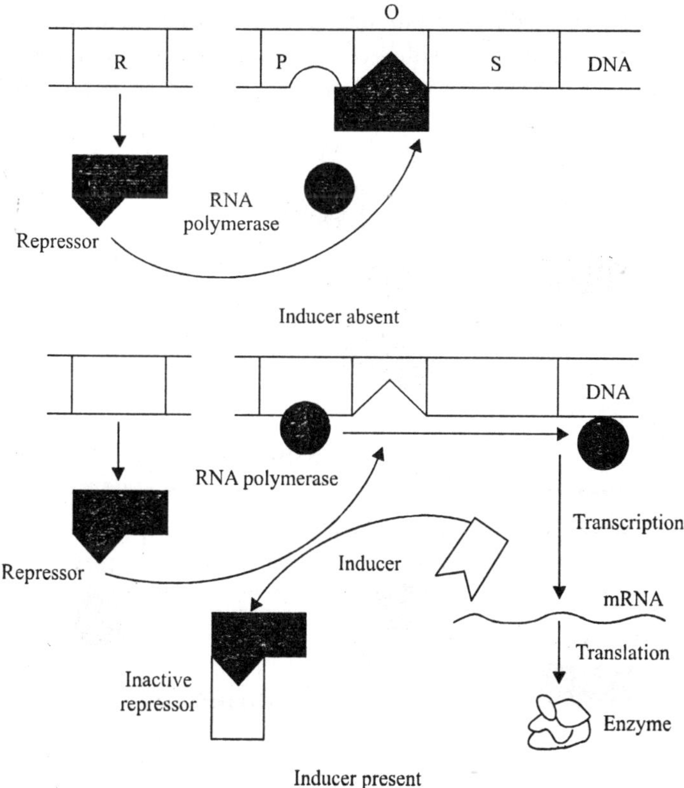


**Fig. 7.11.** Diagrammatic representation of an inducible enzyme system.

### *Catabolite repression*

This phenomenon frequently occurs when the cell is grown in a medium containing more than one utilisable growth substrate. In carbon catabolite repression, enzymes are synthesised which catabolise the best substrate, usually glucose and only after exhaustion of this substrate are enzymes produced to break down the poorer growth substrate. Inhibition of cyclic 3′,5′-adenosine monophosphate (cAMP) formation appears to be the key factor in catabolite repression in *E. coli.* cAMP is necessary for synthesis of mRNA. Glucose metabolism reduces the cAMP content of cells by 1000-fold. cAMP level is a function of adenyl cyclase and cAMP phosphodiesterase which make and degrade cAMP, respectively. Adenyl cyclase appears to be inhibited by transport of readily utilisable carbon sources into the cell. While glucose actabolite repression is widely observed in fungi, yeasts and *Bacillus* species, the regulatory mechanisms are not well understood.

Many enzymes, including proteases are repressed by rapidly utilisable amino acids or ammonia (nitrogen catabolite repression). Limitation of ammonia in fermentation growth media usually de-represses synthesis of these enzymes.

### *Feedback repression*

While synthesis of catabolic enzymes is often controlled by induction and catabolite repression, biosynthesis of anabolic enzymes may be regulated by feedback or end-product repression. Feedback repression is widely used in nature to control amino acid, purine, pyrimidine and vitamin synthesis. In the classical model explaining feedback repression (Fig. 7.12) the regulator gene produces an apo-repressor protein, which when combined with a co-repressor (end product), binds to the operator site and blocks transcription. In the absence of the end-product, no repression takes place.

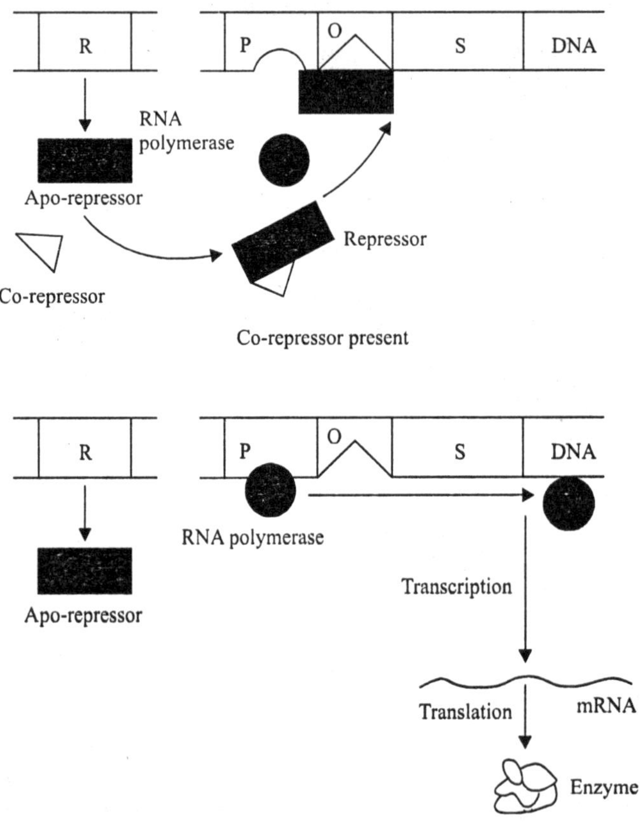


**Fig. 7.12.** Diagrammatic representation of a feedback repressible enzyme system.

## Moderation of Enzyme Activity

Many enzymes have single or multiple sites other than their active site, called allosteric sites, which reversibly bind particular metabolites or effector molecules, causing a change in enzyme conformation and thereby increasing or decreasing its activity. The activities of other enzymes are regulated by covalent modulation, whereby enzymatic formation or degradation of a covalent bond in the enzyme serves to activate or inactivate enzyme activity. Compounds such as ATP, ADP, AMP and nicotinamide nucleotides, which reflect the level of the cell's available energy, also modulate enzyme activity and control metabolism by adjusting the balance of catabolic processes and energy-using anabolic processes.

## Regulation of Branched Metabolic Pathways

Biosynthetic metabolic pathways often have a common enzyme sequence and then branch leading to more than one end-product. Micro-organisms have evolved feedback mechanisms, whereby a build-up of one end-product causes a feedback effect on the first enzyme of the branch leading to that product. In addition, mechanisms exist whereby the end-product of a branched pathway causes partial feedback inhibition of the first enzyme of the common sequence so that the flux of substrate passing through this sequence is proportionately reduced. This effect is achieved by use of isoenzymes, concerted feedback regulation and cumulative feedback regulation (Fig. 7.13). These regulatory effects can be of two types: inhibition of enzyme activity and repression of enzyme synthesis. Where isoenzymes (multiple enzyme forms capable of catalysing the same reaction) are involved, the synthesis or inhibition of each enzyme form may be regulated by a different end-product. With concerted feedback regulation, only one enzyme is involved, but more than one product must be present to inhibit activity or repress enzyme synthesis. With cumulative feedback regulation, each end-product causes partial inhibition or repression and all end-products are required to completely block activity or synthesis.

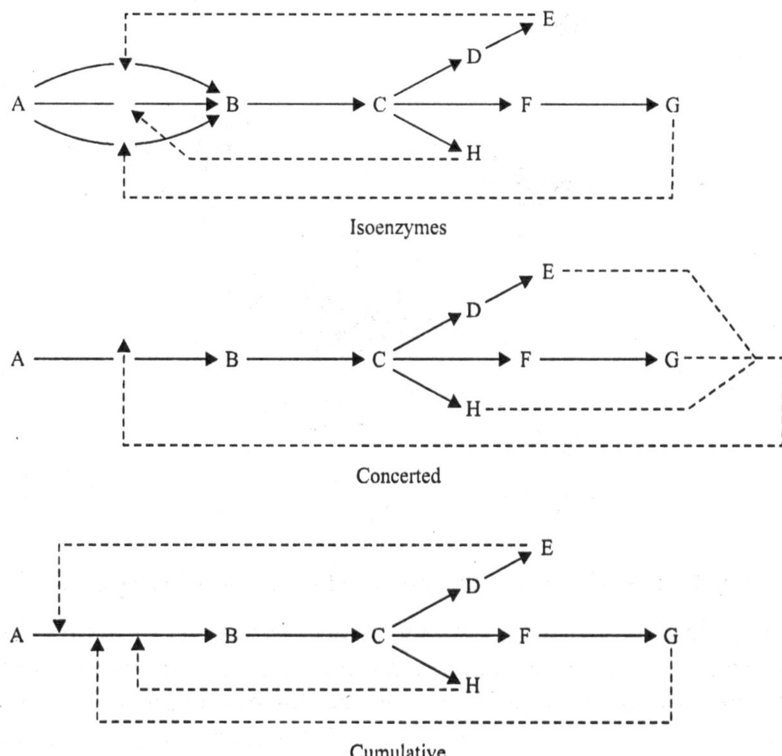


**Fig. 7.13.** Mechanism of feedback regulation of branched metabolic.

## Substrate Assimilation/Product Secretion

Efficient substrate influx and product efflux mechanisms are essential for rapid cell growth, supply of substrate for conversion to primary and secondary metabolites and other products and for product excretion. In prokaryotes, substrates move into and out from the cell directly by traversing membranes associated with the cell envelope. Some bacterial fermentations have been manipulated to alter the cell membrane or wall composition in order to modify uptake mechanisms or facilitate secretion. In addition to having an external plasma membrane, eukaryotes also contain an extensive system of internal membranes and organelles. Some materials are transported into the cell by endocytosis. In this process a region of the plasma membrane invaginates and forms a vesical containing material from outside the cell. The vesicle fuses with another organelle, a primary lysozome, to form a secondary lysozome where much of the external material can be degraded and transported to the cytoplasm. Other degradation products are packaged in vesicles and subjected to exocytosis (Fig. 7.14). Such vesicles are present in large numbers at the hyphal apices in fungi. Consequently, in eukaryotes many of the processes for transport of substrates and products across membranes occur between cell protoplasm and organelles such as vesicles and the endoplasmic reticulum.

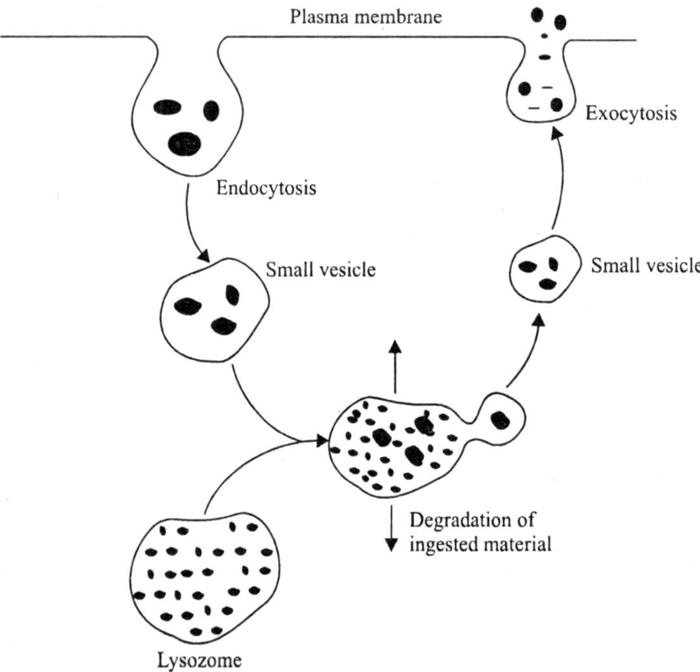


**Fig. 7.14.** Processes of endocytosis and exocytosis in eukaryotes.

Solutes are transported across membranes without modification: (i) by passive diffusion down a concentration gradient; (ii) by facilitated diffusion down a concentration gradient, involving a carrier system, which, like enzymes, displays saturation kinetics; and (iii) by active transport, which also involves a carrier system but allows net movement of the solute against an electrochemical gradient. Solutes may also be modified during translocation by enzymes with vectorial properties.

Specific transport proteins appear to be responsible for transport of many solutes across membranes, some of which are energy-coupled and some of which are inducible. In the biosynthesis of some microbial extracellular products, the final biosynthetic enzyme may be membrane-associated and function to facilitate excretion. For example, translocases have been implicated in the extrusion of exopolysaccharides from cells.

## Enzyme Secretion

Gram-positive prokaryotes and eukaryotes have a single plasma membrane at their surface. In contrast, Gram-negative bacteria have cell walls containing two membranes separated by a periplasmic space. Extracellular enzymes, synthesised by Gram-positive bacteria, are transported through the cytoplasmic membrane, diffuse through the cell wall and accumulate in the extracellular culture fluid. Extracellular enzymes are secretory proteins are synthesised by eukaryotes at the surface of the endoplasmic reticulum

(ER) and transported through the membrane into the ER cisternal space. The protein contained in the cisternal spaces is later processed through the Golgi apparatus and other vesicles and secreted into the cell's external environment by fusion of the vesicles with the plasma membrane (Fig. 7.15). Glycosylation of glycoproteins occurs in the ER cisternal space and in the Golgi bodies. This general secretory pathway appears to be similar in higher eukaryotes and in yeast and probably exists in filamentous fungi as well, although variations in the process may exist. In Gram-negative bacteria, many enzymes are transported across the cytoplasmic membrane and are located in the periplasmic space.

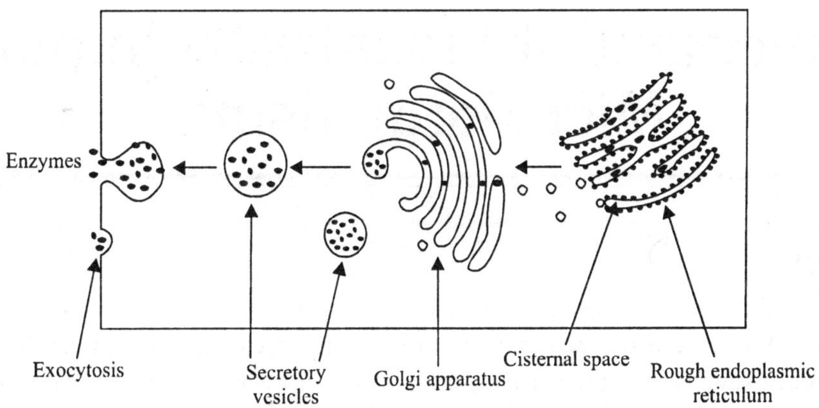


**Fig. 7.15.** Pathway for secretion of proteins in eukaryotes.

Proteins which are secreted across membranes usually contain an amino terminal peptide extension, called the signal sequence. The signal peptide, emerging from the ribosome, interacts with the inner surface of the plasma membrane of bacteria or the ER of eukaryotes, forming a pore or channel. As the polypeptide elongates it passes through the pore in the membrane, in a process known as co-translational secretion, and a signal peptidase located on the external side of the membrane removes the signal peptide, enabling the protein to assume its normal three-dimensional structure. The signal hypothesis for co-translational protein secretion is illustrated diagrammatically in Fig. 7.16. Some extracellular proteins are secreted after they have been synthesised in a process (also involving a signal peptide) known as post-translational secretion.

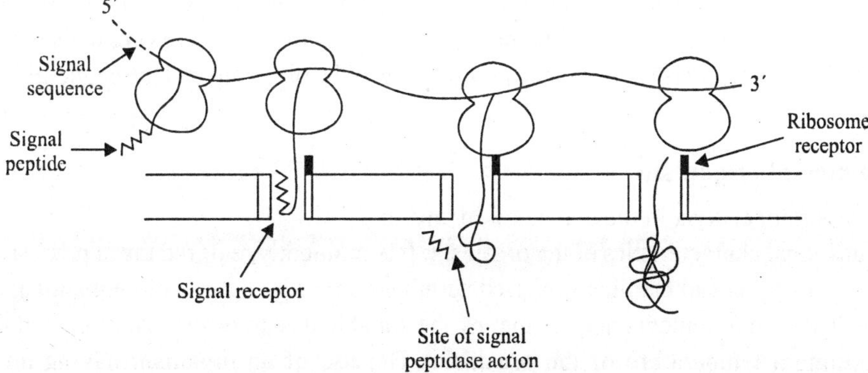


**Fig. 7.16.** Diagrammatic illustration of co-translational secretion.

Chapter 8

# Isolation, Preservation and Improvement of Industrially Important Micro-organisms

## INTRODUCTION

The most publicised advances in biotechnology over the last twenty years have been those in recombinant DNA technology. Whilst these advances have resulted in the development of extremely valuable new commercial processes and have improved many others, it is worthwhile reiterating Buckland's comment that the combined sales of four new microbial secondary metabolites introduced in the 1980s was greater than the sales of all the recombinant products added together. Thus, the diversity of micro-organisms may be exploited still by searching for strains from the natural environment able to produce products of commercial value.

The first stage in the screening for micro-organisms of potential industrial application is their isolation. Isolation involves obtaining either pure or mixed cultures followed by their assessment to determine which carry out the desired reaction or produce the desired product.

## ISOLATION OF INDUSTRIALLY IMPORTANT MICRO-ORGANISMS

In some cases it is possible to design the isolation procedure in such a way that the growth of producers is encouraged or that they may be recognised at the isolation stage, whereas in other cases organisms must be isolated and producers recognised at a subsequent stage. However, it should be remembered that the isolate must eventually carry out the process economically and, therefore, the selection of the culture to be used is a compromise between the productivity of the organism and the economic constraints of the process.

### Criteria in Choice of Organism

There are six important criteria in the choice of organism:

1. The nutritional characteristics of the organism: It is frequently required that a process be carried out using a very cheap medium or a predetermined one, e.g., the use of methanol as an energy source. These requirements may be met by the suitable design of the isolation medium.
2. The optimum temperature of the organism: The use of an organism having an optimum temperature above 40°C considerably reduces the cooling costs of a large-scale fermentation and therefore, the use of such a temperature in the isolation procedure may be beneficial.

3. The reaction of the organism with the equipment to be employed and the suitability of the organism to the type of process to be used.
4. The stability of the organism and its amenability to genetic manipulation.
5. The productivity of the organism, measured in its ability to convert substrate into product and to give a high yield of product per unit time.
6. The ease of product recovery from the culture.

Points 3, 4 and 6 would have to be assessed in detailed tests subsequent to isolation and the organism most well suited to an economic process chosen on the basis of these results. However, before the process may be put into commercial operation the toxicity of the product and the organism has to be assessed.

The above account implies that cultures must be isolated, in some way, from natural environment. However, the industrial microbiologist may also 'isolate' micro-organisms from culture collections. Such collections may provide organisms of known characteristics but may not contain those possessing the most desirable features, whereas the environment contains a myriad of organisms, very few of which may be satisfactory. It is certainly cheaper to buy a culture than to isolate from nature, but it is also true that a superior organism may be found after an exhaustive search of a range of natural environment. However, it is always worthwhile to purchase cultures demonstrating the desired characteristics, however weakly, as they may be used as model systems to develop culture and assay techniques which may then be applied to the assessment of natural isolates.

The ideal isolation procedure commences with an environmental source (frequently soil) which is highly probable to be rich in the desired types, is so designed as to favour the growth of those organisms possessing the industrially important characteristic (i.e., the industrially useful characteristic is used as a selective factor) and incorporates a simple test to distinguish the most desirable types. Selective pressure may be used in the isolation of organisms which will grow on particular substrates, in the presence of certain compounds or under cultural conditions adverse to other types. However, if it is not possible to apply selective pressure for the desired character it may be possible to design a procedure to select for a microbial taxon which is known to show the characteristic at a relatively high frequency, e.g., the production of antibiotics by streptomycetes. Alternatively, the isolation procedure may be designed to exclude certain microbial 'weeds' and to encourage the growth of more novel types. Indeed, as pointed out by Bull for screening programmes to continue to generate new products it is becoming increasingly more important to concentrate on lesser known microbial taxa or to utilise very specific screening tests to identify the desired activity. During the last 30 years significant advances have been made in the establishment of taxonomic databases describing the properties of microbial groups and these databases have been used to predict the cultural conditions which would select for the growth of particular taxa. Thus, the advances in the taxonomic description of taxa have allowed the rational design of procedures for the isolation of strains which may have a high probability of being productive or are representatives of unusual groups. The advances in pharmacology and molecular biology have also enabled the design of more effective screening tests to identify productive strains amongst the isolated organisms.

### Isolation Methods Utilising Selection of the Desired Characteristic

In this section we consider methods which take direct advantage of the industrially relevant property exhibited by an organism to isolate that organism from an environmental source. Isolation methods depending on the use of desirable characteristics as selective factors are essentially types of enrichment culture. Enrichment culture is a technique resulting in an increase in the number of a given organism relative to the numbers of other types in the original inoculum. The process involves taking a mixed population and providing conditions either suitable for the growth of the desired type or unsuitable for the growth of the other organisms, e.g., by the provision of particular substrates or the inclusion of certain inhibitors. Prior to the culture stage it is often advantageous to subject the environmental source (normally soil) to conditions which favour the survival of the organisms in question. For example, air-drying the soil will favour the survival of actinomycetes.

### *Enrichment liquid culture*

Enrichment liquid culture is frequently carried out in shake flasks. However, the growth of the desired type from a mixed inoculum will result in the modification of the medium and therefore changes the selective force which may allow the growth of other organisms, still viable from the initial inoculum, resulting in a succession. The selective force may be re-established by inoculating the enriched culture into identical fresh medium. Such sub-culturing may be repeated several times before the dominant organism is isolated by spreading a small inoculum of the enriched culture onto solid medium. The time of sub-culture in an enrichment process is critical and should correspond to the point at which the desired organism is dominant.

The prevalence of an organism in a batch enrichment culture will depend on its maximum specific growth rate compared with the maximum specific growth rates of the other organisms capable of growth in the inoculum. Thus, provided that the enrichment broth is sub-cultured at the correct times, the dominant organism will be the fastest growing of those capable of growth. However, it is not necessarily true that the organism with the highest specific growth rate is the most useful, for it may be desirable to isolate the organism with the highest affinity for the limiting substrate.

The problems of time of transfer and selection on the basis of maximum specific growth rate may be overcome by the use of a continuous process where fresh medium is added to the culture at a constant rate. Under such conditions the selective force is maintained at a constant level and the dominant organism will be selected on the basis of its affinity for the limiting substrate rather than its maximum growth rate. From the basic principles of continuous culture it may be seen that the growth rate in continuous culture is controlled by the dilution rate and is related to the limiting substrate concentration by the equation:

$$\mu = \mu_{max} s/(K_s + s) \qquad \text{... (8.1)}$$

Equation (8.1) is represented graphically in Fig. 8.1. A model of the competition between two organisms capable of growth in a continuous enrichment culture is represented in Fig. 8.2. Consider the behaviour of the two organisms, A and B, in Fig. 8.2. In continuous culture the specific growth rate is determined by the substrate concentration and is equal to the dilution rate, so that at dilution rates below point Y in Fig. 8.2 strain B would be able to maintain a higher growth rate than strain A, whereas at dilution rates above Y strain A would be able to maintain a higher growth rate.

Thus, if A and B were present in a continuous enrichment culture, limited by the substrate depicted in Fig. 8.2, strain A would be selected at dilution rates above Y and strain B would be selected at dilution rates below Y. Thus, the organisms which are isolated by continuous enrichment culture will depend on the dilution rate employed which may result in the isolation of organisms not so readily recovered by batch techniques.

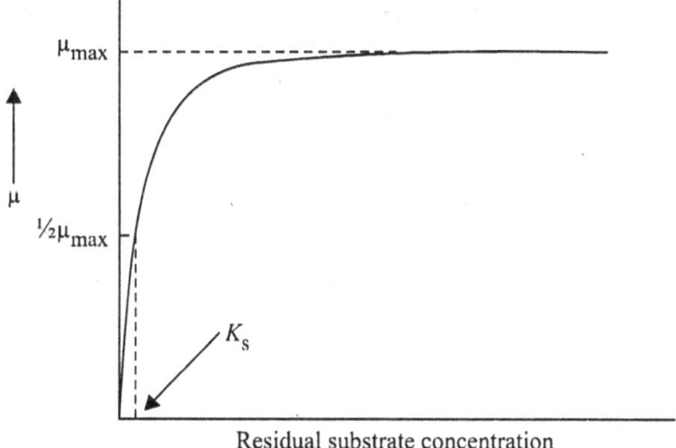


**Fig. 8.1.** The effect of substrate concentration on the specific growth rate of a micro-organism.

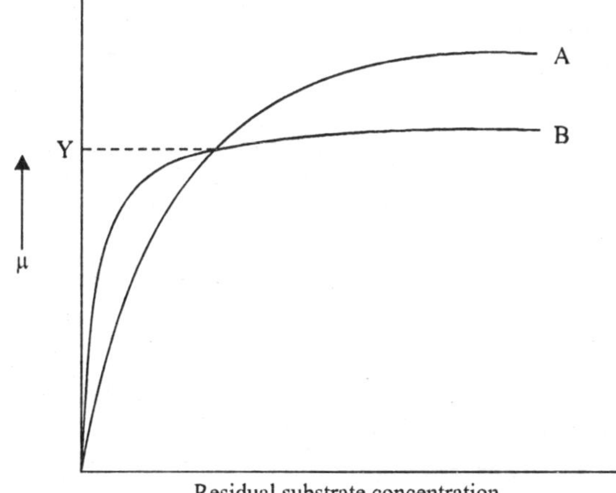


**Fig. 8.2.** The effect of substrate concentration on the specific growth rates of two micro-organisms A and B.

Continuous enrichment techniques are especially valuable in isolating organisms to be used in a continuous-flow commercial process. Organisms isolated by batch enrichment and purification on solid media frequently perform poorly in continuous culture, whereas continuous enrichment provides an organism or mixture of organisms, adapted to continuous culture. The enrichment procedure should be designed such that the predicted isolate meets as many of the criteria of the proposed process as possible. Johnson emphasised the importance of using the carbon source to be employed in the subsequent commercial process as the sole source of organic carbon in the enrichment medium and that the medium should be carbon limited. The inclusion of other organic carbon sources, such as vitamins or yeast extract, may result in the isolation of strains adapted to using these, rather than the principal carbon source, as energy sources. The isolation of an organism capable of growth on a simple medium should also form the basis of a cheaper commercial process and should be more resistant to contamination—a major consideration in the design of a commercial continuous process. The use of as high as possible an isolation temperature should also result in the isolation of a strain presenting minimal cooling problems in the subsequent process.

The main difficulty in using a continuous-enrichment process is the washout of the inoculum before an adapted culture is established. Johnson suggested that the isolation process should be started in batch culture using a 20 per cent inoculum and as soon as growth is observed, the culture should be transferred to fresh medium and the subsequent purification and stabilisation of the enrichment performed in continuous culture. The continuous system should be periodically inoculated with soil or sewage which may not only be a source of potential isolates but should also ensure that the dominant flora is extremely resistant to contamination.

Harrison proposed two solutions to the problem of early washout in continuous isolation processes: The first uses a turbidostat and the second uses a two-stage chemostat. A turbidostat is a continuous-flow system provided with a photoelectric cell to determine the turbidity of the culture and maintain the turbidity between set points by initiating or terminating the addition of medium. Thus, washout is avoided as the medium supply will be switched off if the biomass falls below the lower fixed point. The use of a turbidostat will result in selection on the basis of maximum specific growth rate as it operates at high levels of limiting substrate. Thus, although the use of the turbidostat removes the danger of washout it is not as flexible a system as the chemostat which may be used at a range of dilution rates. The two-stage chemostat described by Harrison is very similar to Johnson's procedure. The first stage of the system was used as a continuous inoculum for the second stage and consisted of a large bottle containing a basic medium inoculated with a soil infusion. Continuous inoculation was employed until an increasing absorbance was observed in the second stage. Bull advocated the use of feedback continuous systems for the isolation of strains with particularly high affinity for substrate and this approach would also guard against premature washout.

The use of continuous enrichment culture has frequently resulted in the selection of stable, mixed cultures presumably based on some form of symbiotic relationship. It is extremely unlikely that such mixed, stable systems could be isolated by batch techniques so that the adoption of continuous enrichment may result in the development of novel, mixed culture fermentations. Harrison isolated a mixed culture using methane as the carbon source in a continuous enrichment and demonstrated that the mixture contained one methylotroph and a number of non-methylotrophic symbionts. The performance of the methylotroph in pure culture was invariably poorer than the mixture in terms of growth rate, yield and culture stability.

Continuous enrichment has also been used for the isolation of organisms to be used in systems other than biomass production; Rowley and Bull used the technique to isolate an *Arthrobacter* sp. producing a yeast lysing enzyme complex.

### *Enrichment cultures using solidified media*

Solidified media have been used for the isolation of certain enzyme producers and these techniques usually involve the use of a selective medium incorporating the substrate of the enzyme which encourages the growth of the producing types. Aunstrup isolated species of *Bacillus* producing alkaline proteases. Soils of various pHs were used as the initial inoculum and, to a certain extent, the number of producers isolated correlated with the alkalinity of the soil sample. The soil samples were pasteurised to eliminate non-sporulating organisms and then spread onto the surface of agar media at pH 9–10, containing a dispersion of an insoluble protein. Colonies which produced a clear zone due to the digestion of the insoluble protein were taken to be alkaline protease producers. The size of the clearing zone could not be used quantitatively to select high producers as there was not an absolute correlation between the size of the clearing zone and the production of alkaline protease in submerged culture. However, this example demonstrates the importance of choice of starting material, the use of a selective force in the isolation and the incorporation of a preliminary diagnostic test, albeit of limited use.

## Isolation Methods not Utilising Selection of the Desired Characteristic

The synthesis of some products does not give the producing organism any selective advantage which may be exploited directly in an isolation procedure. Examples include the production of antibiotics and growth promoters. Therefore, a pool of organisms has to be isolated and subsequently tested for the desired characteristic. The major problem faced by industrial microbiologists in this situation is the re-isolation of strains which have already been screened many times before. However, this 're-invention of the wheel' syndrome may be minimised in two major ways:

1. Developing procedures to favour the isolation of unusual taxa which are less likely to have been screened previously.
2. Identifying selectable features correlated with the unselectable industrial trait thus enabling an enrichment process to be developed.

Much use has been made of numerical taxonomic databases to design media selective for particular taxa. For example, some scientists used such databases to design isolation media to either encourage the growth of uncommon streptomycetes or to discourage the growth of *Streptomyces albidoflavus*. Several groups of workers have taken advantage of the antibiotic sensitivity information stored in taxonomic databases to design media selective for particular taxa, as shown in Table 8.1. The incorporation of particular antibiotics into isolation media may result in the selection of the resistant taxa. Such techniques are reviewed by Bull. Bull emphasised that the taxonomic approach may be optimised and developed according to Fig. 8.1. The isolates from an isolation procedure would be screened for activity and then the growth requirements of the positive cultures determined. This knowledge can then be used to optimise the isolation medium and the cycle begins again.

**Table 8.1.** Antibacterial compounds used in selective media for the isolation of actinomycetes.

| *Selective agent* | *Target organism* |
|---|---|
| Bruneomycin | *Micromonospora* |
| Dihydroxymethylfuratriazone | *Microtetraspora* |
| Gentamycine | *Micromonospora* |
| Kanamycin | *Actinomadura* |
| Nitrofurazone | *Streptomyces* |
| Novobiocin | *Micromonospora* |
| Tellurite | *Actinoplanes* |
| Tunicamycin | *Micromonospora* |

Whilst taxonomic databases are a convenient source of information, it is important to appreciate Huck observation that these systems were designed to provide information for taxonomic differentiation within a group. Thus, some of the diagnostic data may not be applicable to isolation systems. More significantly, the reactions of organisms outside the taxon in question would not be listed. Of course, an environmental sample contains a vast variety of organisms and the design of isolation media based on knowledge of only one taxon may inadvertently result in the preferential isolation of undesirable types. Huck used the statistical stepwise discrimination analysis (SDA) technique to design media for the positive selection of antibiotic producing soil isolates. This was achieved by characterising a collection of eubacterial and actinomycete soil isolates according to 43 physiological and nutritional tests. Certain features were identified which, when used as selective factors, enhanced the probability of either isolating actinomycetes or antibiotic producing actinomycetes. These features are shown in Table 8.2. Using this approach several media were developed which enhanced the isolation of antibiotic producers.

**Table 8.2.** Selective substrates for the isolation of actinomycetes and antibiotic-producing actinomycetes.

| *Substrates selective for Actinomycetes* | *Substrates selective for antibiotic-producing actinomycetes* |
|---|---|
| Proline | Proline |
| Glucose (1.0%) | Glucose (1.0%) |
| Glycerol | Glycerol |
| Starch | Starch |
| Humic acid (0.1%) | Humic acid (0.1%) |
| Propionate (0.1%) | Zinc |
| Methanol | Alanine |
| Nitrate | Potassium |
| Calcium | Vitamins |
| | Cobalt (0.05%) |
| | Phenol (0.01%) |
| | Asparagine |

The most desirable isolation medium would be one which selects for the desired types and also allows maximum genetic expression. Cultures grown on such media could then be used directly in a screen. However, it is more common that, once isolated, the organisms are grown on a range of media designed to enhance productivity. Nisbet put forward some guidelines for the design of such media and these are summarised in Table 8.3.

**Table 8.3.** Guidelines for 'overproduction media'.

1. Prepare a range of media in which different types of nutrients become growth-limiting e.g., C, N, P, O
2. For each type of nutrient depletion use different forms of the growth-sufficient nutrient
3. Use a polymeric or complexed form of the growth-limiting nutrient
4. Avoid the use of readily assimilated forms of carbon (glucose) or nitrogen ($NH_4^+$) that may cause catabolite repression
5. Ensure that known co-factors are present ($Co^{3+}$, $Mg^{2+}$, $Mn^{2+}$, $Fe^{2+}$).
6. Buffer to minimise pH changes.

## Screening Methods

Early screening strategies tended to be empirical, labour intensive and had relatively low success rates. As the number of commercially important compounds isolated increased, the success rates of such screens decreased further. Thus, new screening methods have been developed which are more precisely targeted to identify the desired activity. The evolution of antibiotic screens serves as an excellent illustration of the development of more precise, targeted systems.

Antibiotics were initially detected by growing the potential producer on an agar plate in the presence of an organism (or organisms) against which antimicrobial action was required. Production of the antibiotic was detected by inhibition of the test organism(s). Alternatively, the microbial isolate could be grown in liquid culture and the cell-free broth tested for activity. This approach was extended by using a range of organisms to detect antibiotics with a defined antibacterial spectrum. For example, Zahner discussed the use of test organisms to detect the production of antibiotics with confined action spectra. The use of *Bacillus subtilis* and *Streptomyces viridochromogenes* or *Clostridium pasteurianum* allows the identification of antibiotics with a low activity against *B. subtilis* and a high activity against the other test organisms. Such antibiotics may be new because they would not be isolated by the more common tests using *B. subtilis* alone. The kirromycine group of antibiotics were discovered using methods of this type.

However, the advances made in the understanding of cell-wall biosynthesis and the mode of action of antibiotics allowed the development of mode of action screens in the 1970s which resulted in a very significant increase in the discoveries of new β-lactam antibiotics. Nagarajan discovery of the cephamycins was based on the detection of compounds which induced morphological changes in susceptible bacteria. The appreciation that the mode of action of penicillins was the inhibition of the transpeptidase enzyme cross-linking mucopeptide molecules led to the development of enzyme inhibitor assays, which were particularly attractive because they could be automated. Fleming described the development of an automated screen for the detection of carboxypeptidase inhibitors which has led to the detection of several novel cephamycin and carbapenem compounds.

The increasing frequency of penicillin and cephalosporin resistance amongst clinical bacteria led to the development of mechanism based screens for the isolation of more effective antibacterials. The logic of the Beechams Pharmaceuticals group of Brown was to search for a compound which would inhibit β-lactamase and could be incorporated with ampicillin as a combination therapeutic agent. Samples were tested for their ability to increase the inhibitory effect of ampicillin on a β-lactamase producing *Klebsiella aerogenes* and this strategy resulted in the discovery of clavulanic acid.

The concept of using the inhibition of enzymes as a screening mechanism was pioneered by Umezawa in his search for microbial products inhibiting key enzymes of human metabolism. His approach was based on the logic that, if a compound inhibits a key human enzyme *in vitro*, it may have a pharmacological action *in vivo*. Such screens have been applied to a wide range of pharmacological targets and have resulted in the isolation of several important drugs. From *Aspergillus*, Alberts and others isolated mevinolin which is an inhibitor of hydroxy-methyl-glutaryl reductase, the rate limiting step in the biosynthesis of sterols. It is an agent to lower high cholesterol levels. Bull list the following clinical situations in which microbial products have been shown to inhibit key enzymes: hypercholesterolaemia, hypertension, gastric inflammation, muscular dystrophy, benign prostate hyperplasia and systematic lupus erythemosus. The activity of carbapenem antibiotics is lost in therapy due to renal dehydropeptidase activity. These isolated microbial products are capable of inhibiting the enzyme which could then be administered along with the antibiotic to maintain its clinical activity.

The detection of pharmacological agents by receptor-ligand binding assays has been developed rapidly by pharmaceutical companies. These are extensions of the enzyme inhibitor approach but agents which block receptor sites are likely to be more effective at very low concentrations. The gastrointestinal hormone cholecystokinin (CCK) controls a range of digestive activities such as pancreatic secretion and gall bladder contraction. Receptor screening identified a fungal metabolite, aperlicin (from *Aspergillus alliaceus*) that had a very high affinity for CCK receptors. Although the fungal product did not prove to be a suitable drug it was used as a model for the design of analogues which were receptor binders and pharmacologically acceptable.

The progress in molecular biology, genetics and immunology has also contributed extensively to the development of innovative screens, by enabling the construction of specific detector strains, increasing the availability of enzymes and receptors and constructing extremely sensitive assays. Bull summarised the major contributions as follows:

1. The provision of test organisms that have increased sensitivities or resistances, to known agents. For example, the use of super-sensitive strains for the detection of β-lactam antibiotics.
2. The cloning of genes coding for enzymes or receptors that may be used in inhibitor or binding screens makes such materials more accessible and available in much larger amounts.
3. The development of reporter gene assays. A reporter gene is one which codes for an easily assayable product so that it can be used to detect the activation of a control sequence to which it is fused. Such systems have been used in the search for metabolites that disrupt viral replication.
4. Molecular probes for particular gene sequences may enable the detection of organisms capable of producing certain product groups. This information may be used to focus the search on these organisms in an attempt to find novel representatives of an already known commercially attractive chemical family.
5. The development of immunologically based assays such as ELISA.

It is important to appreciate at this stage that the advances in the biological sciences which have enabled the design of sophisticated screening tests have been paralleled by the development of robotic automation systems. This means that the engineering now exists to automate such tests, resulting in an enormous throughput. Indeed, Nisbet and Porter claimed that 'the modern microbial discovery programme must achieve rates of $10^5$ tests per year, for as many as 20 different assay systems, to compete in the race for novel therapeutics'.

## PRESERVATION OF INDUSTRIALLY IMPORTANT MICRO-ORGANISMS

The isolation of a suitable organism for a commercial process may be a long and very expensive procedure and it is, therefore, essential that it retains the desirable characteristics that led to its selection. Also, the culture used to initiate an industrial fermentation must be viable and free from contamination. Thus, industrial cultures must be stored in such way as to eliminate genetic change, protect against contamination and retain viability. An organism may be kept viable by repeated sub-culture into fresh medium, but, at each cell division, there is a small probability of mutations occurring and because repeated sub-culture involves very many such divisions, there is a high probability that strain degeneration would occur. Also, repeated sub-culture carries with it the risk of contamination. Thus, preservation techniques have been developed to maintain cultures in a state of 'suspended animation' by storing either at reduced temperature or in a dehydrated form.

### Storage at Reduced Temperature

#### *Storage on agar slopes*

Cultures grown on agar slopes may be stored in a refrigerator (5°) or a freezer (–20°) and sub-cultured at approximately 6-monthly intervals. The time of sub-culture may be extended to one year if the slopes are covered with sterile medicinal grade mineral oil.

#### *Storage under liquid nitrogen*

The metabolic activities of micro-organisms may be reduced considerably by storage at the very low temperatures (–150° to –196°) which may be achieved using a liquid nitrogen refrigerator. Snell claimed that this approach is the most universally applicable of all preservation methods. Fungi, bacteriophage, viruses, algae, yeasts, animal and plant cells and tissue cultures have all been successfully preserved.

The technique involves growing a culture to the maximum stationary phase, re-suspending the cells in a cryoprotective agent (such as 10 per cent glycerol) and freezing the suspension in sealed ampoules before storage under liquid nitrogen. Some loss of viability is suffered during the freezing and thawing stages but there is virtually no loss during the storage period. Thus, viability may be predictable even after a period of many years. Snell suggested that liquid nitrogen is the method of choice for the preservation of valuable stock cultures and may be the only suitable method for the long-term preservation of cells that do not survive freeze-drying. Although the equipment is expensive the process is economical on labour. However, the method has the major disadvantage that liquid nitrogen evaporates and must be replenished regularly. If this is not done or the apparatus fails, then the consequences are the loss of the collection.

### Storage in a Dehydrated Form

#### *Dried cultures*

Dried soil cultures have been used widely for culture preservation, particularly for sporulating mycelial organisms. Moist, sterile soil may be inoculated with a culture and incubated for several days for some growth to occur and then allowed to dry at room temperature for approximately 2 weeks. The dry soil may be stored in a dry atmosphere or preferably, in a refrigerator. The technique has been used extensively for the storage of fungi and actinomycetes and Pridham observed that of 1800 actinomycetes dried on soil about 50 per cent were viable after 20-year storage. Malik described methods which extend the approach using substrates other than soil. Silica gel and porcelain beads are suggested alternatives.

#### *Lyophilisation*

Lyophilisation or freeze-drying involves the freezing of a culture followed by its drying under vacuum, which results in the sublimation of the cell water. The technique involves growing the culture to the maximum stationary phase and re-suspending the cells in a protective medium such as milk, serum or sodium glutamate. A few drops of the suspension are transferred to an ampoule, which is then frozen and subjected to a high vacuum until sublimation is complete, after which the ampoule is sealed. The ampoules may be stored in a refrigerator and the cells may remain viable for 10 years or more.

Lyophilisation is very convenient for service culture collections because, once dried, the cultures need no further attention and the storage equipment (a refrigerator) is cheap and reliable. Also, the freeze dried ampoules may be dispatched as such, still in a state of 'suspended animation' whereas liquid nitrogen stored cultures begin to deteriorate. However, freeze-dried cultures are tedious to open and revitalise and several sub-cultures may be needed before the cells regain their typical characteristics. Overall, the technique appears to be second only to liquid nitrogen storage and even when liquid nitrogen is used makes an excellent insurance against the possibility of the breakdown of the nitrogen freezer.

### Quality Control of Preserved Stock Cultures

Whichever technique is used for the preservation of an industrial culture it is essential to be certain of the quality of the stocks. Each batch of newly preserved cultures should be routinely checked to ensure their quality and such a procedure has been outlined by Lincoln: a single colony of the culture to be preserved is inoculated into a shake flask and the growth of the culture observed to ensure a typical growth pattern; after a further shake-flask sub-culture the broth is used to prepare a large number of storage ampoules. At least 3 per cent of the ampoules are reconstituted and the cultures assessed for purity, viability and productivity. If the samples fail any one of these tests the entire batch should be destroyed. Thus, by the use of such a quality-controlled system stock cultures may be retained and used, with confidence.

## IMPROVEMENT OF INDUSTRIAL MICRO-ORGANISMS

Natural isolates usually produce commercially important products in very low concentrations and therefore every attempt is made to increase the productivity of the chosen organism. Increased yields may be achieved by optimising the culture medium and growth conditions, but this approach will be limited by the organism's maximum ability to synthesise the product. The potential productivity of the organism is controlled by its genome and therefore, the genome must be modified to increase the potential yield. The cultural requirements of the modified organism would then be examined to provide conditions that would fully exploit the increased potential of the culture, while further attempts are made to beneficially change the genome of the already improved strain. Thus, the process of strain improvement involves the continual genetic modification of the culture, followed by reappraisals of its cultural requirements.

Genetic modification may be achieved by selecting natural variants, by selecting induced mutants and by selecting recombinants. There is a small probability of a genetic change occurring each time a cell divides and when it is considered that a microbial culture will undergo a vast number of such divisions it is not surprising that the culture will become more heterogeneous. The heterogeneity of some cultures can present serious problems of yield degeneration because the variants are usually inferior producers compared with the original culture. However, variants have been isolated which are superior producers and this has been observed frequently in the early stages in the development of a natural product from a newly isolated organism. An explanation of this phenomenon for mycelial organisms may be that most new isolates are probably heterokaryons (contain more than one type of nucleus) and the selection of the progeny of uninucleate spores results in the production of homokaryons (contain only one type of nucleus) which may be superior producers. However, the phenomenon is also observed with unicellular isolates which are certainly not heterokaryons. Therefore, it is worthwhile to periodically plate out the producing culture and screen a proportion of the progeny for productivity; this practice has the added advantage that the operator tends to become familiar with morphological characteristics associated with high productivity and by selecting 'typical' colonies, a strain subject to yield degeneration may still be used with consistent results. Therefore, selection of natural variants may result in increased yields but it is not possible to rely on such improvements and techniques must be employed to increase the chances of improving the culture. These techniques are the isolation of induced mutants and recombination. The most dramatic examples of strain improvement come from the applications of recombinant DNA technology which has resulted in organisms producing compounds which they were not able to produce previously. Furthermore, the advances in these techniques have resulted in very significant improvements in the production of conventional fermentation products. However, it should be remembered that these methods have not replaced mutant isolation but have made an invaluable addition to an impressive repertoire. The techniques for mutant isolation have contributed enormously to the development of present-day industrial strains and these techniques will be considered first. The applications of recombination systems will then be discussed along with the interactions between mutant isolation and recombination in strain improvement advances.

Dulaney compared the spread in productivity of chlortetracycline of natural variants of *Streptomyces viridifaciens* with the spread in productivity of the survivors of an ultraviolet treatment. The results of their comparison are shown in Figs. 8.3 and 8.4, from which it may be seen that although there are more inferior producers amongst the survivors of the ultraviolet treatment there are also strains producing more than twice the parental level, far greater than the best of the natural variants. The use of ultraviolet light is only one of the large number of physical or chemical agents which increase the mutation-rate— such agents are termed mutagens. The vast majority of induced mutations are deleterious to the yield of the desired product but, as shown in Fig. 8.3, a minority are more productive than the parent. The problem of obtaining the high-yielding mutants may be approached from two theoretical standpoints;

the number of desirable mutants may be increased by 'directed mutation', i.e., the use of a technique which will preferentially produce particular mutants at a high rate or techniques may be developed to improve the separation of the few desirable types from the large number of mediocre producers.

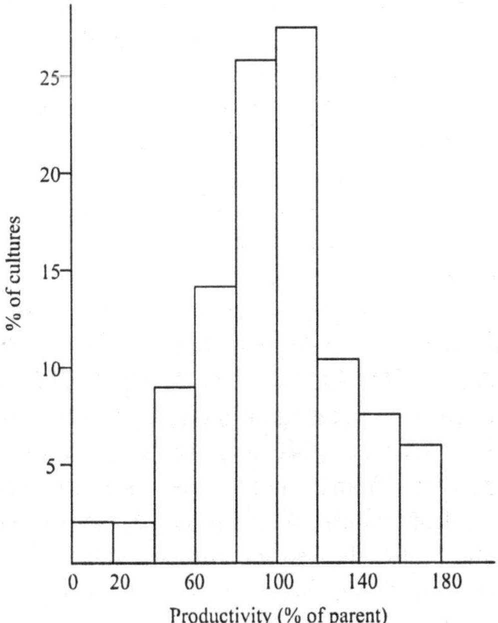


**Fig. 8.3.** The spread in productivity of chlortetracycline of natural variants of *Streptomyces viridifaciens*.

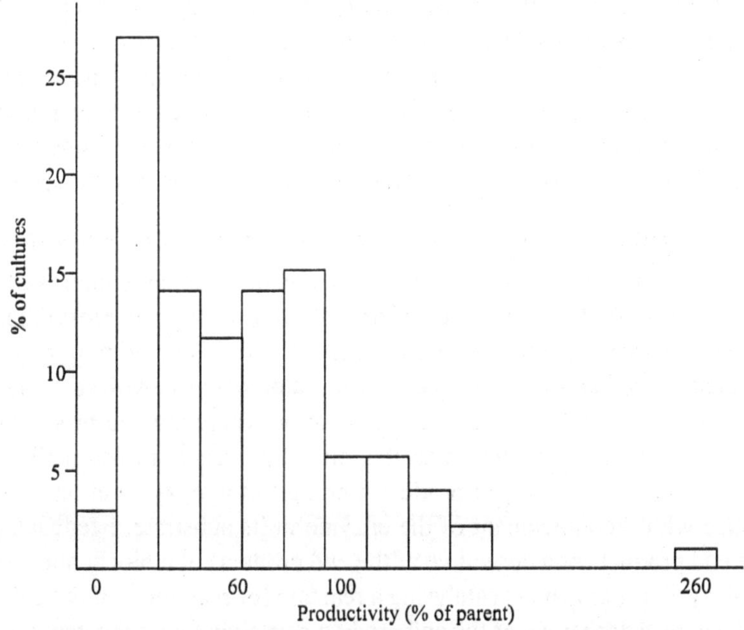


**Fig. 8.4.** The spread in chlortetracycline productivity of the survivors of a UV-treated population of *Streptomyces viridifaciens*.

Inherent in the concept of directed mutation is the assumption that a mutation programme can be optimised to produce mutants of a particular kind. The choice of mutagen was demonstrated to affect the success of mutation programmes early in the history of strain improvement schemes. For example, Hostalek claimed that ultraviolet radiation was the most effective mutagen for increasing the yield of tetracyline by strains of *Streptomyces aureofaciens*. DeWitt emphasised that certain mutagens being more beneficial, the dose will affect the generation of the desired types. Despite these observations it is frequently the case that it is difficult to predict what type of mutation is required at the molecular level to improve a strain and, therefore, it is extremely unlikely that the concept of directed mutation can be applied in these circumstances. Thus, it is the second approach specified above that is likely to provide the solution to this type of problem, i.e., the development of selection techniques.

However, the concept of directed mutation is of considerable relevance when the genes to be modified are known and the organism is genetically well documented. In these systems a cloned DNA molecule may be subjected to *in vitro* enzymatic cleavage and manipulation such that the mutation is present at a particular site. Thus, re-introduction of the modified DNA into the cell should result in the production of a cell mutated at a particular site. The incorporation of a selectable marker into the cloned DNA would assist in the recovery of cells incorporating the modified gene(s). Thus, such systems still depend on the development of methods to recover cells displaying the desirable feature.

The separation of desirable mutants from the other survivors of a mutation treatment is similar to the isolation of desirable organisms from nature. Where possible, the mutant isolation procedure should use the improved characteristic of the desired mutant as a selective factor. Presumably, superior productivity is a result of a diversion of precursors into the product and/or a modification of the control mechanisms limiting the level of production. Thus, a knowledge of the biosynthetic route and the mechanisms of control of the biosynthesis of the product should enable the prediction of a 'blueprint' of the desirable mutant. Such a 'blueprint' might then enable the design of isolation techniques which would give the desired mutants a selective advantage over the other types present. Knowledge of biosynthetic routes and control mechanisms are more detailed for primary metabolites and therefore, the use of selective pressure in mutant isolation is more common in the fields of amino acid, nucleotide and enzyme production than in secondary metabolite production. However, considerable progress has also been made in the design of such procedures for the isolation of secondary metabolite over-producers.

### Selection of Induced Mutants Synthesising Improved Levels of Primary Metabolites

Before considering the methods used for the selection of mutants producing improved levels of primary metabolites it is necessary to study the mechanisms of control of their biosynthesis such that the 'blueprints', referred to above, may be drawn accurately. The levels of primary metabolites in micro-organisms are regulated by feedback control systems. The major systems involved are feedback inhibition and feedback repression. Feedback inhibition is the situation where the end product of a biochemical pathway inhibits the activity of an enzyme catalysing one of the reactions (normally the first reaction) of the pathway. Inhibition acts by the end product binding to the enzyme at an allosteric site which results in interference with the attachment of the enzyme to its substrate. Feedback repression is the situation where the end product (or a derivative of the end product) of a biochemical pathway prevents the synthesis of an enzyme (or enzymes) catalysing a reaction (or reactions) of the pathway. Repression occurs at the gene level by a derivative of the end product combining with the genome in such a way as to prevent the transcription of the gene into messenger RNA, thus resulting in the prevention of enzyme synthesis.

Feedback inhibition and repression frequently act in concert in the control of biosynthetic pathways, where inhibition may be visualised as a rapid control which switches off the biosynthesis of an end product and repression as a mechanism to then switch off the synthesis of temporarily redundant enzymes. The control of pathways giving rise to only one product (i.e., unbranched pathways) is normally achieved by the first enzyme in the sequence being susceptible to inhibition by the end product and the synthesis of all the enzymes being susceptible to repression by the end product, as shown in Fig. 8.5.

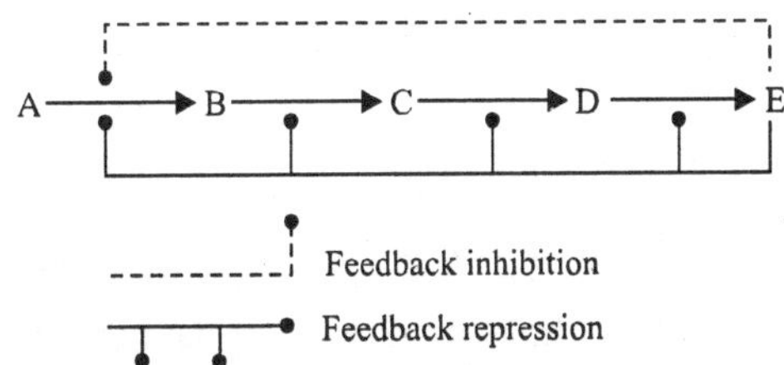


**Fig. 8.5.** Control of a biosynthetic pathway converting precursor A to end product E via the intermediates B, C and D.

The control of biosynthetic pathways giving rise to a number of end products (branched pathways) is more complex than the control of simple, unbranched sequences. The end products of the same, branched biosynthetic pathways are rarely required by the micro-organism to the same extent, so that if an end product exerts control over a part of the pathway common to two or more, end products then the organism may suffer deprivation of the products not participating in the control. Thus, mechanisms have evolved which enable the level of end products of branched pathways to be controlled without depriving the cell of essential intermediates. The following descriptions of these mechanisms are based on the effect of the control, which may be arrived at by inhibition, repression or a combination of both systems.

Concerted or multivalent feedback control: This control system involves the control of the pathway by more than one end product—the first enzyme of the pathway is inhibited or repressed only when all end products are in excess, as shown in Fig. 8.6.

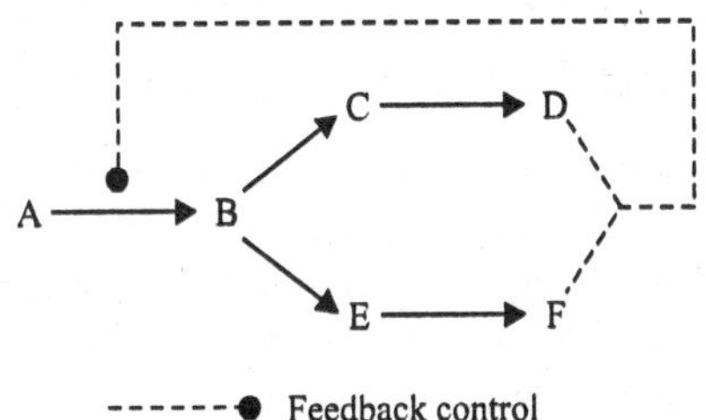


**Fig. 8.6.** Control of a biosynthetic pathway by the concerted effects of products D and F on the first enzyme of the pathway.

Co-operative feedback control: The system is similar to concerted control except that weak control may be effected by each end product independently. Thus, the presence of all end products in excess results in a synergistic repression or inhibition. The system is illustrated in Fig. 8.7 and it may be seen that for efficient control to occur when one product is in excess there should be a further control operational immediately after the branch point to the excess product. Thus, the reduced flow of intermediates will be diverted to the product which is still required.

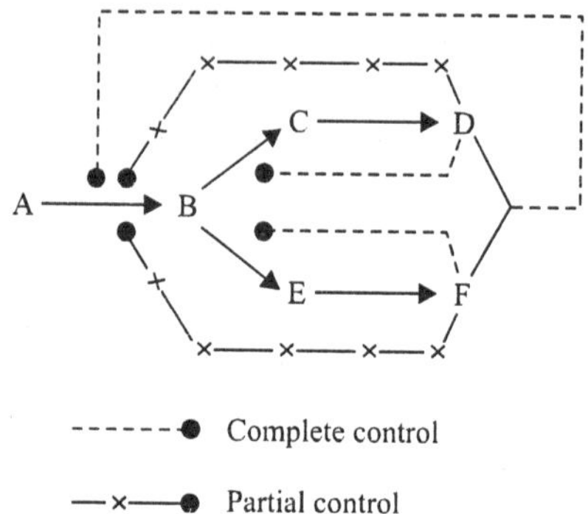


Fig. 8.7. Control of a biosynthetic pathway by the co-operative control by end products D and F.

Cumulative feedback control: Each of the end products of the pathway inhibits the first enzyme by a certain percentage independently of the other end products. In Fig. 8.8 both D and F independently reduce the activity of the first enzyme by 50 per cent, resulting in total inhibition when both products are in excess. As in the case of co-operative control, each end product must exert control immediately after the branch point so that the common intermediate, B, is diverted away from the pathway of the product in excess.

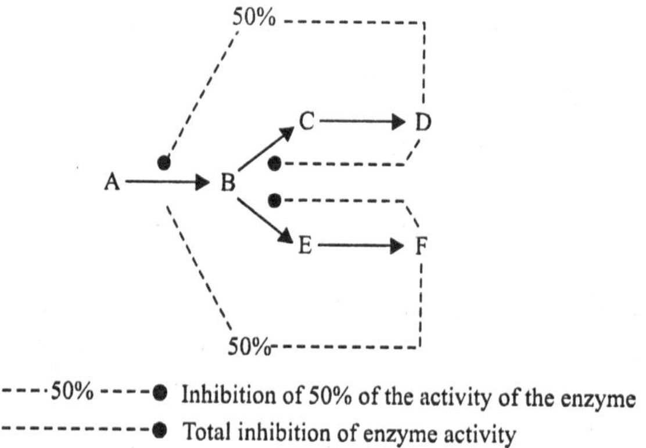


Fig. 8.8. Control of a biosynthetic pathway by the cumulative control of products D and F.

Sequential feedback control: Each end product of the pathway controls the enzyme immediately after the branch point to the product. The intermediates which then build up as a result of this control earlier enzymes in the pathway. Thus, in Fig. 8.9, D inhibits the conversion of B to C and F inhibits the conversion of B to E. The inhibitory action of D, F or both, would result in an accumulation of B which, in turn, would inhibit the conversion of A to B.

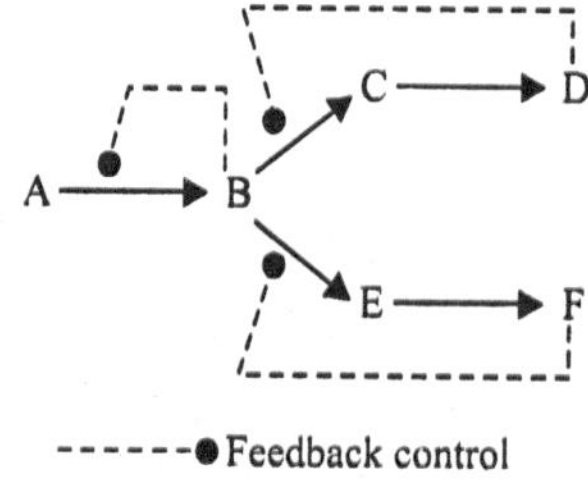


Fig. 8.9. Control of a biosynthetic pathway by sequential feedback control.

Isoenzyme control: Isoenzymes are enzymes which catalyse the same reaction but differ in their control characteristics. Thus, if a critical control reaction of a pathway is catalysed by more than one isoenzyme, then the different isoenzymes may be controlled by the different end products. Such a control system should be very efficient, provided that control exists immediately after the branch point so that the reduced flow of intermediates is diverted away from the product in excess. An example of the system is shown in Fig. 8.10.

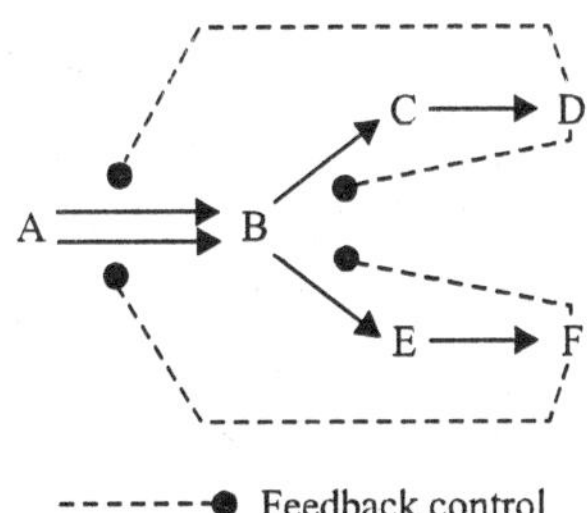


**Fig. 8.10.** Control of two isoenzymes (catalysing the conversion of A to B) by end products D and F.

Thus, the levels of microbial metabolites may be controlled by a variety of mechanisms, such that end products are synthesised in amounts not greater than those required for growth. However, the ideal industrial micro-organism should produce amounts far greater than those required for growth and, as suggested earlier, an understanding of the control of production of a metabolite may enable the construction of a 'blueprint' of the most useful industrial mutant, i.e., one where the production of the metabolite is not restricted by the organism's control system. Such postulated mutants may be modified in three ways:

1. The organism may be modified such that the end products, which control the key enzymes of the pathway, are lost from the cell due to some abnormality in the permeability of the cell membrane.
2. The organism may be modified such that it does not produce the end products which control the key enzymes of the pathway.
3. The organism may be modified such that it does not recognise the presence of inhibiting or repressing levels of the normal control metabolites.

### *Modification of the permeability*

The best example of the modification of the permeability of a micro-organism is provided by the glutamic acid fermentation. Kinoshita isolated a biotin-requiring, glutamate-producing organism, subsequently named *Corynebacterium glutamicum*, the permeability of which could be modified by the level of biotin. Provided that the level of biotin in the production medium was below 5 μg $dm^{-3}$ then the organism would excrete glutamate, but at concentrations of biotin optimum for growth the organism produced lactate. Thus, the permeability of *C. glutamicum* may be controlled by the composition of the culture medium and as such, is not an example of strain improvement in the sense that the term has been used in this chapter. However, the isolation of *C. glutamicum* was a major advance in the microbial production of amino acids and the demonstration of the role of permeability in the production of glutamate suggested the possibility of the genetic modification of permeability to achieve high levels of productivity. Thus, it is relevant at this stage to consider the physiology of glutamate production by biotin-limited culture.

Kinoshita's isolate was not only deficient in its ability to produce biotin but also in the level of α-ketoglutarate dehydrogenase, which normally converts α-ketoglutarate to succinate in the tricarboxylic acid cycle. Thus, a metabolic block results in the accumulation of a large concentration of α-ketoglutarate which the organism converts to glutamic acid. Oxaloacetate is regenerated in glutamate producers by the activity of the glyoxylate pathway as shown in Fig. 8.11.

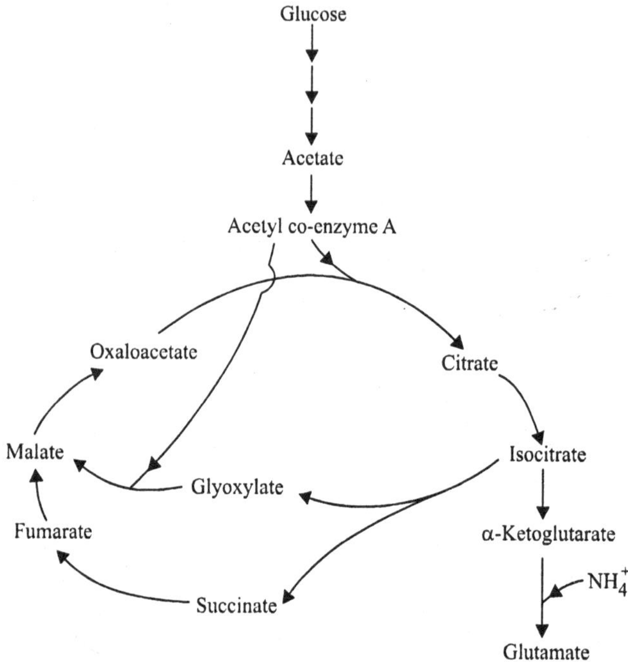


**Fig. 8.11.** Biosynthesis of glutamate by *C. glutamicum*. Heavy lines indicate the route to glutamate and the light lines indicate the route in the regeneration of oxaloacetate via the glyoxylate cycle.

When *C. glutamicum* was grown in a medium containing a high concentration of biotin, the organism synthesised glutamate at a level of 25–36 μg $mg^{-1}$ dry weight of cells, further production being assumed to be prevented by some form of feedback control by glutamate of its own synthesis. Under conditions of biotin limitation, glutamate was released from the cells and accumulated to a level of up to 50 g $dm^{-3}$. The increased permeability of the biotin-limited cells to glutamate corresponds with a change in the fatty acid and phospholipid content of the cell envelopes. The crucial factor in the control of permeability appears to be the synthesis of membranes deficient in phospholipids.

The role of biotin in the glutamate fermentation obviously necessitates the inclusion of limiting levels of biotin in the medium. Thus, the use of crude carbon sources such as cane molasses, which are rich in biotin, presented difficulties which have been overcome by the modification of the permeability by means other than biotin limitation. The inclusion of penicillin or fatty acid derivatives such as polyoxyethylene sorbitan mono-oleatic during logarithmic growth in biotin-rich medium caused an aberration of the cell envelope permeability resulting in the release of glutamate. The production of glutamate from hydrocarbons (as carbon sources) has also presented difficulties in the control of permeability, in that the assimilation of hydrocarbons results in the production of fatty acids which effectively bypasses the site of biotin control. The permeability of the producing organism may be controlled by the addition of penicillin but a genetic solution to this problem has also been found. Nakao isolated a glycerol requiring auxotrophic mutant of *Corynebacterium alkanolyticum* in which phospholipid synthesis was controlled by the supply of glycerol. The mutant produced about 40 g $dm^{-3}$ glutamate when grown on *n*-paraffins in the presence of 0.01 per cent glycerol. Thus, an understanding of the mode of action of biotin limitation in the glutamate fermentation has led to the use of genetic modification of permeability as a method to overcome the normal control mechanisms of the producing organism.

### *Isolation of mutants which do not produce feedback inhibitors or repressors*

Mutants which do not produce certain feedback inhibitors or repressors may be useful for the production of intermediates of unbranched pathways and intermediates and end products of branched pathways. Demain presented several 'blue-prints' of hypothetical mutants producing intermediates and end products of biosynthetic pathways and these are illustrated in Fig. 8.12. The mutants illustrated in Fig. 8.12 do not produce some of the inhibitors or repressors of the pathways considered and thus, the control of the pathway is lifted, but, because the control factors are also essential for growth, they must be incorporated into the medium at concentrations which will allow growth to proceed but will not evoke the normal control reactions.

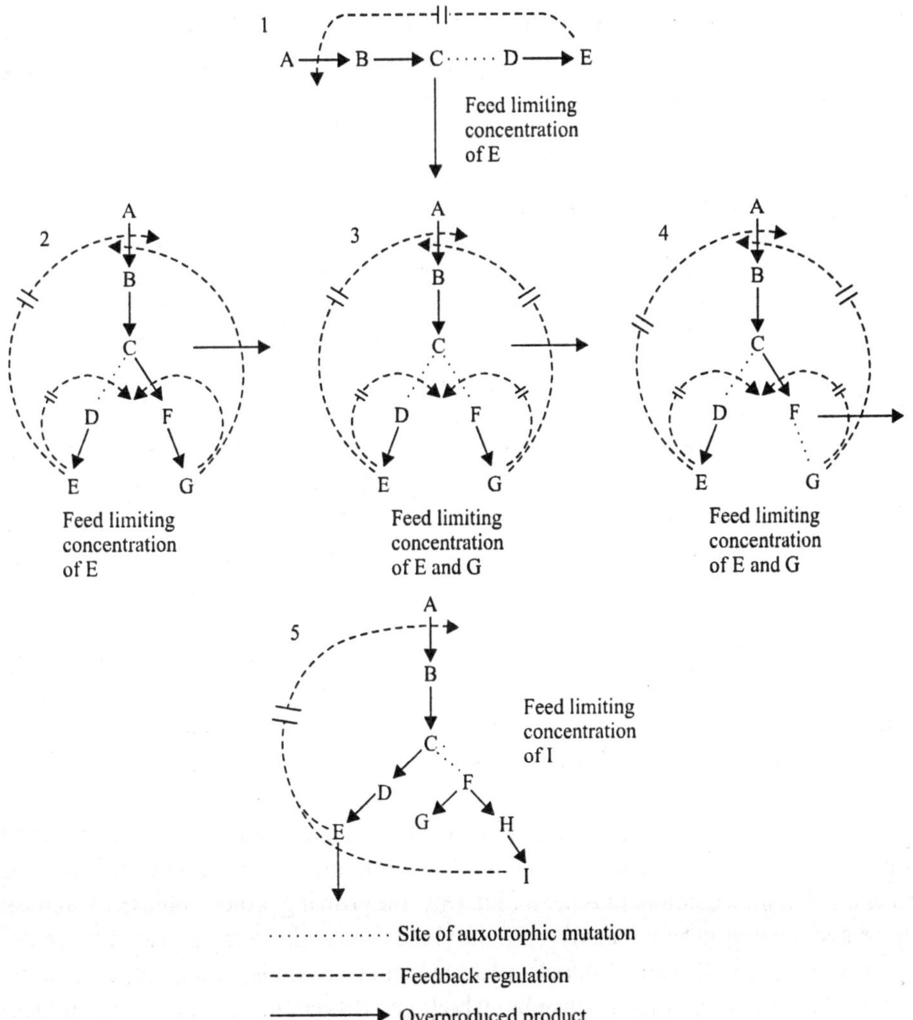


**Fig. 8.12.** Overproduction of primary metabolites by decreasing the concentration of a repressing or inhibiting end product.

In the case of Fig. 8.12(1) the unbranched pathway is normally controlled by feedback inhibition or repression of the first enzyme of the pathway by the end product, E. However, the organism represented in Fig. 8.12(1) is auxotrophic for E due to the inability to convert C to D so that control of the pathway is lifted and C will be accumulated provided that E is included in the medium at a level sufficient to maintain growth but insufficient to cause inhibition or repression.

Figure 8.12(2) is a branched pathway controlled by the concerted inhibition of the first enzyme in the pathway by the combined effects of E and G. The mutant illustrated is auxotrophic for E due to an inability to convert C to D, resulting in the removal of the concerted control of the first enzyme. Provided that E is included in the medium at a level sufficient to allow growth but insufficient to cause inhibition then C will be accumulated due to the control of the end product G on the conversion of C to F. The example shown in Fig. 8.12(3) is similar to that in Fig. 8.12(2) except that it is a double auxotroph and requires the feeding of both E and G. Figure 8.12(4) is, again, the same pathway and illustrates another double mutant with the deletion for the production of G occurring between F and G, resulting in the accumulation of F. Fig. 8.12(5) illustrates the accumulation of an end product of a branched pathway which is normally controlled by the feedback inhibition of the first enzyme in the pathway by the concerted effects of E and I. The mutant illustrated is auxotrophic for I and G due to an inability to convert C to F and thus, provided G and I are supplied in quantities which will satisfy growth requirements without causing inhibition, the end product, E, will be accumulated.

All the hypothetical examples discussed above are auxotrophic mutants and under certain circumstances, may accumulate relatively high concentration of intermediates or end products. Therefore, the isolation of auxotrophic mutants may result in the isolation of high-producing strains, provided that the mutation for auxotrophy occurs at the correct site, e.g., between C and D in Figs. 8.12(1) and 8.12(2). The recovery of auxotrophs is a simpler process than is the recovery of high producers, as such, so that the best approach is to design a procedure to select relevant auxotrophs from the survivors of a mutation and subsequently screen the selected auxotrophs for productivity. Productive strains amongst the auxotrophs may be detected by overlayering colonies of the mutants with agar suspensions of bacteria auxotrophic for the required product. The high-producing mutants may be identified by the growth of the overlay around the producer. The most commonly used methods for the recovery of auxotrophic mutants are the use of some form of enrichment culture or the use of a technique to visually identify the mutants. The enrichment processes employed are based on the provision of conditions which adversely affect the prototrophic cells but do not damage the auxotrophs. Such conditions may be achieved by exposing the population, in minimal medium, to an antimicrobial agent which only affects dividing cells which should result in the death of the growing prototrophs but the survival of the non-growing auxotrophs. Several techniques have been developed using different antimicrobials suitable for use with a range of micro-organisms.

Davis developed an enrichment technique utilising penicillin as the inhibitory agent. The survivors of a mutation treatment were first cultured in complete medium, harvested by centrifugation, washed and re-suspended in minimal medium plus penicillin. Only the growing prototrophic cells were susceptible to the penicillin and the non-growing auxotrophs survived. The cells were harvested by centrifugation, washed (to remove the penicillin and products released from lysed cells) and re-suspended in complete medium to allow the growth of the auxotrophs, which could then be purified on solid medium. The nature of the auxotrophs isolated may be determined by the design of the so-called complete medium; if only one addition is made to the minimal medium; then mutants auxotrophic for the additive should be isolated.

Abe described the use of Davis' technique to isolate auxotrophic mutants of the glutamic acid producing organism *C. glutamicum*. The procedure is outlined in Fig. 8.13.

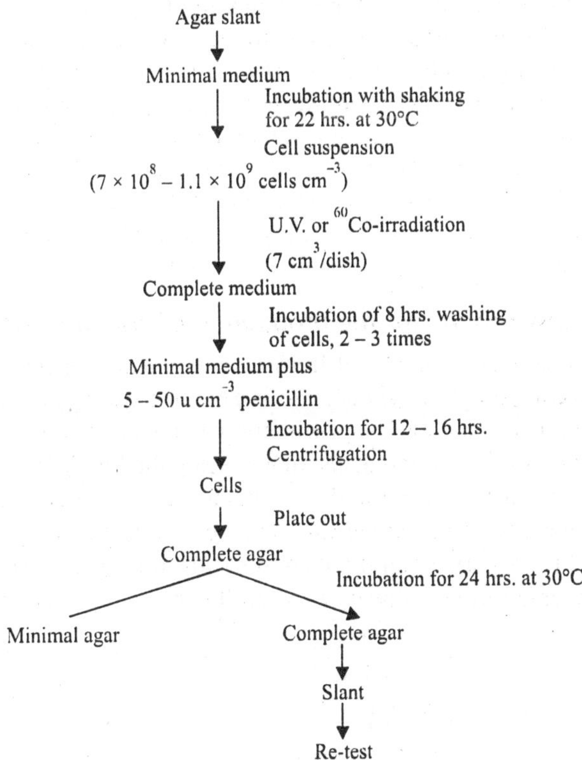


**Fig. 8.13.** The use of the penicillin selection method for the isolation of auxotrophic mutants of *C. glutamicum*.

Advantage has also been taken of the fact that the ungerminated spores of some organisms are more resistant to certain compounds than are the germinated spores. Thus, by culturing mutated spores in minimal medium only the prototrophs will germinate and subsequent treatment of the spore suspension with a suitable compound would kill the germinated prototrophic spores but leave the ungerminated auxotrophic spores unharmed. The auxotrophic spores may then be isolated by washing, to remove the inhibitor and cultured on supplemented medium. Ganju and Iyengar developed a technique of this type using sodium pentachlorophenate against the spores of *Penicillium chrysogenum, Streptomyces aureofaciens, S. olivaceus* and *Bacillus subtilis.*

The mechanical separation of auxotrophic and prototrophic spores of filamentous organisms has been achieved by the 'filtration enrichment method'. Liquid minimal medium is inoculated with mutated spores and shaken for a few hours, during which time the prototrophs will germinate but the auxotrophs will not. The suspension may then be filtered through a suitable medium, such as sintered glass, which will tend to retain the germinated spores resulting in a concentration of auxotrophic spores in the filtrate.

The visual identification of auxotrophs is based on the alternating exposure of suspected colonies to supplemented and minimal media. Colonies which grow on supplemented media, but not on minimal, are auxotrophic. The alternating exposure of colonies to supplemented and minimal medium has been achieved by replica plating. The technique consists of allowing the survivors of a mutation treatment to develop colonies on petri dishes of supplemented medium and then transferring a portion of each colony to minimal medium. The transfer process may be 'mechanised' by using some form of replicator. For bacteria the replicator is a sterile velvet pad attached to a circular support and replication is achieved by inverting the petri dish on to the pad, thus leaving an imprint of the colonies on the pad which may be used to inoculate new plates by pressing the plates on to the pad. It may be possible to replicate fungal and streptomycete cultures using a velvet pad, but, if unsatisfactory results are obtained, a steel pin replicator may be more appropriate. Visual identification of auxotrophs has also been achieved by the so-called 'sandwich technique'. The survivors of a mutation treatment are seeded in a layer of minimal agar in a petri dish and covered with a layer of sterile minimal agar. The plate is incubated for 1 or 2 days and the colonies developed are marked on the base of the plate, after which a layer of supplemented agar is poured over the surface. The colonies which then appear after a further incubation period are auxotrophic, as they were unable to grow on the minimal medium.

### *Examples of the use of auxotrophs for the production of primary metabolites*

Many auxotrophic mutants have been produced from *C. glutamicum* for the synthesis of both amino acids and nucleotide related compounds. *C. glutamicum* is a biotin-requiring organism which will produce glutamic acid under biotin-limited conditions but it is important to remember that mutants of this organism, employed for the production of other amino acids, must be supplied with levels of biotin optimum for growth. Biotin-limited conditions will result in these mutants producing glutamate and not the desired amino acid. Auxotrophic mutants of *C. glutamicum* have been used for the production of lysine. The control of the production of the aspartate family of amino acids in *C. glutamicum* is shown in Fig. 8.14. Aspartokinase, the first enzyme in the pathway, is controlled by the concerted feedback inhibition of lysine and threonine.

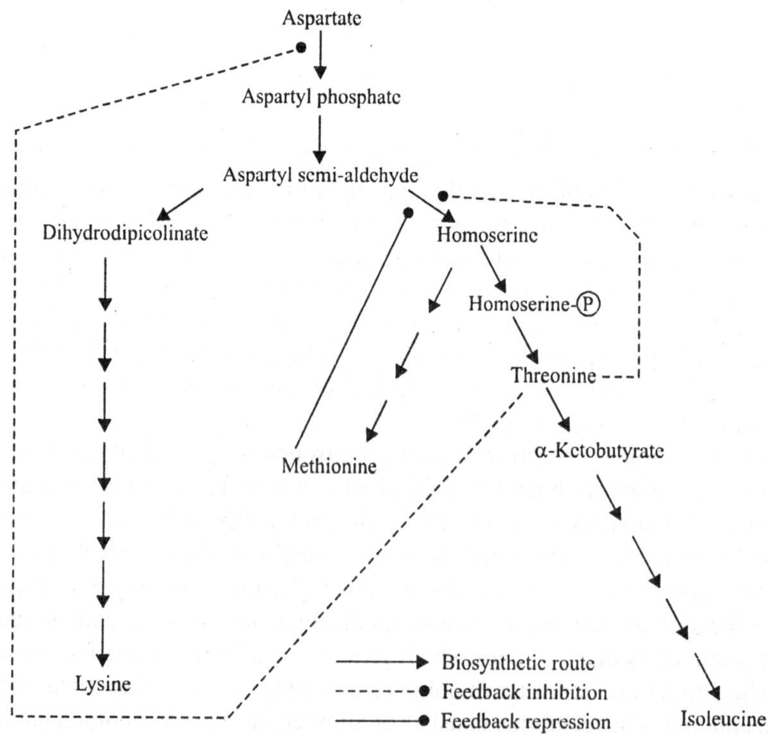


**Fig. 8.14.** The control of the aspartate family of amino acids in *C. glutamicum*.

Homoserine dehydrogenase is subject to feedback inhibition by threonine and repression by methionine. The first enzyme in the route from aspartate semi-aldehyde to lysine is not subject to feedback control. Thus, the control system found in *C. glutamicum* is a relatively simple one. Nakayama selected a homoserine auxotroph of *C. glutamicum*, by the penicillin selection and replica plating method, which produced lysine in a medium containing a low level of homoserine or threonine plus methionine. The mutant lacked homoserine dehydrogenase which allowed aspartic semi-aldehyde to be converted solely to lysine and the resulting lack of threonine removed the concerted feedback inhibition of aspartokinase. Kinoshita quoted the homoserine auxotroph, *C. glutamicum 901*, as producing 44 g $dm^{-3}$ lysine.

The control of the production of arginine in *C. glutamicum* is shown in Fig. 8.15. The major control of the pathway is the feedback inhibition of the second enzyme in the sequence, acetylglutamic acid phosphorylating enzyme, although the first enzyme may also be subject to regulation.

Kinoshita isolated a citrulline requiring auxotroph of *C. glutamicum* which would accumulate ornithine at a molar yield of 36 per cent from glucose, in the presence of limiting arginine and excess biotin. The mutant lacked the enzyme converting ornithine to citrulline which resulted in the cessation of arginine synthesis and therefore, the removal of the control of the pathway.

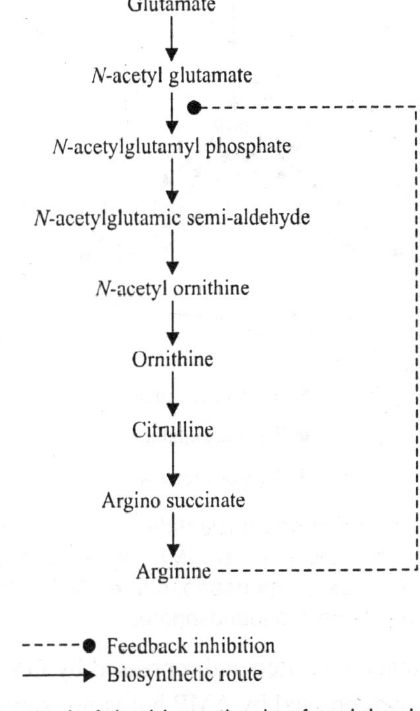


**Fig. 8.15.** The control of the biosynthesis of arginine in *C. glutamicum.*

Inosine monophosphate has been shown to demonstrate flavour-enhancing qualities and is produced commercially by the chemical phosphorylation of inosine which is produced from auxotrophic strains of *Bacillus subtilis*. The control of the production of purine nucleotides is shown in Fig. 8.16. The main sites of control shown in Fig. 8.16 are:

1. Phosphoribosyl pyrophosphate (PRPP) amidinotransferase (the first enzyme in the sequence) is feedback inhibited by AMP but only very slightly by GMP.
2. The synthesis of PRPP amidinotransferase is repressed by the co-operative action of AMP and GMP, as are the syntheses of the other enzymes indicated in Fig. 8.16.

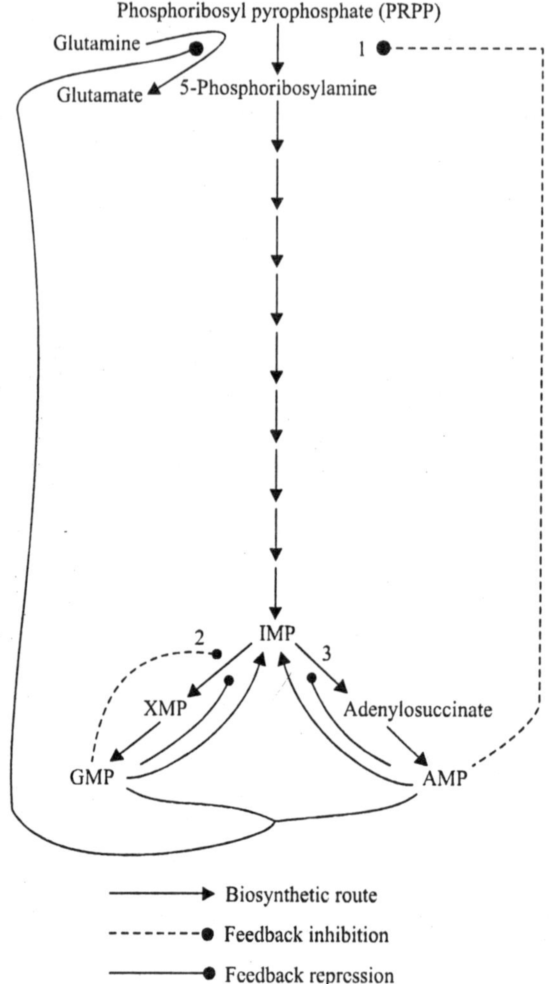


**Fig. 8.16.** Control of the biosynthesis of the purine nucleotides in *Bacillus subtilis*. 1. Reaction catalysed by PRPP amidinotransferase. 2. Reaction catalysed by IMP dehydrogenase. 3. Reaction catalysed by adenylosuccinate synthase. AMP = Adenosine monophosphate. IMP = Inosine monophosphate. XMP = Xanthine monophosphate. GMP = Guanosine monophosphate.

3. IMP dehydrogenase is feedback inhibited and repressed by GMP.
4. Adenylosuccinate synthase is repressed by AMP but is not significantly inhibited.

Mutants which are auxotrophic for AMP or doubly auxotrophic for AMP and GMP have been isolated which will excrete inosine at levels of up to 15 g $dm^{-3}$. AMP auxotrophs, lacking adenylosuccinate synthase activity, require the addition of small quantities of adenosine but will accumulate inosine due to the removal of the inhibition and co-operative repression of PRPP amidinotransferase, as shown in Fig. 8.17.

AMP and GMP double auxotrophs will produce inosine due to the removal of the controls normally imposed by the two end products, as illustrated in Fig. 8.18. Such double auxotrophs require the addition of both adenosine and guanosine, in small concentrations.

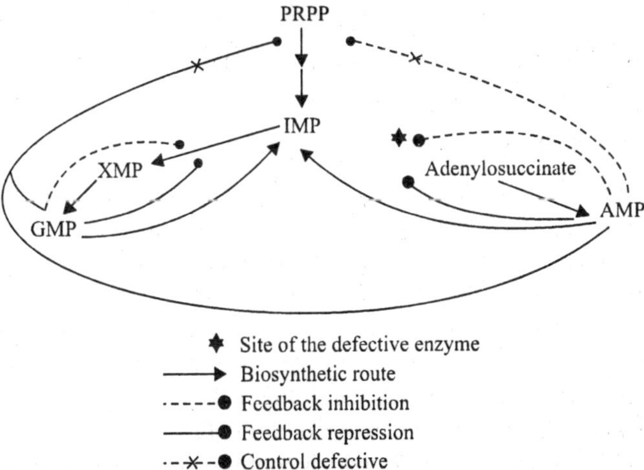


**Fig. 8.17.** The control of the synthesis of purine nucleotides in a mutant with a defective adenylosuccinate synthase.

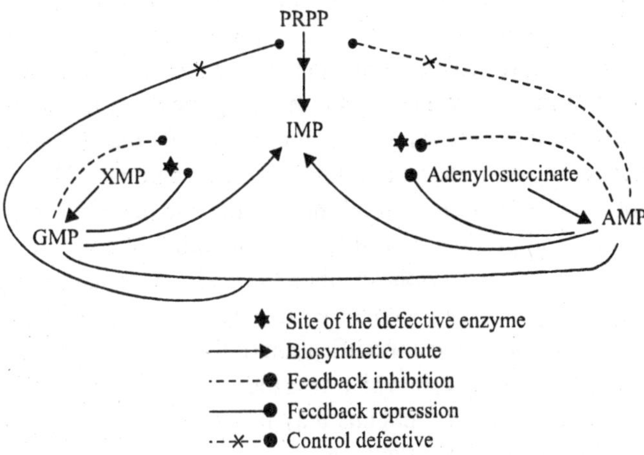


**Fig. 8.18.** The control of the synthesis of purine nucleotides in a mutant defective in adenylosuccinate synthase and IMP dehydrogenase activities.

### *Isolation of mutants that do not recognise the presence of inhibitors and repressors*

The use of auxotrophic mutants has resulted in the production of many microbial products in large concentrations, but, obviously, such mutants are not suitable for the synthesis of products which control their own synthesis independently. A hypothetical example is shown in Fig. 8.19 where the end product P controls its own biosynthesis by feedback inhibition of the first enzyme in the pathway. If it is required to produce the intermediate F in large concentrations then this may be achieved by the isolation of a mutant auxotrophic for P, blocked between F and P. However, if P is required to be synthesised in large concentrations it is quite useless to produce an auxotrophic mutant. The solution to this problem is to modify the organism such that the first enzyme in the pathway no longer recognises the presence of inhibiting levels of P. The isolation of mutants altered in the recognition of control factors has been achieved principally by the use of two techniques:

1. The isolation OB of analogue resistant mutants.
2. The isolation of revertants.

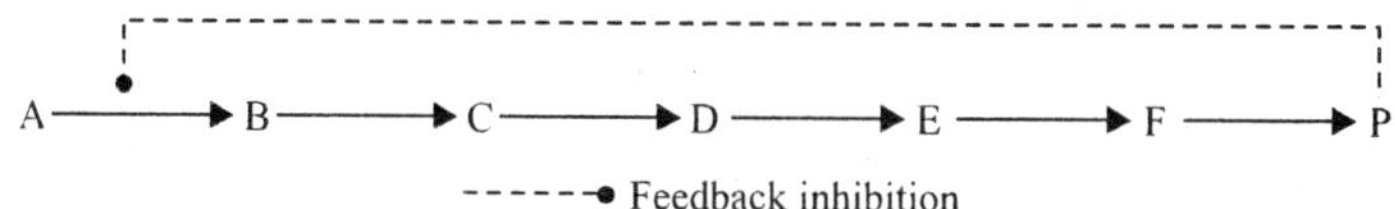


**Fig. 8.19.** The control of the production of an end product P.

An analogue is a compound which is very similar in structure to another compound. Analogues of amino acids and nucleotides are frequently growth inhibitory and their inhibitory properties may be due to a number of possible mechanisms. For example, the analogue may be used in the biosynthesis of macromolecules resulting in the production of defective cellular components. In some circumstances the analogue is not incorporated in place of the natural product but interferes with its biosynthesis by mimicing its control properties. For example, consider the pathway illustrated in Fig. 8.19 where the end product, P, feedback inhibits the first enzyme in the pathway. If P* were an analogue of P (which could not substitute for P in biosynthesis) and were to inhibit the first enzyme in a similar way to P, then the biosynthesis of P may be prevented by P* which could result in the inhibition of the growth of the organism.

Mutants may be isolated which are resistant to the inhibitory effects of the analogue and, if the site of toxicity of the analogue is the mimicing of the control properties of the natural product, such mutants may overproduce the compound to which the analogue is analogous. To return to the example of the biosynthesis of P where P* is inhibitory due to its mimicking the control properties of P; a mutant may be isolated which may be capable of growing in the presence of P* due to the fact that the first enzyme in the pathway is no longer susceptible to inhibition by the analogue. The modified enzyme of the resistant mutant may not only be resistant to inhibition by the analogue but may also be resistant to the control effects of the natural end product, P, resulting in the uninhibited production of P. If the control system were the repression of enzyme synthesis, then the resistant mutant may be modified such that the enzyme synthesis machinery does not recognise the presence of the analogue. However, the site of resistance of the mutant may not be due to a modification of the control system; for example, the mutant may be capable of degrading the analogue, in which case the mutant would not be expected to overproduce the end product. Thus, analogue resistant mutants may be expected to overproduce the end product to which the analogue is analogous provided that:

1. The toxicity of the analogue is due to its mimicking the control properties of the natural product.
2. The site of resistance of the resistant mutant is the site of control by the end product.

Resistant mutants may be isolated by exposing the survivors of a mutation treatment to a suitable concentration of the analogue in growth medium and purifying any colonies which develop. Sermonti described a method to determine the suitable concentration. The organism was exposed to a range of concentrations of the toxic analogue by inoculating each of a number of agar plates containing increasing levels of the analogue with $10^6$ to $10^9$ cells. The plates were incubated for several days and examined to determine the lowest concentration of analogue which allowed only a very few isolated colonies to grow or completely inhibited growth. The survivors of a mutation treatment may then be challenged with the predetermined concentration of the analogue on solid medium. Colonies which develop in the presence of the analogue may be resistant mutants.

Szybalski constructed a method of exposing the survivors of a mutation to a range of analogue concentrations on a single plate. Known as the gradient plate technique, it consists of pouring 20 $cm^3$ of molten agar medium, containing the analogue, into a slightly slanted petri dish and allowing the agar to set at an angle. After the agar has set, a layer of medium not containing the analogue is added and allowed to set with the plate level. The analogue will diffuse into the upper layer giving a concentration gradient across the plate and the survivors of a mutation treatment may be spread over the surface of the plate and incubated. Resistant mutants should be detected as isolated colonies appearing beyond a zone of confluent growth, as indicated in Fig. 8.20. Whichever method is used for the isolation of analogue-resistant mutants, great care should be taken to ensure that the isolates are genuinely resistant to the analogue by streaking them, together with analogue-sensitive controls, on both analogue-supplemented and analogue-free media. The resistant isolates should then be screened for the production of the desired compound by overlayering them with a bacterial strain requiring the compound; producers may then be recognised by a halo of growth of the indicator strain.

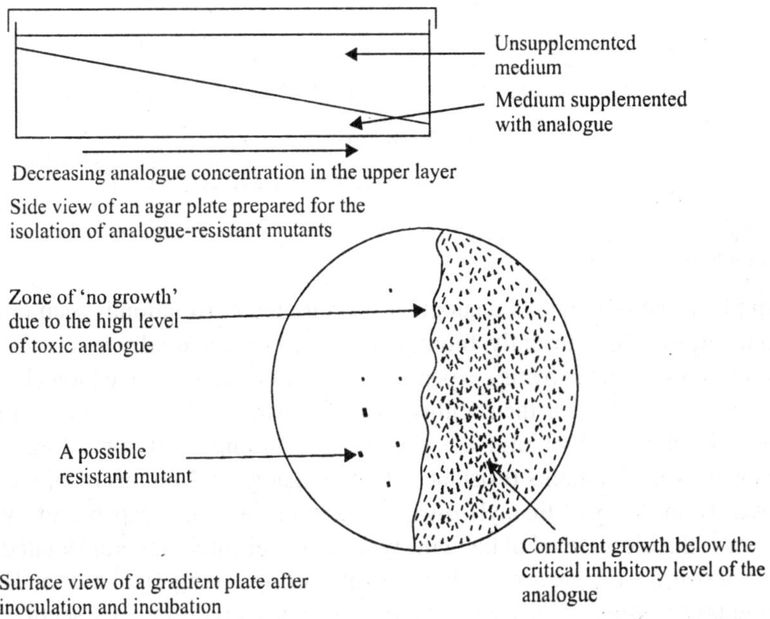


**Fig. 8.20.** The gradient plate technique for the isolation of analogue-resistant mutants.

Sano and Shiio investigated the use of lysine analogue-resistant mutants of *Brevibacterium flavum* for the production of lysine. The control of the biosynthesis of the aspartate family of amino acids in *B. flavum* is illustrated for *C. glutamicum* in Fig. 8.14. The main control of lysine synthesis is the concerted feedback inhibition of aspartokinase by lysine and threonine. Sano and Shiio demonstrated that *S*-(2-aminoethyl) cysteine (AEC) completely inhibited the growth of *B. flavum* in the presence of threonine, but only partially in its absence. Also, the inhibition by AEC and threonine could be reversed by the addition of lysine. This evidence suggested that the inhibitory effect of AEC was due to its mimicing lysine in the concerted inhibition of aspartokinase. AEC-resistant mutants were isolated by plating the survivors of a mutation treatment on minimal agar containing 1 mg $cm^{-3}$ of both AEC and threonine. A relatively large number of the resistant isolates accumulated lysine, the best producers synthesising more than 30 g $dm^{-3}$. Investigation of the lysine producers indicated that their aspartokinases had been desensitised to the concerted inhibition by lysine and threonine.

The development of an arginine-producing strain of *B. flavum* by Kubota provides an excellent example of the selection of a series of mutants resistant to increasing levels of an analogue. The control of the biosynthesis of arginine in *B. flavum* is similar to that shown for *C. glutamicum* in Fig. 8.15.

Kubota selected mutants resistant to the arginine analogue, 2-thiazolealanine and the genealogy of the mutants is shown in Fig. 8.21. Strain number 352 produced 25.3 g dm$^{-3}$ arginine. Presumably, the mutants were altered in the susceptibility of the second enzyme in the pathway to inhibition by arginine.

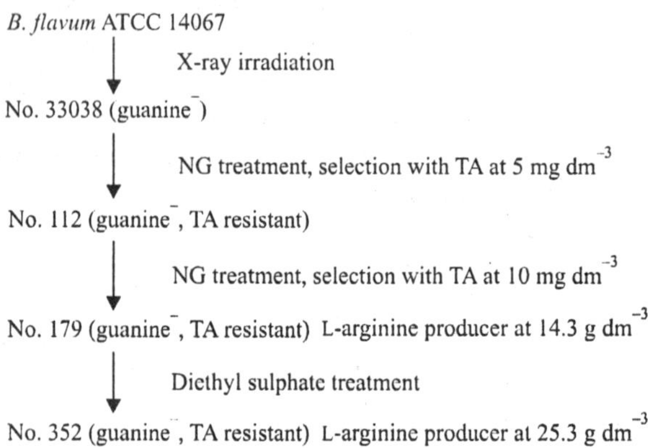


**Fig. 8.21.** The genealogy of L-arginine-producing mutants of *B. flavum.* TA, thiazolealanine; NG, *N*-methyl-*N'*-nitro-*N*-nitroso-guanidine.

A classic example of the rationale of analogue resistant mutant isolation is seen in the development of biotin overproducing strains. Currently, biotin is produced commercially by chemical synthesis, but considerable effort has been (and is being) expended to develop a competitive biotechnological process. The screening of natural isolates for their ability to accumulate biotin-vitamers led to the isolation of *Bacillus sphaericus*. Repression by biotin was shown to be an important control mechanism and thus, attempts were made to isolate mutants resistant to biotin analogues. Mutants of *B. sphaericus* resistant to acidomycin (ACM) and/or 5-(2-thienyl)-*n*-valeric acid (TVA) were capable of synthesising up to 11 times (0.4 mg cm$^{-3}$) the biotin level of the wild-type. More recently, Sakurai isolated dual ACM/TVA resistant strains of *Serratia marcescens* capable of synthesising 20 mg cm$^{-3}$ biotin. The use of genetic engineering techniques to further develop these strains is considered in a later section of this chapter.

The second technique used for the isolation of mutants altered in the recognition of control factors is the isolation of revertant mutants. Auxotrophic mutants may revert to the phenotype of the mutant 'parent'. Consider the hypothetical pathway illustrated in Fig. 8.19 where P controls its own production by feedback inhibiting the first enzyme (*a*) of the pathway. A mutant does not produce the enzyme, *a* and is, therefore, auxotrophic for P. However, a revertant of the mutant produces large concentrations of P. The explanation of the behaviour of the revertant is that, with two mutations having occurred at loci concerned with the production of enzyme *a*, the enzyme of the revertant is different from the enzyme of the original prototrophic strain and is not susceptible to the control by P. Revertants may occur spontaneously or mutagenic agents may be used to increase the frequency of occurrence, but the recognition of the revertants would be achieved by plating millions of cells on medium which would allow the growth of only the revertants, i.e., in the above example, on medium lacking P.

Shiio and Sano investigated the use of prototrophic revertants of *B. flavum* for the production of lysine. These workers isolated prototrophic revertants from a homoserine dehydrogenase-defective mutant. The revertants were obtained as small-colony forming strains and produced up to 23 g dm$^{-3}$ lysine. The overproduction of lysine was shown to be due to the very low level of homoserine dehydrogenase in the revertants which, presumably, resulted in the synthesis of threonine and methionine in quantities sufficient for some growth, but insufficient to cause inhibition or repression.

Mutant isolation programmes for the improvement of strains producing primary metabolites have not relied on the use of only one selection technique. Most projects employed a number of methods including the selection of natural variants and the selection of induced mutants by a variety of means. The selection of bacteria overproducing threonine provides a good example of the use of a variety of selection techniques. Attempts to isolate auxotrophic mutants of *C. glutamicum* producing threonine were unsuccessful despite the fact productive auxotrophic strains of *Escherichia coli* had been isolated. Huang demonstrated threonine production at a level of 2–4 g $dm^{-3}$ by a diaminopimelate and methionine double auxotroph of *E. coli*. Kase isolated a triple auxotrophic mutant of *E. coli* which required diaminopimelate, methionine and isoleucine and produced between 15 and 20 g $dm^{-3}$ threonine. The control of the production of the aspartate family of amino acids in *E. coli* is shown in Fig. 8.22 and that in *C. glutamicum* in Fig. 8.14. The mechanism of control in *E. coli* involves a system of isoenzymes, three isoenzymic forms of aspartokinase and two of homoserine dehydrogenase, under the influence of different end products. However, in *C. glutamicum* control is effected by the concerted inhibition of a single aspartokinase by threonine and lysine; by the inhibition of homoserine dehydrogenase by threonine and the repression of homoserine dehydrogenase by methionine. Thus, the control of homoserine dehydrogenase may not be removed by auxotrophy without the loss of threonine production. However, in *E. coli* methionine auxotrophy would remove control of the methionine-sensitive homoserine dehydrogenase and aspartokinase which would allow threonine production, despite the control of the threonine-sensitive isoenzymes by threonine. *E. coli* mutants also lacking lysine and isoleucine would be relieved of the control of the lysine-sensitive aspartokinase and the degradation of threonine to isoleucine.

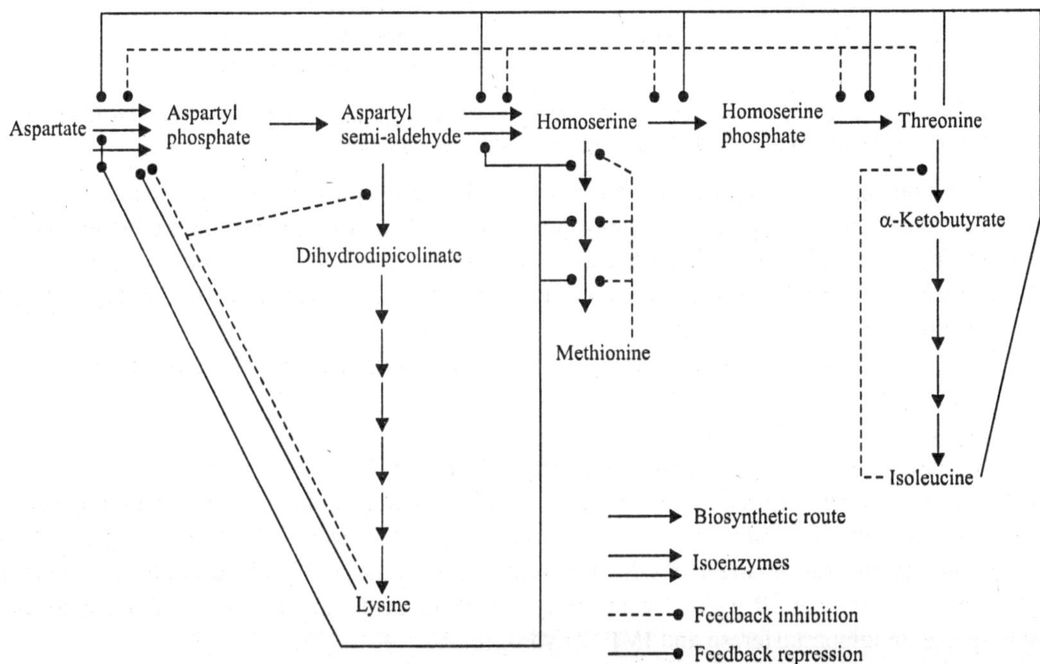


**Fig. 8.22.** Control of the aspartate family of amino acids in *Escherichia coli*.

The production of threonine by *C. glutamicum* has been achieved by the use of combined auxotrophic and analogue resistant mutants. A good example of the approach is given by Kase who obtained stepwise improvements in productivity by the imposition of resistance to α-amino-β-hydroxy valeric acid (a threonine analogue) and *S*-(β-aminoethyl)-L-cysteine (a lysine analogue) on a methionine auxotroph of *C. glutamicum*. The genealogy of the mutants is shown in Fig. 8.23. The analogue-resistant strains were shown to be altered in the susceptibility of aspartokinase and homoserine dehydrogenase to control and the lack of methionine removed the repression control of homoserine dehydrogenase. The use of recombinant DNA technology has resulted in the construction of far more effective threonine producers and these strains are considered in a later section of this chapter.

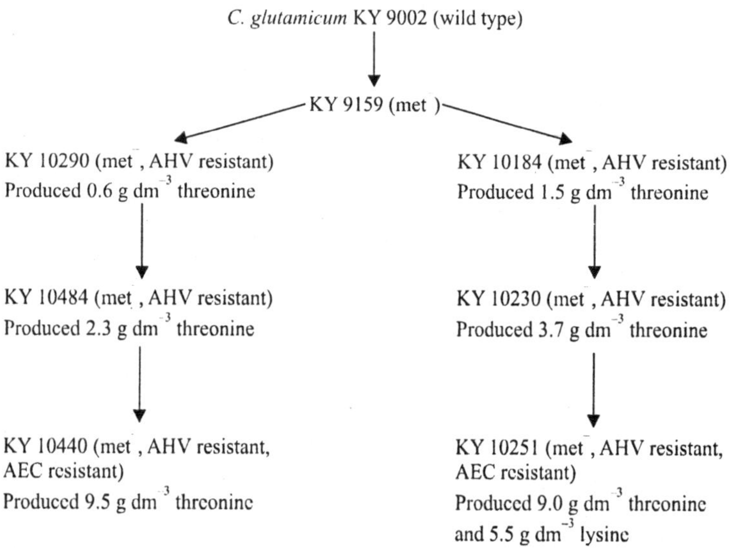


**Fig. 8.23.** The genealogy of mutants of *C. glutamicum* producing L-threonine or L-threonine plus L-lysine. AHV, α-amino-β-hydroxy valeric acid: AEC, *S*-(β-aminoethyl)-L-cysteine.

The development of strains producing guanosine in high levels provides a further example of the use of auxotrophs and analogue-resistant mutants in the same programme. Guanosine production has been achieved using auxotrophic and analogue resistant mutants of *B. subtilis*. The control of the production of purine nucleotides in *B. subtilis* is shown in Fig. 8.16. The major points of control to be modified to achieve guanosine overproduction are:

1. Removal of AMP inhibition of phosphoribosyl pyrophosphate amidino transferase by adenine auxotrophy.
2. Elimination of GMP reductase.
3. By-passing the inhibition of IMP dehydrogenase by GMP.

This approach is demonstrated in the work of Shiio who obtained analogue-resistant mutants from an adenine auxotroph of *B. subtilis*. Mutants were isolated in two stages—first, for resistance to low levels of 8-azaguanine (an analogue of GMP) and then for resistance to high levels of the analogue. The resulting mutant produced 9 g $dm^{-3}$ guanosine due to the removal of control of phosphoribosyl pyrophosphate amidinotransferase and IMP dehydrogenase.

### Isolation of Induced Mutants Producing Improved Yields of Secondary Metabolites Where Directed Selection is Difficult to Apply

The discussion so far has considered the isolation of mutants producing products whose biosynthesis and control have been sufficiently understood to prepare 'blueprints' of the desirable mutants which have enabled the construction of suitable selection procedures. In contrast, important secondary metabolites were being produced long before their biosynthetic pathways and certainly the control of those pathways, had been elucidated. Thus, strain improvement programmes had to be developed without this fundamental knowledge which meant that they depended on the random selection of the survivors of mutagen exposure. Elander and Vournakis described these techniques as 'hit or miss methods that require brute force, persistence and skill in the art of microbiology'. However, despite the limited knowledge underlying these approaches they were extremely effective in increasing the yields of antibiotics. More rational approaches have been developed which reduce the empirical nature of strain-improvement programmes. These developments include the streamlining of the empirical techniques and the use of more directed selection methods. Before discussing the attempts at directed selection the empirical approach will be considered along with the attempts made to improve this approach, including miniaturised programmes. The empirical approach to strain improvement involves subjecting a population of the micro-organism to a mutation treatment and then screening a proportion of the survivors of the treatment for improved productivity. The assessment of the chosen survivors was usually carried out in shake flasks, resulting in the procedure being costly, both in terms of time and personnel. According to Fantini the two questions which arise in the design of such programmes are:

1. How many colonies from the survivors of a mutation treatment should be isolated for testing?
2. Which colonies should be isolated?

In attempting to answer the first question Fantini dismissed statistical approaches as impractical and claimed that the number of colonies isolated for testing is determined by the practical limiting factors of personnel, incubator and shaker space and time. However, Davies demonstrated that a statistical approach could give valuable guidelines for the efficient utilisation of physical resources in strain-development programmes. Davies based a computer simulation of a mutation and screening programme on the availability of 200 shaker spaces and practical results of error variance and the distribution of yield likely to occur amongst the mutants. He assumed that the majority of the progeny of a mutation would give a small, rather than a large, increase in productivity and it would, therefore, be more feasible to screen a small number in the hope of obtaining a small increase rather than screen a large number in the hope of obtaining a large increase. The major difficulty inherent in this approach is the error involved in determining small increases which implies the replication of screening tests and therefore, the use of more facilities. Davies used the computer simulation to investigate the merits of replication in screening programmes and finally proposed the use of a two-stage scheme where mutants were tested singly in the first stage and then the better producers were tested in quadruplicate in the second stage. Davies concluded that such two-stage schemes were adequate over a wide range of conditions, although a one-stage screen could be used if the testing error were small and the frequency of occurrence of favourable types high and a three-stage screen if the testing error were high and the frequency of occurrence of favourable types low. A screening programme based on Davies' proposals is shown in Fig. 8.24.

The answer to Fantini's second question (which colonies should be isolated?) is extremely difficult in the field of secondary metabolites and in Davies' scheme the colonies were chosen at random. The selection of colonies on the basis of changed morphology has been considered by a number of workers and it appears that this is an undesirable technique. Elander demonstrated that it was preferable to isolate normal morphological types as, although a morphological variant may be a superior producer, it might require considerable fermentation development to materialise the increased production. Also, Alikhanian claimed that the vast majority of morphological mutants of the antibiotic producing actinomycetes tested were inferior producers. The most common type of shake flask programme quoted is similar to that of Davies where the choice of colonies is random. However, there are now many reports of screens which miniaturise the procedure for the improvement of secondary metabolite producing strains. Such miniaturised systems are designed to enable the productivity of all (or a significant proportion) of the survivors to be assessed which should eliminate (or reduce) the problem of choice of colonies for assessment and increase the throughput of cultures.

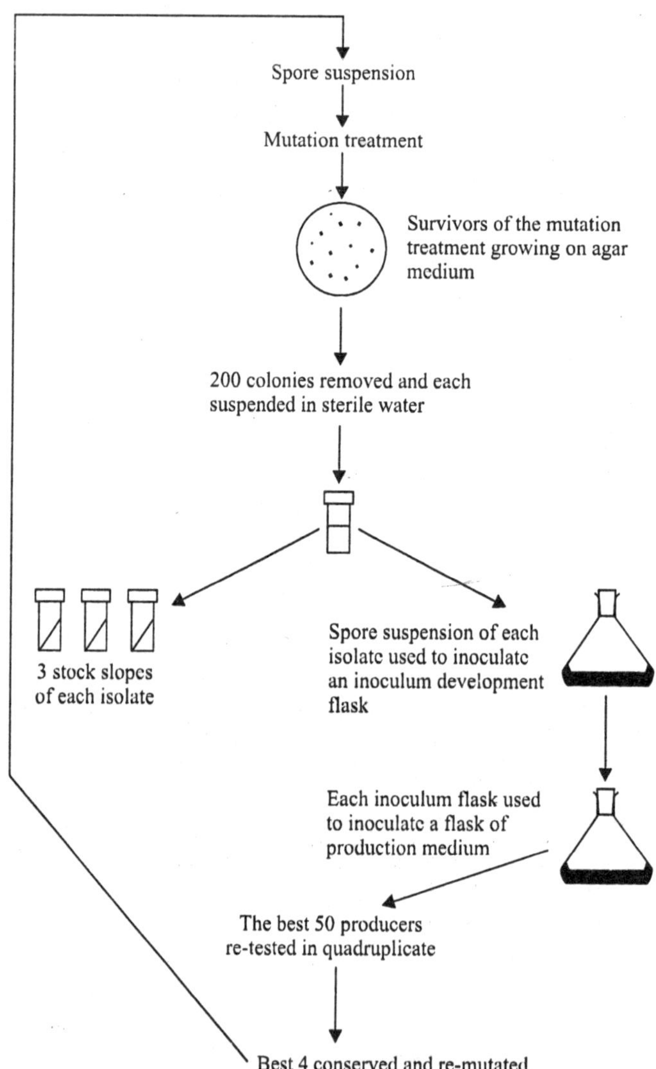


**Fig. 8.24.** A strain-improvement programme for a secondary metabolite producing culture.

The basis of the miniaturised techniques is to grow the survivors of the mutation treatment either in a very low volume of liquid medium or on solidified (agar) medium. If the product is an antibiotic, the agar-grown colonies may be overlayed with an indicator organism sensitive to the antibiotic produced, allowing assay to be done *in situ*. The level of antibiotic is assessed by the degree of inhibition of the overlayed indicator. The system is simple to apply to strains producing low levels of antibiotic but must be modified to allow the screening of high producers where very large zones of inhibition would be obtained. Also a system should be used to free the superior producers from contaminating indicator organisms. Dulaney used overlay techniques in the isolation of mutants producing chlortetracycline. Mutated spores were cultured on an agar medium in petri dishes for 6 days and then covered with pieces of sterile cellophane. An overlay of agar containing the indicator organism was then added and the plates incubated overnight. The mutant colonies were kept free from contamination by the cellophane and the size of the inhibition zone could be controlled by the depth of the base layer, the age of the colonies when overlayed, the depth of the overlay and the temperature of incubation. The system was calibrated initially such that a single colony of a non-mutated strain would not produce an inhibition zone but that two such adjacent colonies would. In practice, the size of the inhibition zone was controlled by the depth of the overlay. Dulaney obtained a far greater enrichment in the number of desired phenotypes by the overlay technique than by random selection and testing in liquid medium.

Ichikawa screened for the increased production of the antibiotic kasugamycin by a *Streptomyces* sp., using a miniaturised technique termed the agar piece method. In order to prevent interference between colonies, mutated spores were grown on plugs of agar which were then placed on assay plates containing agar seeded with the indicator organism, levels of the antibiotic being determined by the size of inhibition zones. By combining this technique with a medium improvement programme the authors obtained a ten-fold increase in productivity. Ichikawa's method was modified by Ditchburn for the isolation of *Aspergillus nidulans* mutants synthesising improved levels of penicillin. These workers obtained promising results using the technique and claimed that its potential for the recovery of higher-yielding mutants, for a given expenditure of labour time, was greater than the shake-flask method. The level of production of penicillin by *A. nidulans* is very small compared with *P. chrysogenum* and the technique would have to be modified considerably for commercial use.

Ball and McConagle developed a miniaturised technique suitable for the assay of penicillin production by 'high-yielding' strains of *P. chrysogenum* (producing up to 6000 units $cm^{-3}$). These workers highlighted the design of the solidified medium as a critical factor in the optimisation of the method. They claimed that the growth of a colony on agar-solidified medium would be unlikely to modify the medium to the same extent as the growth of the organism on the same medium in submerged liquid culture. Thus, the nutrient-limiting conditions that favour the onset of antibiotic production might not be achieved by the growth of the culture on solid medium which would not allow the full production potential of the culture to be detected. However, nutrient-limiting conditions were achieved in the solidified medium by omitting the main carbon source and reducing the corn steep-liquor content. Mutated spores were plated over the surface of petri dishes of the nutrient limited medium such that ten to twenty colonies developed per plate. The time of incubation of the plates was not quoted but 'when the colonies were of a size suitable for accurate measurement' a pasteurised spore suspension of *B. subtilis* containing 0.16 units $cm^{-3}$ of penicillinase (to limit the size of the inhibition zones) was dispersed over the surface of the dish. The plates were then incubated for 18 to 24 hours and the inhibition zones examined. Suitable colonies were freed of contaminating *B. subtilis* by culturing on nutrient agar containing sodium penicillin and streptomycin sulphate. The use of this technique enabled three operators to scan 15000 survivors from ultraviolet irradiated spore population in 3 months.

The major disadvantages of the miniaturised solidified medium technique approach is that productivity expressed on the solid medium may not be expressed in subsequent liquid culture and conversely, colonies not showing activity on solid media may be highly productive in liquid medium. Despite these limitations the above workers have demonstrated that the approach has considerable merit and Ball claimed that the increase in throughput may be as much as 20 times that of a conventional shake-flask programme. Some recent work by Bushell's group provides an interesting insight into the problem of certain streptomycete isolates producing a secondary metabolite on agar but not in liquid culture. It is important to appreciate that the strains were natural isolates and not the survivors of mutation treatments but the conclusions do have relevance to mutant development. Working on the premise that non-production in liquid medium may be due to a more fragmented morphology these strains were subjected to a filtration enrichment procedure. Liquid cultures were filtered through a linen filter and the filtrate and retentate mycelium re-cultured in fresh medium. The procedure was then repeated sequentially. The enriched retentate mycelium of several isolates which were previously unable to produce antibiotic in liquid culture gave rise to stable filamentous types synthesising in liquid medium. Although this approach could not be used routinely in a high throughput mutant screen it may be applied to a few high-producing strains which have not fulfilled their promise (detected in an agar screen) in liquid media.

An excellent example of a miniaturised screening programme is given by the work of Dunn-Coleman of Genencor International. These workers developed a low-volume liquid medium system to isolate mutants of *Aspergillus niger* var. *awamori* capable of improved secretion of bovine chymosin. The *Aspergillus* strain had been constructed using recombinant DNA technology and this construction will be considered in a later section of this chapter. Mutated spore suspensions were diluted and inoculated into 96-well microtitre plates using robots. The dilutions were such that each well should have contained one viable spore. The plates were incubated in a static incubator and then harvested using a robotic pipetting station and assayed for product. Typically, 50–60,000 mutated viable spores were assessed in each screen. The most promising 10–50 strains from the miniaturised screen were then tested in shake flask culture. The best producer from the shake-flask screen was then used as the starter for a further round of mutagenesis. The results of seven rounds of mutation and selection are shown in Table 8.4. Investigation of the improved strains showed that they were capable of secreting a range of enzymes and thus, the strains were presumed to be modified in their secretion mechanisms. It is interesting to note that an improved secreter of a heterologous protein has been isolated using a traditional mutation and selection approach which has been miniaturised and automated.

**Table 8.4.** Chymosin production by NTG mutated strains of *Aspergillus niger* var *awamori.*

| *Rounds of NTG mutagenesis and recurrent selection* | *Chymosin concentration* ($\mu g\ cm^{-3}$) | |
|---|---|---|
| | *Microtitre plate* | *Shake flask* |
| Parent | 42 | 286 |
| First | 50 | 273 |
| Second | 64 | 280 |
| Third | 136 | 377 |
| Fourth | 170 | 475 |
| Fifth | 200 | 510 |
| Sixth | 256 | 646 |

In several cases the examination of randomly selected high-yielding mutants has indicated that their superiority may have been due to modifications of their control systems. Goulden and Chataway demonstrated that a mutant of *P. chrysogenum*, producing high levels of penicillin, was less sensitive to the control of acetohydroxyacid synthase by valine. Valine is one of the precursors of penicillin and its synthesis is controlled by its feedback inhibition of acetohydroxyacid synthase. Thus, removal of the control of valine synthesis may result in the production of higher valine levels and hence, greater production of penicillin. Pruess and Johnson demonstrated that higher-yielding strains of *P. chrysogenum* also contained higher levels of acyltransferase, the enzyme which catalyses the addition of phenylacetic acid to 6-aminopenicillanic acid. Dulaney reported that the best producer of streptomycin amongst *Streptomyces griseus* mutants was auxotrophic for vitamin $B_{12}$ and Demain claimed that this auxotrophic mutation was still a characteristic of production strains. Thus, it appears that some strains isolated by random selection and overproducing secondary metabolites, are altered in ways similar to those strains isolated by directed selection techniques and overproducing primary metabolites.

There are many examples in the literature where a more directed selection approach has been adopted for the improvement of secondary metabolite producers. The techniques used include the isolation of auxotrophs, revertants and analogue-resistant mutants.

### *Isolation of auxotrophic mutants*

Although mutation of secondary metabolite producers to auxotrophy has resulted frequently in their producing lower yields, cases of improved productivity have also been demonstrated. For example, Alikhanian investigated the tetracycline producing abilities of fifty-three auxotrophic mutants, all of which produced significantly less tetracycline than the parent strain in normal production medium. Supplementation of the medium with the growth requirement resulted in one mutant expressing productivity superior to that of the parent.

It is sometimes difficult to explain in the precise reason for the effect of mutation to auxotrophy on the production of secondary metabolites, but in the majority of cases it has been demonstrated to be an effect on the secondary metabolic system rather than, simply, an effect on the growth of the organism. The simplest explanation for the deleterious effect on secondary metabolite yield is that the auxotroph is blocked in the biosynthesis of a precursor of the end product, for example, Polsinelli demonstrated that auxotrophs of *Streptomyces antibioticus* which required any of the precursors of actinomycin (isoleucine, valine or threonine) were poor producers of the antibiotic.

Many secondary metabolites may be considered as end products of branched pathways which also give rise to primary metabolites. Thus, a mutation to auxotrophy for the primary end product may also influence the production of the secondary product. In *P. chrysogenum*, lysine and penicillin share the same common biosynthetic route to α-aminoadipic acid, as shown in Fig. 8.25. This biosynthetic route may explain Bonner's observation that 25 per cent of the lysine auxotrophs of *P. chrysogenum* he isolated could not produce penicillin. The role of lysine in the penicillin fermentation has also led to the investigation of lysine auxotrophs as potential superior penicillin producers. Demain demonstrated that lysine was inhibitory to the biosynthesis of penicillin. The explanation of this phenomenon is considered to be the inhibition of homocitrate synthase by lysine resulting in the depletion of α-aminoadipic acid required for penicillin synthesis. It may be postulated that lysine auxotrophs blocked immediately after α-aminoadipic acid would produce higher levels of penicillin due to the diversion of the intermediate towards penicillin synthesis and the removal of any control of homocitrate synthase by endogenous lysine. O'Sullivan and Pirt investigated the production of penicillin by lysine auxotrophs of *P. chrysogenum* in continuous culture but were unable to demonstrate improved productivity in a range of lysine-supplemented media. Luengo examined the lysine regulation of penicillin biosynthesis in low-producing and industrial strains of *P. chrysogenum*. The industrial strain was capable of producing up to 12000 units $cm^{-3}$ penicillin. These workers demonstrated that although the onset of penicillin synthesis in the industrial strain was less sensitive to lysine than in the low-producing strain, the empirical industrial selection procedures had not completely removed the mechanisms of lysine regulation of penicillin biosynthesis. Luengo expressed the possibility that overproducers of penicillin may be obtained by the selection of mutants to lysine regulation. However, it should be noted that the industrial strain employed was a representative of only one series of penicillin producers and that other series may have already been modified with respect to the regulatory effects of lysine. Nevertheless, this study does indicate the possibility that selection of strains resistant to lysine control may be overproducers of penicillin.

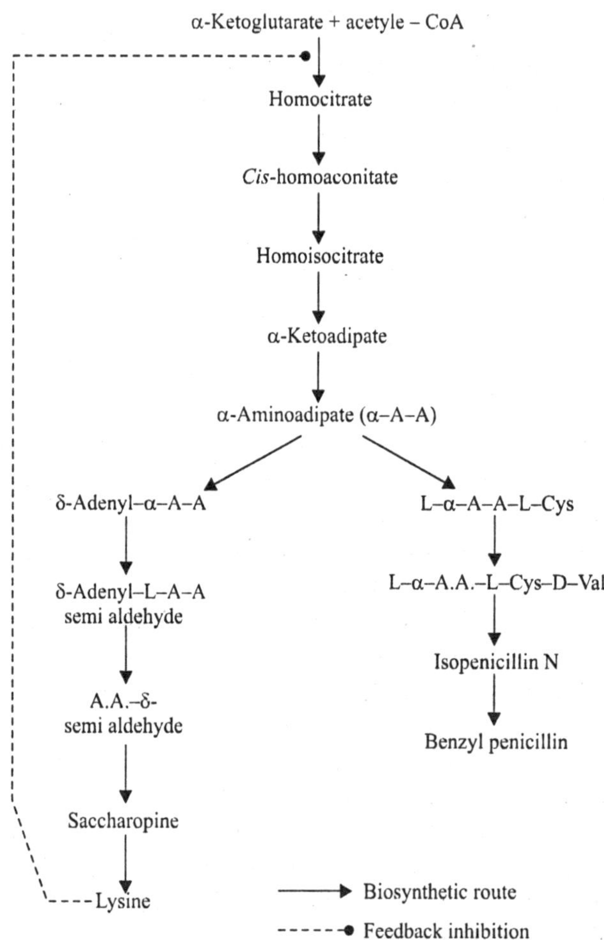


**Fig. 8.25.** Biosynthesis of benzyl penicillin and lysine in *Penicillium chrysogenum.*

It is far more difficult to explain the effect of auxotrophy for factors not associated with the biosynthesis of the secondary metabolite. Polsinelli demonstrated that seven out of twenty-seven auxotrophs of *S. griseus* produced more actinomycin than did the prototrophic parent. Dulaney demonstrated increased chlortetracycline yields in eight out of eleven auxotrophs of *S. viridifaciens* when grown in supplemented media. In neither of these cases were the auxotrophic requirements directly involved in the biosynthesis of the secondary metabolites. Demain put forward two possible explanations to attempt to account for the behaviour of such auxotrophs. The first explanation is that the auxotrophic factors were involved in 'cross-pathway' regulation with the secondary metabolite or its precursors. Demain quoted several examples of cross-pathway regulation in primary metabolism where the activity of one pathway is affected by the product of an apparently unrelated sequence. The alternative explanation is that the effect on secondary metabolism is not due to the auxotrophy but to a second mutation accompanying the auxotrophy, i.e., a double mutation. Demain cited two attempts to determine whether the effects of auxotrophy on secondary metabolism were due to double mutations or to the auxotrophy. MacDonald reverted a low-producing thiosulphate-requiring mutant of *P. chrysogenum* to thiosulphate independence and examined penicillin productivity by the revertants. Approximately half of the revertants re-acquired their 'grandparents' production level, whereas the other half retained their poor productivity. Polsinelli reverted five isoleucine-valine auxotrophs of *S. antibioticus* (which also produced low levels of actinomycin compared with the parent strain) to prototrophy and discovered that some were returned to normal production and others to higher production levels than the 'grandparent'. Thus, in the case of Polsinelli mutants it is unlikely that the effect on secondary metabolism was due to a double mutation, but it is possible that this was the case for some of MacDonald strains. However, it should be remembered that both these groups of auxotrophs were poor secondary metabolite producers blocked in routes directly involved with the secondary biosynthetic pathway. It may be more relevant to examine the nature of auxotrophic strains blocked in apparently unrelated pathways and produce improved levels of the secondary metabolite. It is not possible to say whether any of the auxotrophic mutants previously discussed produced superior levels of the secondary metabolite as a result of double mutations, but it may be possible to exploit this possibility in the future.

The treatment of bacterial cells with nitrosoguanidine (NTG) has been demonstrated to result in clusters of mutations around the replicating fork of the chromosome. Thus, if one of the mutations were selectable (for example, auxotrophy) it may be possible to isolate a strain containing the selectable mutation along with other non-selectable mutations which map close by. The efficient application of this technique would depend on the accurate mapping of the gene involved in producing the secondary metabolite so that neighbouring mutations may be selected. This technique may be valuable for the selection of mutants of the bacilli and streptomycetes producing high levels of antibiotics where mutations affecting synthesis may be mapped for each biosynthetic step. Co-mutation by NTG may then be followed by selection for changes to genes adjacent to those loci known to be involved in production of the particular secondary metabolite.

### *Isolation of resistant mutants*

The isolation of analogue-resistant mutants has already been discussed in the field of primary metabolism, where the rationale was that a mutant resistant to the inhibitory effects of an analogue, which mimics the control characteristics of the natural metabolite, may overproduce the natural metabolite. This approach has been adopted or may be adopted, in a number of guises in the field of secondary metabolism.

1. Mutants may be isolated which are resistant to the analogues of primary metabolite precursors of the secondary metabolite, thus increasing the availability of the precursor.
2. Mutants may be isolated which are resistant to the feedback effects of the secondary metabolite.
3. Mutants may be selected which are resistant to the toxic effects of the secondary metabolite when added to the trophophase of the producing organism.
4. Mutants may be isolated which are resistant to the toxic effects of the compound due to the production of the secondary metabolite.

*Mutants resistant to the analogues of primary metabolic precursors of the secondary metabolite*

This approach has been adopted by Elander in the isolation of mutants of *Pseudomonas aureofaciens* overproducing the antibiotic pyrrolnitrin. Tryptophan is a precursor of pyrrolnitrin and although it is stimulatory to production it is uneconomic to use as an additive in an industrial process. Thus, Elander isolated mutants resistant to tryptophan analogues using the gradient plate technique described previously. A strain was eventually isolated which produced two to three times more antibiotic than the parent and was resistant to feedback inhibition by tryptophan. Addition of tryptophan to the improved strain would not result in higher pyrrolnitrin synthesis, indicating that tryptophan supply was no longer the limiting factor and the organism was producing sufficient endogenous tryptophan for antibiotic synthesis.

The role of tryptophan in the control of the candicidin fermentation has been investigated by Martin and co-workers. Candicidin is a polyene macrolide antibiotic containing an aromatic *p*-aminoacetophenone moiety derived from chorismic acid, a precursor of the aromatic amino acids, so that tryptophan and candicidin may be considered to be end products of a branched biosynthetic pathway. Tryptophan has been demonstrated to inhibit the biosynthesis of the antibiotic which led Martin to isolate mutants resistant to tryptophan analogues. Such mutants produced more candicidin than the parent strain, apparently due to the removal of tryptophan control.

Godfrey isolated mutants of *Streptomyces lipmanii* resistant to the valine analogue, trifluoroleucine. These mutants produced higher levels of cephamycin than the parent strain and appeared to be deregulated for the isoleucine, leucine, valine biosynthetic pathway, indicating that valine may have been a rate-limiting step in cephamycin synthesis.

Methionine has been demonstrated to stimulate the biosynthesis of cephalosporin by *Acremonium chrysogenum* and superior producers have been isolated in the form of methionine analogue-resistant mutants. Lysine analogue resistant mutants have yielded a greater frequency of superior β-lactam antibiotic producers compared with random selection.

*Mutants resistant to the feedback effects of the secondary metabolite*

There are many cases in the literature of a secondary metabolite preventing its own synthesis. Martin cited the following examples: chloramphenicol, aurodox, cycloheximide, staphylomycin, ristomycin, puromycin, fungicidin, candihexin, mycophenolic acid and penicillin. The precise mechanisms of these controls is not clear but they appear to be specific against the synthesis of the secondary metabolite and not against the general metabolism of the cells. The mechanism of the control of its own synthesis by chloramphenicol appears to be the repression of arylamine synthetase (the first enzyme in the pathway from chorismic acid to chloramphenicol) by chloramphenicol. Jones and Vining demonstrated that arylamine synthetase was fully repressed by 100 mg $dm^{-3}$ chloramphenicol, a level of antibiotic which neither affected cell growth nor the activities of the other enzymes of the chloramphenicol pathway.

Martin quoted several examples of correlations between the level of secondary metabolite accumulation and the level which causes 'feedback' control, which may imply that the factor limiting the yields of some secondary metabolites is the feedback inhibition by the end product. The selection of mutants resistant to feedback inhibition by a secondary metabolite is a far more difficult task than the isolation of strains resistant to primary metabolic control. It is extremely unlikely that a toxic analogue of the secondary metabolite could be found where the toxicity lay in the mimicking of the feedback control of the secondary metabolite, a compound which would not be necessary for growth. However, the detection of mutants resistant to feedback inhibition by antibiotics may be achieved by the use of solidified media screening techniques, similar to the miniaturised screening techniques previously described. The technique would involve culturing the survivors of a mutation treatment on solidified medium containing hitherto repressing levels of the antibiotic and detecting improved producers by overlaying the colonies with an indicator organism. The difficulty inherent in this technique is that the incorporated antibiotic, itself, will inhibit the development of the indicator organism. This problem may be overcome by adjusting the depth of the overlay or the concentration of the indicator such that an inhibition zone could be produced only by a level of antibiotic greater than that incorporated in the original medium. However, it is unlikely that this would be a satisfactory solution for the selection of a high-producing commercial strain. Another approach would be to utilise an analogue of the antibiotic which mimicked the feedback control by the natural product but which did not have antimicrobial properties. Inhibition of antibiotic synthesis by analogues has been demonstrated in the cases of aurodox and penicillin. An alternative may be to use a mutant indicator bacterium for the overlay which is only sensitive to levels of the antibiotic in excess of that originally incorporated into the medium. In cases where it has been demonstrated that feedback control by the end product plays an important role in limiting productivity it would probably be worthwhile to design such procedures.

*Isolation of mutants resistant to the toxic effects of the secondary metabolite in the trophophase*

It has been demonstrated for many secondary metabolite producing organisms that the secondary metabolite is toxic to the producing cell when it is present in the trophophase (growth phase). Thus, it appears that a 'switch' in the metabolism of the organism in the idiophase enables it to produce an otherwise 'autotoxic' product. Furthermore, it has been demonstrated that, in some cases, the higher the resistance to the secondary metabolite in the growth phase the higher is productivity in the production phase. Dolezilova demonstrated that the level of production of nystatin by various strains of *S. noursei* was related to the resistance of the strain to the antibiotic in the growth phase; a non-producing mutant was inhibited by 20 units $cm^{-3}$, the parent strain produced 6000 units $cm^{-3}$ and was inhibited by 2000 units $cm^{-3}$ in the growth phase and a mutant producing 15,000 units $cm^{-3}$ was found to be resistant to 20,000 units $cm^{-3}$.

The possible relationship between antibiotic resistance and productivity may be used to advantage in the selection of high-producing mutants by culturing the survivors of a mutation treatment in the presence of a high level of the end product. Those strains capable of growth in the presence of a high level of the antibiotic may also be capable of high productivity in the idiophase. This approach has been used successfully for antifungal agesterols, streptomycin and ristomycin but without success for novobiocin.

A similar rationale was used by McGuire for the improvement of daunorubicin production, an anthracycline antitumour agent synthesised by the red pigmented streptomycete *Streptomyces peuceticus*. The red pigment is presumably anthracycline which shares its oligoketide origin with daunorubicin. The directed selection was based on the ability of the antibiotic cerulenin to suppress oligoketide synthesis. This was indicated in *S. peuceticus* by the lack of the red pigment on cerulenin agar. Thus, mutants which were still capable of oligoketide synthesis in the presence of the inhibitor would remain red. Approximately a third of the resistant mutants were superior daunorubicin producers.

*Isolation of mutants in which secondary metabolite synthesis gives resistance to toxic compounds*

A potentially toxic compound may be rendered harmless by an organism converting it to a secondary metabolite or a secondary metabolite complexing with it. Ions of heavy metals such as $Hg^{2+}$, $Cu^{2+}$ and related organometallic ions are known to complex with β-lactam antibiotics and such agents have been used to select resistant mutants. The logic of this selection process is that overproduction of a β-lactam may be the mechanism for increased resistance to the heavy metal. Elander and Vournakis reported that the frequency of superior cephalosporin C producers was greater amongst mutants resistant to mercuric chloride than amongst random samples of the survivors of ultraviolet treatment.

The conversion of the penicillin precursor, phenylacetic acid, to penicillin is thought to be an example of detoxification by conversion to a secondary metabolite. It appears that strains capable of withstanding higher concentrations of the precursor may be able to synthesis higher levels of the end product. Polya and Nyiri applied this hypothesis in selecting phenylacetic acid-resistant mutants of *P. chrysogenum* and demonstrated that 7 per cent of the isolates showed enhanced penicillin production. Barrios-Gonzalez investigated the same phenomenon and developed methods for the enrichment of both spores and early idiophase mycelium resistant to phenyl acetic acid. Of the resistant spore population, 16.7 per cent were superior penicillin producers as compared with 50 per cent of the resistant idiophase population. This suggests that the selective force may be more 'directed' by using a population already committed to penicillin synthesis. Although the best mutants contained elevated levels of acyltransferase (the enzyme that directly detoxifies PAA) they also showed higher levels of isopenicillin N synthetase (cyclase) which the authors claimed to be the limiting step. Thus, it was claimed that the screening method was not specific to selecting for acyltransferase elevation but could select for strains having a faster carbon flow through the pathway enabling them to use PAA faster.

Ball stressed that the major difficulty in the selection of toxic precursor resistant mutants is that the site of resistance of a mutant may not result in the organism overproducing the end product—for example, the resistance may be due to an alteration in the permeability of the mutant or due to the mutants' ability to degrade the precursor to a harmless metabolite unrelated to the desired end product. Barrios-Gonzalez's approach of using mycelial fragments already committed to penicillin synthesis decreases the likelihood of such 'false selection' occurring.

### *Isolation of revertant mutants*

As discussed in the section on primary metabolites, a mutant may revert to the phenotype of its 'parent', but the genotype of the revertant may not, necessarily, be the same as the original 'parent'. Some revertant auxotrophs have been demonstrated to accumulate primary metabolites and attempts have been made to apply the technique to the isolation of mutants overproducing secondary metabolites. Two approaches have been used in the field of secondary metabolites with respect to the isolation of revertants:

1. The isolation of revertants of mutants auxotrophic for primary metabolites which may influence the production of a secondary metabolite.
2. The reversion of mutants which have lost the ability to produce the secondary metabolite.

#### *Isolation of revertants of mutants auxotrophic for primary metabolites which may influence the production of a secondary metabolite*

As previously discussed, it is difficult to account for some of the effects of auxotrophy on secondary metabolism and it appears that some may be due to as yet unresolved cross-pathway phenomena and others due to the expression of other mutations associated with the auxotrophy. Similarly, revertant mutants may affect secondary metabolism in a number of ways—direct effects on the pathway, cross-pathway effects and effects due to mutations other than the detected reversion.

Dulaney investigated the tetracycline productivities of a population of prototroph revertants of *S. viridifaciens* derived from each of five auxotrophs. These workers predicted that some revertants may be productive due to direct influence of the mutations on tetracycline biosynthesis but that others may be so because they contained other lesions. Superior producers were obtained in all the prototroph-revertant population apart from those derived from a homocysteine auxotroph. However, the frequency of the occurrence of the superior producers was similar to that obtained by the random selection of the survivors of a mutation treatment in all but one of the population. The exceptional population was the revertants of a methionine auxotroph, 98 per cent of which produced between 1.2 and 3.2 times as much tetracycline as the original prototrophic culture. A possible explanation of the very favourable titres of the population may be the role of methionine as the methyl donor in tetracycline biosynthesis and that methionine availability limited the production of the secondary metabolite in the original prototroph. However, addition of exogenous methionine to the prototroph did not result in superior productivity.

Polsinelli also demonstrated that reversion of five mutants of an actinomycin producing strain of *S. antibioticus* blocked in the isoleucine-valine pathway resulted in the isolation of some superior mutants. Godfrey reported that the reversion of a cysteine auxotrophic mutant of *S. lipmanii* resulted in improved production of cephamycin.

These studies provide promising evidence that the selection of revertants of auxotrophs of primary metabolites involved in secondary metabolism may yield a high number of productive mutants.

#### *Isolation of revertants of mutants which have lost the ability to produce the secondary metabolite*

The reversion of non-producing strains may result in the detection of a high-producing mutant as that mutant would have undergone at least two mutations associated with the production of the secondary metabolite. Dulaney plated the progeny of a mutation of a non-producing strain of *S. viridifaciens* onto solidified production medium and screened for superior tetracycline producers by an overlay technique. A mutant was isolated which produced nine times the tetracycline yield of the original 'parent'. The major difficulty inherent in this technique is that non-producing mutants of high-yielding strains may be incapable of being reverted due to extreme deficiencies in their metabolism. Indeed, Rowlands suggested that strains which are non (or low) producers and illustrate other effects such as poor growth and sporulation are best discarded, but that revertant mutants not showing pleiotrophic effects are perhaps more likely to possess genuine increases in biosynthetic activity than those produced by any other technique, having been mutated twice in genes directly affecting product formation.

### Use of Recombination Systems for the Improvement of Industrial Micro-organisms

Hopwood defined recombination, in its broadest sense, as 'any process which helps to generate new combinations of genes that were originally present in different individuals'. The use of recombination mechanisms for the improvement of industrial strains has increased significantly due to the developments in recombinant DNA technology and the necessity to develop new methods of strain improvement as the returns generated from mutation and selection programmes decreased. However, it should be appreciated that mutation and selection techniques are frequently used in association with recombination systems in a strain improvement programme. The parasexual cycle in the filamentous fungi has been applied to strain development as have protoplast fusion techniques in a wide range of micro-organisms.

#### *Application of the parasexual cycle*

Many industrially important fungi do not posse a sexual stage and therefore it would appear difficult to achieve recombination in these organisms. However, Pontecorvo demonstrated that nuclear fusion and gene segregation could take place outside or in the absence of the sexual organs. The process was termed the parasexual cycle and has been demonstrated in the imperfect fungi. *A. niger* and *P. chrysogenum*, as well as the sexual fungus *A. nidulans*. In order for parasexual recombination to take place in an imperfect fungus, nuclear fusion must occur between unlike nuclei in the vegetative hyphae of the organism. Thus, 'recombination may be achieved only in an organism in which at least two different types of nuclei co-exist, i.e., a heterokaryon'. The heterozygous diploid nucleus resulting from the fusion of the two different haploid nuclei may give rise to a diploid clone and in rare cases, a diploid nucleus in the clone may undergo an abnormal mitosis resulting in mitotic segregation and the development of recombinant clones which may be either diploid or haploid.

Recombinant clones may be detected by their display of recessive characteristics not expressed in the heterokaryon. Analysis of the recombinants normally demonstrates them to be segregant for only one, or a few linked, markers and culture of the segregants results in the development of clones displaying more recessive characters than the initial segregant. The process of recombination during the growth of the heterozygous diploid may occur in two ways: mitotic crossing over, which results in diploid recombinants and haploidisation, which results in haploid recombinants.

Mitotic crossing over is the result of an abnormal mitosis. The normal mitosis of a heterozygous diploid cell is shown in Fig. 8.26. During mitosis, each pair of homologous chromosomes replicate to produce two pairs of chromatids and a chromatid of one pair migrates to a pole of the cell with a chromatid of the other pair. Division of the cell at the equator results in the production of two cells, both of which are heterozygous for all the genes on the chromosome. Mitotic crossing over involves the exchange of distal segments between chromatids of homologous chromosomes as shown in Fig. 8.27. This process may result in the production of daughter nuclei homozygous for a portion of one pair of chromosomes and in the expression of any recessive alleles contained in that portion. Thus, the clone arising from the partial homozygote will be recombinant and further mitotic crossing over in the recombinant will result in the expression of more recessive alleles.

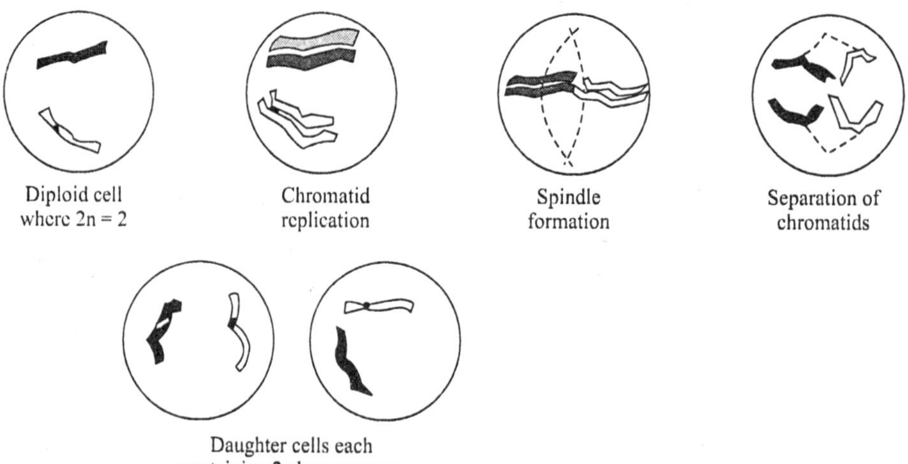


**Fig. 8.26.** Diagrammatic representation of the mitotic division of a eukaryotic cell containing two chromosomes. The nuclear membrane has not been portrayed in the figure.

Haploidisation is a process which results in the unequal distribution of chromatids between the progeny of a mitosis. Thus, of the four chromatids of an homologous chromosome pair, three may migrate to one pole and one to another resulting in the formation of two nuclei, one containing 2n + 1 chromosomes and the other containing 2n – 1 chromosomes. The 2n – 1 nuclei tend towards the haploid state by the progressive random loss of further chromosomes. Thus, the resulting haploid nucleus will contain a

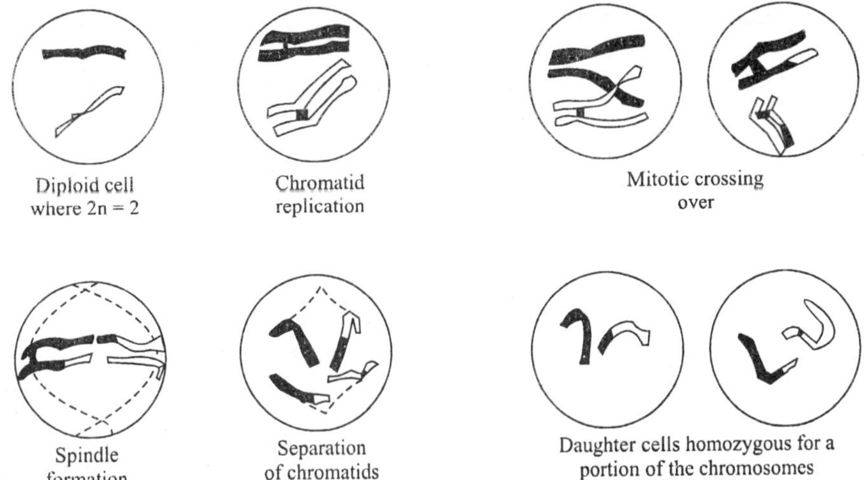


**Fig. 8.27.** Diagrammatic representation of mitosis including mitotic crossing over.

random assortment of the homologues of the chromosomes of the organism. A representation of the process is shown in Fig. 8.28.

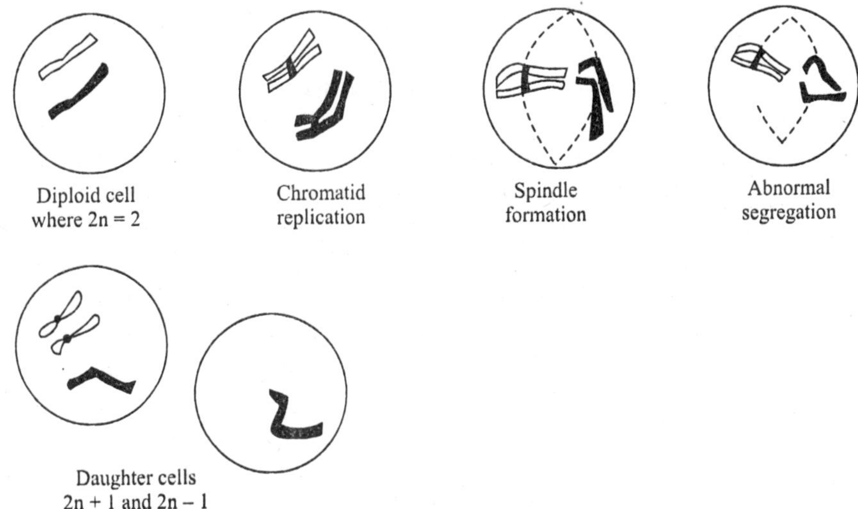


**Fig. 8.28.** Diagrammatic representation of mitosis involving haploidisation.

Therefore, the major components of the parasexual cycle are the establishment of a heterokaryon, vegetative nuclear fusion and mitotic crossing over or haploidisation resulting in the formation of a recombinant. In practice, the occurrence and detection of these stages may be enhanced by the use of auxotrophic markers. The parents of the cross are made auxotrophic for different requirements and cultured together on minimal medium. The auxotrophs will grow very slightly due to the carry over of their growth requirements from the previous media, but if a heterokaryon is produced, by anastomoses forming between the two parents, then it will grow rapidly. The frequency of vegetative nuclear fusion in the heterokaryon may be enhanced by the use of agents such as camphor vapour or ultraviolet light;

mitotic segregation may be enhanced by the use of agents such as X-rays, nitrogen mustard, *p*-fluorophenylalanine and ultraviolet light.

The application of the parasexual cycle to industrially important fungi has been hindered by a number of problems. A major difficulty is the influence of the auxotrophic markers (used for the selection of the heterokaryon) on the synthesis of the desired product. As discussed earlier, auxotrophic mutations have been observed to have quite unpredictable results on the production of some secondary metabolites. Even conidial colour markers have been shown to have deleterious effects on product synthesis. Even when suitable markers are available, the induction of heterokaryons in some industrial fungi has been demonstrated to be a difficult process and specialised techniques had to be employed to increase the probabilities of heterokaryon formation. However, the development of protoplast fusion methods in the late 1970s enabled efficient heterokaryon formation to be achieved and removed the major barrier to the application of the parasexual cycle to strain improvement. These developments are considered in more detail in the next section on protoplast fusion.

Despite the early difficulties of inducing the parasexual cycle in some industrial fungi it was used in two ways to study these organisms. The cycle was used to investigate the genetics of the producing strains as well as to develop recombinant superior producers. Information obtained on the basic genetics of the industrial fungi using the parasexual cycle included the number of chromosomes (or linkage groups), the allocation of genes, important in product synthesis, to particular chromosomes and the mapping of important genes on a chromosome.

Initial studies on haploid strains of *P. chrysogenum* derived from parasexual crosses demonstrated that most of the progeny exhibited the genotype of one of the parents, i.e., no recombination had occurred. MacDonald suggested that one of the reasons for this lack of success may have been the differences in gross chromosomal morphology between the parents caused by certain mutagens used in the development of the strains. Thus, one precaution to be adopted in the development of strains which may be used in a parasexual cross is the avoidance of mutagens which may cause gross changes in chromosomal morphology. Subsequently, Ball suggested that the careful choice of markers and the use of strains giving similar titres and being 'not too divergent' may result in achieving more recombinants.

Although diploids produced by the parasexual cycle are frequently unstable, stable diploids have been reported to have been used for the industrial production of penicillin. Elander isolated a diploid from a sister cross of *P. chrysogenum* which was shown to be a better penicillin producer than the parents and was morphologically more stable (a sister cross is one where the two strains differ only in the markers used to induce the heterokaryon). One explanation of the superior performance of the diploid may have been its resistance to strain degeneration caused by deleterious recessive mutations. Such mutations would only have been expressed in the diploid if both alleles had been modified. Calam demonstrated that a diploid strain of *P. chrysogenum* was more stable than haploid mutants and mutation and selection of the diploid gave rise to a diploid strain producing higher levels of penicillin than the parents. Ball claimed that the usefulness of diploids may only be short-term, presumably implying problems of degeneration and may best be used as stepping stones to a recombinant haploid.

The advantages to be gained from the industrial use of parasexual recombinants are not confined to the amalgamation of different yield improving mutations. Equally advantageous would be the introduction of characteristics which would make the process more economic, for example low viscosity, sporulation and the elimination of unwanted products. The development of these issues will be considered in the next section on protoplast fusion.

### *Application of protoplast fusion techniques*

Protoplasts are cells devoid of their cell walls and may be prepared by subjecting cells to the action of wall degrading enzymes in isotonic solutions. Protoplasts may regenerate their cell walls and are then capable of growth as normal cells. Cell fusion, followed by nuclear fusion, may occur between protoplasts of strains which would otherwise not fuse and the resulting fused protoplast may regenerate a cell wall and grow as a normal cell. Thus, protoplasts may be used to overcome some recombination barriers. Protoplast fusion has been demonstrated in a large number of industrially important organisms including *Streptomyces* spp., *Bacillus* spp., corynebacteria, filamentous fungi and yeasts.

Fusion of fungal protoplasts appears to be an excellent technique to obtain heterokaryons between strains where conventional techniques have failed or, indeed, as the method of choice. Thus, this approach has allowed the use of the parasexual cycle for breeding purposes in situations where it had not been previously possible. This situation is illustrated by the work of Peberdy who succeeded in obtaining heterokaryons between *P. chrysogenum* and *P. cyaneofulvum* and demonstrated the formation of diploids which gave rise to recombinants after treatment with *p*-fluorophenylalanine or benomyl. Although it has been claimed that *P. chrysogenum* and *P. cyaneofulvum* are not different species of *Penicillium*, Peberdy still demonstrated that protoplast fusion could be successful where conventional techniques had failed.

A demonstration of the use of protoplast fusion for an industrial fungus is provided by the work of Hamlyn and Ball on the cephalosporin producer, *C.acremonium*. These workers compared the effectiveness of conventional techniques of obtaining nuclear fusion between strains of *C. acremonium* with the protoplast fusion technique. The results from conventional techniques suggested that nuclear fusion was difficult to achieve. Electron microscopic examination of fused protoplasts indicated that up to 1 per cent underwent immediate nuclear fusion. Recombinants were obtained in both sister and divergent crosses. A cross between an asporulating, slow-growing strain with a sporulating fast-growing strain which only produced one-third of the cephalosporin level of the first strain eventually resulted in the isolation of a recombinant which combined the desirable properties of both strains, i.e., a strain which demonstrated good sporulation, a high growth rate and produced 40 per cent more antibiotic than the higher-yielding parent. Chang utilised protoplast fusion to combine the desirable qualities of two strains of *Penicillium chrysogenum*. Protoplasts from two strains, differing in colony morphology and the abilities to produce penicillin V (the desired product) and OH-V penicillin (an undesirable product), were fused, followed by plating on a non-selective medium. Out of 100 stable colonies which were scored, two possessed the desirable morphology, high penicillin V and low OH-V penicillin productivities.

Lein reported the penicillin strain improvement programme which included random and directed selection as well as protoplast fusion. Table 8.5 illustrates the properties of the two strains used in a protoplast fusion and one of the recombinants selected. To avoid any adverse effects no selective genetic markers were used and the regenerated colonies were screened on the basis of colony morphology and spore colour. A total of 238 colonies judged to be recombinants were screened for penicillin V production and the culture with the best combination of properties is shown in the Table 8.5. Thus, the desirable characteristics of each strain were combined in the recombinant.

Protoplasts are also useful in the filamentous fungi for manipulations other than cell fusion. Rowlands suggested that they may be used in mutagenesis of non-sporulating fungi. Spores are the cells of choice for the mutagenesis of filamentous fungi but this is obviously impossible for non-sporulating strains. Mycelial fragments may be used but these will be multinucleate and very high mutagen doses are required. Although some protoplasts will be non-nucleate or multi-nucleate at least some will be uninucleate which will express any modified genes after mutation. Also, protoplasts will take up DNA in *in vitro* genetic manipulation experiments and this aspect will be discussed in a latter section of this chapter.

**Table 8.5.** The use of protoplast fusion for the improvement of a penicillin V producer.

| *Characteristic* | *Parent A* | *Parent B* | *Best recombinant* |
|---|---|---|---|
| Spores per slant ($\times 10^8$) | 2.2 | 2.5 | 7.5 |
| Germination frequency (%) | 99 | 40 | 49 |
| Colour of sporulating colonies | Green | Pale green | Deep green |
| Seed growth | Good | Poor | Good |
| Penicillin V yield (mg $cm^{-3}$) | 11.7 | 18.5 | 18.0 |
| Phenylacetic oxidation | Yes | No | No |

Recombinant can take place between actinomycetes by conjugation and phage transduction. However, both these mechanisms involve the transfer of only small regions of the bacterial chromosome. Furthermore, the low frequencies at which recombination occurs necessitate the use of selectable genetic markers such as auxotrophy or antibiotic resistance. The introduction of such markers is time-consuming but also they can detrimentally affect the synthetic capacity of the organism. Protoplast fusion has particular advantages over conjugation in that the technique involves the participation of the entire genome in recombination. Also, Hopwood has developed techniques which have resulted in the recovery of a very high proportion of recombinants from the fusion products of *Streptomyces coelicolor* protoplasts. By subjecting protoplasts to an exposure of ultraviolet light, sufficient to kill 99 per cent of them prior to fusion, Hopwood has claimed a ten-fold increase in recombinant detection for strains normally giving a low yield of recombinants (1 per cent) and a doubling of the recombination of frequency for a cross normally yielding 20 per cent recombinants. Such yields of recombinants means that they would be detectable by simply screening a random proportion of the progeny of a protoplast fusion and the use of selectable markers to 'force' out the recombinants would not be necessary. Examples of the application of the technique to actinomycete strain improvement include cephamycin C yield enhancement in *Nocardia lactamdurans* and the improvement of lignin degradation in *Streptomyces viridosporus*.

Protoplast fusion has also been applied to the improvement of amino acid producing strains. Karasawa used the technique to improve the fermentation rates of lysine producers developed using repeated mutation and directed selection. Such strains were good lysine producers but showed low glucose consumption and growth rates, undesirable features which had been inadvertently introduced during the selection programme. A protoplast fusion was performed between the lysine producer and a fast growing strain and a fusant was isolated displaying the desirable characteristics of high lysine production and high glucose consumption rate resulting in a much faster fermentation. The same authors used protoplast fusion to produce a superior threonine producing *Brevibacterium lactofermentum* strain. Lysine auxotrophy was introduced into a threonine and lysine overproducer by fusing it with a lysine auxotroph—the recombinant produced higher levels of threonine due to its lysine auxotrophy.

### *Application of recombinant DNA techniques*

The transfer of DNA between different species of bacteria has been achieved experimentally using both *in vivo* and *in vitro* techniques. Thus, genetic material derived from one species may be incorporated into another where it may be expressed. *In vivo* techniques make use of phage particles which will pick up genetic information from the chromosome of one bacterial species, infect another bacterial species and in so doing introduce the genetic information from the first host. The information from the first host may then be expressed in the second host. Whereas, the *in vivo* techniques depend on vectors collecting information from one cell and incorporating it into another, the *in vitro* techniques involve the insertion of the information into the vector by *in vitro* manipulation followed by the insertion of the carrier and its associated 'extra' DNA into the recipient cell. Because the DNA is incorporated into the vector by *in vitro* methods the source of the DNA is not limited to that of the host organism of the vector. Thus, DNA from human or animal cells may be introduced into the recipient cell. Atherton listed the basic requirements for the *in vitro* transfer and expression of foreign DNA in a host micro-organism as follows:

1. A 'vector' DNA molecule (plasmid or phage) capable of entering the host cell and replicating within it. Ideally the vector should be small, easily prepared and must contain at least one site where integration of foreign DNA will not destroy an essential function.
2. A method of splicing foreign genetic information into the vector.
3. A method of introducing the vector/foreign DNA recombinants into the host cell and selecting for their presence. Commonly used simple characteristics include drug resistance, immunity, plaque formation or an inserted gene recognisable by its ability to complement a known auxotroph.
4. A method of assaying for the 'foreign' gene product of choice from the population of recombinants created.

The initial work focused on *E. coli* but subsequently techniques have been developed for the insertion of foreign DNA into a range of bacteria, yeasts, filamentous fungi and animal cells. The range of vectors has been discussed by Gingold for bacteria, Curran and Bugeja for yeasts, Hopwood for streptomycetes, Elander for filamentous fungi and Murray for animal cells. The insertion of information into the vector molecule is achieved by the action of restriction endonucleases and DNA ligase. Site-specific endonucleases produce specific DNA fragments which may be joined to another similarly treated DNA molecule using DNA ligase. The modified vector is then normally introduced into the recipient cell by transformation. Because the transformation process is an inefficient one, selectable genes must be incorporated into the vector DNA so that the transformed cells may be cultured preferentially from the mixture of transformed and parental cells. This is normally accomplished by the use of drug-resistant markers so that those cells containing the vector will be capable of growth in the presence of a certain antimicrobial agent. The process is shown diagrammatically in Fig. 8.29.

Once the desired gene has been introduced into the recipient cell the problem of expression of the gene arises. This is particularly difficult when a eukaryotic gene is introduced into a bacterium. In the late 1970s a large number of mammalian genes were successfully introduced into bacterial cells, but there was little evidence of any gene expression. The problem of eukaryotic gene expression in prokaryotic cells is due to the different structure of eukaryote genes which contain non-coding segments of DNA. Thus, the production of a eukaryotic product by a prokaryotic cell necessitates the incorporation of the genes coding for the product in a form that may be translated by the recipient cell. Two approaches have been adopted to construct eukaryotic genes in a prokaryotic form: the first is to synthesise DNA corresponding to the primary structure of the protein product of the gene, although this method is suitable only for the construction of genes coding for small peptides of known structure. Itakura synthesised the gene coding for the human hormone, somatostatin and succeeded in incorporating it in *E. coli* where it was expressed. The alternative technique is to synthesise DNA from the messenger RNA, corresponding to the gene, using the enzyme, reverse transcriptase.

Eukaryote messenger RNA is similar to bacterial RNA in that it does not contain non-coding segments so that DNA synthesised from an eukaryotic messenger RNA template should be in a form which is transcribable by a prokaryotic cell. Nagata used the reverse transcriptase method to produce the genes coding for human interferon. Complementary single-stranded DNA was prepared from a mixture of messenger RNA extracted from virus-induced human lymphocytes. The DNA was then introduced into *E. coli* at random using a plasmid vector and the recipient cells screened for the presence of the interferon gene. Those colonies that contained the gene were then examined for interferon production. Using this method, Nagata's group succeeded in demonstrating the expression of the interferon gene in *E. coli*.

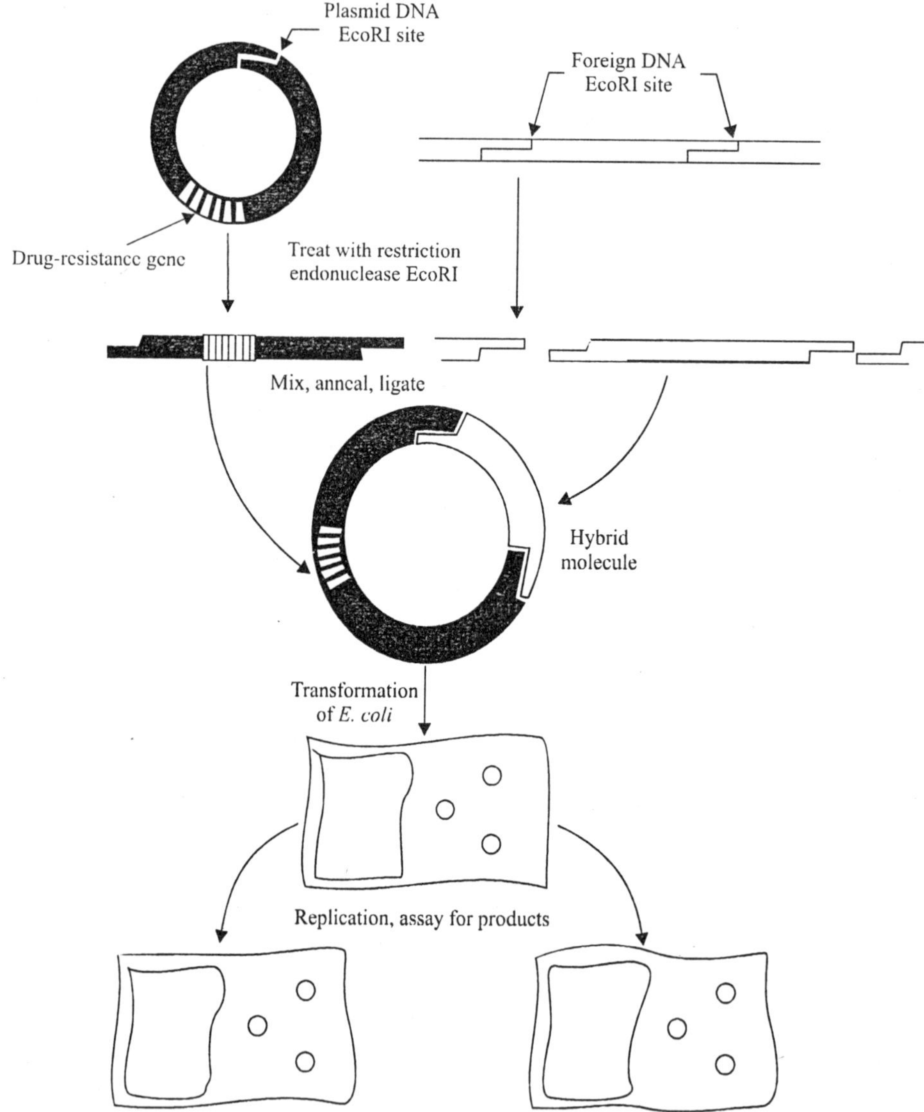


**Fig. 8.29.** A summary of the steps in *in vitro* genetic recombination. Both plasmid vector and foreign DNA are cut by the restriction endonuclease, EcoRI, producing linear double-stranded DNA fragments with single-stranded cohesive projections. EcoRI recognises the oligonucleotide sequence GAATIC/GTTAAG and will cut any double-stranded DNA molecule to yield fragments with the same cohesive ends GAATT/C, C/TTAAG. On mixing vector and foreign DNA, hybrids form into circular molecules which can be covalently joined using DNA ligase. Transformation of *E. coli* results in the low-frequency uptake of hybrid molecules whose presence can be detected by the ability of the plasmid to confer drug resistant on the host.

The most publicised application of recombinant DNA technology in the context of fermentation is the construction of strains capable of synthesising foreign proteins. Although this chapter is concerned with strain improvement it is the obvious place to consider such chimeric strains. The use of the technique for the improvement of microbial product synthesis has also been very successful and this will be discussed in a later in this chapter.

*Production of heterologous proteins*

The rationale for the commercial production of foreign proteins in micro-organisms depends on the protein under consideration. The first commercial heterologous protein to be produced was human growth hormone (hGH) which is used to treat hypopituitary dwarfism and prior to its manufacture by fermentation, was extracted from the brains of human cadavers. Naturally, this source was not readily available and carried the additional disadvantage of the risk of contamination with human pathogens. The successful production of recombinant hGH from *E. coli* both satisfied the demand for the compound and eliminated the risks associated with the human source. Factor VIII is a blood clotting agent used in the treatment of haemophilia. Prior to its production as a recombinant protein it was extracted from human blood with the associated risk of contamination with HIV, so that the logic behind its production by fermentation is very similar to that of hGH—availability and safety.

The logic for the development of recombinant human insulin is not quite as clear because diabetes had been treated successfully for many years with animal insulin. However, it was assumed that the recombinant human product would cause fewer immunological difficulties and it would be pure and not contaminated with such pancreatic peptides as proinsulin, glucagon, somatostatin, pancreatic polypeptides and vaso-active intestinal peptides. Furthermore, the incidence of diabetes was also expected to increase due to changes in diet, the improved care of pregnant diabetic women (resulting in an increase of diabetes in the gene pool) and the increased life expectancy of diabetics.

Many other human proteins are synthesised at very low levels and the only practical way to produce them in sufficient quantities for use as therapeutic agents is as recombinant proteins. Examples of such products include the interferons and erythropoietin. Table 8.6 lists the recombinant proteins which have been licensed for therapeutic use. According to Dykes although insulin and hGH have been two of the most successful products some newer products have greater potential. For example, erythropoietin stimulates the production of erythrocytes from immature erythroid progenitor cells and has been used for the successful treatment of renal failure-induced anaemia.

**Table 8.6.** Recombinant proteins licensed for therapeutic use.

| *Protein* | *Clinical use* |
|---|---|
| Insulin | Diabetes |
| Growth hormone | Hypopituitary dwarfism |
| Tissue plasminogen activator | Clot lysis |
| Erythropoietin | Anaemia |
| G-CSF | Cancer chemotherapy |
| GM-CSF | Bone marrow transplantation |
| Factor VIII | Haemophilia |
| Interferon $\alpha$ | Cancers, hepatitis B, leukaemia |
| Interferon $\beta$ | Cancers, amyotrophic lateral sclerosis, genital warts |
| Interferon | Cancers, AIDS-related complex, osteopetrosis |
| Hepatitis B surface antigen | Hepatitis vaccine |

Hepatitis B virus causes an infection of the liver giving rise to chronic viral hepatitis which may lead to progressive liver disease. There is no effective cure of the disease and thus, vaccination is critical. The first hepatitis B vaccine was prepared from hepatitis B surface antigen (HBsAg) purified from the plasma of human carriers of the disease. Although the vaccine was successful its origin obviously presented the same difficulties of availability and contamination as discussed for hGH and Factor VIII. These difficulties provided the impetus to develop a recombinant vaccine which was achieved by expressing the HBsAg gene in *Saccharomyces cerevisiae*.

The vast majority of the early work on recombinant DNA technology concerned the transfer of genetic material into *E. coli* and this led to the adoption of the organism as the host for the production of several heterologous proteins. The bacterium had several advantages in its favour, primarily:

1. The availability of genetic knowledge.
2. The very wide range of vectors available.
3. A simple fermentation process using cheap media.
4. Promising protein yields in the range of 2–5 g $dm^{-3}$.

Despite these advantages *E. coli* also presents several problems for the production of heterologous proteins:

1. Proteins are formed as insoluble aggregates.
2. The proteins are not secreted.
3. Lack of post-translational modification.

By the mid-1980s the use of *E. coli* had declined considerably and yeasts, filamentous fungi and animal cells in culture were being investigated as alternative hosts. However, by 1992 *E. coli* was once again in favour. The protein yields obtained from animal cells were disappointing and the understanding of the secretion and protein folding had progressed such that soluble proteins could be secreted by engineered *E. coli*. However, mammalian cells are still the preferred host when the protein activity depends upon post-translational modification. Whichever organism is used for the synthesis of an heterologous protein it is important that the strain is constructed such that expression of the gene may be controlled during the fermentation process. This may be achieved by inserting the gene into an inducible system, such as β-galactosidase, so that expression of the gene may be initiated by the addition of the inducer.

As an example of the production of an heterologous protein the development of the Genencor process for the production of recombinant chymosin will be considered. Bovine chymosin is an aspartyl protease extracted from the abomasum of unweaned calves and is used as a milk-clotting agent in the manufacture of cheese. Its natural source means that the enzyme is available only in limited quantities which made the possibility of producing it as an heterologous protein an attractive proposition. This process utilises the filamentous fungus *Aspergillus niger* var. *awamori* which was claimed to have the following advantages as a potential host organism:

1. It is capable of secreting large amounts of protein.
2. The organism is regarded as safe.
3. Transformation systems were available.

Secretion from a eukaryotic cell is achieved via the endoplasmic reticulum and the Golgi apparatus. Entry into the secretory pathway is determined by the presence of a short, hydrophobic 'signal' sequence on the *N*-terminal end of secreted proteins. The signal sequence directs the protein to sites on the endoplasmic reticulum (ER) from where transport into the lumen of the ER occurs. The signal sequence is cleaved by a luminal signal peptidase and the protein passes through the ER and Golgi body, before being packaged into secretory vesicles and exported beyond the cell membrane.

Several expression vectors were constructed incorporating the organism's glucoamylase gene and prochymosin cDNA. The inclusion of the prochymosin cDNA between the glucoamylase promoter and terminator resulted in a strain which accumulated only 15 mg $dm^{-3}$ chymosin in the medium despite high intracellular mRNA and chymosin levels. This implied that the enzyme was not being secreted. To facilitate secretion, the prochymosin cDNA was incorporated between the glucoamylase coding region and the glucoamylase terminator. In this case, the organism should synthesise a 'fusion protein' consisting of chymosin attached to the carboxyl terminus of glucoamylase. The fusion protein was secreted and cleaved autocatalytically liberating the chymosin which was produced at 150 mg $dm^{-3}$.

It is fascinating to appreciate that further improvement of this strain has been achieved by mutant induction and selection. Mutagenesis followed by random selection resulted in the isolation of a superior producer which was subsequently shown to produce very low amounts of the native aspartyl protease. Armed with this information the Genencor group cloned the aspartyl protease and then deleted the gene from the strain. Thus, the insight gained from the random screen allowed the directed *in vitro* approach to be used. The improved strain produced 250 mg $dm^{-3}$ chymosin. This strain was then used in further mutation and selection programmes. An automated microtitre plate screen was developed which has already been described in an earlier section of this chapter and the results are as shown in Table 8.4.

The strains developed using the automated screen were then mutated and subjected to a directed screen. Spores which survived nitrosoguanidine exposure were plated onto glycerol media containing 1 per cent deoxyglucose. Deoxyglucose is a toxic compound which Allen used to isolate resistant mutants of *Neurospora crassa* capable of producing higher levels of extracellular enzymes normally repressed by glucose. The best resistant strains (termed dgr) produced in excess of 1200 mg $dm^{-3}$. Parasexual crosses between improved strains showed that two different loci were involved in resistance to deoxyglucose. The chymosin story is an excellent example of the integration of recombinant DNA technology, mutant induction and automated random selection, directed selection and parasexual genetics in the development of a commercial strain.

### *Use of recombinant DNA technology for the improvement of native microbial products*

Recombinant DNA technology has been used widely for the improvement of native microbial products. Frequently, this has involved 'self cloning' work where a chromosomal gene is inserted into a plasmid and the plasmid incorporated into the original strain and maintained at a high copy number. Thus, this is not an example of recombination because the engineered strain is altered only in the number of copies of the gene and does not contain genes which were present originally in a different organism. However, the techniques employed in the construction of these strains are the same as those used in the construction of chimeric strains, so it is logical to consider this aspect here.

The first application of gene amplification to industrial strains was for the improvement of enzyme production. Indeed, some regulatory mutants isolated by conventional means owed their productivity to their containing multiple copies of the relevant gene as well as the regulatory lesion. For example, the *E. coli* β-galactosidase constitutive mutants isolated by Horiuchi in chemostat culture also contained up to four copies of the *lacZ* gene. According to Demain, during the 1960s and early 1970s the number of gene copies were increased by using plasmids or transducing phage in the same species. The production of β-galactosidase, penicillinase, chloramphenicol transacetylase and aspartate transcarbamylase were all increased by transferring plasmids containing the structural gene into recipient cultures, especially when the plasmid replicated faster than the host chromosome. The advent of recombinant DNA technology increased the applicability of this approach by allowing the construction of vectors containing the desired gene and enabling the transfer of DNA to other species. Table 8.7 includes examples of gene amplification giving rise to improved enzyme yields.

**Table 8.7.** Examples of the enhancement of enzyme production by gene amplification.

| *Enzyme* | *Gene donor* | *Gene recipient* | *Increase (fold)* |
|---|---|---|---|
| α-Amylase | *Bacillus amyloliquefaciens* | *Bacillus subtilis* | 10 |
| α-Amylase | *Bacillus stearothermophilus* | *Bacillus stearothermophilus* | 5 |
| α-Amylase | *Bacillus stearothermophilus* | *Bacillus brevis* | 100 |
| *Eco*RI restriction endonuclease | *E. coli* | *E. coli* | 50–100 |
| DNA polymerase | *E. coli* | *E. coli* | 100 |

The development of the process for the production of a lipase for use in washing powders is an excellent example of the application of recombinant DNA technology to enzyme fermentations. A lipase was isolated from a fungus which was unsuitable for commercial development. A cDNA clone coding for the lipase was prepared and transformed into an industrial *Aspergillus* strain used for enzyme production. The recombinant *Aspergillus* produced the lipase at high levels. Only 8 months elapsed between cloning the gene and the first commercial fermentation. Thus, recombinant DNA technology allowed the exploitation of a valuable microbial enzyme by facilitating its production in a well-established commercially acceptable strain.

The first successful application of genetic engineering techniques to the production of amino acids was obtained in threonine production with *E. coli*. Debabov investigated the production of threonine by a threonine analogue resistant mutant of *E. coli* K12. The entire threonine operon was introduced into a plasmid which was then incorporated into the organism by transformation. The plasmid copy number in the cell was approximately twenty and the activity of the threonine enzymes (measured as homoserine dehydrogenase activity) was increased 40–50 times. The manipulated organism produced 30 g $dm^{-3}$ threonine, compared with 2–3 g $dm^{-3}$ by the non-manipulated strain. Miwa utilised similar techniques in constructing an *E. coli* strain capable of synthesising 65 g $dm^{-3}$ threonine. It is important to appreciate that the genes which were amplified in these production strains were already resistant to feedback repression so that the multi-copies present in the modified organism were expressed and not subject to control. Thus, the recombinant DNA techniques built on the achievements made with directed mutant isolation.

The application of genetic engineering to the industrially important corynebacteria was hampered for some years by the lack of suitable vectors. However, vectors have been constructed from corynebacterial plasmids and transformation and selective systems developed. Transformation has been achieved using protoplasts and more recently, electroporation has been used successfully to introduce the required DNA. These systems have enabled not only the use of recombinant DNA technology for strain improvement but have also facilitated the detailed investigation of the molecular biology of these important amino acid and nucleotide producers; for example the molecular organisation of the pathway to the aspartate family of amino acids in *C. glutamicum* has been elucidated.

It is not surprising that the improvement of threonine production was the first reported use of recombinant DNA technology with amino acid producing corynebacteria. It may be recalled from the discussion of the development of the early threonine producers that *C. glutamicum* was not particularly amenable for threonine over-production using auxotrophs and analogue resistant mutants. Over-production was achieved by incorporating a DNA fragment coding for homoserine dehydrogenase from a *Brevibacterium lactofermentum* threonine producer into a plasmid and introducing the modified plasmid back into the *Brevibacterium*. A similar approach was used for homoserine kinase and a strain was developed with remarkably increased homoserine kinase and dehydrogenase activities which produced 33 g $dm^{-3}$ threonine. The application of recombinant DNA technology to the development of processes for the production of phenylalanine is an excellent illustration of the interrelationship between mutant development and genetic engineering. Phenylalanine is a precursor of the sweetener, aspartame and is thus an exceptionally important fermentation product. Backman described the rationale used in the construction of an *E. coli* strain capable of synthesising commercial levels of phenylalanine. *E. coli* was chosen as the producer because of its rapid growth, the availability of recombinant DNA techniques and the extensive genetic database.

The control of the biosynthesis of the aromatic family of amino acids in *E. coli* is shown in Fig. 8.30. The first step in the pathway is catalysed by three isoenzymes of dihydroxyacetone phosphate (DAHP) synthase, each being susceptible to one of the three end products of the aromatic pathway, phenylalanine, tyrosine or tryptophan. Control is achieved by both repression of enzyme synthesis and inhibition of enzyme activity. Within the common pathway to chorismic acid the production of shikimate kinase is also susceptible to repression. The conversion of chorismic acid to prephenic acid is catalysed by two isoenzymes of chorismate mutase, each being susceptible to feedback inhibition and repression by one of either tyrosine or phenylalanine. Each isoenzyme also carries an additional activity associated with either the phenylalanine or tyrosine branch. The tyrosine sensitive isoenzyme carries pre-phenate dehydrogenase activity (the next enzyme in the route to tyrosine) whilst the phenylalanine sensitive enzyme carries pre-phenate dehydratase (the next enzyme in the route to phenylalanine).

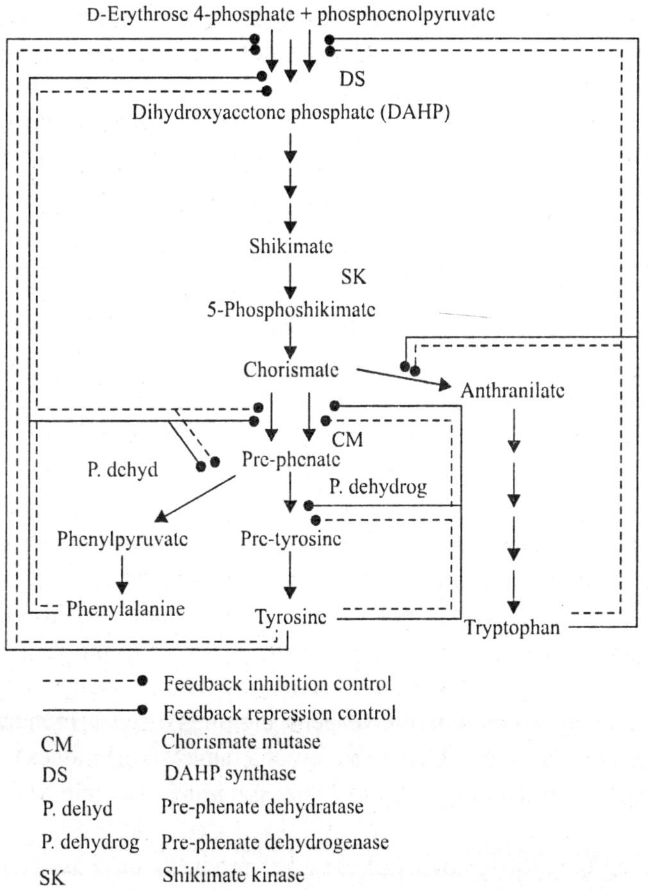


**Fig. 8.30.** Control of the aromatic amino acid family in *Escherichia coli*.

The regulation of gene expression was modified by:

1. Both tyrosine sensitive and phenylalanine sensitive DAHP synthase and shikimic kinase are regulated by the repressor protein coded by the *tyr*R gene. The *tyr*R gene had been cloned and was subjected to *in vitro* mutagenesis and the wild type gene replaced. This one mutation then resulted in the derepression of both tyrosine and phenylalanine DAHP synthase as well as shikimic kinase.
2. The chorismic mutase/pre-phenate dehydratase protein is under both repression and attenuation control by phenylalanine. These controls were eliminated by replacing the normal promoter with one not containing the regulatory sequences, thus giving a ten times increase in gene expression.

The regulation of enzyme inhibition was modified by:

1. Chorismic mutase/pre-phenate dehydratase is subject to feedback inhibition by phenylalanine. It was shown that a certain tryptophan residue was particularly important in the manifestation of feedback inhibition. *In vitro* mutagenesis was used to delete the tryptophan codon and the modified gene introduced along with the substituted promoter referred to earlier. The enzyme produced by the modified gene was no longer susceptible to phenylalanine.
2. The rationale was to limitOA the availability of tyrosine such that the tyrosine-sensitive DAHP synthase isoenzyme would not be inhibited. Although the other two isoenzymes would still be susceptible to phenylalanine and tryptophan inhibition, sufficient DAHP would be synthesised by the third isoenzyme (in the absence of tyrosine) to facilitate overproduction of phenylalanine. Traditionally, this objective would have been achieved by using a tyrosine auxotroph fed with limiting tyrosine. However, these workers developed an excision vector system. An excision vector is a genetic element that can carry a cloned gene and can both integrate into and be excised from, the bacterial chromosome. The vector is based on the bacteriophage lambda and the technology is explained in Fig. 8.31. The excision of the vector can be induced by a temperature shock. Thus, when the vector is excised the progeny of the cell will lose the inserted DNA. The *tyr*A gene was deleted from the production strain, inserted into the excision vector and transformed back into the organism where the vector became integrated into the chromosome. The fermentation could then be conducted using a cheap medium (the organism was not auxotrophic at inoculation) and allowing growth to an acceptable density. A heat shock may then be used to initiate vector excision, tyrosine auxotrophy and hence, phenylalanine synthesis.

These efforts should have generated a high-producing strain. However, the tyrosine-sensitive DAHP synthase was susceptible to inhibition by high concentrations of phenylalanine. It will be recalled that the flow of DAHP was intended to come from the deregulated tyrosine sensitive isoenzyme. Thus, the final step in the development of the strain was to render this isoenzyme resistant to phenylalanine inhibition. This was achieved by the selection of mutants resistant to phenylalanine analogues. Thus, the strain was improved using a combination of gene cloning, *in vitro* mutagenesis and analogue resistance, indicating the importance of the contribution of a range of techniques to strain development. The final strain was capable of producing 50 g $dm^{-3}$ phenylalanine at a yield of 0.23 g $g^{-1}$ glucose and 2 g $g^{-1}$ biomass.

The organism produced very low amounts of the other products and intermediates of the pathway which the workers claimed was due to the very precise manipulation of the strain which avoided the concomitant adverse characteristics associated with many highly mutated organisms.

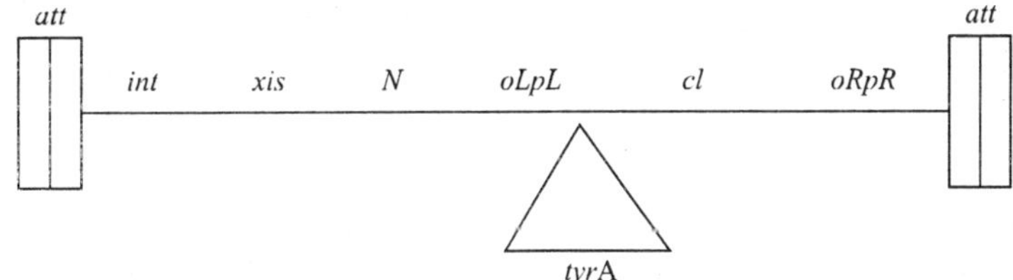


**Fig. 8.31.** Excision vector technology. An excision vector is represented as a line between two boxes. Above the line and boxes are indicated those genes from bacteriophage lambda that are carried on the vector. oRpR is required for the expression of a repressor determined by *cl*. This repressor binds to *oRpR* and *oLpL* to prevent the expression of other lambda genes. The other genes collectively act to form a recombination activity that allows the vector to integrate into or excise from the bacterial chromosome at the sites (indicated by the boxes) named *att*. Other genes, such as *tyr*A (indicated below the excision vector), can be cloned into the excision vector prior to its introduction into a target cell and can thereby be present or absent in the cell in co-ordination with the vector. Upon entering a cell, the genes for recombination (*N, xis* and *int*) are expressed because there is not yet any repressor. As repressor accumulates, it shuts off the expression of those genes. In a fraction of the recipient cells, the recombination enzymes cause the vector to integrate into the cell chromosome before those enzymes decay away. That cell and all of its progeny inherit the vector and any gene it might carry (such as *tyr*A). If the repressor is inactivated, such as by high temperature, new recombination enzymes are formed that excise the vector from the chromosome. In such a cell and all of its progeny, the vector and the gene(s) it might carry are lost.

Ikeda and Katsumata re-designed a tryptophan producing *C.glutamicum* strain such that it overproduced either phenylalanine or tyrosine. The regulation of the aromatic pathway in *C. glutamicum* is shown in Fig. 8.32. Phenylalanine producers which were resistant to phenylalanine analogues were used as sources of genes coding for enzymes resistant to control. Thus, a plasmid was constructed containing genes coding for the deregulated forms of DAHP synthase, chorismate mutase and prephenate dehydratase. The vector was introduced into a tryptophan overproducer having the following features:

1. Chorismate mutase deficient and therefore, auxotrophic for both tyrosine and phenylalanine.
2. Wild type DAHP synthase.
3. Anthranilate synthase partially desensitised to inhibition by tryptophan.

The transformed strain was capable of producing 28 g $dm^{-3}$ phenylalanine and the levels of all three enzymes coded for by the vector were amplified approximately seven-fold. If the original tryptophan producing strain was transformed with a plasmid containing only DAHP synthase and chorismate mutase, then tyrosine was overproduced (26 g $dm^{-3}$). Thus, genetic engineering techniques allowed a tryptophan producer to be re-designed into either a phenylalanine or tyrosine producer.

The development of *B. sphaericus* strains overproducing biotin was discussed in our consideration of the use analogue resistant mutants. Although *B. sphaericus* is an intrinsically good biotin producer one of the limitations to its commercial exploitation is its complex nutritional requirements. It uses carbon sources such as glucose inefficiently and requires complex organic nitrogen sources. Thus, workers have cloned the biotin genes from *B. sphaericus* into *E. coli*. Again, it is important to appreciate that it was the analogue resistant genes that were cloned allowing deregulated production in a more commercially amenable host.

The two gene cassettes which were cloned were under the control of inducible promoters so that production could be initiated after the growth phase and therefore maintain plasmid stability. Yields of up to 45 mg $dm^{-3}$ were achieved.

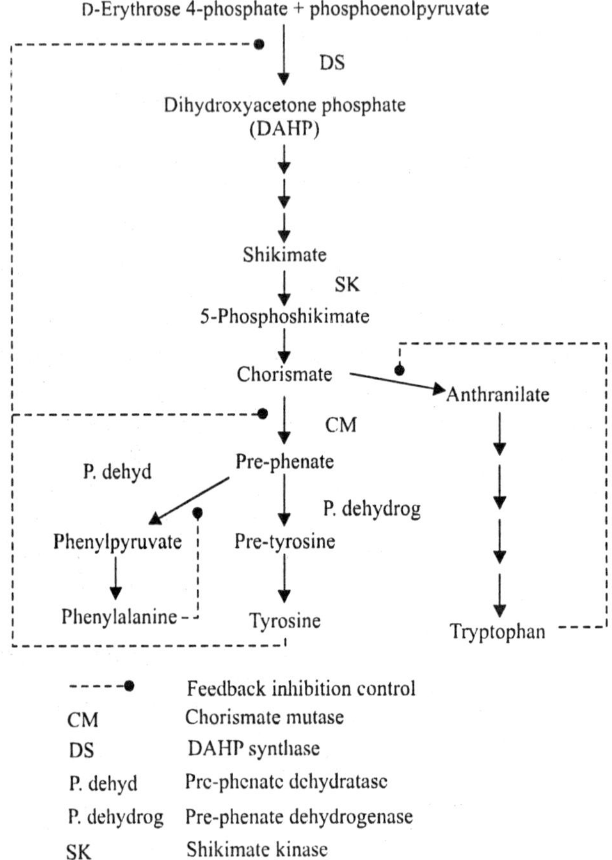


**Fig. 8.32.** Control of the aromatic amino acid family in *Corynebacterium glutamicum.*

A different application of recombinant DNA technology is seen in the modification of the ICI plc Pruteen organism, *Methylomonas methylotrophus*. The efficiency of the organism's ammonia utilisation was improved by the incorporation of a plasmid containing the glutamate dehydrogenase gene from *E. coli*. The wild type *M. methylotrophus* contained only the glutamine synthetase/glutamate synthase system which, although having a lower $K_m$ value than glutamate dehydrogenase, consumes a mole of ATP for every mole of $NH_3$ incorporated. Glutamate dehydrogenase, on the other hand, has a lower affinity for ammonia but does not consume ATP. In the commercial process ammonia was in excess because methanol was the limiting substrate so the expenditure of ATP in the utilisation of ammonia was wasteful of energy. The manipulated organism was capable of more efficient $NH_3$ metabolism, which resulted in a 5 per cent yield improvement in carbon conversion. However, the strain was not used in the industrial process due to problems of scale-up.

The application of *in vitro* recombinant DNA technology to the improvement of secondary metabolite production may not be as advanced as it is for primary metabolites, but it has made a very significant contribution. Techniques have been developed for the genetic manipulation of streptomycetes and the filamentous fungi and a number of different strategies have been devised for cloning secondary metabolism genes. In all the streptomycete systems so far studied the genes for the biosynthesis of a secondary metabolite are clustered. Furthermore, these clusters also contain the genes for regulation, resistance, export and extracellular processing. Work of this type has not only increased the basic understanding of the molecular genetics of secondary metabolism, but it has also facilitated strain improvement.

An excellent example of the application of recombinant DNA technology to the improvement of secondary metabolite production is provided by the work of the Lilly Research Laboratories group. These workers attempted to increase cephalosporin C synthesis by *Cephalosporium acremonium* by increasing the gene dosage at limiting steps in the pathway. Four critical steps were involved:

1. Identifying the biochemical rate limiting step in the cephalosporin C industrial fermentation.
2. Cloning the gene coding for the enzyme catalysing the rate limiting step.
3. Constructing a vector containing the cloned gene and introducing it into the production strain.
4. Screening the transformants for increased productivity.

The biosynthetic route to cephalosporin C is shown in Fig. 8.33. Analysis of fermentation broths showed an accumulation of penicillin N which indicated that there was a bottleneck at the next reaction, i.e., the ring expansion step where the 5-membered penicillin N ring is expanded to the 6-membered ring of deacetoxycephalosporin C. This step is catalysed by deacetoxycephalosporin C synthase (DAOCS, commonly called expandase) coded for by the gene *cef* EF. *cef* EF also codes for the next enzyme is the route, deacetylcephalosporin C synthase (DACS, commonly called hydroxylase). Thus, if the expandase step were rate limiting, introduction of extra copies of *cef* EF should relieve the limitation and eliminate the accumulation of penicillin N.

The production strain of *C. acremonium* was transformed with a plasmid containing an exact copy of *cef* EF and the transformants screened for increased productivity. Approximately one in four transformants were superior producers and one produced almost 50 per cent more antibiotic in a laboratory scale fermentation. Analysis of the transformant showed that a single copy of the transforming DNA had integrated into chromosome III, whereas native *cef* EF resides in chromosome II. In pilot scale fed-batch fermentations the transformant showed a 15 per cent increase in yield which is still a very significant increase for an industrial strain. As predicted, the transformant did not accumulate penicillin N and the bottleneck appeared to have been relieved. The superior strain was among the first eight transformants examined, whereas 10000 survivors of a mutagen exposure rendered no improved types. This work illustrates the enormous potential of recombinant DNA technology for secondary metabolite yield improvement, but it must be appreciated that such a programme involves a very considerable financial investment and the product must be sufficiently valuable to enable a good return to be realised on that investment.

The major difficulty to be overcome in the use of genetically manipulated cultures is their potential instability in large-scale culture, especially continuous process. The genetically manipulated strain of *Methylomonas methylotrophus* appears to be stable in continuous culture, but this is not surprising since the modification renders the cell more efficient and thus, selection force in the chemostat would tend to work to its advantage. However, strains which have been manipulated to produce products which give the cell no selective advantage would be selected against in long-term culture, especially in chemostat systems. Thus, the development of techniques to stabilise manipulated cultures and the incorporation of selectable marker genes consistent with the use of cheap fermentation media are critical to the successful fulfilment of much of the promise of genetic-manipulation techniques. The improved cephalosporin C producer described earlier was transformed with an integrating vector which means that the transformant is stable even in the absence of selective pressure.

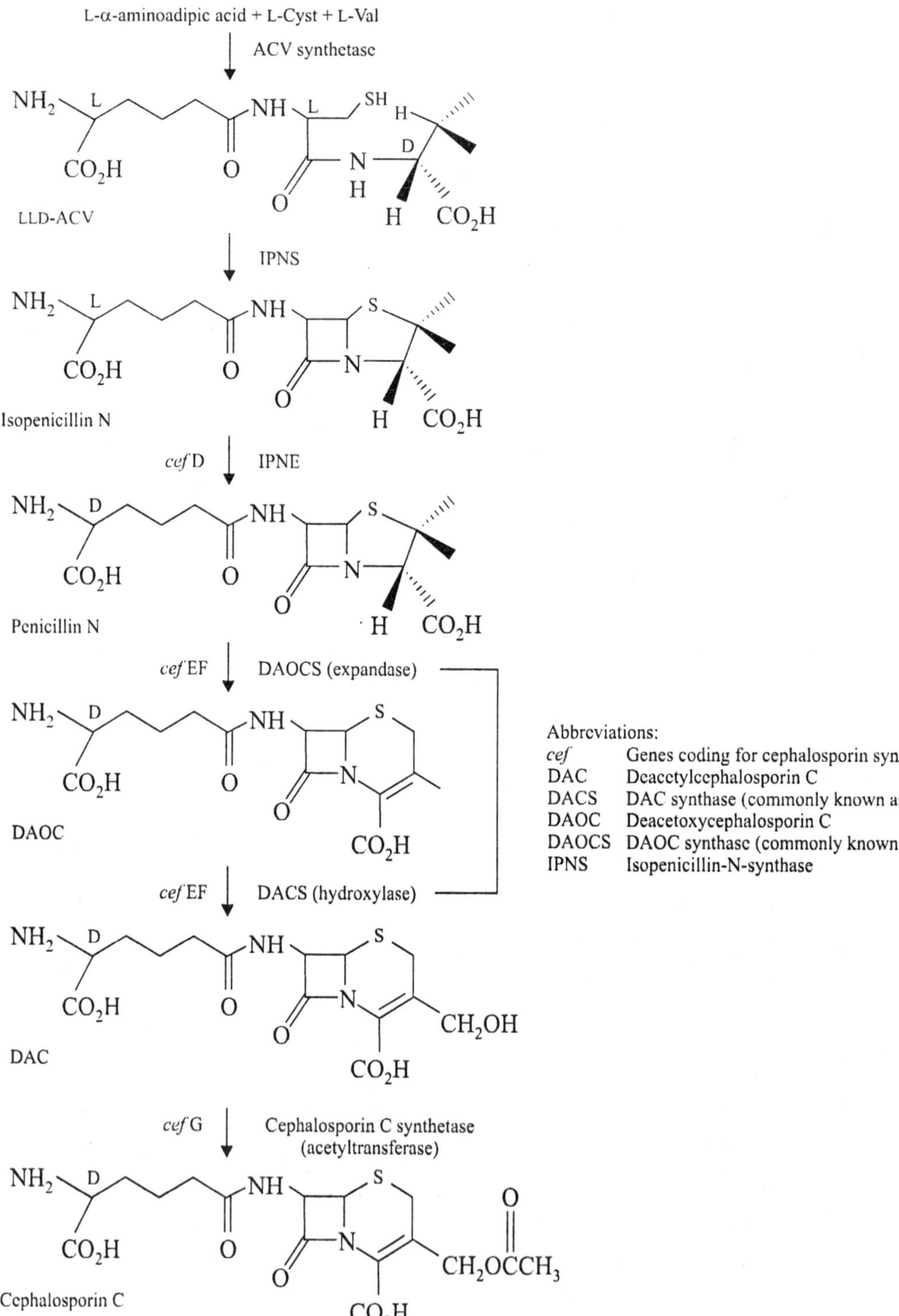


Abbreviations:

| | |
|---|---|
| *cef* | Genes coding for cephalosporin synthesis |
| DAC | Deacetylcephalosporin C |
| DACS | DAC synthase (commonly known as hydroxylase) |
| DAOC | Deacetoxycephalosporin C |
| DAOCS | DAOC synthase (commonly known as expandase) |
| IPNS | Isopenicillin-N-synthase |

**Fig. 8.33.** The biosynthetic route to cephalosporin C in *Cephalosporium acremonium*.

## Improvement of Industrial Strains by Modifying Properties other than the Yield of Product

The previous sections of this Chapter have considered ways of increasing the yields of metabolites produced by industrial micro-organisms. However, the design and economics of a commercial process are influenced by properties of the organism other than its productivity. For example, although a strain may produce a very high level of a metabolite it would be unsuitable for a commercial process if its productivity were extremely unstable or if the organism's oxygen demand were such that it could not be satisfied in the industrial fermenter available for the process. Therefore, characteristics of the producing organism which affect the process may be critical to its commercial success. Thus, it may be desirable to modify such characteristics of the producing organism which may be achieved by selecting natural and induced variants and recombinants. Naturally, it is crucial that the modified strain retains its desirable productivity so that the screening involved in these procedures should include assay of the yield as well as the characteristics being selected. Some examples of the characteristics which may be important in the context are: strain stability, resistance to phage infection, response to dissolved oxygen, tolerance of medium components, the production of foam, the production of undesirable by-products and the morphological form of the organism.

### *Selection of stable strains*

The ability of the producing strain to maintain its high productivity during both culture maintenance and a fermentation is a very important quality. Yield decay during culture storage may be avoided by the use of maintenance techniques such as those discussed earlier in this Chapter, but loss of productivity during the fermentation is far more difficult to control. A decrease in the productivity of a commercial strain is normally due to the occurrence of lower-yielding, spontaneous revertant mutants which frequently have a higher growth rate than the high-producing parent, so that yield decay is especially problematical in long-term fermentations such as fed-batch or continuous culture where the faster-growing, lower producer may predominate or even replace, the high-producing original strain. This situation is illustrated by the commercial, amino acid-producing organisms, many of which are insufficiently stable to be used in continuous-culture processes. Many workers have attempted to control the stability of amino acid-producing strains. As may be seen from the previous section on the isolation of mutants overproducing primary metabolites the amino acid producers tend to be auxotrophs, analogue-resistant mutants or revertants; such mutations removing the normal mechanisms controlling the production of amino acids. The introduction of more than one mutation giving the same phenotype may give a more stable strain since all the mutations would have to revert for the strain to lost its productivity.

Woodruff and Johnson selected a double auxotrophic mutant of *Micrococcus glutamicus* requiring both homoserine and threonine and compared its lysine-producing properties with those of a homoserine auxotroph. The authors claimed that the double mutant had a two-fold advantage in that it produced higher levels of lysine compared with the single auxotroph and was also less susceptible to reversion to low productivity, the stability of the double auxotroph being such that it was a suitable organism for the production of lysine by continuous culture.

Nakayama discussed the problem of strain reversion in the lysine fermentation and cited an example of a fermentation where 87 per cent of the cells from a 60–88-hour culture were revertants. This very high level of revertants was probably due to the faster growth rate of the revertant compared with the single auxotroph in homoserine-limited culture. This situation was controlled, to a certain extent, by the incorporation of antibiotics, such as erythromycin, which are more inhibitory to the rapidly growing revertants than to the homoserine limited auxotrophs. However, the use of mutants which have multiple markers appears to be a better solution, as this considerably reduces the probability of reversion to the wild type as reversion of all the markers must occur, which should be an extremely rare event. Sano and Shiio cited the use of a *C. glutamicum* mutant for the industrial production of lysine which was auxotrophic for homoserine and leucine and was resistant to *S*-(2-aminoethyl)-L-cysteine and produced 39 g $dm^{-3}$ lysine. This strain would have had to have several reversion before it was restored to anything approaching the wild-type.

The stability of fungal diploids used in commercial fermentations may be controlled by a technique discussed by Ball. Ball claimed that it may be possible to control the degeneration of a diploid into haploids by incorporating non-homologous recessive lethal mutations on separate chromosomes of an homologous pair in the diploid. The deleterious effects of the mutations would be repressed by the dominant alleles in the diploid but would be expressed in a haploid derived from the diploid, resulting in any haploids being non-viable.

A very simple but effective technique for selecting stable strains of *P. chrysogenum* was used in the Panlabs strains development programme. Final evaluation of a culture was made using the second slant of a slant-to-slant transfer. If the culture was unstable then the yield of a fermentation from the second slant would be poor, resulting in it being rejected. This procedure was followed sequentially through the programme with the result that the later strains showed less tendency to degenerate after sub-culturing.

Recombinant plasmids used as vectors in genetic manipulation are susceptible to two types of instability. Segregational instability is due to uneven partitioning of the plasmids at replication resulting in the production of plasmid-free daughter cells. Structural instability is a result of recombination events occurring within the vector, sometimes resulting in disruption of the desired gene. It is essential for acceptable stability that a plasmid vector contains a partitioning locus and that segregation of the plasmid is not simply a random event. The problem has been overcome in some systems by the integration of the vector into the chromosomal DNA, as already seen for filamentous fungal vectors. The disadvantage of this approach is that only one (or a few) copy of the desired gene is present whereas the use of an autonomously replicating plasmid would result in many gene copies.

### *Selection of strains resistant to infection*

Bacterial fermentations may be affected very seriously by phage infections, which may result in the lysis of the bacteria. A possible method for reducing the risk of failure due to phage contamination is to select bacterial strains which are resistant to the phages isolated in the fermentation plant. It is important that the apparent resistant strains isolated are not lysogenic as the carrying of a population of phages in the fermentation is a source of potential lytic phage mutants. The use of phage-resistant mutants does not ensure immunity from phage infections because new host range phages may be introduced on to the plant or phage mutants may appear. Plant hygiene is essential to minimise the risk of contamination and it is also possible to utilise chemical agents in the fermentation which selectively inhibit phage replication.

It may be possible to design the host organism of a recombinant fermentation such that it is more resistant to phage infection. Primrose suggested that the inclusion of one or more host-restriction and modification systems (HRM) may achieve this objective. HRM is a mechanism whereby foreign nucleic acids which enter a cell may be destroyed. Most HRM systems reduce the infectivity of bacteriophage DNA by a factor of $10^2$ to $10^4$. The genes for many HRM systems have been cloned and they can be introduced into host strains provided, of course, that they do not degrade the foreign DNA deliberately introduced into production strain. Also, recombinant fermentations may be made phage-resistant by changing the host organism. Although this may seem a drastic step it would be feasible if the rationale were built into the development of the process.

Infection of some fermentations with 'wild' micro-organisms may be made more easily controlled by selecting commercial strains which are resistant to various antibiotics. The antibiotics to which the commercial strain is resistant may then be used to control the level of contaminants. The danger inherent in this technique is that resistant contaminants will also tend to be selected, but reasonable protection should be given provided that stringent sterilisation of a contaminated fermenter is carried out before the vessel is used again so that any antibiotic resistant contaminants are removed.

### *Selection of non-foaming strains*

Foaming during a fermentation may result in the loss of broth, cells and product via the air outlet as well as putting the fermentation at risk from contamination. Thus, foaming is normally controlled by either chemical or mechanical means, but this task may be made easier if a non-foaming strain of the commercial organism can be developed. Foaming which occurs early in the fermentation is usually due to a component in the medium whereas foaming late in the fermentation is normally a property of the growing organism and therefore, it is only this latter type of foam which may be controlled by strain selection. Both mutant screening and recombination may be used to develop non-foaming types but, obviously, the organisms must be tested for productivity as well. Recombination appears to be a particularly attractive proposition especially if non-foaming mutants have been isolated which are poor producers or good producers of foam. Ancestral crosses using protoplast fusion techniques may then render a strain combining the desirable features of both strains.

### *Selection of strains which are resistant to components in the medium*

Some media components which are required for product formation may interfere with the growth of the organism and therefore, it may be desirable to select for strains which are resistant to the medium component. Polya and Nyiri applied this approach to the isolation of mutants of *P. chrysogenum* resistant to phenylacetic acid, a precursor of penicillin and toxic to the organism at high concentrations. Barrios-Gonzalez used a very similar approach in isolating superior penicillin producers. Alikhanian selected strains of the oxytetracycline producer, *S. rimosus*, resistant to high levels of phosphate which prevents product synthesis. This enabled the strain to be used in media containing high levels of corn steep liquor (which is rich in phosphate).

Analogues of repressing media components have been used to select resistant mutants. For example, arsenate and vanadate have been used to isolate phosphate resistant strains and deoxyglucose to isolate glucose catabolite repression resistant strains.

### *Selection of morphologically favourable strains*

The morphology of a micro-organism in submerged culture frequently has an effect on the economics or ease of operation of a process. The morphological form of a filamentous micro-organism will affect both the aeration of the system and the ease of filtration of the fermentation broth. As already discussed, the morphology of a fungus in submerged culture may be controlled by the level of a spore inoculum and by the medium, but it is also possible to influence the organism's morphology by altering its genotype. Backus and Stauffer recognised the influence of the genetics of a strain on the morphology of *P. chrysogenum* in submerged culture and its role in controlling foaming and broth filtration characteristics.

Bartholomew selected strains of *P. chrysogenum* (for penicillin production) which gave a lower viscosity broth which increased the oxygen-transport ability of their fermentation plant equipment. The use of mutation and screening compared with recombination for the production of strains of a particular morphological type has been discussed by Ball. Ball claimed that recombination techniques would be at an advantage over mutation and selection because of the large number of mutants that would have to be screened. A recombination programme would involve the crossing of a strain of the required morphology with a commercial producer, which should involve the screening of relatively few progeny. Hamlyn and Ball applied the technique of protoplast fusion to the construction of a desirable stain of *C. acremonium* for the production of cephalosporin, as discussed in an earlier section. A cross between an asporulating, slow-growing strain with a sporulating, fast-growing strain which only produced one-third of the cephalosporin level of the first strain eventually resulted in the isolation of a recombinant which displayed the desirable properties of both strains, i.e., a strain which demonstrated good sporulation, a high growth rate and produced 40 per cent more antibiotic than the higher-yielding parent. The ability of a strain to sporulate profusely is a very useful characteristic in the development of inoculum.

Rowlands reported that genetically altering *P. chrysogenum* strains to produce a pelleted rather than a filamentous broth was an important feature in the Panlabs penicillin improvement programme. The pelletted form gives rise to a much lower broth viscosity, resulting in lower power consumption. Although Rowlands did not indicate which genetic technique was used (mutant selection or protoplast fusion), protoplast fusion was certainly used to improve the sporulation and growth of the Panlabs strain, as discussed in a previous section (Table 8.5). Flocculation of yeasts is the adherence of cells in clumps resulting in the separation of the cells from the liquid in which they were suspended. Thus, the flocculating property of yeasts may be described as a morphological characteristic. When flocculation occurs in a beer fermentation the yeast will rise to the top of the vessel if it is a 'top-fermenting' yeast or drop to the bottom of the vessel if it is a 'bottom-fermenting' yeast. Thus, the flocculence of a yeast will determine the time of contact of the yeast with the wort (and hence the conversion of the wort to alcohol) and the ease of clarification of the beer. Although mutant selection is rarely applied, the selection of natural variants is a very common practice in many breweries. As already discussed, the yeast produced from one beer fermentation may be used to inoculate a new fermentation, a practice which is very rarely employed in the rest of the fermentation industry. By the careful selection of the yeast produced during a fermentation, strains showing the desired degree of flocculence may be selected. The first cells to rise to the top of a top fermentation or drop to the bottom of a bottom fermentation, are the most flocculent cells whereas the last to rise or drop (depending on the type of fermentation) are the least flocculent cells. Thus, by selecting those cells the flocculate at an intermediate time the brewer isolates the natural variants which have the most desirable morphological properties for the inoculation of the next fermentation.

### *Selection of strains which are tolerant of low oxygen tension*

The provision of oxygen is frequently the limiting factor of many fermentations and it would, therefore, be desirable to select an organism which was capable of producing the product at a lower oxygen tension than normal. This may be achieved by screening for increased production under oxygen limited conditions. For example, Mindlin and Zaitseva isolated a lysine-producing strain which maintained its productivity under aeration conditions which decreased the parental strain productivity by almost a half.

### *Elimination of undesirable products from a production strain*

Although an industrial micro-organism may produce large quantities of a desirable metabolite it may also produce a large amount of a metabolite which is not required, is toxic or may interfere with the extraction process. Thus, in these circumstances it would be an advantage to alter the strain such that the undesirable product is no longer produced. An example in the penicillin-producing strains is the elimination of the production of the yellow pigment, chrysogenein, by the selection of non-pigmented mutants which made the extraction of the antibiotic much simpler.

Dolezilova considered the production of fungicidin (nystatin) and cycloheximide by mutants of *S. noursei*. These workers demonstrated that mutants could be isolated which produced increased levels of fungicidin but produced no cycloheximide. The use of protoplast fusion techniques for the elimination of *p*-OH penicillin V has been considered earlier in the chapter.

### *Development of strains producing new fermentation products*

The isolation of organisms from the natural environment synthesising commercially useful metabolites is an expensive and laborious process. Therefore, other means of producing novel compounds which may be of some industrial significance have been attempted. Probably the most successful alternative approach has been the semi-synthetic one where microbial products have been chemically modified, for example, the semi-synthetic penicillins. Precursor feeding has also met with some success in this context; by incorporating a precursor of a natural product into the fermentation medium the level of the end product may be increased, for example, the use of phenylacetic acid in the production of penicillin G. The feeding of an analogue of the normal precursor of a natural product frequently results in the production of an analogue of the natural product. Hamill demonstrated that if 5, 6 or 7-substituted tryptophan replaced tryptophan, the normal precursor of the antifungal agent, pyrrolnitrin, in cultures of *Pseudomonas aureofaciens*, then a series of substituted pyrrolnitrins were obtained, some of which had improved antifungal activity. However, the disadvantages of the analogue-precursor technique are that the end product tends to be very similar to the natural product and the new product will be contaminated with the normal product which the organism may still synthesise from its self-produced natural precursors. Birch suggested that the problem of mixed products may be overcome by isolating mutants which would not produce the normal precursor but could convert it into the end product. Thus, the analogue precursor could be converted into the novel end product without competition from the normal endogenous precursor. Shier succeeded in applying Birch's idea in the study of neomycin production by *Streptomyces fradiae* from the precursor, deoxystreptamine. By using a replica plating technique, the survivors of a mutation treatment were screened for the ability to inhibit the growth of the test organism, *Bacillus subtilis*, only in the presence of deoxystreptamine. By feeding the mutant isolated by this method, different antibiotics were synthesised. Nagaoka and Demain described this technique for the isolation of new products as 'mutational biosynthesis' and the mutants isolated as 'idiotrophs'. Besides the aminoglycoside-aminocyclitol antibiotics (of which neomycin is an example), mutational biosynthesis has been applied to the macrolide antibiotics, the novobiocin antibiotics and the β-lactam antibiotics.

The advent of readily available recombination techniques resulted in attempts to produce novel compounds from recombinants, particularly streptomycetes. The rationale behind these experiments was that by mixing the genotypes of two organisms synthesising different metabolites then new combinations of biosynthetic genes and hence pathways, may be produced. Little progress was achieved using protoplast fusion but the exploitation of recombinant DNA technology has yielded some significant successes. *Streptomyces coelicolor* produces the polyketide actinorhodin (Fig. 8.34) whilst *Streptomyces* sp. AM 7161 produces medermycin. Hopwood transformed some of the cloned genes coding for actinorhodin into *Streptomyces* sp. AM7161. The recombinant produced another antibiotic, mederrhodin A (Fig. 8.34). The modified strain contained the *act*V gene from *S. coelicolor* coding for the *p*-hydroxylation of actinorhodin; the enzyme was also capable of hydroxylating medermycin. Hopwood also introduced the entire actinorhodin gene cluster into *Streptomyces violaceoruber* which produces granaticin or dihydrogranaticin (Fig. 8.35). The recombinant synthesised the novel antibiotic, dihydrogranatirhodin which has the same structure as dihydrogranaticin apart from the stereochemistry at one of its chiral centers (Fig. 8.35).

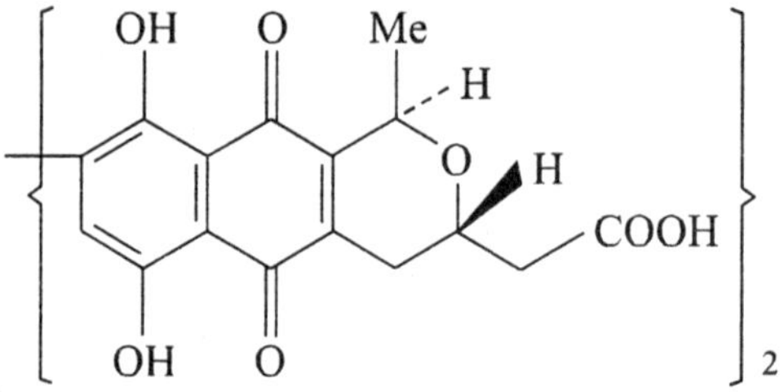

Actinorhodin
(a)

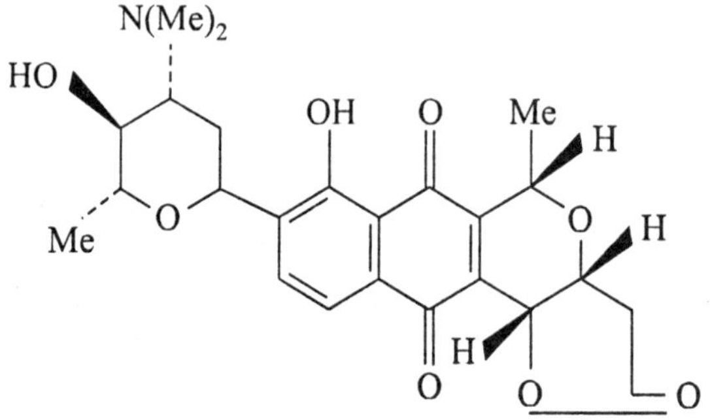

Medermycin
(b)

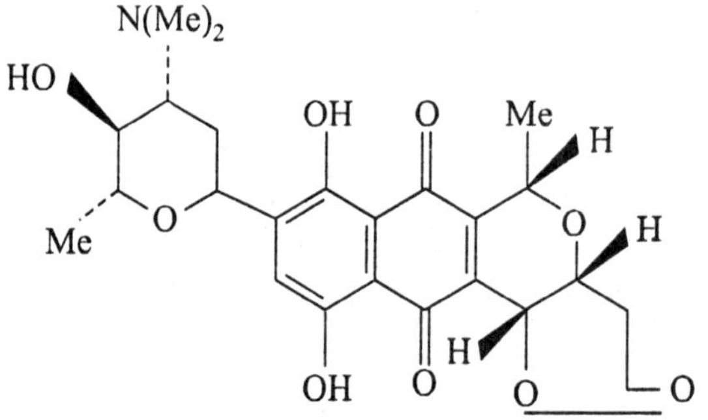

Mederrhodin
(c)

**Fig. 8.34.** (a) Structure of actinorhodin; (b) structure of medermycin; and (c) structure of mederrhodin.

The enzyme, responsible for polyketide biosynthesis have been extensively studied in recent years and significant advances have been made in the understanding of both their biochemistry and genetics. Polyketide synthases (PKSs) are multifunctional enzymes which catalyse repeated decarboxylative condensations between co-enzyme A thioesters and are very similar to the fatty acid synthases. An enormous range of microbial polyketides are known and the variation is due to the chain length, the nature of the precursors and the subsequent modification reactions which occur after cyclisation. McDaniel have constructed plasmids coding for recombinant PKSs and achieved expression in *Streptomyces coelicolor.*

Five novel polyketides were synthesised and the system shows great promise for the production of new, potentially valuable, compounds as well as providing carbon skeletons amenable to derivatisation by organic chemists.

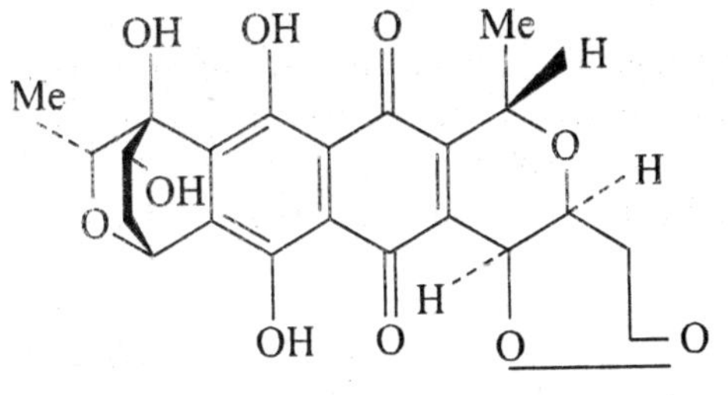

Granaticin

(a)

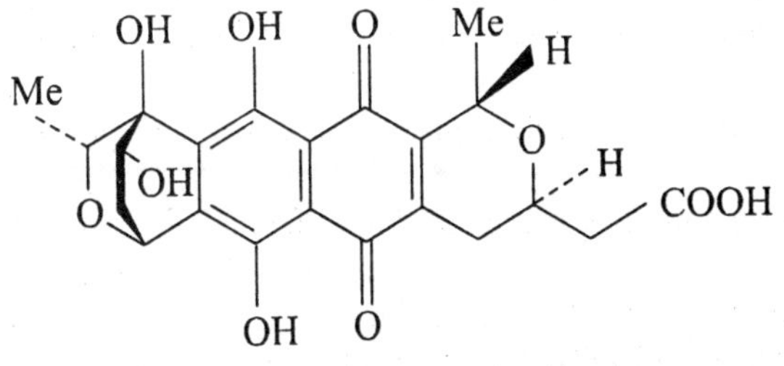

Dihydrogranaticin

(b)

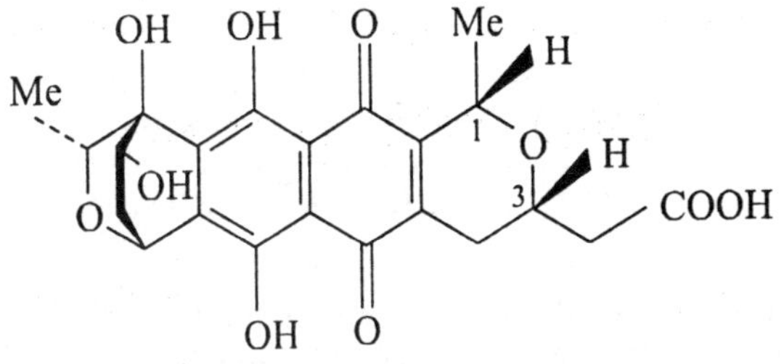

Dihydrogranatirhodin

(c)

**Fig. 8.35.** (a) Structure of granaticin; (b) structure of dihydrogranaticin; and (c) structure of dihydrogranatirhodin.

A commercially relevant example is provided by the Lilly Research Laboratories group who produced a strain of *Penicillium chrysogenum* capable of accumulating deacetoxycephalosporin C (DAOC), a compound which can be biotransformed into 7-aminodesacetoxycephalosporanic acid (7ADAOC) which is a precursor of at least three chemically synthesised clinically important cephalosporins. The conventional route to 7ADAOC is by the chemical ring expansion of benzylpenicillin which is far more complex than the biotransformation of DAOC.

DAOC is produced as an intermediate in the synthesis of cephalosporin C by *Cephalosporium acremonium*, as shown in Fig. 8.33. However, it would be very difficult to manufacture the compound from *C. acremonium* by blocking the conversion of DAOC to deacetylcephalosporin C because a single bifunctional enzyme catalyses both the conversion of penicillin N to DAOC (expandase activity) and the hydroxylation of DAOC to deacetylcephalosporin C (hydroxylase activity). The bifunctional protein is encoded by the gene *cef* EF. In *Streptomyces clavuligerus* the expandase and hydroxylase enzymes are separate proteins encoded by the genes *cef* E and *cef* F respectively. Thus, *S. clavuligerus* could be modified for the commercial production of DAOC. However, the Lilly group did not have a strain capable of producing high cephalosporin levels available to them, but they did have a *Penicillium chrysogenum* that overproduced isopenicillin N (IPN) as an intermediate in phenoxymethylpenicillin synthesis as shown in Fig. 8.36. Thus, to enable the *Penicillium* to produce DAOC the insertion of two genes was necessary, i.e., those encoding isopenicillin epimerase and expandase. These workers had already cloned *cef* E (coding for expandase) from *S. clavuligerus* and *cef* D (coding for epimerase) from *S. lipmanii*. Thus, using sophisticated vector technology a *P. chrysogenum* was transformed with a plasmid containing both *cef* D and *cef* E. One of the isolated transformants was capable of synthesising up to 2.5 g $dm^{-3}$ DAOC and both the *cef* D and *cef* E genes were integrated into the organism's genome.

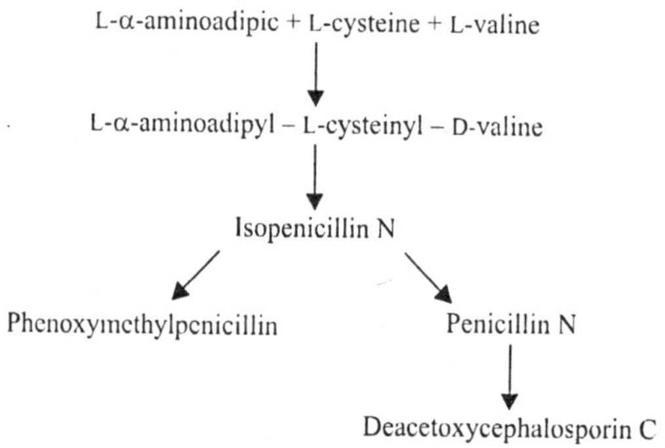


**Fig. 8.36.** β-Lactam biosynthetic pathway in wild-type *P. chrysogenum* (pathway in medium type ) and in *P. chrysogenum* transformed with pBOB13 and pZAZ6 (branched pathway which includes steps in boldface type). Enzyme steps (medium type, top to bottom): ACV synthetase, IPN synthase, acyl co-enzyme A:IPN acyltransferase. Enzyme steps (boldface type, top to bottom): IPN epimerase, penicillin N expandase (= DAOC synthase).

Thus, the recombinant produced by the Lilly group contained a branched biosynthetic pathway giving rise to the two end products, DAOC and phenoxymethylpenicillin. However, the conversion of IPN to DAOC in the recombinant represented less that 10 per cent of the conversion of IPN to phenoxymethylpenicillin in the parent. Thus, the recombinant may still be developed to improve the productivity. These developments could involve the screening of a larger population of transformants to find one where integration of the inserted genes occurred at sites not associated with penicillin synthesis. Also, the isolation of a strain blocked in the conversion of IPN to phenoxymethylpenicillin should give improved DAOC production. The tasks of both discovering new microbial compounds and improving the synthesis of known ones have become more and more challenging. Early work on isolation and improvement relied on a 'blunderbus' approach yet, due to the ingenuity and resourcefulness of the individuals involved, resulted in the establishment of a wide range of highly successful processes. The rational improvement programmes developed in the amino-acid industry pointed the way to the adoption of such approaches for both secondary metabolite discovery and improvement and the development of miniaturised screening systems allowed the industry to take full advantage of robotic systems to revolutionise screening throughput. However, it is the gigantic developments in molecular biology which have allowed the industry to enter the next stage in its evolution. Recombinant DNA technology has enabled the production of heterologous products and has built on the achievements of directed selection to increase yields of conventional products still further.

Chapter 9

# Media for Industrial Fermentations

## INTRODUCTION

Detailed investigation is needed to establish the most suitable medium for an individual fermentation process, but certain basic requirements must be met by any such medium. All micro-organisms require water, sources of energy, carbon, nitrogen, mineral elements and possibly vitamins plus oxygen if aerobic. On a small scale it is relatively simple to devise a medium containing pure compounds, but the resulting medium, although supporting satisfactory growth, may be unsuitable for use in a large scale process.

On a large scale one must normally use sources of nutrients to create a medium which will meet as many as possible of the following criteria:

1. It will produce the maximum yield of product or biomass per gram of substrate used.
2. It will produce the maximum concentration of product or biomass.
3. It will permit the maximum rate of product formation.
4. There will be the minimum yield of undesired products.
5. It will be of a consistent quality and be readily available throughout the year.
6. It will cause minimal problems during media making and sterilisation.
7. It will cause minimal problems in other aspects of the production process particularly aeration and agitation, extraction, purification and waste treatment.

The use of cane molasses, beet molasses, cereal grains, starch, glucose, sucrose and lactose as carbon sources and ammonium salts, urea, nitrates, corn steep liquor, soyabean meal, slaughter-house waste and fermentation residues as nitrogen sources, have tended to meet most of the above criteria for production media because they are cheap substrates. However, other more expensive pure substrates may be chosen if the overall cost of the complete process can be reduced because it is possible to use simpler procedures. Other criteria are used to select suitable sporulation and inoculation media.

It must be remembered that the medium selected will affect the design of fermenter to be used. For example, the decision to use methanol and ammonia in the single cell protein process necessitated the design of a novel fermenter design. The microbial oxidation of hydrocarbons is a highly aerobic and exothermic process.

Thus, the production fermenter had to have a very high oxygen transfer capacity coupled with excellent cooling facilities. These problems can be solved by developing an air lift fermenter. Equally, if a fermenter is already available this will obviously influence the composition of the medium. Rhodes observed that the optimum concentrations of available nitrogen for griseofulvin production showed some variation with the type of fermenter used.

The problem of developing a process from the laboratory to the pilot scale and subsequently to the industrial scale, must also be considered. A laboratory medium may not be ideal in a large fermenter with a low gas-transfer pattern. A medium with a high viscosity will also need a higher power input for effective stirring. Besides meeting requirements for growth and product formation, the medium may also influence pH variation, foam formation, the oxidation-reduction potential and the morphological form of the organism. It may also be necessary to provide precursors or metabolic inhibitors. The medium will also affect product recovery and effluent treatment.

Historically, undefined complex natural materials have been used in fermentation processes because they are much cheaper than pure substrates. However, there is often considerable batch variation because of variable concentrations of the component parts and impurities in natural materials which cause unpredictable biomass and/or product yields. As a consequence of these variations in composition small yield improvements are difficult to detect. Undefined media often make product recovery and effluent treatment more problematical because not all the components of a complex nutrient source will be consumed by the organism. The residual components may interfere with recovery and contribute to the BOD of the effluent.

Thus, although manufacturers have been reluctant to use defined media components because they are more expensive, pure substrates give more predictable yields from batch to batch and recovery, purification and effluent treatment are much simpler and therefore cheaper. Process improvements are also easier to detect when pure substrates are used.

Collins has given an excellent example of a process producing recombinant protein from *S. cerevisiae* instead of just biomass. The range of growth conditions which can be used is restricted because of factors affecting the stability of the recombinant protein. The control of pH and foam during growth in a fermenter were identified as two important parameters. Molasses would normally be used as the cheapest carbohydrate to grow yeast biomass in a large scale process. However, this is not acceptable for the recombinant protein production because of the difficulties and incurred costs caused in subsequent purification which result from using crude undefined media components. Collins and co-workers therefore used a defined medium with glucose, sucrose or another suitable carbon source of reasonable purity plus minimal salts, trace elements, pure vitamins and ammonia as the main nitrogen source and for pH control. Other impurities in molasses might have helped to stabilise foams and led to the need to use antifoams. Media for culture of animal cells will be discussed later in this chapter.

## TYPICAL MEDIA

Table 9.1 gives the recipes for some typical media for submerged culture fermentations. These examples are used to illustrate the range of media in use, but are not necessarily the best media in current use.

**Table 9.1.** Some examples of fermentation media.

| *Itaconic acid* | | *Clavulanic acid* | |
|---|---|---|---|
| Cane molasses (as sugar) | 150 g $dm^{-3}$ | Glycerol | 1% |
| $ZnSO_4$ | 1.0 g $dm^{-3}$ | Soyabean flour | 1.5% |
| $ZnSO_4 \cdot 7H_2O$ | 3.0 g $dm^{-3}$ | $KH_2PO_4$ | 0.1% |
| $CuSO_4 \cdot 5H_2O$ | 0.01 g $gm^{-3}$ | 10% Pluronic L81 antifoam in soyabean oil | 0.2% (v/v) |

*(Cont'd ...)*

| | |
|---|---|
| *Amylase* | |
| Ground soyabean meal | 1.85% |
| Autolysed brewers yeast fractions | 1.50% |
| Distillers dried solubles | 0.76% |
| NZ-amine (enzymatic casein hydrolysate) | 0.65% |
| Lactose | 4.75% |
| $MgSO_4 \cdot 7H_2O$ | 0.04% |
| Hodag KG-1 antifoam | 0.05% |
| *Avermectin* | |
| Cerelose | 45 g |
| Peptonised milk | 24 g |
| Autolysed yeast | 2.5 g |
| Polyglycol P-2000 | 2.5 $cm^3$ |
| Distilled water | 1 $dm^3$ |
| pH | 7.0 |
| Endotoxin from *Bacillus thuringiensis* | |
| Molasses | 0–4% |
| Soya flour | 2–6% |
| $KH_2PO_4$ | 0.5% |
| $KH_2PO_4$ | 0.5% |
| $MgSO_4 \cdot 7H_2O$ | 0.005% |
| $MnSO_4 \cdot 4H_2O$ | 0.003% |
| $FeSO_4 \cdot 7H_2O$ | 0.001% |
| $CaCl_2$ | 0.005% |
| $Na(NH_4)_2PO_4 \cdot 4H_2O$ | 0.15% |
| *Lysine* | |
| Cane blackstrap molasses | 20% |

| | |
|---|---|
| *Oxytetracycline* | |
| Starch | 12% + 4% (Additional feeding) |
| Technical amylase | 0.1% |
| Yeast (dry wt.) | 1.5% |
| $CaCO_3$ | 2% |
| Ammonium sulphate | 1.5% |
| Lactic acid | 0.13% |
| Lard oil | 2% |
| Total inorganic salts | 0.01% |
| *Gibberellic acid* | |
| Glucose monohydrate | 20 g $dm^{-3}$ |
| $MgSO_4$ | 1 g $dm^{-3}$ |
| $NH_4H_2HPO_4$ | 2 g $dm^{-3}$ |
| $KH_2PO_4$ | 5 g $dm^{-4}$ |
| $FeSO_4 \cdot 7H_2O$ | 0.01 g $dm^{-3}$ |
| $MnSO_4 \cdot 4H_2O$ | 0.01 g $dm^{-3}$ |
| $ZnSO_4 \cdot 7H_2O$ | 0.01 g $dm^{-3}$ |
| $CuSO_4 \cdot 5H_2O$ | 0.01 g $dm^{-3}$ |
| Corn steep liquor (as dry solids) | 7.5 g $dm^{-3}$ |
| *Glutamic acid* | |
| Dextrose | 270 g $dm^{-3}$ |
| $NH_4H_2PO_4$ | 2 g $dm^{-3}$ |
| $(NH_4)_2HPO_4$ | 2 g $dm^{-3}$ |
| $K_2SO_4$ | 2 g $dm^{-3}$ |
| $MgSO_4 \cdot 7H_2O$ | 0.5 g $dm^{-3}$ |
| $MnSO_4 \cdot H_2O$ | 0.04 g $dm^{-3}$ |
| $FeSO_4 \cdot 7H_2O$ | 0.02 g $dm^{-3}$ |
| Polyglycol 2000 | 0.3 g $dm^{-3}$ |
| Biotin | 12 µg $dm^{-3}$ |
| Penicillin | 11 µg $dm^{-3}$ |
| Penicillin | |
| Glucose or molasses (by continuous feed) | 10% of total |
| Corn-steep liquor | 4–5% of total |

*(Cont'd ...)*

| Soyabean meal hydrosylate (as weight of meal before hydrolysis with 6N $H_2SO_4$ and neutralised with ammonia water) | 1.8% | Phenylacetic acid (by continuous feed) | 0.5–0.8% of total |
|---|---|---|---|
| $CaCO_3$ or $MgSO_4$ added to buffer medium | | Lard oil (or vegetable oil) antifoam by continuous addition pH to 6.5 to 7.5 by acid or alkali addition | 0.5% of total |
| Antifoam agent | | | |

## MEDIUM FORMULATION

Medium formulation is an essential stage in the design of successful laboratory experiments, pilot-scale development and manufacturing processes. The constituents of a medium must satisfy the elemental requirements for cell biomass and metabolite production and there must be an adequate supply of energy for biosynthesis and cell maintenance. The first step to consider is an equation based on the stoichiometry for growth and product formation. Thus, for an aerobic fermentation:

$$\text{Carbon and energy source} + \text{nitrogen source} + O_2 + \text{other requirements} \rightarrow \text{biomass} + \text{products} + CO_2 + H_2O + \text{heat}$$

This equation should be expressed in quantitative terms, which is important in the economical design of media if component wastage is to be minimal. Thus, it should be possible to calculate the minimal quantities of nutrients which will be needed to produce a specific amount of biomass. Knowing that a certain amount of biomass is necessary to produce a defined amount of product, it should be possible to calculate substrate concentrations necessary to produce required product yields. There may be medium components which are needed for product formation which are not required for biomass production. Unfortunately, it is not always easy to quantify all the factors very precisely.

A knowledge of the elemental composition of a process micro-organism is required for the solution of the elemental balance equation. This information may not be available so that data which is given in Table 9.2 will serve as a guide to the absolute minimum quantities of N, S, P, Mg and K to include in an initial medium recipe.

**Table 9.2.** Element composition of bacteria, yeasts and fungi (% by dry weight).

| *Element* | *Bacteria* | *Yeasts* | *Fungi* |
|---|---|---|---|
| Carbon | 50–53 | 45–50 | 40–63 |
| Hydrogen | 7 | 7 | – |
| Nitrogen | 12–15 | 7.5–11 | 7–10 |
| Phosphorus | 2.0–3.0 | 0.8–2.6 | 0.4–4.5 |
| Sulphur | 0.2–1.0 | 0.01–0.24 | 0.1–0.5 |
| Potassium | 1.0–4.5 | 1.0–4.0 | 0.2–2.5 |
| Sodium | 0.5–1.0 | 0.01–0.1 | 0.02–0.5 |
| Calcium | 0.01–1.1 | 0.1–0.3 | 0.1–1.4 |
| Magnesium | 0.1–0.5 | 0.1–0.5 | 0.1–0.5 |
| Chloride | 0.5 | – | – |
| Iron | 0.02–0.2 | 0.01–0.5 | 0.1–0.2 |

Trace elements (Fe, Zn, Cu, Mn, Co, Mo, B) may also be needed in smaller quantities. An analysis of relative concentrations of individual elements in bacterial cells and commonly used cultivation media quoted by Cooney showed that some nutrients are frequently added in substantial excess of that required, e.g., P, K; however, others are often near limiting values, e.g., Zn, Cu. The concentration of P is deliberately raised in many media to increase the buffering capacity. These points emphasise the need for considerable attention to be given to medium design.

Some micro-organisms cannot synthesise specific nutrients, e.g., amino acids, vitamins or nucleotides. Once a specific growth factor has been identified it can be incorporated into a medium in adequate amounts as a pure compound or as a component of a complex mixture.

The carbon substrate has a dual role in biosynthesis and energy generation. The carbon requirement for biomass production under aerobic conditions may be estimated from the cellular yield coefficient (Y) which is defined as:

$$\frac{\text{Quantity of cell dry matter produced}}{\text{Quantity of carbon substrate utilised}}$$

Some values are given in Table 9.3. Thus for bacteria with a Y for glucose of 0.5, which is 0.5 gram cells $g^{-1}$ glucose, the concentration of glucose needed to obtain 30 gram $dm^{-3}$ cells will be 30/0.5 = 60 gram $dm^{-3}$ glucose. One liter of this medium would also need to contain approximately 3.0 gram N, 1.0 gram P, 1.0 gram K, 0.3 gram S and 0.1 gram Mg. More details of Y values for different micro-organisms and substrates are given by Atkinson and Mavituna.

**Table 9.3.** Cellular yield coefficients (Y) of bacteria on different carbon substrates.

| *Substrate* | *Cellular yield coefficient (g biomass dry wt. $g^{-1}$ substrate)* |
|---|---|
| Methane | 0.62 |
| n-Alkanes | 1.03 |
| Methanol | 0.40 |
| Ethanol | 0.68 |
| Acetate | 0.34 |
| Malate | 0.36 |
| Glucose (molasses) | 0.51 |

An adequate supply of the carbon source is also essential for a product-forming fermentation process. In a critical study, analyses are made to determine how the observed conversion of the carbon source to the product compares with the theoretical maximum yield. This may be difficult because of limited knowledge of the biosynthetic pathways. Cooney has calculated theoretical yields for penicillin G biosynthesis on the basis of material and energy balances using a biosynthetic pathway based on reaction stoichiometry. The stoichiometry equation for the overall synthesis is:

$$a_2C_6H_{12}O_6 + b_2NH_3 + c_2O_2 + d_2H_2SO_4 + e_2PAA \rightarrow n_2\text{Pen G} + p_2CO_2 + q_2H_2O$$

where $a_2$, $b_2$, $c_2$, $d_2$, $e_2$, $n_2$, $p_2$ and $q_2$ are the stoichiometric coefficients and PAA is phenylacetic acid. Solution of this equation yields:

$$\tfrac{10}{6}\,C_6H_{12}O_6 + 2NH_3 + \tfrac{1}{2}\,O_2 + H_2SO_4 + C_8H_8O_2 \rightarrow C_{16}H_{18}O_4N_2S + 2CO_2 + 9H_2O$$

In this instance it was calculated that the theoretical yield was 1.1 gram penicillin G $g^{-1}$ glucose (1837 units $mg^{-1}$).

Using a simple model for a batch-culture penicillin fermentation it was estimated that 28, 61 and 11 per cent of the glucose consumed was used for cell mass, maintenance and penicillin respectively. When experimental results of fed-batch penicillin fermentation were analysed, 26 per cent of the glucose has been used for growth, 70 per cent for maintenance and 6 per cent for penicillin. The maximum experimental conversion yield for penicillin was calculated to be 0.053 gram $g^{-1}$ glucose (88.5 units $mg^{-1}$). Thus, the theoretical conversion value is many times higher than the experimental value. Hersbach and others concluded that there were six possible biosynthetic pathways for penicillin production and two possible mechanisms for ATP production from NADH and $FADH_2$. They calculated that conversion yields by different pathways varied from 638 to 1544 units of penicillin per mg glucose. At that time the best quoted yields were 200 units penicillin per mg glucose. This gives a production of 13 to 29 per cent of the maximum theoretical yield.

The other major nutrient which will be required is oxygen which is provided by aerating the culture. The design of a medium will influence the oxygen demand of a culture in that the more reduced carbon sources will result in a higher oxygen demand. The amount of oxygen required may be determined stoichiometrically.

## WATER

Water is the major component of all fermentation media and is needed in many of the ancillary services such as heating, cooling, cleaning and rinsing. Clean water of consistent composition is therefore required in large quantities from reliable permanent sources. When assessing the suitability of a water supply it is important to consider pH, dissolved salts and effluent contamination.

The mineral content of water is very important in brewing and most critical in the mashing process and historically influenced and siting of breweries and the types of beer produced. Hard water containing high $CaSO_4$ concentrations are better for the English Burton bitter bears and Pilsen type lagers, while water with a high carbonate content are better for the darker beers such as stouts. Now-a-days, the water may be treated by deionisation or other techniques and salts added or the pH adjusted, to favour different beers so that breweries are not so dependent on the local water source.

The re-use or efficient use of water is normally of high priority. When ICI developed a continuous-culture single cell protein (SCP) process at a production scale of 60,000 tonnes per year it was realised that very high costs would be incurred if fresh purified water was used on a once-through basis, since operating at a cell concentration of 30 gram biomass (dw) $dm^{-3}$ would require $2700 \times 10^6$ $dm^3$ of water per annum. Laboratory tests to simulate the process showed that the *Methylophilus methylotrophus* could be grown successfully with 86 per cent continuous recycling of supernatant with additions to make up depleted nutrients. This approach was therefore adopted in the full scale process to reduce capital and operating costs and it was estimated that water used on a once-through basis without any recycling would have increased water costs by 50 per cent and effluent treatment costs 10-fold.

## ENERGY SOURCES

Energy for growth comes from either the oxidation of medium components or from light. Most industrial micro-organisms are chemo-organotrophs, therefore the commonest source of energy will be the carbon source such as carbohydrates, lipids and proteins. Some micro-organisms can also use hydrocarbons or methanol as carbon and energy sources.

## CARBON SOURCES

### Factors Influencing the Choice of Carbon Source

It is now recognised that the rate at which the carbon source is metabolised can often influence the formation of biomass or production of primary or secondary metabolites. Fast growth due to high concentrations of rapidly metabolised sugars is often associated with low productivity of secondary metabolites. This has been demonstrated for a number of processes (Table 9.4). At one time the problem was overcome by using the less readily metabolised sugars such as lactose, but many processes now use semi-continuous or continuous feed of glucose or sucrose (Table 9.15). Alternatively, carbon catabolite regulation might be overcome by genetic modification of the producer organism.

**Table 9.4.** Carbon catabolite regulation of metabolite biosynthesis.

| *Metabolite* | *Micro-organism* | *Interfering carbon source* |
|---|---|---|
| Griseofulvin | *Penicillium griseofulvin* | Glucose |
| Penicillin | *P. chrysogenum* | Glucose |
| Cephalosporin | *Cephalosporium acremonium* | Glucose |
| Aurantin | *Bacillus aurantinus* | Glycerol |
| α-Amylase | *B. licheniformis* | Glucose |
| Bacitracin | *B. licheniformis* | Glucose |
| Puromycin | *Streptomyces alboniger* | Glucose |
| Actinomycin | *S. antibioticus* | Glucose |
| Cephamycin C | *S. clavuligerus* | Glycerol |
| Neomycin | *S. fradiae* | Glucose |
| Cycloserine | *S. graphalus* | Glycerol |
| Streptomycin | *S. griseus* | Glucose |
| Kanamycin | *S. kanamyceticus* | Glucose |
| Novobiocin | *S. niveus* | Citrate |
| Siomycin | *S. sioyaensis* | Glucose |

The main product of a fermentation process will often determine the choice of carbon source, particularly if the product results from the direct dissimilation of it. In fermentations such as ethanol or single-cell protein production where raw materials are 60 to 77 per cent of the production cost, the selling price of the product will be determined largely by the cost of the carbon source. It is often part of a company development programme to test a range of alternative carbon sources to determine the yield of product and its influence on the process and the cost of producing biomass and/or metabolite. This enables a company to use alternative substrates, depending on price and availability in different locations and remain competitive.

The purity of the carbon source may also affect the choice of substrate. For example, metallic ions must be removed from carbohydrate sources used in some citric acid processes.

The method of media preparation, particularly sterilisation, may affect the suitability of carbohydrates for individual fermentation processes. It is often best to sterilise sugars separately because they may react with ammonium ions and amino acids to form black nitrogen containing compounds which will partially inhibit the growth of many micro-organisms. Starch suffers from the handicap that when heated in the sterilisation process it gelatinises, giving rise to very viscous liquids, so that only concentrations of up to 2 per cent can be used without modification.

The choice of substrate may also be influenced by government legislation. Within the European Economic Community (EEC), the use of beet sugar and molasses is encouraged and the minimum price controlled. The quantity of imported cane sugar and molasses is carefully monitored and their imported prices set so that they will not be competitive with beet sugar. If the world market sugar price is very low then the EEC fermentation industry will be at a disadvantage unless it receives realistic subsidies. Refunds for a defined list of products are available in the EEC when sugar and starch are used as substrates. Legislation for recognition of new products is time consuming and manufacturers may be uncertain as to whether they would benefit from carbon substrate refunds. This uncertainty has meant that some manufacturers might prefer to site factories for new products outside the EEC.

Local laws may also dictate the substrates which may be used to make a number of beverages. There are similar laws applying to beer production in Germany. Scotch malt whisky may be made only from barley malt, water and yeast. Within France, many wines may be called by a certain name only if the producing vineyard is within a limited geographical locality.

## Examples of Commonly Used Carbon Sources

### *Carbohydrates*

It is common practice to use carbohydrates as the carbon source in microbial fermentation processes. The most widely available carbohydrates is starch obtained from maize grain. It is also obtained from other cereals, potatoes and cassava. Maize and other cereals may also be used directly in a partially ground state, e.g., maize chips. Starch may also be readily hydrolysed by dilute acids and enzymes to give a variety of glucose preparations (solids and syrups). Hydrolysed cassava starch is used as a major carbon source for glutamic acid production in Japan. Syrups produced by acid hydrolysis may also contain toxic products which may make them unsuitable for particular processes.

Barley grains may be partially germinated and heat treated to give the material known as malt, which contains a variety of sugars besides starch (Table 9.5). Malt is the main substrate for brewing beer and lager in many countries. Malt extracts may be prepared from malted grain.

**Table 9.5.** Carbohydrate composition of barley malt (expressed as % dry weight of total).

| | |
|---|---|
| Starch | 58–60 |
| Sucrose | 3–5 |
| Reducing sugars | 3–4 |
| Other sugars | 2 |
| Hemicellulose | 6–8 |
| Cellulose | 5 |

Sucrose is obtained from sugar cane and sugar beet. It is commonly used in fermentation media in a very impure form as beet or cane molasses (Table 9.6), which are the residues left after crystallisation of sugar solutions in sugar refining. Molasses is used in the production of high-volume/low-value products such as ethanol, SCP, organic and amino acids and some microbial gums. Molasses or sucrose also may be used for production of higher value/low-bulk products such as antibiotics, speciality enzymes, vaccines and fine chemicals. The cost of molasses will be very competitive when compared with pure carbohydrates.

However, molasses contains many impurities and molasses-based fermentations will often need a more expensive and complicated extraction/purification stage to remove the impurities and effluent treatment will be more expensive because of the unutilised waste materials which are still present. Some new processes may require critical evaluation before the final decision is made to use molasses as the main carbon substrate.

**Table 9.6.** Analysis of beet and cane molasses (expressed as % of total w/v).

| | *Beet* | *Cane* |
|---|---|---|
| Sucrose | 48.5 | 33.4 |
| Raffinose | 1.0 | 0 |
| Invert sugar | 1.0 | 21.2 |

Remainder is non-sugar.

The use of lactose and crude lactose (milk whey powder) in media formulations is now extremely limited since the introduction of continuous-feeding processes utilising glucose, discussed in a later section of this chapter.

Corn steep liquor (Table 9.7) is a by-product after starch extraction from maize. Although primarily used as a nitrogen source, it does contain lactic acid, small amounts of reducing sugars and complex polysaccharides. Certain other materials of plant origin, usually included as nitrogen sources, such as soyabean meal and Pharmamedia, contain small but significant amounts of carbohydrate.

**Table 9.7.** Partial analysis of corn-steep liquor.

| | |
|---|---|
| Total solids | 51% w/v |
| Acidity as lactic acid | 15% w/v |
| Free reducing sugars | 5.6% w/v |
| Free reducing sugars after hydrolysis | 6.8% w/v |
| Total nitrogen | 4% w/v |
| Amino acids as % of nitrogen | |
| Alanine | 25 |
| Arginine | 8 |
| Glutamic acid | 8 |
| Leucine | 6 |
| Proline | 5 |
| Isoleucine | 3.5 |
| Threonine | 3.5 |
| Valine | 3.5 |
| Phenylalanine | 2.0 |
| Methionine | 1.0 |
| Cystine | 1.0 |
| Ash | 1.25% w/v |
| Potassium | 20% |
| Phosphorus | 1–5% |

*(Cont'd ...)*

| | |
|---|---|
| Sodium | 0.3–1% |
| Magnesium | 0.003–0.3% |
| Iron | 0.01–0.3% |
| Copper | 0.01–0.03% |
| Calcium | 0.01–0.03% |
| Zinc | 0.003–0.08% |
| Lead | 0.001–0.003% |
| Silver | 0.001–0.003% |
| Chromium | 0.001–0.003% |
| B Vitamins | |
| Aneurine | 41–49 μg $g^{-1}$ |
| Biotin | 0.34–0.38 μg $g^{-1}$ |
| Calcium pantothenate | 14.5–21.5 μg $g^{-1}$ |
| Folic acid | 0.26–0.6 μg $g^{-1}$ |
| Nicotinamide | 30–40 μg $g^{-1}$ |
| Riboflavine | 3.9–4.7 μg $g^{-1}$ |
| Also niacin and pyridoxine | |

### *Oils and fats*

Oils were first used as carriers for antifoams in antibiotic processes. Vegetable oils (olive, maize, cotton-seed, linseed, soyabean, etc.) may also be used as carbon substrates, particularly for their content of the fatty acids, oleic, linoleic and linolenic acid, because costs are competitive with those of carbohydrates. Bader discussed factors favouring the use of oils instead of carbohydrates. A typical oil contains approximately 2.4 times the energy of glucose on a per weight basis. Oils also have a volume advantage as it would take 1.24 $dm^3$ of soyabean oil to add 10 kcal of energy to a fermenter, whereas it would take 5 $dm^3$ of glucose or sucrose assuming that they are being added as 50 per cent w/w solutions. Ideally, in any fermentation process, the maximum working capacity of a vessel should be used. Oil based fed-batch fermentations permit this procedure to operate more successfully than those using carbohydrate feeds where a larger spare capacity must be catered for to allow for responses to a sudden reduction in the residual nutrient level. Oils also have antifoam properties which may make downstream processing simpler, but normally they are not used solely for this purpose.

Stowell reported the results of a Pfizer antibiotic process operated with a range of oils and fats on a laboratory scale. On a purely technical basis glycerol trioleate was the most suitable substrate. In the UK however, when both technical and economic factors are considered, soyabean oil or rapeseed oil are the preferred substrates. Glycerol trioleate is known to be used in some fermentations where substrate purity is an important consideration. Methyl oleate has been used as the sole carbon substrate in cephalosporin production.

### *Hydrocarbons and their derivatives*

There has been considerable interest in hydrocarbons. Development work has been done using n-alkanes for production of organic acids, amino acids, vitamins and co-factors, nucleic acids, antibiotics, enzymes and proteins. Methane, methanol and n-alkanes have all been used as substrates for biomass production.

Drozd discussed the advantages and disadvantages of hydrocarbons and their derivatives as fermentation substrates, particularly with reference to cost, process aspects and purity. In processes where the feedstock costs are an appreciable fraction of the total manufacturing cost, cheap carbon sources are important. In the 1960s and early 1970s there was an incentive to consider using oil or natural gas derivatives as carbon substrates as costs were low and sugar prices were high. On a weight basis n-alkanes have approximately twice the carbon and three times the energy content of the same weight of sugar. Although petroleum-type products are initially impure they can be refined to obtain very pure products in bulk quantities which would reduce the amount of effluent treatment and downstream processing. At this time the view was also held that hydrocarbons would not be subject to the same fluctuations in cost as agriculturally derived feedstocks because it would be a stable priced commodity and might be used to provide a substrate for conversion to microbial protein (SCP) for economic animal and/or human consumption. Sharp gives a very good account of market considerations of changes in price and how this would affect the price of SCP. The SCP would have had to have been cheaper or as cheap as soya meal to be marketed as an animal feed supplement.

Only BP and ICI plc eventually developed SCP at a production scale as an animal feed supplement.

Drozd has made a detailed study of hydrocarbon feedstocks and concluded that the cost of hydrocarbons does not make them economically attractive bulk feedstocks for the production of established products or potential new products where feedstock costs are an appreciable fraction of manufacturing costs of low-value bulk products. In SCP production, raw materials account for three quarters of the operating or variable costs and about half of the costs of manufacture. It was considered that hydrocarbons and their derivatives might have a potential role as feedstocks in the microbial production of higher value products such as intermediates, pharmaceuticals, fine chemicals and agricultural chemicals.

## NITROGEN SOURCES

### Examples of Commonly Used Nitrogen Sources

Most industrially used micro-organisms can utilise inorganic or organic sources of nitrogen. Inorganic nitrogen may be supplied as ammonia gas, ammonium salts or nitrates. Ammonia has been used for pH control and as the major nitrogen source in a defined medium for the commercial production of human serum albumin by *Saccharomyces cerivisiae*. Ammonium salts such as ammonium sulphate will usually produce acid conditions as the ammonium ion is utilised and the free acid will be liberated. On the other hand nitrates will normally cause an alkaline drift as they are metabolised. Ammonium nitrate will first cause an acid drift as the ammonium ion is utilised and nitrate assimilation is repressed. When the ammonium ion has been exhausted, there is an alkaline drift as the nitrate is used as an alternative nitrogen source. One exception to this pattern is the metabolism of *Gibberella fujikuroi*. In the presence of nitrate the assimilation of ammonia is inhibited at pH 2.8–3.0. Nitrate assimilation continues until the pH has increased enough to allow the ammonia assimilation mechanism to re-start.

Organic nitrogen may be supplied as amino acid, protein or urea. In many instances growth will be faster with a supply of organic nitrogen and a few micro-organisms have an absolute requirement for amino acids. It might be thought that the main industrial need for pure amino acids would be in the deliberate addition to amino acid requiring mutants used in amino acid production. However, amino acids are more commonly added as complex organic nitrogen sources which are non-homogeneous, cheaper and readily available. In lysine production, methionine and threonine are obtained from soyabean hydrolysate since it would be too expensive to use the pure amino acids.

Other proteinaceous nitrogen compounds serving as sources of amino acids include corn-steep liquor, soya meal, peanut meal, cotton-seed meal (Pharmamedia, Table 9.8 and Proflo), Distiller's solubles, meal and yeast extract. In storage these products may be affected by moisture, temperature changes and ageing.

**Table 9.8.** The composition of Pharmamedia.

| *Component* | *Quantity* |
|---|---|
| Total solids | 99% |
| Carbohydrate | 24.1% |
| Reducing sugars | 1.2% |
| Non-reducing sugars | 1.2% |
| Protein | 57% |
| Amino nitrogen | 4.7% |
| Components of amino nitrogen | |
| Lysine | 4.5% |
| Leucine | 6.1% |
| Isoleucine | 3.3% |
| Threonine | 3.3% |
| Valine | 4.6% |
| Phenylalanine | 5.9% |
| Tryptophan | 1.0% |
| Methionine | 1.5% |
| Cystine | 1.5% |
| Aspartic acid | 9.7% |
| Serine | 4.6% |
| Proline | 3.9% |
| Glycine | 3.8% |
| Alanine | 3.9% |
| Tyrosine | 3.4% |
| Histidine | 3.0% |
| Arginine | 12.3% |
| Mineral components | |
| Calcium | 2530 ppm |
| Chloride | 685 ppm |
| Phosphorus | 13100 ppm |
| Iron | 94 ppm |
| Sulphate | 18000 ppm |
| Magnesium | 7360 ppm |
| Potassium | 17200 ppm |
| Fat | 4.5% |

*(Cont'd ...)*

| *Component* | *Quantity* |
| --- | --- |
| Vitamins | |
| Ascorbic acid | 32.0 mg $kg^{-1}$ |
| Thiamine | 4.0 mg $kg^{-1}$ |
| Riboflavin | 4.8 mg $kg^{-1}$ |
| Niacin | 83.3 mg $kg^{-1}$ |
| Pantothenic acid | 12.4 mg $kg^{-1}$ |
| Choline | 3270 mg $kg^{-1}$ |
| Pyidoxine | 16.4 mg $kg^{-1}$ |
| Biotin | 1.5 mg $kg^{-1}$ |
| Folic acid | 1.6 mg $kg^{-1}$ |
| Inositol | 10,800 mg $kg^{-1}$ |

Chemically defined amino acid media devoid of protein are necessary in the production of certain vaccines when they are intended for human use.

### Factors Influencing the Choice of Nitrogen Source

Control mechanisms exist by which nitrate reductase, an enzyme involved in the conversion of nitrate to ammonium ion, is repressed in the presence of ammonia. For this reason ammonia or ammonium ion is the preferred nitrogen source. In fungi that have been investigated, ammonium ion represses uptake of amino acid by general and specific amino acid permeases. In *Aspergillus nidulans*, ammonia also regulates the production of alkaline and neutral proteases. Therefore, in mixtures of nitrogen sources, individual nitrogen components may influence metabolic regulation so that there is preferential assimilation of one component until its concentration has diminished.

It has been shown that antibiotic production by many micro-organisms is influenced by the type and concentration of the nitrogen source in the culture medium. Antibiotic production may be inhibited by a rapidly utilised nitrogen source ($NH_4^+$, $NO_3^-$, certain amino acids). The antibiotic production only begins to increase in the culture broth after most of the nitrogen source has been consumed.

In shake flask media experiments, salts of weak acids (e.g., ammonium succinate) may be used to serve as a nitrogen source and eradicate the source of a strong acid pH change due to chloride or sulphate ions which would be present if ammonium chloride or sulphate were used as the nitrogen source. This procedure makes it possible to use lower concentrations of phosphate to buffer the medium. High phosphate concentrations inhibit production of many secondary metabolites.

The use of complex nitrogen sources for antibiotic production has been common practice. They are thought to help create physiological conditions in the trophophase which favour antibiotic production in the idiophase. For example, in the production of polyene antibiotics, soyabean meal has been considered a good nitrogen source because of the balance of nutrients, the low phosphorus content and slow hydrolysis. It has been suggested that this gradual breakdown prevents the accumulation of ammonium ions and repressive amino acids. These are probably some of the reasons for the selection of ideal nitrogen sources for some secondary metabolites (Table 9.9).

In gibberellin production the nitrogen source has been shown to have an influence on directing the production of different gibberellins and the relative proportions of each type.

**Table 9.9.** Best nitrogen sources for some secondary metabolites.

| *Product* | *Main nitrogen source(s)* |
|---|---|
| Penicillin | Corn-steep liquor |
| Bacitracin | Peanut granules |
| Riboflavin | Pancreatic digest of gelatine |
| Novobiocin | Distiller's solubles |
| Rifomycin | Pharmamedia |
| | Soyabean meal, $(NH_4)_2SO_4$ |
| Gibberellins | Ammonium salt and natural plant nitrogen source |
| Butirosin | Dried beef blood or haemoglobin with $(NH_4)_2SO_4$ |
| Polyenes | Soyabean meal |

Other predetermined aspects of the process can also influence the choice of nitrogen source. Rhodes has shown that the optimum concentration of available nitrogen for griseofulvin production showed some variation depending on the form of inoculum and the type of fermenter being used. Obviously these factors must be borne in mind in the interpretation of results in media-development programmes. Some of the complex nitrogenous material may not be utilised by a micro-organism and create problems in downstream processing and effluent treatment. This can be an important factor in the final choice of substrate.

## MINERALS

All micro-organisms require certain mineral elements for growth and metabolism. In many media, magnesium, phosphorus, potassium, sulphur, calcium and chlorine are essential components and because of the concentrations required, they must be added as distinct components. Others such as cobalt, copper, iron, manganese, molybdenum and zinc are also essential but are usually present as impurities in other major ingredients. There is obviously a need for batch analysis of media components to ensure that this assumption can be justified, otherwise there may be deficiencies or excesses in different batches of media. See Tables 9.7 and 9.8 for analysis of corn steep liquor and Pharmamedia. When synthetic media are used the minor elements will have to be added deliberately. The form in which the minerals are usually supplied and the concentration ranges are given in Table 9.10. As a consequence of product composition analysis, as outlined earlier in this chapter, it is possible to estimate the amount of a specific mineral for medium design, e.g., sulphur in penicillins and cephalosporins, chlorine in chlortetracycline.

The concentration of phosphate in a medium, particularly laboratory media in shake flasks, is often much higher than that of other mineral components. Part of this phosphate is being used as a buffer to minimise pH changes when external control of the pH is not being used.

In specific processes the concentration of certain minerals may be very critical. Some secondary metabolic processes have a lower tolerance range to inorganic phosphate than vegetative growth. This phosphate should be sufficiently low as to be assimilated by the end of trophophase. Garner suggested that an important function of calcium salts in fermentation media was to precipitate excess inorganic phosphates and suggested that the calcium indirectly improved the yield of streptomycin. The inorganic phosphate concentration also influences production of bacitracins, citric acid (surface culture), ergot, monomycin, novobiocin, oxytetracycline, polyenes, ristomycin, rifamycin Y, streptomycin, vancomycin and viomycin. However, pyrrolnitrin, bicyclomycin, thiopeptin and methylenomycin are produced in a medium containing a high concentration of phosphate. Two monomycin antibiotics are selectively produced by *Streptomyces jamaicensis* when the phosphate is 0.1 mM or 0.4 mM.

**Table 9.10.** The range of typical concentrations of mineral components (g $dm^{-3}$).

| *Component* | *Range* |
|---|---|
| *$KH_2PO_4$ | 1.0–4.0 (part may be as buffer) |
| $MgSO_4 \cdot 7H_2O$ | 0.25–3.0 |
| KCl | 0.5–12.0 |
| $CaCO_3$ | 5.0–17.0 |
| $FeSO_4 \cdot 4H_2O$ | 0.01–0.1 |
| $ZnSO_4 \cdot 8H_2O$ | 0.1–1.0 |
| $MnSO_4 \cdot H_2O$ | 0.01–0.1 |
| $CuSO_4 \cdot 5H_2O$ | 0.003–0.01 |
| $Na_2MoO_4 \cdot 2H_2O$ | 0.01–0.1 |

*Complex media derived from plant and animal materials normally contain a considerable concentration of inorganic phosphate.

In a recent review of antibiotic biosynthesis, Liras recognised target enzymes which are: (i) repressed by phosphate; (ii) inhibited by phosphate; or (iii) repression of an enzyme occurs but phosphate repression is not clearly proved. A phosphate control sequence has also been isolated and characterised from the phosphate regulated promoter that controls biosynthesis of candicidin.

Weinberg has reviewed the nine trace elements of biological interest (atomic numbers 23–30, 42). Of these nine, the concentrations of manganese, iron and zinc are the most critical in secondary metabolism.

In every secondary metabolic system in which sufficient data has been reported, the yield of the product varies linearly with the logarithmic concentration of the 'key' metal. The linear relationship does not apply at concentrations of the metal which are either insufficient or toxic, to cell growth. Some of the primary and secondary microbial products whose yields are affected by concentrations of trace metals greater than those required for maximum growth are given in Table 9.11.

**Table 9.11.** Trace elements influencing primary and secondary metabolism.

| *Product* | *Trace element(s)* |
|---|---|
| Bacitracin | Mn |
| Protease | Mn |
| Gentamicin | Co |
| Riboflavin | Fe, Co |
| Mitomycin | Fe |
| Monensin | Fe |
| Actinomycin | Fe, Zn |
| Candicidin | Fe, Zn |

*(Cont'd ...)*

| *Product* | *Trace element(s)* |
|---|---|
| Chloramphenicol | Fe, Zn |
| Neomycin | Fe, Zn |
| Patulin | Fe, Zn |
| Streptomycin | Fe, Zn |
| Citric acid | Fe, Zn, Cu |
| Penicillin | Fe, Zn, Cu |
| Griseofulvin | Zn |

Chlorine does not appear to play a nutritional role in the metabolism of fungi. It is, however, required by some of the halophilic bacteria. Obviously, in those fermentations where a chlorine-containing metabolite is to be produced the synthesis will have to be directed to ensure that the non-chloro-derivative is not formed. The most important compounds are chlortetracycline and griseofulvin. In griseofulvin production, adequate available chloride is provided by the inclusion of at least 0.1 per cent KCl, as well as the chloride provided by the complex organic materials included as nitrogen sources. Other chlorine containing metabolites are caldriomycin, nornidulin and mollisin.

### Chelators

Many media cannot be prepared or autoclaved without the formation of a visible precipitate of insoluble metal phosphates. Gaunt demonstrated that when the medium of Mandels and Weber was autoclaved, a white precipitate of metal ions formed, containing all the iron and most of the calcium, manganese and zinc present in the medium.

The problem of insoluble metal phosphate(s) may be eliminated by incorporating low concentrations of chelating agents such as ethylene diamine tetraacetic acid (EDTA), citric acid, polyphosphates, etc., into the medium. These chelating agents preferentially form complexes with the metal ions in a medium. The metal ions then may be gradually utilised by the micro-organism. Gaunt showed that the precipitate was eliminated from Mandel and Weber's medium by the addition of EDTA at 25 mg $dm^{-3}$. It is important to check that a chelating agent does not cause inhibition of growth of the micro-organism which is being cultured. In many media, particularly those commonly used in large scale processes, there may not be a need to add a chelating agent as complex ingredients such as yeast extracts or proteose peptones will complex with metal ions and ensure gradual release of them during growth.

## GROWTH FACTORS

Some micro-organisms cannot synthesise a full complement of cell components and therefore require preformed compounds called growth factors. The growth factors most commonly required are vitamins, but there may also be a need for specific amino acids, fatty acids or sterols. Many of the natural carbon and nitrogen sources used in media formulations contain all or some of the required growth factors. When there is a vitamin deficiency it can often be eliminated by careful blending of materials. It is important to remember that if only one vitamin is required it may be occasionally more economical to add the pure vitamin, instead of using a larger bulk of a cheaper multiple vitamin source. Calcium pantothenate has been used in one medium formulation for vinegar production. In processes used for the production of glutamic acid, limited concentrations of biotin must be present in the medium. Some production strains may also require thiamine.

## NUTRIENT RECYCLE

It was shown that *M. methylotrophus* could be grown in a medium containing 86 per cent recycled supernatant plus additional fresh nutrients to make up losses. This approach made it possible to reduce the costs of media components, media preparation and storage facilities.

## BUFFERS

The control of pH may be extremely important if optimal productivity is to be achieved. A compound may be added to the medium to serve specifically as a buffer or may also be used as nutrient source. Many media are buffered at about pH 7.0 by the incorporation of calcium carbonate (as chalk). If the pH decreases the carbonate is decomposed. Obviously, phosphates which are part of many media also play an important role in buffering. However, high phosphate concentrations are critical in the production of many secondary metabolites.

The balanced use of the carbon and nitrogen sources will also form a basis for pH control as buffering capacity can be provided by the proteins, peptides and amino acids, such as in corn-steep liquor. The pH may also be controlled externally by addition of ammonia or sodium hydroxide and sulphuric acid.

## ADDITION OF PRECURSORS AND METABOLIC REGULATORS TO MEDIA

Some components of a fermentation medium help to regulate the production of the product rather than support the growth of the micro-organism. Such additives include precursors, inhibitors and inducers, all of which may be used to manipulate the progress of the fermentation.

### Precursors

Some chemicals, when added to certain fermentations, are directly incorporated into the desired product. Probably the earliest example is that of improving penicillin yields. A range of different side chains can be incorporated into the penicillin molecule. The significance of the different side chains was first appreciated when it was noted that the addition of corn-steep liquor increased the yield of penicillin from 20 units $cm^{-3}$ to 100 units $cm^{-3}$. Corn-steep liquor was found to contain phenylethylamine which was preferentially incorporated into the penicillin molecule to yield benzyl penicillin (Penicillin G). Having established that the activity of penicillin lay in the side chain and that the limiting factor was the synthesis of the side chain, it became standard practice to add side-chain precursors to the medium, in particular phenylacetic acid. Smith and Bide showed that addition of phenylacetic acid and its derivatives to the medium were capable of both increasing penicillin production three-fold and to directing biosynthesis towards increasing the proportion of benzyl penicillin from 0 to 93 per cent at the expense of other penicillins. Phenylacetic acid is still the most widely used precursor in penicillin production. Some important examples of precursors are given in Table 9.12.

### Inhibitors

When certain inhibitors are added to fermentations, more of a specific product may be produced or a metabolic intermediate which is normally metabolised is accumulated. One of the earliest examples is the microbial production of glycerol. Glycerol production depends on modifying the ethanol fermentation by removing acetaldehyde. The addition of sodium bisulphite to the broth leads to the formation of the acetaldehyde bisulphite addition compound (sodium hydroxy ethyl sulphite). Since acetaldehyde is no longer available for re-oxidation of $NADH_2$, its place as hydrogen acceptor is taken by dihydroacetone phosphate, produced during glycolysis. The product of this reaction is glycerol-3-phosphate, which is converted to glycerol.

**Table 9.12.** Precursors used in fermentation processes.

| *Precursor* | *Product* | *Micro-organism* |
|---|---|---|
| Phenylacetic-acid related compounds | Penicillin G | *Penicillium chrysogenum* |
| Phenoxy acetic acid | Penicillin V | *Penicillium chrysogenum* |
| Chloride | Chlortetracycline | *Streptomyces aureofaciens* |
| Chloride | Griseofulvin | *Penicillium griseofulvin* |
| *Propionate | Riboflavin | *Lactobacillus bulgaricus* |
| Cyanides | Vitamin B12 | *Proprianobacterium, Streptomyces* spp. |
| β-Iononones | Carotenoids | *Phycomyces blakesleenus* |
| α-Amino butyric acid | L-Isoleucine | *Bacillus subtilis* |
| D-Threonine | L-Isoleucine | *Serratia marcescens* |
| Anthranilic acid | L-Tryptophan | *Hansenula anomala* |
| Nucleosides and bases | Nikkomycins | *Streptomyces tendae* |
| Dihydronovobionic acid | Dihydronovobiocin | *Streptomyces* sp. |
| *p*-Hydroxycinnamate | Organomycin A and B | *Streptomyces organonensis* |
| DL-α-Amino butyric acid | Cyclosporin A | *Tolypocladium inflatum* |
| L-Threonine | Cyclosporin C | |
| Tyrosine or *p*-hydroxyphenylglycine | Dimethylvancomycin | *Nocardia orientalis* |

*Yields are not so high as by other techniques.

The application of general and specific inhibitors are illustrated in Table 9.13. In most cases the inhibitor is effective in increasing the yield of the desired product and reducing the yield of undesirable related products. A number of studies have been made with potential chlorination inhibitors, e.g., bromide, to minimise chlortetracycline production during a tetracycline fermentation.

**Table 9.13.** Specific and general inhibitors used in fermentations.

| *Product* | *Inhibitor* | *Main effect* | *Micro-organism* |
|---|---|---|---|
| Glycerol | Sodium bisulphite | Acetaldehyde production repressed | *Saccharomyces cerivisiae* |
| Tetracycline | Bromide | Chlortetracycline formation repressed | *Streptomyces aureofaciens* |
| Glutamic acid | Penicillin | Cell wall permeability | *Micrococcus glutamicus* |
| Citric acid | Alkali metal/phosphate, pH below 2.0 | Oxalic acid repressed | *Aspergillus niger* |
| Valine | Various inhibitors | Various effects with different inhibitors | *Brevibacterium roseum* |
| Rifamycin B | Di-ethyl barbiturate | Other rifamycins inhibited | *Nocardia mediterranei* |
| 7-Chloro-6 demethyltetracycline | Ethionine | Affects one-carbon transfer reactions | *Streptomyces aureofaciens* |

Inhibitors have also been used to affect cell-wall structure and increase the permeability for release of metabolites. The best example is the use of penicillin and surfactants in glutamic acid production.

### Inducers

The majority of enzymes which are of industrial interest are inducible. Induced enzymes are synthesised only in response to the presence in the environment of an inducer. Inducers are often substrates such as starch or dextrins for amylases, maltose for pullulanase and pectin for pectinases. Some inducers are very potent, such as isovaleronitrile inducing nitralase. Substrate analogues that are not attacked by the enzyme may also serve as enzyme inducers. Most inducers which are included in microbial enzyme media (Table 9.14) are substrates or substrate analogues, but intermediates and products may sometimes be used as inducers. For example, maltodextrins will induce amylase and fatty acids induce lipase. However, the cost may prohibit their use as inducers in commercial process.

One unusual application of an inducer is the use of yeast mannan in streptomycin production. During the fermentation varying amounts of streptomycin and mannosidostreptomycin are produced. Since mannosidostreptomycin has only 20 per cent of the biological activity of streptomycin, the former is an undesirable product. The production organism *Streptomyces griseus* can be induced by yeast mannan to produce β-mannosidase which will convert mannosidostreptomycin to streptomycin.

**Table 9.14.** Some examples of industrially important enzyme inducers.

| *Enzyme* | *Inducer* | *Micro-organism* |
|---|---|---|
| α-Amylase | Starch | *Aspergillus* spp. |
| | Maltose | *Bacillus subtilis* |
| Pullulanase | Maltose | *Aerobacter aerogenes* |
| α-Mannosidase | Yeast mannans | *Streptomyces griseus* |
| Penicillin acylase | Phenylacetic acid | *Escherichia coli* |
| Proteases | Various proteins | *Bacillus* spp. |
| | | *Streptococcus* spp. |
| | | *Streptomyces* spp. |
| | | *Aspergillus* spp. |
| | | *Mucor* spp. |
| Cellulase | Cellulose | *Trichoderma viride* |
| Pectinases | Pectin (beet pulp, apple pomace, citrus peel) | *Aspergillus* spp. |
| Nitralase | Isovaleronitrile | *Rhodococcus rhodochrous* |

It is now possible to produce a number of heterologous proteins in yeasts, fungi and bacteria. These include proteins of viral, human, animal, plant and microbial origin. However, heterologous proteins may show some degree of toxicity to the host and have a major influence on the stability of heterologous protein expression. As well as restricting cell growth as biomass the toxicity will provide selective conditions for segregant cells which no longer synthesise the protein at such a high level. Therefore, optimum growth conditions may be achieved by not synthesising a heterologous protein continuously and only inducing it after the host culture has grown up in a vessel to produce sufficient biomass. In cells of *S. cerevisiae* where the *Gal*1 promoter is part of the gene expression system, product formation may be induced by galactose addition to the growth medium which contains glycerol or low non-repressing levels of glucose as a carban source.

One commercial system that has been developed is based on the *alc*A promoter in *Aspergillus nidulans* to express human interferon α2. This can be induced by volatile chemicals, such as ethylmethyl ketone, which are added when biomass has increased to an adequate level and the growth medium contains a non-repressing carbon source or low non-repressing levels of glucose.

Methylotrophic yeasts such as *Hansenula polymorpha* and *Pichia pastoris* may be used as alternative systems because of the presence of an alcohol oxidase promoter. During growth on methanol, which also acts as an inducer, the promoter is induced to produce about 30 per cent of the cell protein. In the presence of glucose or ethanol it is undetectable. Expression systems have been developed with *P. pastoris* for tumour necrosis factor, hepatitis B surface antigen and α-galactosidase. Hepatitis B surface antigen and other heterologous proteins can also be expressed by *H. polymorpha*.

## OXYGEN REQUIREMENTS

It is sometimes forgotten that oxygen, although not added to an initial medium as such, is nevertheless a very important component of the medium in many processes and its availability can be extremely important in controlling growth rate and metabolite production.

The medium may influence the oxygen availability in a number of ways including the following:

1. *Fast metabolism*: The culture may become oxygen limited because sufficient oxygen cannot be made available in the fermenter if certain substrates, such as rapidly metabolised sugars which lead to a high oxygen demand, are available in high concentrations.
2. *Rheology*: The individual components of the medium can influence the viscosity of the final medium and its subsequent behaviour with respect to aeration and agitation.
3. *Antifoams*: Many of the antifoams in use will act as surface active agents and reduce the oxygen transfer rate.

### Fast Metabolism

Nutritional factors can alter the oxygen demand of the culture. *Penicillium chrysogenum* will utilise glucose more rapidly than lactose or sucrose and it therefore has a higher specific oxygen uptake rate when glucose is the main carbon source. Therefore, when there is the possibility of oxygen limitation due to fast metabolism, it may be overcome by reducing the initial concentration of key substrates in the medium and adding additional quantities of these substrates as a continuous or semi-continuous feed during the fermentation (Tables 9.1 and 9.15). It can also be overcome by changing the composition of the medium, incorporating higher carbohydrates (lactase, starch, etc.) and proteins which are not very rapidly metabolised and do not support such a large specific oxygen uptake rate.

**Table 9.15.** Some processes using batch feed or continuous feed or in which they have been tried.

| *Product* | *Additions* |
|---|---|
| Yeast | Molasses, nitrogen sources, P and Mg |
| Glycerol | Sugar, $Na_2CO_3$ |
| Acetone-butyl alcohol | Additions and withdrawals of wort |
| Riboflavin | Carbohydrate |
| Penicillin | Glucose and $NH_3$ |
| Novobiocin | Various carbon and nitrogen sources |

*(Cont'd ...)*

| *Product* | *Additions* |
|---|---|
| Griseofulvin | Carbohydrate |
| Rifamycin | Glucose, fatty acids |
| Gibberellins | Glucose |
| Vitamin $B_{12}$ | Glucosc |
| Tetracyclines | Glucose |
| Citric acid | Carbohydrates, $NH_3$ |
| Single-cell protein | Methanol |
| Candicidin | Glucose |
| Streptomycin | Glucose, ammonium sulphate |
| Cephalosporin | Fresh medium addition |

### Rheology

Deindoerfer and West reported that there can be considerable variation in the viscosity of compounds that may be included in fermentation media. Polymers in solution, particularly starch and other polysaccharides, may contribute to the rheological behaviour of the fermentation broth. As the polysaccharide is degraded, the effects on rheological properties will change. Allowances may also have to the made for polysaccharides being produced by the micro-organism.

## ANTIFOAMS

In most microbiological processes, foaming is a problem. It may be due to component in the medium or some factor produced by the micro-organism. The most common cause of foaming is due to proteins in the medium, such as corn-steep liquor, Pharmamedia, peanut meal, soyabean meal, yeast extract or meat extract. These proteins may denature at the air-broth interface and form a skin which does not rupture readily. The foaming can cause removal of cells from the medium which will lead to autolysis and the further release of microbial cell proteins will probably increase the stability of the foam. If uncontrolled, then numerous changes may occur and physical and biological problems may be created. These include reduction in the working volume of the fermenter due to oxygen-exhausted gas bubbles circulating in the system, changes in bubble size, lower mass and heat transfer rates, invalid process data due to interference at sensing electrodes and incorrect monitoring and control. The biological problems include deposition of cells in upper parts of the fermenter, problems of sterile operation with the air filter exits of the fermenter becoming wet and there is danger of microbial infection and the possibility of siphoning leading to loss of product.

Hall have recognised five patterns of foaming in fermentations:

1. Foaming remains at a constant level throughout the fermentation. Initially it is due to the medium and later due to microbial activity.
2. A steady fall in foaming during the early part of the fermentation, after which it remains constant. Initially it is due to the medium but there are no later effects caused by the micro-organism.
3. The foaming falls slightly in the early stages of the fermentation then rises. There are very slight effects caused by the medium but the major effects are due to microbial activity.
4. The fermentation has a low initial foaming capacity which rises. These effects are due solely to microbial activity.

5. A more complex foaming pattern during the fermentation which may be a combination of two or more of the previously described patterns.

If excessive foaming is encountered there are three ways of approaching the problem:

1. To try and avoid foam formation by using a defined medium and a modification of some of the physical parameters (pH, temperature, aeration and agitation). This assumes that the foam is due to a component in the medium and not a metabolite.
2. The foam is unavoidable and antifoam should be used. This is the more standard approach.
3. To use a mechanical foam breaker.

Antifoams are surface active agents, reducing the surface tension in the foams and destabilising protein films by (i) hydrophobic bridges between two surfaces; (ii) displacement of the absorbed protein; and (iii) rapid spreading on the surface of the film.

An ideal antifoam should have the following properties:

1. Should disperse readily and have fast action on an existing foam.
2. Should be active at low concentrations.
3. Should be long acting in preventing new foam formation.
4. Should not be metabolised by the micro-organism.
5. Should be non-toxic to the micro-organism.
6. Should be non-toxic to human and animals.
7. Should not cause any problems in the extraction and purification of the product.
8. Should not cause any handling hazards.
9. Should be cheap.
10. Should have no effect on oxygen transfer.
11. Should be heat sterilisable.

The following compounds which meet most of these requirements have been found to be most suitable in different fermentation processes.

1. Alcohols; stearyl and octyl decanol.
2. Esters.
3. Fatty acids and derivatives, particularly glycerides, which include cottonseed oil, linseed oil, soyabean oil, olive oil, castor oil, sunflower oil, rapeseed oil and cod liver oil.
4. Silicones.
5. Sulphonates.
6. Miscellaneous: Alkaterge C, oxazaline, polypropylene glycol.

These antifoams are generally added when foaming occurs during the fermentation. Because many antifoams are of low solubility they need a carrier such as lard oil, liquid paraffin or castor oil, which may be metabolised and affect the fermentation process.

Unfortunately, the concentrations of many antifoams which are necessary to control fermentations will reduce the oxygen-transfer rate by as much as 50 per cent; therefore antifoam additions must be kept to an absolute minimum. There are also other antifoams which will increase the oxygen-transfer rate. If the oxygen-transfer rate is severely affected by antifoam addition then mechanical foam breakers may have to be considered as a possible alternative. Vardar-Sukan concluded that foam control in industry is still an empirical art. The best method for a particular process in one factory is not necessarily the best for the same process on another site. The design and operating parameters of a fermenter may affect the properties and quantity of foam formed.

## MEDIUM OPTIMISATION

At this stage it is important to consider the optimisation of a medium such that it meets as many as possible of the seven criteria given in the introduction to this chapter. The meaning of optimisation in this context does need careful consideration. When considering the biomass growth phase in isolation it must be recognised that efficiently grown biomass produced by an 'optimised' high productivity growth phase is not necessarily best suited for its ultimate purpose, such as synthesising the desired product. Different combinations and sequences of process conditions need to be investigated to determine the growth conditions which produce the biomass with the physiological state best constituted for product formation. There may be a sequence of phases each with a specific set of optimal conditions.

Medium optimisation by the classical method of changing one independent variable (nutrient, antifoam, pH, temperature, etc.) while fixing all the others at a certain level can be extremely time consuming and expensive for a large number of variables. To make a full factorial search which would examine each possible combination of independent variable at appropriate levels could require a large number of experiments, $x^n$, where x is the number of levels and n is the number of variables. This may be quite appropriate for three nutrients at two concentrations ($2^3$ trials) but not for six nutrients at three concentrations. In this instance $3^6$ (729) trials would be needed. Industrially the aim is to perform the minimum number of experiments to determine optimal conditions. Other alternative strategies must therefore be considered which allow more than one variable to be changed at a time.

When more than five independent variables are to be investigated, the Plackett-Burman design may be used to find the most important variables in a system, which are then optimised in further studies. These authors give a series of designs for up to one hundred experiments using an experimental rationale known as balanced incomplete blocks. This technique allows for the evaluation of X – 1 variables by X experiments. X must be a multiple of 4, e.g., 8, 12, 16, 20, 24, etc. Normally one determines how many experimental variables need to be included in an investigation and then selects the Plackett-Burman design which meets that requirement most closely in multiples of 4. Any factors not assigned to a variable can be designated as a dummy variable. Alternatively, factors known do not have any effect may be included and designated as dummy variables. As will be shown shortly in a worked example (Table 9.16), the incorporation of dummy variables into an experiment makes it possible to estimate the variance of an effect (experimental error).

Table 9.16 shows a Plackett-Burman design for seven variables (A–G) at high and low levels in which two factors, E and G, are designated as 'dummy' variables. These can then be used in the design to obtain an estimate of error. Normally three dummy variables will provide an adequate estimate of the error. However, more can be used if fewer real variables need to be studied in an investigation. Each horizontal row represents a trial and each vertical column represents the H (high) and L (low) values of one variable in all the trials. This design (Table 9.16) requires that the frequency of each level of a variable in a given column should be equal and that in each test (horizontal row) the number of high and low variables should be equal. Consider the variable A; for the trials in which A is high, B is high in two of the trials and low in the other two. Similarly, C will be high in two trials and low in two, as will all the remaining variables. For those trials in which A is low, B will be high two times and low two times. This will also apply to all the other variables. Thus, the effects of changing the other variables cancel out when determining the effect of A. The same logic then applies to each variable. However, no changes are made to the high and low values for the E and G columns. Greasham and Inamine state that although the difference between the levels of each variable must be large enough to ensure that the optimum response will be included, caution must be taken when setting the level differential for sensitive variables, since a

**Table 9.16.** Plackett-Burman design for seven variables.

| *Trial* | *Variables* | | | | | | | *Yield* |
|---|---|---|---|---|---|---|---|---|
| | *A* | *B* | *C* | *D* | *E* | *F* | *G* | |
| 1 | H | H | H | L | H | L | H | 1.1 |
| 2 | L | H | H | H | L | H | L | 6.3 |
| 3 | L | L | H | H | H | L | H | 1.2 |
| 4 | H | L | L | H | H | H | L | 0.8 |
| 5 | L | H | L | L | H | H | H | 6.0 |
| 6 | H | L | H | L | L | H | H | 0.9 |
| 7 | H | H | L | H | L | L | H | 1.1 |
| 8 | L | L | L | L | L | L | L | 1.4 |

H denotes a high level value; and L denotes a low level value.

differential that is too large could mask the other variables. The trials are carried out in a randomised sequence.

The effects of the dummy variables are calculated in the same way as the effects of the experimental variables. If there are no interactions and no errors in measuring the response, the effect shown by a dummy variable should be 0. If the effect is not equal to 0, it is assumed to be a measure of the lack of experimental precision plus any analytical error in measuring the response.

This procedure will identify the important variables and allow them to be ranked in order of importance to decide which to investigate in a more detailed study to determine the optimum values to use.

The stages in analysing the data (Tables 9.16 and 9.17) using Nelson's example are as follows:

1. Determine the difference between the average of the H (high) and L (low) responses for each independent and dummy variable.

   Therefore the difference = $\Sigma A(H) - \Sigma A(L)$.

   The effect of an independent variable on the response is the difference between the average response for the four experiments at the high level and the average value for four experiments at the low level.

   Thus, the effect of

$$A = \frac{\Sigma A(H)}{4} - \frac{\Sigma A(L)}{4}$$

$$= \frac{2(\Sigma A(H) - \Sigma A(L))}{8}$$

   This value should be near zero for the dummy variables.

2. Estimate the mean square of each variable (the variance of effect).

   For A the mean square will be $= \dfrac{\left(\Sigma A(H) \quad \Sigma A(L)^2\right)}{8}$

3. The experimental error can be calculated by averaging the mean squares of the dummy effects of E and G.

**Table 9.17.** Analysis of the yields shown in Table 4.16.

| | Factors | | | | | | |
|---|---|---|---|---|---|---|---|
| | *A* | *B* | *C* | *D* | *E* | *F* | *G* |
| Σ(H) | 3.9 | 14.5 | 9.5 | 9.4 | 9.1 | 14.0 | 9.2 |
| Σ(L) | 14.9 | 4.3 | 9.3 | 9.4 | 9.7 | 4.8 | 9.6 |
| Difference | –11.0 | 10.2 | 0.2 | 0.0 | –0.6 | 9.2 | –0.4 |
| Effect | –2.75 | 2.55 | 0.05 | 0.00 | –0.15 | 2.30 | –0.10 |
| Mean square | 15.125 | 13.005 | 0.005 | 0.000 | 0.045 | 10.580 | 0.020 |

Mean square for 'error' $= \dfrac{0.045 + 0.020}{2} = 0.0325$

Thus, the mean square for error $= \dfrac{0.045 + 0.020}{2} = 0.0325$

This experimental error is not significant.

4. The final stage is to identify the factors which are showing large effects. In the example this was done using an F-test for

$$\frac{\text{Factor mean square}}{\text{Error mean square}}$$

This gives the following values:

$$A = \frac{15.125}{0.0325} = 465.4$$

$$B = \frac{13.005}{0.0325} = 400.2$$

$$C = \frac{0.0500}{0.0325} = 3.255$$

$$D = \frac{0.0000}{0.0325} = 0.00$$

$$F = \frac{10.580}{0.0325} = 325.6$$

When probability Tables are examined it is found that factors *A*, *B* and *F* show large effects which are very significant, whereas *C* shows a very low effect which is not significant and *D* shows no effect. *A*, *B* and *F* have been identified as the most important factors. The next stage would then be the optimisation of the concentration of each factor, which will be discussed later.

Nelson has also referred to the possibility of two-factor interactions which might occur when designing Table 9.16.

The next stage in medium optimisation would be to determine the optimum level of each key independent variable which has been identified by the Plackett-Burman design. This may be done using response surface optimisation techniques which were introduced by Box and Wilson. Hendrix has given a very readable account of this technique and the way in which it may be applied. Response surfaces are similar to contour plots or topographical maps. Whilst topographical maps show lines of constant elevation, contour plots show lines of constant value. Thus, the contours of a response surface optimisation plot show lines of identical response. In this context, response means the result of an experiment carried out at particular values of the variables being investigated.

The axes of the contour plot are the experimental variables and the area within the axes is termed the response surface. To construct a contour plot, the results (responses) of a series of experiments employing different combinations of the variables are inserted on the surface of the plot at the points delineated by the experimental conditions. Points giving the same results (equal responses) are then joined together to make a contour line. In its simplest form two variables are examined and the plot is two-dimensional. It is important to appreciate that both variables are changed in the experimental series, rather than one being maintained constant, to ensure that the data are distributed over the response surface. In Fig. 9.1, the profile generated by fixing $X_1$ and changing $X_2$ and then using the best $X_2$ value and changing $X_1$ constitutes a cross which may not encroach upon the area in which the optimum resides.

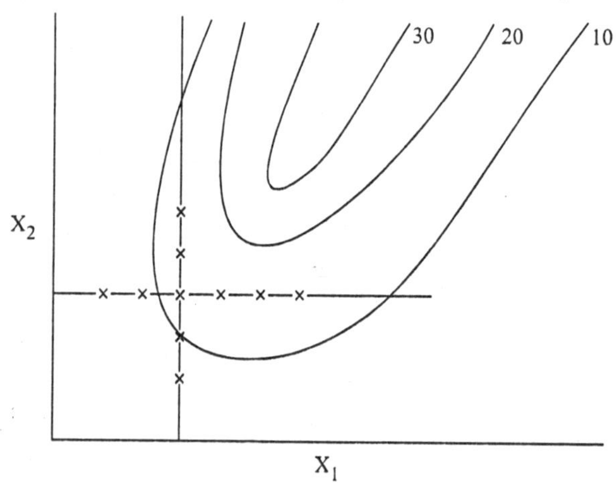


**Fig. 9.1.** Optimal point of a response surface by one factor at a time.

The technique may be applied at different levels of sophistication. Hendrix applied the technique at its simplest level to predict the optimum combination of two variables. The values of the variables for the initial experiments are chosen randomly or with the guidance of previous experience of the process. There is little to be gained from using more than 15–20 experiments. The resulting contour map gives an indication of the area of the optimum combination of variables resides. A new set of experiments may then be designed within the indicated zone. Hendrix proposed the following strategy to arrive at the optimum in an incremental fashion:

1. Define the space on the plot to be explored.
2. Run five random experiments in this space.
3. Define a new space centred upon the best of the five experiments and make the new space smaller than the previous one, perhaps by cutting each dimension by one-half.
4. Run five more random experiments in this new space.
5. Continue doing this until no further improvement is observed or until you cannot afford any more experiments.

The more sophisticated applications of the response surface technique use mathematical models to analyse the first round of experimental data and to predict the relationship between the response and the variables. These calculations then allow predictive contours to be drawn and facilitate a more rapid optimisation with fewer experiments. If three or more variables are to be examined then several contour maps will have to be constructed. Hicks gives an excellent account of the development of equations to model the different interactions which may take place between the variables. Several computer software packages are now available which allow the operator to determine the equations underlying the responses and thus, to determine the likely area on the surface in which the optimum resides. Some examples of the types of response surface profiles that may be generated are illustrated in Fig. 9.2.

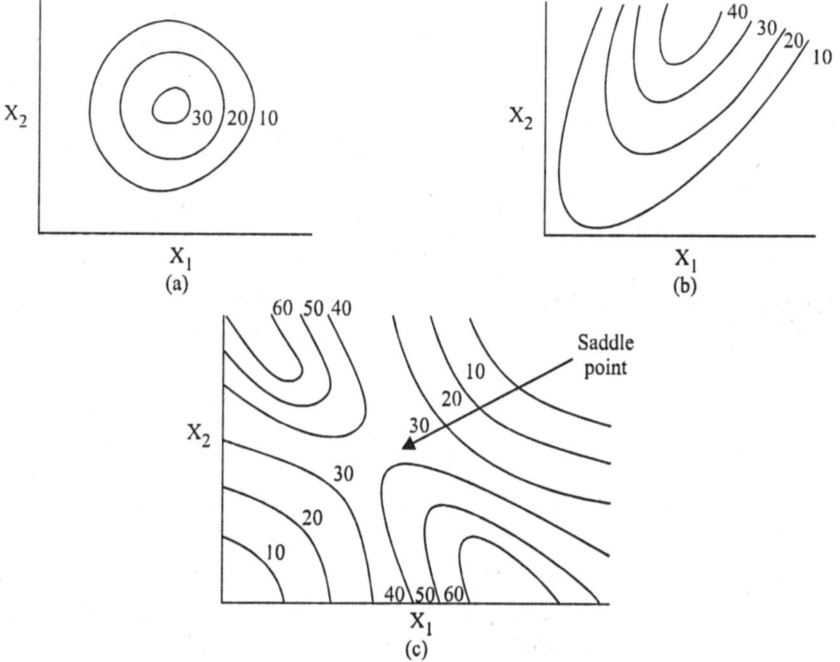


**Fig. 9.2.** Typical response surfaces in two dimensions; (a) mound, (b) rising ridge, (c) saddle.

The following examples illustrate the application of the technique:

1. The variables under investigation were cerelose and soyabean level, with the analysis indicating the optimum to be 6.2 per cent cerelose and 3.2 per cent soyabean (Fig. 9.3).
2. The medium for streptomycin production was optimised for four components resulting in a 52 per cent increase in streptomycin yield, a 10 per cent increase in mycelial dry weight and a 48 per cent increase in specific growth rate (Table 9.18).

**Table 9.18.** Concentration of nutrients in an original and optimised medium for streptomycin production.

| *Nutrient ($g\ dm^{-3}$)* | *Original medium* | *Optimised medium* |
|---|---|---|
| Glucose | 10 | 23 |
| Beer-yeast autolysate | 25 | 27 |
| NaCl | 10 | 8 |
| $K_2HPO_4$ | 1 | 1 |

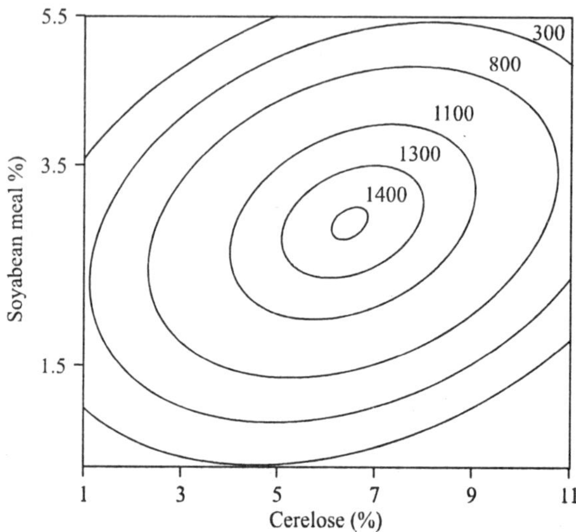


**Fig. 9.3.** Contour plot of two independent variables, cerelose and soyabean meal, for optimisation of the candidin fermentation.

When further optimisation experiments are necessary for medium development in large vessels, the number of experiments will normally be restricted because of the cost and the lack of spare large vessels. The simplex search method attempts to optimise n variables by initially performing n + 1 experimental trials. The results of this initial set of trials are then used to predict the conditions of the next experiment and the situation is repeated until the optimum combination is attained. Thus, after the first set of trials the optimisation proceeds as individual experiments. The prediction is achieved using a graphical representation of the trials where the experimental variables are the axes. Using this procedure, the experimental variables are plotted and not the results of the experiments. The initial experimental conditions are chosen such that the points on the graph are equidistant from one another and form the vertices of a polyhedron described as the simplex. Thus, with two variables the simplex will be an equilateral triangle. The results of the initial set of three experiments are then used to predict the next experiment enabling a new simplex to be constructed. The procedure will be explained using an example to optimise the concentrations of carbon and nitrogen sources in a medium for antibiotic production.

In our example a graph is constructed in which the x-axis represents the concentration range of the carbon source (the first variable) and the y-axis represents the concentration range of the nitrogen source (the second variable). The first vertex A (experimental point) of the simplex represents the current concentrations of the two variables which are producing the best yield of the antibiotic. The experiment for the second vertex B is planned using a new carbon-nitrogen mixture and the position of the third vertex C can now be plotted on the graph using lengths AC and BC equal to AB (the simplex equilateral triangle, Fig. 9.4a). The concentrations of the carbon and nitrogen sources to use in the third experiment can now be determined graphically and the experiment can be undertaken to determine the yield of antibiotic. The results of the three experiments are assessed and the worst response to antibiotic production identified. In our example, experiment A was the worst and B the best. The simplex design is now used to design the next experiment. A new simplex (equilateral triangle) BCD is constructed opposite the worst response (i.e., A) using the existing vertices B and C. A line is drawn from A through the centroid (mid-point) of BC. D (the next experiment) will be on this line and the sides BD and CD will be the same length as BC. This process of constructing the new simplex is described as reflection. Once the position of D is known, the concentrations of the carbon and nitrogen sources can be determined graphically, the experiment performed and the production of antibiotic assayed. Thus, a series of simplexes can be constructed moving in a crabwise way. The procedure is continued until the optimum is located. At this point the simplex begins to circle on its self, indicating the optimum concentration (Fig. 9.4b). However, if a new vertex exhibits the lowest response, the simplex would reflect back on to the previous one, halting movement towards the optimum. In this case the new simplex is constructed opposite the second least desirable response using the method previously described.

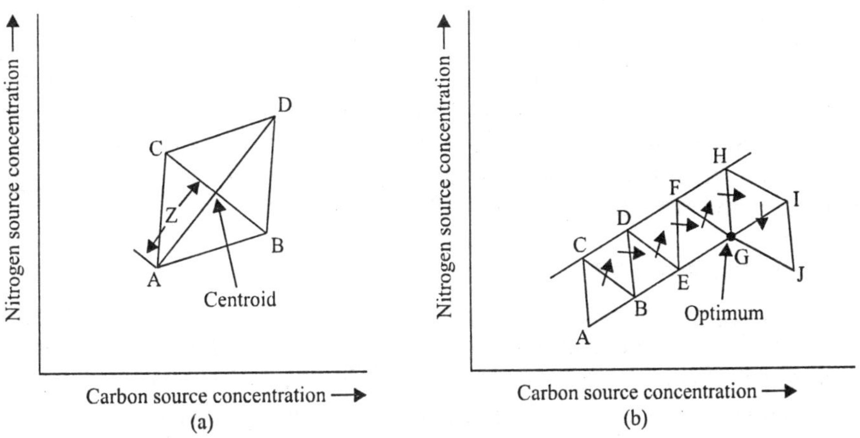


**Fig. 9.4.** (a) Simplex optimisation for a pair of independent variables (with reflection); and (b) Simplex optimisation of pair of independent variables which has reached the optimum.

If it is decided that the supposed optimum should be reached more rapidly then the distance $z$ between the centroid and D may be increased (expanded) by a factor which is often 2. If the optimum is thought to have nearly been reached then the distance $z$ may be decreased by a factor of 0.5 (contraction). This modified simplex optimisation was first proposed by Nelder and Mead and has been discussed by Greasham and Inamine.

The simplex method may also be used in small scale media development experiments to help identify the possible optimum concentration ranges to test in more extensive multifactorial experiments.

## Animal Cell Media

Mammalian cell lines have been cultured *in vitro* for 40 years. Initially, animal cells were required for vaccine manufacture but they are now also used in the production of monoclonal antibodies, interferon, etc. The media initially used for this purpose contained about 10 per cent serum (foetal calf or calf) plus other organic and inorganic components. Since by this pioneering work it has been possible to develop a range of serum-free media. These media contain carbohydrates, amino acids, vitamins, nucleic acids, etc., dissolved in high purity water. Media costs are therefore considerably higher than those for microbial cells. At a 1000 $dm^3$ scale the medium costs may account for 40 per cent of the unit costs and serum may be 80 per cent of the medium cost.

## Serum

The serum is a very complex mixture containing approximately 1000 components including inorganic salts, amino acids, vitamins, carbon sources, hormones, growth factors, haemoglobin, albumin and other compounds. However, most of them do not appear to be needed for growth and differentiation of cell lines which have been tested.

Serum is used extensively in production media for animal cell culture to produce recombinant proteins and antibody based products *in vivo* use in human. At present regulations governing the quality of serum which can be used for manufacturing processes vary considerably from country to country. However, FDA approval of a process will be essential to market a product in the USA and therefore regulate the quality of serum which can be used. Serum tested by approved laboratories should be free of bacterial, viral or BSE (bovine sporangiphorm encephalitis) contamination and other components should be within strictly defined limits. Serum of this standard is needed for the cell culture media which is used to maintain the cell culture stocks as well as the production media.

The cost of foetal calf serum, $dm^{-3}$ in Europe, makes serum free media attractive economic alternatives, but it would take a number of years to develop suitable serum free media. The absence of the many unutilised components in serum will also simplify purification of potential products produced in such media. However, these process changes would need approval by the FDA or other regulatory bodies before a product could be marketed using a modified process.

### Serum-Free Media Supplements

The development of serum-free media was initiated by Ham who reduced the amount of serum in media and optimised other medium components and Sato who investigated a range of components to promote cell growth and differentiation. Some of the more important replacements in serum-free media are albumin, insulin, transferrin, ethanolamine, selenium and β-mercaptoethanol.

The advantages of removing serum from media include:

1. More consistent and definable medium composition to reduce batch variation.
2. Reduction in potential contamination to make sterility easier to achieve.
3. Potential cost savings because of cheaper replacement components.
4. Simplifying downstream processing because the total protein content of the medium has been reduced.

### Protein-Free Media

The elimination of proteins seems an attractive objective. However, the design of such media is difficult and their use may be very limited and not very cost effective. Hamilton and Ham demonstrated the growth of Chinese hamster cell lines in a protein-free medium formulated from amino acids, vitamins, organic compounds and inorganic salts. Other media have been developed by Cleveland and Shive.

### Trace Elements

The role of trace elements in medium formulation can be significant. Cultured cells normally require Fe, Zn, Cu, Se, Mn, Mo and V. These are often present as impurities in other media components. Cleveland found that if the number of trace elements were increased, insulin, transferrin, albumin and liposomes were not needed in a serum-free hybridoma medium. They included Al, Ag, Ba, Br, Cd, Co, Cr, F, Ge, J, Rb, Zr, Si, Ni and Sn as well as those previously mentioned.

### Osmolality

The optimum range of the osmotic pressure for growth is often quite narrow and varies with the type of cell and the species from which it was isolated. It may be necessary to adjust the concentration of NaCl when major additions are made to a medium.

## pH

The normal buffer system in tissue culture media is the $CO_2$-bicarbonate system. This is a weak buffering system and can be improved by the use of a zwitterionic buffer such as Hepes, either in addition to or instead of the $CO_2$-bicarbonate buffer. Continuous pH control is achieved by the addition of sodium bicarbonate or sodium hydroxide (with fast mixing) when too acidic. The pH does not normally become too alkaline so acid additions are not required but provision may be made for $CO_2$ additions.

## Non-Nutritional Media Supplements

Sodium carboxy methyl cellulose may be added to media at 0.1 per cent to help to minimise mechanical damage caused by the shear force generated by the stirrer impeller. The problems of foam formation and subsequent cell damage and losses can affect animal cell growth. Pluronic F-68 (polyglycol) can provide a protective effect to animal cells in stirred and sparged vessels. In media which are devoid of Pluronic F-68, cells may become more sensitive to direct bubble formation in the presence of an antifoam agent being used to suppress foam formation.

Chapter 10

# Sterilisation

## INTRODUCTION

A fermentation product is produced by the culture of a certain organism or organisms, in a nutrient medium. If the fermentation is invaded by a foreign micro-organism then the following consequences may occur:

1. The medium would have to support the growth of both the production organism and the contaminant, resulting in a loss of productivity.
2. If the fermentation is a continuous one then the contaminant may 'outgrow' the production organism and displace it from the fermentation.
3. The foreign organism may contaminate the final product, e.g., single-cell protein where the cells, separated from the broth, constitute the product.
4. The contaminant may produce compounds which make subsequent extraction of the final product difficult.
5. The contaminant may degrade the desired product; this is common in bacterial contamination of antibiotic fermentations where the contaminant would have to be resistant to the normal inhibitory effects of the antibiotic and degradation of the antibiotic is a common resistance mechanism, e.g., the degradation of β-lactam antibiotics by β-lactamase-producing bacteria.
6. Contamination of a bacterial fermentation with phage could result in the lysis of the culture.

Avoidance of contamination may be achieved by:

1. Using a pure inoculum to start the fermentation.
2. Sterilising the medium to be employed.
3. Sterilising the fermenter vessel.
4. Sterilising all materials to be added to the fermentation during the process.
5. Maintaining aseptic conditions during the fermentation.

The extent to which these procedures are adopted is determined by the likely probability of contamination and the nature of its consequences. Some fermentations are described as 'protected'—that is, the medium may be utilised by only a very limited range of micro-organisms or the growth of the process organism may result in the development of selective growth conditions, such as a reduction in pH. The brewing of beer falls into this category; hop resins tend to inhibit the growth of many micro-organisms and the growth of brewing yeast tends to decrease the pH of the medium. Thus, brewing worts are boiled, but not necessarily sterilised and the fermenters are thoroughly cleaned with disinfectant solution but are not necessarily sterile. Also, the precautions used in the development of inoculum for brewing are far less stringent than, for example, in an antibiotic fermentation. However, the vast majority of fermentations are not 'protected' and if contaminated, would suffer some of the consequences previously listed.

## MEDIUM STERILISATION

As pointed out by Corbett, media may be sterilised by filtration, radiation, ultrasonic treatment, chemical treatment or heat. However, for practical reasons, steam is used almost universally for the sterilisation of fermentation media. The major exception is the use of filtration for the sterilisation of media for animal-cell culture—such media are completely soluble and contain heat labile components making filtration the method of choice. Filtration techniques will be considered later in this chapter. Before the techniques which are used for the steam sterilisation of culture media are discussed, it is necessary to discuss the kinetics of sterilisation. The destruction of micro-organisms by steam (moist heat) may be described as a first-order chemical reaction and thus, may be represented by the following equation:

$$-dN/dt = kN \quad \text{... (10.1)}$$

where $N$ is the number of viable organisms present,

$t$ is the time of the sterilisation treatment,

$k$ is the reaction rate constant of the reaction or the specific death rate.

It is important at this stage to appreciate that we are considering the total number of organisms present in the volume of medium to be sterilised, not the concentration—the minimum number of organisms to contaminate a batch is one, regardless of the volume of the batch. On integration of equation (10.1) the following expression is obtained:

$$N_t/N_0 = e^{-kt} \quad \text{... (10.2)}$$

where $N_0$ is the number of viable organisms present at the start of the sterilisation treatment,

$N_t$ is the number of viable organisms present after a treatment period, t.

On taking natural logarithms, equation (10.2) is reduced to:

$$\ln(N_t/N_0) = -kt \quad \text{... (10.3)}$$

The graphical representations of equations (10.1) and (10.3) are illustrated in Fig. 10.1, from which it may be seen that viable organism number declines exponentially over the treatment period. A plot of the natural logarithm of $N_t/N_0$ against time yields a straight line, the slope of which equals $-k$. This kinetic description makes two predictions which appear anomalous:

1. An infinite time is required to achieve sterile conditions (i.e. $N_t = 0$).
2. After a certain time there will be less than one viable cell present.

Thus, in this context, a value of $N_t$ of less than one is considered in terms of the probability of an organism surviving the treatment. For example, if it were predicted that a particular treatment period reduced the population to 0.1 of a viable organism, this implies that the probability of one organism surviving the treatment is one in ten. This may be better expressed in practical terms as a risk of one batch in ten becoming contaminated. This aspect of contamination will be considered later.

The relationship displayed in Fig. 10.1 would be observed only with the sterilisation of pure culture in one physiological form, under ideal sterilisation conditions. The value of $k$ is not only species dependent, but dependent on the physiological form of the cell; for example, the endospores of the genus *Bacillus* are far more heat resistant than the vegetative cells. Richards produced a series of graphs illustrating the deviation from theory which may be experienced in practice. Figures 10.2a, 10.2b and 10.2c illustrate the effect of the time of heat treatment on the survival of population of bacterial endospores. The deviation from an immediate exponential decline in viable spore number is due to the heat activation of the spores, that is the induction of spore germination by the heat and moisture of the initial period of the sterilisation process. In Fig. 10.2a the activation of spores is significantly more than their destruction during the early stages of the process and therefore, viable numbers increase before the observation of exponential decline. In Fig. 10.2b activation is balanced by spore death and in Fig. 10.2c activation is less than spore death.

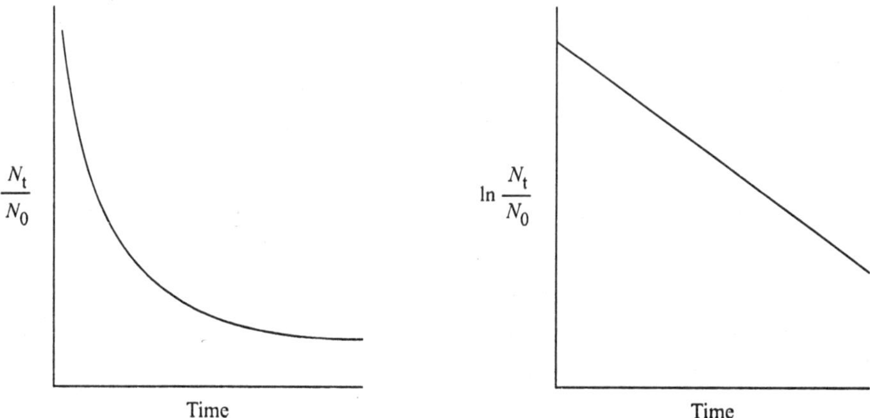


**Fig. 10.1.** Plots of the proportion of survivors and the natural logarithm of the proportion of survivors in a population of micro-organisms subjected to a lethal temperature over a time period.

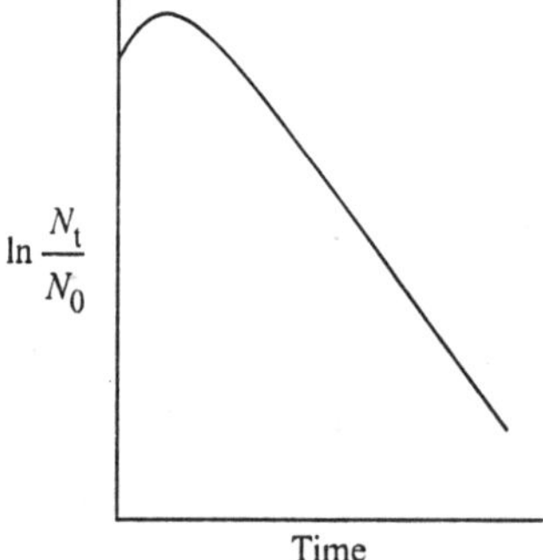


**Fig. 10.2a.** Initial population increase resulting from the heat activation of spores in the early stages of a sterilisation process.

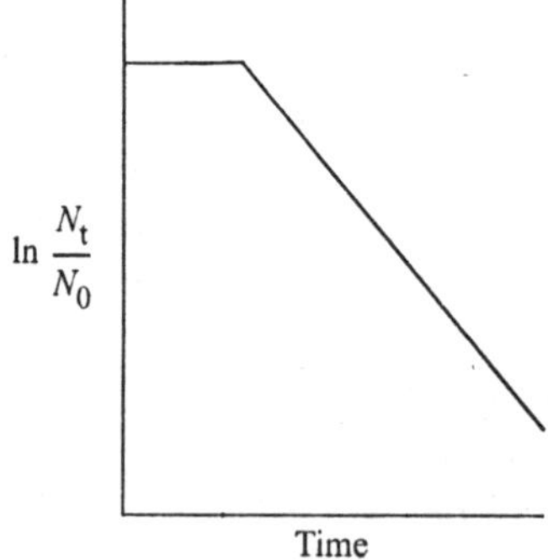


**Fig. 10.2b.** An initial stationary period observed during a sterilisation treatment due to the death of spores being completely compensated by the heat activation of spores.

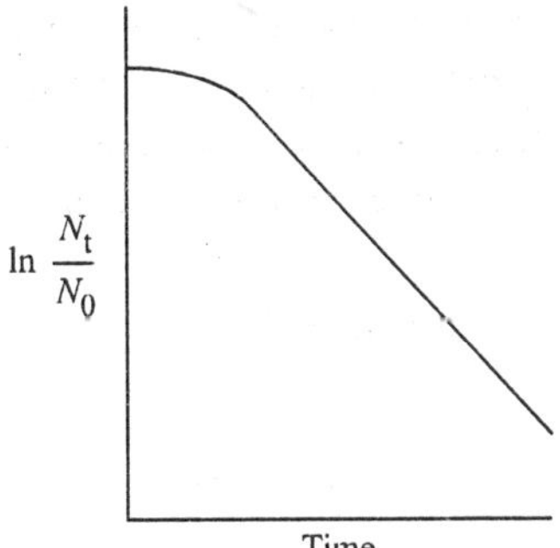


**Fig. 10.2c.** Initial population decline at a sub-maximum rate during a sterilisation treatment due to the death of spores being compensated by the heat activation of spores.

Figures 10.3a and 10.3b illustrate typical results of the sterilisation of mixed cultures containing two species with different heat sensitivities. In Fig. 10.3a the population consists mainly of the less-resistant type where the initial decline is due principally to the destruction of the less-resistant cell population and the later, less rapid decline, is due principally to the destruction of the more resistant cell population. Fig. 10.3b represents the reverse situation where the more resistant type predominates and its presence disguises the decrease in the number of the less resistant type.

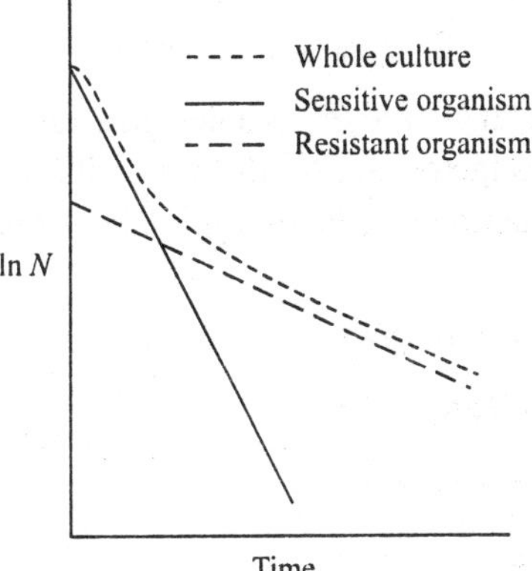


**Fig. 10.3a.** The effect of a sterilisation treatment on a mixed culture consisting of a high proportion of a very sensitive organism.

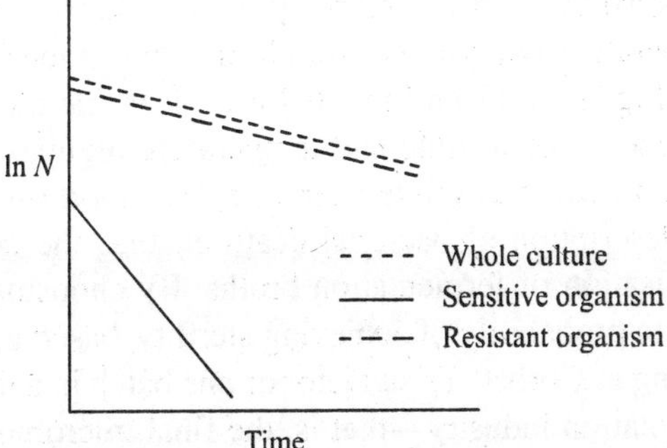


**Fig. 10.3b.** The effect of a sterilisation treatment on a mixed culture consisting of a high proportion of a relatively resistant organism.

As with any first-order reaction, the reaction rate increases with increase in temperature due to an increase in the reaction rate constant, which, in the case of the destruction of micro-organisms, is the specific death rate (k). Thus, k is a true constant only under constant temperature conditions. The relationship between temperature and the reaction rate constant was demonstrated by Arrhenius and may be represented by the equation:

$$d \ln k/dT = E/RT^2 \qquad \text{... (10.4)}$$

where $E$ is the activation energy,
$R$ is the gas constant,
$T$ is the absolute temperature.
On integration equation (10.4) gives:

$$k = Ae^{-E/RT} \qquad \text{... (10.5)}$$

where $A$ is the Arrhenius constant. On taking natural logarithms, equation (10.5) becomes:

$$\ln k = \ln A - E/RT \qquad \text{... (10.6)}$$

From equation (10.6) it may be seen that a plot of ln k against the reciprocal of the absolute temperature will give a straight line. Such a plot is termed an Arrhenius plot and enables the calculation of the activation energy and the prediction of the reaction rate for any temperature. By combining together equations (10.3) and (10.5), the following expression may be derived for the heat sterilisation of a pure culture at a constant temperature:

$$\ln N_0/N_t = A{\cdot}t{\cdot}e^{-E/RT} \qquad \text{... (10.7)}$$

Deindoerfer and Humphrey used the term ln $N_0/N_t$ as a design criterion for sterilisation, which has been variously called the Del factor, Nabla factor and sterilisation criterion represented by the term $\nabla$. Thus, the Del factor is a measure of the fractional reduction in viable organism count produced by a certain heat and time regime. Therefore:

$$\nabla = \ln (N_0/N_t)$$

but $$\ln (N_0/N_t) = kt$$

and $$kt = A{\cdot}t{\cdot}e^{-(E/RT)}$$

thus $$\nabla = A{\cdot}t{\cdot}e^{-(E/RT)} \qquad \text{... (10.8)}$$

On rearranging, equation (10.8) becomes:

$$\ln t = E/RT + \ln (\nabla/A) \qquad \text{... (10.9)}$$

Thus, a plot of the natural logarithm of the time required to achieve a certain $\nabla$ value against the reciprocal of the absolute temperature will yield a straight line, the slope of which is dependent on the activation energy, as shown in Fig. 10.4. From Fig. 10.4 it is clear that the same degree of sterilisation ($\nabla$) may be obtained over a wide range of time and temperature regimes; that is, the same degree of sterilisation may result from treatment at a high temperature for a short time as from a low temperature for a long time. This kinetic description of bacterial death enables the design of procedures (giving certain $\nabla$ factors) for the sterilisation of fermentation broths. By choosing a value for $N_t$, procedures may be designed having a certain probability of achieving sterility, based upon the degree of risk that is considered acceptable. According to Corbett a risk factor of one batch in a thousand being contaminated is frequently used in the fermentation industry—that is, the final microbial count in the medium after sterilisation should be $10^{-3}$ viable cells. However, to apply these kinetics it is necessary to know the thermal death.

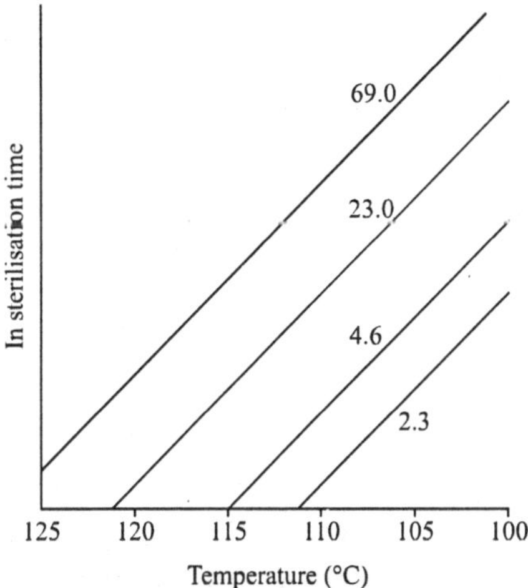


**Fig. 10.4.** The effect of sterilisation and temperature on the Del factor achieved in the process. The figures on the graph indicate the Del factors for each straight line.

characteristics of all the taxa contaminating the fermenter and unsterile medium. This is an impossibility and therefore, the assumption may be made that the only microbial contaminants present are spores of *Bacillus stearothermophilus*—that is, one of the most heat-resistant microbial types known. Thus, by adopting *B. stearothermophilus* as the design organism a considerable safety factor should be built into the calculations. It should be remembered that *B. stearothermophilus* is not always adopted as the design organism. If the most heat-resistant organism contaminating the medium ingredients is known, then it may be advantageous to base the sterilisation process on this organism. Deindoerfer and Humphrey determined the thermal death characteristics of *B. stearothermophilus* spores as:

$$\text{Activation energy} = 67.7 \text{ kcal mole}^{-1}$$

$$\text{Arrhenius constant} = 1 \times 10^{36.2} \text{ second}^{-1}$$

However, it should be remembered that these kinetic values will vary according to the medium in which the spores are suspended and this is particularly relevant when considering the sterilisation of fats and oils (which are common fermentation substrates) where the relative humidity may be quite low. Bader demonstrated that spores of *Bacillus macerans* suspended in oil were ten times more resistant to sterilisation if they were dry than if they were wet.

A regime of time and temperature may now be determined to achieve the desired Del factor. However, a fermentation medium is not an inert mixture of components and deleterious reactions may occur in the medium during the sterilisation process, resulting in a loss of nutritive quality. Thus, the choice of regime is dictated by the requirement to achieve the desired reduction in microbial content with the least detrimental effect on the medium. Figure 10.5 illustrates the deleterious effect of increasing medium sterilisation time on the yield of product of subsequent fermentations. The initial rise in yield is due to some components of the medium being made more available to the process micro-organisms by the 'cooking effect' of a brief sterilisation period.

## Types of Reaction

Two types of reaction contribute to the loss of nutrient quality during sterilisation:

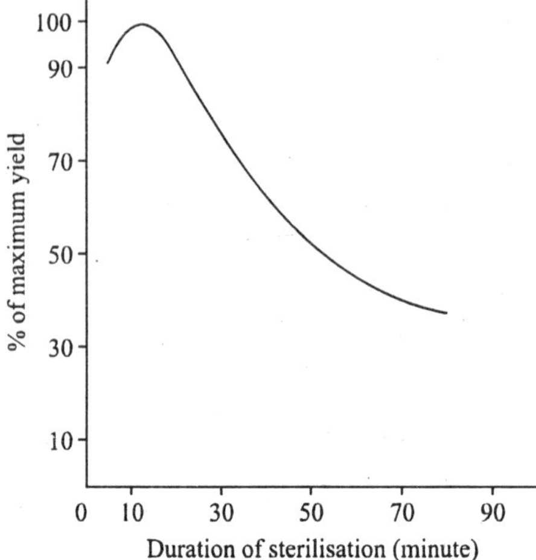


**Fig. 10.5.** The effect of the time of sterilisation on the yield of a subsequent fermentation.

### *Interactions between nutrient components of the medium*

A common occurrence during sterilisation is the Maillard-type browning reaction which results in discolouration of the medium as well as loss of nutrient quality. These reactions are normally caused by the reaction of carbonyl groups, usually from reducing sugars, with the amino groups of amino acids and proteins. An example of the effect of sterilisation time on the availability of glucose in a corn-steep liquor medium is shown in Table 10.1. Problems of this type are normally resolved by sterilising the sugar separately from the rest of the medium and re-combining the two after cooling.

**Table 10.1.** The effect of sterilisation time on glucose concentration and product accretion rate in an antibiotic fermentation.

| *Time at 121° (min.)* | *Amount of added glucose remaining (%)* | *Relative accretion rate* |
|---|---|---|
| 60 | 35 | 90 |
| 40 | 46 | 92 |
| 30 | 64 | 100 |

### *Degradation of heat labile components*

Certain vitamins, amino acids and proteins may be degraded during a steam sterilisation regime. In extreme cases, such as the preparation of media for animal-cell culture, filtration may be used and this aspect will be discussed later in the chapter. However, for the vast majority of fermentations these problems may be resolved by the judicious choice of steam sterilisation regime.

The thermal destruction of essential media components conforms approximately with first order reaction kinetics and therefore, may be described by equations similar to those derived for the destruction of bacteria:

$$x_t/x_0 = e^{-kt} \qquad \text{... (10.10)}$$

where $x_t$ is the concentration of nutrient after a heat treatment period, t,

$x_0$ is the original concentration of nutrient at the onset of sterilisation,

k is the reaction rate constant.

It is important to appreciate that we are considering the decline in the concentration of the nutrient component, whereas we consider the decline in the number of contaminants. The effect of temperature on the reaction rate constant may be expressed by the Arrhenius equation:

$$\ln k = \ln A - E/RT.$$

Therefore, a plot of the natural logarithm of the reaction rate against $1/T$ will give a straight line, slope—$(E/R)$. As the value of $R$, the gas constant, is fixed the slope of the graph is determined by the value of the activation energy ($E$). The activation energy for the thermal destruction of *B. stearothermophilus* spores has been cited as 67.7 kcal $mol^{-1}$, whereas that for thermal destruction of nutrients is 10 to 30 kcal $mole^{-1}$. Figure 10.6 is an Arrhenius plot for two reactions—one with a lower activation energy than the other. From this plot it may be seen that as temperature is increased, the reaction rate rises more rapidly for the reaction with the higher activation energy. Thus, considering the difference between activation energies for spore destruction and nutrient degradation, an increase in temperature would accelerate spore destruction more than medium denaturation.

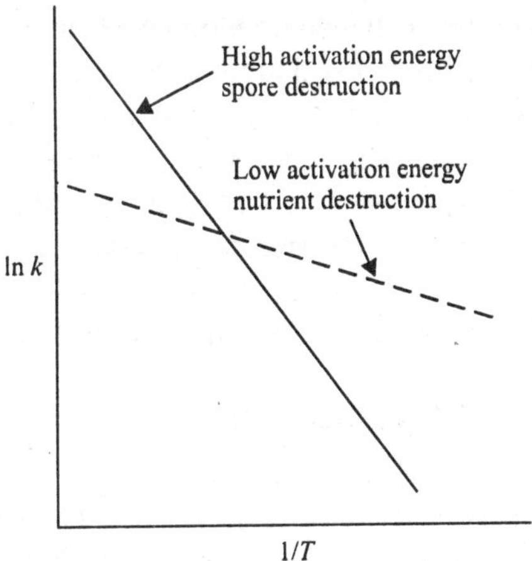


**Fig. 10.6.** The effect of activation energy on spore and nutrient destruction.

In the consideration of Del factors it was evident that the same Del factor could be achieved over a range of temperature/time regimes. Thus, it would appear to be advantageous to employ a high temperature for a short time to achieve the desired probability of sterility, yet causing minimum nutrient degradation. Thus, the ideal technique would be to heat the fermentation medium to a high temperature, at which it is held for a short period, before being cooled rapidly to the fermentation temperature. However, it is obviously impossible to heat a batch of many thousands of litres of broth in a tank to a high temperature, hold for a short period and cool without the heating and cooling periods contributing considerably to the total sterilisation time. The only practical method of materialising the objective of a short-time, high-temperature treatment is to sterilise the medium in a continuous stream. In the past the fermentation industry was reluctant to adopt continuous sterilisation due to a number of disadvantages outweighing the advantage of nutrient quality. The relative merits of batch and continuous sterilisation may be summarised as follows:

### Advantages of Continuous Sterilisation over Batch Sterilisation

1. Superior maintenance of medium quality.
2. Ease of scale-up—discussed later.

3. Easier automatic control.
4. The reduction of surge capacity for steam.
5. The reduction of sterilisation cycle time.
6. Under certain circumstances, the reduction of fermenter corrosion.

### Advantages of Batch Sterilisation over Continuous Sterilisation

1. Lower capital equipment costs.
2. Lower risk of contamination—continuous processes require the aseptic transfer of the sterile broth to the sterile vessel.
3. Easier manual control.
4. Easier to use with media containing a high proportion of solid matter.

The early continuous sterilisers were constructed as plate heat exchangers and these were unsuitable on two accounts:

1. Failure of the gaskets between the plates resulted in the mixing of sterile and unsterile streams.
2. Particulate components in the media would block the heat exchangers.

However, modern continuous sterilisers use double spiral heat exchangers in which the two streams are separated by a continuous steel division. Also, the spiral exchangers are far less susceptible to blockage. However, a major limitation to the adoption of continuous sterilisation was the precision of control necessary for its success. This precision has been achieved with the development of sophisticated computerised monitoring and control systems resulting in continuous sterilisation being very widely used and it is now the method of choice. However, batch sterilisation is still used in many fermentation plants and thus, it will be considered here before continuous sterilisation is discussed in detail.

## DESIGN OF BATCH STERILISATION PROCESSES

Although a batch sterilisation process is less successful in avoiding the destruction of nutrients than a continuous one, the objective in designing a batch process is still to achieve the required probability of obtaining sterility with the minimum loss of nutritive quality. The highest temperature which appears to be feasible for batch sterilisation is 121°C so the procedure should be designed such that exposure of the medium to this temperature is kept to a minimum.

This is achieved by taking into account the contribution made to the sterilisation by the heating and cooling periods of the batch treatment. Deindoerfer and Humphrey presented a method to assess the contribution made by the heating and cooling periods. The following information must be available for the design of a batch sterilisation process:

1. A profile of the increase and decrease in the temperature of the fermentation medium during the heating and cooling periods of the sterilisation cycle.
2. The number of micro-organisms originally present in the medium.
3. The thermal death characteristics of the 'design' organism. As explained earlier this may be *Bacillus stearothermophilus* or an alternative organism relevant to the particular fermentation.

Knowing the original number of organisms present in the fermenter and the risk of contamination considered acceptable, the required Del factor may be calculated. A frequently adopted risk of contamination is 1 in 1000, which indicates that $N_t$ should equal $10^{-3}$ of a viable cell. It is worth reinforcing at this stage that we are considering the total number of organisms present in the medium and not the concentration. If a specific case is considered where the unsterile broth was shown to contain $10^{11}$ viable organisms, then the Del factor may be calculated, thus:

$$\nabla = \ln (10^{11}/10^{-3})$$

$$\nabla = \ln 10^{14}$$

$$= 32.2 \quad \text{... (10.11)}$$

Therefore, the overall Del factor required is 32.2. However, the destruction of cells occurs during the heating and cooling of the broth as well as during the period at 121°C, thus, the overall Del factor may be represented as:

$$\nabla_{\text{overall}} = \nabla_{\text{heating}} + \nabla_{\text{holding}} + \nabla_{\text{cooling}}$$

Knowing the temperature-time profile for the heating and cooling of the broth (prescribed by the characteristics of the available equipment) it is possible to determine the contribution made to the overall Del factor by these periods. Thus, knowing the Del factors contributed by heating and cooling, the holding time may be calculated to give the required overall Del factor.

### Calculation of the Del Factor during Heating and Cooling

The relationship between Del factor, the temperature and time is given by equation (10.8):

$$\nabla = A \cdot t \cdot e^{-(E/RT)}$$

However, during the heating and cooling periods the temperature is not constant and therefore, the calculation of $\nabla$ would require the integration of equation (10.8) for the time-temperature regime observed. Deindoerfer and Humphrey produced integrated forms of the equation for a variety of temperature-time profiles, including linear, exponential and hyperbolic. However, the regime observed in practice is frequently difficult to classify, making the application of these complex equations problematical. Richards demonstrated the use of a graphical method of integration and this is illustrated in Fig. 10.7. The time axis is divided into a number of equal increments, $t_1$, $t_2$, $t_3$, etc. Richards suggesting 30 as a reasonable number.

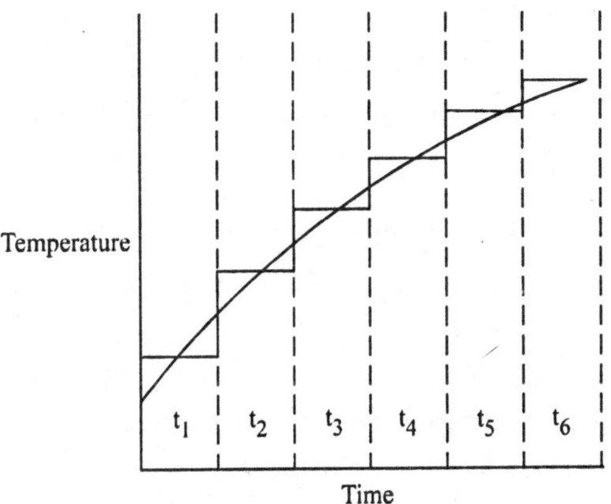


**Fig. 10.7.** The graphical integration method applied to the increase in temperature over a time period. $t_1$, $t_2$, etc., represent equal time intervals.

For each increment, the temperature corresponding to the mid-point time is recorded. It may now be approximated that the total Del factor of the heating-up period is equivalent to the sum of the Del factors of the mid-point temperature for each time increment. The value of the specific death rate of *B. stearothermophilus* spores at each mid-point temperature may be deduced from the Arrhenius equation using the thermal death characteristic. The value of the Del factor corresponding to each time increment may then be calculated from the equations:

$$\nabla_1 = k_1 t$$

$$\nabla_2 = k_2 t$$

$$\nabla_3 = k_3 t$$

The sum of the Del factors for all the increments will then equal the Del factor for the heating-up period. The Del factor for the cooling-down period may be calculated in a similar fashion.

### Calculation of the Holding Time at Constant Temperature

From the previous calculations the overall Del factor, as well as the Del factors of the heating and cooling parts of the cycle, have been determined. Therefore, the Del factor to be achieved during the holding time may be calculated by difference:

$$\nabla_{\text{holding}} = \nabla_{\text{overall}} - \nabla_{\text{heating}} - \nabla_{\text{cooling}}$$

Using our example where the overall Del factor is 32.2 (equation 10.11) and if it is taken that the heating Del factor was 9.8 and the cooling Del factor 10.1, the holding Del factor may be calculated:

$$\nabla_{\text{holding}} = 32.2 - 9.8 - 10.1$$

$$\nabla_{\text{holding}} = 12.3$$

But $\nabla = kt$ and from the data of Deindoerfer and Humphrey the specific death rate of *B. stearothermophilus* spores at 121°C is 2.54 $\text{min}^{-1}$.

Therefore, $t = \nabla/k$ or $t = 12.3/2.54 = 4.84$ min.

If the contribution made by the heating and cooling parts of the cycle were ignored then the holding time would be given by the equation:

$$t = \nabla_{\text{overall}}/k = 32.2/2.54 = 12.68 \text{ min.}$$

Thus, by considering the contribution made to the sterilisation process by the heating and cooling parts of the cycle a considerable reduction in exposure time is achieved.

### Richards' Rapid Method for the Design of Sterilisation Cycles

Richards proposed a rapid method for the design of sterilisation cycles avoiding the time-consuming graphical integrations. The method assumes that all spore destruction occurs at temperatures above 100°C and that those parts of the heating and cooling cycle above 100°C are linear. Both these assumptions appear reasonably valid and the technique loses very little in accuracy and gains considerably in simplicity. Furthermore, based on these assumptions, Richards has presented a table of Del factors for *B. stearothermophilus* spores which would be obtained in heating and cooling a broth up to (and down from) holding temperatures of 101–130°C, based on a temperature change of 1°C per minute. This information is presented in Table 10.2, together with the specific death rates for *B. stearothermophilus* spores over the temperature range.

If the rate of temperature change is 1°C per minute, the Del factors for heating and cooling may be read directly from the table; if the temperature change deviates from 1°C per minute, the Del factors may be altered by simple proportion. For example, if a fermentation broth were heated from 100° to

121°C in 30 minutes and cooled from 121° to 100°C in 17 minutes, the Del factors for the heating and cooling cycles may be determined as follows:

From Table 10.2, if the change in temperature had been 1°C per minute, the Del factor for both the heating and cooling cycles would be 12.549. But the temperature change in the heating cycle was 21°C in 30 minutes; therefore,

$$\text{Del}_{\text{heating}} = (12.549 \times 30)/21 = 17.93$$

and the temperature change in the cooling cycle was 21°C in 17 minutes, therefore,

$$\text{Del}_{\text{cooling}} = (12.549 \times 17)/21 = 10.16$$

Having calculated the Del factors for the heating and cooling periods the holding time at the constant temperature may be calculated as before.

**Table 10.2.** Del values for *B. stearothermophilus* spores for the heating-up period over a temperature range of 100 to 130°C, assuming a rate of temperature change of 1°C $min^{-1}$ and negligible spore destruction at temperatures below 100°C.

| *T(°C)* | *k* $(min^{-1})$ | ∇ |
|---|---|---|
| 100 | 0.019 | – |
| 101 | 0.025 | 0.044 |
| 102 | 0.032 | 0.076 |
| 103 | 0.040 | 0.116 |
| 104 | 0.051 | 0.168 |
| 105 | 0.065 | 0.233 |
| 106 | 0.083 | 0.316 |
| 107 | 0.105 | 0.420 |
| 108 | 0.133 | 0.553 |
| 109 | 0.168 | 0.720 |
| 110 | 0.212 | 0.932 |
| 111 | 0.267 | 1.199 |
| 112 | 0.336 | 1.535 |
| 113 | 0.423 | 1.957 |
| 114 | 0.531 | 2.488 |
| 115 | 0.666 | 3.154 |
| 116 | 0.835 | 3.989 |
| 117 | 1.045 | 5.034 |
| 118 | 1.307 | 6.341 |
| 119 | 1.633 | 7.973 |
| 120 | 2.037 | 10.010 |
| 121 | 2.538 | 12.549 |
| 122 | 3.160 | 15.708 |
| 123 | 3.929 | 19.638 |

*(Cont'd ...)*

| $T(°C)$ | $k\ (min^{-1})$ | $\nabla$ |
|---|---|---|
| 124 | 4.881 | 24.518 |
| 125 | 6.056 | 30.574 |
| 126 | 7.506 | 38.080 |
| 127 | 9.293 | 47.373 |
| 128 | 11.494 | 58.867 |
| 129 | 14.200 | 73.067 |
| 130 | 17.524 | 90.591 |

## Scale-up of Batch Sterilisation Processes

The use of the Del factor in the scale-up of batch sterilisation processes has been discussed by Banks. It should be appreciated by this stage that the Del factor does not include a volume term, i.e., absolute numbers of contaminants and survivors are considered, not their concentration. Thus, if the size of a fermenter is increased the initial number of spores in the medium will also be increased, but if the same, probability of achieving sterility is required the final spore number should remain the same resulting in an increase in the Del factor. For example, if a pilot sterilisation were carried out in a 1000 $dm^3$ vessel with a medium containing $10^6$ organisms $cm^{-3}$ requiring a probability of contamination of 1 in 1000, the Del factor would be:

$$\begin{aligned}\nabla &= \ln \{(10^6 \times 10^3 \times 10^3)/10^{-3}\}\\ &= \ln (10^{12}/10^{-3})\\ &= \ln 10^{15} = 34.5\end{aligned}$$

If the same probability of contamination were required in a 10,000 $dm^3$ vessel using the same medium the Del factor would be:

$$\begin{aligned}\nabla &= \ln \{(10^6 \times 10^3 \times 10^4)/10^{-3}\}\\ &= \ln (10^{13}/10^{-3})\\ &= \ln 10^{16} = 36.8\end{aligned}$$

Thus, the Del factor increases with an increase in the size of the fermenter volume. The holding time in the large vessel may be calculated by the graphical integration method or by the rapid method of Richards, as discussed earlier, based on the temperature-time profile of the sterilisation cycle in the large vessel. However, it must be appreciated that extending the holding time on the larger scale (to achieve the increased $\nabla$ factor) will result in increased nutrient degradation. Also, the contribution of the heating-up and cooling-down periods to nutrient destruction will be greater as scale increases. Maintaining the same nutrient quality on a small and a large scale can be achieved in batch sterilisation only by compromising the sterility of the vessel, which is totally unacceptable. Thus, the decrease in the yield of a fermentation when it is scaled-up is often due to problems of nutrient degradation during batch sterilisation and the only way to eradicate the problem is to sterilise the medium continuously.

## Methods of Batch Sterilisation

The batch sterilisation of the medium for a fermentation may be achieved either in the fermentation vessel or in a separate mash cooker. Richards considered the relative merits of *in situ* medium sterilisation and the use of a special vessel. The major advantages of a separate medium sterilisation vessel may be summarised as:

1. One cooker may be used to serve several fermenters and the medium may be sterilised as the fermenters are being cleaned and prepared for the next fermentation, thus saving time between fermentations.
2. The medium may be sterilised in a cooker in a more concentrated form than would be used in the fermentation and then diluted in the fermenter with sterile water prior to inoculation. This would allow the construction of smaller cookers.
3. In some fermentations, the medium is at its most viscous during sterilisation and the power requirement for agitation is not alleviated by aeration as it would be during the fermentation period. Thus, if the requirement for agitation during *in situ* sterilisation were removed, the fermenter could be equipped with a less powerful motor. Obviously, the sterilisation kettle would have to be equipped with a powerful motor, but this would provide sterile medium for several fermenters.
4. The fermenter would be spared of corrosion which may occur with medium at high temperature.

The major disadvantages of a separate medium sterilisation vessel may be summarised as:

1. The cost of constructing a batch medium steriliser is much the same as that for the fermenter.
2. If a cooker serves a large number of fermenters complex pipework would be necessary to transport the sterile medium, with the inherent dangers of contamination.
3. Mechanical failure in a cooker supplying medium to several fermenters would render all the fermenters temporarily redundant. The provision of contingency equipment may be prohibitively costly.

Overall, the pressure to decrease the 'down time' between fermentations has tended to outweigh the perceived disadvantages of using separate sterilisation vessels. Thus, sterilisation in dedicated vessels is the method of choice for batch sterilisation. However, as pointed out by Corbett, the fact that separate batch sterilisers are used lends further weight to the argument that continuous sterilisation should be adopted in preference to batch. The capital cost of a separate batch steriliser is similar to that of a continuous one and the problems of transfer of sterile media are then the same for both batch and continuous sterilisation. Thus, two of the major objections to continuous systems (capital cost and aseptic transfer) may be considered as no longer relevant.

## DESIGN OF CONTINUOUS STERILISATION PROCESSES

The design of continuous sterilisation cycles may be approached in exactly the same way as for batch sterilisation systems. The continuous system includes a time period during which the medium is heated to the sterilisation temperature, a holding time at the desired temperature and a cooling period to restore the medium to the fermentation temperature. The temperature of the medium is elevated in a continuous heat exchanger and is then maintained in an insulated serpentine holding coil for the holding period. The length of the holding period is dictated by the length of the coil and the flow rate of the medium. The hot medium is then cooled to the fermentation temperature using two sequential heat exchangers—the first utilising the incoming medium as the cooling source (thus conserving heat by heating-up the incoming medium) and the second using cooling water. The major advantage of the continuous process is that a much higher temperature may be utilised, thus reducing the holding time and reducing the degree of nutrient degradation. The required Del factor may be achieved by the combination of temperature and holding time which gives an acceptably small degree of nutrient decay. Richards quoted the following example to illustrate the range of temperature-time regimes which may be employed to achieve the same probability of obtaining sterility. The Del factor for the example sterilisation was 45.7 and the following temperature-time regimes were calculated to give the same Del factor:

| *Temperature* | *Holding time* |
|---|---|
| 130° | 2.44 minutes |
| 135° | 51.9 seconds |
| 140° | 18.9 seconds |
| 150° | 2.7 seconds |

Furthermore, because a continuous process involves treating small increments of medium the heating-up and cooling-down periods are very small compared with those in a batch system. There are two types of continuous steriliser which may be used for the treatment of fermentation media: the indirect heat exchanger and the direct heat exchanger (steam injector).

The most suitable indirect heat exchangers are of the double-spiral type which consists of two sheets of high-grade stainless steel which have been curved around a central axis to form a double spiral, as shown in Fig. 10.8. The ends of the spiral are sealed by covers. To achieve sterilisation temperatures steam is passed through one spiral and medium through the other in countercurrent streams. Spiral heat exchangers are also used to cool the medium after passing through the holding coil. Incoming unsterile medium is used as the cooling agent in the first cooler so that the incoming medium is partially heated before it reaches the steriliser and thus, heat is conserved.

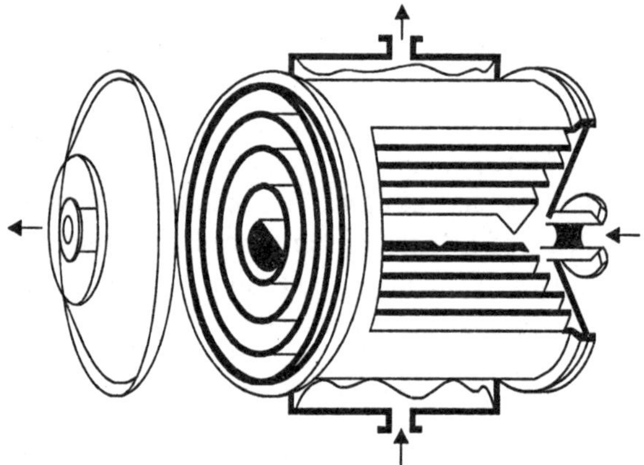

**Fig. 10.8.** A schematic representation of a spiral heat exchanger.

The major advantages of the spiral heat exchanger are:

1. The two streams of medium and cooling liquid or medium and steam, are separated by a continuous stainless steel barrier with gasket seals being confined to the joints with the end plates. This makes cross-contamination between the two streams unlikely.
2. The spiral route traversed by the medium allows sufficient clearances to be incorporated for the system to cope with suspended solids. The exchanger tends to be self-cleaning which reduces the risk of sedimentation, fouling and 'burning-on'.

Indirect plate heat exchangers consist of alternating plates through which the countercurrent streams are circulated. The plates are separated by gaskets and failure of these gaskets can cause cross-contamination between the two streams. Also, the clearances between the plates are such that suspended solids in the medium may block the exchanger and thus, the system is only useful in sterilising completely soluble media. However, the plate exchanger is more adaptable than the spiral system in that extra plates may be added to increase its capacity. The continuous steam injector injects steam directly into the unsterile broth.

### Advantages

1. Very short (almost instantaneous) heating up times.
2. It may be used for media containing suspended solids.
3. Low capital cost.
4. Easy cleaning and maintenance.
5. High steam utilisation efficiency.

### Disadvantages

1. Foaming may occur during heating.
2. Direct contact of the medium with steam requires that allowance be made for condense dilution and requires 'clean' steam, free from anticorrosion additives.

In some cases the injection system is combined with flash cooling, where the sterilised medium is cooled by passing it through an expansion valve into a vacuum chamber. Cooling then occurs virtually instantly. A flow chart of a continuous sterilisation system using direct steam injection is shown in Fig. 10.9. In some cases a combination of direct and indirect heat exchangers may be used. This is especially true for starch-containing broths when steam injection is used for the pre-heating step. By raising the temperature virtually instantaneously the critical gelatinisation temperature of the starch is passed through very quickly and the increase in viscosity normally associated with heated starch colloids can be reduced.

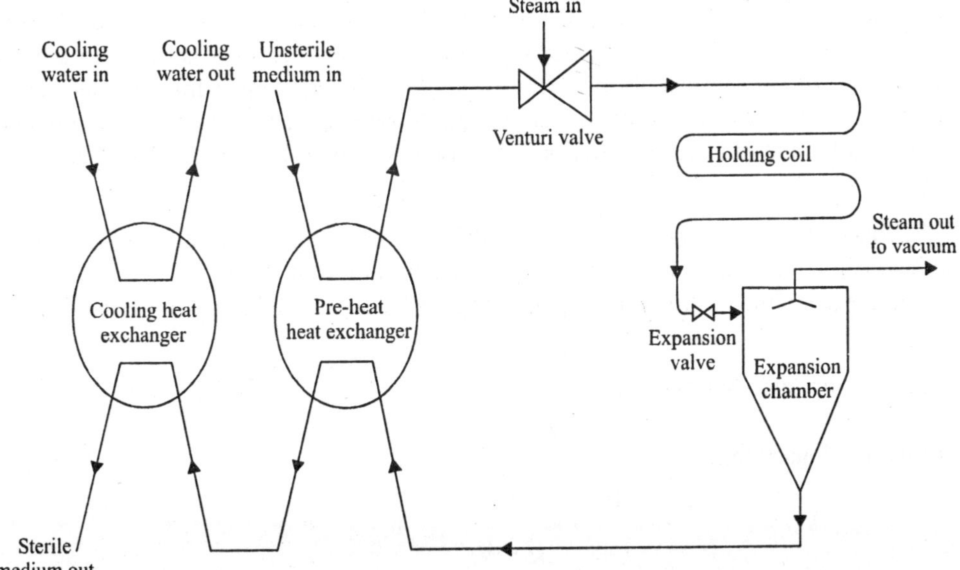


**Fig. 10.9.** Flow diagram of a typical continuous injector-flash cooler steriliser.

The most widely used continuous sterilisation system is that based on the spiral heat exchangers and a typical layout is shown in Fig. 10.10. The plant is sterilised prior to sterilisation of the medium by circulating hot water through the plant in a closed circuit. At the same time, the fermenter and the pipework between the fermenter and the steriliser are steam sterilised. Heat conservation is achieved by cooling the sterile medium against cold, incoming unsterile medium which will then be partially heated before it reaches the steriliser.

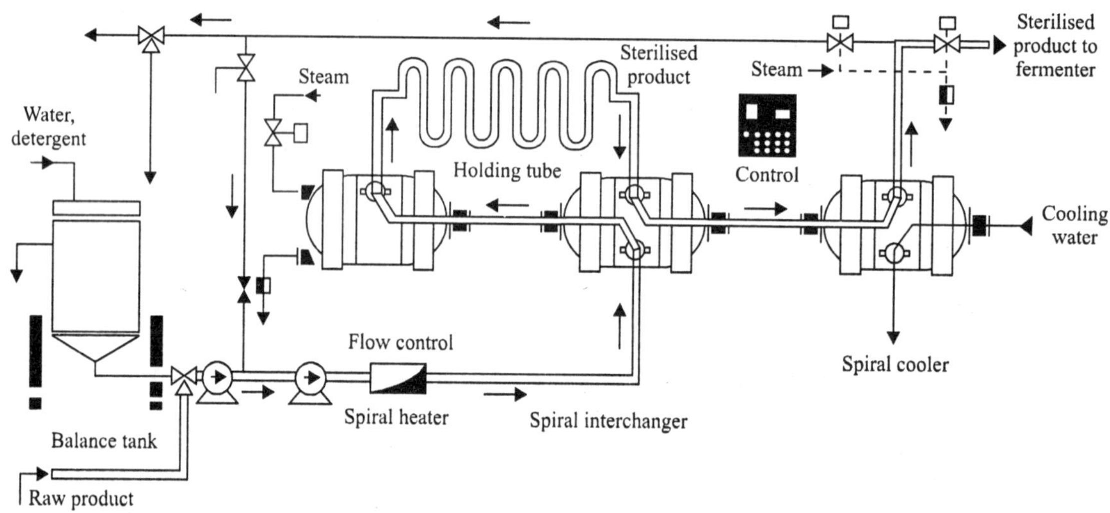


**Fig. 10.10.** Flow diagram of a typical continuous sterilisation system employing spiral heat exchangers.

The Del factor to be achieved in a continuous sterilisation process has to be increased with an increase in scale and this is calculated exactly as described in the consideration of the scale up of batch regimes. Thus, if the volume to be sterilised is increased from 1000 $dm^3$ to 10,000 $dm^3$ and the risk of failure is to remain at 1 in 1000 then the Del factor must be increased from 34.5 to 36.8. However, the advantage of the continuous process is that temperature may be used as a variable in scaling up a continuous process so that the increased $\nabla$ factor may be achieved whilst maintaining the nutrient quality constant. Deindoerfer and Humphrey attempted to rationalise the choice of time-temperature regime by the use of a nutrient quality criterion ($Q$), based on similar logic to the Del factor:

$$Q = \ln (x_0/x_t) \quad \text{... (10.12)}$$

where $x_0$ is the concentration of essential heat labile nutrient in the original medium. $x_t$ is the concentration of essential heat labile nutrient in the medium after a sterilisation time, t.

As considered earlier, the destruction of a nutrient may be considered a first-order reaction:

$$x_t/x_0 = e^{-kt}$$

where k is the reaction rate constant

or

$$x_0/x_t = e^{kt}$$

Thus, taking natural logarithms,

$$\ln (x_0/x_t) = kt$$

Therefore,

$$Q = kt$$

The relationship between k and absolute temperature is described by the Arrhenius equation:

$$k = A \cdot e^{-(E/RT)}$$

therefore, substituting for k:

$$Q = A \cdot t \cdot e^{-(E/RT)}$$

Therefore, as for the Del factor equation, by taking natural logarithms and re-arranging, the following equation is obtained

$$\ln t = E/RT + \ln Q/A \qquad \text{... (10.13)}$$

Thus, a plot of the natural logarithms of the time required to achieve a certain $Q$ value against the reciprocal of the absolute temperature will yield a straight line, the slope of which is dependent on the activation energy; that is, a very similar plot to Fig. 10.5 for the Del factor relationship. If both plots were superimposed on the same figure, then a continuous sterilisation performance chart is obtained. The example put forward by Deindoerfer and Humphrey is shown in Fig. 10.11. Thus, in Fig. 10.11 each line of a constant Del factor specifies temperature-time regimes giving the same fractional reduction in spore number and each line of a constant nutrient quality criterion specifies temperature-time regimes giving the same destruction of nutrient. By considering the effect of nutrient destruction on product yield, limits may be imposed on Fig. 10.11 indicating the nutrient quality criterion below which no further increase in yield is achieved (i.e., the nutrient is in excess) and the nutrient quality criterion at which the product yield is at its lowest (i.e., there is no nutrient remaining). Thus, from such a plot a temperature-time regime may be chosen which gives the required Del factor and does not adversely affect the yield of the process.

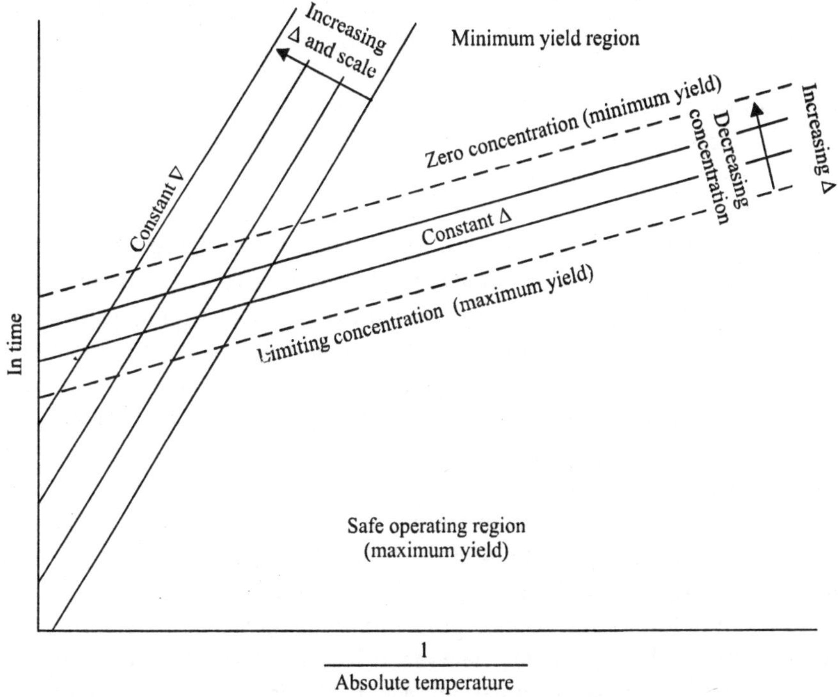


**Fig. 10.11.** Continuous sterilisation performance chart.

The adoption of Deindoerfer and Humphrey's approach is possible only if the limiting heat-labile nutrient is identified and the Arrhenius constant and activation energy for its thermal destruction experimentally determined. As pointed out by Banks, this information may not be available for a complex fermentation medium and therefore, the technique is fairly limited in its practical application. However, provided this information is available the technique may be used in the scale-up of continuous-sterilisation processes. As discussed earlier, the Del factor is scale dependent and therefore as the volume to be sterilised is increased so the Del factor should be increased if the probability of achieving sterility is to remain the same. However, the nutrient-quality criterion is not scale dependent so that by changing the temperature-time regime to accommodate the attainment of sterility the nutrient quality may be adversely affected. Examination of Fig. 10.11 indicates that the only way in which the Del factor may be increased without any change in the nutrient quality criterion is to increase the temperature and to decrease the holding time. Although this scale-up could be performed exactly only if the thermal degradation kinetics of the medium were known, the analysis indicates that even without this information the approach would be to reduce the holding time and increase the temperature.

When designing a continuous sterilisation process based on spiral heat exchangers it is important to consider the effect of suspended solids on the sterilisation process. Micro-organisms contained within solid particles are given considerable protection against the sterilisation treatment. If the residence time in the steriliser is insufficient for heat to penetrate the particle then the fermentation medium may not be rendered sterile. The routine solution to this problem is to 'overdesign' the process and expose the medium to a far more severe regime than may be necessary. Armenante and Li discussed this problem in considerable detail and produced a model to predict the behaviour of a continuous system. Their analysis suggested that the temperature of the particle cores is significantly less than that of the bulk liquid. Furthermore, there is a considerable time lag in heat penetrating to the particle cores, resulting in a very different time-temperature profile for the particles as compared with the liquid medium. Thus, the temperature of the particles may not reach the critical point before they leave the steriliser and heat penetration into the particles will continue downstream of the steriliser. Armenante and Li's conclusion is that it is the steriliser and/or the first cooling exchanger that should be 'overdesigned' rather than the length of the holding coil. Remember that the first cooling exchanger transfers a significant amount of heat from the sterile medium to the incoming medium and increasing its surface area would give more opportunity for the heat to penetrate the particles. This, coupled with increasing the temperature or residence time in the steriliser, would ensure that the particle cores are up to temperature before the holding coil is reached. Also, this work suggests that it is unwise to use the direct steam injection method to heat a particulate medium because, again, there will be insufficient time for the heat to penetrate the particles.

An example of the scale-up of sterilisation regimes is given by Buckland on the production of efrotomycin by *Nocardia lactamdurans*. In this case a beneficial interaction appeared to be occurring between the protein nitrogen source and glucose during sterilisation, thus making the protein less available but resulting in a more controlled fermentation. When glucose was sterilised separately the oxygen demand of the subsequent fermentation was excessive and the fermentation terminated prematurely with very poor product formation. On scaling-up the fermentation it was very difficult to attain the correct sterilisation conditions using a batch regime. However, continuous sterilisation using direct steam injection allowed the design of a precise process producing sterile medium with the required degree of interaction between the ingredients. The identification of this phenomenon was dependent upon careful monitoring of the small scale fermentation and consideration being given to sterilisation as an important scale-up factor.

When a fermentation is scaled-up it is important to appreciate that the inoculum development process is also increased in scale and a larger seed fermenter may have to be employed to generate sufficient inoculum to start the production scale. Thus, the sterilisation regime of the seed fermenter (and its medium) will also have to be scaled-up. Therefore, the performance of the seed fermentation should be assessed carefully to ensure that the quality of the inoculum is maintained on the larger scale and that it has not been adversely affected by any increase in the severity of the sterilisation regime.

## STERILISATION OF THE FERMENTER

If the medium is sterilised in a separate batch cooker or is sterilised continuously, then the fermenter has to be sterilised separately before the sterile medium is added to it. This is normally achieved by heating the jacket or coils of the fermenter with steam and sparging steam into the vessel through all entries, apart from the air outlet from which steam is allowed to exit slowly. Steam pressure is held at 15 psi in the vessel for approximately 20 minutes. It is essential that sterile air is sparged into thc fermenter after the cycle is complete and a positive pressure is maintained; otherwise a vacuum may develop and unsterile air be drawn into the vessel.

## STERILISATION OF THE FEEDS

A variety of additives may be administered to a fermentation during the process and it is essential that these materials are sterile. The sterilisation method depends on the nature of the additive and the volume and feed rate at which it is administered. If the additive is fed in large quantities then continuous sterilisation may be desirable, for example, Aunstrup cited the use of continuous heat sterilisation for the feed medium of microbial enzyme fermentations. Aunstrup also referred to the use of filtration for the sterilisation of certain feeds. Stowell described the use of a continuous steriliser for the addition of oil feeds to industrial scale fermenters, but stressed that each oil has its own characteristics which makes it impossible to predict the performance of a steriliser for different oil feeds. Batch sterilisation of feed liquids normally involves steam injection into the material held in storage vessels. Whatever the sterilisation system employed it is essential that all ancillary equipment and feed pipework associated with the additions are sterilisable.

## STERILISATION OF LIQUID WASTES

Process organisms which have been engineered to produce 'foreign' products and therefore contain heterologous genes are subject to strict containment regulations. Thus, waste biomass of such organisms must be sterilised before disposal. Sterilisation may be achieved by either batch or continuous means but the whole process must be carried out under contained conditions. Batch sterilisation involves the sparging of steam into holding tanks, whereas continuous processes would employ the type of heat exchangers which have been discussed in the previous section. A holding vessel for the batch sterilisation of waste is shown in Fig. 10.12. Whichever method is employed the effluent must be cooled to below 60°C before it is discharged to waste. The sterilisation processes have to be validated and are designed using the Del factor approach considered in the previous sections. However, the kinetic characteristics used in the calculation would be those of the process organism rather than of *B. stearothermophilus*. Also, the $N_t$ value used in the design calculations would be smaller than $10^{-3}$ which is used for medium sterilisation and would depend on the assessment of the hazard involved should the organism survive the decontamination process. Thus, the sterilisation regime used for destruction of the process organism will be different from that used in sterilising the medium.

## FILTER STERILISATION

Suspended solids may be separated from a fluid during filtration by the following mechanisms:

1. Inertial impaction.
2. Diffusion.
3. Electrostatic attraction.
4. Interception.

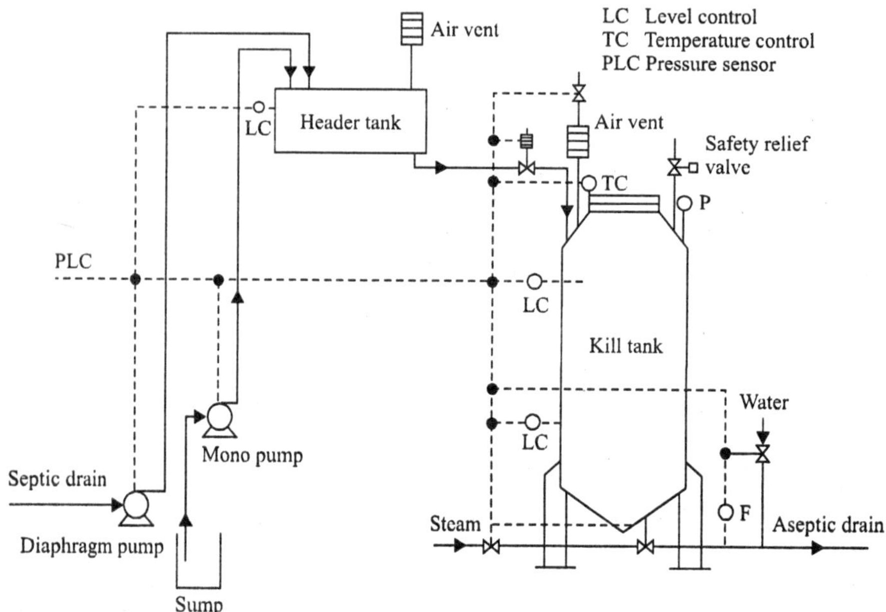


**Fig. 10.12.** A vessel for the batch sterilisation of liquid waste from a contained fermentation.

### Inertial Impaction

Suspended particles in a fluid stream have momentum. The fluid in which the particles are suspended will flow through the filter by the route of least resistance. However, the particles, because of their momentum, tend to travel in straight lines and may therefore become impacted upon the fibres where they may then remain. Inertial impaction is more significant in the filtration of gases than in the filtration of liquids.

### Diffusion

Extremely small particles suspended in a fluid are subject to Brownian motion which is random movement due to collisions with fluid molecules. Thus, such small particles tend to deviate from the fluid flow pattern and may become impacted upon the filter fibres. Diffusion is more significant in the filtration of gases than in the filtration of liquids.

### Electrostatic Attraction

Charged particles may be attracted by opposite charges on the surface of the filtration medium.

### Interception

The fibres comprising a filter are interwoven to define openings of various sizes. Particles which are larger than the filter pores are removed by direct interception. However, a significant number of particles which are smaller than the filter pores are also retained by interception. This may occur by several mechanisms—more than one particle may arrive at a pore simultaneously, an irregularly shaped particle may bridge a pore, once a particle has been trapped by a mechanism other than interception the pore may be partially occluded enabling the entrapment of smaller particles. Interception is equally important a mechanism in the filtration of gases and liquids.

Filters have been classified into two types—those in which the pores in the filter are smaller than the particles which are to be removed and those in which the pores are larger than the particles which are to be removed. The former type may be regarded as an absolute filter, so that filters of this type (provided they are not physically damaged) are claimed to be 100 per cent efficient in removing micro-organisms. Filters of the latter type are frequently referred to as depth filters and are composed of felts, woven yarns, asbestos pads and loosely packed fibreglass. The terms absolute and depth can be misleading as they imply that absolute filtration only occurs at the surface of the filter, whereas absolute filters also have depth and filtration occurs within the filter as well as at the surface. Terms which bear more relationship to the construction of filters are 'non-fixed pore filters' (corresponding with depth filters) and 'fixed pore filters' (corresponding with absolute filters).

Non-fixed pore filters rely on the removal of particles by inertial impaction, diffusion and electrostatic attraction rather than interception. The packing material contains innumerable tortuous routes through the filter but removal is a statistical phenomenon and thus, sterility of the product is predicted in terms of the probability of failure (similar to the situation for steam sterilisation). Thus, in theory, the removal of micro-organisms by a fibrous filter cannot be absolute as there is always the possibility of an organism passing through the filter, regardless of the filter's depth. Also, because the fibres are not cemented into position an increase in the pressure to which the filter is subjected may result in movement of the material, producing larger channels through the filter. Increased pressure may also result in the displacement of previously trapped particles.

Fixed pore filters are constructed so that the filtration medium will not be distorted during operation so that the flow patterns through the filter will not change due to disruption of the material. Pore size is controlled during manufacture so that an absolute rating can be quoted for the filter—i.e., the removal of particles above a certain size can be guaranteed. Thus, interception is the major mechanism by which particles are removed. Because fixed pore filters have depth they are also capable of removing particles which are smaller than the pores by the mechanisms of inertial impaction, diffusion and attraction and these mechanisms do play significant roles in the filtration of gases. Fixed pore filters are superior for most purposes in that they have absolute ratings, are less susceptible to changes in pressure and are less likely to release trapped particles. The major disadvantage associated with absolute filters was the resistance to flow they present and hence, the large pressure drop across the filters which represents a major operating cost. However, modern absolute filter cartridges which have been developed by many filtration companies contain pleated membranes with very large surface areas, thereby minimising the pressure drop across the filter.

It is important to realise that filters should be steam sterilised before and after operation. Thus, the materials must be stable at high temperatures and the steam must be free of particulate matter because the filter modules are particularly vulnerable to damage at high temperatures. Thus, the steam itself is filtered through stainless steel mesh filters rated at 1 μm.

### Filter Sterilisation of Fermentation Media

Media for animal-cell culture cannot be sterilised by steam because they contain heat-labile proteins. Thus, filtration is the method of choice and as may be appreciated from the previous discussion, fixed pore or absolute filtration is the better system to use. An ideal filtration system for the sterilisation of animal cell culture media must fulfil the following criteria:

1. The filtered medium must be free of fungal, bacterial and mycoplasma contamination.
2. There should be minimal adsorption of protein to the filter surface.

3. The filtered medium should be free of viruses.
4. The filtered medium should be free of endotoxins.

Several filter manufacturers supply absolute filtration systems for the sterilisation of animal cell culture medium. Such systems consist of membrane cartridges which are fitted into stainless steel, steam sterilisable modules. The membranes for media filtration are constructed from steam sterilisable hydrophilic material and are treated to produce a filtrate of particular quality. For example, if minimal protein adsorption is a major criterion then a specially coated filter membrane is used. It would be very difficult to construct a single filtration membrane which would fulfil all four criteria cited above. Thus, a series of filters are used to achieve the desired result. The example shown in Fig. 10.13 illustrates a system to produce sterile, mycoplasma free serum and consists of four filters arranged in sequence. The first filter is a positively charged polypropylene pre-filter with an absolute rating of 5 μm for the removal of coarse precipitates, clot-like material and other gross contaminants; the second filter is also positively charged polypropylene but with an absolute rating of 0.5 μm for bulk microbial removal, deformable gels, lipid-based materials and endotoxin reduction; the third filter is a single layered, nylon/polyester positively charged filter with a 0.1 μm absolute rating for further microbial and endotoxin removal and optimum protection of the final filter; the fourth filter is similar to the third and has the same rating, but is double layered and removes mycoplasmas, gives absolute sterility and final endotoxin control. Thus, the combination of four filters gives a sequential removal of decreasingly small particles and prolongs the life of the final filter. If it is necessary to remove viral contamination then a final 0.04 μm nylon/polyester filter would be added.

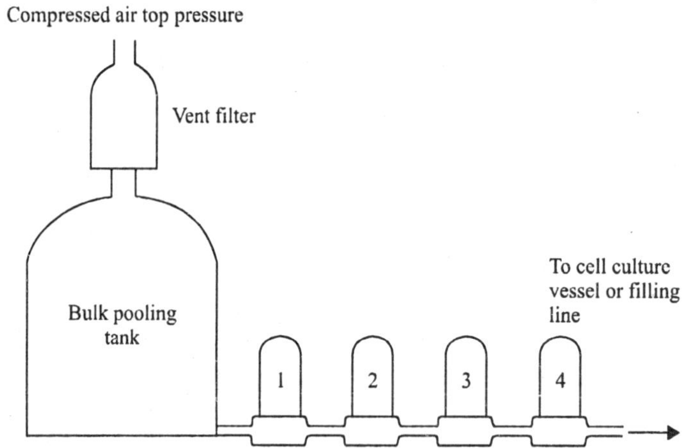


Filter 1: 5 μm absolute rated pre-filter for removal of coarse precipitates.
Filter 2: 0.5 μm absolute rated pre-filter for bulk bioburden removal.
Filter 3: 0.1. μm absolute rated single layer pre-filter for further bioburden and endotoxin removal.
Filter 4: 0.1 μm absolute rated double layer final filter for absolute sterility, mycoplasma removal and further endotoxin control.

**Fig. 10.13.** Filtration system for the provision of sterile, mycoplasma free serum.

Similar systems may be used in downstream processing of animal-cell products where the rating and properties of the filters would be optimised for the particular process. Figure 10.14 illustrates a system for the removal of cells and cell debris from an animal-cell fermentation broth. The pre-filter is a polypropylene 1.0 μm rated filter to remove the bulk of the cells and debris and the second filter is an hydroxyl modified nylon/polyester 0.2 μm rated filter giving absolute cell removal with minimal protein adsorption.

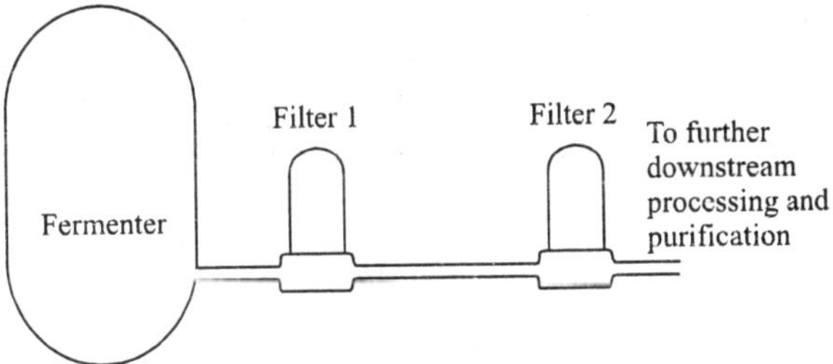


Filter 1:1.0 μm absolute rated pre-filter for bulk cell and cell debris removal.
Filter 2: 0.2 μm absolute rated single layer 'bio-inert' filter for final bioburden removal.

**Fig. 10.14.** Filtration system for the removal of cells and cell debris from an animal cell culture fermentation.

## Filter Sterilisation of Air

Aerobic fermentations require the continuous addition of considerable quantities of sterile air. Although it is possible to sterilise air by heat treatment, the most commonly used sterilisation process is filtration. Fixed pore filters (which have an absolute rating) are very widely used in the fermentation industry and several manufacturers produce filtration systems for air sterilisation. These systems, like those for the sterilisation of liquids, consist of pleated membrane cartridges designed to be accommodated in stainless steel modules. A sectioned filter unit is shown in Fig. 10.15. The most common construction material used for the pleated membranes for air sterilisation is PTFE, which is hydrophobic and is therefore resistant to wetting. Also, PTFE filters may be steam sterilised and are resistant to ammonia which may be injected into the air stream, prior to the filter, for pH control. As was seen for the filter sterilisation of liquids it is essential that a pre-filter is incorporated up-stream of the absolute filter. The pre-filter traps large particles such as dust, oil and carbon (from the compressor) and pipescale and rust (from the pipework). The use of a coalescing pre-filter also ensures the removal of water from the air; entrained water is coalesced in the filter (air flow being from the inside of the filter to the outside) and is discharged via an automatic drain. Figure 10.16 illustrates the layout of such a filtration unit showing the steam sterilisation system and Fig. 10.17 is a photograph of the air sterilisation plant for an 85 $m^3$ fermenter.

**Fig. 10.15.** An absolute membrane filter sectioned to show the pleated membrane structure.

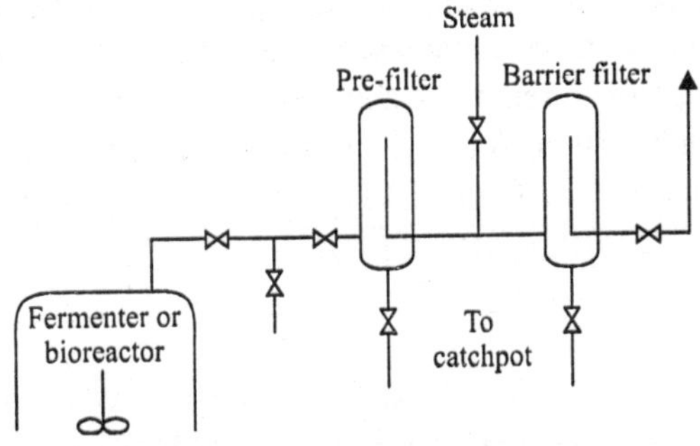


**Fig. 10.16.** Dual hydrophobic filter system for the sterilisation of off-gas from a fermenter.

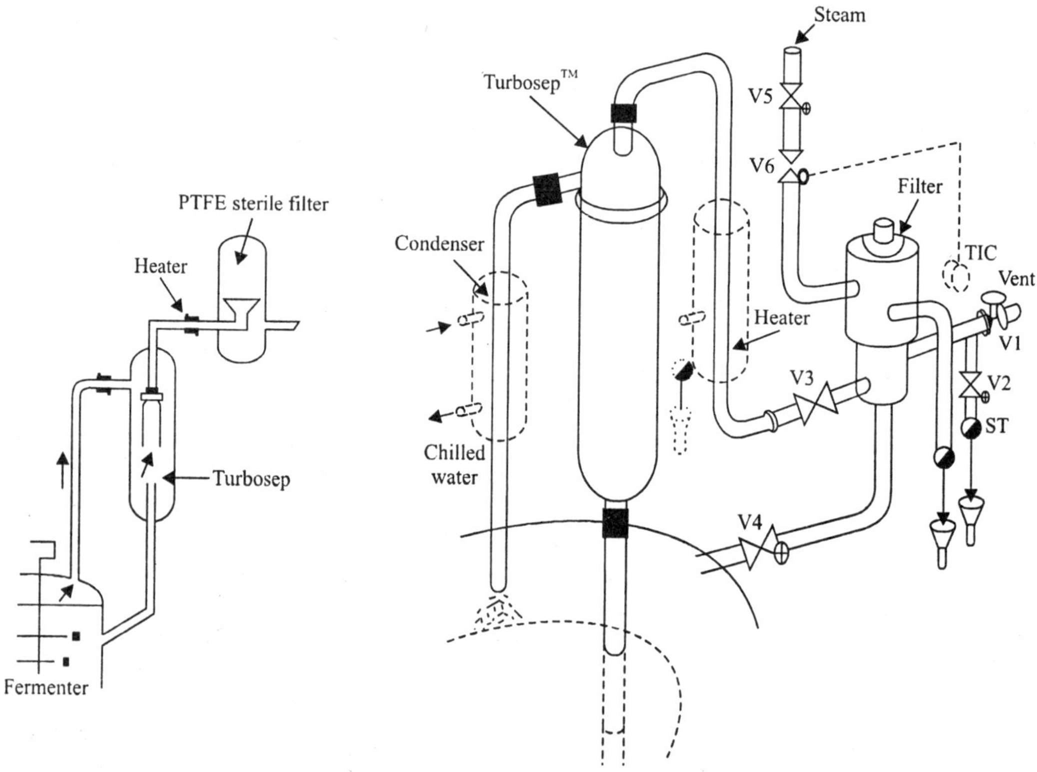


**Fig. 10.17.** A mechanical separator and hydrophobic filter system for the sterilisation of off-gas from a fermenter. Left. Cut-away diagram. Right, equipment arrangement, showing steam supply. V1–V6, valves; and O, steam, traps.

## Sterilisation of Fermenter Exhaust Air

In many traditional fermentations the exhaust gas from the fermenter was vented without sterilisation or vented through relatively inefficient depth filters. With the advent of the use of recombinant organisms and a greater awareness of safety and emission levels of allergic compounds the containment of exhaust air is more common (and in the case of recombinant organisms, compulsory). Fixed pore membrane modules are also used for this application but the system must be able to cope with the sterilisation of water saturated air, at a relatively high temperature and carrying a large contamination level. Also, foam may overflow from the fermenter into the air exhaust line. Thus, some form of pre-treatment of the exhaust gas is necessary before it enters the absolute filter. This pre-treatment may be a hydrophobic pre-filter or a mechanical separator to remove water, aerosol particles and foam. The pre-treated air is then fed to a 0.2 μm hydrophobic filter. Again, it is important to appreciate that the filtration system must be steam sterilisable. Figures 10.16 and 10.17 illustrate the pre-filter and mechanical separator systems respectively.

Chapter 11

# Development of Inocula for Industrial Fermentations

## INTRODUCTION

It is essential that the culture used to inoculate a fermentation satisfies the following criteria:

1. It must be in a healthy, active state thus minimising the length of the lag phase in the subsequent fermentation.
2. It must be available in sufficiently large volumes to provide an inoculum of optimum size.
3. It must be in a suitable morphological form.
4. It must be free of contamination.
5. It must retain its product-forming capabilities.

The process adopted to produce an inoculum meeting these criteria is called inoculum development. Hockenhull is credited with the quotation 'once a fermentation has been started it can be made worse but not better'. Whereas this is an over-statement it does illustrate the importance of inoculum development. Much of the variation observed in small-scale laboratory fermentations is due to poor inocula being used and thus, it is essential to appreciate that the establishment of an effective inoculum development programme is equally important regardless of the scale of the fermentation. Such a programme not only aids consistency on a small scale but is invaluable in scaling up the fermentation and forms an essential part in progressing a new process.

A critical factor in obtaining a suitable inoculum is the choice of the culture medium. It must be stressed that the suitability of an inoculum medium is determined by the subsequent performance of the inoculum in the production stage. As already discussed, the design of a production medium is determined not only by the nutritional requirements of the organism, but also by the requirements for maximum product formation. The formation of product in the seed culture is not an objective during inoculum development so that the seed medium may be of a different composition from the production medium. However, Lincoln stated that the lag time in a fermentation is minimised by growing the culture in the 'final-type' medium. Lincoln's argument is an important one, so the inoculum development medium should be sufficiently similar to the production medium to minimise any period of adaptation of the culture to the production medium, thus reducing the lag phase and the fermentation time. Furthermore, Hockenhull pointed out the dangers of using very different media in consecutive stages. Major differences in pH, osmotic pressure and anion composition may result in very sudden changes in uptake rates which, in turn, may affect viability. Hockenhull also emphasised that for antibiotic fermentations the inoculum medium should contain sufficient carbon and nitrogen to support maximum growth until transfer, so that secondary metabolism remains repressed during growth of the inoculum. If secondary metabolism is derepressed in the seed fermentation, then selection may enrich the culture with non-producing variants having a growth advantage over high-producing types. Hockenhull drew attention to Righelato's work in which it was shown that chemostat culture of *Penicillium chrysogenum* under carbohydrate-limited conditions led to a loss of penicillin synthesising ability and an increase in the proportion of non-conidiated variants whereas this did not occur in ammonia, phosphate or sulphate-limited conditions. The relevance of this phenomenon is supported by Hockenhull's observation that *P. chrysogenum* inocula produced under non-limiting conditions are remarkably free from variants whereas variants arise relatively frequently during the carbon-limited production phase. Examples of inoculum and production phase. Examples of inoculum and production media are given in Table 11.1, from which it may be seen that inoculum media are, generally, less nutritious than production media and contain a lower level of carbon.

**Table 11.1.** Inoculum development and production media for a range of processes.

| *Process* | *Inoculum development medium* | | *Production medium* | |
|---|---|---|---|---|
| Griseofulvin | Whey powder and lactose together to give: | | Lactose | 7% |
| | Lactose | 3.5% | Corn-steep liquor solids to give: | |
| | Nitrogen | 0.05% | Nitrogen | 0.2% |
| | Corn-steep liquor solids | 0.38% | Limestone | 0.8% |
| | (to give approximately 0.04% N) | | $KH_2PO_4$ | 0.4% |
| | $KH_2PO_4$ | 0.4% | KCl | 0.1% |
| | KCl | 0.05% | | |
| Clavulanic acid | Soyabean flour | 1.0% | Soyabean flour | 1.5% |
| | Dextrin | 2.0% | Oil | 1.0% |
| | Pluronic L81 (Antifoam) | 0.03% | $KH_2PO_4$ | 0.1% |
| | | (g $dm^{-3}$) | | (g $dm^{-3}$) |
| Vitamin $B_{12}$ | Sugar beet | | | |
| | Molasses | 70 | | 105 |
| | Sucrose | – | | 15 |
| | Betaine | – | | 3 |
| | $NH_4H_2PO_4$ | 0.8 | | – |
| | $(NH_4)_2SO_4$ | 2 | | 2.5 |
| | $MgSO_4$ | 0.2 | | 0.2 |
| | $ZnSO_4$ | 0.02 | | 0.08 |
| | 5–6 Dimethyl-benzimidazole | 0.005 | | 0.025 |
| Lysine | Cane molasses | 5% | | 20% |
| | Corn-steep liquor | 1% | | – |
| | $CaCO_3$ | 1% | | – |
| | Soyabean meal hydrolysate | – | | 1.8% |

The quantity of inoculum normally used is between 3 and 10 per cent of the medium volume. A relatively large inoculum volume is used to minimise the length of the lag phase and to generate the maximum biomass in the production fermenter in as short a time as possible, thus increasing vessel productivity. Thus, starting from a stock culture, the inoculum must be built up in a number of stages to produce sufficient biomass to inoculate the production-stage fermenter. This may involve two or three stages in shake flasks and one to three stages in fermenters, depending on the size of the ultimate vessel. Throughout this procedure there is a risk of contamination and strain degeneration necessitating stringent quality-control procedures. The greater the number of stages between the master culture and the production fermenter the greater the risk of contamination and strain degeneration. Therefore, a compromise may be reached regarding the size of the inoculum to be used and the risk of contamination and strain degeneration. Another factor to be considered in the determination of the inoculum volume is the economics of the process. A seed fermenter 10 per cent of the size of the production fermenter represents a considerable financial investment and must be justified in terms of productivity. A large-scale continuous fermentation for the production of biomass would be expected to operate at steady state in excess of 100 days. Thus, the number of times that the fermenter is inoculated should be very few compared with batch or fed-batch systems. In such circumstances it may be more economic to compromise on the size of the inoculum and to tolerate a relatively lengthy period of growth up to maximum biomass than to invest a large seed vessel which would be used on very few occasions. This is particularly relevant for biomass continuous processes where one very large fermenter may be used and thus, any seed vessel would only be servicing the one production vessel.

A typical inoculum-development programme will now be described in detail. The master culture is reconstituted and plated on to solid medium; approximately ten colonies of typical morphology of high producers are selected and inoculated on to slopes as the sub-master cultures, each sub-master culture being used for a new production run. At this stage, shake flasks may be inoculated to check the productivity of these cultures, the results of such tests being known before the developing inoculum eventually reaches the production plant. A sub-master culture is used to inoculate a shake flask (250 or 500 $cm^3$ containing 50 or 100 $cm^3$ medium) which, in turn, is used as inoculum for a larger flask or a laboratory fermenter, which may then be used to inoculate a pilot-scale fermenter. Culture purity checks are carried out at each stage to detect contamination as early as possible. Although the results of these tests may not be available before the culture has reached the production plant, at least it is known at which stage in the procedure contamination occurred. For a sporulating organism the process may be modified to facilitate the use of a spore suspension as inoculum and this will be discussed in more detail later in this chapter.

Lincoln suggested a more elaborate procedure for the development of inoculum for bacterial fermentations which, with minor modifications, is applicable to any type of culture. The procedure involved the use of one sub-master culture to develop a bulk inoculum which was sub-divided, stored in a frozen state and used as inocula for several months. A single colony, derived from a sub-master culture, was inoculated into liquid medium and grown to maximum log phase. This culture was then transferred into nineteen times its volume of medium and incubated again to the maximum log phase, at which point it was dispensed in 20 $cm^3$ volumes, plug frozen and stored at below $-20°C$. At least 3 per cent of the samples were tested for purity and productivity in subsequent fermentation and provided these were suitable, the remaining samples could be used as initial inocula for subsequent fermentations. To use one of the stored samples as inoculum it was thawed and used as a 5 per cent inoculum for a seed culture which, in turn, was used as a 5 per cent inoculum for the next stage in the programme. This procedure ensured that a proven inoculum was used for the penultimate stage in inoculum development.

## CRITERIA FOR THE TRANSFER OF INOCULUM

The physiological conditions of the inoculum when it is transferred to the next culture stage can have a major effect on the performance of the fermentation. The optimum time of transfer must be determined experimentally and then procedures established so that inoculation with an ideal culture may be achieved routinely. These procedures include the standardisation of cultural conditions and monitoring the state of an inoculum culture so that it is transferred at the optimum time, i.e., in the correct physiological state. The most widely used criterion for the transfer of vegetative inocula is biomass and such parameters as packed cell volume, dry weight, wet weight, turbidity, respiration, residual nutrient concentration and morphological form have been used. Ettler demonstrated that the rheological behaviour of *Streptomyces noursei* could be used as the transfer criterion in the nystatin fermentation. The rheology of the seed fermentation changed from Newtonian to non-Newtonian behaviour and the optimum inoculum transfer time corresponded with this transformation.

Criteria which may be monitored on-line are the most convenient parameters to use as indicators of inoculum quality and these would include dissolved oxygen, pH (although pH would normally be controlled in seed fermentations) and oxygen or carbon dioxide in the effluent gas. Parton and Willis advocated the use of the carbon dioxide production rate (CPR) as a transfer criterion which requires analysis of the fermenter effluent air. This approach is suitable only when transfer is being made from a fermenter, but Parton and Willis stressed the importance of adopting this strategy even for the inoculation of laboratory-scale fermentations despite the fact that an adequate inoculum volume could be produced in shake flask. These workers provide an excellent example of the effect of inoculum transfer time on the productivity of the streptomycete secondary metabolite, as shown in Fig. 11.1(a), (b) and (c). The CPR of the inoculum fermentation and the points at which inoculum was transferred are shown in Fig. 11.1(a). The CPR of the subsequent production fermentations are shown in Fig. 11.1(b), from which it may be seen that the three fermentations performed similarly. However, Fig. 11.1(c) illustrates the very different secondary metabolite production of the three fermentations. Thus, although the time of transfer had only a marginal influence on biomass in the production fermentation the effect on product formation was critical. It should be emphasised that the amount of biomass transferred was standardised for the three fermentations and thus, the differences in performance were due to the physiological states of the inocula.

In recent years, probes have been developed for on-line assessment of biomass and these could be invaluable in estimating the time of inoculum transfer. Boulton reported the use of a biomass sensor (the Bugmeter) to control the yeast pitching rate (inoculum level) in brewing. The probe measures the dielectric permittivity of viable yeast cells and is unaffected by the presence of dead cells, air bubbles or detritus, making it ideal for the routine monitoring of yeast inoculum. Using the probe, these workers developed an automatic inoculum dispenser allowing a preset viable yeast mass to be transferred from a yeast mass to be transferred from a yeast storage vessel to the brewery fermentation.

Alford reported the use of a real-time expert computer system to predict the time of inoculum transfer for industrial-scale fermentations. The system involves the comparison of on-line fermentation data with detailed historical data of the process. A problem with the interpretation of carbon dioxide production rate figures is that the data are not available continuously because the analyser is not dedicated to any one fermenter, but is analysing process streams from a large number of vessels via a multiplexer system. Thus, a fermentation may have passed a critical stage between monitoring times. Also, occasional false readings may be generated. The expert system enabled the verification of data points as well as prediction of the outcome of the fermentation from early information. Data from seed fermentations were analysed by the expert system and the transfer time predicted. As a result of this approach operators were able to plan their work more effectively, the need for manual sampling was reduced and early warning of contamination was provided if the seed-culture profile predicted from early readings was abnormal.

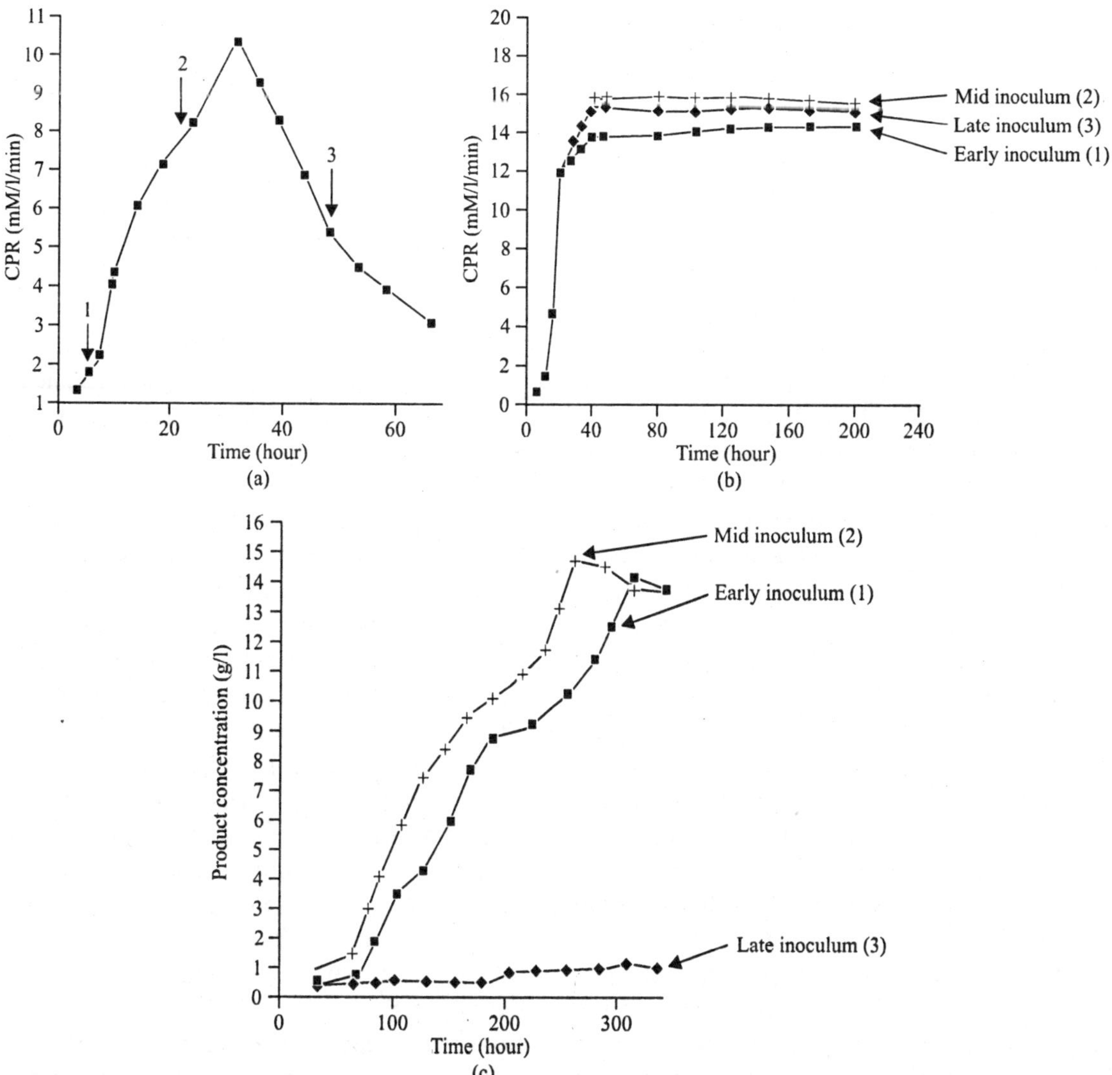


**Fig. 11.1.** The effect of inoculum age on growth and productivity in a streptomycete fermentation. (a) The carbon dioxide production rate (CPR) profile of the inoculum culture showing the points (1, 2 and 3) at which inocula were removed. (b) The effect of inoculum age on the CPR of the production fermentation. (c) The effect of inoculum age on productivity in the production fermentation.

Smith and Calam compared the quality and enzymic profile of differently prepared inocula *Penicillium patulum* (producing griseofulvin) and demonstrated that a low level of glucose 6-phosphate dehydrogenase was indicative of a good quality inoculum. The enzyme profile of good quality inoculum was established early in the growth of the seed culture. Thus, this approach could be used to assess the cultural conditions giving rise to satisfactory inoculum, but would be of less value in determining the time of transfer.

Yeast, unicellular bacterial, fungal and streptomycete fermentations have different requirements for inoculum development and these are dealt with separately.

## DEVELOPMENT OF INOCULA FOR YEAST PROCESSES

Whilst the largest industrial fermentations utilising yeasts are the brewing of beer and the production of biomass, recent processes have also been established for the production of recombinant products.

### Brewing

It is common practice in the British brewing industry to use the yeast from the previous fermentation to inoculate a fresh batch of wort. The brewing terms used to describe this process are 'crop', referring to the harvested yeast from the previous fermentation and 'pitch', meaning to inoculate. One of the major factors contributing to the continuation of this practice is the wort-based excise laws in the United Kingdom where duty is charged on the sugar consumed rather than the alcohol produced. Thus, dedicated yeast propagation systems are expensive to operate because duty is charged on the sugar consumed by the yeast during growth. It can then be appreciated that the reduced cost of using yeast from a previous fermentation is an attractive proposition. The dangers inherent in this practice are the introduction of contaminants and the degeneration of the strain, the most common degenerations being a change in the degree of flocculence and attenuating abilities of the yeast. In breweries employing top fermentations in open fermenters these dangers are minimised by collecting yeast to be used for future pitching from 'middle skimmings'. During the fermentation the yeast cells flocculate and float to the surface, the first cells to do this being the most flocculent and the last cells the least flocculent. As the head of yeast develops, the surface layer (the most flocculent and highly contaminated yeasts) is removed and discarded and the underlying cells (the 'middle skimmings') are harvested and used for subsequent pitching. Therefore, the 'middle skimmings' contain cells which have the desired flocculence and which have been protected from contamination by the surface layer of the yeast head. The pitching yeast may be treated to reduce the level of contaminating bacteria and remove protein and dead yeast cells by such treatments as reducing the pH of the slurry to 2.5 to 3, washing with water, washing with ammonium persulphate and treatment with antibiotics such as polymixin, penicillin and neomycin.

However, traditional open vessels are becoming increasingly rare and the bulk of beer is brewed using cylindro-conical fermenters. In these systems the yeast flocculates and collects in the cone at the bottom of the fermenter where it is subject to the stresses of nutrient starvation, high ethanol concentration, low water activity, high carbon dioxide concentration and high pressure. Thus, the viability and physiological state of the yeast crop would not be ideal for an inoculum. The viability of the crop may be assessed using a biomass probe of the type described earlier, thus ensuring that at least the correct amount of viable biomass is used to start the next fermentation. However, the physiological state of the biomass will not have been influenced by such monitoring procedures. The situation is further complicated by the fact that the harvested yeast is stored before it is used as inoculum. Metabolic activity is minimised during this time by chilling rapidly to about 1°C, suspending in beer and storing in the absence of oxygen. If oxygen is present during the storage period then the yeast cells consume their stored glycogen which renders them very much less active at the start of the fermentation.

One of the key physiological features of yeast inoculum is the level of sterol in the cells. Sterols are required for membrane synthesis but they are only produced in the presence of oxygen. Thus, we have the anomaly of oxygen being required for sterol synthesis, yet anaerobic conditions are required for ethanol production. This anomaly is resolved traditionally by aerating the wort before inoculation. This oxygen allows sufficient sterol synthesis early in the fermentation to support growth of the cells throughout the process, that is after the oxygen is exhausted and the process is anaerobic. Boulton developed an alternative approach where the pitching yeast was vigorously aerated prior to inoculation. The yeast was then sterol rich and had no requirement for oxygen during the alcohol fermentation.

The difficulties outlined above and the likelihood of strain degeneration and contamination mean that yeasts are rarely used for more than five to ten consecutive fermentations which necessitates the periodical production of a pure inoculum. This would involve developing sufficient biomass from a single colony to pitch a fermentation at a level of approximately 2 grams of pressed yeast per litre. Hansen pioneered the use of pure inocula and devised a yeast propagation scheme utilising a 10 per cent inoculum volume at each stage in the programme and employing conditions similar to those used during brewing. However, modern propagation schemes use inoculum volumes of 1 per cent or even lower and may use conditions different from those used during brewing.

Therefore, continuous aeration may be used during the propagation stage which seems to have little effect on the beer produced in the subsequent fermentation. Yeast inoculum produced in this way would also be sterol rich, obviating the need for aerated wort.

A number of yeast propagators (which are basically closed, aerated vessels) have been described in the literature. The simplest type of propagator is a single-stage system resembling an unstirred, aerated fermenter which is inoculated with a shake-flask culture developed from a single colony. Clark and Thorne described two-stage systems which could be operated semi-continuously. Thorne's propagator consisted of two linked vessels, 1.5 and 150 $dm^3$ respectively. The smaller vessel was filled with wort, sterilised, cooled, aerated and inoculated with a flask-grown culture.

After growth for 3 to 4 days the culture was forced by air pressure into the second vessel which had been filled with sterilised, cooled wort and aerated. An aliquot of 1.5 $dm^3$ was forced back into the first vessel after mixing. In a further 3 to 4 days the larger vessel contained sufficient biomass to pitch a 1000 $dm^3$ fermenter and the first vessel contained sufficient inoculum for another second stage. However, although this procedure should produce a pure inoculum there is a danger of strain degeneration occurring in such a semi-continuous system.

Boulton suggested that the bakers' yeast aerobic fed-batch inoculum programme could be adopted to produce sterol-rich catabolite-derepressed cells. Such an inoculum would remove the necessity for aerating the wort prior to inoculation because sterol synthesis would not be necessary due to the high sterol content of the cells.

## Bakers' Yeast

The commercial production of bakers' yeast involves the development of an inoculum through a large number of aerobic stages. Although the production stages of the process may not be operated under strictly aseptic conditions a pure culture is used for the initial inoculum, thereby keeping contamination to a minimum in the early stages of growth. Reed and Nagodawithana discussed the development of inoculum for the production of bakers' yeast and quoted a process involving eight stages, the first three being aseptic while the remaining stages were carried out in open vessels. The yeast may be pumped from one stage to the next or the seed cultures may be centrifuged and washed before transfer, which reduces the level of contamination.

The yields obtained in the first five stages are relatively low because they are not fed-batch systems, whereas the last three stages are fed-batch. A summary of a typical inoculum development programme for the production of bakers' yeast is given in Fig. 11.2.

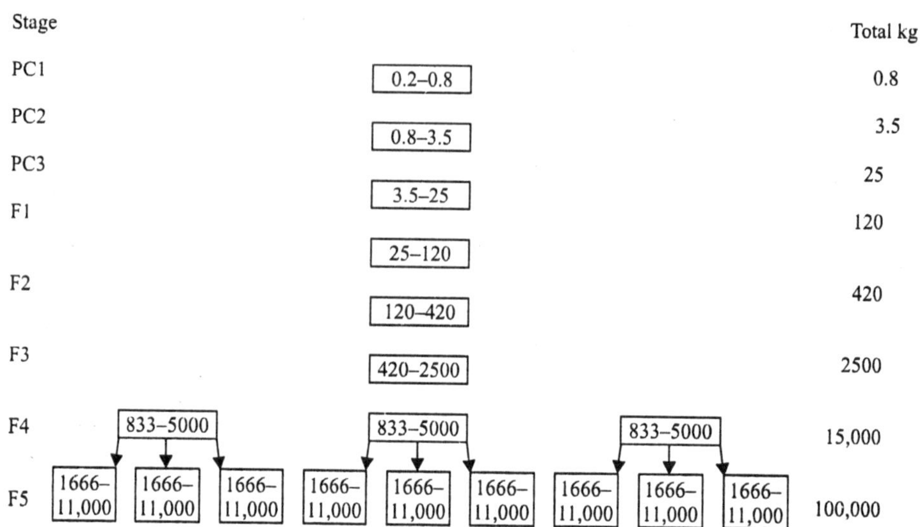


**Fig. 11.2.** The development of inoculum for the commercial production of bakers' yeast. PC 1, 2 and 3 are pure culture batch fermentations. F1 and F2 are non-aseptic batch fermentations. F3 and F4 are fed-batch fermentations and F5 is the final fed-batch fermentation.

## DEVELOPMENT OF INOCULA FOR BACTERIAL PROCESSES

The main objective of inoculum development for traditional bacterial fermentations is to produce an active inoculum which will give as short a lag phase as possible in subsequent culture. A long lag phase is disadvantageous in that not only is time wasted but also medium is consumed in maintaining a viable culture prior to growth. The length of the lag phase is affected by the size of the inoculum and its physiological condition. As already stated, the inoculum size normally ranges between 3 and 10 per cent of the culture volume. Lincoln stressed that bacterial inocula should be transferred in the logarithmic phase of growth, when the cells are still metabolically active. The age of the inoculum is particularly important in the growth of sporulating bacteria, for sporulation is induced at the end of the logarithmic phase and the use of an inoculum containing a high percentage of spores would result in a long lag phase in a subsequent fermentation.

Keay quoted the use of a 5 per cent inoculum of a logarithmically growing culture of a thermophilic *Bacillus* for the production of proteases. Aunstrup described a two-stage inoculum development programme for the production of proteases by *Bacillus subtilis*. Inoculum for a seed fermenter was grown for 1 to 2 days on a solid or liquid medium and then transferred to a seed vessel where the organism was allowed to grow for a further ten generations before transfer to the production stage. Priest and Sharp cited the use of a 5 per cent inoculum, still in the exponential phase, for the commercial production of *Bacillus* α-amylase. Underkofler emphasised that, in the production of bacterial enzymes, the lag phase in plant fermenters could be almost completely eliminated by using inoculum medium of the same composition as used in the production fermenter and employing large inocula of actively growing seed cultures. The inoculum development programme for a pilot-plant scale process for the production of vitamin $B_{12}$ from *Pseudomonas denitrificans* is shown in Fig. 11.3.

The necessity to use an inoculum in an active physiological state is taken to its extreme in the production of vinegar. The acetic-acid bacteria used in the vinegar process are extremely sensitive to oxygen starvation.

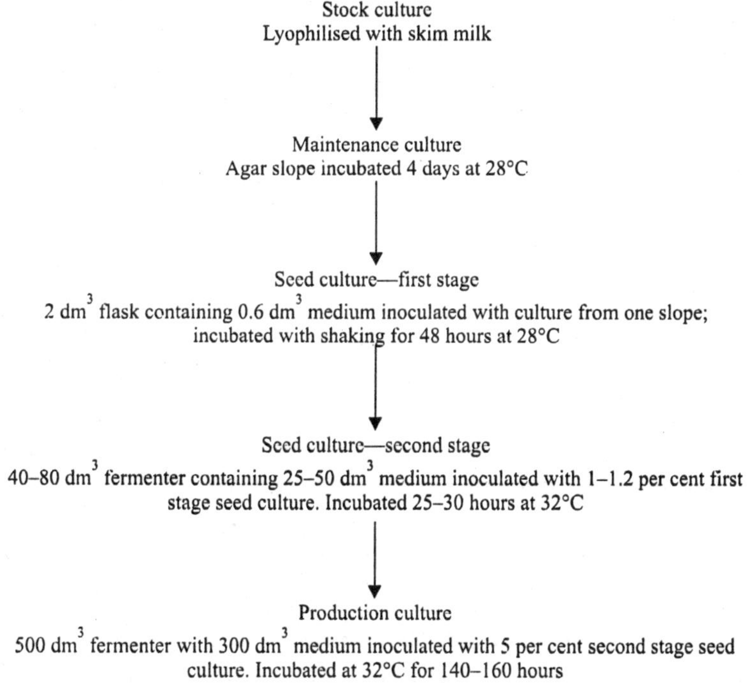


**Fig. 11.3.** The inoculum development programme for a vitamin $B_{12}$ pilot scale fermentation using *Pseudomonas denitrificans.*

Therefore, to avoid disturbing the system, the cells at the end of a fermentation are used as inoculum for the next batch by removing approximately 60 per cent of the culture and restoring the original level with fresh medium. The advantage of a highly active inoculum apparently outweighs the disadvantages of possible strain degeneration and contaminant accumulation. However, strain stability is a major concern in inoculum development for fermentations employing recombinant bacteria. Sabatie demonstrated that plasmid stability and productivity in an *E. coli* biotin fermentation was greatly improved if stationary, rather than exponential, phase cells were used as inoculum. They postulated that the plasmid copy number may be higher in stationary cells than in exponential ones, resulting in a lower plasmid loss in the subsequent fermentation when a stationary culture is used as inoculum. A stationary phase inoculum would result in a lag phase, but this disadvantage was more than compensated for by the considerable improvement in plasmid retention and biotin production compared with that obtained using an exponential inoculum.

In the lactic-acid fermentation the producing organism may be inhibited by lactic acid. Thus, production of lactic acid in the seed fermentation may result in the generation of poor quality inoculum. Yamamoto generated high quality inoculum of *Lactococcus lactis* IO-1 on a laboratory scale using electrodialysis seed culture which reduced the lactate in the inoculum and reduced the length of the lag phase in the production fermentation.

An example of the development of inoculum for an anaerobic bacterial process is provided by the clostridial acetone-butanol fermentation. The process was out-competed by the petrochemical industry but there is still considerable interest in re-establishing the fermentation. The inoculum development programme described by McNeil and Kristiansen is given in Table 11.2.

**Table 11.2.** The inoculum development programme for the clostridial acetone-butanol fermentation.

| *Stage* | *Cultural conditions* | *Incubation time (hours)* |
|---|---|---|
| 1 | Heat-shocked spore suspension inoculated into 150 $cm^3$ of potato glucose medium | 12 |
| 2 | Stage 1 culture used as inoculum for 500 $cm^3$ molasses medium | 6 |
| 3 | Stage 2 culture used as inoculum for 9 $dm^3$ molasses medium | 9 |
| 4 | Stage 3 culture used as inoculum for 90,000 $dm^3$ molasses medium | – |

The stock culture is heat shocked to stimulate spore germination and to eliminate the weaker spores. The production stage is inoculated with a very low volume and this corresponds with Lurie's description of the South African acetone-butanol fermentation in which a 1,00,000 $dm^3$ fermenter was inoculated with only 10 $dm^3$ of seed.

The use of such small inocula necessitates the achievement of as near perfect conditions as possible to prevent contamination and to avoid an abnormally long lag phase.

## DEVELOPMENT OF INOCULA FOR MYCELIAL PROCESSES

The preparation of inocula for fermentations employing mycelial (filamentous) organisms is more involved than that for unicellular bacterial and yeast processes. The majority of industrially important fungi and streptomycetes are capable of asexual sporulation, so it is common practice to use a spore suspension as seed during an inoculum development programme.

A major advantage of a spore inoculum is that it contains far more 'propagules' than a vegetative culture. Three basic techniques have been developed to produce a high concentration of spores for use as an inoculum.

### Sporulation on Solidified Media

Most fungi and streptomycetes will sporulate on suitable agar media but a large surface area must be employed to produce sufficient spores. Parker described the 'roll-bottle' technique for the production of spores of *Penicillium chrysogenum*. Quantities of medium (300 $cm^3$) containing 3 per cent agar were sterilised in 1 $dm^3$ cylindrical bottles, which were then cooled to 45°C and rotated on a roller mill so that the agar set as a cylindrical shell inside the bottle. The bottles were inoculated with a spore suspension from a sub-master slope and incubated at 24°C for 6 to 7 days. Parker claimed that although the use of the 'roll-bottle' involved some sacrifice in ease of visual examination, it provided a large surface area for cultivation of spores in a vessel of a convenient size for handling in the laboratory.

Hockenhull described the production of $10^{10}$ spores of *Penicillium chrysogenum* on a 300 $cm^2$ agar layer in a Roux bottle and El Sayed quoted the use of spore suspensions derived from agar media containing between $10^7$ and $10^8$ $cm^{-3}$.

Butterworth described the use of a Roux bottle for the production of a spore inoculum of *Streptomyces clavuligerus* for the production of clavulanic acid. The spores produced from one bottle containing 200 $cm^2$ agar surface could be used to inoculate a 75 $dm^3$ seed fermenter which, in turn, was used to inoculate a 1500 $dm^3$ fermenter. The clavulanic acid inoculum development programme is illustrated in Fig. 11.4. Some representative solidified media for the production of streptomycete and fungal spores are given in Tables 11.3 and 11.4 respectively.

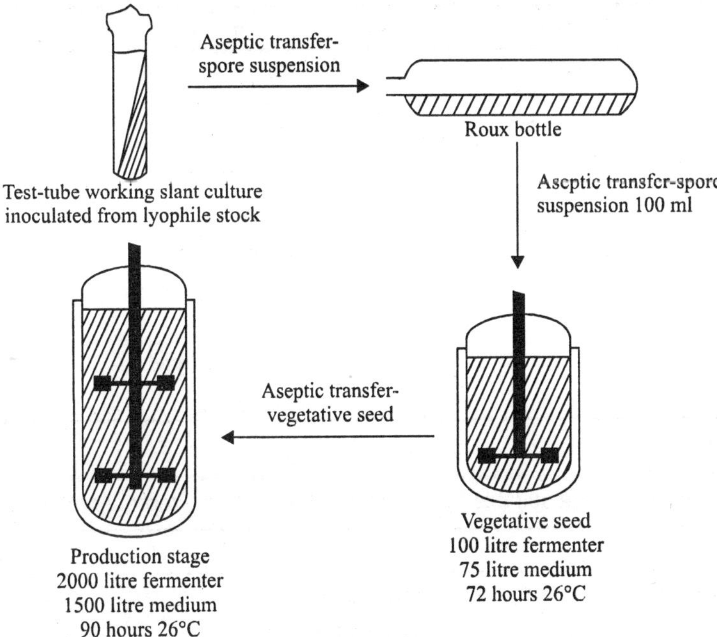


**Fig. 11.4.** The inoculum development programme for the production of clavulanic acid from *Streptomyces clavuligerus*.

**Table 11.3.** Solidified media suitable for the sporulation of some representative *streptomycetes*.

| *Organism* | *Product* | *Medium* | |
|---|---|---|---|
| *S. aureofaciens* | Tetracycline* | Malt extract (Difco) | 1.0% |
| | | Yeast extract (Difco) | 0.4% |
| | | Glucose | 0.4% |
| *S. erythreus* | Erythromycin* | Beef extract (Difco) | 0.1% |
| | | Yeast extract (Difco) | 0.1% |
| | | Casamino acids (Difco) | 0.2% |
| | | Glucose | 0.2% |
| *S. vinaceus* | Viomycin* | Corn-steep liquor (50% dry matter) | 1.0% |
| | | Starch | 1.0% |
| | | $(NH_4)_2SO_4$ | 0.3% |
| | | NaCl | 0.3% |
| | | $CaCO_3$ | 0.3% |
| *S. clavuligerus* | Clavulanic acid* | Soluble starch | 1.0% |
| | | $K_2HPO_4$ | 0.1% |
| | | $MgSO_4{\cdot}7H_2O$ | 0.1% |
| | | NaCl | 0.1% |
| | | $(NH_4)_2SO_4$ | 0.2% |

*(Cont'd ...)*

| *Organism* | *Product* | *Medium* | |
|---|---|---|---|
| | | $CaCO_3$ | 0.2% |
| | | $FeSO_4 \cdot 7H_2O$ | 0.001% |
| | | $MnCl_2 \cdot 4H_2O$ | 0.001% |
| | | $ZnSO_4 \cdot 7H_2O$ | 0.001% |
| *S. hygroscopicus* | Maridomycin | Soluble starch | 1.0% |
| | | Peptone | 0.04% |
| | | Meat extract | 0.02% |
| | | Yeast extract | 0.02% |
| | | N-Z amine (type A) | 0.02% |
| | | Agar | 2.0% |

*Agar content not quoted.

**Table 11.4.** Solidified media suitable for the sporulation of some representative fungi.

| *Fungus* | *Medium* | ($g\ dm^{-3}$) |
|---|---|---|
| *Penicillium chrysogenum* | Glycerol | 7.5 |
| | Cane molasses | 7.5 |
| | Curbay BG | 2.5 |
| | $MgSO_4 \cdot 7H_2O$ | 0.05 |
| | $KH_2PO_4$ | 0.06 |
| | Peptone | 5.0 |
| | NaCl | 4.0 |
| | Agar | 20 |
| *Aspergillus niger* | Molasses | 300 |
| | $KH_2PO_4$ | 0.5 |
| | Agar | 20 |

## Sporulation on Solid Media

Many filamentous organisms will sporulate profusely on the surface of cereal grains from which the spores may be harvested. Substrates such as barley, hard wheat bran, ground maize and rice are all suitable for the sporulation of a wide range of fungi. The sporulation of a given fungus is particularly affected by the amount of water added to the cereal before sterilisation and the relative humidity of the atmosphere, which should be as high as possible during sporulation. Singh has described a system for the sporulation of *Aspergillus ochraces* in which a 2.8 $dm^3$ Fernbach flask containing 200 grams of 'pot' barley or 100 grams of moistened wheat bran produced $5 \times 10^{11}$ conidia after six days at 28°C and 98 per cent relative humidity. This was 5 times the number obtainable from a Roux bottle batched with Sabouraud agar and 50 times the number obtainable from such a vessel batched with Difco Nutrient Agar, incubated for the same time period. Vezina has published a list of fungi which are capable of sporulating heavily on cereal grains. El-Sayed quoted the use of cooked rice for the production of spores of *Penicillium* and *Cephalosporium* in penicillin and cephalosporin inoculum development. Sansing and Cieglem described the mass production of spores of several *Aspergillus* and *Penicillium* species on whole loaves of white bread and Podojil quoted the use of millet for the sporulation of *Streptomyces aureofaciens* in the development of inoculum for the chlortetracycline fermentation (Fig. 11.5).

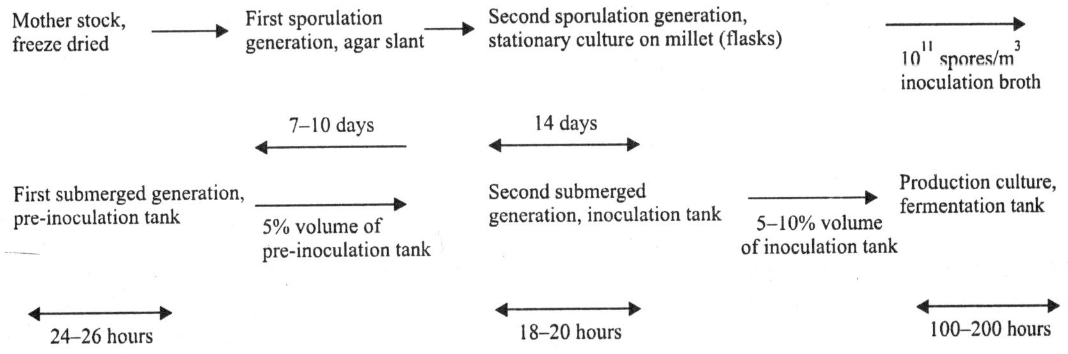


**Fig. 11.5.** The inoculum development programme for the production of chlortetracycline by *Streptomyces aureofaciens.*

## Sporulation in Submerged Culture

Many fungi will sporulate in submerged culture provided a suitable medium is employed. This technique is more convenient than the use of solid or solidified media because it is easier to operate aseptically and it may be applied on a large scale. The technique was first adopted by Foster who induced submerged sporulation in *Penicillium notatum* by including 2.5 per cent calcium chloride in a defined nitrate-sucrose medium. An example of the use of this technique for the production of inoculum for an industrial fermentation is provided by the griseofulvin process. Rhodes described the conditions necessary for the submerged sporulation of the griseofulvin-producing fungus, *Penicillium patulum* and the medium utilised is given in Table 11.5. These authors found that for prolific sporulation the nitrogen level had to be limited to between 0.05 and 0.1 per cent w/v and that good aeration had to be maintained. Also, an interaction was demonstrated between the nitrogen level and aeration in that the lower the degree of aeration the lower the concentration of nitrogen needed to induce sporulation. Submerged sporulation was induced by inoculating 600 cm$^3$ of the above medium, in a 2 dm$^3$ shake flask, with spores from a well-sporulated Czapek-Dox agar culture and incubating at 25°C for 7 days. The resulting suspension of spores was then used as a 10 per cent inoculum for a vegetative seed stage in a stirred fermenter, the seed culture subsequently providing a 10 per cent inoculum for the production fermentation. The vegetative seed and production media are given in Table 11.1.

**Table 11.5.** Media for the submerged sporulation of selected fungi.

| | | |
|---|---|---|
| Rhodes: *Penicillium patulum* | | |
| Whey powder, to give | Lactose 3.5% | |
| | Nitrogen 0.05% | |
| $KH_2PO_4$ | | 0.4% |
| KCl | | 0.05% |
| Corn-steep liquor solids to give approximately 0.04% N | | 0.38% |
| Foster: *Penicillium notatum* | | |
| Sucrose | | 2.0% |

*(Cont'd ...)*

| | |
|---|---|
| $NaNO_3$ | 0.6% |
| $KH_2PO_4$ | 0.15% |
| $MgSO_4 \cdot 7H_2O$ | 0.05% |
| $CaCl_2$ | 2.5% |
| Vezina: *Aspergillus ochraceus* | |
| Glucose | 2.5% |
| NaCl | 2.5% |
| Corn-steep liquor | 0.5% |
| Molasses | 5.0% |

Most actinomycetes do not sporulate in submerged culture and thus, solid or solidified media tend to be used for the production of spore inocula.

### Use of the Spore Inoculum

The stage in an inoculum development programme at which a large-scale spore inoculum is used varies according to the process; it appears to be common practice that the penultimate stage is so inoculated, but this will depend on the scale of the production fermentation. In the inoculum development programme for the early penicillin fermentation described by Parker the penultimate stage was inoculated with a spore suspension (from a 'roll-bottle') and this stage may have produced either a vegetative or a submerged spore inoculum for the final fermentation. For the griseofulvin process, Rhodes stated that the spore suspension obtained from the submerged sporulation stage could either be used for direct inoculation of the production fermentation or it could be germinated in an inoculum development medium to yield a vegetative inoculum for the final fermentation. The latter course was preferred and an inoculum volume of 7–10 per cent was used. From Figs. 11.4 to 11.6, it can be seen that in the clavulanic acid process the spore inoculum is used to inoculate the final seed stage, in the chlortetracycline process a vegetative stage is interspersed between the spore inoculated batch and the production fermentation and in the sagamicin process the spore inoculum is used at a very early stage followed by the vegetative growth.

When considering the production of gluconic acid by *Aspergillus niger*, Lockwood discussed the merits of inoculating the final fermentation directly with a spore suspension as compared with germinating the spores in a seed tank to give a vegetative inoculum. Direct spore inoculation would avoid the cost of installation and operation of the seed tanks whereas the use of germinated spores would reduce the fermentation time of the final stage, thus allowing a greater number of fermentations to be carried out per year. However, labour costs for the production of the vegetative inoculum could be almost as high as for the final fermentation although some of these costs may be recovered, in that gluconic acid produced in the penultimate stage would be recoverable from the final fermentation broth and would contribute to the buffering capacity throughout the fermentation. Thus, Lockwood claimed that the choice of inoculum for the production stage depends on the length of the cycle of the fermentation process, plant size and the availability and cost of labour.

### Inoculum Development for Vegetative Fungi

Some fungi will not produce asexual spores and therefore, an inoculum of vegetative mycelium must be used. *Gibberella fujikuroi* is such a fungus and is used for the commercial production of gibberellin. Hansen described an inoculum development programme for the gibberellin fermentation. Cultures were

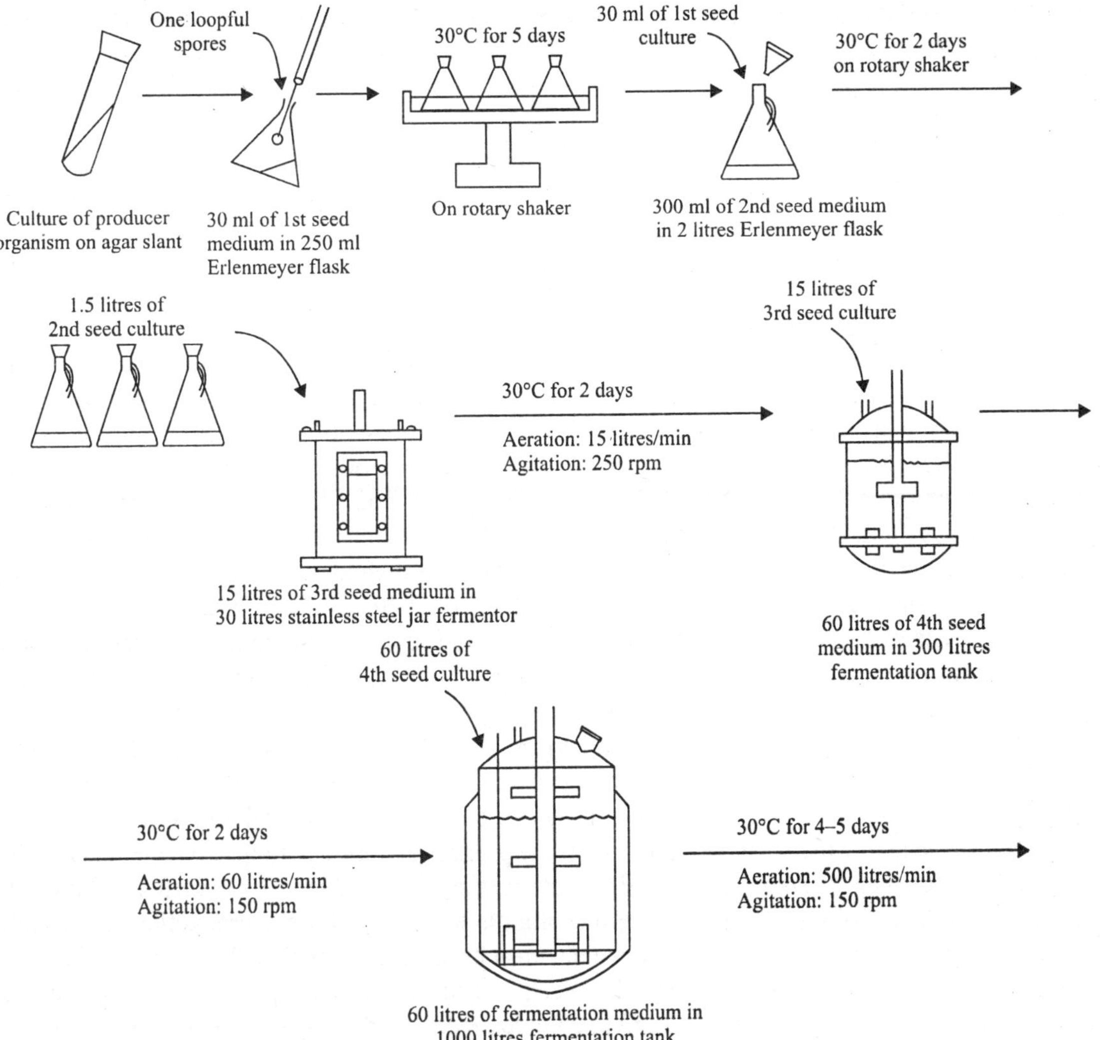


**Fig. 11.6.** The inoculum development programme for the production of sagamicin by *Micromonospora sagamiensis.*

grown on long slants (25 × 10 mm test tubes) of potato dextrose agar for 1 week at 24°C. Growth from three slants were scraped off and transferred to a 9 $dm^3$ carboy containing 4 $dm^3$ of a liquid medium composed of 2 per cent glucose, 0.3 per cent $MgSO_4 \cdot 7H_2O$, 0.3 per cent $NH_4Cl$ and 0.3 per cent $KH_2PO_4$. The medium was aerated for 75 hours at 28°C before transfer to a 100 $dm^3$ seed fermenter containing the same medium.

The major problem in using vegetative mycelium as initial seed is the difficulty of obtaining a uniform, standard inoculum. The procedure may be improved by fragmenting the mycelium in a homogeniser, such as a Waring blender, prior to use as inoculum. This method provides a large number of mycelial particles and therefore a large number of growing points. Worgan has given a detailed account of the use of this technique in the preparation of inocula for the submerged culture of the higher fungi.

### Effect of the Inoculum on the Morphology of Filamentous Organisms in Submerged Culture

When filamentous fungi are grown in submerged culture the type of growth varies from the 'pellet' form, consisting of compact discrete masses of hyphae, to the filamentous form in which the hyphae form a homogenous suspension dispersed through the medium. The filamentous type of habit gives rise to an extremely viscous broth which may be very difficult to aerate adequately, whereas the pellet type of habit gives rise to a far less viscous, but also less homogenous, broth. In a pelleted culture there is a danger that the mycelium at the center of the pellet may be starved of nutrients and oxygen due to diffusion limitations. Also, there is considerable evidence that the morphological form of the organism influences the productivity of the culture, but whether this is due to the phenomena already mentioned or to some form of metabolic control is far from clear. Thus, some fermentations are carried out with the fungus in a filamentous habit, whereas others are carried out with the organism growing as pellets. For example, filamentous growth has been claimed to be optimum for penicillin production by *P. chrysogenum*, whereas pelleted growth has been claimed to be optimum for citric acid production from *Aspergillus niger* and lovastatin from *Aspergillus terreus*. The necessity for filamentous growth is taken to the extreme in the ICI-Rank Hovis McDougal mycoprotein process where *Fusarium graminearium* is produced for human consumption. A highly filamentous morphology is required to produce the desired texture in the product which resembles the strength and eating texture of white and soft, red meats. Thus, in this process a median hyphal length of 400 μm is required.

The relevance of this consideration of mycelial morphology to inoculum development is that the morphology may be influenced considerably by both the concentration of spores in a spore inoculum and the inoculum development medium. Usually, a high spore inoculum will tend to produce a dispersed form of growth whilst a low one will favour pellet formation. The effect of the concentration of a spore inoculum on the morphology of *P. chrysogenum* is given in Table 11.6. Thus, in the commercial production of fungal products it is critical to grow the organism in the desired morphological form which necessitates the use of an inoculum which achieves this end. If the production fermentation is to be inoculated with a spore suspension then the spore concentration must be such as to produce the production culture in the desired morphological form; if a vegetative inoculum is to be used for the production fermentation then, again, the concentration of its spore inoculum must be such as to produce the vegetative inoculum in the desired morphological form. Although the effects of media on morphological form can be extremely varied dispersed growth is more likely in rich, complex media and pelleted growth tends to occur in chemically defined media. Thus, the medium used in the spore germination stage must be optimised in terms of the morphology of the inoculum.

An interesting series of experiments on the effects of inoculum conditions on the morphology of *Penicillium citrinum* were reported by Hosobuchi. This *Penicillium* species synthesises compound ML-236B, a precursor of pravastatin which is a cholesterol-lowering drug. Optimum productivity was achieved when the organism grew as compact pellets in the production fermentation. The vegetative inoculum for the production fermentation had to contain an optimum number of short, filamentous propagules in order to initiate pellet formation in the final culture. This was achieved by using a four-stage inoculum development programme (initiated by a spore-inoculated shake flask) with very rich media in the third and fourth cultures. Thus, this system required a dispersed vegetative inoculum to generate a pelleted production fermentation.

The information available on the morphology of actinomycetes in submerged culture is very limited compared with that on fungi. However, Whitaker has reviewed the area and it is obvious that actinomycetes are capable of producing a wide range of morphological types. Also, it appears to be accepted that a dispersed mycelial morphology is desirable for most industrial actinomycete fermentations. Mycelial forms have been shown to be desirable for the production of streptomycin by *Streptomyces griseus* and turimycin by *S. hygroscopicus*, whereas the pelleted form of *S. nigrificans* was better for glucose isomerase production. As already discussed for the fungi, the concentration of spores in the inoculum has also been shown to influence the morphology of certain streptomycetes. These workers also demonstrated that medium composition and the shear forces operating during culture also affect morphological form. Thus, the principles applied to the optimisation of fungal inoculum development regimes are also relevant to actinomycete processes. Hunt and Stieber described the optimisation of the inoculum regime of a small-scale streptomycete cephamycin C fermentation. Pellet formation was observed to be detrimental to product formation and the key factor in establishing the correct form appeared to be the concentration of iron in the seed medium, a higher iron concentration giving the optimum inoculum.

**Table 11.6.** The effect of spore concentration and medium on the morphology and penicillin productivity of *Penicillium chrysogenum*.

| *Medium* | *Spore concentration in the medium* | *Morphology* |
|---|---|---|
| Camici: | | |
| Corn-steep dextrin | More than $10 \times 10^5$ dm$^{-3}$ | Filamentous |
| | Less than $10 \times 10^5$ dm$^{-3}$ | Pellets |
| Czapek-dox | More than $3.0 \times 10^5$ dm$^{-3}$ | Filamentous |
| | Less than $3.0 \times 10^5$ dm$^{-3}$ | Pellets |
| Glucose, lactose and | More than $2.0 \times 10^5$ dm$^{-3}$ | Filamentous |
| ammonium lactate | Less than $2.0 \times 10^5$ dm$^{-3}$ | Pellets |
| *Spore concentration in the inoculum (cm$^{-3}$)* | *Penicillin yield (units cm$^{-3}$)* | *Morphology* |
| Calam: | | |
| $10^2$ | 500 | Dense pellets |
| $10^3$ | 1800 | Dense pellets |
| $2 \times 10^3$ | 4000 | Open pellets |
| $10^4$ | 5000 | Filamentous |

## ASEPTIC INOCULATION OF PLANT FERMENTERS

The inoculation of plant-scale fermenters may involve the transfer of culture from a laboratory fermenter or spore suspension vessel, to a plant fermenter or the transfer from one plant fermenter to another. Obviously, it is extremely important that the fermentation is not contaminated during inoculation but if the process is a contained one then it is equally important that the process organism does not escape. Thus, the nature of the inoculation system will be dictated by the containment category of the process.

At Containment levels 1 and B2, the addition of inoculum must be carried out in such a way that release of micro-organism is restricted. This should be done by aseptic piercing of membranes or connections with steam locks. At Containment level 2 and B3/4, no micro-organisms must be released during inoculation or other additions. In order to meet these stringent requirements, all connections must be screwed or clamped and all pipelines must be steam sterilisable.

### Inoculation from a Laboratory Fermenter or a Spore Suspension Vessel

Several systems have been described in the literature which are suitable for inoculating fermentations requiring only Good Industrial Large Scale Practice (GILSP). To prevent contamination during the transfer process it is essential that both vessels be maintained under a positive pressure and the inoculation port be equipped with a steam supply. Meyrath and Suchanek described a system for the inoculation of a plant fermenter from a laboratory vessel. The apparatus is shown in Fig. 11.7. The connecting point A is normally covered with the blank plug P and prior to inoculation this plug is slightly loosened, valve E closed and valve F opened to allow steam to exit at A. Valve F is then closed and E opened so that when P is removed sterile air will be released from the vessel. After removal, plug P is placed in strong disinfectant. Blank plug Q is then removed and a coupling made between B and A. Valve E is closed and an air line attached to point C, establishing a pressure inside the inoculum fermenter greater than in the plant vessel. Valve E is then opened and the inoculum will be forced into the plant fermenter. After closing valve E the inoculum fermenter may be removed, plug P replaced and the line steamed out by opening valve F. This system may be modified by using the quick connection devices which are now available.

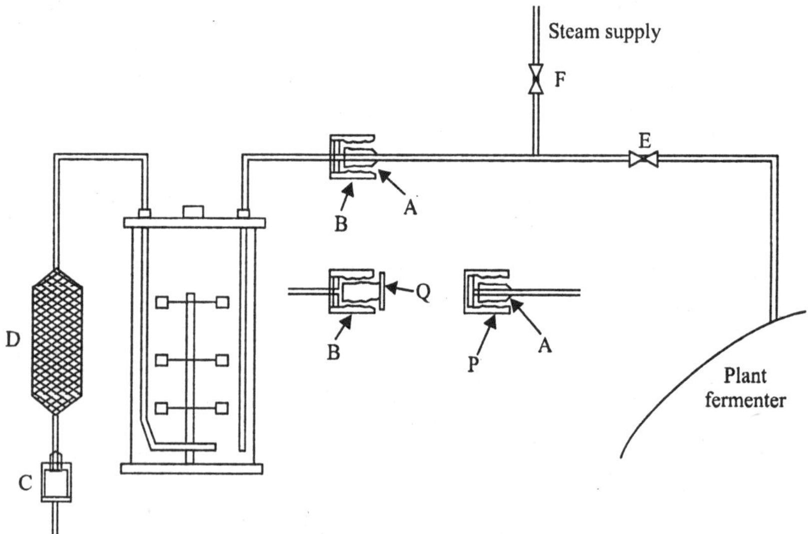


**Fig. 11.7.** Inoculation of a plant fermenter from a laboratory fermenter.

Jackson has described a very similar system for the introduction of a spore suspension into a plant fermenter. The apparatus is shown in Fig. 11.8 and its operation is identical to that described by Meyrath and Suchanek.

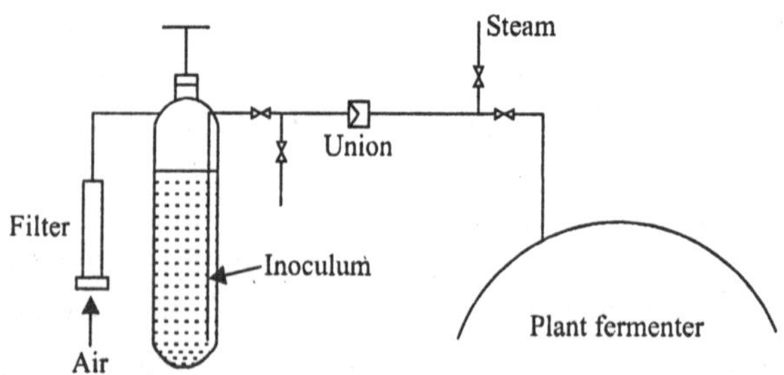


**Fig. 11.8.** Inoculation of a plant fermenter from a spore suspension vessel.

Parker described a more complex system for inoculation of a plant fermenter from a spore-suspension vessel. The apparatus is shown in Fig. 11.9. The sterile spore-suspension vessel is batched with the spore suspension in the sterile room. The plant vessel, containing the medium and with blank plugs screwed on at A and B, is sterilised by steam injection. The plugs A and B are slightly loosened to allow steam to emit for 20 minutes when the whole system is under steam pressure. The blanks are then tightened, the valves E and G shut and sterile air is allowed into the plant fermenter by opening valves D, F and C. The spore suspension vessel is loosely connected at A and B and valves D, H, I and C closed and E, F and G opened. After 20 minutes steaming A and B are tightened and G and E are closed. D is then opened to establish a positive pressure in the pipework. When the pipework has cooled, the pressure in the plant vessel is reduced to approximately 5 psi, valve F is closed and H, I and C are opened. This procedure allows the spore suspension to be forced into the fermenter. Valves D, H, I and C are closed and the suspension vessel replaced by blank plugs at A and B and the pipework steamed by opening valves E and G.

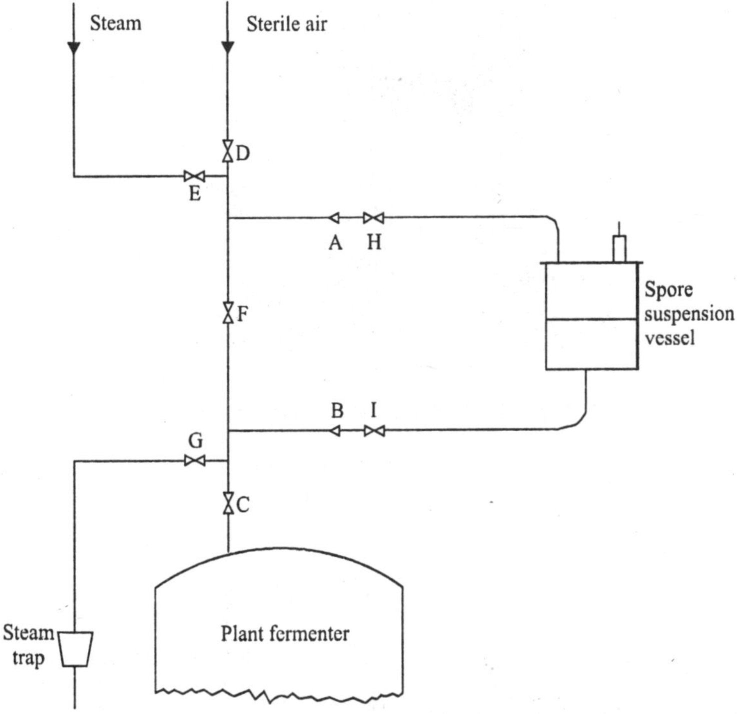


**Fig. 11.9.** The inoculation of a plant fermenter from a spore suspension vessel.

GILSP and category 1 fermentations may also be inoculated by aseptic piercing of a membrane. In this system the inoculum vessel is connected to an inoculating needle assembly (as shown in Fig. 11.10) and the sterile needle pierces a rubber septum set into a fermenter port. However, the use of a needle presents safety problems and many companies prohibit such systems. Also, aerosols may be created on the removal of a needle.

The inoculation of level 2 or B3/4 contained fermentations requires the use of modified systems. It must be possible to steam sterilise all the pipework after the inoculation and the condensate from the sterilisation must be collected in a kill tank. The inoculation flask is then removed after inoculation and sterilised in an autoclave. One such system is shown in Fig. 11.11.

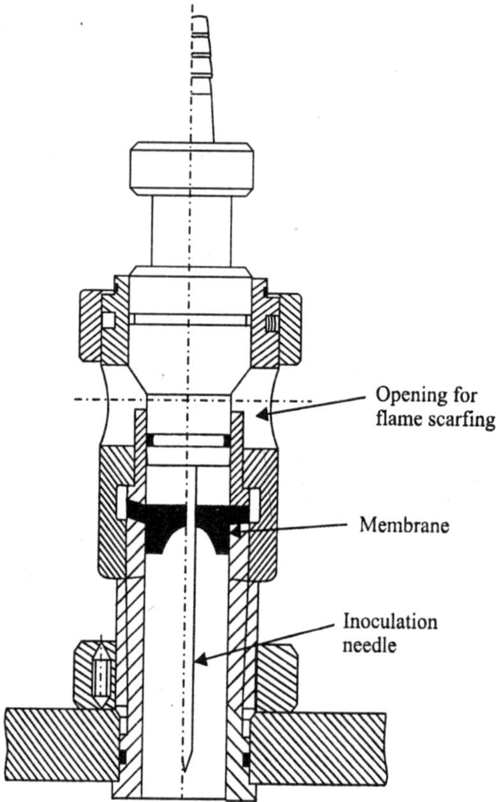


Fig. 11.10. Needle inoculation device.

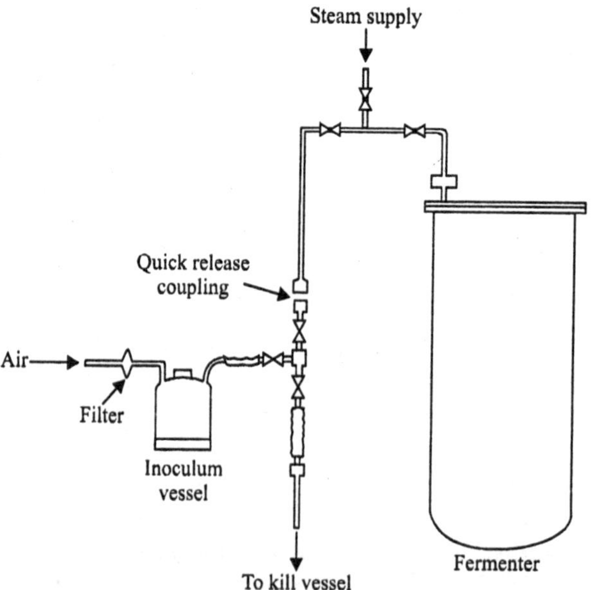


Fig. 11.11. Inoculum system suitable for contained fermentations.

### Inoculation from a Plant Fermenter

Figure 11.12 illustrates the system described by Parker. The two vessels are connected by a flexible pipeline A–B. The batched fermenter is sterilised by steam injection via valves G and J, valves D, I, A, B, H, E and F being open and valve C closed. Valves H and I lead to steam traps for the removal of condensate. After 20 minutes at the desired pressure the steam supply is switched off at J and G and the steam-trap valves I and H are closed. F, E and D are left open so that the connecting pipeline fills with sterile medium.

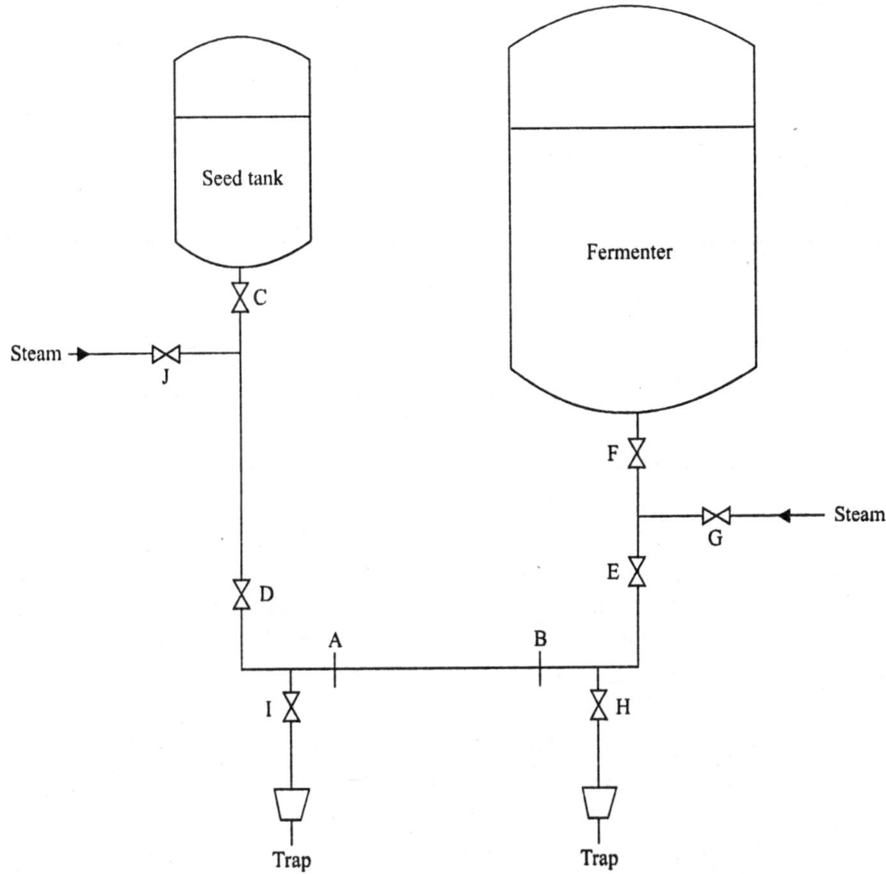


**Fig. 11.12.** Inoculation of a plant fermenter from another plant fermenter.

The medium in the fermenter is sparged with sterile air and when it has cooled to the desired temperature the pressure in the seed tank is increased to at least 10 psi whilst the pressure in the fermenter is reduced to about 2 psi. Valve C is opened and the inoculum is forced into the production vessel. After inoculation is complete the connecting pipeline is re-sterilised before it is removed. For a contained fermentation the condensate from the two steam traps attached to valves I and H would be directed to a kill tank.

Chapter 12

# Design of a Fermenter

## INTRODUCTION

The first true large-scale aerobic fermenters were used in Central Europe for the production of compressed yeast. The fermenters consisted of large cylindrical tanks with air introduced at the base via networks of perforated pipes.

In later modifications mechanical impellers were used to increase the rate of mixing and to break up and disperse the air bubbles. This procedure led to the compressed-air requirements being reduced by a factor of 5.

Baffles on the walls of the vessels prevented a vortex forming in the liquid. Even at this time it was recognised that the cost of energy necessary to compress air could be 10 to 20 per cent of the total production cost.

Prior to 1940, the other important fermentation products besides bakers' yeast were ethanol, glycerol, acetic acid, citric acid, other organic acids, enzymes and sorbose. These processes used highly selective environment such as acidic or anaerobic conditions or the use of an unusual substrate, resulting in contamination being a relatively minor problem compared with the acetone fermentation or the subsequent aerobic antibiotic fermentations.

The decision to use submerged culture techniques for penicillin production, where aseptic conditions, good aeration and agitation were essential, was a very important factor in forcing the development of carefully designed and purpose-built fermentation vessels.

The purpose of this chapter is to give an introduction to the methodology of batch fermentation using a small, autoclavable, bench-top fermentor system. Specifically, this chapter will provide an understanding of the following:

1. The engineering and biological concepts of fermenters.
2. The principles behind the instrumentation used with a modern fermenter.
3. The assembly and preparation of a fermenter vessel and ancillary equipment for autoclaving.
4. How samples are taken from the fermenter vessel during operation.
5. How simple continuous culture can be accomplished.

## BASIC FUNCTIONS OF A FERMENTER FOR MICROBIAL OR ANIMAL CELL CULTURE

The main function of a fermenter is to provide a controlled environment for the growth of micro-organisms or animal cells, to obtain a desired product.

In designing and constructing a fermenter a number of points must be considered:

1. The vessel should be capable of being operated aseptically for a number of days and should be reliable in long-term operation and meet the requirements of containment regulations.
2. Adequate aeration and agitation should be provided to meet the metabolic requirements of the micro-organism. However, the mixing should not cause damage to the organism.
3. Power consumption should be as low as possible.
4. A system of temperature control should be provided.
5. A system of pH control should be provided.
6. Sampling facilities should be provided.
7. Evaporation losses from the fermenter should not be excessive.
8. The vessel should be designed to require the minimal use of labour in operation, harvesting, cleaning and maintenance.
9. Ideally the vessel should be suitable for a range of processes, but this may be restricted because of containment regulations.
10. The vessel should be constructed to ensure smooth internal surfaces, using welds instead of flange joints whenever possible.
11. The vessel should be of similar geometry to both smaller and larger vessels in the pilot plant or plant to facilitate scale-up.
12. The cheapest materials which enable satisfactory results to be achieved should be used.
13. There should be adequate service provisions for individual plants (Table 12.1).

**Table 12.1.** Service provisions for a fermentation plant.

| |
|---|
| Compressed air |
| Sterile compressed air (at 1.5 to 3.0 atmospheres) |
| Chilled water (12 to 15°F) |
| Cold water (4°F) |
| Hot water |
| Steam (high pressure) |
| Steam condensate |
| Electricity |
| Stand-by generator |
| Drainage of effluents |
| Motors |
| Storage facilities for media components |
| Control and monitoring equipment for fermenters |
| Maintenance facilities |
| Extraction and recovery equipment |
| Accessibility for delivery of materials |
| Appropriate containment facilities |

The first two points are probably the most critical. It is obvious from the above points that the design of a fermenter will involve co-operation between experts in microbiology, biochemistry, chemical engineering, mechanical engineering and costing. Although many different types of fermenter have been described in the literature, very few have proved to be satisfactory for industrial aerobic fermentations. The most commonly used ones are based on a stirred upright cylinder with sparger aeration. This type of vessel can be produced in a range of sizes from one $dm^3$ to thousands of $dm^3$. Figures 12.1 and 12.2 are diagrams of typical mechanically agitated and aerated fermenters with one and three multi-bladed impellers respectively. Tables 12.2 and 12.3 give geometrical ratios of various dimensions for a variety of sizes of vessel.

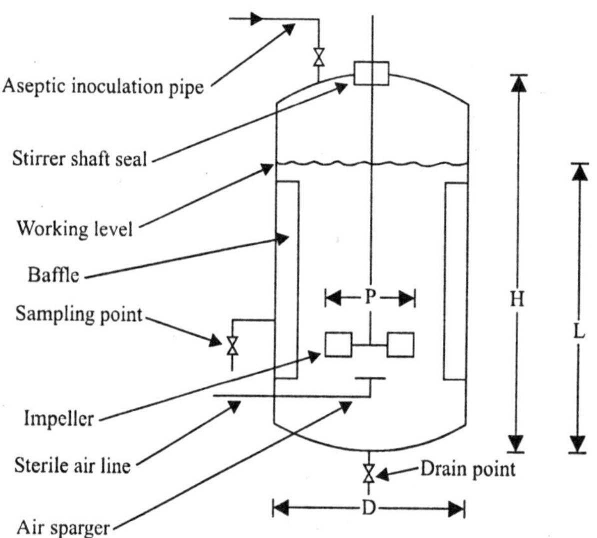


**Fig. 12.1.** Diagram of a fermenter with one multi-bladed impeller.

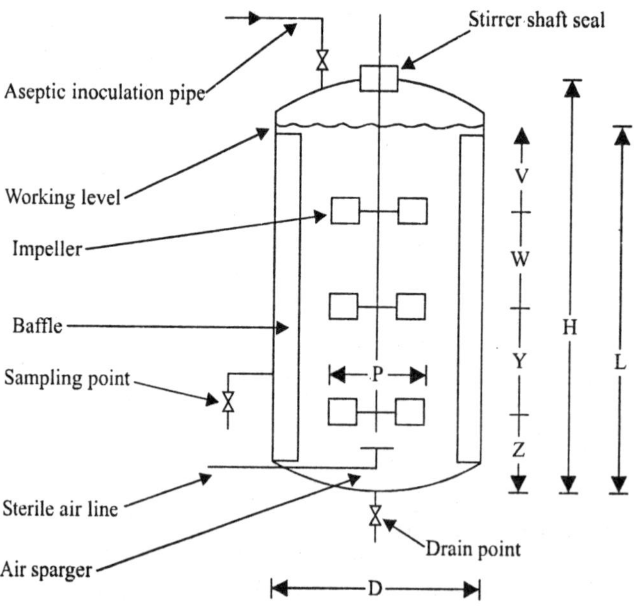


**Fig. 12.2.** Diagram of a fermenter with three multi-bladed impellers.

**Table 12.2.** Details of geometrical ratios of fermenters with single multi-blade impellers.

| *Dimension* | *Steel and Maxon* | *Wegrich and Shurter* | *Blakeborough* |
|---|---|---|---|
| Operating volume | 250 $dm^3$ | 12 $dm^3$ | – |
| Liquid height (L) | 55 cm | 27 cm | – |
| L/D (tank diameter) | 0.72 | 1.1 | 1.0–1.5 |
| Impeller diameter (P/D) | 0.4 | 0.5 | 0.33 |
| Baffle width/D | 0.10 | 0.08 | 0.08–0.10 |
| Impeller height/D | – | – | 0.33 |

**Table 12.3.** Details of geometrical ratios of fermenters with three multi-bladed impellers.

| *Dimension* | *Jackson* | *Aiba* | *Paca* |
|---|---|---|---|
| Operating volume | – | 1,00,000 $dm^3$ (total) | 170 $dm^3$ |
| Liquid height (L) | – | – | 150 cm |
| L/D (tank diameter) | – | – | 1.7 |
| Impeller diameter | 0.34–0.5 | 0.4 | 0.33 |
| Baffle width/D | 0.08–0.1 | 0.095 | 0.098 |
| Impeller height/D | 0.5 | 0.24 | 0.37 |
| P/V | 0.5–1.0 | – | 0.74 |
| P/W | 0.5–1.0 | 0.85 | 0.77 |
| P/Y | 0.5–1.0 | 0.85 | 0.77 |
| P/Z | – | 2.1 | 0.91 |
| H/D | 1.0–1.6 | 2.2 | 2.95 |

At this stage the discussion will be concerned with stirred, aerated vessels for microbial cell culture. More varied shapes are commonly used for alcohol, biomass production, animal cell culture and effluent treatment and will be dealt with later in this section.

## ASEPTIC OPERATION AND CONTAINMENT

Aseptic operation involves protection against contamination and it is a well established and understood concept in the fermentation industries, whereas containment involves prevention of escape of viable cells from a fermenter or downstream equipment and is much more recent in origin.

To establish the appropriate degree of containment which will be necessary to grow a micro-organism, it and in fact the entire process, must be carefully assessed for potential hazards that could occur should there the accidental release. Different assessment procedures are used depending on whether or not the organism contains foreign DNA (genetically engineered). Once the hazards are assessed, an organism can be classified into a hazard group for which there is an appropriate level of containment. The procedure which has been adopted within the European Community is outlined in Fig. 12.3. Non-genetically engineered organisms may be placed into a hazard group (1 to 4) using criteria to assess risk such as those given by Collins:

1. The known pathogenicity of the micro-organism.
2. The virulence or level of pathogenicity of the micro-organism—are the diseases it causes mild or serious?

3. The number of organisms required to initiate an infection.
4. The routes of infection.
5. The known incidence of infection in the community and the existence locally of vectors and potential reserves.
6. The amounts or volumes of organisms used in the fermentation process.
7. The techniques or processes used.
8. Ease of prophylaxis and treatment.

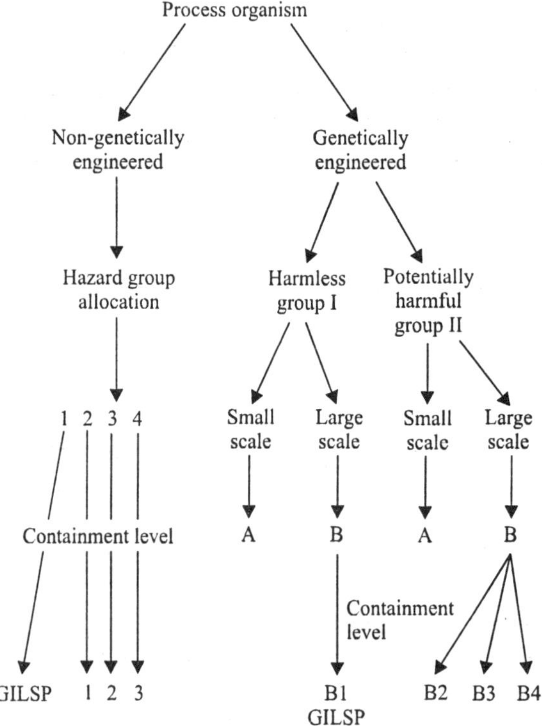


**Fig. 12.3.** Categorisation of process micro-organism and designation of its appropriate level of containment (GILSP = Good Industrial Large Scale Practice).

Once the organism has been allocated to a hazard group, the appropriate containment requirements can be applied (Table 12.4). Hazard group 1 organisms used on a large scale only require Good Industrial Large Scale Practice (GILSP). Processes in this category need to be operated aseptically but no containment steps are necessary, including prevention of escape of organisms. If the organism is placed in Hazard group 4 the stringent requirements of level 3 will have to be met before the process can be operated.

Genetically engineered organism are classified as either harmless (Group I) or potentially harmful (Group II). The process is then classified as either small scale (A: less than 10 $dm^3$) or large scale (B: more than 10 $dm^3$).

Therefore large scale processes fall into two categories, IB or IIB. IB processes require containment level B1 and are subject to GILSP, whereas IIB processes are further assessed to determine the most suitable containment level, ranging from B2 to B4 as outlined in Table 12.4. Levels B2 to B4 correspond to levels 1 to 3 for non-genetically engineered organisms.

**Table 12.4.** Summary of safety precautions for biotechnological operations in the Europen Federation for Biotechnology (EFB).

| *Procedures* | *GILSP** | *Containment category* | | |
|---|---|---|---|---|
| | | *1* | *2* | *3* |
| Written instruction and code of practice | + | + | + | + |
| Biosafety manual | – | + | + | + |
| Good occupational hygiene | + | + | + | + |
| Good microbiological techniques (GMT) | – | + | + | + |
| Biohazard sign | – | + | + | + |
| Restricted access | – | + | + | + |
| Accident reporting | + | + | + | + |
| Medical surveillance | – | + | + | + |
| *Primary containment:* | | | | |
| *Operation and equipment* | | | | |
| Work with viable micro-organisms should take place in closed systems (CS), which minimise (m) or prevent (p) the release of cultivated micro-organisms | – | m | p | p |
| Treatment of exhaust air or gas from CS | – | m | p | p |
| Sampling from CS | – | m | p | p |
| Addition of materials to CS, transfer of cultivated cells | – | m | p | p |
| Removal of material, products and effluents from CS | – | m | p | p |
| Penetration of CS by agitator shaft and measuring devices | – | m | p | p |
| Foam-out control | – | m | p | p |
| *Secondary containment:* | | | | |
| *Facilities* | | | | |
| Protective clothing appropriate to the risk category | + | + | + | + |
| Changing/washing facility | + | + | + | + |
| Disinfection facility | – | + | + | + |
| Emergency shower facility | – | – | + | + |
| Airlock and compulsory shower facilities | – | – | – | + |
| Effluents decontaminated | – | – | + | + |
| Controlled negative pressure | – | – | – | + |
| HEPA filters in air ducts | – | – | + | + |
| Tank for spilled fluids | – | – | + | + |
| Area hemetically sealable | – | – | – | + |

* Unless required for product quality; –, not required; +, required.

m, minimise release. The level of contamination of air, working surface and personnel shall not exceed the level found during microbiological work applying good microbiological techniques.

p, prevent release. No detectable contamination during work should be found in the air, working surfaces and personnel.

In future it is possible, under new legislation, that no distinction will be made between non-genetically engineered and genetically engineered organism. The key factor will be whether the organism is harmless or potentially harmful, regardless of its genetic constitution. Containment would then be decided using the scheme which is currently being used for genetically engineered organism.

Other hazard-assessment systems for classifying organisms have been introduced in many other countries. Production and research workers must abide by appropriate local official hazard lists. Problems can occur when different official bodies place the same organism in different hazard categories.

Most micro-organisms used in industrial processes are in the lowest hazard group which only require GILSP, although some organisms used in bacterial and viral vaccine production and other processes are categorised in higher groups. There is an obvious incentive for industry to use an organism which poses a low risk as this minimises regulatory restrictions and reduces the need for expensive equipment and associated containment facilities. In this chapter, where appropriate, a distinction will be made between equipment which is used to maintain aseptic conditions for GILSP and those modifications which are needed to meet physical containment requirements in order to grow specific cultures which have been classified in higher hazard groups.

### Overall Containment Categorisation

In this chapter the emphasis will be on containment levels which might be obtained from particular designs of fermenters and associated equipment. However, this is only one aspect of assessment. To meet the standards of the specific level of containment it will also be necessary to consider the procedures to bo used, staff training, the facilities in the laboratory and factory, downstream processing, effluent treatment, work practice, maintenance, etc. It will be necessary to ensure that all these aspects are of a sufficiently high standard to meet the levels of containment deemed necessary for a particular process by a government regulatory body. If these are met, then the process can be operated.

## BODY CONSTRUCTION

### Construction Materials

In fermentations with strict aseptic requirements it is important to select materials that can withstand repeated steam sterilisation cycles. On a small scale (1 to 30 $dm^3$) it is possible to use glass and/or stainless steel. Glass is useful because it gives smooth surfaces, is non-toxic, corrosion proof and it is usually easy to examine the interior of the vessel. Two basic types of fermenter are used:

1. A glass vessel with a round or flat bottom and a top flanged carrying plate. The large glass containers originally used were borosilicate battery jars. All vessels of this type have to be sterilised by autoclaving. The largest practical diameter for glass fermenters is 60 cm.
2. A glass cylinder with stainless-steel top and bottom plates. These fermenters may be sterilised *in situ*, but 30 cm diameter is the upper size limit to safely withstand working pressures. Vessels with two stainless steel plates cost approximately 50 per cent more than those with just a top plate.

At pilot and large scale, when all fermenters are sterilised *in situ*, any materials used will have to be assessed on their ability to withstand pressure sterilisation and corrosion and on their potential toxicity and cost. Pilot-scale and industrial scale vessels are normally constructed of stainless steel or at least have a stainless-steel cladding to limit corrosion. The American Iron and Steel Institute (AISI) states that steels containing less than 4 per cent chromium are classified as steel alloys and those containing more than 4 per cent are classified as stainless steels. Mild steel coated with glass or phenolic epoxy materials has occasionally been used. Wood, plastic and concrete has been used when contaminations was not a problem in a process.

Walker and Holdsworth stated that the extent of vessel corrosion varied considerably and did not appear to be entirely predicable. Although stainless steel is often quoted as the only satisfactory material, it has been reported that mild-steel vessels were very satisfactory after 12 years use for penicillin fermentations and mild steel clad with stainless steel has been used for at least 35 years for acetone-butanol production. Pitting to a depth of 7 mm was found in a mild-steel fermenter after 7 years use for streptomycin production.

The corrosion resistance of stainless steel is thought to depend on the existence of a thin hydrous oxide film on the surface of the metal. The composition of this film varies with different steel alloys and different manufacturing process treatments such as rolling, pickling or heat treatment. The film is stabilised by chromium and is considered to be continuous, non-porous, insoluble and self-healing. If damaged, the film will repair itself when exposed to air or an oxidising agent.

The minimum amount of chromium needed to resist corrosion will depend on the corroding agent in a particular environment, such as acids, alkalies, gases, soil, salt or fresh water. Increasing the chromium content enhances resistance to corrosion, but only grades of steel containing at least 10 to 13 per cent chromium develop an effective film. The inclusion of nickel in high per cent chromium steels enhances their resistance and improves their engineering properties. The presence of molybdenum improves the resistance of stainless steels to solution of halogen salts and pitting by chloride ions in brine or sea water. Corrosion resistance can also be improved by tungsten, silicone and other elements.

AISI grade 316 steels which contain 18 per cent chromium, 10 per cent nickel and 2–2.5 per cent molybdenum are now commonly used in fermenter construction. The thickness of the construction material will increase with scale. At 3,00,000 to 4,00,000 dm$^3$ capacity, 7-mm plate may be used for the side of a vessel and 10 mm plate for the top and bottom, which should be hemispherical to withstand pressures. At this stage it is important to consider the ways in which a reliable aseptic seal is made between glass and glass, glass and metal or metal and metal joints such as between a fermenter vessel and a detachable top or base plate. With glass and metal, a seal can be made with a compressible gasket, a lip seal or an 'O' ring (Fig. 12.4). With metal to metal joints only an 'O' ring is suitable.

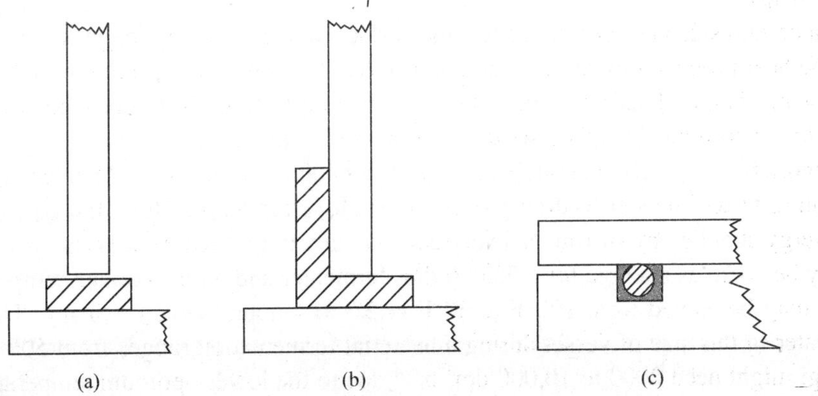


**Fig. 12.4.** Joint seals for glass-glass, glass-metal and metal-metal; (a) gasket; (b) lip seal; and (c) 'O' ring in groove.

This is placed in a groove, machined in either the end plate, the fermenter body or both. This seal ensures that a good liquid and/or gas-tight joint is maintained in spite of the glass or metal expanding or contracting at different rates with changes in temperature during a sterilisation cycle or an incubation cycle. Nitryl or butyl rubbers are normally used for these seals as they will withstand fermentation process conditions. These rubber seals have a finite life and should be checked regularly for damage or perishing. Before purchasing a fermenter it is important to check that standard sized 'O' rings can be purchased cheaply from local suppliers.

A single 'O' ring seal is adequate for GILSP and levels 1 and B2, a double 'O' ring seal is required for levels 2 and B3 and a double 'O' ring seal with steam between the seals (steam tracing) is necessary for levels 3 and B4. In the USA, however, simple seals are used at containment levels comparable with the UK B3.

Titchener-Hooker criticised Chapman's proposal to use two seals without a steam trace for a number of reasons including:

1. Double seals are more difficult to assemble correctly.
2. It is difficult to detect failure of one seal of a pair during operation or assembly.
3. Neither of the two seals can be tested independently.
4. Dead space between two seals must be considered to be contaminated.

Leaver and Titchener-Hooker consider that correctly fitted single static seals can provide adequate containment for most processes and double static seals with a steam trace should be strictly limited to the small number of processes for which an extreme level of protection may be required.

## Temperature Control

Normally in the design and construction of a fermenter there must be adequate provision for temperature control which will affect the design of the vessel body. Heat will be produced by microbial activity and mechanical agitation and if the heat generated by these two processes is not ideal for the particular manufacturing process then heat may have to be added to or removed from, the system. On a laboratory scale little heat is normally generated and extra heat has to be provided by placing the fermenter in a thermostatically controlled bath or by the use of internal heating coils or a heating jacket through which water is circulated or by a silicone heating jacket. The silicone jacket consists of a double silicone rubber mat with heating wires between the two mats; it is wrapped around the vessel and held in place by Velcro strips.

Once a certain size has been exceeded, the surface area covered by the jacket becomes too small to remove the heat produced by the fermentation. When this situation occurs internal coils must be used and cold water is circulated to achieve the correct temperature. Different types of fermentation will influence the maximum size of vessel that can be used with jackets alone.

It is impossible to specify accurately the necessary cooling surface of a fermenter since the temperature of the cooling water, the sterilisation process, the cultivation temperature, the type of micro-organism and the energy supplied by stirring can vary considerably in different processes. A cooling area of 50 to 70 $m^2$ may be taken as average for a 55,000 $dm^3$ fermenter and with a coolant temperature of 14°F the fermenter may be cooled from 120°F to 30°F in 2.5 to 4 hours without stirring. The consumption of cooling water in this size of vessel during a bacterial fermentation ranges from 500 to 2000 $dm^3$ $hr.^{-1}$, while fungi might need 2000 to 10,000 $dm^3$ $hr.^{-1}$ due to the lower optimum temperature for growth.

To make an accurate estimate of heating/cooling requirements for a specific process it is important to consider the contributing factors. An overal energy balance for a fermenter during normal operation can be written as:

$$Q_{met} + Q_{ag} + Q_{gas} = Q_{acc} + Q_{exch} + Q_{evap} + Q_{sen} \quad ...(12.1)$$

where,

$Q_{met}$ = heat generation rate due to microbial metabolism
$Q_{ag}$ = heat generation rate due to mechanical agitation
$Q_{gas}$ = heat generation rate due to aeration power input
$Q_{acc}$ = heat accumulation rate by the system
$Q_{exch}$ = heat transfer rate to the surroundings and/or heat exchanger
$Q_{evap}$ = heat loss rate by evaporation
$Q_{sen}$ = rate of sensible enthalpy gain by the flow streams (exit–inlet)

This equation can be rearranged as:

$$Q_{exch} = Q_{met} + Q_{ag} + Q_{gas} - Q_{acc} - Q_{sen} - Q_{evap} \quad \text{... (12.2)}$$

$Q_{exch}$ is the heat which will have to be removed by a cooling system.

Atkinson and Mavituna have presented data to estimate $Q_{met}$ for a range of substrates, methods to calculate power input for $Q_{ag}$ and $Q_{gas}$, a formula to calculate the sensible heat loss for flow streams ($Q_{sen}$) and a method to calculate the heat loss due to evaporation ($Q_{evap}$). Cooney determined representative low and high heats of fermentation values for *Bacillus subtilis* grown on molasses at 37°F (Table 12.5). They concluded that $Q_{evap}$ and $Q_{sen}$ are small contributory factors and $Q_{acc}$ = 0 in a steady-state system. $Q_{evap}$ can also be eliminated by using a saturated air stream at the temperature of the broth.

**Table 12.5.** Representative low and high values of calculated heats (kcal dm$^{-3}$ hr.$^{-1}$) or fermentation for *Bacillus subtilis* on molasses at 37°C.

| | *Heats of fermentation* | | | | | |
|---|---|---|---|---|---|---|
| | $Q_{acc}$ | $Q_{ag}$ | $Q_{evap}$ | $Q_{sen}$ | $Q_{exch}$ | $Q_{met}$ |
| Low | 3.81 | 3.32 | 0.023 | 0.005 | 0.61 | 1.12 |
| High | 11.3 | 3.31 | 0.045 | 0.010 | 0.65 | 8.65 |

When designing a large fermenter, the operating temperature and flow conditions will determine $Q_{evap}$ and $Q_{sen}$, the choice of agitator, its speed and the aeration rate will determine $Q_{ag}$ and the sparger design and aeration rate will determine $Q_{gas}$.

The cooling requirements (jacket and/or pipes) to remove the excess heat from a fermenter may be determined by the following formula:

$$Q_{exch} = U \cdot A \cdot \Delta T \quad \text{... (12.3)}$$

where,

$A$ = the heat transfer surface available, m$^2$
$Q$ = the heat transferred, W
$U$ = the overall heat transfer coefficient, W/m$^2$K
$\Delta T$ = the temperature difference between the heating or cooling agent and the mass itself, K.

The coefficient $U$ represents the conductivity of the system and it depends on the vessel geometry, fluid properties, flow velocity, wall material and thickness. $1/U$ is the overall resistance to heat transfer. It is the reciprocal of the overall heat-transfer coefficient. It is defined as the sum of the individual resistances to heat transfer as heat passes from one fluid to another and can be expressed as:

$$\frac{1}{U} = \frac{1}{h_o} + \frac{1}{h_i} + \frac{1}{h_{of}} + \frac{1}{h_{if}} + \frac{1}{h_w} \quad \text{... (12.4)}$$

where,

$h_o$ = outside film coefficient, W/m$^2$K
$h_i$ = inside film coefficient, W/m$^2$K
$h_{of}$ = outside fouling film coefficient, W/m$^2$K
$h_{if}$ = inside fouling film coefficient, W/m$^2$K
$h_w$ = wall heat transfer coefficient = k/x, W/m$^2$K
$k$ = thermal conductivity of the wall, W/mK
$x$ = wall thickness, m

Atkinson and Mavituna have given three methods to determine ΔT (the temperature driving force) depending on the operating circumstances. If one side of the wall is at a constant temperature, as is often the case in a stirred fermenter and the coolant temperature rises in the direction of the coolant flow along a cooling coil, an arithmetic mean is appropriate:

$$\Delta T_{am} = \frac{(T_f - T_e) + (T_f - T_i)}{2} \quad \text{... (12.5)}$$

$$= \frac{T_f \quad (T_e \quad T_i)}{2} \quad \text{... (12.6)}$$

where,

$T_f$ = the bulk liquid temperature in the vessel
$T_e$ = the temperature of the coolant entering the system
$T_i$ = the temperature of the coolant leaving the system

If the fluids are in counter- or co-current flow and the temperature varies in both fluids then a log mean temperature difference is appropriate:

$$\Delta T_m = \Delta T_e - \Delta T_i / \ln (\Delta T_e / \Delta T_i) \quad \text{... (12.7)}$$

where,

$\Delta T_e$ = the temperature of the coolant entering
$\Delta T_i$ = the temperature of the coolant leaving

If the flow pattern is more complex than either of the two previous situations then the log mean temperature difference defined in equation (12.7) is multiplied by an appropriate dimensionless factor which has been evaluated for a number of heat-exchanger systems by McAdams.

Appropriate techniques have just been discussed to obtain values for $Q_{exch}$, $U$ and $\Delta T$ (or $\Delta T_{am}$ or $\Delta T_m$). If equation (12.3) is now rearranged:

$$A = \frac{Q_{cxch}}{U\Delta T} \quad \text{... (12.8)}$$

Substituting in this equation makes it possible to calculate the heat-transfer surface necessary to obtain adequate temperature control.

## AERATION AND AGITATION

The primary purpose of aeration is to provide micro-organisms in submerged culture with sufficient oxygen for metabolic requirements, while agitation should ensure that a uniform suspension of microbial cells is achieved in a homogeneous nutrient medium. The type of aeration-agitation system used in a particular fermenter depends on the characteristics of the fermentation process under consideration.

Although fine bubble aerators without mechanical agitation have the advantage of lower equipment and power costs, agitation may be dispensed with only when aeration provides sufficient agitation, i.e., in processes where broths of low viscosity and low total solids are used. Thus, mechanical agitation is usually required in fungal and actinomycete fermentations. Non-agitated fermentations are normally carried out in vessels of a height/diameter ratio of 5:1. In such vessels aeration is sufficient to produce high turbulence, but a tall column of liquid does require greater energy input in the production of the compressed air.

The structural components of the fermenter involved in aeration and agitation are:

1. Agitator (impeller).
2. Stirrer glands and bearings.
3. Baffles.
4. Aeration system (sparger).

## Agitator (Impeller)

The agitator is required to achieve a number of mixing objectives, e.g., bulk fluid and gas-phase mixing, air dispersion, oxygen transfer, heat transfer, suspension of solid particles and maintaining a uniform environment throughout the vessel contents. It should be possible to design a fermenter to achieve these conditions; this will require knowledge of the most appropriate agitator, air sparger, baffles, the best positions for nutrient feeds, acid or alkali for pH control and antifoam addition. There will also be a need to specify agitator size and number, speed and power imput.

Agitators may be classified as disc turbines, vaned discs, open turbines of variable pitch and propellers and are illustrated in Fig. 12.5. The disc turbine consists of a disc with a series of rectangular vanes set in a vertical plane around the circumference and the vaned disc has a series of rectangular vanes attached vertically to the undersides. Air from the sparger hits the underside of the disc and is displaced towards the vanes where the air bubbles are broken up into smaller bubbles. The vanes of a variable pitch open turbine and the blades of a marine propeller are attached directly to a boss on the agitator shaft. In this case the air bubbles do not initially hit any surface before dispersion by the vanes or blades.

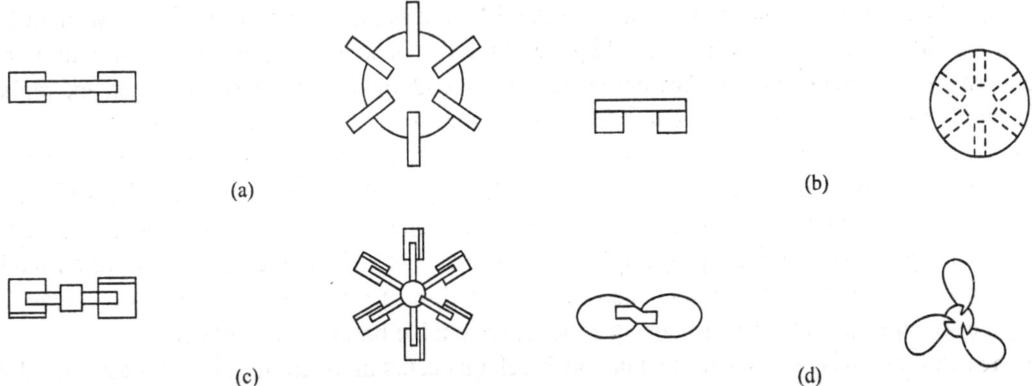


**Fig. 12.5.** Types of agitators: (a) disc turbine; (b) vaned disc; (c) open turbine, variable pitch; and (d) marine propeller.

Four other modern agitator developments, the Scaba 6SRGT, the Prochem Maxflo T, the Lightnin A315 and the Ekato Intermig (Figs. 12.6 and 12.7), which are derived from open turbines, will also be discussed for energy conservation and use in high-viscosity broths.

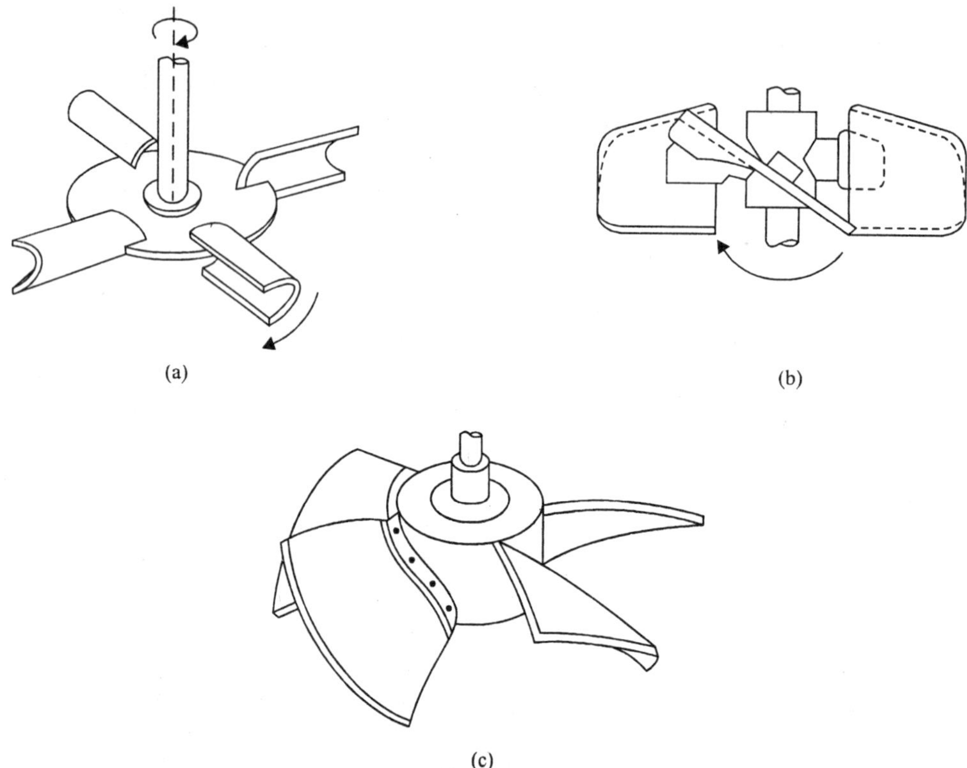


**Fig. 12.6.** Diagrams of: (a) Scaba agitator; (b) Lightnin' A315 agitator (four blades); and (c) Prochem Maxflo T agitator (four, five or six blades).

Since the 1940s a Rushton disc turbine of one-third the fermenter diameter has been considered the optimum design for use in many fermentation processes. It had been established experimentally that the disc turbine was most suitable in a fermenter since it could break up a fast air stream without itself becoming flooded in air bubbles. This flooding condition is indicated when the bulk flow pattern in the vessel normally associated with the agitator design (radial with the Rushton turbine) is lost and replaced by a centrally flowing air-broth plume up the middle of the vessel with a liquid flow as an annulus. The propeller and the open turbine flood when $V_s$ (superficial velocity, i.e. volumetric air flow rate/cross-sectional area of fermenter) exceeds 21 m hr.$^{-1}$, whereas the flat blade turbine can tolerate a $V_s$ of up to 120 m hr.$^{-1}$ before being flooded, when two sets are used on the same shaft. Besides being flooded at a lower $V_s$ than the disc turbine, the propeller is also less efficient in breaking up a stream of air bubbles and the flow it produce is axial rather than radial. The disc turbine was thought to be essential for forcing the sparaged air into the agitator tip zone where bubble break up would occur.

In other studies it has been shown that bubble break up occurs in the trailing vortices associated with all agitator types which give rise to gas-filled cavities and provided the agitator speed is high enough, good gas dispersion will occur in low-viscosity broths. It has been also shown that similar oxygen-transfer efficiencies are obtained at the same power input per unit volume, regardless of the agitator type.

In high-viscosity broths, gas dispersion also occurs from gas filled cavities trailing behind the rotating blades, but this is not sufficient to ensure satisfactory bulk blending of all the vessel contents. When the cavities are of maximum size, the impeller appears to be rotating in a pocket of gas from which little actual dispersion occurs into the rest of the vessel.

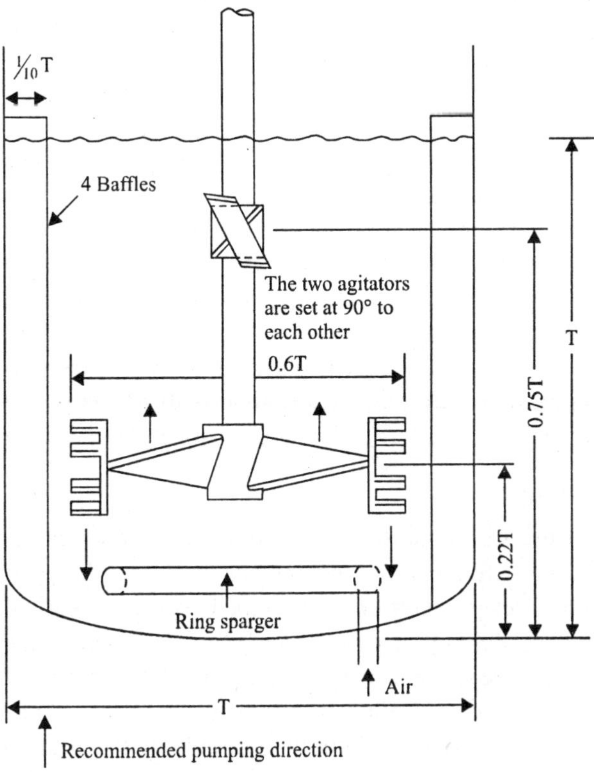


**Fig. 12.7.** Arrangement for a pair of intermig agitators. Relative dimensions are given as a proportion of the fermenter vessel diameter (T).

Recently, a number of agitators have been developed to overcome problems associated with efficient bulk blending in high-viscosity fermentations (Figs. 12.6 and 12.7). The Scaba 6SRGT agitator is one which, at a given power input, can handle a high air flow rate before flooding. This radial-flow agitator is also better for bulk blending than a Rushton turbine, but does not give good top to bottom blending in a large fermenter which leads to lower concentrations of oxygen in broth away from the agitators and higher concentration of nutrients, acid or alkali or antifoam near to the feed points.

Another is the Prochem Maxflo agitator. It consists of four, five or six hydrofoil blades set at a critical angle on a central hollow hub. A high hydrodynamic thrust is created during rotation, increasing the downwards pumping capacity of the blades. This design minimises the drag forces associated with rotation of the agitator such that the energy losses due to drag are low. This leads to a low power number. The recommended agitator to vessel diameter ratio is greater than 0.4. When the agitator was used with a 800 $dm^3$ *Streptomyces* fermentation, the maximum power requirement at the most viscous stage was about 66 per cent of that with Rushton turbines. The fall in power was also less in a 14000 $dm^3$ fermentation. The oxygen-transfer efficiency was also significantly improved. It was thought that in improvement in bulk mixing was another contributing factor.

Intermig agitators (Fig. 12.7) are more complex in design. Two units are used (with agitator diameter to vessel diameter ratios of 0.6 to 0.7) instead of a single Rushton turbine because their power number is so low. A large diameter air sparger is used to optimise air dispersion (see later section on spargers). The loss in power is less than when aerating with a Rushton turbine. Air dispersion starts from the air cavities which form on the wing tips of the agitator blades. In spite of the downwards pumping direction of the wings, the cavities extend horizontally from the back of the agitator blades, reducing the effectiveness of top to bottom mixing in a vessel.

Cooke and Nienow give comparisons of performance of Rushton turbines with some of the newer designs. These new turbine designs make it possible to replace Rushton turbines by larger low power agitators which do not lose as much power when aerated, are able to handle higher air volumes without flooding and give better bulk blending and heat transfer characteristics in more viscous media. However, there can be mechanical problems which are mostly of a vibrational nature.

Good mixing and aeration in high viscosity broths may also be achieved by a dual impeller combination, where the lower impeller acts as the gas disperser and the upper impeller acts primarily as a device for aiding circulation of vessel contents.

Steel and Maxon tested a multi-rod mixing impeller. In a 15000 $dm^3$ vessel, the same novobiocin yield and oxygen availability rate were obtained at about half of the power required by a standard turbine-stirred fermenter, but this type of impeller does not appear to have come into general use.

## Stirrer Glands and Bearings

The satisfactory sealing of the stirrer shaft assembly has been one of the most difficult problems to overcome in the construction of fermentation equipment which can be operated aseptically for long periods. A number of different designs have been developed to obtain aseptic seals. The stirrer shaft can enter the vessel from the top, side or bottom of the vessel. Top entry is most commonly used, but bottom entry may be advantageous if more space is needed on the top plate for entry ports and the shorter shaft permits higher stirrer speeds to be used by eliminating the problem of the shaft whipping at high speeds. Originally, bottom entry stirrers were considered undesirable as the bearings would be submerged. Chain and others successfully operated vessels of this type and they have since been used by many other workers. Mechanical seals can be used for bottom entry provided that they are routinely maintained and replaced at recommended intervals.

One of the earliest stirrer seals described was that used by Rivett, Johnson and Peterson in a laboratory fermenter (Fig. 12.8). A porous bronze bearing for a 13 mm shaft was fitted in the center of the fermenter top and another in a yoke directly above it. The bearings were pressed into steel housings, which screwed into position in the yoke and the fermenter top. The lower bearing and housing were covered with a skirt-like shield having a 6.5 mm overhang which rotated with the shaft and prevented air-borne contaminants from settling on the bearing and working their way through it into the fermenter.

Four basic types of seal assembly have been used: the stuffing box (packed-gland seal), the simple bush seal, the mechanical seal and the magnetic drive. Most modern fermenter stirrer mechanisms now incorporate mechanical seals instead of stuffing boxes and packed glands. Mechanical seals are more expensive, but are more durable and less likely to be an entry point for contaminants or a leakage point for organisms or products which should be contained. Magnetic drives, which are also quite expensive, are being used in some animal cell culture vessels.

### *Stuffing box (Packed-gland seal)*

In stuffing box the shaft is sealed by several layers of packing rings of asbestos or cotton yarn, pressed against the shaft by a gland follower. At high stirrer speeds the packing wears quickly and excessive pressure may be needed to ensure tightness of fit.

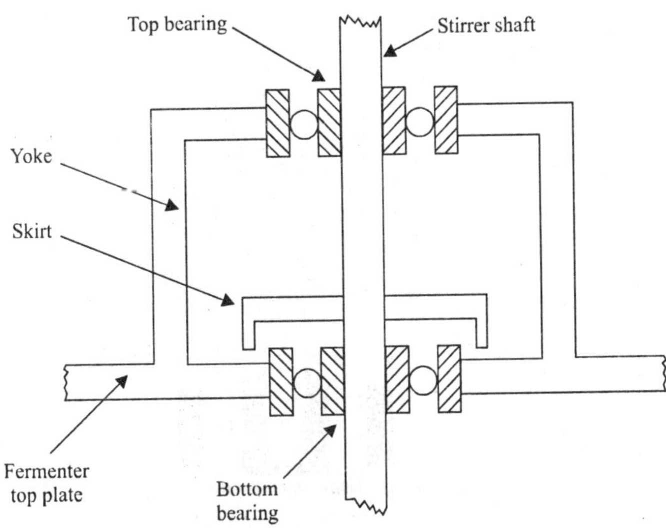


**Fig. 12.8.** A simple stirrer seal.

### *Mechanical seal*

The mechanical seal assembly is now commonly used in both small and large fermenters. The seal is composed of two parts, one part is stationary in the bearing housing, the other rotates on the shaft and the two components are pressed together by springs or expanding bellows. The two meeting surfaces have to be precision machined, the moving surface normally consists of a carbon-faced unit while the stationary unit is of satellite faced stainless steel. Steam condensate is used to lubricate and cool the seals during operation and serve as a containment barrier. Single mechanical seals are used with a steam barrier in fermenters whereas double mechanical seals are typically used in vessels with the outer seal as a backup for the inner seal for primary containment. Monitoring of the steam condensate flowing out of the seal is an effective way for checking for seal failure. Disinfectants are alternatives for flushing the seals.

### *Magnetic drives*

The problems of providing a satisfactory seal when the impeller shaft passes through the top or bottom plate of the fermenter may be solved by the use of a magnetic drive in which the impeller shaft does not pierce the vessel. A magnetic drive (Fig. 12.9) consists of two magnets: one driving and one driven. The driving magnet is held in bearings in a housing on the outside of the head plate and connected to a drive shaft. The internal driven magnet is placed on one end of the impeller shaft and held in bearings in a suitable housing on the inner surface of the headplate. The stirring mechanism is ideal for animal cell culture to minimise the chances of potential contamination. This application is discussed later in this chapter.

### Baffles

Four baffles are normally incorporated into agitated vessels of all sizes to prevent a vortex and to improve aeration efficiency. In vessels over 3 dm$^3$ diameter six or eight baffles may be used. Baffles are metal strips roughly one-tenth of the vessel diameter and attached radially to the wall (Figs. 12.1 and 12.2 and Tables 12.2 and 12.3). The agitation effect is only slightly increased with wider baffles, but drops sharply

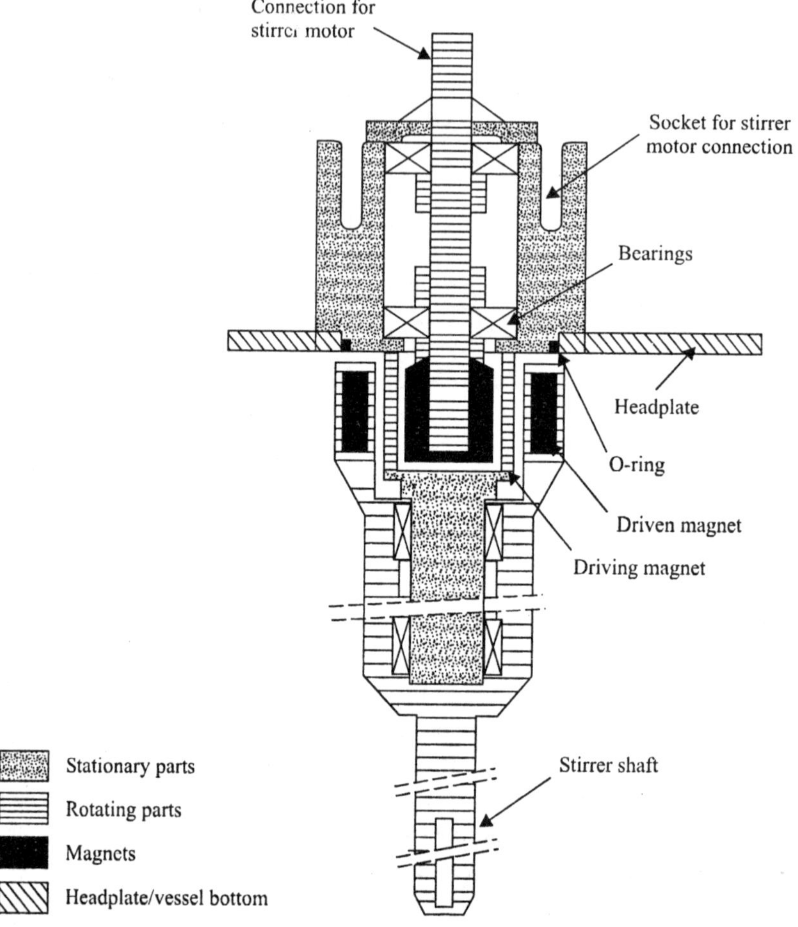


**Fig. 12.9.** Diagram of magnetically coupled top stirrer assembly.

with narrower baffles. Walker and Holdsworth recommended that baffles should be installed so that a gap existed between them and the vessel wall, so that there was a scouring action around and behind the baffles thus minimising microbial growth on the baffles and the fermenter walls. Extra cooling coils may be attached to baffles to improve the cooling capacity of a fermenter without unduly affecting the geometry.

## Aeration System (Sparger)

A sparger may be defined as a device for introducing air into the liquid in a fermenter. Three basic types of sparger have been used and may be described as the porous sparger, the orifice sparger (a perforated pipe) and the nozzle sparger (an open or partially closed pipe). A combined sparger-agitator may be used in laboratory fermenters.

### *Porous sparger*

The porous sparger of sintered glass, ceramics or metal, has been used primarily on a laboratory scale in non-agitated vessels. The bubble size produced from such spargers is always 10 to 100 times larger than the pore size of the aerator block. The throughput of air is low because of the pressure drop across the sparger and there is also the problem of the fine holes becoming blocked by growth of the microbial culture.

### *Orifice sparger*

Various arrangements of perforated pipes have been tried in different types of fermentation vessel with or without impellers. In small stirred fermenters the perforated pipes were arranged below the impeller in the form of crosses or rings (ring sparger), approximately three-quarters of the impeller diameter. In most designs the air holes were drilled on the under surfaces of the tubes making up the ring or cross.

Orifice spargers without agitation have been used to a limited extent in yeast manufacture effluent treatment and later in the production of single-cell protein in the air-lift fermenter.

### *Nozzle sparger*

Most modern mechanically stirred fermenter designs from laboratory to industrial scale have a single open or partially closed pipe as a sparger to provide the stream of air bubbles. Ideally the pipe should be positioned centrally below the impeller and as for away as possible from it to ensure that the impeller is not flooded by the air stream. The single-nozzle sparger causes a lower pressure loss than any other sparger and normally does not get blocked.

### *Combined sparger-agitator*

On a small scale (1 $dm^3$), Herbert developed the combined sparger-agitator design, introducing the air via a hollow agitator shaft and emitting it through holes drilled in the disc between the blades and connected to the base of the main shaft.

## ACHIEVEMENT AND MAINTENANCE OF ASEPTIC CONDITIONS

Once the design problems of aeration and agitation have been solved, it is essential that the design meets the requirements of the degree of asepsis and containment demanded by the particular process being considered. It will be necessary to be able to sterilise and keep sterile, a fermenter and its contents throughout a complete growth cycle. There may be also a need to protect workers and the environment from exposure to hazardous micro-organisms or animal cells. As has been mentioned earlier, the containment requirements depend on the size of the fermentation vessel.

The following operations may have to be performed according to certain specifications to achieve and maintain aseptic conditions and containment during a fermentation:

1. Sterilisation of the fermenter.
2. Sterilisation of the air supply and the exhaust gas.
3. Aeration and agitation.
4. The addition of inoculum, nutrients and other supplements.
5. Sampling.
6. Foam control.
7. Monitoring and control of various parameters.

On a small scale, below 10 $dm^3$, the biohazard risk can be controlled by a combination of containment cabinets and work practices. When the volume of culture exceeds 10 $dm^3$, GILSP is required for those non-pathogenic non-toxigenic agents which have an extended history of large scale use. For this category there should be prevention of contamination of the product, control of aerosols and minimisation of the release of micro-organisms during sampling, addition of material, transfer of cells and removal of materials, products and effluents. It should be appreciated that the majority of fermentations fall into this category. At level 1, B2 or at higher containment levels, the following points need to be considered when designing a fermenter or other vessel, so that it can operate as a contained system:

1. All vessels containing live organisms should be suitable for steam sterilisation and have sterile vent filters.
2. Exhaust gases from vessels should pass through sterile filters.
3. Seals on flange joints should be fitted with a single 'O'-ring at the lower levels of containment. Flange joints on vessels for containment levels 3 and B3/4 need double 'O'-rings of double 'O'-rings plus a steam barrier. This has been discussed in an earlier section of this chapter.
4. Appropriate seals should be provided for entry ports for sensor probes, inoculum, sampling, medium addition, acid, alkali and antifoam. This will be discussed later in this chapter.
5. Rotating shafts into a closed system should be sealed with double acting mechanical seal with steam or condensate between the seals. This has been discussed in an earlier section in this chapter.
6. During operation a steam barrier should be maintained in all fixed piping leading to the 'contained' vessels.
7. Provision of appropriate pressure relief facilities.

Some of the issues discussed in points 1, 3, 4, 5 and 7 will need to be considered for aseptic operation of GILSP processes.

### Sterilisation of the Fermenter

The fermenter should be so designed that it may be steam sterilised under pressure. The medium may be sterilised in the vessel or separately and subsequently added aseptically. If the medium is sterilised *in situ* its temperature should be raised proper to the injection of live steam to prevent the formation of large amounts of condensate. This may be achieved by steam being introduced into the fermenter coils or jacket. As every point of entry to and exit from the fermenter is a potential source of contamination, steam should be introduced through all the entry and exit points except the air outlet from which steam should be allowed to leave.

All pipes should be constructed as simply as possible and slope towards drainage points to make sure that steam reaches all parts of the equipment and is not excluded by siphons or pockets of condensate or mash. Each drainage point in the pipework should be fitted with a steam trap. Parker, Chain and Muller have all stressed the need to eliminate fine fissures or gaps such as flange seals which might be filled with nutrient solutions and micro-organisms. Hambleton described a high specification pilot scale fermenter with surfaces free of crevices greater than 0.05 mm depth, which is needed if the vessel was to be used for animal cells in suspension culture or on micro-carriers. For long-term aseptic operation welded joints should be used wherever possible, even though sections may have to be cut and re-welded during maintenance and repair.

### Sterilisation of the Air Supply

Sterile air will be required in very large volumes in many aerobic fermentation processes. Although there are a number of ways of sterilising air, only two have found permanent application. These are heat and filtration. Heat is generally too costly for full-scale operation.

Historically, glass wool, glass fibre or mineral slag wool have been used as filter material, but currently most fermenters are fitted with cartridge-type filters. However, before the filter may be used, it too must be sterilised in association with the fermenter. Two procedures are commonly followed depending on the construction of the filter unit.

Fig. 12.10 shows the simple unit. During sterilisation the main non-sterile air-inlet valve A is shut and initially the sterile air valve B is closed. Steam is applied at valve C and air is purged downwards through the filter to a bleed valve at the base. When the steam is issuing freely through the bleed valve, the valve B is opened to allow steam to pass into the fermenter as well as the filter. It is essential to adjust the bleed valve to ensure that the correct sterilisation pressure is maintained in the fermenter and filter for the remainder of the sterilisation cycle.

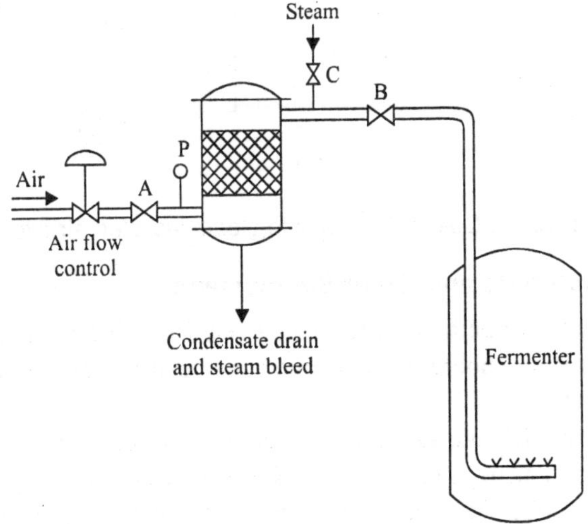


**Fig. 12.10.** An arrangement of packed air filter and fermenter.

An alternative approach is to use a steam-jacketed air filter (Fig. 12.11). At the beginning of a sterilisation cycle valve A will be closed and steam passed through valves B and C and bled out of D. Simultaneously steam will be passed into the steam jacket through valve F and out of G. When steam is issuing freely from valve D, valve F, may be opened and steam circulated into the fermenter. The bleed valve D will have to be adjusted to ensure that the correct pressure is maintained.

Once the sterilisation cycle is complete, valves B and E are closed and valve A is opened to allow air to pass through the heated filter and out of valve D to dry the filter. Finally the steam supply to the steam jacket is stopped. Valve D is closed and valve E opened, thus introducing sterile air into the fermenter to achieve a slight positive pressure in the vessel.

## Sterilisation of the Exhaust Gas from a Fermenter

Sterilisation of the exhaust gas can be achieved by 0.2 μm filters on the outlet pipe. Under normal operation aerosol formation may occur in the fermenter and moisture and solid matter may then plug the filter. To ensure satisfactory operation a cyclone separator (for solids) and a coalescer (for liquids) would be included upstream of two filters in series. The filters should be checked regularly to ensure that no viable cells are escaping. A test procedure to ensure integrity has been described by Hesselink.

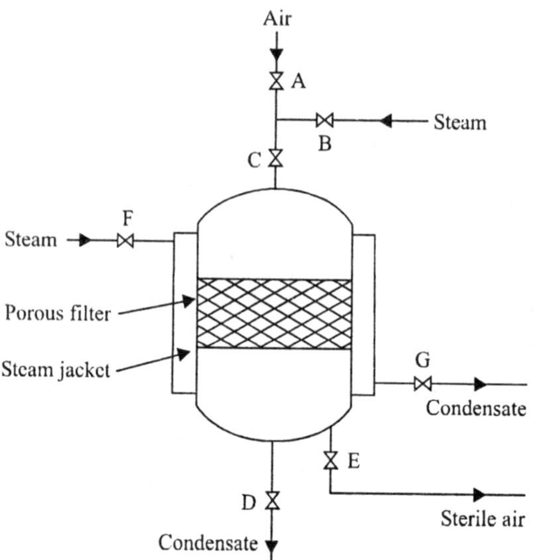


**Fig. 12.11.** Design for a simple steam-jacketed air-filter.

## Addition of Inoculum, Nutrients and Other Supplements

To prevent contamination when operating a fermenter requiring GILSP, it is essential that both the addition vessel and the fermenter should be maintained at a positive pressure and that the addition port is equipped with a steam supply.

At containment levels 1 and B2, the addition of inoculum, nutrients, etc., must be carried out in such a way that release of micro-organisms is restricted. This should be done by aseptic piercing of membranes or connections with steam locks. At Containment levels 2 and B 3/4, no micro-organisms must be released during inoculation or other additions. In order to meet these stringent requirements all connections must be screwed or clamped and all pipelines must be steam sterilisable.

## Sampling

The sampling points fitted to larger fermenters also illustrate the principles for maintaining sterility. A sterile barrier must be maintained between the fermenter contents and the exterior when the sample port is not being used and it must be sterilisable after use. A simple design is shown in Fig. 12.12.

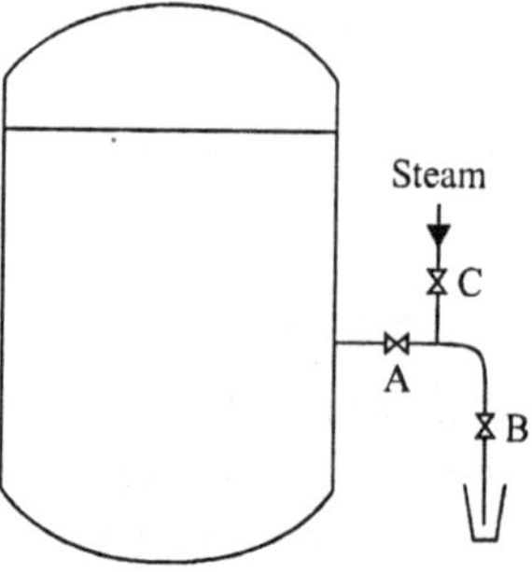


**Fig. 12.12.** Simple design for a sampling port.

In normal operation valves A, B and C are closed and a barrier is formed by submerging the end of the sampling port in 40 per cent formalin or a suitable substitute. A sample is obtained by removing the container of formalin and closing valve A. Valves B and C are then opened until the piping has been sterilised by steam. Valves B and C are then partially closed to allow a slow stream of steam and condensate out of the sampling port. Valve A is then opened slightly to cool the piping. The broth is discarded. Valve C is then closed and a sample is collected. Valve A is then closed and the piping is resterilised and left in the out of use arrangement.

An alternative arrangement for a sampling port is illustrated in Fig. 12.13 where the sterile barrier between the sample port and the exterior is a condensed flow of steam between valves C and D, valves A and B being closed. Valve D is connected to a steam trap to avoid condensate accumulation. To sterilise valve B prior to sampling, valve C is partially closed, valve D completely closed and valve B partially opened to allow a slow stream of steam and condensate out of the sampling port. Valve A is opened briefly to cool the pipe and the broth is discarded. Valve C is then closed and a sample is collected. Valve A is then closed and the piping is re-sterilised. In between collecting samples valves C and D are left partially open.

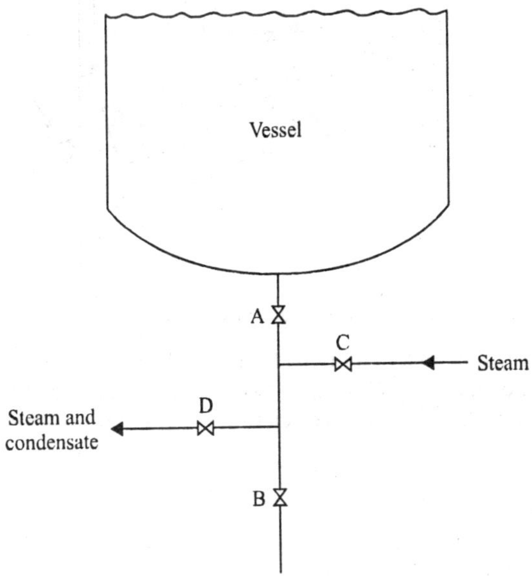


**Fig. 12.13.** An alternative simple sample port.

A more modern sterilisable system, as illustrated in Fig. 12.14, has been described by Werner. In the closed position steam enters through the entry point 6 and passes through the sampling hole 11 and into the sampling hole 2. When sampling, the handle is pushed so that the hole in the sampling tube is in the fermenter and the vessel contents may be sampled. When the sampler is closed the unit can be re-sterilised.

Marshall recognised that sampling requires good mechanical design and good operator practice to ensure sterility. Complex valve sequencing can lead to operator error. The automatic MX-3 rotary valve is used (Fig. 12.15). In this device, broth is continuously recycled to and from the fermenter and up to 12 samples can be taken automatically in sample bottles according to a pre-programmed sequence. The storage of the bottles in an integral refrigeration block reduces spoilage until they can be removed for analysis. This sampler can be used for GILSP.

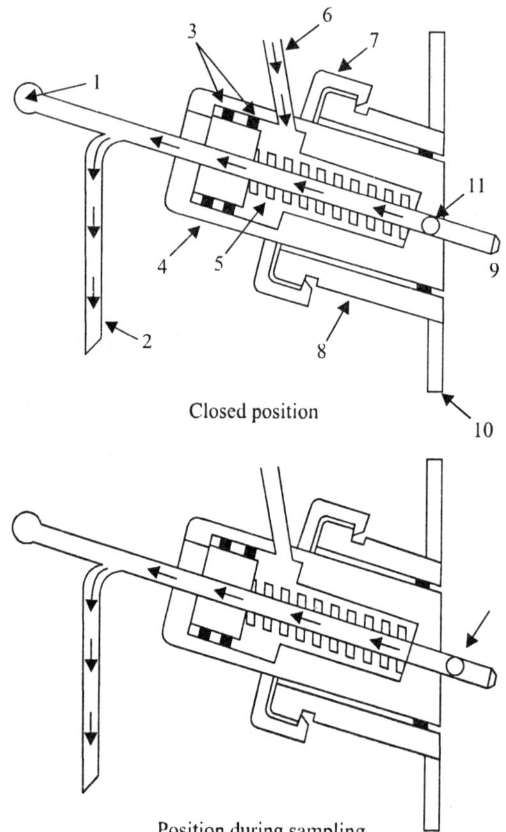


**Fig. 12.14.** Sterilisable sampling system for category 1. Key: (1) Handle for opening and closing; (2) piping for sample or condensate, respectively; (3) 'O'-ring; (4) housing; (5) spring; (6) steam inlet; (7) union nut; (8) welding socket; (9) product; (10) wall of the bioreactor; and (11) port of sampling tube.

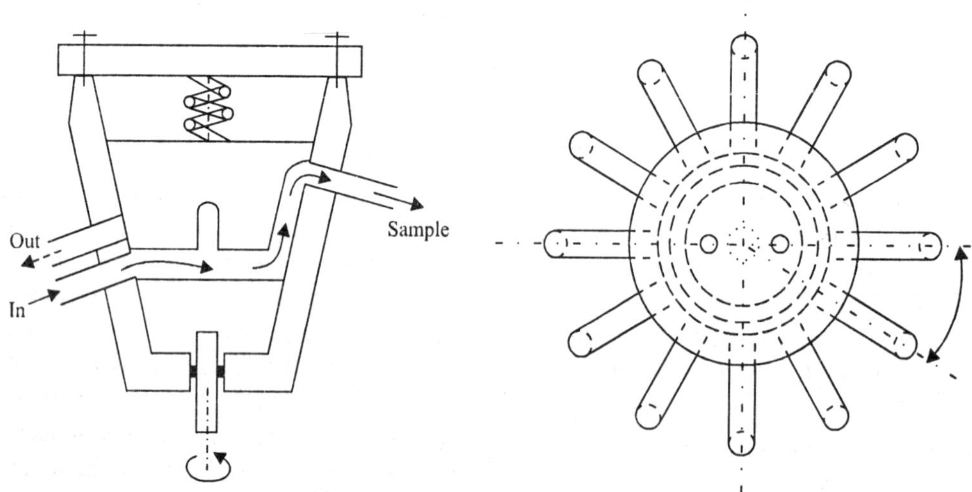


**Fig. 12.15.** Schematic representation of the MX-3 Bio-sampler rotary valve system.

In all the sampling devices described above, steam is used merely to maintain aseptic conditions. If the process organism has to be contained then the sample vessel must be designed accordingly. At level 1 or B2, removal of broth should be carried out in such a way that release of micro-organisms is minimised. Such a system (Fig. 12.16) is described by Janssen. The sampling bottle assembly is attached to the fermenter by a double-end shut-off quickconnect B and air is vented from the vessel during sampling by a suitable membrane filter. The pipe work is steamed for 15 minutes using 1.3 bar steam with the valve E on the sample bottle closed. With valves C and D closed and valves A and B open a sample can be taken, then the pipe is re-sterilised before removing the sample bottle.

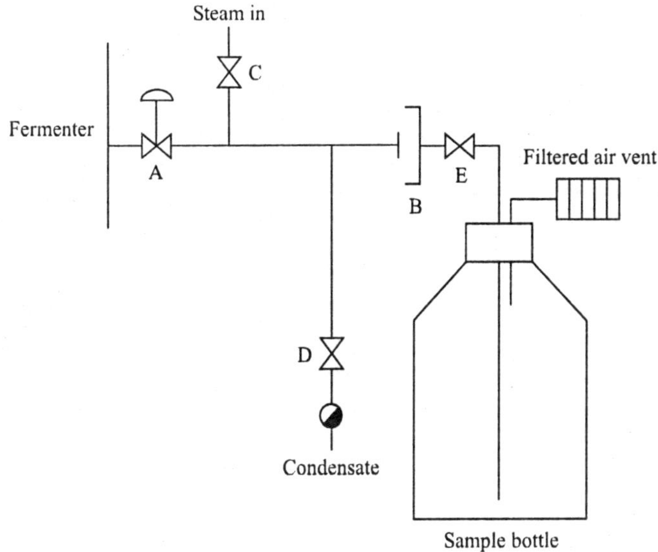


**Fig. 12.16.** Sample system for level 1.

At level 2 or B3, sampling should be done in a closed system (Figs. 12.17 and 12.18) and all piping coming into contact with a sample must be sterilisable with steam. The sampling vessel should be closed during transport and samples should be examined under containment conditions corresponding to those specified for the process.

## Feed Ports

Additions of nutrients and acid/alkali to small fermenters are normally made via silicone tubes which are autoclaved separately and pumped by a peristaltic pump after aseptic connection. In large fermenter units, the nutrient reservoirs and associated piping are usually an integrated part which can be sterilised with the vessel. However, there may also be ports which are used intermittently. These can be sterilised *in situ* with steam after connection has been completed and before any additions are made.

## Sensor Probes

Double 'O' ring seals have been used for many years to provide an aseptic seal for glass electrodes in stainless steel housings in fermenters using GILSP. This system is also suitable for levels 1 and B2, provided that release of micro-organisms is minimised and there are adequate disinfection procedures for dealing with leakages.

At containment levels 2 and B3/4, probes are fitted with triple 'O'-ring seals. Although double 'O'-ring seals with steam tracing have been described they are not normally considered to be feasible. The use of pre-inserted back-up probes is recommended as a means for dealing with probe failure rather than using a retractable electrode housing during a fermentation cycle because of the danger of leakage of broth.

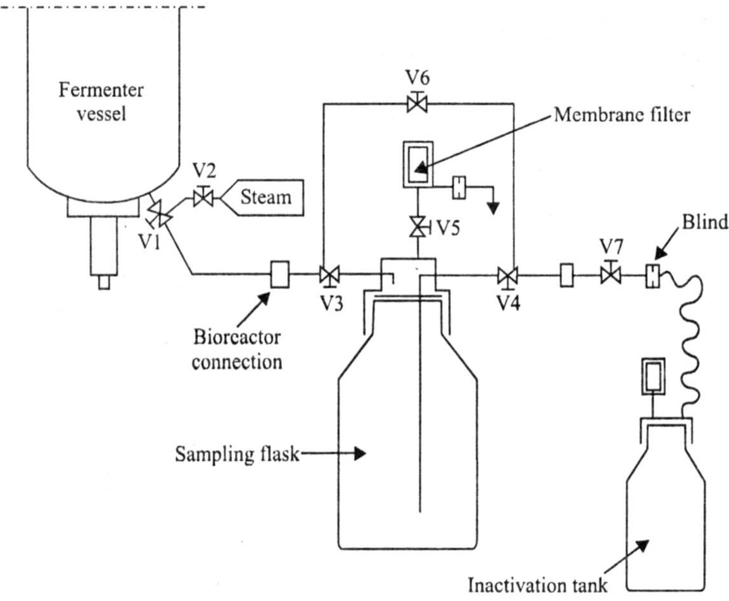


Valve sequences during sampling

| *Valve* | *Sterilisation sampling flask* | *Sampling* | *Sterilisation after sampling of sampling flask* | *Removal of sampling flask* |
|---|---|---|---|---|
| V1 | Closed | Open | Closed | Closed |
| V2 | Open | Closed | Open | Closed |
| V3 | Open | Open | Closed | Closed |
| V4 | Open | Closed | Closed | Closed |
| V5 | Open | Open | Open | Closed |
| V6 | Closed | Closed | Open | Closed |
| V7 | Open | Open | Open | Closed |

**Fig. 12.17.** Containment sampling unit.

## Foam Control

In any fermentation it is important to minimise foaming. When foaming becomes excessive, there is a danger that filters become wet resulting in contamination. There is also the possibility that siphoning will develop, leading to the loss of all or part of the contents of the fermenter. In certain circumstances antifoams may cause problems with aeration or downstream processing. Foam breakers break down foam by an impact mechanism created by some type of rotating mechanisms inside the fermenter. Mechanical foam breakers used in Acetators for vinegar production are described later in this chapter.

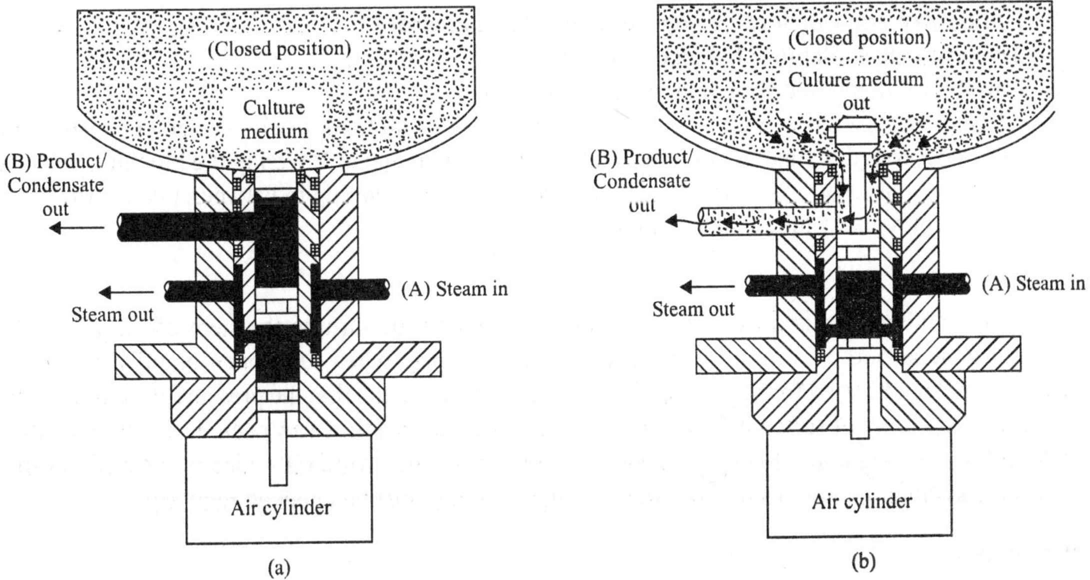


**Fig. 12.18.** Re-sterilisable harvest-sampling valve for level 2 containment. This spool-type valve is connected to the bottom of the fermenter vessel in the closed position (a), pressurised steam circulates throughout the entire valve body and through the product condensate line (B) via a steam inlet line (A). Aseptic withdrawal of samples is achieved with the valve in the open position (b). To prevent possible contamination when the plunger is raised, steam is circulated to the lower valve area. Action of the plunger is controlled by an air cyclinder.

At containment level 1 or B2, the exhaust air system should shut-down automatically if the foam control system fails.

## VALVES AND STEAM TRAPS

Valves attached to fermenters and ancillary equipment are used for controlling the flow of liquids and gases in a variety of ways. The valves may be:

1. Simple ON/OFF valves which are either fully open or fully closed.
2. Valves which provide coarse control of flow rates.
3. Valves which may be adjusted very precisely so that flow rates may be accurately controlled.
4. Safety valves which are constructed in such a way that liquids or gases will flow in only one direction.

When making the decision as to which valves to use in the design and construction of a fermenter it is essential to consider the follwoing points:

1. Will the valve serve its chosen purpose? Is it suitable for aseptic operation or contained processes and of the correct dimensions? Is the pressure drop across the valve tolerable?
2. Will the valve withstand the rigours of the process? The materials used to construct the valve should be suited to the process. It is also important to know whether corrosive liquids are used or synthesised during the process. The maximum operating temperature and pressure of the process should be known.
3. Are there welds or flanges in the valve?

4. Is the valve one which can be operated by remote control?
5. The cost and availability of suitable valves.
6. Can the valve be used for containment purposes?

A wide range of valves is available, but not all of them are suitable for use in fermenter construction.

The valves described in this section open and close by: (i) raising or lowering the blocking unit with a screw thread (rising stem); (ii) a drilled sphere or plug or a disc rotating in between two bearings; and (iii) a rubber diaphragm or tube which is pinched.

### Gate Valves

In this valve a sliding disc is moved in or out of the flow path by turning the stem of the valve. It is suitable for general purposes on a steam or a water line for use when fully open or fully closed and therefore should not be used for regulating flow. The flow path is such that the pressure drop is minimal, but unfortunately it is not suitable for aseptic conditions as mash solids can pack in the groove where the gate slides and there may be leakage round the stem of the valve which is sealed by a simple stuffing box. This means that the nut around the stem and the packing must be checked regularly.

### Globe Valves

In this valve, a horizontal disc or plug is raised or lowered in its seating to control the rate of flow. This type of valve is very commonly used for regulating the flow of water or steam since it may be adjusted rapidly. It is not suitable for aseptic operation because of potential leakage round the valve stem, which is similar in design to that of the gate valve. There is a high-pressure drop across the valve because of the flow path. In both the gate and globe valves it is possible to incorporate a flexible metallic membrane around the steam of the valve, to replace the standard packing. This modified type of valve can be operated aseptically, but is bigger and more expensive. Valves with non-rising stems have been used, but they are still potential sources of contamination.

### Piston Valves

The piston valve is similar to a globe valve except that flow is controlled by a piston passing between two packing rings. This design has proved in practice to be very efficient under aseptic operation. It is important to sterilise them partly open so that steam can reach as far as possible into the valve body. There may be blockage problems with mycelial cultures. The pressure drop is similar to that of a globe valve.

### Needle Valves

The needle valve is similar to the globe valve, except that the disc is replaced by a tapered plug or needle fitting into a tapered valve seat. The valve can be used to give fine control of steam or liquid flow. Accurate control of flow is possible because of the variable orifice formed between the tapered plug and the tapered seat. The aseptic applications are very limited.

### Plug Valves

In this valve there is a parallel or tapered plug sitting in a housing through which an orifice, has been machined. This type of valve has a tendency to leak or seize up, but the use of lubricants and/or sealants may overcome these problems. If suitable packing sleeves (e.g., compressed asbestos) are incorporated into the valve it will be suitable for use in a steam line as it is quick to operate, has protected seals, a minimal pressure drop and a positive closure. It can also provide good flow control.

### Ball Valves

This valve has been developed from the plug valve. The valve element is a stainless-steel ball through which an orifice is machined. The ball is sealed between two wiping surfaces which wipe the surface and prevent deposition of matter at this point. The orifice in the ball can be of the same diameter as the pipeline, giving an excellent flow path. The valve is suitable for aseptic operation, can handle mycelial broths and can be operated under high temperatures and pressures. The pressure and temperature range is normally limited by the PTFE seat and stem seals.

### Butterfly Valves

The butterfly valve consists of a disc which rotates about a shaft in a housing. The disc closes against a seal to stop the flow of liquid. This type of valve is normally used in large diameter pipes operating under low pressure where absolute closure is not essential. It is not suitable for aseptic operation.

### Pinch Valves

In the pinch valve a flexible sleeve is closed by a pair of pinch bars or some other mechanism which can be operated by compressed air remotely or automatically. The flow rate can be controlled from 10 to 95 per cent of rated flow capacity. The valve is suitable for aseptic operation with fermentation broths, even when mycelial, as there are no dead spaces in the valve structure and the closing mechanism is isolated from the contents of the piping. Obviously, the sleeve of rubber, neoprene, etc., must be checked regularly for signs of wear.

### Diaphragm Valves

Like the pinch valve, the diaphragm valve makes use of a flexible closure, with or without a weir. It may also be fitted with a quick action lever. This valve is very suitable for aseptic operation provided that the diaphragm is of a material which will withstand repeated sterilisation. The valve can be used for ON/OFF, flow regulation and for steam services within pressure limits. Diaphragm failure, which is often due to excessive handling, is the primary fault of the valve. Ethylene propylene diene modified (EPDM) is now the preferred material.

### Most Suitable Valve

Among these groups of valves which have just been described, globe and butterfly valves are most commonly used for ON/OFF applications, gate valves for crude flow control, needle valves for accurate flow control and ball, pinch or diaphragm valves for all sterile uses.

Ball and diaphragm valves are now the most widely used designs in fermenter and other biotechnology equipment. Although ball valves are more robust, they contain crevices which make sterilisation more difficult.

### Check Valves

The purpose of the check valve is to prevent accidental reversal of flow of liquid or gas in a pipe due to breakdown in some part of the equipment. There are three basic types of valve: swing check, lift check and combined stop and check with a number of variants. The swing check valve is most commonly used in fermenter designs. The functional part is a hinged disc which closes against a seat ring when the intended direction of flow is accidentally reversed.

### Pressure-Control Valves

When planning the design of a plant for a specific process, the water, steam and air should be at different, but specified pressures and flow rates in different parts of the equipment. For this reason it is essential to control pressures precisely and this can be done using reduction or retaining valves.

#### *Pressure-reduction valves*

Pressure-reduction valves are incorporated into pipelines when it is necessary to reduce from a higher to a lower pressure and be able to maintain the lower pressure in the downstream side within defined limits irrespective of changes in the inlet pressure or changes in demand for gas, steam or water.

#### *Pressure-retaining valves*

A pressure-retaining valve will maintain pressure in the pipeline upstream of itself and the valve is designed to open with a rising upstream pressure. It is constructed with a reverse action of the pressure-reducing valve.

### Safety Valves

Safety valves must be incorporated into every air or steam line and vessel which is subject to pressure to ensure that the pressure will never exceed the safe upper limit recommended by the manufacturer or a code of practice and to satisfy government legislation and insurance companies. They must also be of the correct type and size to suit the operating conditions and be in sufficient numbers to protect the plant. The reliability of such valves is crucial.

In the simplest valves a spindle is lifted from its seating against the pressure of gas, steam or liquid. Once the pressure falls below the value set by the tensioned spring, the spindle should return to its original position. However, the valve may stick open if waste material lodges on the valve seat and plant operators may interfere with the release pressure setting. Bursting/rupture discs may be used as an alternative and are of a more hygienic design than some valves.

It is also important to ensure that vessels do not collapse when vacuums occur. Vacuum release valves have been designed to cope with cold rinsing of a hot vessel or absorption of $CO_2$ when cleaning with caustic solutions. Shuttlewood has discussed design pressure and vacuum pressure in relation to vessel thickness requirements for design codes for American and British standards of safety.

### Steam Traps

In all steam lines it is essential to remove any steam condensate which accumulates in the piping to ensure optimum process conditions. This may be achieved by incorporating steam traps, which will collect and remove automatically any condensate at appropriate points in steam lines. A steam trap has two elements. One is a valve and seat assembly which provides an opening, which may be of variable size, to ensure effective removal of any condensate. This opening may operate on an open/close basis so that the average discharge rate matches the steam condensation rate or the dimensions of this opening may be varied continually to provide a continuous flow of condensate. The second element is a device which will open or close the valve by measuring some parameter of the condensate reaching it to determine whether it should be discharged.

The steam trap may be designed to operate automatically on the basis of:

1. The density of the fluid by using a float (ball or bucket) which will float in water or sink in steam.

2. By measuring the temperature of the fluid, closing the valve at or near steam temperature and opening it when the fluid has cooled to temperature, say 8°F below the saturated steam temperature of the steam. The sensing element is often a stainless-steel capsule filled with a water-alcohol mixture which expands when steam is present and presses a ball into a valve seat or contracts when cooler condensate is present and lifts the ball from the seating. These are used in thermostatic and balanced pressure steam traps.
3. By measuring the kinetic effects of the fluid in motion. At a given pressure drop, low-density steam will move at a much greater velocity than higher density condensate. The conversion of pressure energy into kinetic energy can be used to control the degree of opening of a valve. This type of steam trap is not used very widely.

Armer thinks that balanced pressure thermostatic traps are the most suitable for autoclaves and sterilisers. Some can operate close to steam-saturation temperature with little back-up of condensate. Any steam condensate may be: (i) returned to the boiler; (ii) used in a steam condensate line; (iii) vented to waste; and (iv) piped to a kill tank in a containment location.

### Complete Loss of Contents from a Fermenter

Leaver and Hambleton have suggested building a wall or dyke around a fermenter which would retain the entire contents were they to leak out. The escaping fluid would then be pumped into a suitable vessel for sterilisation. The floor should be covered with an impervious epoxy based surface and coved up the walls.

### Testing New Fermenters

When a new fermenter (10 $dm^3$ or larger) has finally been assembled in a factory it must be hydraulically pressure tested and checked by independent inspectors using nationally approved test procedures. A vessel which has passed the approved tests is given a certificate which is recognised by insurance companies which will allow it to be operated in a laboratory or factory. If any subsequent modifications are made to a vessel in this size range it must be re-tested and certificated before it can be legally used. Thus a number of fermenter manufacturer cut extra 'O'-ring grooves and steam traces in head plates or ports to satisfy higher containment-level needs so that the vessels will not have to be modified subsequently and to avoid extra pressure checks.

## OTHER FERMENTATION VESSELS

Some of the other forms of fermentation vesels will now be considered. These vessels have more limited applications and have been developed for specific purposes or closely related processes. Some are historical developments, such as the earliest forms of packed tower, others were being developed in parallel with the standard mechanically stirred fermenter during the 1940s, while other approaches are more recent.

### Waldhof-type Fermenter

The investigations on yeast growth in sulphite waste liquor in Germany, Japan and the United States of America led to the development of the Waldhof-type fermenter. The fermenter was of carbon steel, clad in stainless steel, 7.9 m in diameter and 4.3 meter high with a center draught tube 1.2 meter in diameter. A draught tube was held by tie rods attached to the fermenter walls. The operating volume was 2,25,000 $dm^3$ of emulsion (broth and air) or 100,000 $dm^3$ of broth without air. Non-sterile air was introduced into the fermenter through a rotating pin-wheel type of aerator, composed of open-ended tubes rotating at 300 rpm

(Fig. 12.19). The broth passed down the draught tube from the outer compartment and reduced the foaming.

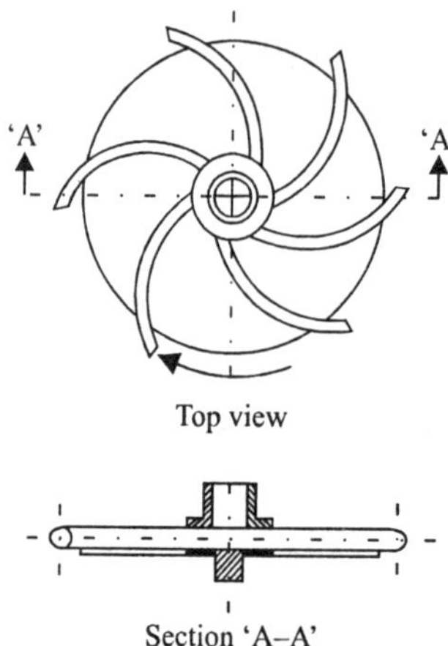


**Fig. 12.19.** Top view and section of a Waldhof aeration wheel.

## Acetators and Cavitators

Fundamental studies by Hromatka and Ebner on vinegar production showed that if *Acetobacter* cells were to remain active in stirred aerated fermenter, the distribution of air had to be almost perfect within the entire contents of the vessel. They solved the full-scale problem by the use of a self-aspirating rotor. In this design (Fig. 12.20), the turning rotor sucked in air and broth and dispersed the mixture through the rotating stator (d). The aerator also worked without a compressor and was self-priming.

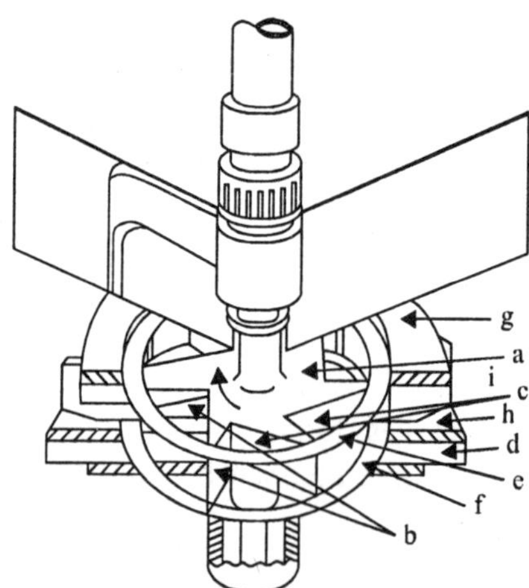


**Fig. 12.20.** Axonometric view of the self-priming aerator used with the Frings generator. The turbine is designed as a hollow body (a) with openings which are arranged radially and open against the direction of rotation (b). The openings are shielded by vertical sheets (c). The turbine sucks liquid from above and below and mixes it with air sucked in through the openings. The suspension is thrown through the stator (d) towards the circumference of the tank. An upper and lower ring on the turbine (e,f) helps to direct and regulate the air-liquid suspension. The stator (d) consists of an upper and lower ring (g,h) which are connected by vertical sheets (i) inclined at about 30° towards the radius.

Vinegar fermentations often foam and chemical antifoams were not thought feasible because they would decrease aeration efficiency and additives were not desirable in vinegar. A mechanical defoamer therefore had to be incorporated into the vessel and as foam builds up it is forced into a chamber in which a rotor turns at 1000 to 1450 rpm. The centrifugal force breaks the foam and separates it into gas and liquid. The liquid is pumped back into the fermenter and the gas escapes by a venting mechanism. An illustration of the basic components is given in Fig. 12.21.

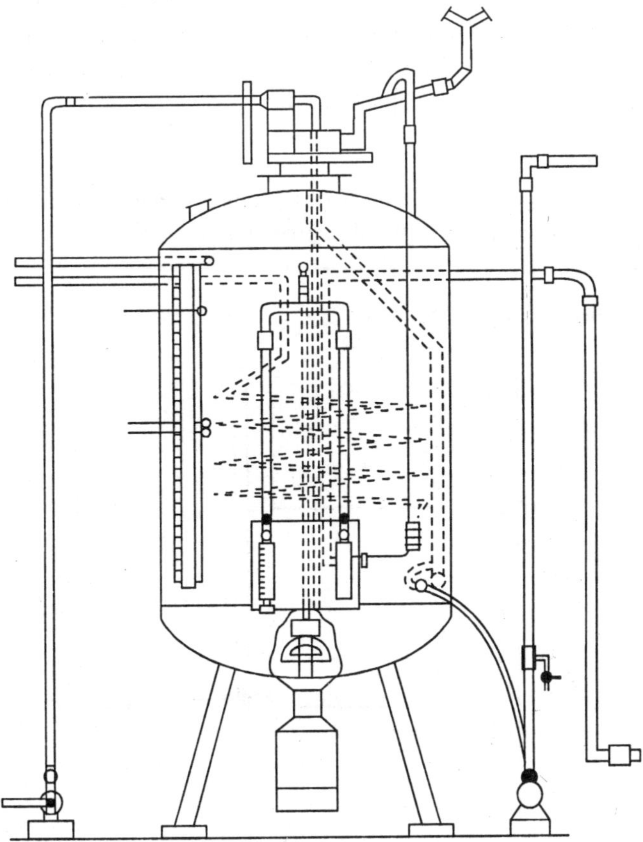

**Fig. 12.21.** Diagram of a section through a Frings generator fermenter used for the manufacture of vinegar. The fermenter, which can be used semi-continuously or continuously, employs vortex stirring.

The fermenter had an agitator of different design, but similar operating principles to the acetator. Uniform distribution of air bubbles was obtained by means of the circulation pattern created by the centrally located draught (draft) tube. The agitator withdrew liquid from the draught tube and pushed liquid into the main part of the vessel. The outer level rose and overflow occurred back into the top of the draught tube.

## Tower Fermenter

It is difficult to formulate a single definition which encompasses all the types of tower fermenter. Their main common feature appears to be their height: diameter ratio or aspect ratio. Several different types of tower fermenter exist and these will be examined in broad groups based on their design.

The simplest types of fermenter are those that consist of a tube which is air sparged at the base (bubble columns). This type of fermenter was first described for citric acid production on a laboratory scale. This batch fermenter was in the form of glass column having a height: diameter ratio of 16:1 with a volume of 3 $dm^3$. Humid sterile air was supplied through a sinter at the base.

The next group of tower fermenters are the multistage systems, for brewing beer, although these systems were not used on an industrial scale. Later work reported using these systems includes continous cultures of *E. coli*, bakers' yeast and activated sludge. The fermenters used by all these workers were basically similar. Each consisted of a column forming the body of the vessel and a number of perforated plates which were positioned across the fermenter, dividing it into compartments. Approximately 10 per cent of the horizontal plate area was perforated.

### Cylindro-Conical Vessel

The cylindro-conical vessel (Fig. 12.22) consists of a stainless-steel vertical tube with hemispherical top and a conical base with an included angle of approximately 70°.

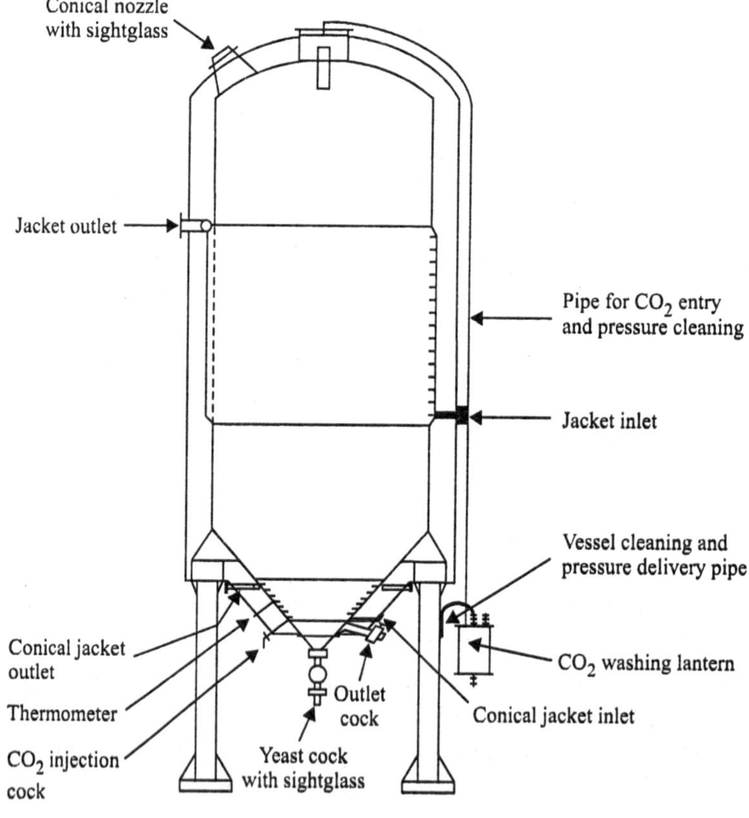


**Fig. 12.22.** Cylindro-conical fermentation vessel.

Aspect ratios are usually 3:1 and fermenter heights are 10 to 20 m. Operating volumes are chosen to suit the individual brewery requirements, but are often 150,000 to 200,000 $dm^3$. Vessels are not normally agitated unless a particularly flocculant yeast is used, but small impellers may be used to ensure homogeneity when filling with wort. In the vessel, the wort is pitched (inoculated) with yeast and the fermentation proceeds for 40 to 48 hours. Mixing is achieved by the generation of carbon dioxide bubbles that rise rapidly in the vessel. Temperature control is monitored by probes positioned at suitable points within the vessel. A number of cooling jackets are fitted to the vessel wall to regulate and cause flocculation and settling of the yeast. The fermentation is terminated by the circulation of chilled water via the cooling jackets which results in yeast floculation. Thus, it is necessary to select a yeast strain which will flocculate readily in the period of chilling. Part of this yeast may be withdrawn and used for re-pitching another vessel. The partially cleared beer may be left to allow a secondary fermentation and conditioning. Some of the adantages of this vessel in brewing are:

1. Reduced process times may be achieved due to increased movement within the vessel.
2. Primary fermentation and conditioning may be carried out in the same vessel.
3. The sedimented yeast may be easily removed since yeast separation is good.
4. The maturing time may be reduced by gas washing with carbon dioxide.

### Air-lift Fermenters

An air-lift fermenter (Fig. 12.22) is essentially a gas-tight baffled riser tube (liquid ascending) connected to a downcomer tube (liquid descending). Figure 12.23a shows an external riser and Fig. 12.23b an internal riser. Air or gas mixtures are introduced into the base of the riser by a sparger during normal operating conditions. The driving force for circulation of medium in the vessel is produced by the difference in density between the liquid column in the riser (excess air bubbles in the medium) and the liquid column in the downcomer (depleted in air bubbles after release at the top of the loop). Circulation times in loops of 45 m height may be 120 seconds. This type of vessel can be used for continuous culture.

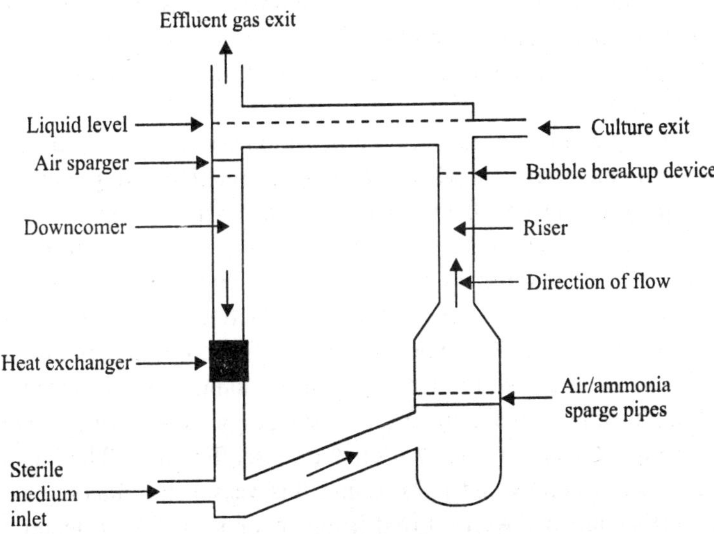


**Fig. 12.23a.** Air-lift fermenter with outer loop.

It would be uneconomical to use a mechanically stirred fermenter to produce SCP (single-cell protein) from methanol as a carbon substrate, as heat removal would be needed in external cooling loops because of the high rate of aeration and agitation required to operate the process. To overcome these problems, particularly that of cooling the medium when mechanical agitation is used, air-lift fermenters with outer or inner loops (Fig. 12.23) were chosen.

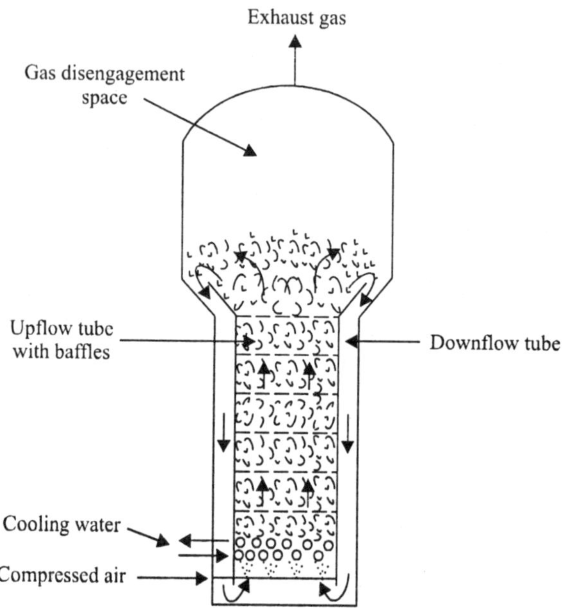


**Fig. 12.23b.** Air-lift fermenter with inner loop.

In the continuous process, air and gaseous ammonia were introduced at the base of the fermenter. Sterilised methanol, other nutrients and recycled spent medium were also introduced into the downcomer. Heat from this exothermic fermentation was removed by surrounding part of the downcomer with a cooling jacket in the pilot plant, while at full scale ($2.3 \times 10^6$ $dm^3$) it was found necessary to insert cooling coils at the base of the riser.

Rank Hovis and McDougall developed a protein-rich food primarily for human consumption. They have grown *Fusarium graminareum* on a wheat starch based medium using a 40 $m^3$ air-lift fermenter (Fig. 12.24) to produce the myco-protein Quorn. The use of an air-lift fermenter for culture of a mycelial fungus would seem unusual as lower rates of oxygen transfer occur in a viscous culture which give rise to lower biomass yields. Because low shear conditions are present in the vessel, long fungal hyphae can be cultured (the preferred product form) even though production yields are only 20 g $dm^{-3}$. At the present time a fermenter to produce 10,000 tonnes per annum of myco-protein is considered economically feasible. Okabe modified a 3 $dm^3$ air-lift fermenter by putting stainless steel four-mesh sieves at the top and bottom of the draught tube to manipulate the morphology of *Aspergillus terreus* for optimum production of itaconic acid as the culture circulates in the vessel flow path. The fungal morphology was an intermediate state between pellets and pulp. Using this vessel, the itaconic acid production rate (g $dm^{-1}$ $hr.^{-1}$) was double that obtained with a stirred fermenter or an air-lift fermenter with a conventional draught tube. Work has also begun to examine oxygen transfer rates with modified draught tubes. Carrington used a 20 $m^3$ pilot-scale bubble column fermenter fitted with an internal helical cooling coil (Fig. 12.25) or a solid draught tube. The fermentation study was a commercial *Streptomyces* antibiotic fermentation in a complex medium which produced a viscous non-Newtonian broth. Tracer studies indicated that the vessel fitted with only the cooling coil behaved like an air-lift fermenter with a region of good mixing in the zone above the cooling coil. The coil acted as a leaky draught tube with back mixing taking place between the coils into the riser section. No poorly oxygenated zones were observed.

Liquid velocities of 1 m $sec^{-1}$ were measured giving circulation times of 9 to 12 seconds and mixing times of 14 to 18 seconds. The $K_La$ at different power inputs and viscosities was found to increase almost linearly with increasing power input and decreased exponentially with increasing viscosity. When a solid draught tube was installed inside the cooling coil the circulation time was similar, but the mixing time increased to 18 to 24 seconds. The $K_La$ was also determined at different power inputs and viscosities. This gave a reduction of 5 to 25 per cent in $K_La$ with the biggest reduction at a high viscosity.

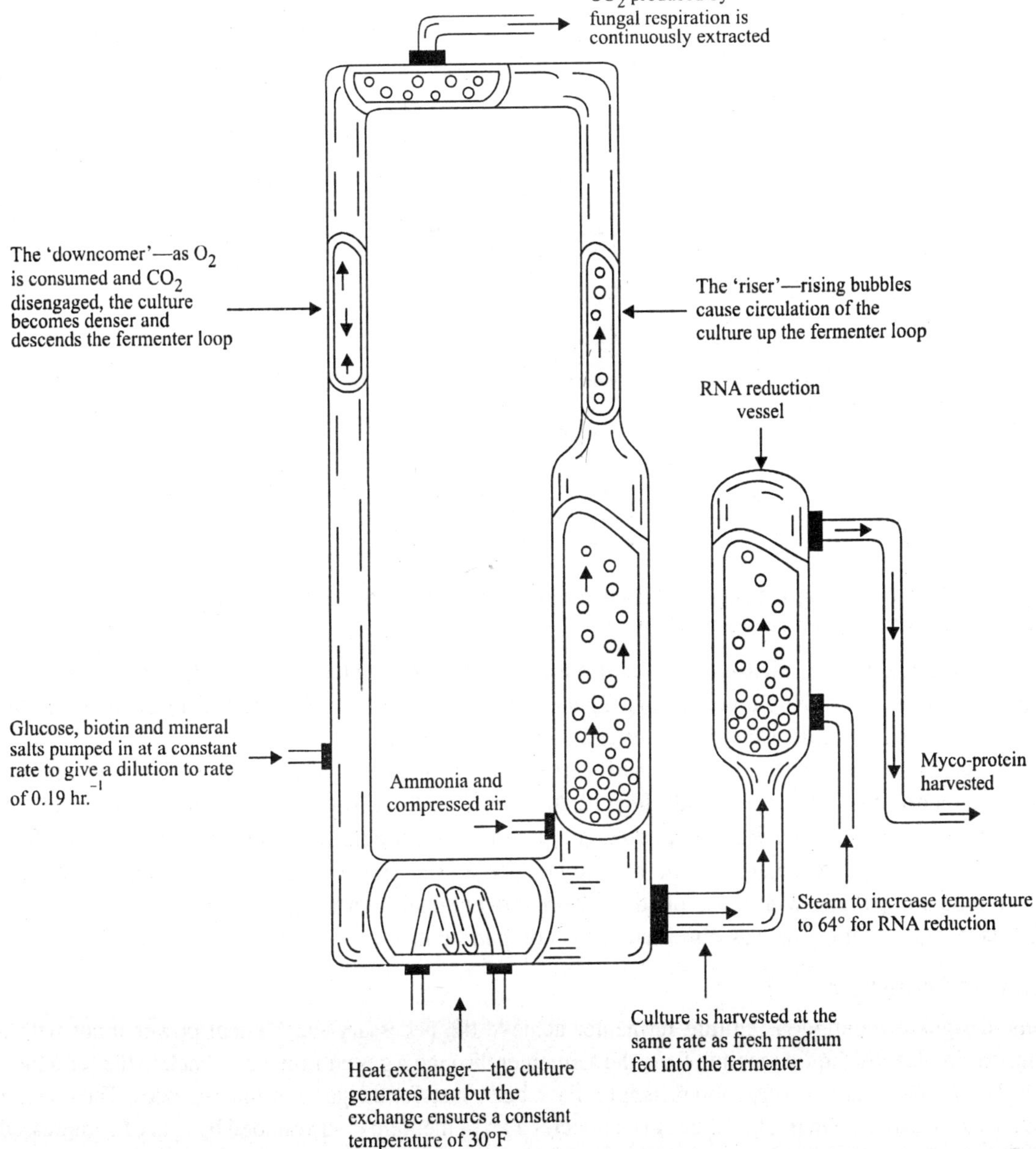


**Fig. 12.24.** Schematic diagram of the air-lift fermenter for the production of myco-protein in continuous flow culture.

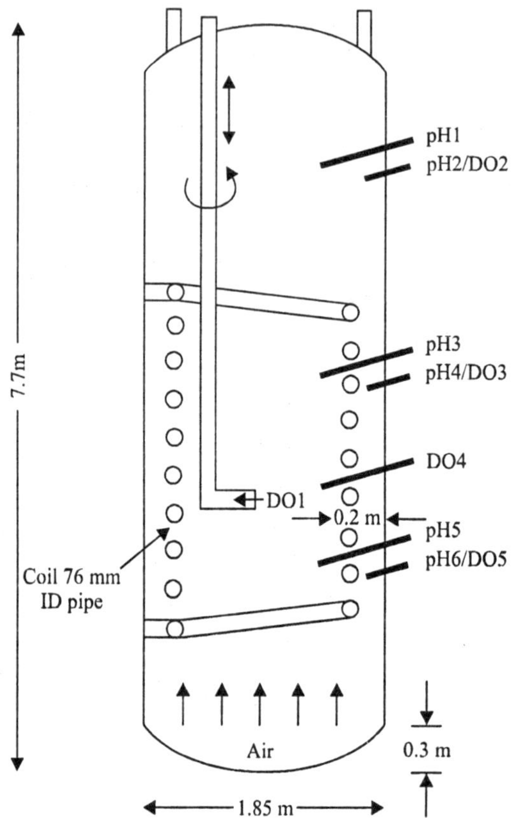


**Fig. 12.25.** Schematic diagram of 20 m$^3$ bubble column fermenter fitted with internal cooling coil (pH and DO indicate positions of pH and oxygen electrodes).

Wu and Wu compared $K_L$as in a range of mesh draught tubes and a solid draught tube in an air-lift vessel of 15 dm$^3$ working volume. When a 24-mesh tube was used at high superficial gas velocities the $K_La$ was double that of the same vessel with a solid draught tube.

Bakker has developed a multiple air-lift fermenter in which tree air-lift fermenters with internal loops are incorporated into one vessel (Fig. 12.26). Fresh medium is fed into the central compartment, depleted medium overflows into the middle compartment, from here to the outer compartment where medium is eventually discharged. The hydrodynamics and mixing in the middle compartment were found to be comparable with those obtained with conventional internal loop air-lift vessels, although some tangential liquid flow was observed in the compartment.

## Deep-Jet Fermenter

Some designs of continuous culture fermenter achieve the necessary mechanical power input with a pump to circulate the liquid medium from the fermenter through a gas entrainer and back to the fermenter (Fig. 12.27). Two basic construction principles have been used for the gas entrainer nozzles. The injector and the ejector are shown in Fig. 12.28. In an injector a jet of medium is surrounded by a jet of compressed air. The gas from the outlet enters the larger tube with a nozzle velocity of 5 to 100 m s$^{-1}$ and expands in the tube to form large air bubbles which are dispersed by the shear of the water jet. In an ejector the liquid jet enters into a larger converging-diverging nozzle and entrains the gas around the jet.

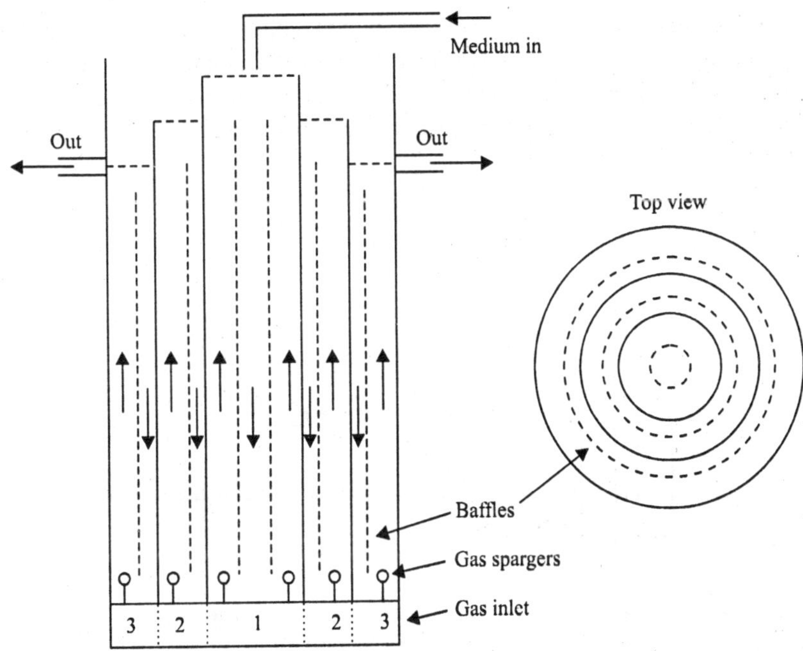


**Fig. 12.26.** Three-compartment multiple air-lift loop fermenter. Cross-sectional side and top view.

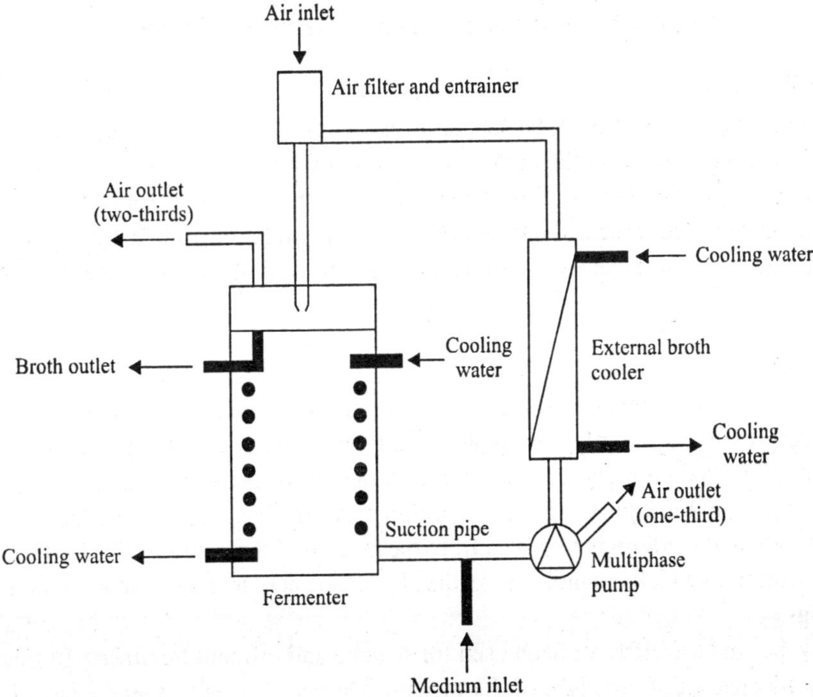


**Fig. 12.27.** Diagram of the Vogelbusch deep-jet fermenter system.

The gas which is sucked into the converging-diverging jet is dispersed in that zone. Partially aerated medium is pumped by a multiphase pump through a broth cooler to an air entrainer above the fermenter.

The air-medium mixture falls down a slightly conical shaft at a high velocity and creates a turbulence in the fermenter. Two-thirds of the exhaust gas is vented from the fermenter headspace and the remainder via the multiphase pump. Oxygen-transfer rates of 4.5 gram $dm^{-3}$ $hr.^{-1}$ with an energy consumption of 1 kW $hr.^{-1}$ $kg^{-1}$ have been achieved for industrial-scale yeast production from whey using such a fermentation system.

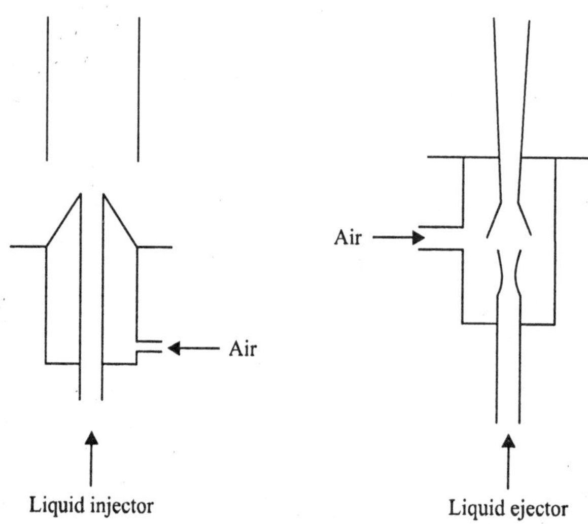


**Fig. 12.28.** Gas entrainer nozzles of deep-jet fermenters.

## Cyclone Column

Dawson developed the cyclone column, particularly for the growth of filamentous cultures (Fig. 12.29). The culture liquid was pumped from the bottom to the top of the cyclone column through a closed loop. The descending liquid ran down the wall of the column in a relatively thin film.

Nutrients and air were fed in near the base of the column whilst the exhaust gases left at the top of the column. Good gas exchange, lack of foaming and limited wall growth have been claimed with this fermenter.

## Packed Tower

The packed tower is a well established application of immobilised cells. A vertical cylindrical column is packed with pieces of some relatively inert material, e.g., wood shavings, twigs, coke, an aggregate or polythene. Initially both medium and cells are fed into the top of the packed bed. Once the cells have adhered to the support and are growing well as a thin film, fresh medium is added at the top of the column and the fermented medium is removed from the bottom of the column. The best known example is the vinegar generator, in which ethanol was oxidised to acetic acid by strains of *Acetobacter* supported on beech shavings.

More recently, packed towers have been used for sewage and effluent treatment. In treatment of gas liquor, a column was packed to a height of 7.9 m with 'Dowpac', a polystyrene derivative. The main advantages compared with other methods of effluent treatment being its simplicity of operation and a saving in land because of the increased surface areas within the column. Other possible applications with immobilised cells are now being investigated.

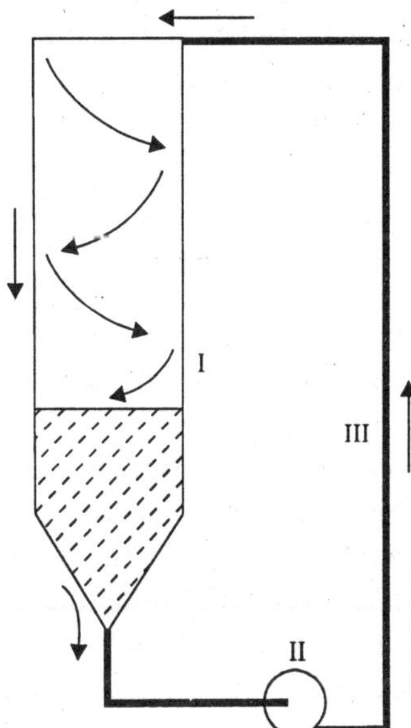


**Fig. 12.29.** Schematic diagram of a cyclone column fermenter: (I) cyclone column; (II) circulating pump; and (III) recirculating limb.

### Rotating-Disc Fermenters

Rotating-disc contactors have been used in effluent treatment. They utilise a growing microbial film on slow rotating discs to oxidise the effluent. Anderson and Blain have used the same principle to construct small fermenters of up to 40 $dm^3$ working volume. A range of filamentous fungi, including species of *Aspergillus, Rhizopus, Mucor* and *Penicillium*, could be grown on the polypropylene discs. It has been possible to obtain yields of 80 gram $dm^{-3}$ of citric acid from *A. niger* using this design of fermenter.

## ANIMAL CELL CULTURE

Interest in the *in vitro* cultivation of animal cells has developed because of the need for large scale production of monoclonal antibodies, hormones, vaccines and other products which are difficult or impossible to produce synthetically or by using other culture techniques.

Animal cells are usually more nutritionally demanding than microbial cells. They lack cell walls which makes them sensitive to shear and extremes of osmolarity. The doubling times are normally 12 to 48 hours and cell densities in suspension cultures rarely exceed $10^6$ to $10^7$ cells $cm^{-3}$. Two distinct modes of growth can be recognised:

1. Anchorage dependent cells: These cells require a solid support for their replication. They produce cellular protrusions (pseudopodia) which allow them to adhere to positively charged surfaces and often grow as monolayers.
2. Anchorage independent cells: These cells do not require a support and can grow as a suspension in submerged culture. Established and transformed cell lines are normally in this category.

There are also intermediate categories of cells which may be grown as anchorage dependent or suspension cells. It is also possible to grow some anchorage dependent cells in suspension, provided the cells can be grown on suitable microcarriers.

A range of free and immobilised culture systems have therefore been developed for the culture of different lines of animal cells and their products. The range of available equipment and techniques may be confusing to scientists not familiar with animal cells who may have to decide on the most appropriate culture systems for laboratory, pilot or production scale. Because of the relatively small quantities of product required, the volume at production scale may only be 100 to 10000 $dm^3$. Shear is a phenomenon recognised as being critical to the scale up of animal cell culture processes, irrespective of the cell line or reactor configuration.

Shear may influence the cell culture causing damage which may result in cell death or metabolic changes. The shear sensitivity of animal cells may vary between cell lines, the phase of growth or with a change to fresh medium.

Mijnbeek has reviewed research on shear stress of free and immobilised animal cells in stirred fermenters and air-lift fermenters. It was concluded that the predominant damage mechanism in both types of vessel was due to sparging and the break up of bubbles on the medium surface. This type of damage causing cell death might be reduced by increasing the height to diameter ratio in the vessel, increasing the bubble size, decreasing the gas flow rate or by adding protective agents. Other damage mechanisms in stirred fermenters are caused by cell-microcarrier eddies and microcarrier-microcarrier interactions. Damage of this type may be reduced by reducing the impeller speed, impeller diameter, microcarrier size and concentration or by increasing the viscosity of the medium.

The maximum cell densities obtainable in stirred and air-lift fermenters are often only $10^6$ cells $cm^{-3}$. This is not ideal if secreted proteins are present only in very low concentration and mixed with other proteins present in the original medium. A number of modified fermenters and reactors have been developed to grow cells at higher concentrations using microcarriers, encapsulation, perfusion, glass beads or hollow fibres to obtain the required product at higher concentrations.

### Stirred Fermenters

Unmodified stirred fermenters have been used for the batch production of some virus vaccines but modified vessels are used for most cultures. The modifications made to fermenters are to reduce the possibility of cell damage due to shear, heat or contamination. Marine propellers revolving at a slow speed (10 to 100 rpm) will normally provide adequate mixing. Hemispherical bottoms on the vessels will ensure better mixing of the broth at slow stirrer speeds. Water jacket heating is often preferred since heating probes may give rise to localised zones of high temperature which might damage some of the cells. Magnetic driven stirrers may be used to reduce the risk of contamination. A novel sparger-impeller design is shown in Fig. 12.30. When the impeller rotates the swept-back ports produce a negative pressure inside the hollow impeller complex. This creates a suction lift that produces highly efficient circulation and gas transfer at low rpm without damaging the cells. Gases are introduced through a ring sparger into the medium as it circulates through a fine stainless steel mesh jacket which excludes cells and microcarriers. Because the gas sparging is restricted to a relatively small zone, foaming is reduced and any foam formed is broken up by the mesh in the foam-eliminator chamber.

### Air-lift Fermenters

Air-lift fermenters have proved ideal for growth of some cell lines because of the gentle mixing action and reduced shear forces when compared with those in stirred vessels. The absence of a stirrer and associated seals excludes a potential source of contamination.

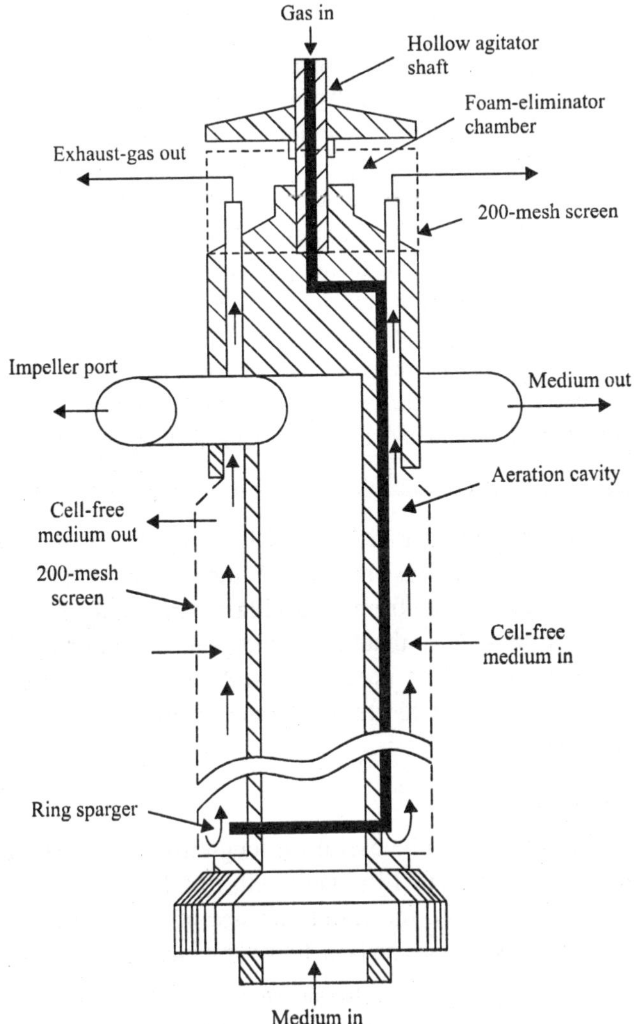


**Fig. 12.30.** Sparger-impeller in the Celligen cell-culture fermenter.

## Microcarriers

Microcarriers may provide a solution to the problem of growing anchorage-dependent cultures in suspension culture in fermenters, by providing the necessary surface for attachment. Animal cells normally have a net negative surface charge and will attach to a positively charged surface by electrostatic forces. Van Wezel made use of this property and attached anchorage dependent cell to chromatographic grade DEAE Sephadex A-50 resin beads and suspended the coated beads in a slowly stirred liquid medium. The density of the electrostatic charges on the microcarrier surface is critical if cell growth at high bead concentrations is to be achieved. If the net charge density on the bead surface is too low, cell attachment will be restricted. When the charge density is too high, apparent toxic effects will limit cell growth. A number of microcarrier beads, manufactured from dextran, cellulose, gelatine, plastic or glass, are now commercially available. Dextran microcarriers have been used for large scale production of viral vaccines and interferon.

### Encapsulation

At least three methods of encapsulation have been developed. Encapsel, a technique developed by Damon Biotechnology, USA, traps the animal cells in sodium alginate spheres which are then coated with polylysine to form a semi-permeable membrane. The enclosed alginate gel is solubilised with sodium citrate to release the cells into free suspension within the capsules which are usually 50 to 500 μm diameter. After few weeks growth it is possible to obtain cell concentrations of $5 \times 10^8$ $cm^{-3}$ and product levels 100 times higher than with free cells can be achieved. The high molecular weight products are retained within the capsules. This technique has been used commercially for monoclonal antibody production. In a second method the cells are entrapped in calcium alginate which will allow high molecular weight products to diffuse into the medium. Unfortunately, the spheres tend to be 0.5 to 1.0 mm diameter and slow diffusion into the spheres may cause nutrient limitations. Alternatively the cells can be entrapped in agarose beads in which the cells are contained in a honeycomb matrix within the gel. These capsules have a wide size distribution and a low mechanical strength compared with alginate ones.

### Hollow Fiber Chambers

Anchorage dependent cells can be cultured at densities of $10^8$ cells $cm^{-3}$ using bundles of hollow fibres held together in cartridge chambers. The cells are grown in the extra capillary spaces (ECS) within the cartridge. Medium and gases diffuse through from the capillary lumea to the ECS. The molecular weight cut-off of the fibre walls may be selected so that the product is retained in the ECS or released into the perfusing medium. Many chambers will be needed for scale-up because of diffusion limitations in larger chambers. These have been used to study production of monoclonal antibodies, viruses, gonadotropin, insulin and antigens.

### Packed Glass Bead Reactors

Packed glass bead reactors have proved to be useful for long-term culture of attached dependent cell lines. It is possible to obtain cell densities of $10^{10}$ viable cells in a 1 $dm^3$ vessel with moderate medium flow rates through the vessel (Fig. 12.31). Increasing the size of vessels causes problems with mass transfer of oxygen and nutrients and scale-up can be achieved by increasing the number of small vessels.

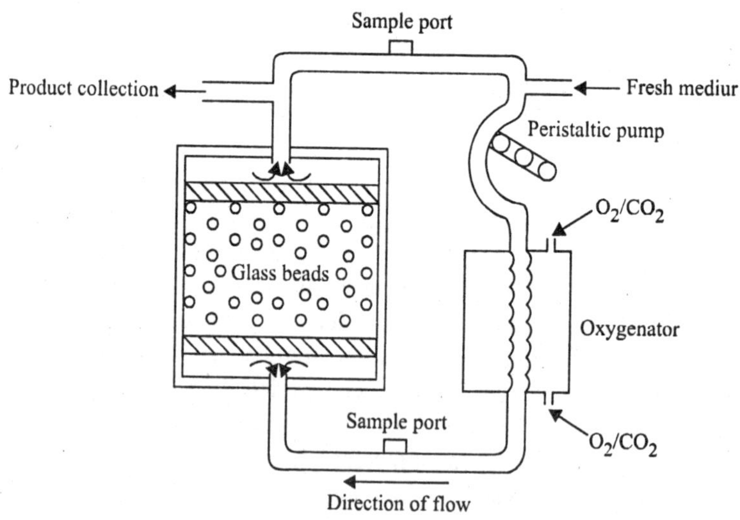


**Fig. 12.31.** Schematic diagram of a glass bead reactor.

### Perfusion Cultures

Perfusion culture is a technique where modified fermenters of up to 100 $dm^3$ are gently stirred and broth is withdrawn continuously from the vessel and passed through a stainless steel or ceramic filter. This type of culture is sometimes referred to as spin culture, since the filter is spun to prevent blocking with cells. The filtered medium is pumped to a product reservoir and fresh medium is pumped into the culture vessel. The rate of addition and removal can be regulated depending on the cell concentration in the vessel. With this method it is possible to obtain cell densities 10 to 30 times higher than the maximum cell density in an unmodified fermenter.

Chapter 13

# Aeration and Agitation in Bioprocess

## INTRODUCTION

The majority of fermentation processes are aerobic and therefore, require the provision of oxygen. If the stoichiometry of respiration is considered, then the oxidation of glucose may be represented as:

$$C_6H_{12}0_6 + 60_2 = 6H_2O + 6CO_2$$

Thus, 192 grams of oxygen are required for the complete oxidation of 180 grams of glucose. However, both components must be in solution before they are available to a micro-organisms and oxygen is approximately 6000 times less soluble in water than is glucose (a fermentation medium saturated with oxygen contains approximately 7.6 mg $dm^{-3}$ of oxygen at 30°C). Thus, it is not possible to provide a microbial culture with all the oxygen it will need for the complete oxidation of the glucose (or any other carbon source) in one addition. Therefore, a microbial culture must be supplied with oxygen during growth at a rate sufficient to satisfy the organism's demand.

The oxygen demand of an industrial fermentation process is normally satisfied by aerating and agitating the fermentation broth. However, the productivity of many fermentations is limited by oxygen availability and therefore, it is important to consider the factors which affect a fermenter's efficiency in supplying microbial cells with oxygen. This chapter considers the requirement for oxygen in fermentation processes, the quatification of oxygen transfer and the factors which will influence the rate of oxygen transfer into solution.

## OXYGEN REQUIREMENTS OF INDUSTRIAL FERMENTATIONS

Although a consideration of the stoichiometry of respiration gives an appreciation of the problem of oxygen supply, it gives no indication of an organism's true oxygen demand as it does not take into account the carbon that is converted into biomass and products. A number of workers have considered the overall stoichiometry of the conversion of oxygen, a source of carbon and a source of nitrogen into biomass and have used such relationships to predict the oxygen demand of a fermentation. A selection of such equations is shown in Table 13.1. From these determinations it may be seen that a culture's demand for oxygen is very much dependent on the source of carbon in the medium.

Thus, the more reduced the carbon source, the greater will be the oxygen demand. From Darlington's and Johnson's equations (Table 13.1) it may be seen that the production of 100 grams of biomass from hydrocarbon requires approximately three times the amount of oxygen to produce the same amount of biomass from carbohydrate. This point is also illustrated in Table 13.2. However, it must be remembered that the high carbon content of hydrocarbon substrates means that high yield factors (g biomass $g^{-1}$ substrate consumed) are obtained and the decision to use such substrates is based on the balance between the advantage of high biomass yield and the disadvantage of high oxygen demand and heat generation.

**Table 13.1.** Stoichiometric equations describing oxygen demand in a fermentation.

| *Equation* | *Terms used* | *Reference* |
|---|---|---|
| $6.67CH_2O + 2.1O_2 = C_{3.92}H_{6.5}O_{1.94} + 2.75CO_2 + 3.42H_2O$ | $C_{3.92}H_{6.5}O_{1.94}$ is 100 g (dry weight) of yeast cells; $CH_2O$ is carbohydrate | Darlington |
| $7.14CH_2 + 6.135O_2 = C_{3.92}H_{6.5}O_{1.94} + 3.22CO_2 + 3.89H_2O$ | $CH_2$ is hydrocarbon | Darlington |
| $(A/Y) - B = C$ | $A$ = Amount of oxygen for combustion of 1 g of substrate to $CO_2$, $H_2O$ and $NH_3$, if nitrogen is present in the substrate<br>$B$ = Amount of oxygen required for the combustion of 1 g cells to $CO_2$, $H_2O$ and $NH_3$<br>$Y$ = Cell yield (g cells $g^{-1}$ substrate)<br>$C$ = g oxygen consumed for the production of 1 gram of cells | Johnson |
| $(dx/dt)/Y + mx + p = (dCO_2/dt) = -(dO_2/dt)RQ$ | $x$ = biomass concentration<br>$t$ = time<br>$Y$ = g biomass $g^{-1}$ carbon substrate<br>$m$ = maintenance<br>$p$ = allowance for antibiotic production | Righelato |
| $Y_{O/P} = (0.53/Y_{P/G}) - (0.6X/P) - 0.43$ | $Y_{O/P}$ = g oxygen consumed $g^{-1}$ glucose<br>$Y_{P/G}$ = g sodium penicillin G produced $g^{-1}$ glucose<br>$X$ = g cells (dry weight) produced<br>$P$ = g sodium penicillin G produced | Cooney |
| $Y_O = \{(32C + 8H - 16O)/YM\} - 1.58$ | $Y_O$ = g oxygen consumed $g^{-1}$ cells produced<br>$Y$ = Cell yield (g cells $g^{-1}$ substrate)<br>$M$ = Molecular weight of the carbon source<br>C, H and O = Number of atoms of carbon, hydrogen and oxygen per molecule of carbon source | Mateles |

**Table 13.2.** Oxygen requirements of a range of micro-organisms grown on a range of substrates.

| *Substrate* | *Organism* | *Oxygen requirement ($g\ O_2\ g^{-1}$ dry wt.)* |
|---|---|---|
| Glucose | *Escherichia coli* | 0.4 |
| Methanol | *Pseudomonas C* | 1.2 |
| Octane | *Pseudomonas* sp. | 1.7 |

Darlington's, Johnson's and Matele's equations only include biomass production and do not consider product formation, whereas Cooney's and Righelato's equations consider product formation. Ryu and

Hospodka used Righelato's approach to calculate that the production of 1 gram penicillin consumes 2.2 gram of oxygen.

However, it is inadequate to base the provision of oxygen for a fermentation simply on an estimation of overall demand, because the metabolism of the culture is affected by the concentration of dissolved oxygen in the broth. The effect of dissolved oxygen concentration on the specific oxygen uptake rate ($Q_{O_2}$, mmoles of oxygen consumed per gram dry weight of cells per hour) has been shown to be of the Michaelis-Menten type, as shown in Fig. 13.1.

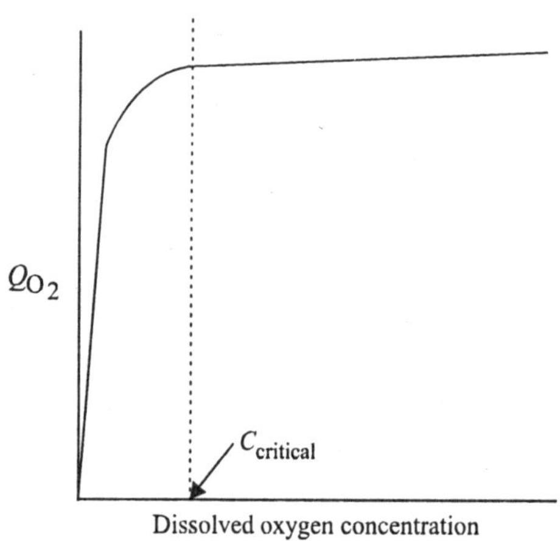


**Fig. 13.1.** The effect of dissolved oxygen concentration on the $Q_{O_2}$ of a micro-organism.

From Fig. 13.1 it may be seen that the specific oxygen uptake rate increases with increase in the dissolved oxygen concentration up to a certain point (referred to as $C_{crit}$) above which no further increase in oxygen uptake rate occurs. Some examples of the critical oxygen levels for a range of micro-organisms are given in Table 13.3. Thus, maximum biomass, production may be achieved by satisfying the organism's maximum specific oxygen demand by maintaining the dissolved oxygen concentration greater than the critical level. If the dissolved oxygen concentration were to fall below the critical level then the cells may be metabolically disturbed. However, it must be remembered that it is frequently the objective of the fermentation technologist to produce a product of the micro-organism rather than the organism itself and that metabolic disturbance of the cell by oxygen starvation may be advantageous to the formation of certain products. Equally, provision of a dissolved oxygen concentration far greater than the critical level may have no influence on biomass production, but may stimulate product formation. Thus, the aeration conditions necessary for the optimum production of a product may be different from those favouring biomass production.

**Table 13.3.** Critical dissolved oxygen concentrations for a range of micro-organisms.

| *Organism* | *Temperature* | *Critical dissolved oxygen concentration (mmoles dm$^{-3}$)* |
|---|---|---|
| *Azotobacter* sp. | 30 | 0.018 |
| *Escherichia coli* | 37 | 0.008 |
| *Saccharomyces* sp. | 30 | 0.004 |
| *Penicillium chrysogenum* | 24 | 0.022 |

Hirose and Shibai's investigations of amino acid biosynthesis by *Brevibacterium flavum* provide an excellent example of the effects of the dissolved oxygen concentration on the production of a range of closely related metabolites. These workers demonstrated the critical dissolved oxygen concentration for *B. flavum* to be 0.01 mg $dm^{-3}$ and considered the extent of oxygen supply to the culture in terms of the degree of 'oxygen satisfaction', that is the respiratory rate of the culture expressed as a fraction of the maximum respiratory rate. Thus, a value of oxygen satisfaction below unity implied that the dissolved oxygen concentration was below the critical level. The effect of the degree of oxygen satisfaction on the production of a range of amino acids is shown in Fig. 13.2. From Fig. 13.2 it may be seen that the production of members of the glutamate and aspartate families of amino acids was affected detrimentally by levels of oxygen satisfaction below 1.0, whereas optimum production of phenylalanine, valine and leucine occurred at oxygen satisfaction levels of 0.55, 0.60 and 0.85, respectively. The biosynthetic routes of the amino acids are shown in Fig. 13.3, from which it may be seen that the glutamate and aspartate families are all produced from tricarboxylic acid (TCA) cycle intermediates, whereas phenylalanine, valine and leucine are produced from the glycolysis intermediates, pyruvate and phosphoenol pyruvate. Oxygen excess should give rise to abundant TCA cycle intermediates, whereas oxygen limitation should result in less glucose being oxidised via the TCA cycle, allowing more intermediates to be available for phenylalanine, valine and leucine biosynthesis. Thus, some degree of metabolic disruption results in greater production of pyruvate derived amino acids.

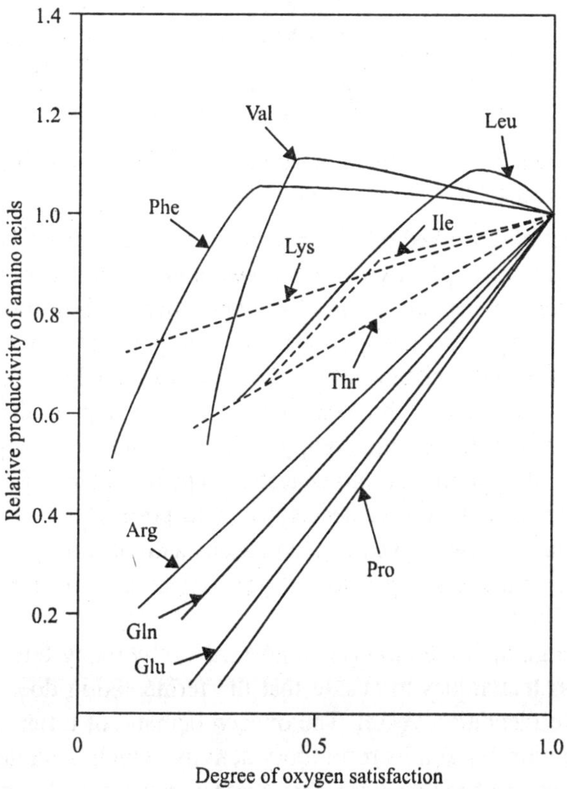


**Fig. 13.2.** The effect of dissolved oxygen on the production of amino acids by *Brevibacterium flavum.*

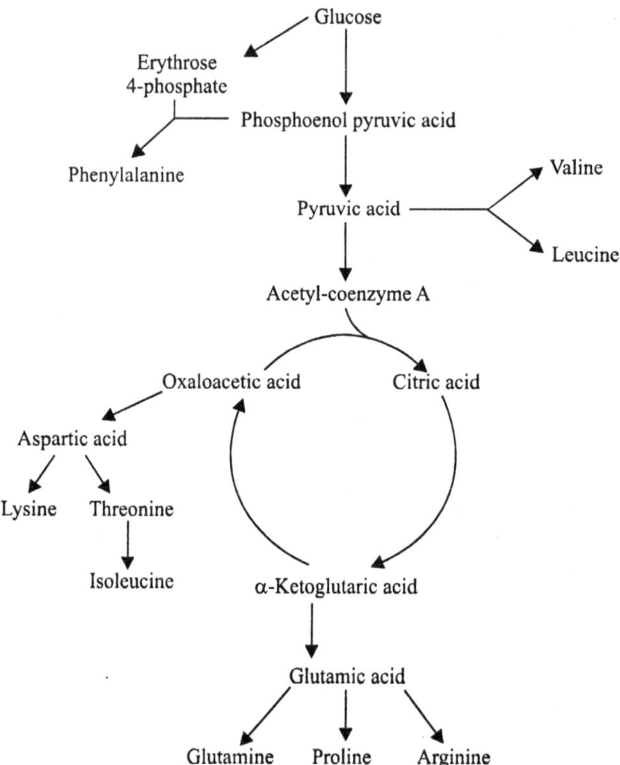


**Fig. 13.3.** The biosynthetic routes to the amino acids phenylalanine, valine, leucine, lysine, threonine, *I*-leucine, glutamic acid, proline, glutamine and arginine in *B. flavum*.

An example of the effect of dissolved oxygen on secondary metabolism is provided by Zhou work on cephalosporin C synthesis by *Cephalosporium acremonium*. These workers demonstrated that the critical oxygen concentration for cephalosporin C synthesis during the production phase was 20 per cent saturation. At dissolved oxygen concentrations below 20 per cent cephalosporin C concentration declined and penicillin N increased. The biosynthetic pathway to cephalosporin C is shown in Fig. 13.4, from which it may be seen that there are three oxygen-consuming steps in the pathway:

1. Cyclisation of the tripeptide, α-amino-adipylcysteinyl-valine into isopenicillin N.
2. The ring expansion of penicillin N into deacetoxycephalosporin C (DAOC).
3. The hydroxylation of DAOC to give deacetylcephalosporin C.

DAOC did not accumulate at low oxygen concentrations and thus, it appears that the most oxygen sensitive step in the pathway is the ring expansion enzyme (expandase) resulting in the accumulation of penicillin N under oxygen limitation.

The requirement for a high dissolved oxygen concentration by many fermentations has resulted in the development of process techniques to ensure that the fermentation does not exceed the oxygen-supply capabilities of the fermentation vessel. The oxygen demand of a fermentation largely depends on the concentration of the biomass and its respiratory activity, which is related to the growth rate. By limiting the initial concentration of the medium, the biomass in the vessel may be kept at a reasonable level and by supplying some nutrient component as a feed, the rate of growth and hence the respiratory rate, may be controlled.

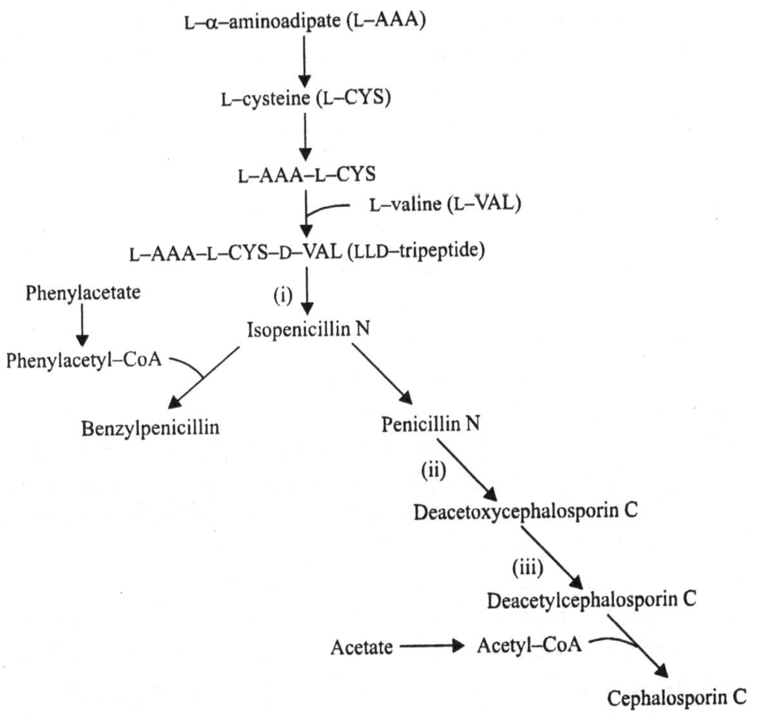


**Fig. 13.4.** The biosynthesis of cephalosporin C, indicating the oxygen consuming steps: (i) isopenicillin-N-synthase; (ii) deacetoxycephalosporin C synthase (commonly called expandase); (iii) deacetylcephalosporin C synthase (commonly called hydroxylase).

## OXYGEN SUPPLY

Oxygen is normally supplied to microbial cultures in the form of air, this being the cheapest available source of the gas. The method for provision of a culture with a supply of air varies with the scale of the process:

1. Laboratory-scale cultures may be aerated by means of the shake-flask technique where the culture (50 to 100 $cm^3$) is grown in a conical flask (250 to 500 $cm^3$) shaken on a platform contained in a controlled environment chamber.
2. Pilot and industrial-scale fermentations are normally carried out in stirred, aerated vessels, termed fermenters. However, it is often advantageous to culture relatively small volumes (1 $dm^3$) in a stirred, aerated vessel as this enables the cultural conditions to be better monitored and controlled and facilitates the addition of supplements and the removal of samples. Some fermenters are so designed that adequate oxygen transfer is obtained without agitation.

Bartholomew represented the transfer of oxygen from air to the cell, during a fermentation, as occurring in a number of steps:

1. The transfer of oxygen from an air bubble into solution.
2. The transfer of the dissolved oxygen through the fermentation medium to the microbial cell.
3. The uptake of the dissolved oxygen by the cell.

These workers demonstrated that the limiting step in the transfer of oxygen from air to the cell in a *Streptomyces griseus* fermentation was the transfer of oxygen into solution. These findings have been shown to be correct for non-viscous fermentations but it has been demonstrated that transfer may be limited by either of the other two stages in certain highly viscous fermentations. The difficulties inherent in such fermentations are discussed later in this chapter.

The rate of oxygen transfer from air bubble to the liquid phase may be described by the equation:

$$dC_L/dt = K_L a(C^* - C_L) \quad \text{... (13.1)}$$

where $C_L$ = is the concentration of dissolved oxygen in the fermentation broth (mmoles $dm^{-3}$),
$t$ = is time (hours),
$dC_L/dt$ = is the change in oxygen concentration over a time period, i.e., the oxygen transfer rate (mmoles $O_2$ $dm^{-3}$ $hr.^{-1}$),
$K_L$ = is the mass transfer coefficient (cm $hr.^{-1}$),
$a$ = is the gas/liquid interface area per liquid volume ($cm^2$ $cm^{-3}$),
$C^*$ = is the saturated dissolved oxygen concentration (mmoles $dm^{-3}$).

$K_L$ may be considered as the sum of the reciprocals of the resistances to the transfer of oxygen from gas to liquid and $(C^* - C_L)$ may be considered as the 'driving force' across the resistances. It is extremely difficult to measure both $K_L$ and '$a$' in a fermentation and therefore, the two terms are generally combined in the term $K_La$, the volumetric mass-transfer coefficient, the units of which are reciprocal time ($hr.^{-1}$). The volumetric mass-transfer coefficient is used as a measure of the aeration capacity of a fermenter. The larger the $K_La$, the higher the aeration capacity of the system. The $K_La$ value will depend upon the design and operating conditions of the fermenter and will be affected by such variables as aeration rate, agitation rate and impeller design. These variables affect '$K_L$' by reducing the resistances to transfer and affect '$a$' by changing the number, size and residence time of air bubbles. It is convenient to use $K_La$ as a yardstick of fermenter performance because, unlike the oxygen-transfer rate, it is unaffected by dissolved oxygen concentration. However, the oxygen transfer rate is the critical criterion in a fermentation and as may be seen from equation 13.1, it is affected by both $K_La$ and dissolved oxygen concentration. The dissolved oxygen concentration reflects the balance between the supply of dissolved oxygen by the fermenter and the oxygen demand of the organism. If the $K_La$ of the fermenter is such that the oxygen demand of the organism cannot be met, the dissolved oxygen concentration will decrease below the critical level ($C_{crit}$). If the $K_La$ is such that the oxygen demand of the organism can be easily met the dissolved oxygen concentration will be greater than $C_{crit}$ and may be as high as 70 to 80 per cent of the saturation level. Thus, the $K_La$ of the fermenter must be such that the optimum oxygen concentration for product formation can be maintained in solution throughout the fermentation.

## DETERMINATION OF $K_La$ VALUES

The determination of the $K_La$ of a fermenter is essential in order to establish its aeration efficiency and to quantify the effects of operating variables on the provision of oxygen. This section considers the merits and limitations of the methods available for the determination of $K_La$ values. It is important to remember at this stage that dissolved oxygen is usually monitored using a dissolved oxygen electrode which records dissolved oxygen activity or dissolved oxygen tension (DOT) whilst the equations describing oxygen transfer are based on dissolved oxygen concentration. The solubility of oxygen is affected by dissolved solutes so that pure water and a fermentation medium saturated with oxygen would have different dissolved oxygen concentrations yet have the same DOT, i.e., an oxygen electrode would record 100 per cent for both. Thus, to translate DOT into concentration the solubility of oxygen in the fermentation medium must be known and this can present difficulties.

## Sulphite Oxidation Technique

Cooper was the first to describe the determination of oxygen-transfer rates in aerated vessels by the oxidation of sodium sulphite solution. This technique does not require the measurement of dissolved oxygen concentrations but relies on the rate of conversion of a 0.5 M solution of sodium sulphite to sodium sulphate in the presence of copper or cobalt catalyst:

$$Na_2SO_3 + 0.5O_2 = Na_2SO_4$$

The rate of reaction is such that as oxygen enters solution it is immediately consumed in the oxidation of sulphite, so that the sulphite oxidation rate is equivalent to the oxygen-transfer rate. The dissolved oxygen concentration, for all practical purposes, will be zero and the $K_La$ may then be calculated from the equation:

$$\text{OTR} = K_L a \cdot C^* \qquad \text{... (13.2)}$$

(where OTR is the oxygen transfer rate)

The procedure is carried out as follows: the fermenter is batched with a 0.5 M solution of sodium sulphite containing $10^{-3}$ M $Cu^{2+}$ ions and aerated and agitated at fixed rates; samples are removed at set time intervals (depending on the aeration and agitation rates) and added to excess iodine solution which reacts with the unconsumed sulphite, the level of which may be determined by a back titration with standard sodium thiosulphate solution. The volumes of the thiosulphate titrations are plotted against sample time and the oxygen transfer rate may be calculated from the slope of the graph.

The sulphite oxidation method has the advantage of simplicity and also, the technique involves sampling the bulk liquid in the fermenter and therefore, removes some of the problems of conditions varying through the volume of the vessel. However, the method is time consuming (one determination taking up to 3 hours, depending on the aeration and agitation rates) and is notoriously inaccurate. Bell and Gallo demonstrated that minor amounts of surface-active contaminants (such as amino acids, proteins, fatty acids, esters, lipids, etc.) could have a major effect on the accuracy of the technique and apparent differences in aeration efficiency between vessels could be due to differences in the degree of contamination. Also, the rheology of a sodium sulphite solution is completely different from that of a fermentation broth, especially a mycelial one so that it is impossible to relate the results of sodium sulphite determinations to real fermentations.

## Gassing-out Techniques

The estimation of the $K_La$ of a fermentation system by gassing-out techniques depends upon monitoring the increase in dissolved oxygen concentration of a solution during aeration and agitation. The oxygen transfer rate will decrease during the period of aeration as $C_L$ approaches $C^*$ due to the decline in the driving force ($C^* - C_L$). The oxygen transfer rate, at any one time, will be equal to the slope of the tangent to the curve of values of dissolved oxygen concentration against time of aeration, as shown in Fig. 13.5.

To monitor the increase in dissolved oxygen over an adequate range it is necessary first to decrease the oxygen level to a low value. Two methods have been employed to achieve this lowering of the dissolved oxygen concentration—the static method and the dynamic method.

### *Static method of gassing-out*

In this technique, the oxygen concentration of the solution is lowered by gassing the liquid out with nitrogen gas, so that the solution is 'scrubbed' free of oxygen. The deoxygenated liquid is then aerated and agitated and the increase in dissolved oxygen monitored using some form of dissolved oxygen probe. The increase in dissolved oxygen concentration has already been described by equation (13.1), i.e.,:

$$dC_L/dt = K_La\ (C^* - C_L)$$

and depicted in Fig. 13.5. Integration of equation (13.1) yields:

$$\ln (C^* - C_L) = -K_Lat \quad \text{... (13.3)}$$

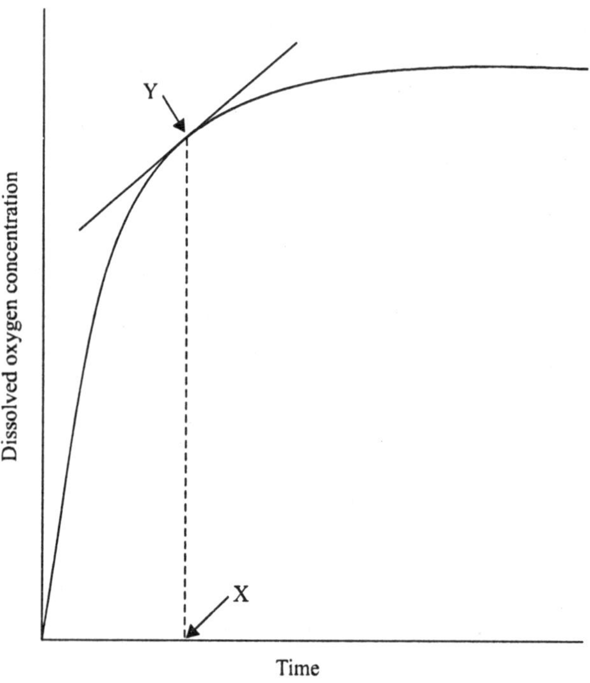


**Fig. 13.5.** The increase in dissolved oxygen concentration of a solution over a period of aeration. The oxygen transfer rate at time X is equal to the slope of the tangent at point Y.

Thus, a plot of ln $(C^* - C_L)$ against time will yield a straight line of slope $-K_La$, as shown in Fig. 13.6. This technique has the advantage over the sulphite oxidation method in that it is very rapid (normally taking up to 15 minutes) and may utilise the fermentation medium, to which may be added dead cells or mycelium at a concentration equal to that produced during the fermentation. However, employing the fermentation medium with or without, killed biomass necessitates the use of a membrane-type electrode, the response time of which may be inadequate to reflect the true change in the rate of oxygenation over a short period of time. The probe response time ($T_p$) is defined as the time needed to record 63 per cent of a stepwise change and this should be much smaller than the mass transfer response time of the system ($1/K_La$). According to Van't Riet, the use of commercially available electrodes, with a response time of 2 to 3 seconds, should enable a $K_La$ of up to 360 hr.$^{-1}$ to be measured with little loss of accuracy. However, for estimations of higher $K_La$ values it would be necessary to incorporate a correction factor into the calculation, as discussed by Taguchi and Humphrey. It is not necessary to know the oxygen solubility in the medium because DOT values may be used directly in order to calculate the rates, i.e., $C^*$ is taken as 100 per cent.

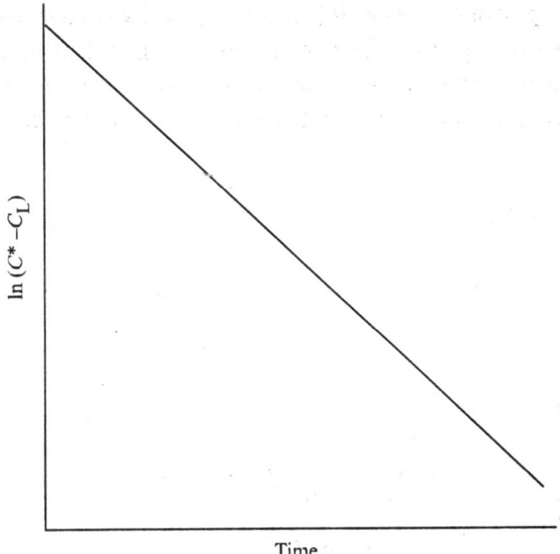


**Fig. 13.6.** A plot of the ln($C^* - C_L$) against time of aeration, the slope of which equals $-K_La$.

Whilst the method is acceptable for small scale vessels, there are severe limitations to its use on large scale fermenters which have high gas residence times. When the air supply to such a vessel is resumed after deoxygenation with nitrogen, the oxygen concentration in the gas phase may change with time as the nitrogen is replaced with air. Thus, $C^*$ will no longer be constant. Although correction factors have been derived to compensate for this phenomenon, Van't Riet and Tramper concluded that the method should not be used for vessels over 1-metre high.

### *Dynamic method of gassing-out*

Taguchi and Humphrey utilised the respiratory activity of a growing culture in the fermenter to lower the oxygen level prior to aeration. Therefore, the estimation has the advantage of being carried out during a fermentation which should give a more realistic assessment of the fermenter's efficiency. Because of the complex nature of fermentation broths the probe used to monitor the change in dissolved oxygen concentration must be of the membrane-covered type which may necessitate the use of the response-correction factors referred to previously. The procedure involves stopping the supply of air to the fermentation which results in a linear decline in the dissolved oxygen concentration due to the respiration of the culture, as shown in Fig. 13.7. The slope of the line AB in Fig. 13.7 is a measure of the respiration rate of the culture. At point B the aeration is resumed and the dissolved oxygen concentration increases until it reaches concentration X. Over the period, BC, the observed increase in dissolved oxygen concentration is the difference between the transfer of oxygen into solution and the uptake of oxygen by the respiring culture as expressed by the equation:

$$dC_L/dt = K_La(C^* - C_L) - xQ_{O_2} \quad \text{... (13.4)}$$

where $x$ = is the concentration of biomass and

$Q_{O_2}$ = is the specific respiration rate (mmoles of oxygen g$^{-1}$ biomass hr.$^{-1}$).

The term $xQ_{O_2}$ is given by the slope of the line AB in Fig. 13.7. Equation (13.4) may be rearranged as:

$$C_L = -1/K_La\ \{(dC_L/dt) + xQ_{O_2}\} + C^* \quad \text{... (13.5)}$$

Thus, from equation (13.5), a plot of $C_L$ versus $dC_L/dt + xQo_2$ will yield a straight line, the slope of which will equal $-1/K_La$, as shown in Fig. 13.8. This technique is convenient in that the equations may be applied using DOT rather than concentration because it is the rates of transfer and uptake that are being monitored so that the percentage saturation readings generated by the electrode may be used directly.

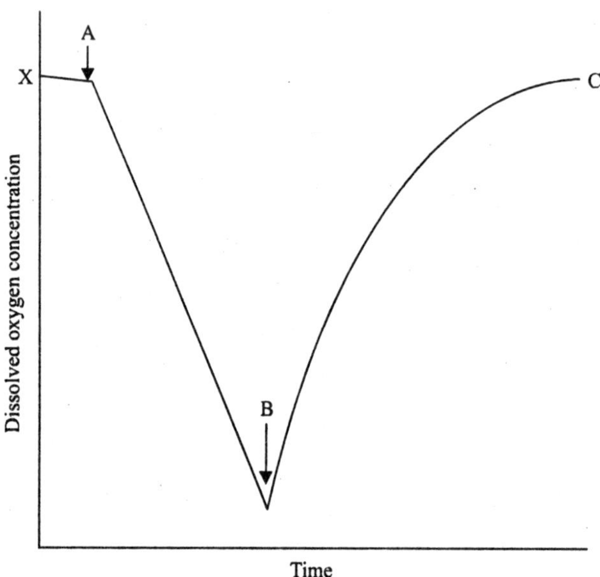


**Fig. 13.7.** Dynamic gassing-out for the determination of $K_La$ values. Aeration was terminated at point A and recommenced at point B.

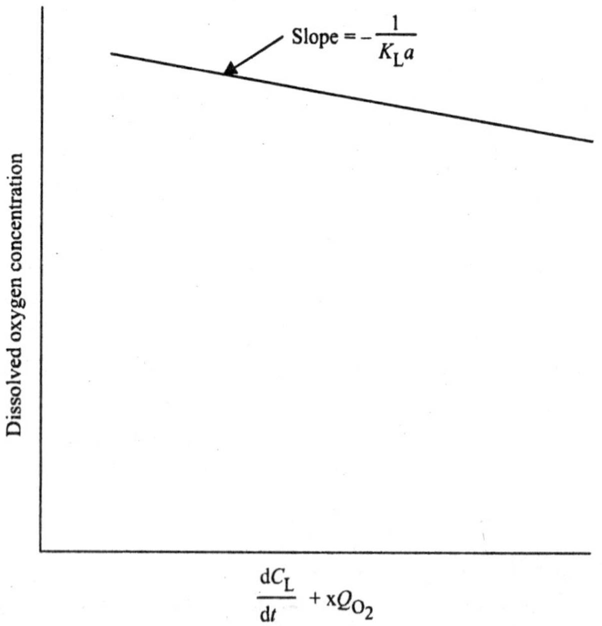


**Fig. 13.8.** The dynamic method for determination of $K_La$ values. The information is gleaned from Fig. 13.7 by taking tangents of the curve, BC, at various values of $C_L$.

The dynamic gassing-out method has the advantage over the previous methods of determining the $K_La$ during an actual fermentation and may be used to determine $K_La$ values at different stages in the process. The technique is also rapid and only requires the use of a dissolved-oxygen probe, of the membrane type. A major limitation in the operation of the technique is the range over which the increase in dissolved oxygen concentration may be measured. It is important not to allow the oxygen concentration to drop below $C_{crit}$ during the deoxygenation step as the specific oxygen uptake rate will then be limited and the term $xQ_{O_2}$ would not be constant on resumption of aeration. The occurrence of oxygen-limited conditions during deoxygenation may be detected by the deviation of the decline in oxygen concentration from a linear relationship with time, as shown in Fig. 13.9.

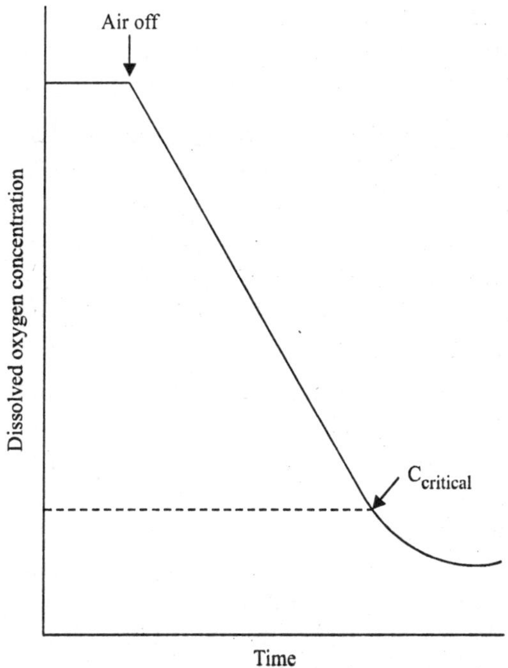


**Fig. 13.9.** The occurrence of oxygen limitation during the dynamic gassing-out of a fermentation.

When the oxygen demand of a culture is very high it may be difficult to maintain the dissolved oxygen concentration significantly above $C_{crit}$ during the fermentation so that the range of measurements which could be used in the $K_La$ determination would be very small. Thus, it may be difficult to apply the technique during a fermentation which has an oxygen demand close to the supply capacity of the fermenter.

Although the difficulty presented by nitrogen degassing does not arise with the dynamic method, it also, is not suitable for use with vessels in excess of one metre high. Van't Riet and Tramper pointed out that in such vessels the time taken to establish an equilibrium population of air bubbles would be significant and the gas-liquid interface area would change over the aeration period resulting in a considerable underestimate of the $K_La$ value achievable under normal operating conditions. Both the dynamic and static methods are also unsuitable for measuring $K_La$ values in viscous systems. This is due to the very small bubbles (< 1 mm diameter) formed in a viscous system which have an extended residence time compared with 'normal' sized bubbles. Thus, the gassing-out techniques are only useful on a small scale with non-viscous systems.

### Oxygen-Balance Technique

The $K_La$ of a fermenter may be measured during a fermentation by the oxygen balance technique which determines, directly, the amount of oxygen transferred into solution in a set time interval. The procedure involves measuring the following parameters:

1. The volume of the broth contained in the vessel, $V_L$ ($dm^3$).
2. The volumetric air flow rates measured at the air inlet and outlet, $Q_i$ and $Q_o$, respectively ($dm^3$ $min^{-1}$).
3. The total pressure measured at the fermenter air inlet and outlet, $P_i$ and $P_o$, respectively (atm. absolute).
4. The temperature of the gases at the inlet and outlet, $T_i$ and $T_o$, respectively (K).
5. The mole fraction of oxygen measured at the inlet and outlet, $y_i$ and $y_o$, respectively.

The oxygen transfer rate may then be determined from the following equation:

$$OTR = (7.32 \times 10^5/V_L)\ (Q_iP_iy_i/T_i - Q_oP_oy_o/T_o) \qquad \text{... (13.6)}$$

where $7.32 \times 10^5$ is the conversion factor equalling (60 minute $hr.^{-1}$)[mole/22.4 $dm^3$ (STP)] (273 K/1 atm).

These measurements require accurate flow meters, pressure gauges and temperature-sensing devices as well as gaseous oxygen analysers. The ideal gaseous oxygen analyser is a mass spectrometer analyser which is sufficiently accurate to detect changes of 1 to 2 per cent.

The $K_La$ may be determined, provided that $C_L$ and $C^*$ are known, from equation (13.1):

$$OTR = K_La\ (C^* - C_L) \text{ or } K_La = OTR/(C^* - C_L)$$

$C_L$ may be determined using a membrane-type dissolved-oxygen electrode and in this case the slow response time is not an important factor because a rate of change is not being measured, simply the steady-state oxygen concentration. However, it should be remembered that an electrode simply measures the oxygen tension at one point and it is, therefore, advisable to monitor the oxygen tension at a number of points in the vessel with a number of electrodes and to use an average value. Also, the DOT reading must be converted to concentration, which necessitates knowing the oxygen solubility in the fermentation medium. The value of $C^*$ is frequently taken as that value which is in equilibrium with the oxygen concentration of the gas outlet. Wang claimed that this approach was adequate for small-scale fermenters but on a large scale there may be a considerable difference between the dissolved oxygen concentration in equilibrium with the inlet and outlet gases. Therefore, these workers suggested that the behaviour of the gas in transit in the fermenter would approximate to plug flow conditions and a logarithmic mean value for the dissolved oxygen concentration should be used.

The oxygen-balance technique appears to be the simplest method for the assessment of $K_La$ and has the advantage of measuring aeration efficiency during a fermentation. The sulphite oxidation and static gassing-out techniques have the disadvantage of being carried out using either a salt solution or an uninoculated, sterile fermentation medium. These techniques are adequate for the comparison of equipment or operating variables, it should not be assumed that the values obtained are those actually operating during a fermentation. This may be the case for bacterial or yeast fermentations where the rheology of the suspended cells in the broth is similar to that in a sterile medium or a salt solution, but it is certainly not true for fungal and streptomycete processes where the rheology is quite different.

Tuffile and Pinho compared a number of methods for the determination of $K_La$ values in viscous streptomycete fermentations. The techniques used were static gassing-out, dynamic gassing-out and the oxygen-balance method. Tuffile and Pinho did not make it clear whether non-respiring mycelium was present during their static gassing-out procedure, but from their results it would appear that it was present in the vessel. Thus, the rheology of the fermenter contents would appear to have been similar for the different determinations. The $K_La$ values, determined by the different techniques, for a 300 $dm^3$ fermenter containing a 90-hour culture of *Streptomyces aureofaciens* are shown in Table 13.4.

**Table 13.4.** $K_La$ values for a 300 $dm^3$ fermenter containing a 90-hour culture of *S. aureofaciens*.

| *Method of $K_La$ determination* | *Measured oxygen uptake rate (mmoles $dm^{-3}$ $hr.^{-1}$)* | *$K_La$ ($hr.^{-1}$)* |
|---|---|---|
| Static gassing-out | – | 58.2 |
| Dynamic gassing-out | 6.6 | 58.2 |
| Oxygen balance | 20.1 | 108.0 |

From Table 13.4 it may be seen that the $K_La$ values for the two gassing-out techniques were very similar but there was a considerable difference between the oxygen-uptake rates and the $K_La$s determined by the dynamic method and the balance method. Tuffile and Pinho claimed that the low oxygen-uptake rate determined by the dynamic method was due to air bubbles remaining in suspension in the mash during the dynamic gassing-out period. Thus, the decline in oxygen concentration after the cessation of aeration was not a measure of the oxygen-uptake rate but the difference between oxygen uptake and the transfer of oxygen from entrapped bubbles. It was demonstrated that a large number of bubbles remained suspended in the medium 15 minutes after aeration had been stopped. The use of the low oxygen-uptake rate in the calculation of the $K_La$ would result in an artificially low $K_La$ being determined. Heijnen also observed anomalies in determining $K_La$ values in viscous systems due to the presence of very small bubbles having a much longer residence time than the more abundant large bubbles in the vessel.

Overall, it would appear that the balance method is the most desirable technique to use and the extra cost of the monitoring equipment involved should be a worthwhile investment.

Before considering the factors which may affect the $K_La$ of a fermenter it is necessary to consider the behaviour of fluids in agitated systems.

## FLUID RHEOLOGY

Fluids may be described as Newtonian or non-Newtonian depending on whether their rheology (flow) characteristics obey Newton's law of viscous flow. Consider a fluid contained between two parallel plates, area $A$ and distance $x$ apart. If the lower plate is moved in one direction at a constant velocity, the fluid adjacent to the moving plate will move in the same direction and impart some of its momentum to the 'layer' of liquid directly above it causing it, also, to move in the same direction at a slightly lower velocity. Newton's law of viscous flow states that the viscous force, $F$, opposing motion at the interface between the two liquid layers, flowing with a velocity gradient of dv/dx, is given by the equation:

$$F = \mu A\,(dv/dx) \qquad \text{... (13.7)}$$

where μ is the fluid viscosity, which may be considered as the resistance of the fluid to flow.

Equation (13.7) may be written as:

$$F/A = \mu\,(dv/dx)$$

$F/A$ is termed the shear stress ($\tau$) and is the applied force per unit area, $dv/dx$ is termed the shear rate ($\gamma$) and is the velocity gradient. Thus:

$$\tau = \mu\gamma \qquad \text{... (13.8)}$$

Equation (13.8) conforms to the general relationship:

$$\tau = K\gamma^n \qquad \text{... (13.9)}$$

where $K$ = is the consistency coefficient and

n = is the flow behaviour index or power law index.

For a Newtonian fluid $n$ is 1 and the consistency coefficient is the viscosity which is the ratio of shear stress to shear rate. Thus, a plot of shear stress against shear rate, for a Newtonian fluid, would produce a straight line, the slope of which would equal the viscosity. Such a plot is termed a rheogram (as shown in Fig. 13.10).

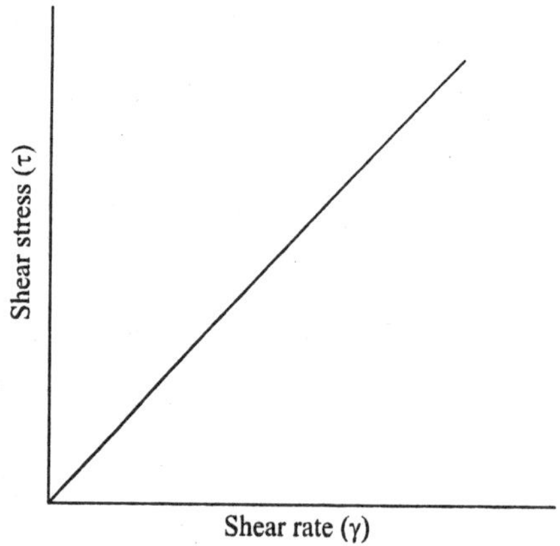


**Fig. 13.10.** A rheogram of a Newtonian fluid.

Thus, a Newtonian liquid has a constant viscosity regardless of shear, so that the viscosity of a Newtonian fermentation broth will not vary with agitation rate. However, a non-Newtonian liquid does not obey Newton's law of viscous flow and does not have a constant viscosity. The value for n [equation (13.9)] of such a fluid deviates from 1 and its behaviour is said to follow a power law model. Thus, the viscosity of a non-Newtonian fermentation broth will vary with agitation rate and is described as an apparent viscosity ($\mu_a$). A plot of shear stress against shear rate for a non-Newtonian liquid will deviate from the relationship depicted in Fig. 13.10, depending on the nature of the liquid. Several types of non-Newtonian liquids are recognised and typical rheograms of types important in the study of culture fluids are given in Fig. 13.11 and their characteristics are discussed below.

## Bingham Plastic Rheology

Bingham plastics are similar to Newtonian liquids apart from the fact that shear rate will not increase until a threshold shear stress is exceeded. The threshold shear stress is termed the yield stress or yield value, $\tau_0$. A linear relationship of shear stress to shear rate is given once the yield stress is exceeded and the slope of this line is termed the coefficient of rigidity or the plastic viscosity. Thus, the flow of a Bingham plastic is described by the equation:

$$\tau = \tau_0 + n\gamma$$

where $n$ = is the coefficient of rigidity and

$\tau_0$ = is the yield stress.

There have been some claims of mycelial fermentation broths displaying Bingham plastic characteristics (Table 13.5). Everyday examples of these fluids include toothpaste and clay.

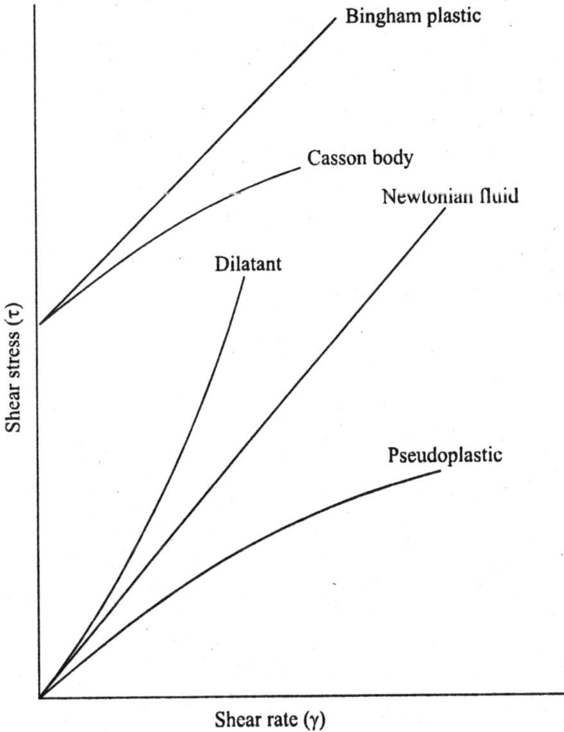


**Fig. 13.11.** Rheogram of fluids of different properties.

**Table 13.5.** Some examples of the rheological nature of fermentation broths.

| *Organism* | *Rheological type* |
|---|---|
| *Penicillium chrysogenum* | Bingham plastic |
| *Streptomyces kanamyceticus* | Bingham plastic |
| *Penicillium chrysogenum* | Pseudoplastic |
| *Endomyces* sp. | Pseudoplastic |
| *Penicillium chrysogenum* | Casson body |

## Pseudoplastic Rheology

The apparent viscosity of a pseudoplastic liquid decreases with increasing shear rate. Most polymer solutions behave as pseudoplastics. The decrease in apparent viscosity is explained by the long chain molecules tending to align with each other at high shear rates resulting in easier flow. The flow of a pseudoplastic liquid may be described by the power law model, equation (13.9), i.e.,:

$$\tau = K(\gamma)^n$$

*K* has the same units as viscosity and may be taken as the apparent viscosity. The flow-behaviour index is less than unity for a pseudoplastic liquid, the smaller the value of *n*, the greater the flow characteristics of the liquid deviate from those of a Newtonian fluid. Equation (13.9) may be converted to the logarithmic form as:

$$\log\tau = \log K + n\log \gamma \qquad \text{... (13.10)}$$

Thus, a plot of log shear stress against log shear rate will produce a straight line, the slope of which will equal the flow-behaviour index and the intercept on the shear stress axis will be equal to the logarithm of the consistency coefficient. Many workers have demonstrated that mycelial fermentation broths display pseudoplastic properties as shown in Table 13.5.

### Dilatant Rheology

The apparent viscosity of a dilatant liquid increases with increasing shear rate. The flow of a dilatant liquid may also be described by equation (13.9) but in this case the value of the flow-behaviour index is greater than 1, the greater the value the greater the flow characteristics deviate from those of a Newtonian fluid. Thus, the values of K and n may be obtained from a plot of log shear stress against log shear rate. Fortunately this type of behaviour is not exhibited by fermentation broths—an everyday example is liquid cement slurry.

### Casson Body Rheology

Casson described a type of non-Newtonian fluid, termed a Casson body, which behaved as a pseudoplastic in that the apparent viscosity decreased with increasing shear rate but displayed a yield stress and therefore, also resembled a Bingham plastic. The flow characteristics of a Casson body may be described by the following equation:

$$\sqrt{\tau} = \sqrt{\tau_0} + K_c \sqrt{\gamma} \qquad \text{... (13.11)}$$

where $K_c$ is the Casson viscosity.

A plot of $\sqrt{\tau}$ against $\sqrt{\gamma}$ will give a straight line, the slope of which will equal the Casson viscosity and the intercept of the $\sqrt{\tau}$ axis will equal $\sqrt{\tau_0}$. Roels claimed that the rheology of a penicillin broth could be best described in terms of a Casson body.

Therefore, to determine the rheological nature of a fluid it is necessary to construct a rheogram which requires the use of a viscometer which is accurate over a wide range of shear rates. Furthermore, the testing of mycelial suspensions may present special difficulties.

## FACTORS AFFECTING $K_La$ VALUES IN FERMENTATION VESSELS

A number of factors have been demonstrated to affect the $K_La$ value achieved in a fermentation vessel. Such factors include the air-flow rate employed, the degree of agitation, the rheological properties of the culture broth and the presence of antifoam agents. If the scale of operation of a fermentation is increased (so-called 'scale-up') it is important that the optimum $K_La$ found on the small scale is employed in the larger scale fermentation. The same $K_La$ value may be achieved in different sized vessels by adjusting the operational conditions on the larger scale and measuring the $K_La$ obtained. However, quantification of the relationship between operating variables and $K_La$ should enable the prediction of conditions necessary to achieve a particular $K_La$ value. Thus, such relationships should be of considerable value in scaling-up a fermentation and in fermenter design.

### Effect of Air-flow Rate on $K_La$

#### *Mechanically agitated reactors*

The effect of air flow rate on $K_La$ values in conventional agitated systems is illustrated in Fig. 13.12. The quantitative relationships between aeration and $K_La$ for agitated vessels are considered in the subsequent

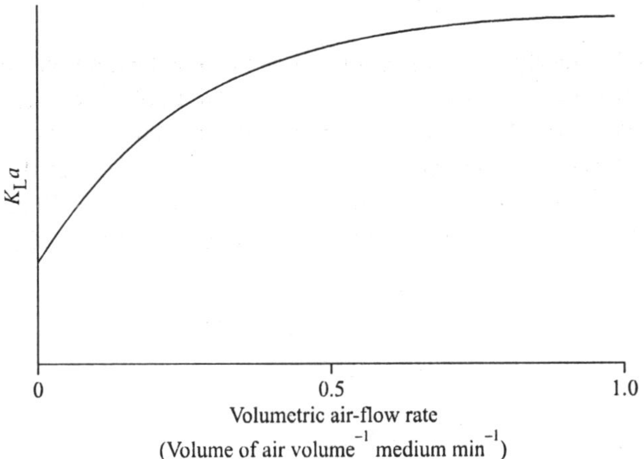


**Fig. 13.12.** The effect of air-flow rate on the $K_La$ of an agitated, aerated vessel.

section on power consumption. The air-flow rate employed rarely falls outside the range of 0.5–1.5 volumes of air per volume of medium per minute and this tends to be maintained constant on scale-up. If the impeller is unable to disperse the incoming air then extremely low oxygen transfer rates may be achieved due to the impeller becoming 'flooded'. Flooding is the phenomenon where the air-flow dominants the flow pattern and is due to an inappropriate combination of air-flow rate and speed of agitation. Nienow categorised the different flow patterns produced by a disc turbine that occur under a range of aeration and agitation conditions (Fig. 13.13). Figure 13.13(a) shows the flow profile of a non-aerated vessel of Figs. 13.13(b) to (f) the profiles with increasing air-flow rate. As air-flow rate increases the flow profile changes from one dominated by agitation [Fig. 13.13(b)] to one dominated by air-flow [Figs. 13.13(d) to (f)] until finally the air flow rate is such that the air escapes without being distributed by the agitator [Fig. 13.13(f)].

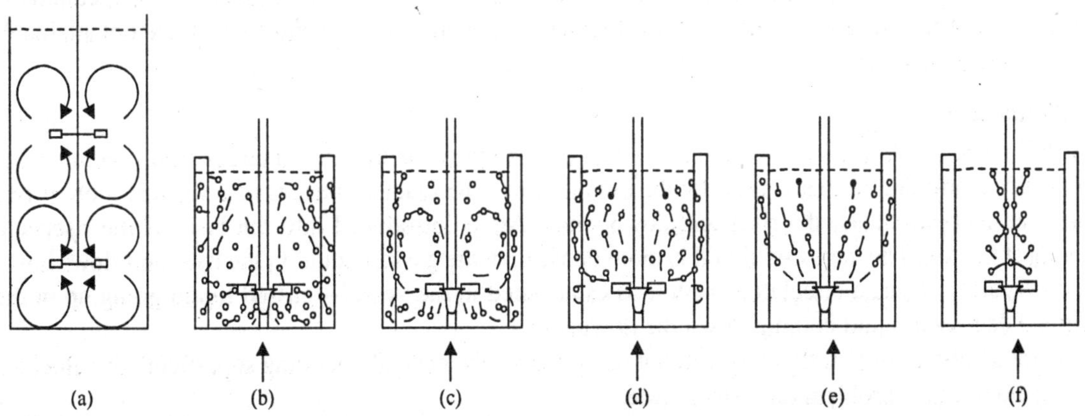


**Fig. 13.13.** The effect of air flow rate on the flow pattern in stirred vessels: (a) Non-aerated; (b) to (f), increasing air-flow rates.

### *Non-mechanically agitated reactors*

Bubble columns and air-lift reactors are not mechanically agitated and therefore, rely on the passage of air to both mix and aerate.

*Bubble columns*

The flow pattern of bubbles through a bubble column reactor is dependent on the gas superficial velocity (cm second$^{-1}$). At gas velocities of below 1–4 cm second$^{-1}$ the bubbles will rise uniformly through the medium and the only mixing will be that created in the bubble wake. This type of flow is referred to as homogeneous. At higher gas velocities bubbles are produced unevenly at the base of the vessel and bubbles coalesce resulting in local differences in fluid density. The differences in fluid density create circulatory currents and flow under these conditions is described as heterogeneous as shown in Fig. 13.14.

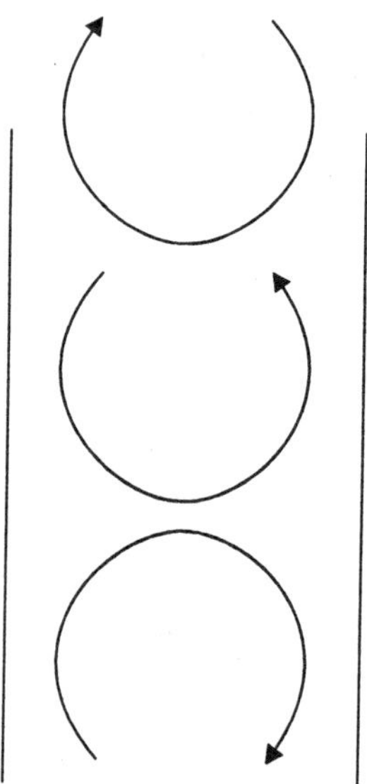

**Fig. 13.14.** Schematic representation of a heterogeneous flow regime in a bubble column.

Flooding in a bubble column is the situation when the air-flow is such that it blows the medium out of the vessel. This requires superficial gas velocities approaching 1 m second$^{-1}$ which are not attainable on commercial scales.

*Air-lift reactors*

The difference between a bubble column and an air-lift reactor is that liquid circulation is achieved in the air-lift in addition to that caused by the bubble flow. The reactor consists of a vertical loop of two connected compartments, the riser and downcomer. Air is introduced into the base of the riser and escapes at the top. The degassed liquid is more dense than the gassed liquid in the riser and flows down the downcomer. Thus, a circulatory pattern is established in the vessel—gassed liquid going up in the riser and degassed liquid coming down the downcomer.

As for a bubble column flooding will not occur within the normal operating superficial air velocities and should not be a problem on a large scale.

## Effect of the Degree of Agitation on $K_La$

The degree of agitation has been demonstrated to have a profound effect on the oxygen-transfer efficiency of an agitated fermenter. The agitation assisted oxygen transfer in the following ways:

1. Agitation increases the area available for oxygen transfer by dispersing the air in the culture fluid in the form of small bubbles.

2. Agitation delays the escape of air bubbles from the liquid.
3. Agitation prevents coalescence of air bubbles.
4. Agitation decreases the thickness of the liquid film at the gas-liquid interface by creating turbulence in the culture fluid.

The degree of agitation may be measured by the amount of power consumed in stirring the vessel contents. The power consumption may be assessed by using a dynamometer, by using strain gauges attached to the agitator shaft and by measuring the electrical power consumption of the agitator motor. The assessment of electrical consumption is suitable only for use with large-scale vessels.

## Effect of Medium and Culture Rheology

The objective of this section is to discuss the effects of medium and culture rheology on oxygen transfer during a fermentation. A fermentation broth consists of the liquid medium in which the organism grows, the microbial biomass and any product which is secreted by the organism. Thus, the rheology of the broth is affected by the composition of the original medium and its modification by the growing culture, the concentration and morphology of the biomass and the concentration and rheological properties of the microbial products. Therefore, it should be apparent that fermentation broths vary widely in their rheological properties and significant changes in broth rheology may occur during a fermentation.

### *Medium rheology*

Fermentation media frequently contain starch as a carbon source which may render the medium non-Newtonian and relatively viscous. However, as the organism grows it will degrade the starch and thus modify the rheology of the medium and reduce its viscosity. The rheological problems presented by the medium are minor compared with those presented by a high mycelial biomass, especially when it is considered that the total oxygen demand is relatively low in the early stages of a fermentation. However, it is worth remembering that in non-viscous unicellular fermentations the highest power draw will occur when the medium is sterilised *in situ* when the vessel is not being aerated and this will correspond with the time when a starch-based medium is at its most viscous.

It was observed that the medium composition could affect the rheological properties of mycelium suspensions by affecting the interactions between the hyphae.

### *Effect of microbial biomass*

#### *Agitator design for non-Newtonian fermentations*

The biomass concentration and its morphological form in a fermentation has been shown to have a profound effect on oxygen transfer. Most bacterial and yeast fermentations tend to give rise to relatively non-viscous Newtonian broths in which conditions of turbulent flow may be achieved. Such fermentations present relatively few oxygen-transfer problems. However, the highly viscous non-Newtonian broths of fungal and streptomycete fermentations present major difficulties in oxygen provision, the productivities of many such fermentations being limited by oxygen availability.

#### *Manipulation of mycelial morphology*

The biomass of mycelial organisms grown in submerged culture may vary from the filamentous type, in which the hyphae form a homogeneous suspension dispersed through the medium, to the 'pellet' type consisting of compact, discrete masses of hyphae. The filamentous form tends to give rise to a highly viscous, non-Newtonian broth whereas the pellet form tends to produce an essentially Newtonian system with a much lower viscosity making oxygen transfer much easier. The morphological form of a mycelial organism in submerged culture has a major effect on the broth rheology and may, therefore, be expected to influence aeration efficiency.

### *Effect of microbial products on aeration*

Generally speaking, the product of a fermentation contributes relatively little to the viscosity of the culture broth. However, the exception is the production of bacterial polysaccharides, where the broths tend to be highly viscous (30000 cp) and non-Newtonian. Charles demonstrated that the bacterial cells in a polysaccharide fermentation made a minimal contribution to the high culture viscosity which was due primarily to the polysaccharide product. Normally, microbial polysaccharides tend to behave as pseudoplastic fluids, although some have also been shown to exhibit a yield stress. The yield stress of a polysaccharide can make the fermentation particularly difficult because, beyond a certain distance from the impeller, the broth will be stagnant and productivity in these regions will be practically zero. Thus, bacterial polysaccharide fermentations present problems of oxygen transfer and bulk mixing similar to those presented by mycelial fermentations. Thus, similar stirrer configurations to those discussed in the previous section have been used in polysaccharide fermentations. Gallindo and Nienow investigated the behaviour of a hydrofoil impeller, the Lightnin' A315, in a simulated xanthan fermentation. These workers adopted Metzner and Otto's approach to construct power curves. Better agitator performance was achieved when its pumping direction was upwards rather than downwards resulting in lower power loss on aeration and less torque fluctuations. It was concluded that such agitators may give improved mixing in a xanthan fermentation provided that the polysaccharide concentration is below 25 kg $m^{-3}$.

## Effect of Foam and Antifoams on Oxygen Transfer

The high degree of aeration and agitation required in a fermentation frequently gives rise to the undesirable phenomenon of foam formation. In extreme circumstances the foam may overflow from the fermenter via the air outlet or sample line resulting in the loss of medium and product, as well as increasing the risk of contamination. The presence of foam may also have an adverse effect on the oxygen-transfer rate. It was pointed out that Waldhof and vortex-type fermenters were particularly affected due to the bubbles becoming entrapped in the continuously recirculating foam, resulting in high bubble residence times and therefore, oxygen-depleted bubbles. The presence of foam in a conventional agitated, baffled fermenter may also increase the residence time of bubbles and therefore result in their being depleted of oxygen. Furthermore, the presence of foam in the region of the impeller may prevent adequate mixing of the fermentation broth. Thus, it is desirable to break down a foam before it causes any process difficulties and this may be achieved by the use of mechanical foam breakers or chemical antifoams. However, mechanical foam control consumes considerable energy and is not completely reliable so that chemical antifoams are preferred.

All antifoams are surfactants and may, themselves, be expected to have some effect on oxygen transfer. The predominant effect observed by most workers is that antifoams tend to decrease the oxygen-transfer rate. Antifoams cause the collapse of bubbles in foam but they may favour the coalescence of bubbles within the liquid phase, resulting in larger bubbles with reduced surface area to volume ratios and hence a reduced rate of oxygen transfer. Thus, a balance must be struck between the necessity for foam control and the deleterious effects of the controlling agent. Foam formation has a particular influence on the liquid height in the fermenter at which it is practical to operate. If inadequate space is provided above the liquid level for foam control, then copious amounts of antifoam must be used to prevent loss of broth from the vessel.

## BALANCE BETWEEN OXYGEN SUPPLY AND DEMAND

Both the demand for oxygen by a micro-organism and the supply to the organism by the fermenter have been considered in this chapter. This section attempts to bring these two aspects together and considers how processes may be designed such that the oxygen uptake rate of the culture does not exceed the oxygen transfer rate of the fermenter.

It will also be recalled that the dissolved oxygen concentration during the fermentation should not fall below the critical dissolved oxygen concentration ($C_{crit}$) or the dissolved oxygen concentration which gives optimum product formation. Thus, it is necessary that the oxygen-transfer rate of the fermenter matches the oxygen uptake rate of the culture whilst maintaining the dissolved oxygen above a particular concentration. To balance supply and demand conditions of fermenter it should be adjusted to match the supply. This may be achieved by:

1. Controlling biomass concentration.
2. Controlling the specific oxygen uptake rate.
3. A combination of 1 and 2.

## SCALE-UP AND SCALE-DOWN

Scale-up means increasing the scale of a fermentation, for example from the laboratory scale to the pilot plant scale or from the pilot plant scale to the production scale. Increase in scale means an increase in volume and the problems of process scale-up are due to the different ways in which process parameters are affected by the size of the unit. It is the task of the fermentation technologist to increase the scale of a fermentation without a decrease in yield or if a yield reduction occurs, to identify the factor which gives rise to the decrease and to rectify it. The major factors involved in scale-up are:

1. Inoculum development. An increase in scale may mean the extra stages have to be incorporated into the inoculum development programme.
2. Sterilisation. Sterilisation is a scale dependent factor because the number of contaminating micro-organisms in a fermenter must be reduced to the same absolute number regardless of scale. Thus, when the scale of a process is increased the sterilisation regime must be adjusted accordingly, which may result in a change in the quality of the medium after sterilisation.
3. Environmental parameters. The increase in scale may result in a changed environment for the organism. These environmental parameters may be summarised as follows:
   (a) Nutrient availability.
   (b) pH.
   (c) Temperature.
   (d) Dissolved oxygen concentration.
   (e) Shear conditions.
   (f) Dissolved carbon dioxide concentration.
   (g) Foam production.

All the above parameters are affected by agitation and aeration, either in terms of bulk mixing or the provision of oxygen. Points a, b, c and e are related to bulk mixing whist d, e, f and g are related to air-flow and oxygen transfer. Thus, agitation and aeration tends to dominate the scale-up literature. However, it should always be remembered that inoculum development and sterilisation difficulties may be the reason for a decrease in yield when a process is scaled up and that achieving the correct aeration/agitation regime is not the only problem to be addressed.

### Scale-up of Aeration/Agitation Regimes in Stirred Tank Reactors

From the list of environmental parameters affected by aeration and agitation it will be appreciated that it is extremely unlikely that the conditions of the small-scale fermentation will be replicated precisely on the large scale. Thus, the most important criteria for a particular fermentation must be established and the scale-up based on reproducing those characteristics. The scale-up window represents the boundaries imposed by the environmental parameters and cost on the aeration/agitation regime and is shown in Fig. 13.15. Suitable conditions of mixing and oxygen transfer can be obtained with a range of aeration/agitation combinations. The two axes of Fig. 13.15 are agitation and aeration and the zone within the hexagon represents suitable aeration/agitation regimes. The boundary of the hexagon is defined by the limits of oxygen supply, carbon dioxide accumulation, shear damage to the cells, cost, foam formation and bulk mixing. For example, the agitation rate must fall between a minimum and maximum value—mixing is inadequate below the minimum level and shear damage to the cells is too great above the maximum value. The limits for aeration are determined at the minimum end by oxygen limitation and carbon dioxide accumulation and at the maximum end by foam formation. The shape of the window will depend on the fermentation—for example, the supply of oxygen would be irrelevant in an anaerobic fermentation, whereas the limitation due to shear would be of major importance in the scale-up of animal cell fermentations.

The solution of the scale-up problem is three-fold:

1. The identification of the principal environmental domain affected by aeration and agitation in the fermentation, e.g., oxygen concentration, shear, bulk mixing.
2. The identification of a process variable (or variables) which affects the identified environmental domain.
3. The calculation of the value of the process variable to be used on the large scale which will result in the replication of the same environmental conditions on both scales.

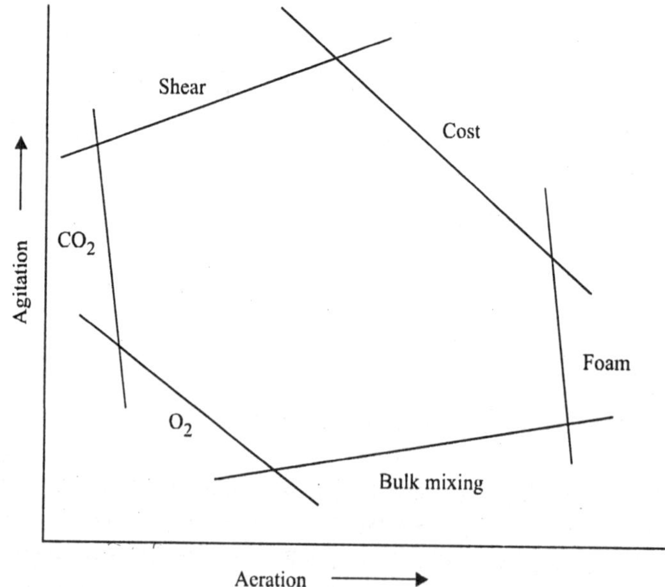


**Fig. 13.15.** The 'scale-up' window defining the operating boundaries for aeration and agitation in the scale-up of a fermentation.

The process variables which affect mixing and mass transfer are summarised in Table 13.6. Thus, if dissolved oxygen concentration is perceived as the overriding environmental condition then power consumption per unit volume and volumetric air-flow rate per unit volume should be maintained constant on scale-up. However, as a result, the other parameters will not be the same in the larger scale and therefore, neither will the environmental factors which they influence. This phenomenon is summarised in Table 13.7 where a 125 fold increase in scale is represented. If power consumption per unit volume is kept constant then impeller tip speed (i.e., shear) increases and flow $min^{-1}$ $vol^{-1}$ (i.e., mixing) decreases. If mixing is kept constant, an enormous (and totally uneconomic) increase in power is required and shear increases 5 fold. If impeller tip speed (shear) is kept constant then power consumption (hence, $K_La$) and mixing decrease. This analysis indicates that it is economically impossible to maintain the same degree of mixing on scale-up and therefore, a decrease in yield may be due to mixing anomalies.

**Table 13.6.** The effect of process variables on mass transfer or mixing characteristics.

| *Process variable* | *Mass transfer or mixing characteristic affected* |
|---|---|
| Power consumption per unit volume | Oxygen-transfer rate |
| Volumetric air flow rate | Oxygen-transfer rate |
| Impeller tip seed | Shear rate |
| Pumping rate | Mixing time |
| Reynolds number | Heat transfer |

**Table 13.7.** The effect of the choice of scale-up criteria on operating conditions in the scaled-up vessel. Based on scale-up from 80 $dm^3$ to $10^4$ $dm^3$.

| *Criterion used in scale-up* | *Effect on the operating conditions on the large scale (large scale value/small scale value)* | | | |
|---|---|---|---|---|
| | *P* | *P/V* | *Flow $min^{-1}$ $vol^{-1}$* | *$ND_i$* |
| P/V | 125.0 | 1.0 | 0.34 | 1.7 |
| Flow $min^{-1}$ $vol^{-1}$ | 3125.0 | 25.0 | 1.0 | 5.0 |
| $ND_i$ (impeller tip speed) | 25.0 | 0.2 | 0.2 | 1.0 |
| Reynolds number | 0.2 | 0.0016 | 0.04 | 0.2 |

The most important environmental domains affected by aeration and agitation for the majority of fermentations are oxygen concentration and shear. Thus, the most widely used scale-up criteria are the maintenance of a constant $K_La$ or constant shear conditions. Constant shear may be achieved by scaling up on the basis of constant impeller tip speed. Constant $K_La$ may be achieved on the basis of constant power consumption per unit volume and constant volumetric air-flow rate. The operating variable dictating constant power consumption in geometrically similar vessels is the agitator speed. The agitator speed on the large scale is then calculated from the correlations between $K_La$ and power consumption and between power consumption and operating variables. The accuracy of these scale-up techniques is only as good as the power and $K_La$ correlations, so it is worth expending some considerable time to test the validity of potential correlations for the fermentation in question.

### Scale-up of Air-lift Reactors

Bubble columns and air-lift vessels tend to be scaled-up on the basis of geometric similarity and constant gas velocity. Under these conditions the $K_La$ and shear rate in the two scales will be similar. The major difference will be the height of the vessels resulting in increased pressure at the base of the larger vessel. This would result in higher oxygen and carbon dioxide solubility which would give a higher $K_La$ but might result in carbon dioxide inhibition. The other problem in the scale-up of air-lift systems is that the organism is exposed to extremes of oxygen levels in the riser and downcomer and the effects of these conditions should be investigated on the laboratory scale.

### Scale-down Methods

Scale-down is the situation where laboratory or pilot-scale experiments are conducted under conditions which mimic the industrial-scale conditions. This approach is important in both the development of a new product and the improvement of an existing full-scale fermentation. Frequently, conditions achievable on a laboratory scale are impractical on an industrial scale, which means that if inappropriate conditions have been used in the laboratory unrealistic yield objectives may be set for the scaled-up process. The aspects to consider in the design of laboratory or pilot-plant experiments in the context of scale-down may be summarised as follows:

#### *Medium design*

Media relevant to the industrial situation should be used in development experiments.

#### *Medium sterilisation*

If the medium is to be batch sterilised on the large scale its exposure time at a high temperature will be much greater than that experienced in the laboratory or pilot plant. Thus, the sterilisation times on the smaller scales should be increased to mimic the industrial situation. Alternatively, medium sterilised in the production fermenter may be used in the laboratory and pilot plant. This highlights the advantage of continuous sterilisation where little loss of medium quality occurs. Furthermore, the same continuous steriliser may be used for both full-scale and pilot scale vessels.

#### *Inoculation procedures*

Due to a range of circumstances, it may not always be possible to inoculate every production fermentation with inoculum in optimum condition. The scale-down approach can be used to predict the consequences of such events by mimicking these situations in the laboratory, for example by storing inoculum or using inocula of different ages.

#### *Number of generations*

An industrial scale fermentation requires a greater number of generations than does a laboratory one; this may place more severe stability criteria on the process strain than may have been appreciated on the small scale. The industrial situation may be modelled in the laboratory by using serial sub-culture to ensure that the strain is sufficiently stable. This approach is particularly pertinent in the development of recombinant fermentations.

#### *Mixing*

As indicated in the previous section it is almost inevitable that the degree of mixing will decrease with an increase in scale. Thus, it is possible to model inadequate mixing in the laboratory by subjecting the organism to pulse medium feeds or fluctuating process conditions such as oxygen concentration, pH and temperature. Such scaled-down experiments then allow predictions to be made about the suitability of new strains for industrial exploitation.

### *Oxygen transfer rate*

Far higher oxygen transfer rates can be achieved in laboratory fermenters than in industrial-scale ones. Thus, unrealistic demands may be made of a fermentation plant if the development work has been done at very high oxygen-transfer rates. Therefore, the laboratory and pilot fermenters should reflect the oxygen transfer rates achievable in the full-scale fermenters.

The adoption of these simple approaches to small scale experimentation can prevent many scale-up problems before they even occur.

Chapter 14

# Optimisation of Fermentation Processes by Quantitative Analysis

## INTRODUCTION

The use of microbial fermentations for the production of 'bulk fine chemicals' is a well-established technology and began in the early 1940s with the development of large-scale processes for the industrial production of products as diverse as organic acids, enzymes and antibiotics.

Since the pioneering work with penicillins, many more antibiotics and related secondary product processes have emerged and been commercialised. Microbial fermentations have also been developed for the production of a wide range of pharmaceutical products and enzymes for use in the food and detergent industries. No accurate figures are generally available for the scale of production of these products, but it is certain that microbial fermentations play an important role in contemporary manufacturing practice; while the expression of 'foreign' proteins with biomedical importance has been achieved in sheep (and the products extracted from their milk), it is unlikely that 1,00,000 litre fermentors will be rapidly replaced by flocks of cloned sheep as 'factories' for the production of such potent biological molecules.

## MICROBIAL FERMENTATION AS A CHEMICAL PROCESS

Today there are new demands and emerging horizons for the rapid development of new processes to a commercial scale. These include the widespread emergence of microbial resistance to older groups of antibiotics, the interest in quite novel biological activities in the microbial fauna of exotic natural sources (such as deep sea sediments, tropical rain forests and high temperature geological sites) and the environmentally (and economically) driven interest in using microbes to produce aromatics and other chemical building blocks from the renewable resource of photosynthetically-derived sugars.

Fermentation processes were initially devised and improved without the aid of modern analytical methods and techniques, mostly by random selection of mutagenised strains and short-term empirical improvements to nutrient media, feeding strategies, fermentor design and process operation. Analytical biochemistry has made great advances in the last 60 years and it is now routine for laboratories to have ready access to powerful analytical facilities such as:

1. High performance liquid chromatography (HPLC).
2. Capillary gas chromatography (GC).

3. Combined chromatography/mass spectrometry (LC-MS, GC-MS).
4. Ion chromatography (IC).
5. Ion-selective electrodes.
6. Enzymatic assays for carbohydrates, carboxylic acids, etc.
7. Capillary electrophoresis (CE).
8. Supercritical fluid extraction and chromatography (SFE/SFC).
9. Two-dimensional polyacrylamide gel electrophoresis (2D-PAGE).

The advent of such techniques provides an alternative to the earlier 'black box' technologies where only the results of changes are assessed without any basis of understanding of the causes. In this chapter, a general method is outlined for the optimisation of fermentation processes using an analytical approach. The framework of this method consists of the following stages:

1. Accurate quantitative analysis of the important chemical and biochemical events which are related to growth and product formation.
2. Identifying target areas for process improvement.
3. Devising practical means to change these key fermentation events.
4. Understanding the metabolic consequences of these changes.

Although biological systems are complex, that complexity is not a barrier to practical intervention. Biochemical teaching laboratories display impressive metabolic charts which include most, if not all, of the major biosynthetic pathways, with all the multiplicity of catabolic and anabolic pathways and their interactions. Direct analysis of what actually happens and from a knowledge of the overall pathways involved, what must occur identifies the key targets for rational intervention and indicates which areas of the biochemical map require further investigation. This view of fermentation processes is derived from systems analysis: inputs (nutrients) and outputs (products) are linked by the generation and functioning of the biomass (the microbial cells). The focus is, therefore, on the process as a series of major events rather than on its fine details (Fig. 14.1). Process optimisation proceeds from what can be measured and does not require a complete understanding of either the molecular biology of the organism or the biosynthetic pathway of the product.

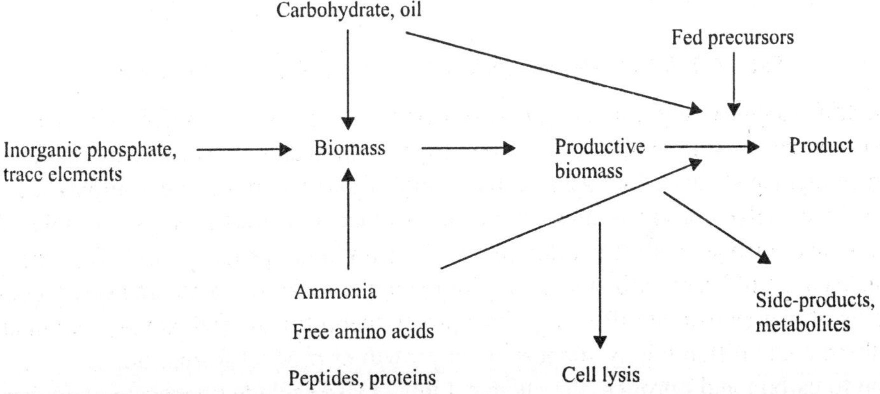


**Fig. 14.1.** Analytical targets in fermentation process improvement.

## UTILISATION OF FERMENTATION INPUTS

The details of industrial manufacturing processes are confidential. For some processes, however, a wealth of practical information has been made available; for the purposes of illustration, therefore, reference will be made, where appropriate, to penicillin processes using data from or quoted in the sources listed at the end of this chapter. Penicillin is unusual in that it has a comparatively simple and well-defined biosynthetic pathway (Fig. 14.2); in most other respects, it is a good example of the older generation of antibiotic-producing fermentations. Progress curves showing input utilisation, growth and product accumulation are computer simulations which exemplify those found in different fermentation processes at scales of production ranging from shake flask cultures to 100 $m^3$ vessels; the length of the hypothetical idealised process (7 days) is chosen to be intermediate between the shorter fermentations with bacterial species and the lengthier processes using yeasts and fungi.

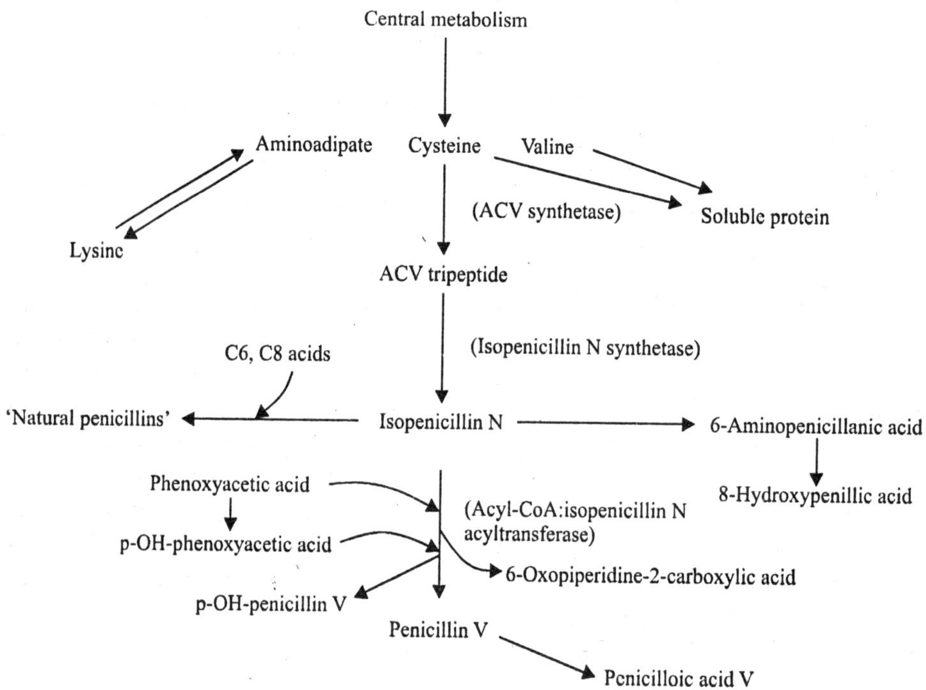


**Fig. 14.2.** Metabolite accumulation in a penicillin V fermentation.

On an industrial manufacturing scale it is seldom economical to use chemically defined media (for example, a sugar carbon source, an ammonium salt as the nitrogen source and a few other inorganic compounds such as a 'trace elements' mix). Inexpensive but complex carbon and nitrogen inputs are commonly used which are themselves products or by-products of other industrial processes (Table 14.1). More than one carbon and nitrogen source is often required for optimum product yield. To some extent, this can be rationalised as different carbon sources (for example, a carbohydrate and oil) feeding into the major pathways of metabolism at different points; protein peptides and free amino acid inputs are also utilised and these could differentially support either growth or product formation.

In addition to carbon and nitrogen, the chemical inputs also include phosphate and other inorganic ions and trace elements (iron, zinc, cobalt, molybdenum, etc.). The necessity for so complex a mix of nutrients can be assessed using this analytical approach which will establish the contribution of each additive to biomass and product formation; this allows the medium to be simplified and optimised. The medium developed for an existing production process is the result of a (probably lengthy) series of empirical trials and it may contain some important nutrients which are poorly utilised and which could

**Table 14.1.** 'Complex' carbon and nitrogen sources for industrial fermentations.

| *Carbon* | *Nitrogen* | |
|---|---|---|
| *Starch* | Plant proteins | Seed meals (20–50% protein) |
| (Corn, potato, barley, wheat, | Plant proteins | Seed flours (>50% protein) |
| oat, rye, artichoke) | Plant proteins | Protein isolates (>70% protein) |
| *Starch hydrolysates* | | |
| 2–30% glucose, | Protein hydrolysates | Enzymic (5.5–13.5% N) |
| 3–30% 'polysaccharide' | Protein hydrolysates | Acid (4–13.5% N) |
| | Yeast extracts | Baker's yeasts |
| *Molasses* | Yeast extracts | Brewer's yeasts |
| Sugar beet | Yeast extracts | Wine yeasts |
| Sugar cane ('blackstrap', refinery, 'high test') | Yeast extracts | 6–11.5% N |
| 20–80% sucrose, 0.1–3% N | | |
| | Corn steep liquors | 3–4% N |
| | Corn steep powders | 7–8.5% N |

be replaced with immediate advantage. To this extent, the medium may, therefore, be the 'message', in the sense of the famous sound-bite from the 1960s. At the very least, the medium is the first focus for analytical scrutiny as heat sterilisation may itself 'damage' some of the nutrients—the best known example is the Maillard reaction between sugars and amino compounds which generates large amounts of unwanted and probably unusable degradation products. The key ability required is that of accurately measuring the utilisation of inputs, both quantitatively and as part of a temporal sequence of events initiated when the medium is inoculated with cells generated in the previous 'seed' stage.

## Carbon Sources

### *Carbohydrates*

Glucose and other monosaccharides and simple disaccharides (such as sucrose and lactose) are readily determined by HPLC or by chemical and enzymic assays. Readily utilisable sugars such as glucose frequently exert 'carbon catabolite' inhibition of secondary product formation and once the sugars in a batched medium are exhausted, must be carefully fed to avoid their accumulation. It is likely that glucose feeds will be routinely monitored on-line in the near future with amperometric sensors based on glucose oxidase.

Starch hydrolysates and polysaccharide inputs present much greater analytical difficulties. Straightforward test-tube chemical and enzymatic assays can measure the overall utilisation of the carbohydrate (Fig. 14.3). This type of carbohydrate input is, however, a population of molecular species with different molecular weights: in starch hydrolysates, for example, the average glucan oligomer may be maltoheptose, but different methods of manufacture can give variable mixtures of oligosaccharides. This is important because micro-organisms utilise the different oligomers at quite different rates (Fig. 14.3). Part of this utilisation is enzymic, i.e., the breakdown of oligoglucans to either free glucose or (if the organism cannot take up glucose) maltose or maltotriose. For example, amyloglucosidase action liberates free glucose from glucans with α-1,4 and (at a slower rate) α-1,6 linkages, while α-amylase hydrolyses larger oligomers but only at α-1,4 linkages and pullulanase liberates large α-1,4 glucans by breaking α-1,6 linkages.

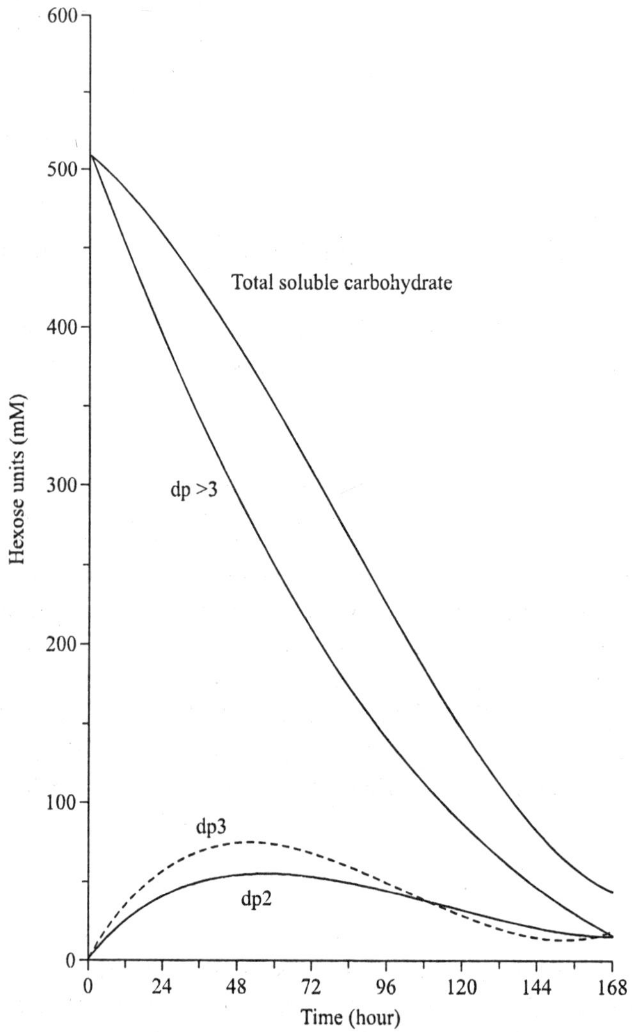


**Fig. 14.3.** Carbohydrate utilisation in a medium containing a starch hydrolysate.

Starch requires some pre-treatment before inclusion in nutrient media otherwise semi-solid preparations could result after heat sterilisation. Amylase action breaks down the polymeric starch but only incompletely, leaving much of the carbohydrate in the form of large oligomers, some of which represent the 'limit dextrins' of classical carbohydrate chemistry and resist further enzymic attack at the α-1,6 branch points. Efficient starch utilisation during the fermentation requires, therefore, three separate enzyme activities—amylase, amyloglucosidase and pullulanase—generated and secreted into the medium by the biomass. The biological entity in the fermentation is thus itself an important part of input processing before the substrates enter the cell.

Given the results of Fig. 14.3, several questions would immediately arise with implications for process development and strain improvement:

1. Is the initial accumulation of small oligosaccharides a sign of a limitation on growth by another input (most likely in the nitrogen sources)?
2. Conversely, would a higher starting content of small oligosaccharides give faster growth in a fermentation with sufficient nitrogen?
3. Would feeding a specific oligosaccharide (for example, maltose) maintain productivity?
4. What carbohydrate 'processing' enzymes can the cells generate and do strains of differing productivities give different spectra of enzymes?

#### *Oils*

Plant-derived oils (from monoculture crops such as soyabean) are cost-effective carbon sources available on a large scale; these also find uses as metabolisable anti-foaming agents. The mixtures of triglycerides in oils can be analysed by chromatographic (HPLC, GC, SFC or gel permeation) or enzymatic methods for the intact triglycerides or the fatty acids liberated from them by lipase action. Oils could be included in the category of complex carbon sources because their exact chemical compositions are variable and reflect their geographical source.

Lipid metabolism has the disadvantage that the bioenergetics of fatty acid breakdown are adverse in comparison with carbohydrate utilisation; the oxygen requirements for fatty acid oxidation are very high and can result in oxygen depletion in the medium. Oleic acid is an alternative to oils but, because of the toxicity of long-chain fatty acids to many microbial species, must be fed and its levels measured accurately to modulate the feed and avoid the accumulation of the unused fatty acid in the fermentation.

#### *Other carbon sources*

Ethanol, some alkanes (chain length 12–18) and simple carboxylic acids such as acetic acid are readily utilised by some micro-organisms. In penicillin processes where corn steep liquor is a major component of the medium, high levels of lactic acid are present (corn steep liquors are products of fermentations in which most of the carbohydrate is converted to lactic acid); the lactate is a minor carbon source but is known to be metabolised by *Penicillium chrysogenum*.

The supply of multiple carbon sources is common in manufacturing practice and both oil and carbohydrates can be metabolised at the same time. This facility to simultaneously utilise different carbon substrates may be of great value because there is a growing family of antibiotics with complex structures derived from clearly different areas of primary metabolism. Polyene macrolides and vancomycins, for example, have backbone structures derived from acetate units (best supplied as an oil or as acetic acid) to which aminated sugars (readily formed from glucose) are attached. Microbial biochemistry 'allows' both acetate and glucose-derived precursors to be generated from either carbohydrate or lipid inputs but maximum productivity and yield might depend on a precise balance of the two types of carbon input. Analysis would show how the utilisation of the inputs proceeds kinetically at different stages of the process.

### Nitrogen Sources

#### *Proteins*

As with oils, plant seed proteins are major fermentation inputs on the industrial scale (in fact, the proteins themselves are crude by-products of the extraction of the more valuable oil from the seeds).

Proteins inputs are supplied in four forms:

1. Macromolecular (undigested) in the starting medium.
2. Digested by a protease in the starting medium.
3. Undigested in a single or combined feed.
4. Proteolytically digested in a single or combined feed.

When used, enzymic digestion may proceed only as far as the generation of peptides of variable molecular weight. Feeds of intact or digested proteins may be with carbon or other nitrogen sources and this is usually determined by the requirement either to maintain optimum C:N ratios or to avoid providing substrates which might encourage excessive growth once product formation has commenced.

Protein utilisation is a deceptively complex topic for analysis. The breakdown of protein during the course of the fermentation is easily followed by a standard macromolecular protein assay (Fig. 14.4).

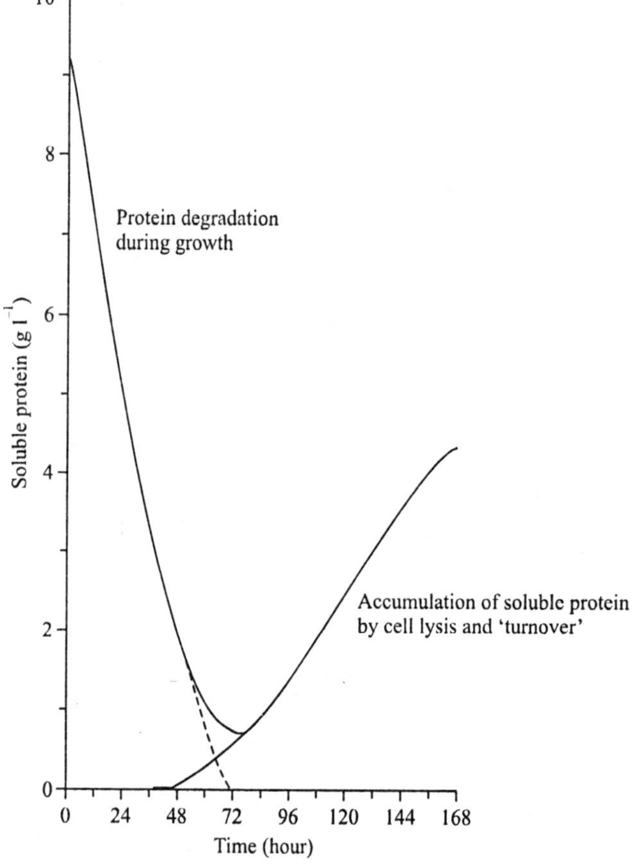


**Fig. 14.4.** Protein utilisation in a fermentation with protein in the starting medium.

Soluble peptides and free amino acids may, however, accumulate and their utilisation might be kinetically quite different from that of a protein (Fig. 14.5). Results such as those contained in Figs. 14.4 and 14.5 suggest various lines for more detailed investigation:

1. Soluble protein was readily broken down to peptides and free amino acids but how much of the protein in the starting medium was insoluble and how much of this insoluble protein became soluble during the process?

2. The end of protein utilisation just after 70 hours was a crucial event in the process—what happened then?
3. If the increase in soluble protein, peptides and free amino acids in the second half of the fermentation was caused by cell lysis, was this a consequence of the protein in the starting medium being fully utilised?

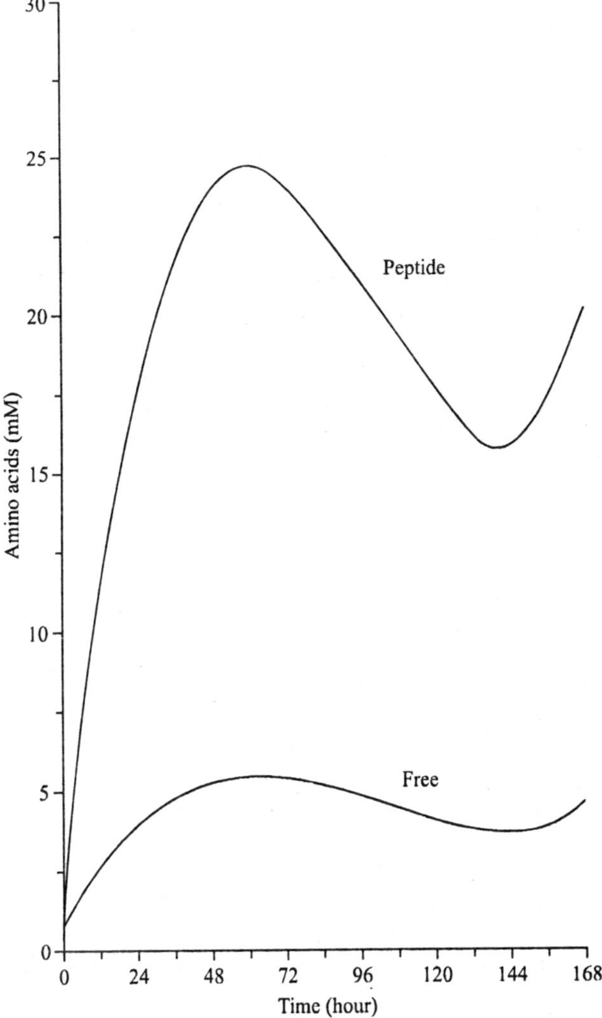


**Fig. 14.5.** Accumulation of soluble amino acids and peptides by degradation in a protein-containing medium.

The extent of the utilisation of a protein in a fermentation is usually measured only indirectly. If the biomass can be grown on a medium lacking protein inputs, the protein level in the cells expressed per unit biomass represents a baseline with which the complex medium fermentation can be compared. Protein contents in the biomass (in practice, the sedimentable material) approaching that found in the 'pure' biomass indicate complete utilisation of that portion of the protein input which was insoluble, while persistently higher values suggest that the protein input is never fully used or, at least, solubilised (Fig. 14.6). The varying degrees of protein utilisation presented in Fig. 14.6 could simply reflect the growth rates in the two fermentations, but the abilities of different strains to utilise macromolecular proteins and differences in the utilisation rates of grades of any protein source would yield similar results.

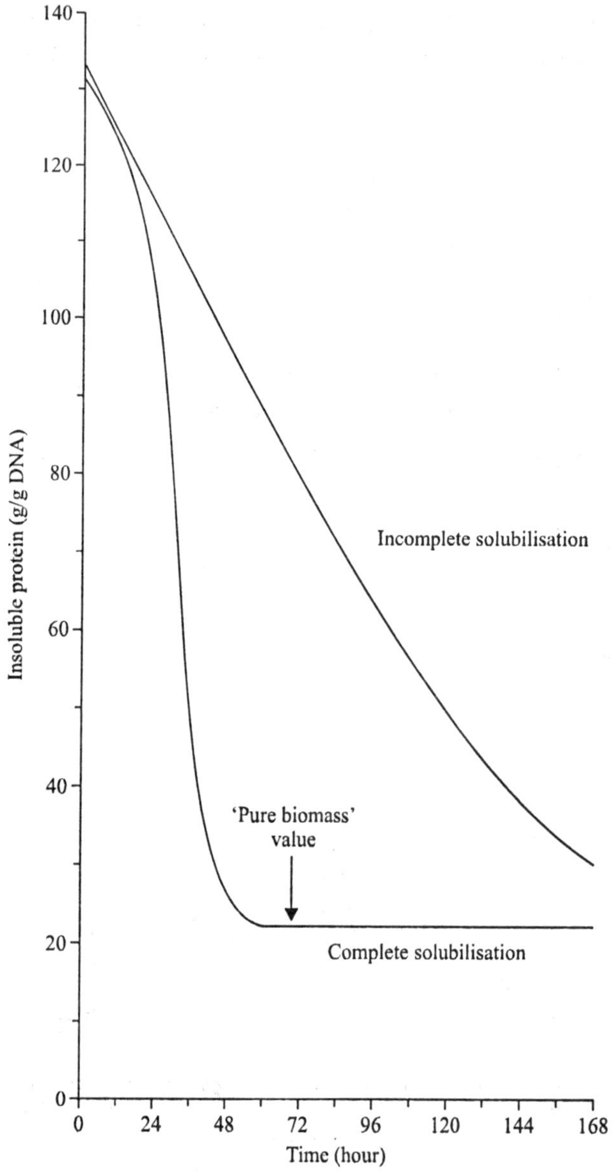


**Fig. 14.6.** Utilisation of insoluble protein from the starting medium in two fermentations.

### *Peptides and free amino acids*

'Peptidic' nitrogen sources such as yeast extracts and corn steep liquors (see Table 14.1) are heterogeneous in that they also contain some macromolecular protein and also low levels of free amino acids (only some casein digests contain most of the nitrogen as free amino acids without any macromolecular protein). Peptidic nitrogen sources are preferable to proteins as feeds because their amino nitrogen is already in a biologically more accessible form. Their quantitative determination requires acid hydrolysis of samples from which the proteins have been precipitated followed by correction of the total amino acid content for the (separately measured) free amino acids (Fig. 14.5). Free amino acids can be used as fermentation inputs but because of their cost are more often supplied in nitrogen sources such as yeast extract and corn steep liquor. The contents of the individual free amino acids are in these, however, quite variable and some may be present in very low concentrations. This is especially true for glutamine (which may have been cyclised to pyroglutamic acid) and the aromatic amino acids phenylalanine, tyrosine and tryptophan. Non-protein amino compounds are also present, some at quite high levels in corn steep liquors (ornithine, 4-aminobutyric acid, β-alanine, amino-alcohols, etc.).

When free amino acids are present in the starting medium, their rates of utilisation are highly variable; some may even be exhausted while others are still present in readily detectable amounts (Fig. 14.7). This is mostly due to the mismatching of the amino acids which are supplied to the inoculum with those required for the generation of new cells (this is a general feature of nitrogen sources for microbial growth and applies equally well to peptides and proteins over the course of the fermentation). Some amino acids are accumulated inside the biomass rather than simply polymerised into proteins (for example, proline and glutamate have roles in osmoprotection and both glutamate and glutamine are required for biosynthetic amination reactions) and the rate of their disappearance from the extracellular phase need not correlate with utilisation rate—this is a major inaccuracy in attempts to measure metabolic flux rates from analytical data obtained solely with culture filtrates.

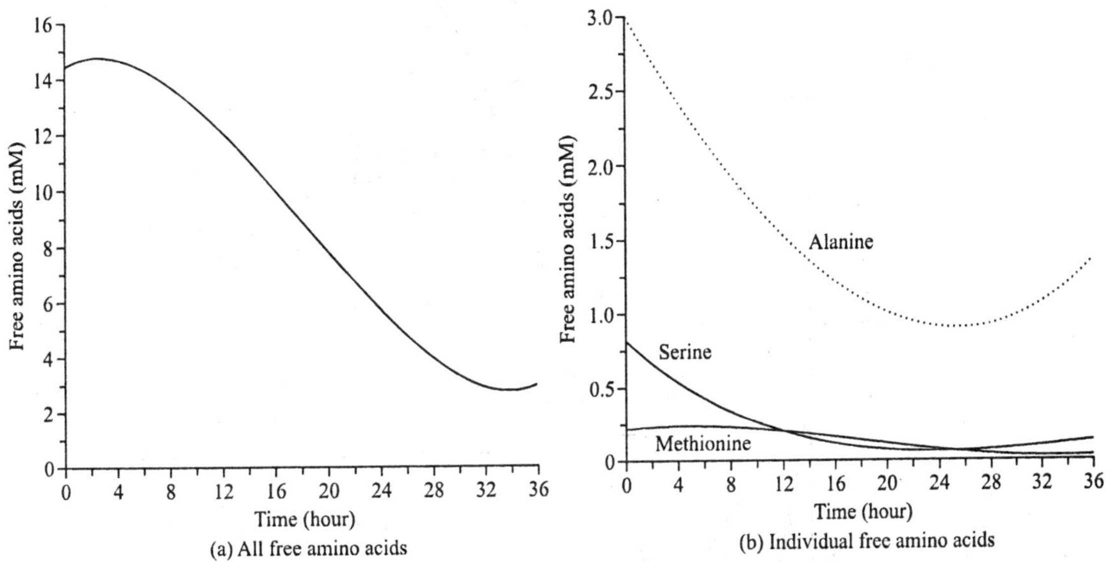


**Fig. 14.7.** Utilisation of free amino acids from a fermentation medium.

### *Ammonia*

Ammonia is the most important nitrogen source for secondary product formation because most of the nitrogen incorporated into organic compounds occurs by amination reactions using glutamate or glutamine as amino donors. Ammonia may be included in the starting medium in the form of ammonium salts (such as ammonium sulphate) and in complex nitrogen sources and some free ammonia is liberated by the partial degradation of the complex nitrogen sources during heat sterilisation. Ammonium salts may also be fed or gaseous ammonia supplied as a buffering agent when the accumulation of acids (or the uptake of ammonium ion in the starting medium) acts to depress the pH.

Industrial micro-organisms vary greatly in their tolerance to high ammonia concentrations and the monitoring of the ammonia level during the course of fermentation is very important. Figure 14.8 shows two contrasting free ammonia profiles. In the first, the level of free ammonia rises as amino acids are deaminated and the carbon skeletons used for biosynthesis and/or catabolism; later the level of accumulated ammonia declines as ammonia is taken up by the growing biomass; in the third stage gaseous ammonia is fed in response to pH and the ammonia level rises greatly as carboxylic acids are formed. The second case shows the rapid utilisation of the ammonia present in the starting medium, after which the level of free ammonia remains low as no further ammonia or ammonium salts are supplied while organic nitrogen sources are used for growth and product formation. Ammonia feeding to the second process may, therefore, be a means of supplying a readily used nitrogen source which would sustain faster rates of product formation. Conversely, in the first process the trend towards higher product formation in improved strains risks generating higher ammonia levels (especially if the ammonia is balancing acid accumulation) which may eventually tax the tolerance of the cells to high ammonia concentrations.

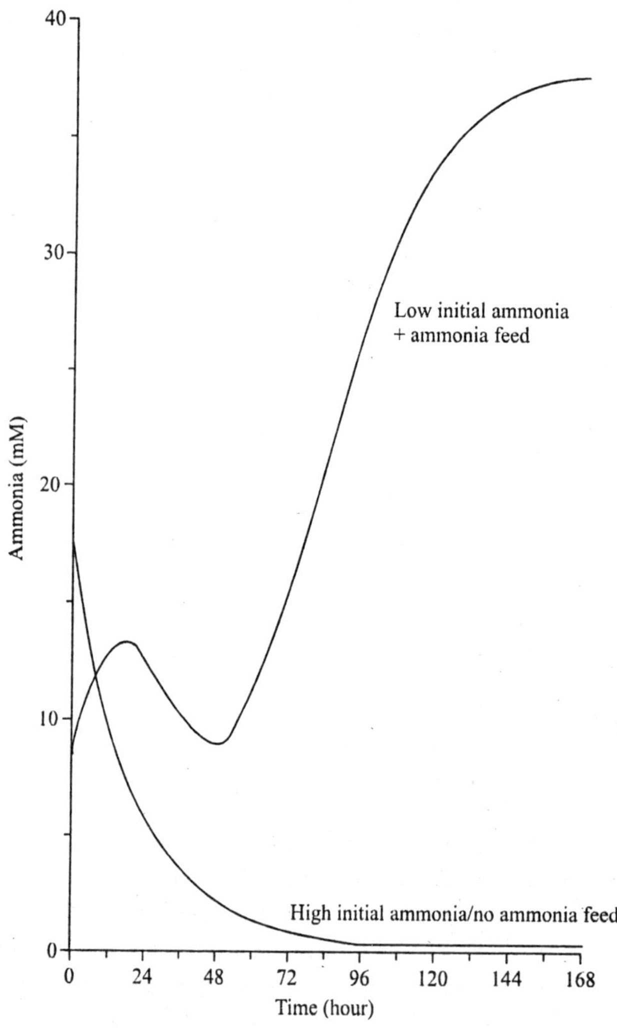


Fig. 14.8. Contrasting ammonia profiles in fermentations.

## Inorganic Components

### *Macro-inorganics*

Ammonia (as ammonium ion) is the most important inorganic cation because of its central role in nitrogen metabolism. Gaseous ammonia is $NH_3$. On dissolving in water, ammonium hydroxide ($NH_4OH$) is formed. The acid dissociation constant, pKa, of the ionic equilibrium is 9.25 at 25°C. The relative amounts of $NH_3$ and $NH_4OH$ in a solution are determined by the pH:

1. At pH 9.25, 50 per cent is in the form of $NH_3$.
2. At pH 8.25 only 10 per cent is present as $NH_3$.
3. At pH 10.25, 90 per cent is present as $NH_3$.

The pH in many fermentations is controlled to 6.0–7.5; in such culture media, therefore, free $NH_3$ is present in very small amounts. Sodium, potassium, magnesium and calcium play different roles in fermentation processes. Sodium may be added as sodium hydroxide to adjust the pH of the initial medium or to control pH after inoculation; it is generally thought to be inert but may add to the osmotic potential of the medium. Potassium salts are used as sources of sulphate and chloride; potassium is accumulated by some microbial species as an osmoprotectant. Magnesium salts are also used as sources of chloride and sulphate while calcium, as calcium carbonate (chalk), is added to increase the pH of the medium; both of these cations participate in precipitation reactions of, for example, ammonia (as magnesium ammonium phosphate) and phosphate (as di and tri-basic calcium phosphates).

Phosphate is the most important inorganic anion because it is essential for biomass formation (DNA, RNA, phospholipids, etc.). Phosphate levels in secondary product fermentations have received much attention due to the inhibitory effect of high phosphate concentrations on product formation. This is usually an indirect effect resulting from the stimulation of growth where rapid growth and high product formation are incompatible (this is discussed later when growth in fermentation processes is considered). Figure 14.9 shows an example of a process with low phosphate content in the medium and where that phosphate was rapidly utilised; such a fermentation may become phosphate-limited at a relatively early stage. This is in contrast to the other examples where much higher levels of phosphate are either incompletely utilised (and where the phosphate concentration is likely to be an important factor in pH regulation during the fermentation) or where the phosphate concentration only became low at an advanced stage in the process. In some cases, phosphate levels must remain low for high productivity (this is often the case with antibiotic-producing processes). Organic phosphates such as phytic acid which are introduced into the medium in complex nitrogen sources are potential sources of inorganic phosphate liberated by enzymic or chemical breakdown; these can greatly affect the phosphate supply during the fermentation.

The unequal utilisation of the acidic and basic ionic species of salts has implications for pH control and regulation. The uptake of ammonia from ammonium sulphate without the parallel uptake of sulphate by the cells would tend to force the pH down because of ionic exchange across the cellular membranes (in effect, leaving sulphuric acid in solution); the uptake of phosphate from potassium phosphate without an equal uptake of potassium would, on the other hand, be a move towards higher pH (leaving potassium hydroxide in the medium). The situation is made more complicated by the subsequent utilisation of the counter-ions and defining the mechanisms causing pH change in complex media can be very difficult. As well as contributing to pH, both chloride and sulphate ions may be directly incorporated into antibiotic production: for example, chloride is an essential precursor of vancomycins, chlortetracycline and avilamycins while sulphate (after reduction to organic sulphur) is a precursor of the cysteine required for the biosynthesis of penicillins, cephalosporins and lincomycins.

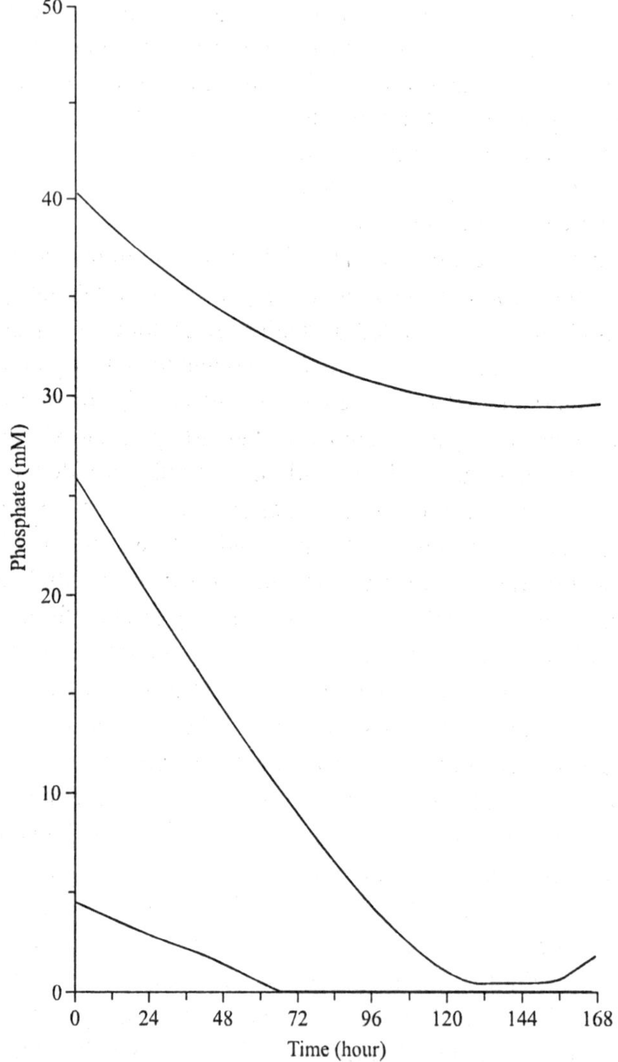


**Fig. 14.9.** Soluble inorganic phosphate in high and low-phosphate media.

### *Trace elements*

Iron, zinc, copper, cobalt and molybdenum are often 'essential' in the sense that in their absence either growth or product formation will be sub-optimal. Industrial media contain them in the 'ash' fraction of complex nitrogen and carbon sources and the unpurified water used in large industrial process is another significant source; supplementation by the addition of salts may be necessary.

Such di and trivalent metals can, however, be predominantly insoluble in fermentation media as a result of precipitation reactions. Phytic acid, contained in corn steep liquor and some grades of plant proteins, forms insoluble complexes with calcium, magnesium and heavy metals. There may be, therefore, a limited bio-availability of trace elements, whether they were added deliberately or gratuitously.

### *Feeding of biosynthetic precursors*

To maximise their efficiency of utilisation, biosynthetic precursors for product formation may be fed rather than included in the starting medium. The phenylacetic acid precursor for penicillin G is unusual in that high concentrations are toxic; a phenylacetate salt is routinely fed to a batch fermentation and its levels in the broth must be continuously monitored to avoid accumulation of the unused acid.

## GROWTH AND BIOMASS PROFILES

### Measurements of Growth and Cellular Contents

The quantitative measurement of growth during the course of a fermentation is probably the single most important parameter for process optimisation and yet is the most difficult and the most controversial. Cell counting methods are slow and difficult to apply except with unicellular organisms. Dry weight estimates and methods based on light scattering are accurate only in simple defined media. More indirect methods have been widely used including viscosity (especially for filamentous organisms) and oxygen uptake and carbon dioxide evolution rates.

Three biochemical measures, DNA, RNA and total cell protein, are generally the most useful if complex media are to be investigated (Fig. 14.10); apart from their intrinsic value in elucidating the amounts of inputs used for biomass formation, they can be used to calibrate indirect measurements of growth for routine use, where this is possible. DNA estimates (either by the classical Burton procedure with acid-hydrolysed DNA or by the more recent and sensitive intercalating dye methods with extracted intact DNA) are the most accurate measures the growth because there are less interferences from media components. This is very useful when investigating the early phases of fast-growing fermentations where plant protein sources exaggerate RNA measurements which rely on the chemical estimation of ribose. Total RNA determinations reflect of ribosomal RNA content rather than the smaller transfer RNA contribution and the yet smaller and more transient messenger RNA in the cell. Cellular protein reflects the gross structural composition and if the utilisation of unsolubilised protein inputs can be quantified, gives an estimate of biomass which can be correlated with those from the nucleic acids. Other chemical methods include biochemical markers for cell wall polymers (for example, diaminopimelic acid) and in oil-less processes, the total lipid content of the biomass.

Growth profiles such as exemplified by Fig. 14.10 yield valuable information, i.e., separating phases of fast and slow growth, indicating later phases of net biomass loss and giving estimates of the specific growth rate and of its change during the course of the fermentation. These are crucial aspects of the fermentation and high-priority targets for process optimisation. The end of the early phase of most rapid growth is signalled by the exhaustion of such limiting factors as inorganic phosphate or one (or more) amino acids with low abundance in the starting medium. The potential auxotrophic factors also relevant to this part of the growth profile include vitamins and vitamin precursors contributed by complex nitrogen and some complex carbon sources, trace metal ions and assorted 'growth factors' (for example, purines and pyrimidines) on which the strain may have become dependent during strain development. This is sometimes apparent if seasonal trends in productivity or fermentation parameters can be related to changes in the source of complex inputs.

A different aspect of growth is that of biomass composition. Expressed on a weight basis, the content of both the macromolecular and the low molecular weight components varies in response to growth rate. Because fermentation processes other than in chemostats have a variable nutrient environment, there will also be an effect exerted by the availability of nutrients, but this is most important when the pool of low molecular weight constituents is considered. Unlike enteric bacteria, secondary product-forming prokaryotes (which are often filamentous) and eukaryotes have large pools of intracellular metabolites. This is, therefore, an important compartment inside the fermentor which is distinct from the soluble phase and changes in the intracellular pools are sensitive indicators of the availability of soluble nutrients for biosynthesis. Phases of increasing intracellular pools indicate an excess of uptake over utilisation while later phases of declining intracellular pools show a failure to maintain uptake (usually the exhaustion of the external solute) with respect to utilisation.

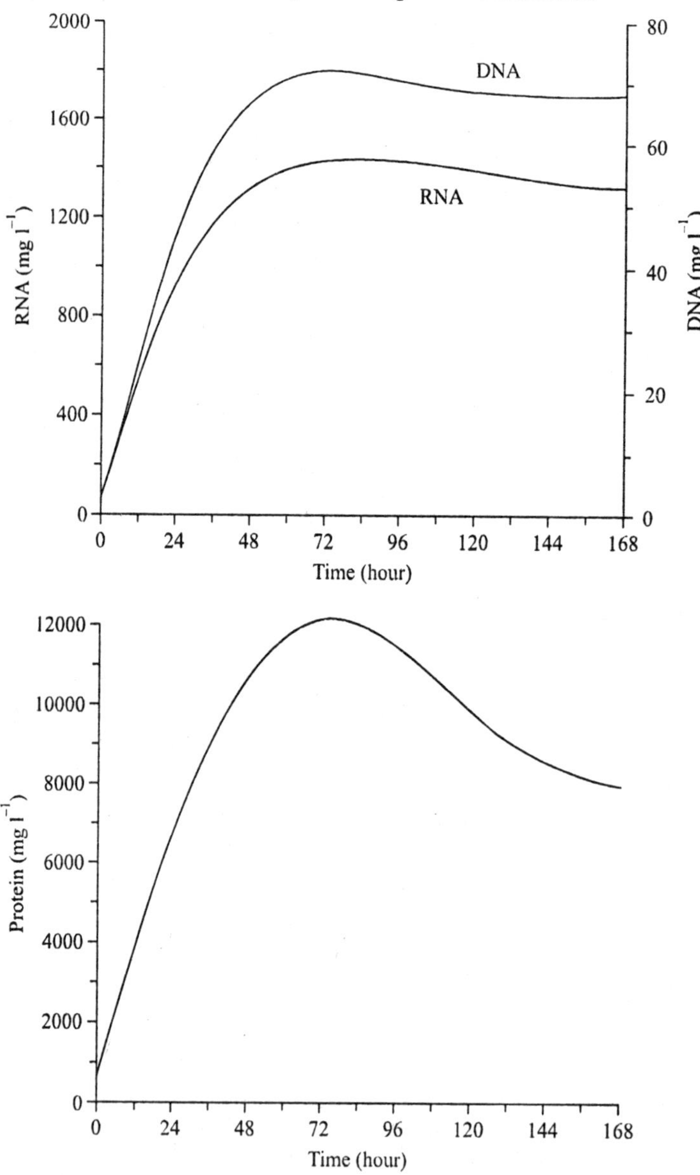


**Fig. 14.10.** Growth profiles by nucleic acids and cell protein measurements.

Many secondary product-forming organisms exhibit a 'life cycle' in which a sequence of morphological changes can be associated with certain phases of growth and varying levels of productivity. These are reflected in major changes in DNA:RNA and DNA:total protein ratios and the levels of other cellular components and these parameters are very useful in assessing the timing of transitions between phases and also the onset of contaminations. Published data on *Penicillium chrysogenum* show indeed that, at a faster growth rate in the early stages of a fed-batch process, both RNA and protein increase with respect to DNA but lipid and carbohydrate contents decrease.

### *Growth and metabolism in the seed stage*

An indispensable but variable 'input' to a fermentation is the inoculum. All microbial fermentations are inoculated with the biomass which is formed after one or more previous 'seed' stages. Media used for seed vessels are frequently quite different from those used in the production vessel. This has many consequences, the most important of which is that the nutritional and environmental conditions experienced by the cells in the inoculum often cause a phase of adaptation after transfer to the production vessel. The metabolic pattern established in the seed stage will need to be 'unlearned' in the production vessel and the risk of adverse effects on initial growth rate and substrate utilisation can be significant—for example, if a seed grown on glucose-containing medium is inoculated into a starch hydrolysate or sucrose-based medium, the enzymes required for carbohydrate utilisation are qualitatively different.

As with the production stage, growth in the seed stage is seldom measured directly and transfer criteria are consequently empirical (this adds a major uncertainty when process optimisation by designed statistical trials of media or operational parameters is undertaken). Traditionally, the seed is optimised by using the production vessel, in effect, as a bioassay for the biological material in the seed tank. This is a seriously flawed methodology because the greatest influence on the outcome of the production stage may not be solely the metabolic competence of the cells in the inoculum. While a poorly grown seed is unlikely to show high productivity, a well-grown one may be adversely affected by operational parameters in the production stage. An accurate measure of biomass in the inoculum (by DNA, RNA, total protein, etc.) is essential; analysis of both what is and is not utilised by the time of transfer also defines potential problems in establishing rapid growth and development after transfer.

## ACCUMULATION OF FERMENTATION OUTPUTS

The accumulation of the product (such as an antibiotic) in an industrial fermentation is carefully monitored so that operational changes can be made to optimise yield in any one production-scale fermentation. This is frequently the only routine chemical assay employed other than pH and gas exchange (oxygen uptake and carbon dioxide evolution) once the fermentation has been inoculated. In terms of the carbon inputs to the process, the product may only represent 10 per cent or less of the carbon supplied. In contrast, carbon dioxide can represent 50 per cent or more and accurate measurement of the carbon dioxide levels in the exit gas are essential if carbon mass balances are to be attempted.

There are, however, several classes of other outputs of quantitative importance, some of which may go unrecognised for many years: side-products, competitive products and accumulated primary metabolites. The biochemical relationships for the various types of metabolites accumulated in penicillin processes are summarised in Fig. 14.2.

Side-products are either breakdown products of biosynthetic intermediates or of the final product itself (resulting from either chemical instability in the fermentation broth or from enzymatic breakdown). These two possibilities can be demonstrated during the course of penicillin production. The commercial product is either penicillin G (in which the aminoadipyl side chain has been exchanged for phenylacetic acid) or penicillin V (with phenoxyacetic acid). The aminoadipyl side chain of isopenicillin N (the final biosynthetic precursor for both penicillin G and V) partially accumulates after side chain exchange in a cyclised form, oxo-piperidine 2-carboxylic acid; the efficiency of recycling of the aminoadipyl side chain may be variable in different fermentations. Similarly, the removal of the aminoadipyl side chain of isopenicillin N leaves 6-aminopenicillanic acid which appears in the extracellular medium where it is carboxylated to 8-hydroxypenillic acid. Penicillins are chemically unstable and the β-lactam ring can be opened to form penicilloic acids. Most biosynthetic pathways for secondary products are much more complex than the route to penicillins and are seldom fully defined; isolation of either major or minor side-products may, in these circumstances, be a useful aid to other biosynthetic studies. Competitive products arise because the organism which was selected from its natural environment on account of one particular biological activity it exhibited may also have synthesised other natural products. Classical strain improvement can amplify as well as lose these other potential outputs, which are competitive in the sense that they represent a waste of nutrients and common substrates of central metabolism.

In the penicillin process there are two types of competitive products: firstly, the amino acid precursors of the penicillins can be used for other biosynthetic purposes; secondly, one of the enzymes of penicillin biosynthesis has a wide substrate selectivity. Cysteine and valine are common amino acids which are polymerised into proteins; once penicillin production has begun, therefore, any generation of soluble protein (including that of enzymes involved in carbohydrate and oil utilisation) must compete with the use of the amino acids for penicillin biosynthesis. Similarly, α-aminoadipate is a biosynthetic precursor not only of penicillin but also of the amino acid lysine (in yeasts but not in bacteria); detailed work has shown that Penicillium strains with higher penicillin production have larger endogenous pools of α-aminoadipate and less of the precursor is used for lysine formation. Lysine can, under some circumstances, stimulate penicillin accumulation; this paradox was resolved when a catabolic pathway was identified which transforms exogenous lysine to α-aminoadipate. The acyltransferase which removes the aminoadipyl group from isopenicillin N and replaces this with a phenylacetyl or phenoxyacetyl group to form penicillin G or V, respectively, is the second source of competitive products. Ring-hydroxylation of pheoxyacetic acid produces p-hydroxyphenoxyacetic acid which can be incorporated by the acyltransferase into p-hydroxypenicillin V. The same acyltransferase will also accept several medium length (C6–C8) fatty acids to form other penicillins (the 'natural' penicillins DF, F and K). The use of oil in penicillin processes is much less common than glucose because triglyceride breakdown results in large amounts of fatty acyl derivatives and increases the levels of the natural penicillins.

Major products of primary metabolism include carboxylic acids such as citric acid which may accumulate as 'overspill' products from carbohydrate metabolism. The most important example of this is the formation of large amounts of acetic acid by recombinant strains of *Escherichia coli*; this has adverse effects on growth and the expression of foreign genes. Acetic acid and glycerol can also accumulate from the breakdown of triglycerides in oils. Free amino acids are formed from the proteolysis of protein and peptide inputs and if in excess of the requirement for growth, can accumulate (Fig. 14.5). A second category of primary metabolite is that of 'storage' products; many micro-organisms accumulate, for example, the polysaccharide glycogen or the disaccharide trehalose, polyhydroxybutyric acid or phosphate polymers ('polyphosphates'), especially under conditions of nutrient limitation.

Enzymes generated and secreted by the biomass could be considered as 'primary' metabolites, but the absolute amounts of protein involved are usually small unless the organism has been selected or designed to produce large amounts of a particular enzyme (for example, the enzymes used in detergents and in the food processing industry). A certain amount of cell lysis ('turnover') occurs in fermentation processes and this has the result of releasing much larger amounts of cellular proteins into the soluble phase; soluble protein can, therefore, accumulate even though protein inputs may still be in the course of utilisation (Fig. 14.5). Similarly, soluble carbohydrate pools may increase despite no further carbohydrate being supplied as cell wall fragments accumulate from cell turnover during the fermentation.

Identifying side-products, competitive products and major metabolites is a key factor in process improvement. Their presence can be recognised by the application of class-selective chromatographic methods for:

1. Primary and secondary amino acids.
2. Organic acids with UV detection at different wavelengths.
3. Carbohydrates using refractive index and electrochemical detection.
4. Reverse-phase chromatography.
5. Isocratic and gradient ion chromatography.

Unexpected peaks in standard chromatograms, i.e., those not identifiable by reference to standard calibration mixtures, are the best sources of fermentation 'unknowns'. In general, the greater the accumulation of the unknown product the more likely it is to be of importance for process optimisation. Once recognised, an unknown metabolite can be isolated by scale-up of conventional analytical HPLC procedures and structural identification undertaken by mass spectroscopy, nuclear magnetic resonance (NMR) spectroscopy and chemical synthesis.

## PROCESS IMPROVEMENT

### Different Kinetic Aspects of Product Formation

The progress curves of product formation are generally sigmoidal. The tangent to the curve is an instantaneous production rate or yield rate (Fig. 14.11). Industrially, a more useful parameter is the amount of product formed per accumulated time after inoculation of the production fermentor (this is sometimes referred to as the productivity). The units of both are product mass (g, kg, tonne) per hour.

The yield rate and its profile during the fermentation, is related to changes in growth rate, utilisation rates of inputs and metabolic events. The productivity measures overall kinetics of product formation and is important for estimating the economics of the process; factors such as the time required to empty, clean and re-fill the production vessel prior to the next inoculation can also be taken into account in the mathematics. The analytical results obtained from a fermentation form the basis for identifying the key events for intervention in process improvement. The first step is to relate the results to the kinetics of product formation. In both batch and fed-batch fermentations product formation follows patterns exemplified in Fig. 14.11. A constant rate of product formation can be achieved over lengthy periods of time in laboratory continuous fermentations but these have not yet been developed to the large production scale. The yield rate in a batch or fed-batch process invariably exhibits a maximum after which the product formation rate then declines. There is a strong interaction between the rates of biomass and product formation. Historically, fermentations were divided into growth-associated and growth-independent; in the former active product formation occurs by actively growing cells, while in the latter the growth phase (trophophase) is followed by product formation under conditions of no net growth (idiophase). More recently, analysis has shown that these are extreme examples of a spectrum where the key parameters are the specific growth rate ($\mu$), the specific productivity ($q_{prod}$) and the requirement for inputs to 'maintain' the biomass in the fermentor in a fully productive state. Mathematical simulations of increasing sophistication have postulated minimum and maximum $\mu$ values outside which maximum $q_{prod}$ is not sustained and have made various assumptions as to how the carbon input is partitioned between growth, product formation and the provision of maintenance energy which is required by the biomass. The general outline of the productivity curves can be described as the generation of productive biomass followed by the loss of productive capacity as the number of competent cells declines; the process has an implicit life cycle in which young cells differentiate to productive cells which themselves only have a limited productive life and it is the reduced rate of generation of the producing cells that underlies the failure to maintain the highest production rates.

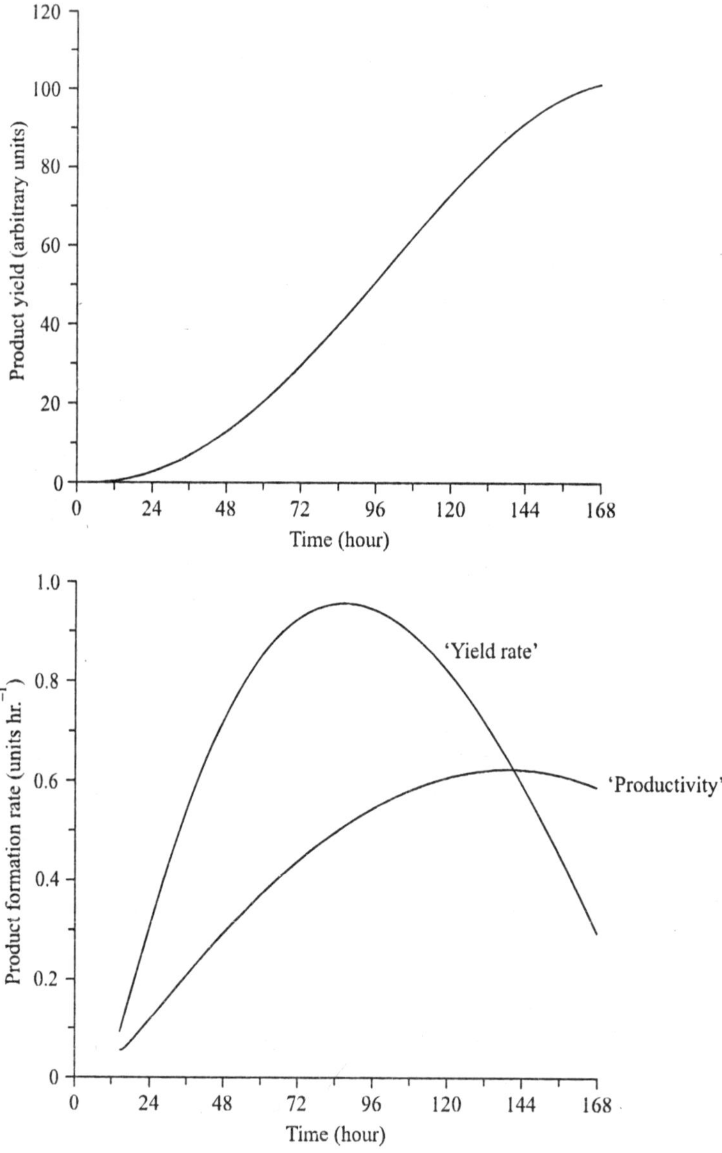


**Fig. 14.11.** Parameters of product accumulation.

Direct measurements of biomass (DNA, RNA, cell protein, lipid, etc.) will guide process improvement to achieve the optimal growth profile. This optimum is a combination of two factors: the fastest generation of productive biomass and the maximum length of time $\mu$ values can be maintained to allow maximum $q_{prod}$. The overall growth profile can then be optimised by the manipulation of fermentation inputs based on the kinetics of their utilisation.

Which input limits the phase of rapid growth (at approximately 60 hours in the idealised process of Fig. 14.10). The inorganic phosphate in the medium became exhausted at this time (lowest curve in Fig. 14.9). The free ammonia may have been the nitrogen substrate for rapid growth (Fig. 14.8, lower curve) or the protein in the starting medium might have fulfilled this role (Fig. 14.4). Alternatively, the free amino acids may have been fully utilised (Fig. 14.7).

Can the levels of the growth-limiting nutrient in the starting medium be increased to give a faster growth rate or higher biomass levels? Poorly growing fermentations are usually also poorly productive but cultures with high cell densities may become oxygen-limited. Simply increasing the concentrations may impose osmotic stress. A controlled feed may be applied once the nutrient in the batched medium becomes depleted, for example adding a phosphate feed to two of the processes illustrated in Fig. 14.9.

Are there successive growth bursts (diauxic transitions)? This may be particularly important when 'rival' carbon sources such as carbohydrate/organic acids or sugars/triglycerides are present in the growth medium. Can nutrient feeds sustain optimum $\mu$ values? The relationship between $\mu$ and specific productivity is well understood for penicillin processes and given accurate growth data, the different stages of growth can be optimised by the choice of feeds and control of feed rates.

Do the macromolecular inputs require enzymic processing? The protein inputs in the example of Fig. 14.5 were degraded during the process to soluble peptides and free amino acids but in other cases a protease digestion before the fermentation is inoculated may be necessary to make the nitrogen supply more biologically accessible. Is the growth erratic? Irreproducible growth in a process can be caused by many factors, including variable precipitation (of phosphate, trace elements, etc.), inconsistent medium preparation and differing amounts of biomass in the inoculum.

Given an improved growth profile, the second key control point is the succession of factors which progressively limit product formation during the course of the fermentation. With a knowledge of the quantitative utilisation of all the inputs which contribute to product formation, it is possible to compute the rates of utilisation required to sustain the maximum rate of product formation; the first utilisation rate to fall below that required for this maximum rate represents the first input (or substrate) limitation. This could be the major carbon source being consumed during product formation, a specific precursor either included in the batched medium or being fed or nitrogen (as amino groups) required for amination reactions. The levels of the limiting substrate can, therefore, either be increased in the medium or fed (if it was only supplied in the batched medium) or its feed rate or timing could be optimised. The procedure is then repeated for each successive limitation, given the constraints of engineering, the volume in the fermentor and the osmotic strength of the starting medium. On a commercial scale, deciding the point in time at which higher time-based productivity could be gained simply by terminating the running fermentation (and setting up a new operating vessel) becomes the final criterion at any one stage in process improvement—in Fig. 14.11 the likely harvest time would be 144 hours.

Definition and measurement of side-products, unwanted and competitive metabolites leads to quite different options for process improvement. In general, higher substrate efficiencies follow the elimination of such undesirable products. Their elimination may, however, require gene technology or the careful scrutiny of strains developed by classical selection methods. The closer the source waste metabolite to the biosynthesis of the fermentation product, the larger may be the potential benefit to productivity.

Consider Fig. 14.12 where the avenues and potential productivity gains by rational improvement of a hypothetical process are visualised as the partitioning of the supplied nitrogen into product, biomass and metabolites. Roughly equal amounts of the total input nitrogen are incorporated into the product and a side-product at the end of the fermentation; in a penicillin process this 'side-product' would represent the sum of the oxopiperidine carboxylic, aminopenicillanic, hydroxypenillic and penicilloic acids together with any natural penicillins formed. If the accumulation of side-product could be eliminated, then the yield increase with this 'cleaner' process would be two-fold. An obvious strategy would be to screen promising improved strains for those which accumulate less side-product and use these as a base for further strain development work.

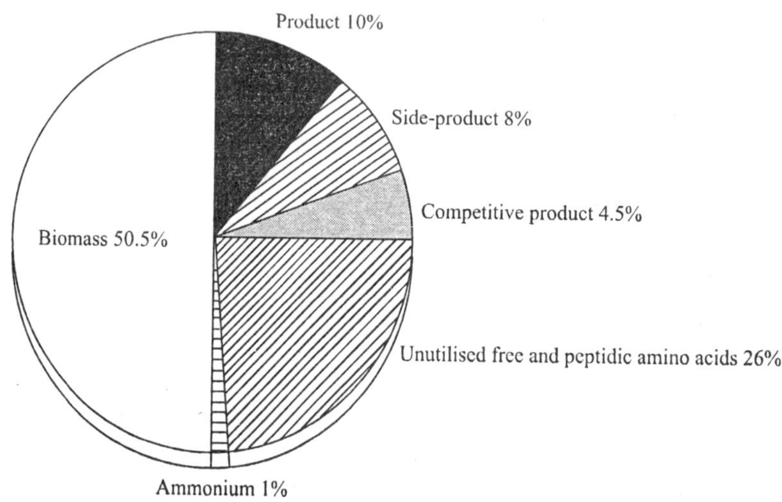


**Fig. 14.12.** Fate of input nitrogen in a secondary product fermentation.

With a competitive product the benefits of eliminating such a product are not as clear because the biosynthetic relationship with that of the main product must first be defined. In the case of penicillin V and p-hydroxypenicillin V, this relationship is very close; mutants could readily be obtained with much reduced levels of the hydroxylated product. Simplified product recovery during downstream processing would also be anticipated if the formation of a competitive product or side-product of similar chemical structure to the product were to be prevented.

A large amount of nitrogen remains unused in amino acids and peptides (Fig. 14.12). This waste of unused nitrogen could represent peptides resistant to breakdown by the proteases secreted by the biomass (in which case a different mix of complex nitrogen inputs could be considered) or remnants of cellular proteins released by cell turnover. A related question is whether the biomass level in the process is unnecessarily high (i.e., of low specific productivity); could a different balance of inputs restrict growth and divert more nitrogen to product? These possibilities should be explored at an early stage in process and strain development to avoid introducing problems which may become acute when large-scale fermentors are used. Scaling up from laboratory and pilot plant fermentors to production vessels may also result in differing quantitative patterns of metabolism and side-product formation which influence the success of the production process. The information gained from growth studies, input utilisation and output accumulation can be used to define targets for rational strain improvement. This can have a very large impact on productivity and does not require a complete knowledge of the genomics of the organism. This illustrates the central theme developed in this chapter: that direct analysis bridges the knowledge gap between the genetic potential of an organism and what portion of this is expressed in a given environment (Fig. 14.13). By identifying what the cells must do (given a nutrient medium from which substrates for growth and product must be taken as building blocks) and what the cells actually do (accumulate side-products, competitive products, 'waste' metabolites, unused portions of inputs, etc.), the use of microbial cells as biocatalysts becomes amenable to precise genetic and metabolic engineering. Finally, the same approach can also be extended to cultures of animal and plant cells with their diverse and mostly unexplored biosynthetic potentials.

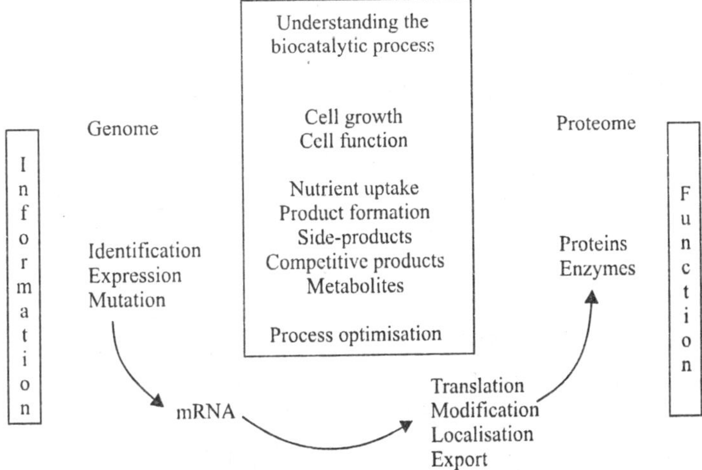


**Fig. 14.13.** Functional genomics, proteomics and direct analysis of fermentation processes.

Thus, product yields in industrial fermentations can be increased by strain selection (both rational and random) and engineering (technological) improvements. Both approaches are limited by the inability to define the metabolic causes and consequences of the changes which are made to the biological system. Modern chemical and biochemical analytical techniques can now give a rapid understanding of the key events in a fermentation process: the utilisation of inputs, growth and the generation of product, side-products and unwanted metabolites. A detailed knowledge of input utilisation, growth and the generation of outputs allows the optimisation of growth profiles and the metabolic engineering of more advanced and efficient fermentation-based manufacturing. The benefits are that key targets can be selected for genetic manipulation and improved process control by optimisation of the media, feeding strategies and physical parameters. Novel strains and processes can be more readily scaled up and the optimal features of strains with increased productivity can be defined more rapidly. This approach benefits both small and large-scale processes, with prokaryotes and eukaryotes and for the full range of products produced by contemporary fermentation biotechnology.

## FUTURE PROSPECTS

Robotic sampling and analysis, on-line sensors for nutrient concentrations and cell density, 'instant' metabolite profiling by high-resolution mass spectroscopy and near-infrared scanning methods, *in situ* NMR probes which can 'see' into cells and sub-cellular compartments and artificial intelligence capable of reacting to the data generated will revolutionise the operation of large-scale fermentors. Most of these technologies are already at an advanced stage of development. Biochemical analysis using two-dimensional gel electrophoresis methods will complement them with rapid analyses of the spectra of individual proteins made at various stages in the process and of how gene expression is quantitatively regulated. With that quality of information, process control will enter the molecular era and the full potential of the biological system represented by a fermentation can be achieved on a real-time basis.

Chapter 15

# Biosensors in Bioprocess Monitoring and Control

## INTRODUCTION

Biosensors belong to a sub-group of chemical sensors in which a biologically based mechanism is used for analyte detection. The International Union of Pure and Applied Chemistry (IUPAC) has defined chemical sensors as miniaturised transducers that reversibly and selectively respond to chemical compounds or ions yielding an analytical signal proportional to the analyte concentration. By definition, a biosensor is an analytical device that combines the specificity of a biological sensing element for the analyte of interest with a transducer to produce a signal proportional to target analyte concentration. This signal can result from: a change in pH, release or uptake of gases, light emission, heat emission, mass change, etc. The transducer converts the biological reaction into a measurable response such as current, potential, temperature change, modulation of light intensity, etc. The signal can be further amplified, processed or stored for later analysis and retrieval. Table 15.1 lists the different components that can be used to construct a biosensor. In principle, any receptor can be combined with any suitable transducer to produce a working probe. Figure 15.1 shows a generalised representation of a biosensor.

**Table 15.1.** Components that can be used in the construction of a biosensor.

| *Biological element* | *Transducer* |
|---|---|
| Whole organisms | Potentiometric |
| Tissues | Amperometric |
| Cells | Conductometric |
| Organelles | Impedimetric |
| Membranes | Optical |
| Enzymes | Calorimetric |
| Receptors | Acoustic |
| Antibodies | Mechanical |
| Genetic material | |
| Organic molecules | |

Fermentation processes by their very nature are complex systems, especially for the purpose of effective bioprocess monitoring. Typically, the biological system (e.g., cell or enzyme) is surrounded by a chemical and physical environment where changes in one of these environment can cause drastic effects on the other parts of the system as a whole. The ultimate aim of bioprocess analysis therefore is a detailed monitoring of the biological system, the chemical and physical environment and how these interact.

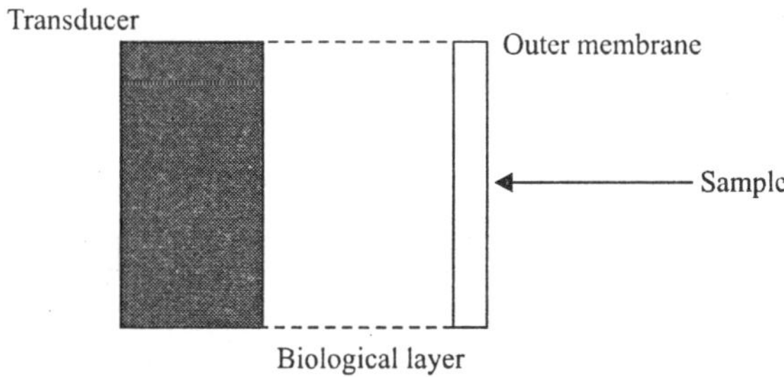


**Fig. 15.1.** Generalised features of a biosensor.

The current state of sensor technology permits the measurement and control of dissolved oxygen, dissolved carbon dioxide, pH, redox potential, temperature, agitation and the level of foam in the fermenter vessel. Recent advances in sensor technologies have led to the on-line determination of biomass through *in situ* optical density or fluorometric probes and capacitance measurements using low radio frequencies. Oxygen uptake and/or carbon dioxide evolution during cell growth have been monitored using piezoelectric mass balances and off-gas analysis. In addition, head space analysis of volatiles such as methanol has been used to monitor and control fermentative processes.

However, no reliable techniques exist to carry out real-time measurements of non-volatile substrates and metabolites in the reactor vessel. In this respect, biosensors have attracted a considerable amount of attention over the last decade because they offer the possibility to monitor single analytes even in the extremely complex and often undefined matrices that occur in fermentation vessels; however, the commercial exploitation of biosensors for fermentation process monitoring and control has not yet been fully realised.

## BASIC COMPONENTS OF ON-LINE PROCESS MONITORING AND CONTROL

Any system for on-line monitoring and control must include three essential elements: (i) a sensor (that is a biosensor); (ii) a suitable analysis manifold employing an *in situ* or *ex situ* arrangement for contacting the fermentation liquor with the bioprobe; and (iii) a control system with the necessary hardware and control algorithms in order to employ a suitable control strategy. In order to use the analysis data efficiently, the time lag between sampling and analysis must be within the timescales of the bioprocess. Thus, in a bacterial cultivation process with short generation times, the time delay should be in the order of a very few minutes. In cultivation processes employing much slower growing mammalian cells, this time delay can be in the order of 1–2 hours. In a reactor employing an immobilised enzyme, the time delay should be of the order of seconds. The time delay is a function of the sampling, the sample handling, the analysis and the data process.

Sensors can be interfaced to a biotechnological process in different ways. A biosensor can be used as an *in situ* probe or can be separated from the fermentation broth by a filtration unit, e.g., tangential flow units in a bypass. An alternative and perhaps more practical way of utilising biosensors in continuous fermentation monitoring and control is to use them in conjunction with flow injection analysis (FIA). In FIA, the liquid sample to be analysed, which forms a discrete zone, is injected into a moving, non-segmented carrier stream flowing continuously past a detector. The many advantages of this technique when applied to process control include:

1. Reduced risk of contamination because the sample is not returned to the bioreactor.
2. Automatic recalibration of the sensor to counter any problems associated with drift.
3. Short response time.
4. Requirement for low sample volumes.
5. Multi-component monitoring if an array of biosensors is used as the detector.

Certain practical considerations need to be taken into account when using biosensors for process monitoring. With the use of *in situ* biosensors, sterilisation of the probe can be a major problem that is usually solved by a complex engineering design. Furthermore, some biocatalysts may require very strict operational conditions that may be different from those in the fermenter. With FIA, the associated problems are different and include: faults associated with in-line membranes used to produce the cell-free samples for analysis, microbial growth on the upstream side of the membrane causing plugging, need for dilution of the sample or adjustment of pH prior to the analysis step, dissolved gas in the sample or flow stream and drift or deactivation of the sensor with time.

In terms of process monitoring and control applications, the emphasis has been on the use of enzymes and in particular in an enzyme electrode configuration. [The enzyme electrode is a combination of any type of electrochemical probe with a thin layer (10–200 μm) of immobilised enzyme.] Typically, the progress of a particular enzymic reaction (which is related to the concentration of analyte) is measured by monitoring the rate of product formation or the rate of disappearance of a reactant. If either the product or reactant is electroactive, the reaction may be monitored directly using amperometry. Consequently, this trend will be reflected in the subject matter of this chapter although other successful strategies will also be discussed.

## ENZYMES

Enzymes are very selective biological catalysts that carry out the conversion of a particular substrate into a product under mild operating conditions, i.e., room temperature, low ionic strength and near neutral pH. An enzyme will usually catalyse a single chemical reaction or a set of closely related chemical reactions. Side reactions leading to wasteful formation of by-products rarely occur. Thus, quantitative assays may be done on crude materials with little or no sample preparation.

Much of the catalytic power of enzymes comes from their ability to bring substrates together in favourable geometric alignments to form enzyme-substrate complexes. The substrates are bound to a specific region of the enzyme known as the active site. Most enzymes are highly specific in the binding of their substrates and indeed, the catalytic specificity of an enzyme is dependent at least in part on the specificity of this binding.

For many enzymes, the rate of catalysis (v), defined as the number of moles of product formed per second, varies with substrate concentration [S] as illustrated in Fig. 15.2. At a fixed concentration of enzyme, v is directly proportional to [S] when the concentration is small. At high substrate concentrations, the rate of reaction becomes independent of [S] and reaches a maximum value. The model proposed, which is the simplest one to account for these kinetic properties is shown below:

$$E + S \underset{k_{-1}}{\overset{k_1}{\rightleftharpoons}} ES \xrightarrow{k_2} E + P$$

According to this model, an enzyme molecule E combines with substrate to form an ES complex, with a rate constant $k_1$. The ES complex then has two possible fates; it can dissociate back to E and S, with a rate constant $k_{-1}$ or it can proceed in an irreversible manner to form a product P, with a rate constant $k_2$. Using this model, it is possible to define a rate equation (Michaelis-Menten equation) for an enzyme catalysed reaction that states:

$$v = v_{\max} \frac{[S]}{[S]+K_M}$$

where $K_M$ (the Michaelis constant) is defined as $\frac{k_{-1}+k_2}{k_1}$ and $v_{\max}$ is the product of the total enzyme concentration and $k_3$.
Thus, when $[S] \ll K_M$ the rate of reaction v is defined as:

$$v = \frac{v_{\max}[S]}{K_M}$$

i.e., the rate is directly proportional to $[S]$.

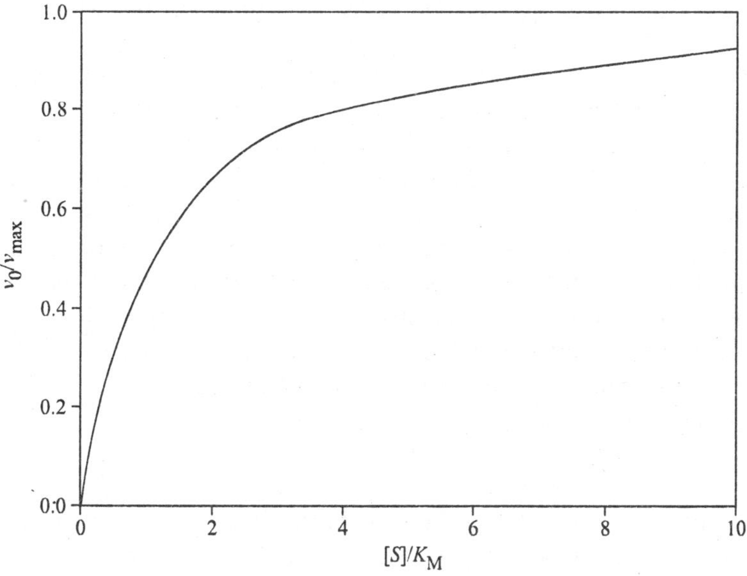


**Fig. 15.2.** Dimensionless plot of the Michaelis-Menten equation.

When using enzymes in biosensor applications, two important considerations have to be taken into account, i.e., operational stability and long-term use. Because both of these factors are to some degree a function of the immobilisation strategy used, correct choice of immobilisation technique is an important criterion. Many methods are used for the immobilisation of enzymes to the surface of a transducer. These can be summarised as follows:

1. Physical methods:
   (a) Entrapment in a gel matrix or polymer.
   (b) Adsorption to the surface of the transducer.
2. Chemical methods:
   (a) Covalent binding to the transducer surface using bifunctional reagents such as cyanuric chloride.

(b) Cross-linking using multifunctional reagents such as glutaraldehyde.
(c) Immobilisation in an electrochemically grown polymer film.

### Preparation of the Immobilised Enzyme Layer

The proper functioning of an enzyme-based sensor is, of course, heavily dependent on both the chemical and physical properties of the immobilised enzyme layer. To this end, there are many possible methods for immobilising an enzyme at the surface of an electrode. For convenience, these can be divided into physical methods and chemical methods. An important consideration when developing an immobilisation procedure is that the process should be applicable to a range of surfaces. This allows the choice of support to be as wide as possible and ensures that no re-fabrication of the support is required. Other advantages sought for in the immobilisation method include: (i) an ability of the biological component to operate at a wider pH range than in solution; (ii) attainment of greater stability resulting from the immobilisation; and (iii) the generation of a defined 'diffusion region' on the surface of the electrode.

#### *Gel entrapment*

Entrapment within a 3-D gel matrix is a common method of enzyme immobilisation. Numerous matrices have been employed but the most favoured have been alginate, cross-linked with linear chains of $Ca^{2+}$ ions, gelatine and polyacrylamide. With due attention to the degree of cross-linking and the nature of the gelling process, i.e., minimising the concentration of free radicals, gel entrapment can be applied to any enzyme. Unfortunately, this process does suffer from three major drawbacks: (i) large diffusional barriers to the transport of substrate or product resulting in reaction retardation and long response times; (ii) continuous loss of enzyme activity because these materials generally do not have a narrow pore size distribution; and (iii) shrinkage and/or swelling of the polymer depending upon the ionic strength of the milieu. A more straightforward method of enzyme entrapment is to retain the protein at the electrode surface behind a thin semi-permeable membrane such as dialysis tubing. Here, a thick paste of the enzyme in a suitable volume of buffer (1 or 2 $mm^3$) is spread over the surface of the electrode. The layer is then covered with a 20–25 μm thick dialysis membrane, of about 10,000 Daltons molecular mass cut-off, held in place by a suitably sized O-ring.

#### *Adsorption*

The great advantage of using adsorption as the immobilisation technique is that usually no additional reagents are required and only a minimum of activation or 'clean up' need be done. Unfortunately however, only weak short-range interactions are involved such as Van der Waal forces, dipole-dipole interactions and hydrogen bonding. Because of this, the reversible nature of the binding equilibrium is highlighted by its susceptibility to changes in ambient conditions, i.e., pH, temperature, ionic strength, polarity, etc. When a protein adsorbs to the surface of a support, the reaction will be influenced by non-covalent interactions between the surface of the protein and the support. Clearly the protein could interact with the surface in several different ways depending on the orientation with which it approaches the surface. In an unperturbed solution this process and the reverse process, is under mass transport control. If every molecule that encounters the surface is adsorbed, a concentration gradient rapidly develops at the surface. The rate of adsorption then becomes proportional to the rate of diffusion such that:

$$\frac{dn}{dt} = C_0\sqrt{\frac{D}{\pi t}}$$

where $n$ is the number of molecules, $C_0$ the bulk concentration of protein and $D$ the diffusion coefficient. The model does not, however, account for the situation where the layer next to the surface is only partially saturated with protein (as happens during sensor use). Under this condition, desorption becomes the dominant process and the biomolecule becomes lost from the electrode surface. Finally, Schmid has shown that the adsorption of glucose oxidase to the surface of a carbon electrode leads to the irreversible denaturation of the protein unless steps are taken to modify the interface. Indeed, recent results in this laboratory have shown that the adsorption of the enzymes glucose oxidase and horseradish peroxidase to carbon does not proceed via a simple Langmuir isotherm. Rather, three equilibrium steps are involved suggesting that the protein exists in three different microcrystalline states on the surface thereby making the immobilised protein difficult to quantify. Despite these apparent shortcomings, the ease of the technique and its general applicability make adsorption particularly attractive as a pilot method or for the production of devices not requiring long-term stability, e.g., one-shot disposable sensors.

### *Covalent immobilisation*

Covalent binding of the enzyme to the surface of the electrode is generally irreversible and as such is the most stable method. Chemical bonding to the surface must be effected by using nucleophilic amino acid (carboxylic acid, hydroxy, thiol, imidazole and phenolic groups) of the protein that are not involved in the biological function of the molecule. In general terms, the attachment of an enzyme to an electrode surface (e.g., a metal or carbon disc) is a two-step process. The first step involves activating the surface of the electrode, i.e., imparting some useful chemical reactivity to the otherwise inert surface. The second step involves the binding of the enzyme to the chemically activated electrode surface.

#### *Immobilisation of enzymes to metal electrodes*

Although metal surfaces are in themselves unreactive, when they are covered by a thin surface oxide film they can be functionalised by reagents such as chloro or alkyl silanes. By analogy with silica surfaces, Pt/PtO, Au/AuO and $SnO_2$ surfaces have many M–OH sites (where M is the metal) and when they are contacted with, for example, a solution of dichlorodimethylsilane $[Cl_2Si(CH_3)_2]$ under anhydrous conditions the organosilane reagent becomes immobilised by formation of chemically stable –MOSi– bonds.

The silanisation reaction can be monitored using surface analysis techniques such as X-ray photoelectron spectroscopy (XPS). Here, the experimenter would detect the silanisation reaction of a 'clean' metal surface by noting the appearance of new XPS Si 2p, Si 2s and O 1s bands in the spectrum. Alternatively, other techniques such as reflectance FTIR may also prove useful if the modifying agent contains a suitable chromophore. Finally, it may also be possible to monitor the surface modification by measuring changes in the double layer capacitance due to the replacement of polar surface groups by non-polar methyl groups. Accurate measurements of the change in the double layer capacitance may allow quantisation of the surface groupings. Although the reaction shown above produces a functionalised electrode surface, the resultant electrode is not suitable for further synthetic modification due to the chemical inertness of the methyl groups. To bond an enzyme to a silanised surface, it is important that the organosilane reagent itself bears chemical functionality such as a primary amine group or a carboxylic acid. A particularly useful reagent in this context is propyl amino silane $[(CH_3CH_2O)_3Si(CH_2)_3NH_2]$ which will not only functionalise a metal electrode but will then allow coupling chemistry to take place with proteins through the attendant amine grouping. A typical synthetic scheme using this reagent is shown in Fig. 15.3., by which an enzyme is coupled to a silanised metal electrode via the formation of an amide bond between the amino group of the silane and a surface carboxylic acid on the protein. Note however, that to make the amide linkage, the carboxylic acid on the protein must itself be activated. This is normally achieved by using carbodiimide chemistry. The reaction uses attractively mild conditions and can draw upon a variety of diimide reagents developed largely for solid phase peptide synthesis. The incorporation of the enzyme onto the electrode surface can be monitored by challenging the electrode with the substrate of the enzyme and detecting the production of an electrochemical product.

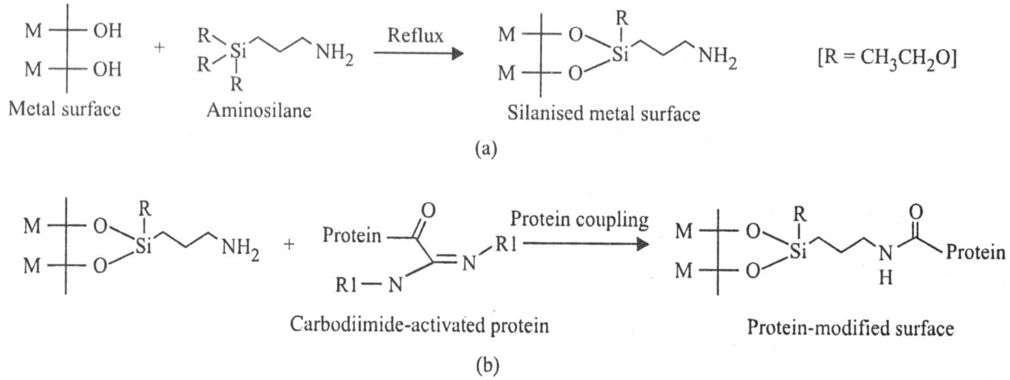

Metal surface + Aminosilane —Reflux→ Silanised metal surface [R = $CH_3CH_2O$]

(a)

Silanised metal surface + Carbodiimide-activated protein —Protein coupling→ Protein-modified surface

(b)

**Fig. 15.3.** Two-step synthetic scheme for the attachment of a protein to a metal electrode. In step (a), the metal oxide surface is functionalised with an aminosilane reagent. In step (b), a carbodiimide-activated protein is attached to the electrode surface via the attendant primary amine groups.

### *Immobilisation of enzymes to carbon electrodes*

Carbon electrodes represent another popular material for the manufacture of enzyme electrodes. The attachment of enzymes to carbon surfaces is achieved through the activation of the carbon surface and the chemical bonding of the enzyme. Graphitic carbon consists of giant sheets of fused aromatic rings, stacked co-planarly. An uninterrupted basal plane surface is non-ionic, of low polarity, hydrophobic and rich in π-electron density. Without alterations, the basal plane surface of carbon is somewhat barren to synthetic coupling reactions. (On the other hand, the high π-electron density is conducive to strong chemisorption interactions and could be used as basis of adsorptive modification particularly where unsaturated compounds and aromatic rings are involved.) The perimeter of the basal plane structure on the other hand is terminated as a graphitic edge plane on which chemical functionalities such as carboxylic, hydroxy, phenolic, quinone, lactone and other carbonyl containing groups abound. Any cleavage of graphite results in reactive edge plane 'dangling' valences that become satisfied by reaction with oxygen and water. Graphitic edge planes are therefore attractive surfaces for chemical coupling and modification procedures. Edge planes are also polar and thus fairly hydrophilic implying that aqueous coupling reactions can be used with a good degree of success. Several chemical pre-treatments have been proposed for enhancing synthetically useful carboxylic acid and hydroxylic groups on the edge planes. Carboxylic acid coverages on spectroscopic carbon rods, highly orientated pyrolytic graphite, glassy carbon and pyrolytic graphite can be enhanced by simply heating in air at 400–500°C. Treatment in a radio frequency (RF) $O_2$ plasma is an equally effective procedure. To enhance hydroxylic groupings on pyrolytic graphite surfaces, Lin and co-workers first thoroughly oxidised the surface with an $O_2$ RF plasma and then reduced it with ethereal $LiAlH_4$. More recently, work in this laboratory has shown that simple oxidation of graphite particles with 1 M nitric acid at room temperature can also greatly increase the surface concentration of hydroxyl functionalities. Hydroxyl-containing surfaces are particularly attractive because they can react with polyfunctional cyanuric chloride (trichloro-*sym*-triazine) producing an ether linkage that is chemically and electrochemically stable in organic solvents and aqueous solutions. The attached cyanuric chloride (CC) can further react with a variety of substances including hydroxyl and amino compounds, alkyl and aryl Grignard reagents and organic hydrazine reagents. This approach has been used to attach the enzymes horseradish peroxidase and glucose oxidase to the surface of graphite particles. The synthetic scheme for the attachment of horseradish peroxidase to the graphite particles is shown in Fig. 15.4. In an analogous fashion to metal oxide electrodes, the derivatisation of the carbon surface with cyanuric chloride can be easily followed by XPS. Here, use is made of the appearance of the Cl 1s peak in the XPS spectrum following reaction with CC. It must be stressed that the actual knowledge of the absolute population and chemical reactivities of the various groupings on carbon surfaces is still limited. Although one of the treatments discussed above may appear to enhance the efficacy of a chemical coupling procedure, e.g., by producing larger electrochemical waves for attached redox species, this in itself is an indirect form of evidence which explicitly pre-supposes that the coupling reaction proceeds as planned.

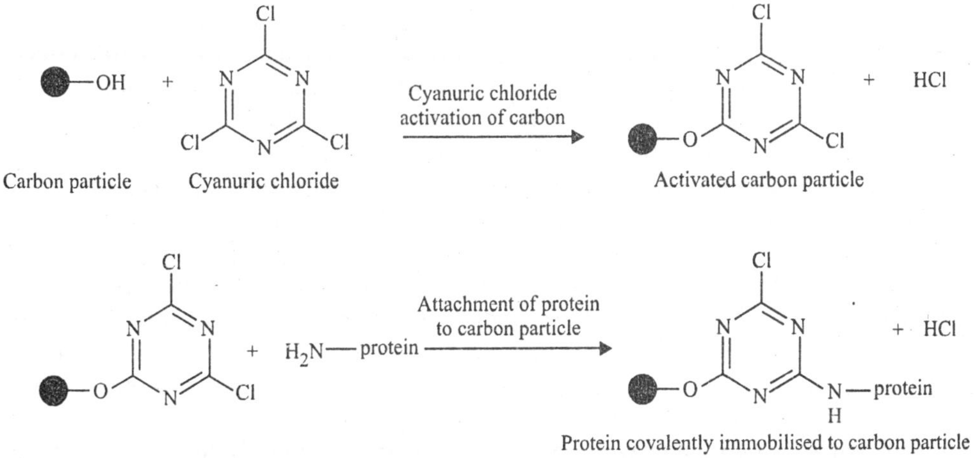

Carbon particle — Cyanuric chloride — Cyanuric chloride activation of carbon → Activated carbon particle + HCl

+ $H_2N$—protein — Attachment of protein to carbon particle → + HCl

Protein covalently immobilised to carbon particle

**Fig. 15.4.** Synthetic scheme showing the usage of cyanuric chloride in the covalent attachment of a protein to the surface of a carbon particle.

### *Immobilisation in an electrochemically grown organic polymer*

Direct *in situ* formation of a polymer film from a solution of monomers can be induced electrochemically. Through electrochemical initiation, a monomer such as pyrrole or thiophene is oxidised to a polymerisable radical as illustrated in Fig. 15.5.

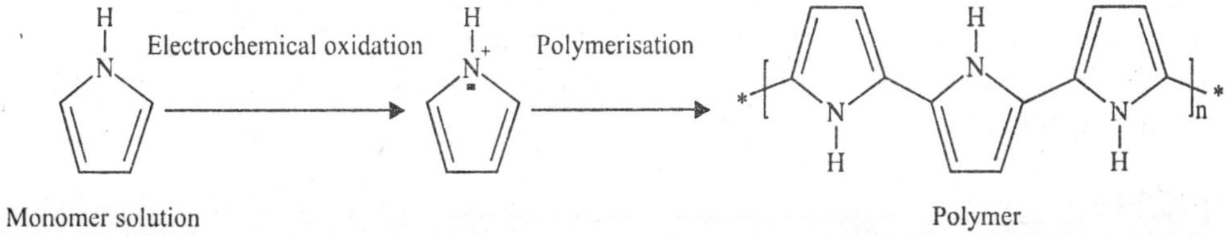

Electrochemical oxidation → Polymerisation →

Monomer solution — Polymer

**Fig. 15.5.** Scheme showing the electrochemical oxidation and subsequent polymerisation of pyrrole onto the surface of an electrode.

This method is applicable to many electrode materials and also many monomer compounds. The technique is often simple to carry out, requiring only basic electrochemical instrumentation. Furthermore, the growth of the polymer film can be controlled by stepping or cycling the electrode potential. The properties and the reproducibility of the surface polymer ultimately depend upon the nature of the polymerisation solution, the electrode material and to some extent, the cell geometry and attendant hydrodynamic conditions. Another material that has proved particularly useful for the immobilisation of enzymes at electrode surfaces is polyaniline.

If an enzyme is present in the aqueous solution containing the monomer, molecules of the enzyme will become physically trapped within the growing matrix during film growth. This approach has several significant advantages as a general method for preparing enzyme electrodes. First, the method is flexible and can be readily controlled. Second, it is simple to carry out and usually results in enzyme loadings with high activity. Third, it is possible to co-immobilise more than one enzyme either in the same film or by growing different layers one on top of each other (note that this is only possible with conducting polymers) and finally, the polymer deposition is localised at the surface of the electrode so that the method is suited to the spatially localised deposition of enzymes onto microelectrode arrays. An example of this approach is the entrapment of glucose oxidase within a conductive matrix of polypyrrole at the surface of a platinum electrode. Film growth occurs because the oxidation of the monomer (pyrrole) results in the formation of a highly reactive radical that reacts with neighbouring pyrrole molecules to give a polymer that is predominantly $\alpha$-$\alpha'$ coupled although some branching of the polymer chains is thought to take place through $\beta$-coupling reactions. The resulting polymer has a net positive charge and incorporates anions from the bulk solution to maintain charge neutrality.

Although the above example makes use of conducting polypyrrole films, conductivity is not a pre-requisite for a suitable immobilisation medium. Indeed, electrode films incorporating enzymes have been constructed from a wide range of materials, either conducting or insulating, which include: polyaniline polyphenols, polypyridine and cobalt tetrakis (*o*-aminophenyl) porphyrin.

Clearly, the successful preparation of the enzyme layer is a primary concern in the design of an enzyme electrode. Unfortunately, this still requires a certain degree of empiricism and an appreciation of the fact that what is a suitable method for one enzyme may not work as well for another. The relevant effects that immobilisation can have are summarised as follows:

1. The apparent activity of an enzyme can be reduced after immobilisation. This in turn may be due to several factors. First, the chosen immobilisation chemistry may not be optimal and may lead to modification of the active site. Even when the immobilisation does not interfere with the active site, the nature of the support may produce diffusional barriers. Also, the very act of immobilisation may lead to unfavourable conformational changes in the protein or reduce the conformational mobility of the enzyme. Finally, changes in the microenvironment resulting from immobilisation may also lead to apparent changes in activity.
2. Although the immobilisation matrix is often seen as solely a support for the enzyme, it may nevertheless introduce partitioning effects. A positively charged matrix for example will exclude protons so an enzyme in this matrix will exhibit a lower optimal pH than usual. Similarly, a hydrophobic substrate will partition into a hydrophobic matrix and thus lower the apparent $K_M$ for that substrate.

## TRANSDUCERS

The physico-chemical changes produced by the biological material must be converted into an electrical output signal by an appropriate transducer. On the one hand, the transducer may be unspecific and indicate general parameters such as reaction enthalpy (thermistor), mass change (piezoelectric crystal oscillator) or layer thickness (reflectometer). On the other hand, a specific indication maybe achieved with ion-selective electrodes or amperometric electrodes responding to species such as hydrogen peroxide or oxygen.

## Electrochemical Transducers

There are three types of electrochemical transducers that have been used in the preparation of enzyme electrode probes: potentiometric, amperometric and conductometric. Conductometric sensors however, are usually non-specific and have a poor signal-to-noise ratio and as yet have not been applied to bioprocess monitoring.

Potentiometric transducers (ion-selective electrodes) measure the potential difference that is generated across an ion-selective membrane separating two solutions (an internal filling solution which contains the ion of interest at a fixed activity and the external sample solution). To ensure that the potential is stable with time, the measurement is made under conditions of virtually zero current flow. In addition, the potential is measured relative to a stable reference electrode of known potential. The equipment needed to make a potentiometric measurement includes an ion-selective electrode (for example pH electrode, $NH_3$ electrode, etc.) and a high input impedance potential measuring device.

Thermodynamic arguments (which will not be elaborated on here) tell us that the potential produced at the membrane corresponds to the free energy difference ($\Delta G$) associated with the gradient of activity of the analyte ions in the inner and outer solutions. Ideally, an ion-selective electrode should obey the following relationship:

$$E = K + \left(2.303\frac{RT}{nF}\right)\log a_i$$

where $E$ is the measured potential, $R$ is the gas constant (8.314 J/K mol), $T$ is the absolute temperature, n is the ionic charge and $a_i$ is the activity of the specified ion. $K$ is a constant comprising of contributions from various sources. Theoretically, the calibration graph obtained from an ion-selective electrode should have a slope of 59/n mV a standard temperature and pressure.

When an enzyme is immobilised close to the sensing head, the enzymic reaction with the analyte of interest generates a change in potential due to ion accumulation or depletion. For example, if one wishes to make potentiometric sensor for the detection of urea, the enzyme urease can be used. Urease, an enzyme isolated from the jack bean, catalyses the following reaction:

$$(H_2N)_2C=O + 2H_2O + H^+ \xrightarrow{\text{Urease}} HCO_3^- + NH_4^+$$

Thus, the reaction can be followed by using either an ammonia or carbon dioxide gas sensing electrode. Example of enzymes/ion-selective electrode combinations are given in Table 15.2.

Amperometric sensors monitor the flow of current at a working (polarised) electrode as a function of the applied potential. The potential of the electrode serves as the driving force for the electrochemical reaction, i.e., it is the controllable parameter in the experiment that causes the electroactive species present in the solution to be electrolysed (oxidised or reduced) at the surface of the electrode. As the potential of the electrode becomes more negative, the electrode becomes a stronger reductant (source of electrons). As the potential becomes more positive, so the electrode becomes a stronger oxidising agent (electron sink). The current resulting from an oxidation or reduction, is termed the faradaic current because it obeys Faraday's laws of electrolysis. The faradaic current is a direct measure of the rate of the electrochemical reaction taking place at the electrode surface and is dependent on two things: (i) the rate at which the species moves from the bulk solution to the electrode surface (mass transport); and (ii) the rate of electron transfer across the charged interface (charge transfer).

**Table 15.2.** Examples of enzyme/ion-selective electrode combinations.

| *Substance detected* | *Enzyme* | *Ion-selective electrode* |
|---|---|---|
| Urea | Urease | pH, $NH_3$, $CO_2$ |
| Glucose | Glucose oxidase | pH, $I^-$ |
| L-Amino acids | L-Amino acid oxidase | $NH_4^+$, $I^-$ |
| L-Tyrosine | L-Tyrosine decarboxylase | $CO_2$ |
| L-Glutamine | Glutaminase | Cation |
| L-Glutamic acid | Glutamate dehydrogenase | Cation |
| L-Asparagine | Asparaginase | Cation |
| D-Amino acids | D-Amino acid oxidase | Cation |
| Penicillin | Penicillinase | pH |
| Amygdalin | β-Glucosidase | $CN^-$ |
| Nitrate | Nitrate reductase | $NH_4^+$ |
| Nitrite | Nitrite reductase | $NH_3$ |

Oxidation involves the loss of electrons. Reduction involves the gain of electrons. Amperometric enzyme electrodes are based upon redox enzymes such as glucose oxidase. This class of enzyme catalyses the oxidation of diverse substrates such as fatty acids, sugars, amino acids, aldehydes and phenols utilising molecular oxygen as the electron sink. Hydrogen peroxide is produced as a by-product of the reaction which can be detected amperometrically. Alternatively, the rate of reaction can be monitored by measuring the depletion in oxygen tension in a sealed reaction vessel.

Oxygen or hydrogen peroxide based electrodes are strongly influenced by local fluctuations in oxygen tension that can result from changes in pH, temperature, ionic strength or partial pressure. In addition, hydrogen peroxide electrodes are prone to interference due to non-specific electrochemical oxidation of endogenous substances like ascorbate, urate, glutathione and cysteine. These problems have led to the development of chemically modified electrodes in which the natural electron acceptor for the enzyme reaction (dioxygen) is replaced by an artificial oxidant (mediators) like ferricinium, methylene blue, ferricyanide, tetrathiafulvalene, etc. Mediators can be incorporated into the electrode by: (i) adsorption using a mediator organic solvent solution that is allowed to evaporate; (ii) entrapment in or behind a polymer film; (iii) covalent binding to polymers that can then be deposited onto the electrode surface; (iv) mixing into a paste of graphite and mineral oil (carbon paste electrodes); (v) covalent attachment to preformed polymers on the electrode surface; and (vi) *in situ* formation during the electropolymerisation of monomer units onto the surface of the electrode.

### Biosensors Based on Thermal Effects

A mediator is a small redox couple which takes the place of the enzyme's natural electron acceptor. Because mediators are mobile, they can interact at the active site of the enzyme, undergo electron exchange and then diffuse to the surface of the electrode where they become re-oxidised.

These types of devices are based on the principle that enzymic reactions are exothermic in nature. This fact can be used to calorimetrically determine the amount of substrate converted to product during the enzyme-catalysed reaction. Molar enthalpies for enzyme-catalysed reactions range from 5 to 100 KJ $mol^{-1}$. Modern thermistors can be used to accurately measure temperature changes as low as $10^{-4}$ °C. As with enzyme electrodes, the enzyme is attached directly to the surface of the temperature transducer (thermistor) either by cross-linking or entrapment. Alternatively, the enzyme can be placed in a temperature controlled 'reactor column' and the heat of reaction measured by recording the difference in temperature between the column outlet and the column inlet. Although a general method for measuring enzyme-catalysed reactions, calorimetry suffers from two major drawbacks: (i) non-specific thermal effects resulting in overestimation of the substrate concentration; and (ii) baseline drift due to heating of the measurement apparatus.

### Biosensors Based on Optical Effects

In sensors based on optical methods of detection, the modulation of electromagnetic radiation such as UV/vis absorption, bio and chemiluminescence, reflectance, fluorescence and surface plasmon oscillation caused by the interaction of the biocatalyst with the target analyte is monitored optically. A key consideration in these types of device is the use of optical fibre waveguide technology. This involves the synthesis of two ideas: (i) the use of optical fibres to bring light of the appropriate wavelength from a spectrometer and back again; and (ii) the use of optical fibres as an immobilisation support for the biocatalyst thereby allowing the reagent to be used on a continuous rather than a 'one-off' basis.

Because most optical sensors that have been developed to date respond to analytes that can also be sensed electrochemically, electrochemical sensors have provided the 'frame of reference' for evaluating the performance of optical sensors. This type of comparison should be treated with caution, however, because optical sensors are based on very different principles than electrical ones. Consequently, the relative merit of each approach will depend on both the particular analyte being measured and the operational demands of the particular application. In general terms, optical sensors offer the following advantages over their electrochemical counterparts.

1. No reference electrode is required.
2. Because the signal is optical it is not subject to electrical interference. This is particularly important for sensors that are operated in a particularly noisy environment such as in a fermenter vessel.
3. The immobilised biocatalyst does not have to be in direct contact with optical fibre.
4. Optical devices are inherently safer than electrical devices.
5. Optical sensors can be highly stable with respect to calibration and baseline drift.

Optical sensors that simultaneously respond to more than one analyte can be prepared using multiple immobilised reagents with different wavelength characteristics. For example, Scheper described a biooptrode (optical biosensor) for the simultaneous detection of glucose, fructose, gluconolactone and sorbitol that could be used for on-line process monitoring. The device made use of the glucose-fructose oxidase complex isolated from *Zymomonas mobilis*. The oxidation and reduction of the co-enzyme $NADP^+/NADPH$ was monitored by measuring the fluorescence intensity at 450 nm. In another type of biooptrode, the pH modulated fluorescent property of fluorescein isothiocyanate was used to measure urea and penicillin after the co-immobilisation of urease and penicillinase. For a diagrammatic representation of the optrode (Fig. 15.6).

## APPLICATIONS

Although perceived as an important area, there have not been many reports of biosensors for fermentation monitoring and control. This is probably due to the fact that the fermenter is not the most ideal environment for biosensors and the need to invest heavy capital towards research and development. In spite of this however, there are groups worldwide who are looking at the development and implementation of biosensors for process control. Their work, in chronological order, will be summarised in this section.

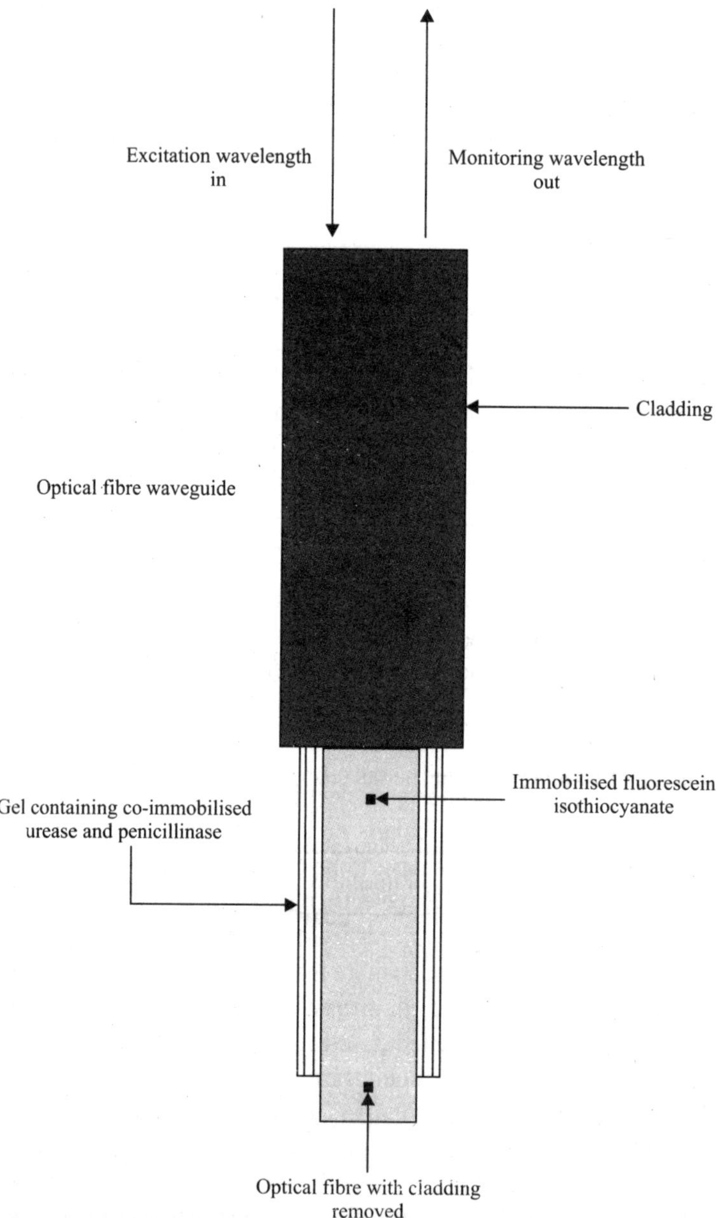


**Fig. 15.6.** Schematic diagram of an optrode for the measurement of urea and penicillin. The immobilised layer contains both urease and penicillinase.

## Electrochemical Probes

### *Potentiometric biosensors*

The biosensors used for process monitoring and control are an on-line fully automated system for the monitoring and control of glucose (carbon source) concentration in a fed-batch yeast fermentation system.

Briefly, cell-free filtrate was withdrawn aseptically from the fermenter using a ceramic membrane cartridge. The filtrate was then passed over a potentiometric glucose biosensor and the glucose concentration determined. The resolution of the analysing system was 1 minute. Using this approach and an 'on–off' control strategy, good feedback control of the glucose concentration in the fermenter was achieved.

Meyerhoff and co-workers utilised a flow injection analysis (FIA) system to monitor glutamine concentration in hybridoma cell cultures. Here, the enzyme glutaminase was immobilised in a reactor column through which samples of filtered growth medium were injected and the concentration of glutamine was determined from the ammonium ions produced in the flow-through glutaminase column.

Brand described a multi-analyte field effect transistor (FET) capable of the simultaneous detection of glucose, urea, penicillin G and cephalosporin in an FIA set-up. This was used successfully to monitor these analytes during the cultivation of a variety of micro-organisms including *E. coli* K12 MF, *E. coli* JM103, *S. cerivisiae* and *Candida boidinii.*

### Amperometric biosensors

Amperometric biosensors for process monitoring and control can be traced back to the late 1980s. For example, Romette and Cooney described a biosensor for monitoring glutamine in mammalian cell cultures. Glutaminase and glutamate oxidase were insolubilised in a thin gelatine membrane that was subsequently fixed over the surface of a polarographic oxygen electrode. Correlation between oxygen consumption and glutamine concentration provided a linear response between 0.2 and 2 mM glutamine.

Kleman reported a computer-assisted on-line glucose analyser for feedback control of *E. coli* cell growth in fed-batch cultures. Glucose concentration was determined after cross-flow removal of particulate matter using a commercially available glucose analyser (YSI glucose analyser). The concentration of glucose in the fermenter was maintained at a state level of $1.5 \pm 0.5$ g/l by feedback control. The device was very stable with the enzyme membrane requiring changing after every 600 to 800 measurements. Holst and co-workers also used a commercial glucose analyser for the monitoring of glucose concentrations in cultures of yeast and *E. coli*. Samples were taken every 90 seconds and measured over the working range 0–5 g/l. This analyser, linked to an automatic injection system, was used to monitor and control glucose concentrations during continuous cultivation of *Schizosaccharomyces pombe* as well as continuous and fed-batch cultures of *S. cerivisiae*. On-line sampling was carried out through a sample loop system in the fermenter vessel itself.

A different approach was employed by Brooks who described a dimethylferrocene-mediated glucose sensor for on-line monitoring of glucose. Despite problems of poor stability and drift with the prototype device, Bradley and Schmid constructed an *in situ* ferrocene-mediated biosensor for glucose consisting of an outer housing of stainless steel closed at one end by a sterilisable polycarbonate membrane. After sterilisation of the membrane the biosensor, which consisted of four graphite electrodes press fitted into cavities on its surface, was introduced and held in place against the polycarbonate membrane. The biosensor performed satisfactorily during 4 hours of continuous use and had a dynamic range up to 20 $g/l^{-1}$ of glucose. Careful placement in the fermenter vessel ensured that fouling of the surface was minimised.

Rishpon described a glucose probe for *in situ* monitoring of glucose concentration during fermentation. The probe had an interesting design in that it consisted of a platinum mesh electrode (onto which glucose oxidase was immobilised) sandwiched between a dialysis membrane on one side and an oxygen-permeable silicone membrane on the other. The function of the dialysis membrane was to prevent fouling from large macromolecules and cells whereas the silicone membrane provided oxygen for the glucose oxidase reaction. This electrode was applied to the continuous monitoring of *S. cerivisiae* cultivation for a period of 28 hours. During this time, however, it was noted that the probe suffered from clogging of the dialysis membrane resulting in unstable outputs.

The use of amperometric oxygen detection was used by Scheper for the on-line monitoring and control of monosaccharides, disaccharides, lactate and amino acids during the production of alkaline protease and penicillin. Immobilised reactor columns of glucose oxidase, lactate oxidase and L-amino acid oxidase were coupled to oxygen electrodes to monitor the respective analytes. Sucrose, lactose and maltose were also monitored by first converting them to glucose using in-line reactor columns containing the enzymes invertase, lactase or maltase.

Finally, a number of groups worldwide have investigated the application of hydrogen peroxide sensors for glucose and glutamine analysis in mammalian cell culture. Renneberg and co-workers for example developed an enzyme sensor based FIA system to monitor glucose, glutamine and lactate in cultures of hybridoma cells producing Mab and interleukin 2. Cattaneo and Luong have also developed a biosensor system for monitoring glutamine concentration in mammalian cell cultures. The system consisted of two electrodes. The first had co-immobilised glutamate oxidase and glutaminase and the second incorporated glutamate oxidase alone. The first electrode responded to both glutamine and glutamate whereas the second electrode responded only to glutamate. The difference reading was used to give a measure of the total glutamine present in the culture.

An alternative method to alleviate the problem of endogenous glutamate interference is to use ion exchange. Here, an anion-exchange column is placed upstream of the enzyme columns in the biosensor-FIA system. The purpose of the anion-exchange column is to remove the negatively charged glutamate ion from the sample slug. This approach is effective not only in removing glutamate but also in removing other potential interferents like ascorbate and urate.

### Thermal biosensors

The approach to process monitoring and control using thermal biosensors as the analytical device is essentially the same as that described for electrochemical sensors with the important difference that in this case, enthalpy changes are measured as opposed to electrical ones. Usage of these devices either as *in situ* devices or in an FIA manifold is however, common.

Mattiasson described the operation of a thermistor type biosensor for monitoring glucose in untreated, unfiltered fermentation samples. Glucose oxidase was immobilised onto glass beads which were packed into a column. A dialysis membrane was used to prevent fouling.

Hundeck and co-workers described a four enzyme column system using a thermistor with a temperature resolution of $10^{-5}$ K. Four different enzyme columns and a blank column were used in conjunction with a calorimeter device. Glucose, fructose, maltose and sucrose were monitored simultaneously in filtered bioreactor liquor.

Rank also described an enzyme-thermistor biosensor system using split flow for monitoring penicillin V, glucose and ethanol in bioreactor vessels. Non-specific heat effects (for example from the movement of the flowing stream) were eliminated by the use of a reference column.

### Optical biosensors

Once again, the ethos of developing sensors based on optical transduction is essentially the same, i.e., an enzyme such as glucose oxidase is used to give the specificity and the enzymic reaction is monitored through an optical change. For example, Huang described a chemiluminescent detection scheme linked to FIA for on-line monitoring of glucose in animal cell cultures. Here, glucose was converted with immobilised glucose oxidase and the hydrogen peroxide produced reacted with luminol and the intensity of the emitted light at 425 nm was measured. By using immobilised glutaminase and glutamate oxidase, Cattaneo and Luong used the same methodology to monitor glutamine concentration in animal cell culture.

Fibre-optic based oxygen sensors have also been used in conjunction with glucose oxidase. Oxygen optrodes are generally based on fibre-optics covered with a rubber film containing decacyclene, a fluorescent dye which responds strongly to oxygen partial pressure. Filtered samples from fermenter vessels can then be passed through immobilised reactor columns and the decrease in oxygen partial pressure detected using the optrode downstream.

Reduced nicotinamide co-enzymes fluoresce strongly. Thus, glucose and lactate can be detected by passing samples through immobilised reactor columns containing the enzymes such as glucose, alcohol and lactate dehydrogenase. $NAD^+$ is provided in the sample buffer and the NADH formed in the biocatalytic reactions is detected via fluorescence. This approach has been investigated by Ogbomo and co-workers for the parallel determination of glucose and ethanol during bakers' yeast fermentation. Finally, the fluorescence properties of NADH have been exploited to monitor ethanol, glucose, gluconolactone and sorbitol in fermentation vessels.

Chapter 16

# Fermentation Process Validation

## INTRODUCTION

Validation is relevant to all stages of process development and is an ongoing activity requiring company-wide input. Validation is an integral component of current Good Manufacturing Practices (cGMP). Validation applies to all the elements involved in production of a therapeutic substance (otherwise known as Active Pharmaceutical Ingredient or API). This includes facility maintenance and operation; equipment maintenance and operation; cleaning; sterilisation; analytical methods; and the process itself. Most of these systems have been well established in the chemical drugs industry and readily transfer to the biotechnology field.

There are only a few sources of guidance which discuss how process validation is actually accomplished. Most of the available documents indicate that validation of processes must be carried out but provide little pragmatic information as to how this might be achieved. The main problems arise when process validation is considered for fermentation-based manufacturing.

It is well understood that biological production systems, as applied for recombinant DNA derived products, are perceived to be inherently variable. The micro-organisms which form the basis of the process will adapt in a transitional manner to changes in the physico-chemical environment. It may equally be argued, from a physiological viewpoint, that biological processes are actually inherently stable. They will tend to do the same things with a substrate over a comparatively broad range of conditions and genetic adaptation through mutation can take a large number of generations to establish. However, the burden of 'proof' remains to demonstration of this stability.

Products derived from biologically-based processes are regulated under the auspices of the FDA's Centre of Biologics Evaluation and Research (CBER). The current philosophy is that this class of drugs cannot rely solely on final product testing for confidence of patient safety. This thinking is rooted in the issues of biological complexity and the limited sensitivity of analytical techniques. The end result is that drugs derived from biological production processes rely on demonstration of process consistency for safety consideration (the 'product by process' doctrine). Process validation thus becomes the primary tool applied to regulate the biotechnology industry.

Validation can be prospective—where it is conducted before a process is used in manufacturing; concurrent—where it is conducted alongside manufacturing; or retrospective—where the manufacturing process is in existence and is ongoing. The ideal is to conduct prospective validation followed by concurrent process monitoring rather than to attempt to establish a manufacturing process first.

## PROCESS VALIDATION

Process validation is defined as: 'Establishing documented evidence which provides a high degree of assurance that a specific process will consistently produce a product meeting its pre-determined specifications and quality attributes.'

It is essential that validation be properly documented and conducted under appropriate levels of control. Consistency within a specified matrix of operational ranges according to a controlled set of manufacturing instructions is the ultimate goal for a validated process. These operational ranges or limits must be demonstrated to encompass parameter values which do not have significant impact on product safety. In terms of fermentation, at least, it is the requirement for 'demonstration' which is the most problematic. There is almost an inconsistency between the concept of supporting validation outcomes by analysis of end product and the 'product by process' approach to drug approval. However, there is a limit to how much clinical testing (the ultimate 'proof' of product safety) that can be done and validation will provide information as to the response surface where risk is minimised. Validation should not be considered as a discrete event, conducted once and subsequently archived. GMP requires ongoing monitoring of processes as an essential component of compliance. This monitoring should be applied using classical trend analysis methods—especially to permit identification of a potential problem before process failure and product rejection. Such monitoring also requires periodic scientific review of data, ideally by members of the Research and Development group responsible for development of the process. Validation thus becomes a company-wide activity.

In classical terms, validation involves manipulation of one variable at a time (be it the preparation of a buffer for an assay, the load to an autoclave, the concentration of a disinfectant or cleaning agent, for example), demonstrating that the outcome of the procedure is the same and showing that there is no adverse impact on the product. This approach is manageable where the number of parameters to be evaluated is small and the interaction between variables inconsequential. In terms of fermentation, such a structured approach rapidly runs into the problem of number of experiments required. The product of a fermentation step in a process is often a relatively crude mixture of components which is difficult to analyse in a meaningful, quantitative manner. In addition, if final product quality becomes the yardstick, the human resources involvement is unsupportable for any but the largest and most profitable manufacturers. This is both because the number of variables is large and because they are interdependent. It is therefore necessary to reduce the number of variables to be tested to a minimum whilst retaining an understanding of the manufacturing plant reality. It is evidently possible to set all operational limits for a process as being the same as the practical limits of control imposed by the process equipment. Such an approach has high safety profile but will also result in a high number of rejected processes. There is thus a trade-off between definition of process ranges which reduce risk of failure and their demonstration as acceptable ('safe').

The preferential entry point to validation for fermentation is the development of the process. During this stage, the important parameters are defined and the preferred values determined. The capacity of the purification process to purify product away from host-derived components is normally determined concurrently with fermentation. Finally, the analytical tools to characterise the product through the process are developed in response to experience with the vagaries of the product itself in context of the production system. The prime importance of good documentation during process development is thus highlighted. The end result is a body of information surrounding a bench-scale process which provides product of a certain quality. The bench-scale process is then taken to pilot or full manufacturing scale prior to testing in animal models for toxicity and subsequently into Phase 1 clinical trial for human safety. This provides the baseline process profile with defined transfer points between stages and a series of in-process control analyses which is the benchmark for validation.

The question of in-process control testing is also an important consideration, since these are included in the quality indicators used in demonstrating successful validation. The process development phase (including scale-up) should be used to gather data to allow elimination of variables from the list of those items requiring validation. During development, there is a tendency to maximise the number of tests in a process—partly on the assumption that this will improve control. These tend to be carried through to scale-up—mainly to provide confidence of successful transfer from bench to pilot or production scale. However, each of these must be considered during validation. Thus the preferred approach is to monitor a large number of variables during process development and to identify those which are critical indicators of process control. The majority of the tests can thus be dropped from validation to monitoring status. In this regard, a good method need only be precise. Variability in an analytical method can only compound the difficulties of validating a potentially variable process.

Once the critical process parameters have been identified through the development programmed a validation protocol is written. The goal of this is to establish a series of experiments to test the limits for these parameters—individually and in combination. There is an important caveat at this point—numerology dictates that the number of combinations to be tested increases dramatically with the number of individual cases involved. A scientific case may be made to eliminate many of these combinations as insignificant, statistically unlikely or physiologically implausible. For those parameters which remain, the validation will ideally, but not necessarily, demonstrate the failure limits of the process as well as the safe limits and the operational limits to form a nested set. That is where the operational (control) limits are encompassed by the range of values shown to have no evident impact on product (safe limits) and which, in turn, lie within those conditions which result in process failure. Thereby providing a high degree of confidence that the process will consistently provide a product which is safe for therapeutic use. The 'good scientific argument' scenario may be coupled with a definition of 'worst case' conditions to further reduce the number of experiments required. Thus, for example, the upper limit of parameter 'A' may have more potential consequence on product quality than the lower limit. It may therefore be possible to define a lower safe limit and both safe and failure limits for the high ends of the range. For other parameters, there may be no discrete 'failure' limits—for example, process temperature, within a relatively broad range, may be considered to simply reduce the rate of the biochemical processes occurring within the cell. Thus, the process will arrive at a certain point at a time which is dependent on the temperature. Assuming that this point is the same (i.e., a target biomass for culture harvest) then the process can be considered not to have failed over this entire range. However, in the real world, process time in a manufacturing plant is costly and so the process will be operated under conditions designed to get the purified product in the shortest reasonable time. Thus the 'failure' limits can be considered to be those outside of which the process takes too long or is too short to fit into the plant schedule. There is no rational reason to define or validate these ranges. However, experience in a manufacturing environment quickly shows that process controllers do malfunction and this can result in 'drift' of a control variable and it is essential, from a pragmatic perspective, to have evidence which shows that this deviation has no adverse effect on the product.

Arguably one of the most important sections of the validation protocol is that which defines the success or failure of the process. The product of a fermentation process is, at best, a process intermediate where the ultimate product exists in a soup of 'background' contaminants. It is often difficult, if not impossible with current methods, to evaluate the quality of the product at this stage of the overall manufacturing process. The quantity or yield of product may be an acceptable surrogate for most experiments. It remains impossible to predict the effects of changes in fermentation on the profile of contaminants through the recovery and purification stages to the final drug substance. That is without actually doing the purification. It may not be necessary to carry all validation experimental processes through to final drug substance since some of them may fail at an early stage and it may be possible to use normal in-process controls and specifications as acceptance criteria. This latter case is, to a certain extent, based on overcapacity built in to the process. Thus, for example, a chromatography step will not generally be conducted at maximum resin capacity—there will be some binding capacity of resin left unused with the normal process stream. Changes in the load to the column, which may arise through changes in the upstream fermentation, are accommodated by this excess capacity and the process runs normally thereafter. However, it is good practice to retain archived samples to permit evaluation of at least the preliminary stages of purification for all validation experiments. It is obvious that the majority of this validation should be conducted at bench-scale insofar as this can be claimed to be representative of the full scale process. For some manufacturing systems, it may be required to conduct some validation at pilot or manufacturing scale. In any case, validation cannot be completed until the process has been conducted successfully in the manufacturing plant on a minimum of three consecutive occasions. It is equally important in this context to consider conditions under which re-validation is required. Thus changes in manufacturing facility (including equipment, environment, procedures and materials) must be evaluated in terms of potential impact on the process and if necessary, tested before implementation.

## FERMENTATION PROCESS VALIDATION

The detailed approach to fermentation process validation is dependent on the nature of the process—whether it is based on intra or extracellular accumulation of product; whether the producing strain has particular sensitivities or characteristics and the characteristics of the product itself. The validation plan must be prepared with an understanding of how the fermentation works. The questions of how and why the system responds in a particular manner should be considered and ideally, understood. This absolutely requires consideration of the physiology of the production strain, since physiological state defines the chemical composition of the cells and therefore the profile of materials transferred to the purification process. From this perspective, the validation approach can be constructed and more importantly, defended on regulatory review.

A representative validation scheme can be developed for a classical recombinant DNA protein product. This assumes a three-stage fermentation from seed flasks through an intermediate fermentation step to a fed-batch production fermentation. The product accumulates inside the cell in response to a specific induction event. Transfer from stage to stage is determined by time and/or biomass-related criteria. The fermentation is operated with dissolved oxygen and pH control. Significant divergence of other common types of process from this model will be discussed where relevant. The issue of scale must also be considered—there are inescapable consequences for increase in volume of the bioreactor. These include mixing times, shear and heating/cooling profiles—all of which can impact the culture physiology and fermentation process. The preference is to conduct as much of the validation at bench scale as possible to minimise cost. However, the approach must also include critical evaluation of process scale to demonstrate why this is appropriate in context of the particular process.

In any industrial fermentation process there are a number of 'control' variables and 'response' variables which are used to direct the process through to the harvest point. The two are not necessarily distinct since a response variable can be used for process control. However, they both must be considered for validation purposes. The control variables are those which are independent of the process— raw materials, process timing, temperature, pH, air flow rate, impeller agitation speed, for example—those which the operator inputs to the system. The response variables would be biomass (whether this be optical density, dry weight, etc.) or culture physiology (carbon dioxide production rate, oxygen consumption rate, respiratory quotient, substrate consumption or metabolite production, for example). It is comparatively trivial to manipulate the control variables but equally difficult to manipulate the response variables—especially in isolation.

In practical terms, the control variables form the place to start validation. Each of these has an effect on the fermentation process at some level and they are discussed in the following sections and in the context of the process stage.

## MAIN PARAMETERS FOR VALIDATION

### Raw Materials

The validation process should include evaluation of the influence of media formulation on the process. This need not extend to a complex array of experiments to establish limits for each component—primarily this should establish that the contributions of the components to growth are understood and that any growth limiting components are identified. Supportive data on media composition can be obtained by determining yield coefficients for the various components (principally the macro-nutrients) during process development. The majority of factors would be expected to be in excess for standard batch processes, but not to the point of being toxic or growth retardant. Hence variability arising from minor changes in concentration (from balance accuracy, for example) is not significant. Where the process involves defined media with well-characterised components, it is arguable that variability is unlikely to arise from the use of different manufacturers lot of raw materials. However this may not be the case where the media formulation includes complex, ill-defined, factors such as protein hydrolysates or crude extracts of plant origin (Fig. 16.1 for example). In this scenario, it is important to evaluate effects of using different batches of the more complex items on the growth of the production organism as early as possible in development. In dealing with raw materials it is important to consider the way in which these are handled prior to introduction of the micro-organism. Thus issues of shelf life (as powder and in solution), hold time prior to sterilisation, the sterilisation process itself and even hold time pre-inoculation should be assessed for potential impact on process. It may also be important to consider the specifications for raw materials release. Typically, these are compendial (against USP or EP standards, for example) and based on chemical identity. There may be specific characteristics which should be introduced to allow for biological systems. These may include such tests as bioburden (very high levels may swamp capacity to sterilise media); endotoxin (which is of real concern for injectable drug products) or growth-affecting substances (with both positive or negative consequence). Sourcing of materials during the development phase should be done in cognizance of the manufacturing reality. When using small quantities of chemicals for bench-scale processes, it is possible to use high purity materials which are uneconomically expensive for production scale. Conversely, bench-scale processes may use water, steam and air of a lower quality than typically encountered in a manufacturing environment. However, there is unlikely to be a radical shift in enzyme complement or in macromolecular composition of the cell. Effects on product quality may be considered to be unlikely—at least in terms of contaminant profile of purified product. This phenomenon may also influence the uptake of an inducer—affecting product expression rate and ultimate process yield. The direct influence of pH on microbial growth is to decrease the specific growth rate on either side of an optimum. The rate of decline is organism specific and should ideally be determined for the production organism and under conditions of temperature and medium formulation as close to those for production as possible. Most often, the rate of decline in growth rate is lowest around the optimum pH value for the organism. This is the preferred condition for validation—where the largest pH range issues may have process impact that is not evident until the process is transferred into manufacturing with consequent costly failure. The optimal situation is to use identical raw materials during development and manufacturing.

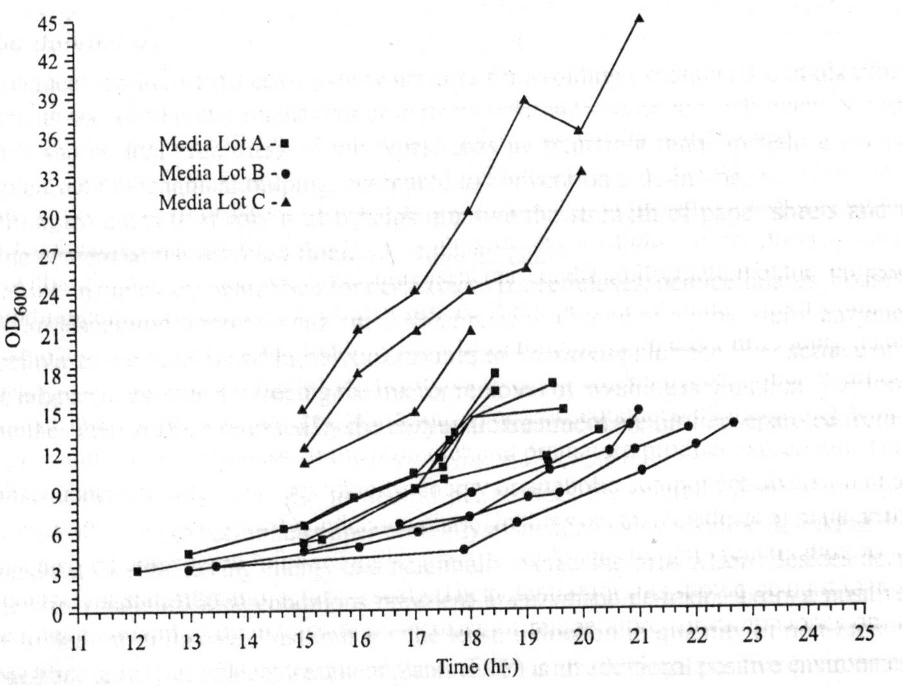


**Fig. 16.1.** Growth profiles of a series of fermentations using media prepared using different lots of casein peptone from a single manufacturer. *Streptomyces lividans* cultures were grown in 15 liters stirred tank bioreactors in glucose-casein peptone-salts medium at 32°C, pH 6.7, >25 per cent dissolved oxygen.

## pH

Micro-organisms react to a change in the pH of the external medium in a manner directed to maintenance of a constant intracellular pH. Transport of substances across the cell membrane is known to be pH sensitive, more specifically by pH gradient between the cytoplasm and the culture medium. This can influence the composition of the cell by altering the balance of tolerated before the process falls out of the biomass/time specification. The culture pH can also have dramatic effects on product where this is secreted from the producing cell (for example through interaction between product and protease). For such systems it may not be possible to conduct the process at optimal pH for growth and a tighter pH specification may result. It is evident, given the multiple possible effects, the pH is a parameter requiring validation.

There are two scenarios with pH—that it is controlled at a defined value by metered addition of acid and/or alkali or that it is not controlled other than by the buffering capacity of the medium. In general and in absence of organism-specific factors, it is arguably better from the validation perspective to control pH at a defined value. Where pH is allowed to drift from the initial value the question for validation is one of ensuring constant profile within specified limits. There are therefore two variables involved—actual pH and time in the process. In this case, validation may consist of adjusting the initial pH of the culture medium and then running the process. Analysis of potential risk is complicated by the fact that effects on the system are not constant—the organism is continually adapting to a new environment. There is also the relationship between pH and buffering capacity—being inversely related to difference between actual pH and $pK_a$ of the system. Hence the pH profile may change significantly through a minor change in initial value. The situation where pH is controlled at a defined value is comparatively simple. The micro-organisms experience a more consistent environment and validation at pH values on either side of the control value is a single parameter case. However, the nature of system controllers is that they sometimes misbehave and step changes in pH can occur. It is not necessarily appropriate to claim that a step change within a validated range is acceptable since the shock effect on cell physiology cannot be ignored. In this case, the suggested approach is to test a range of pH values which is linked to the buffering capacity of the medium. Where this is high, a relatively broad range can be applied on the argument that changes in pH within the range are likely to be slow and hence shock effects minimised. Conversely, if the buffering capacity is low, then a narrow range (but still outside the operating range of the pH control systems in the processing plant) should be applied. Evidently, media formulation development should move towards high buffering capacity around the preferred process pH value. Overall, there is a balance between a better controlled (and validatable) system which is prone to higher failure rate to one which offers lower control confidence but also lower risk of failure.

### Temperature

As for pH, temperature has a primary effect on cell growth rate and most probable cause of process failure is concerned with the biomass/time control parameters. The pattern of response is slightly different from pH in that fall off in growth rate above the optimum is frequently found to be more dramatic than at temperatures below it. Sub-optimal temperatures may be argued to show lower growth rate through thermodynamic effects on rates of reaction. Super-optimal temperatures on the other hand may cause a lowering of growth rate through protein denaturation effects—with consequences for cell composition (primarily a higher content of misfolded protein). It may therefore be argued to be more important to establish the safe and failure limits for temperatures above the process control value. It is recommended to acquire information on the growth rate/temperature relationship for the production strain as part of the development programmed.

Experiments have shown that cellular composition (protein, DNA and RNA) for *Salmonella typhimarium* did not change with temperature of incubation, despite a significant reduction in growth rate. Herbert showed that this is true only for the case in batch culture where the organism is growing at the maximum possible growth rate under the imposed conditions. There must therefore be some caution applied to arguing lack of significant temperature effect on cell composition. In a fed-batch process where growth rate is controlled at a value lower than the maximum by rate of addition of a specific nutrient there may be a change in RNA content with temperature at fixed feed rate (and *vice versa*—with flow rate at fixed temperature). These effects may be small but should not be ignored when planning validation experiments. Of some concern would be a process where feeding is defined to commence at a particular time post-inoculation. When temperature of the process is manipulated, the feed initiation will happen at a different biomass and this may affect the culture growth (by over or under-feeding respectively). Secondly, the feed rate will impose a particular growth rate (in relation to the maximum possible at that temperature, assuming that one component is limiting) and this can alter the physiological status of the organism.

Temperature (and indeed, any parameter which affects growth rate) must also be considered in context of mass transfer phenomena in the bioreactor. As temperature and growth rate increase, metabolic rate increases. The demand for nutrients (and to expel waste) will increase and the capacity of the system to effect transfer from the bulk liquid to and from the cell surface may become growth-limiting. The cells may become energy-starved (or poisoned by accumulation of metabolic by-products) and can significantly shift their composition (and hence potential contaminants to the purified product). The outcome can be process failure at sub-optimal temperature. There is a compounding effect specific to temperature since oxygen solubility decreases with increasing temperature. The capacity of the system to deliver oxygen to cells thus drops off significantly with increasing temperature.

There may be a particular issue with temperature for some commonly used approaches to control of gene product expression. In processes which use the $\gamma_{pl}$ system, temperature is used to control activity of a promoter. In this case, there is a protein (repressor) which binds to the promoter and prevents interaction with RNA polymerase. As temperature is increased, the conformation of the repressor changes to a non-binding competent state, allowing access of RNA polymerase and transcription of the gene product. The issue of concern is that the transition from binding competent to binding incompetent is not a precise value—there is a range of temperatures over which this occurs which is affected by cell composition (primarily pH and ion concentration). The consequences of increasing temperature of incubation may thus be to 'leakiness' at the promoter and premature product expression. This can have process consequences through product toxicity, energy or anabolic component diversion or attenuation of promoter activity during the specific induction phase. In this system, the effects of scale on temperature control must also be considered—for example, it may take longer to ramp temperature to achieve the binding incompetent state at manufacturing scale. The potential for this should be built in to small scale development by evaluating effects of rate of increase of temperature (or ramp slope) on the product yield and quality.

Hence a simplistic approach to validation of process temperature must be tempered by consideration of other components of the process.

### Airflow Rate and Impeller Agitation Speed

These two parameters are considered together since they are inter-related, particularly in aerobic processes. The geometry and operation of the impellers in a stirred tank bioreactor have a major impact on mixing capacity of the system. They also define, to a large extent, the shear forces acting on the micro-organisms in the process. Bulk mixing of a fluid is a product of the energy input to the system—more energy providing a shorter time to homogeneity following a perturbation. The energy input to a stirred tank bioreactor is provided via the impellers which act to accelerate the fluid. The baffles create turbulence in the fluid to enhance mixing. These accelerative and turbulent flow effects produce shear forces acting on the components of the system—including the micro-organisms. In conjunction with airflow rate (and head space pressure) impeller set-up controls gas transfer characteristics of the bioreactor. Micro-organisms vary widely in their sensitivity to shear and their demand for gas exchange and sometimes even over the course of a process. Thus unicellular bacteria are probably the least shear-sensitive and eukaryotic cells the most, with filamentous organisms somewhere in between. It is important to investigate the effects of shear on the process—especially since cell damage and release of intracellular components can have dramatic effects on an extracellular product (both by action of proteases and by increased contaminant concentration) as well as affecting growth rate. Manipulating shear in isolation is difficult because of the effects on oxygen transfer. At high shear there is inevitably better gas exchange between the air/liquid interface and at the bacterial surface. This is both advantageous and deleterious since oxygen is required for energy generation but is toxic at relatively high levels through production of superoxides and free radicals. This can impact both growth rate and cell physiology (production of protective proteins, for example) resulting in process failure. It is possible to manipulate the gas composition to decouple dissolved oxygen and impeller agitation speed but this requires specialist equipment. It is more convenient to consider effects of impeller agitation speed along with variations in dissolved oxygen. In this case, high range dissolved oxygen setpoint also covers high shear (and faster mixing times) and low range with low shear and poor mixing. In processes without oxygen control impeller agitation speed becomes the validation parameter with dissolved oxygen as a secondary (dependent) variable.

### Biomass/Time

Biomass accumulation and time are frequently utilised process control parameters. This is encountered as an instruction to transfer culture from one stage of the process to another at a particular biomass or a particular time. This being on the premise that growth rate and culture physiology are effectively synonymous. The experiments reported by Herbert show that this premise is reasonable where the culture growth rate is defined by rate of supply of a particular nutrient (i.e., in fed-batch or semi-continuous growth). It is also reasonable to assume that, all other parameters being consistent, the micro-organism will grow at a particular rate and have a particular physiology. The important caveat may be the case of temperature or pH where growth rate may be lower or higher but the organism will have essentially the same physiological status since it is growing at the maximum rate possible under the circumstances and neither would be expected to impose a particular metabolic condition. However, this relates to interpretation of validation and not validation *per se*.

Both biomass and time would have ranges associated with them to allow for normal (acceptable) process variability (Fig. 16.2 for example). The rationale is that a process is 'normal' if the organism displays a particular growth profile and anything outside of this is aberrant and should be rejected. These ranges must therefore be validated. The problem is that growth rate is not a factor which can be tightly controlled by the operators. It is generally not possible to run a process to a specified value of either time or biomass independently of the other. Process development involves evaluation of optimum biomass to achieve a particular outcome for product recovery and purification. This is generally within a timeframe convenient to the operator's schedule. The issue with biomass determination is that of precision of the method used. Optical density is less precise because it often involves significant sample dilution to remain within the linear response range of the spectrophotometer, because time taken between sampling and analysis is variable and conditions are uncontrolled and it is sample handling dependent (for example, cells can settle, introducing non-homogeneity of sample). On-line, real time methods such as biomass-specific sensors or metabolite production (especially $CO_2$) are preferred, but may not be universally applied—for example, in the shake flask stages of a process.

Validation of the biomass/time process control parameters is probably only possible concurrently with process development. During this phase, a database of information can be obtained on what is the range of values for time to achieve a specified biomass when the process is conducted normally. Time, by itself, is sometimes used to define when, during a process, to do something—for example, start a feed or induce expression. The preferred option in this case is to substitute a biomass-related parameter for time. In the absence of an appropriate indicator other than time, it is essential to validate a range to demonstrate proper process control. This is in the context of normal growth profile variability and hence biomass at the defined time of process.

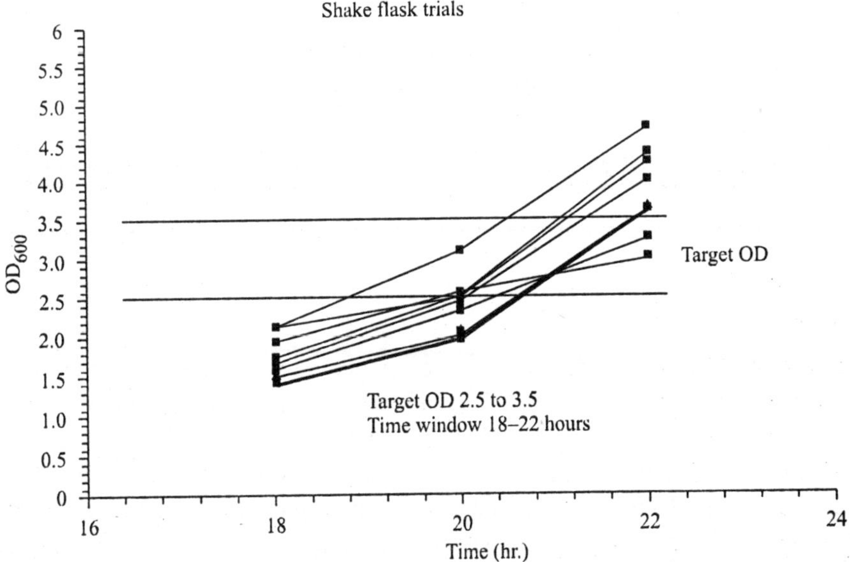


**Fig. 16.2.** Growth profiles for ten 200 ml shake flask cultures of *Streptomyces lividans* grown in 1.0 liter Erlenmeyer flasks with baffles in glucose-casein peptone-salts medium at 32°C and 240 rpm in a new Brunswick series 25 incubator shaker.

To sum up validation of a fermentation process involves consideration of many inter-related parameters. Some of these are operator-specified and some concern the response of the organism to the imposed conditions. The preferred basis for design of a validation protocol is a sound understanding of the physiology of the organism. This understanding is acquired during the development phase and can be applied with good scientific rationale to address the real potential for process failure.

Chapter 17

# Instrumentation and Control

## INTRODUCTION

The success of a fermentation depends upon the existence of defined environmental conditions for biomass and product formation. To achieve this goal it is important to understand what is happening to a fermentation process and how to control it to obtain optimal operating conditions. Thus, temperature, pH, degree of agitation, oxygen concentration in the medium and other factors may have to be kept constant during the process. The provision of such conditions require careful monitoring (data acquisition and analysis) of the fermentation so that any deviation from the specified optimum might be corrected by a control system.

Criteria which are monitored frequently are listed in Table 17.1, along with the control processes with which they are associated. As well as aiding the maintenance of constant conditions, the monitoring of a process may provide information on the progress of the fermentation. Such information may indicate the optimum time to harvest or that the fermentation is progressing abnormally which may be indicative of contamination or strain degeneration. Thus, monitoring equipment produces information indicating fermentation progress as well as being linked to a suitable control system.

In initial studies the number of functions which are to be controlled may be restricted in order to gain more knowledge about a particular fermentation. Thus, the pH may be measured and recorded but not maintained at a specified pH or the dissolved oxygen concentration may be determined but no attempt will be made to prevent oxygen depletion.

Also, it is important to consider the need for a sensor and its associated control system to interface with a computer (to be discussed in a later section). This chapter will consider the general types of control systems which are available, specific monitoring of control systems and the role of computers. It is apparent from Table 17.1 that a considerable number of process variables may need to be monitored during a fermentation. Methods for measuring these variables, the sensors or other equipment available and possible control procedures are outlined below.

There are three main classes of sensor:

1. Sensors which penetrate into the interior of the fermenter, e.g., pH electrodes, dissolved-oxygen electrodes.
2. Sensors which operate on samples which are continuously withdrawn from the fermenter, e.g., exhaust-gas analysers.
3. Sensors which do not come into contact with the fermentation broth or gases, e.g., tachometers, load cells.

**Table 17.1.** Process sensors and their possible control functions.

| *Category* | *Sensor* | *Possible control function* |
|---|---|---|
| Physical | Temperature | Heat/cool |
| | Pressure | |
| | Agitator shaft power | |
| | rpm | |
| | Foam | Foam control |
| | Weight | Change flow rate |
| | Flow rate | Change flow rate |
| Chemical | pH | Acid or alkali addition, carbon source feed rate |
| | Redox | Additives to change redox potential |
| | Oxygen | Change feed rate |
| | Exit-gas analysis | Change feed rate |
| | Medium analysis | Change in medium composition |

It is also possible to characterise a sensor in relation to its application for process control:

1. In-line sensor: The sensor is an integrated part of the fermentation equipment and the measured value obtained from it is used directly for process control.
2. On-line sensor: Although the sensor is an integral part of the fermentation equipment, the measured value cannot be used directly for control. An operator must enter measured values into the control system if the data is to be used in process control.
3. Off-line sensor: The sensor is not part of the fermentation equipment. The measured value cannot be used directly for process control. An operator is needed for the actual measurement (e.g., medium analysis or dry weight sample) and for entering the measured values into the control system for process control.

When evaluating sensors to use in measurement and control it is important to consider response time, gain, sensitivity, accuracy, ease and speed of calibration, stability, reliability, output signal (continuous or discontinuous), materials of construction, robustness, sterilisation, maintenance, availability to purchase and cost.

## METHODS OF MEASURING PROCESS VARIABLES

### Temperature

The temperature in a vessel or pipe is one of the most important parameters to monitor and control in any process. It may be measured by mercury-in-glass thermometers, bimetallic thermometers, pressure bulb thermometers, thermocouples, metal-resistance thermometers or thermistors. Metal-resistance thermometers and thermistors are used in most fermentation applications. Accurate mercury-in-glass thermometers are used to check and calibrate the other forms of temperature sensors, while cheaper thermometers are still used with laboratory fermenters.

#### *Mercury-in-glass thermometers*

A mercury-in-glass thermometer may be used directly in small bench fermenters, but its fragility restricts its use. In larger fermenters it would be necessary to insert it into a thermometer pocket in the vessel, which introduces a time lag in registering the vessel temperature. This type of thermometer can be used solely for indication, not for automatic control or recording.

### *Electrical resistance thermometers*

It is well known that the electrical resistance of metals change with temperature variation. This property has been utilised in the design of resistance thermometers. The bulb of the instrument contains the resistance element, a mica framework (for very accurate measurement) or a ceramic framework (robust, but for less accurate measurement) around which the sensing element is wound. A platinum wire of 100 Ω resistance is normally used. Leads emerging from the bulb are connected to the measuring element. The reading is normally obtained by the use of a Wheat stone bridge circuit and is a measure of the average temperature of the sensing element. This type of thermometer does have a greater accuracy (±0.25 per cent) than some of the other measuring devices and is more sensitive to small temperature changes. There is a fast response to detectable changes (1 to 10 seconds) and there is no restriction on distance between the very compact sensing point (30 × 5 mm) and the display point of reproducible readings. These thermometers are normally enclosed in stainless-steel sheaths if they are to be used in large vessels and ancillary equipment.

### *Thermistors*

Thermistors are semi-conductors made from specific mixtures of pure oxides of iron, nickel and other metals. Their main characteristic is a large change in resistance with a small temperature change. The change in resistance is a function of absolute temperature. The temperature reading is obtained with a Wheatstone bridge or a simpler or more complex circuit depending on the application. Thermistors are relatively cheap and have proved to be very stable, give reproducible readings and can be sited remotely from the read-out point. Their main disadvantages is the marked non-linear temperature versus resistance curve.

### *Temperature control*

In many small systems there is a heating element, 300 to 400 W capacity being adequate for a 10 $dm^3$ fermenter and a cooling water supply; these are on or off depending on the need for heating or cooling. The heating element should be as small as possible to reduce the size of the 'heat sink' and resulting overshoot when heating is no longer required. In some cases it may be better to run the cooling water continuously at a steady rate and to have the heating element only connected to the control unit. This can be an expensive mode of operation if the water flows directly to waste. For small-scale use, Harvard Apparatus Ltd. (Fircroft Way, Edenbridge, Kent, UK) make a unit, the thermocirculator, which will pump recirculating thermostatically heated water through fermenters of up to 10 $dm^3$ capacity and give temperature control of ±0.1°C.

In large fermenters, where heating during the fermentation is not normally required, a regulatory valve at the cooling water inlet may be sufficient to control the temperature. There may be provision for circulation of refrigerated brine if excessive cooling is required. Steam inlet to the coil and jacket must be present if a fermenter is being used for batch sterilisation of media.

Low agitation speeds are often essential in animal culture vessels to minimise shear damage. In these vessels, heating fingers can create local 'hot-spots' which may cause damage to cells very close to them. Heating jackets which have a lower heat output proportional to the surface area (water or silicone rubber covered electrical heating elements) are used to overcome this problem.

## Flow Measurement and Control

Flow measurement and control of both gases and liquids is important in process management.

### *Gases*

One of the simplest methods for measuring gas flow to a fermenter is by means of a variable area meter. The most commonly used example is a rotameter, which consists of a vertically mounted glass tube with an increasing bore and enclosing a free-moving float which may be a ball or a hollow thimble. The position of the float in the graduated glass tube is indicative of flow rate. Different sizes can cater for a wide range of flow rates. The accuracy depends on having the gas at a constant pressure, but errors of up to ±10 per cent of full-scale deflection are quoted. The errors are greatest at low flow rates. Ideally, rotameters should not be sterilised and are therefore normally placed between a gas inlet and a sterile filter. There is no provision for on-line data logging with the simple rotameters. Metal tubes can be used in situations where glass is not satisfactory. In these cases the float position is determined by magnetic or electrical techniques, but this provision has not been normally utilised for fermentation work. Rotameters can also be used to measure liquid flow rates, provided that abrasive particles or fibrous matter are not present.

The use of oxygen and carbon dioxide gas analysers for effluent gas analysis requires the provision of very accurate gas-flow measurement if the analysers are to be used effectively. For this reason thermal mass flowmeters have been utilised for the range 0 to 500 $dm^3\ min^{-1}$. These instruments have a ±1 per cent full-scale accuracy and work on the principle of measuring a temperature difference across a heating device placed in the path of the gas flow (Fig. 17.1). Temperature probes such as thermistors are placed upstream and downstream of the heat source, which may be inside or outside the piping.

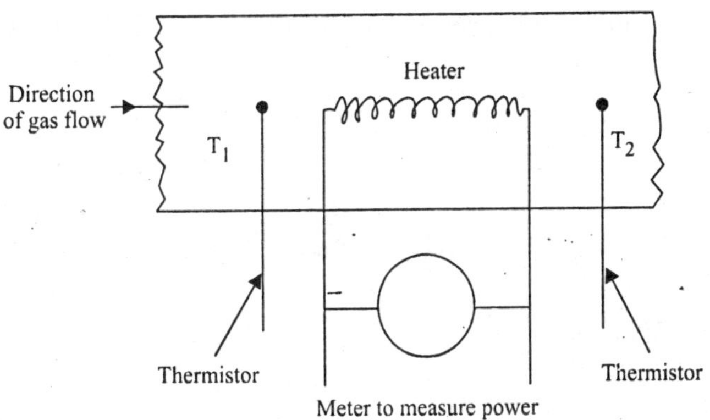


**Fig. 17.1.** Thermal mass flowmeter.

The mass flow rate of the gas, $Q$, can be calculated from the specific heat equation:

$$H = QC_P (T_2 - T_1)$$

where,
$H$ = heat transferred
$Q$ = mass flow rate of the gas
$C_P$ = specific heat of the gas
$T_1$ = temperature of gas before heat is transferred to it
$T_2$ = temperature of gas after heat is transferred to it.

This equation can then rearranged for Q:

$$Q = \frac{H}{C_p(T_2 \quad T_1)}$$

A voltage signal can be obtained by this method of measurement which can be utilised in data logging.

Control of gas flow is usually by needle valves. Often this method of control is not sufficient and it is necessary to incorporate a self-acting flow-control valve. At a small scale, such valves as the 'flowstat' are available. Fluctuations in pressure in a flow-measuring orifice cause a valve or piston pressing against a spring to gradually open or close so that the original, pre-selected flow rate is restored. In a gas 'flowstat', the orifice should be upstream when the gas supply is at a regulated pressure and downstream when the supply pressure fluctuates and the back pressure is constant. Valves operating by a similar mechanism are available for larger scale applications.

### *Liquids*

The flow of non-sterile liquids can be monitored by a number of techniques, but measurement of flow rates of sterile liquids presents a number of problems which have to be overcome. On a laboratory scale flow rates may be measured manually using a sterile burette connected to the feed pipe and timing the exit of a measured volume. The possible use of rotameters has already been mentioned in the previous section. A more expensive method is to use an electrical flow transducer which can cope with particulate matter in suspension and measure a range of flow rates from very low to high (50 $cm^3\ min^{-1}$ to 5,00,000 $dm^3\ min^{-1}$) with an accuracy of ±1 per cent. In this flowmeter (Fig. 17.2) there are two windings outside the tube, supplied with an alternating current to create a magnetic field. The voltage induced in the field is proportional to the relative velocity of the fluid and the magnetic field. The potential difference in the fluid can be measured by a pair of electrodes and is directly proportional to the velocity of the fluid.

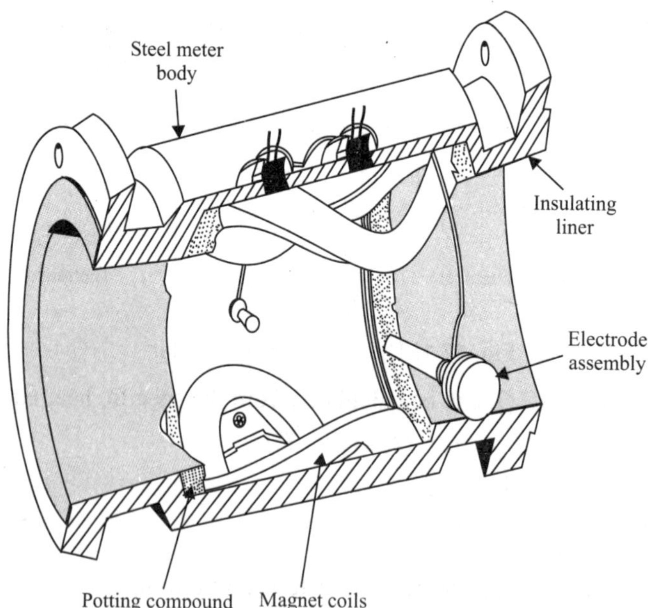


**Fig. 17.2.** A cut-away view of a short-form magnetic flowmeter.

In batch and fed-batch culture fermenters, a cheaper alternative is to measure flow rates indirectly by load cells (see Weight section). The fermenter and all ancillary reservoirs are attached to load cells which monitor the increases and decreases in weight of the various vessels at regular time intervals. Provided the specific gravities of the liquids are known it is possible to estimate flow rates fairly accurately in the different feed pipes. This is another techniques which may be used with particulate suspensions.

Another indirect method of measuring flow rates aseptically is to use a metering pump which pumps liquid continuously at a pre-determined and accurate rate. A variety of metering pumps are commercially available including motorised syringes, peristaltic pumps, piston pumps and diaphragm pumps. Motorised syringes are used only when very small quantities of liquid have to be added slowly to a vessel. In a peristaltic pump, liquid is moved forwards gradually by squeezing a tubing held in a semi-circular housing. A variety of sizes of tubes can be used in different pumps to produce different known flow rates over a very wide range. Suspensions can be handled since the liquid has no direct contact with moving parts. A piston pump contains an accurately machined ceramic or stainless-steel piston moving in a cylinder normally fitted with double ball inlet and outlet valves. The piston is driven by a constant-speed motor. Flow rates can be varied within a defined range by changing the stroke rate, the length of the piston stroke and by using a different piston size. Sizes are available from $cm^3$ $hr.^{-1}$ to thousands of $dm^3$ $hr.^{-1}$ and all can be operated at relatively high working pressure. Unfortunately, they cannot be used to pump fibrous or particulate suspensions. Piston pumps are more expensive than comparable sized peristaltic pumps but do not suffer from tube failure.

Leakage can occur via the shaft housing of a piston pump. The problem can be prevented by the use of a diaphragm pump. This pump uses a flexible diaphragm to pump fluid through a housing (Fig. 17.3) with ball valves to control the direction of flow. The diaphragm may be made of, e.g., teflon, neoprene, stainless steel and is actuated by a piston. A range of sizes of pumps is available for flow rates up to thousands of $dm^3$ $hr.^{-1}$.

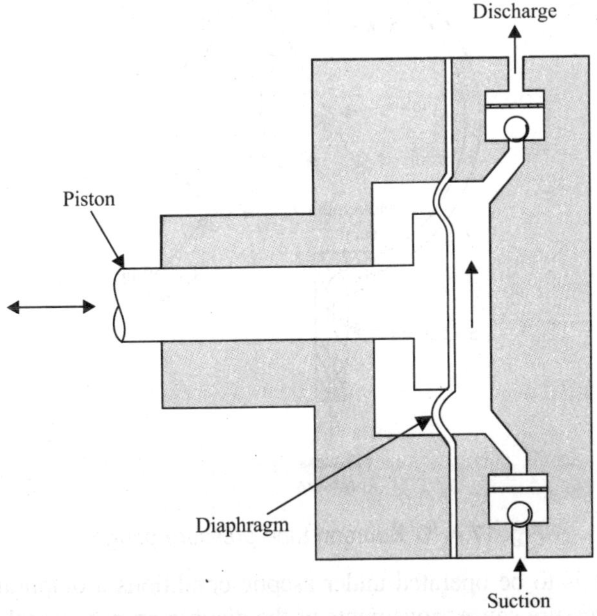


**Fig. 17.3.** A direct-driven diaphragm pump.

Liquid flow from a nutrient feed tank or to or out of a fermenter may be monitored by continuous weighing on a balance or load cell(s). This will be discussed in the Weight section.

### Pressure Measurement

Pressure is one of the crucial measurements that must be made when operating many processes. Pressure measurements may be needed for several reasons, the most important of which is safety. Industrial and laboratory equipment is designed to withstand a specified working pressure plus a factor of safety. It is therefore important to fit the equipment with devices that will sense, indicate, record and control the pressure. The measurement of pressure is also important in media sterilisation. In a fermenter, pressure will influence the solubility of gases and contribute to the maintenance of sterility when a positive pressure is present.

One of the standard pressure measuring sensors is the Bourdon tube pressure gauge (Fig. 17.4), which is used as a direct indicating gauge. The partial coil has an elliptical cross-section (A–A) which tends to become circular with increasing pressure and because of the difference between the internal and external radii, gradually straightens out. The process pressure is connected to the fixed socket end of the tube while the sealed tip of the other end is connected by a geared sector and pinion movement which actuates an indicator pointer to show linear rotational response.

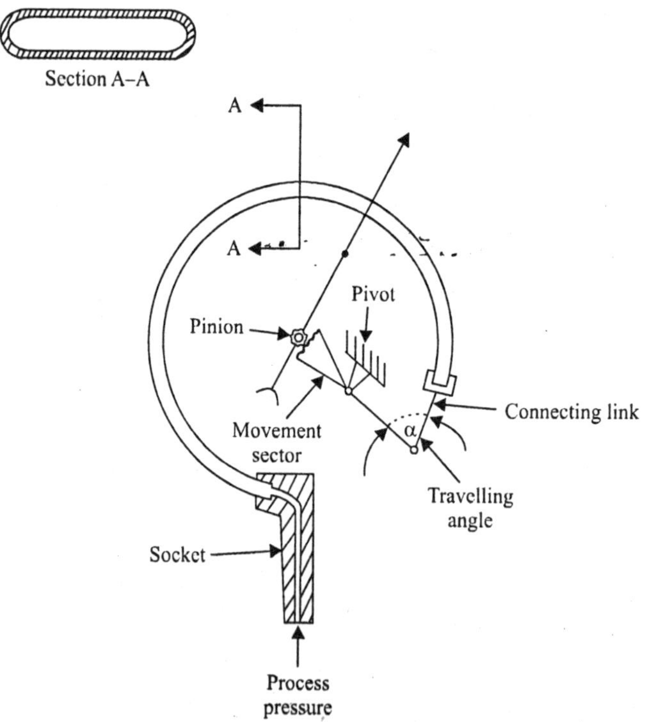


**Fig. 17.4.** 'C' Bourdon tube pressure gauge.

When a vessel or pipe is to be operated under aseptic conditions a diaphragm gauge can be used (Fig. 17.5). Changes in pressure cause movements of the diaphragm capsule which are monitored by a mechanically levered pointer.

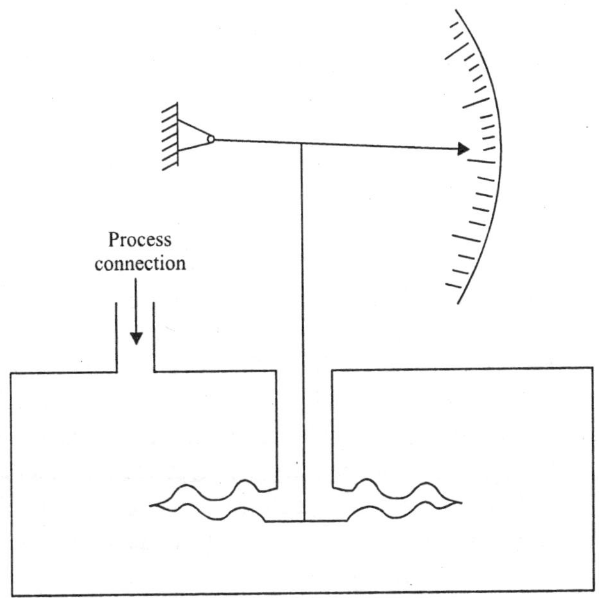


**Fig. 17.5.** Nested diaphragm-type pressure sensor.

Alternatively, the pressure could be measured remotely using pressure bellows connected to the core of a variable transformer. The movement of the core generates a corresponding output. It is also possible to use pressure sensors incorporating strain gauges. If a wire is subject to strain its electrical resistance changes; this is due, in part, to the changed dimensions of the wire and the change in resistivity which occurs due to the stress in the wire. The output can then be measured over long distances. Another electrical method is to use a piezoelectric transducer. Certain solid crystals such as quartz have an asymmetrical electrical charge distribution. Any change in shape of the crystal produces equal, external, unlike electric charges on the opposite faces of the crystal. This is the piezoelectric effect. Pressure can therefore be measured by means of electrodes attached to the opposite surfaces of the crystal. Bioengineering AG (Wald, Switzerland) have made a piezoelectrical transducer with integral temperature compensation, to overcome pyroelectric effects and built into a housing which can be put into a fermenter port.

It will also be necessary to monitor and record atmospheric pressure if oxygen concentrations in inlet and/or exit gases are to be determined using oxygen gas analysers (see later section). Paramagnetic gas analysers are susceptible to changes in barometric pressure. A change of 1 per cent in pressure may cause a 1 per cent change in oxygen concentration reading. This size of error may be very significant in a vessel where the oxygen consumption rate is very low and there is very little difference between the inlet and exit gas compositions. The pressure changes should be constantly monitored to enable the appropriate corrections to be made.

## Pressure Control

Different working pressures are required in different parts of a fermentation point. During normal operation a positive head pressure of 1.2 atmospheres (161 $kN^{-1}$) absolute is maintained in a fermenter to assist in the maintenance of aseptic conditions. This pressure will obviously be raised during a steam-sterilisation cycle. The correct pressure in different components should be maintained by regulatory valves controlled by associated pressure gauges.

### Safety Valves

Safety valves should be incorporated at various suitable places in all vessels and pipe layouts which are likely to be operated under pressure. The valve should be set to release the pressure as soon as it increase markedly above a specified working pressure. Other provisions will be necessary to meet any containment requirements.

### Agitator Shaft Power

A variety of sensors can be used to measure the power consumption of a fermenter. On a large scale, a watt meter attached to the agitator motor will give a fairly good indication of power uptake. This measuring technique becomes less accurate as there is a decrease in scale to pilot scale and finally to laboratory fermenters, the main contributing factor being friction in the agitator shaft bearing. Torsion dynamometers can be used in small-scale applications. Since the dynamometer has to be placed on the shaft outside the fermenter the measurement will once again include the friction in the bearings. For this reason strain gauges mounted on the shaft within the fermenter are the most accurate method of measurement and overcome frictional problems. Aiba mounted four identical strain gauges at 45° to the axis in a hollow shaft. Lead wires from the gauges passed out of the shaft via an axial hole and electrical signals were then picked up by an electrical slip-ring arrangement.

### Rate of Stirring

In all fermenters it is important to monitor the rate of rotation (rpm) of the stirrer shaft. The tachometer used for this purpose may employ electromagnetic induction voltage generation, light sensing or magnetic force as detection mechanisms. Obviously, the final choice of tachometer will be determined by the type of signal which is required for recording and/or process control for regulating the motor speed and other ancillary equipment. Provision is often made on small laboratory fermenters to vary the rate of stirring. In most cases it is now standard practice to use an AC slip motor that has an acceptable torque curve that is coupled to a thyristor control. At pilot or full scale, the need to change rates of stirring is normally reduced. When necessary it can be done using gear boxes, modifying the sizes of wheels and drive belts or by changing the drive motor, the most expensive alternative.

### Foam Sensing and Control

The formation of foam is a difficulty in many types of microbial fermentation which can create serious problems if not controlled. It is common practice to add an antifoam to a fermenter when the culture starts foaming above a certain pre-determined level. The methods used for foam sensing and antifoam additions will depend on process and economic considerations.

A foam sensing and control unit is shown in Fig. 17.6. A probe is inserted through the top plate of the fermenter. Normally the probe is a stainless-steel rod, which is insulated except at the tip and set at a defined level above the broth surface. When the foam rises and touches the probe tip, a current is passed through the circuit of the probe, with the foam acting as an electrolyte and the vessel acting as an earth. The current actuates a pump or valve and antifoam is released into the fermenter for a few seconds. Process timers are routinely included in the circuit to ensure that the antifoam has time to mix into the medium and break down the foam before the probe is programmed after a preset time interval to sense the foam level again and possibly actuate the pump or valve. Alternatively antifoam may be added slowly at a pre-determined rate by a small pump so that foaming never occurs and there is therefore no need for a sensing system.

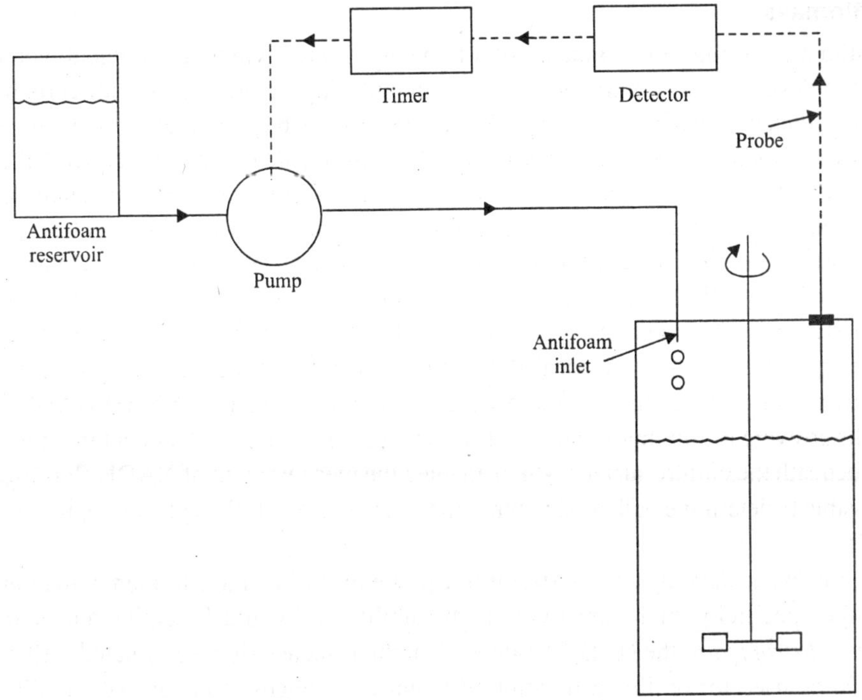


**Fig. 17.6.** Foam sensing and control unit.

A number of mechanical antifoam devices have been described including discs, propellers, brushes or hollow cones attached to the agitator shaft above the surface of the broth. The foam is broken down when it is thrown against the walls of the fermenter. Other devices which have been manufactured include horizontal rotating shafts, centrifugal separators and jets spraying on to deflector plates. Unfortunately most of these devices have to be used in conjunction with an antifoam.

## Weight

A load cell offers a convenient method of determining the weight of a fermenter or feed vessel. This is done by placing compression load cells in or at the foot of the vessel supports. When designing the support system for a fermenter or other vessel, the weight of which is to be measured by load cells, the principle of the three-legged stool should be remembered. Three feet will always rest in stable equilibrium even though the supporting surface is uneven. If more feet are provided, the additional feet must each be fitted with means of adjustment or precision packing to ensure load bearing on all the feet.

A load cell is essentially an elastic body, usually a solid or tubular steel cylinder, the compressive strain of which under axial load may be measured by a series of electrical resistance strain gauges which are cemented to the surface of the cylinder. The load cell is assembled in a suitable housing with electrical cable connecting points. The cell is calibrated by measuring compressive strain over the appropriate range of loading. Changes of resistance with strain which are proportional to load are determined by appropriate electrical apparatus.

It is therefore possible to use appropriately sized load cells to monitor feed rates from medium reservoirs, acid and base utilisation for pH control and the use of antifoam for foam control. The change in weight in a known time interval can be used indirectly as a measure of liquid flow rates.

## Microbial Biomass

Real-time estimation of microbial biomass of a fermenter is an obvious requirement, yet it has proved very difficult to develop a satisfactory sensor. Most monitoring has been done indirectly by dry weight samples (made quicker with microwave ovens), cell density (spectrophotometers), cell numbers (Coulter counters) or by the use of gateway sensors which will be discussed later in this chapter. Other alternative approaches are real-time estimation of a cell component which remains at a constant concentration, such as nicotinamide adenine dinucleotide (NAD), by fluorimetry or measurement of a cell property which is proportional to the concentration of viable cells, such as radio frequency capacitance.

It is well established that fluorimetric measurements are very specific and rapid, but their use in fermentation studies is limited. The measurement of NAD, provided that it remains at a constant concentration in cells, would be an ideal indirect method for continuous measurement of microbial biomass. Harrison and Chance used a fluorescence technique to determine NAD-NADH levels inside microbial cells growing in continuous culture. Einsele mounted a fluorimeter on a fermenter observation port located beneath the culture surface which enabled the measurement of NADH fluorescence *in situ*, making it possible to determine bulk mixing times in the broth and to follow glucose uptake by monitoring NADH levels.

Beyeler were able to develop a small sterilisable probe for fitting into a fermenter to monitor NADH, which had high specificity, high sensitivity, high stability and could be calibrated *in situ*. In batch culture of *Candida tropicalis*, the NAD(P)H-dependent fluorescence signal correlated well with biomass, so that it could be used for on-line estimation of biomass. Changes in the growth conditions, such as substrate exhaustion or the absence of oxygen, were also very quickly detected.

Schneckenburger used this technique to study the growth of methanogenic bacteria in anaerobic fermentations.

Dielectric spectroscopy can be used on-line to monitor biomass. Details of the theory and principles of this technique have been described by Kell. At low radio frequencies (0.1 to 1.0 MHz), a microbial cell membrane will act as a capacitor and become charged by the so-called β-dispersion effect, making it possible to discriminate between microbial cells, gas bubbles and insoluble media particles. The size of this β-dispersion is linearly proportional to the membrane enclosed volume fraction up to high cell densities. Kell were able to show that the capacitance (dielectric permittivity) was linearly proportional to the biomass concentration. The output for unicellular organisms is proportional to the mean cell radius whereas with mycelial suspensions the output remains linear for increases in biomass of a particular cell morphology. The capacitance has been shown to give a linear response with biomass using a number of strains of bacteria, yeasts, mycelial fungi, plant and animal cells.

The sterilisable probe can be inserted directly into a fermenter using a 25 mm diameter port. Fouling of the gold electrodes in the probe can be avoided by the automatic application of electrolytic cleaning pulses. The sensor ('Bug meter') is manufactured by Aber Instruments, Netherland has a capacitance range of 0.1 to 200 pF (picoFarads), which is equivalent to 0.1 to 200 mg dry weight $cm^{-3}$ (approximately $10^6$ to $2 \times 10^9$ cells $cm^{-3}$ of *Saccharomyces cerivisiae*). The resolution depends on the type cells and the conductivity of the medium, but is normally 0.1 mg dry weight $cm^{-3}$. In order for the sensor to work effectively the suspending medium must have a minimum conductance. Yeast slurries after acid washing are satisfactory, but before such washing there may be a need for extra salts in the medium in order to make measurements. This sensor has proved ideal for yeast cells and is now being used by the brewing industry to control yeast pitching rates.

## Measurement and Control of Dissolved Oxygen

In most aerobic fermentations it is essential to ensure that the dissolved oxygen concentration does not fall below a specified minimal level. Since 1970s steam sterilisable oxygen electrodes have become available for this monitoring (Fig. 17.7).

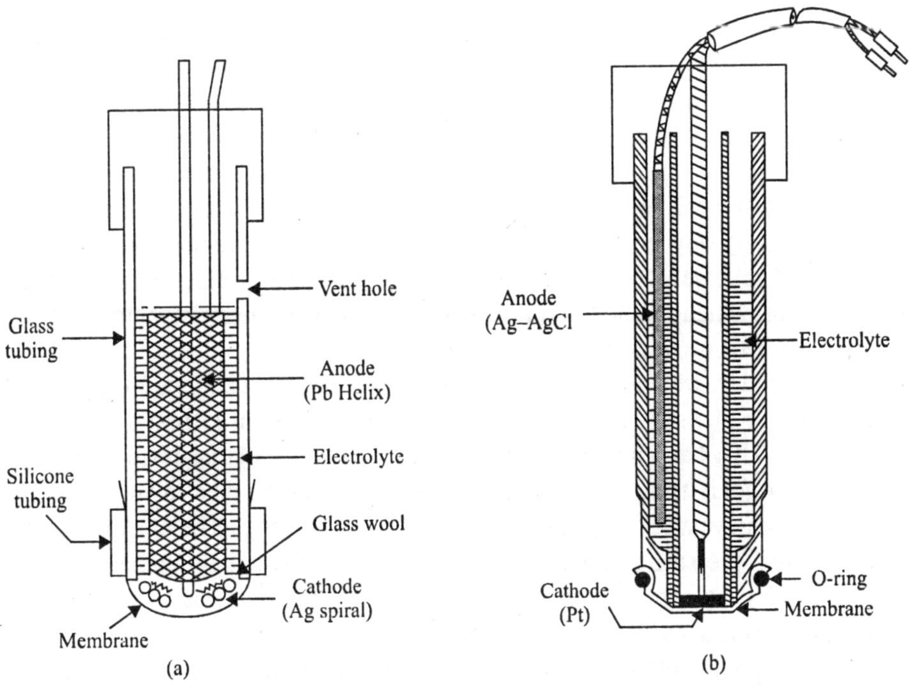


**Fig. 17.7.** Construction of dissolved-oxygen electrodes: (a) galvanic, (b) polarographic.

These electrodes measure the partial pressure of the dissolved oxygen and not the dissolved oxygen concentration. Thus at equilibrium, the probe signal of an electrode will be determined by:

$$P(O_2) = C(O_2) \times P_T$$

where, $P(O_2)$ = partial pressure of dissolved oxygen sensed by the probe
$C(O_2)$ = volume or mole fraction of oxygen in the gas phase
$P_T$ = total pressure.

The actual reading is normally expressed as percentage saturation with air at atmospheric pressure, so that 100 per cent dissolved oxygen means a partial pressure of approximately 160 mmHg.

Pressure changes can have a significant effect on readings. If the total pressure of the gas equilibrating with the fermentation broth varies, the electrode reading will change even though there is no change in the gas composition. Changes in atmospheric pressure can often cause 5 per cent changes and back pressure due to the exit filters can also cause increases in readings. Allowance must also be made for temperature. The output from an electrode increases by approximately 2.5 per cent per °C at a given oxygen tension. This effect is due mainly to increase in permeability in the electrode membrane. Many electrodes have built-in temperature sensors which allow automatic compensation of the output signal. It is also important to remember that the solubility of oxygen in aqueous media is influenced by the composition. Thus, water at 25°C and 760 mmHg pressure saturated with air will contain 8.4 mg $O_2$ $dm^{-3}$, while 25 per cent NaCl in identical conditions will have an oxygen solubility of 2.0 mg $O_2$ $dm^{-3}$. However, the measured partial pressure outputs for $O_2$ would be the same even though the oxygen concentrations would be very different. Therefore, it is best to calibrate the electrode in percentage oxygen saturation.

In small fermenters (1 $dm^3$), the commonest electrodes are galvanic and have a lead anode, silver cathode and employ potassium hydroxide, chloride, biocarbonate or acetate as an electrolyte. The sensing tip of the electrode is a teflon, polyethylene or polystyrene membrane which allows passage of the gas phase so that an equilibrium is established between the gas phases inside and outside the electrode. Because of the relatively slow movement of oxygen across the membrane, this type of electrode has a slow response of the order of 60 seconds to achieve a 90 per cent reading of true value. Buhler and Ingold quote 50 seconds for 98 per cent response for a later version. These electrodes are therefore suitable for monitoring very slow changes in oxygen concentration and are normally chosen because of their compact size and relatively low cost. Unfortunately, this type of electrode is very sensitive to temperature fluctuations, which should be compensated for by using a thermistor circuit. The electrodes also have a limited life because of corrosion of the anode.

Polarographic electrodes, which are bulkier than galvanic electrodes, are more commonly used in pilot and production fermenters, needing instrument ports of 12, 19 or 25 mm diameter. Removable ones need a 25 mm port. They have silver anodes which are negatively polarised with respect to reference cathodes of platinum or gold, using aqueous potassium chloride as the electrolyte. Response times of 0.05 to 15 seconds to achieve a 95 per cent reading have been reported. The electrodes which can be very precise may be both pressure and temperature compensated. Although a polarographic electrode may initially cost 600 per cent more than the galvanic equivalent, the maintenance costs are considerably lower as only the membrane should need replacing.

Prototypes of a fast response phase fluorometric sterilisable oxygen sensor are now being developed. The sensor utilises the differential quenching of a fluorescence lifetime of a chromophore, tris(4,7-diphenyl-1,10-phenanthroline)ruthenium(II) complex, in response to the partial pressure of oxygen. The fluorescence of this complex is quenched by oxygen molecules resulting in a reduction of fluorescence lifetime. Thus, it is possible to obtain a correlation between fluorescence lifetime and the partial pressure of oxygen. However, at room temperature when a Clark-type oxygen electrode shows a linear calibration the optical sensor shows a hyperbolic response. The sensitivity of the optical sensor when compared with an oxygen electrode is significantly higher at low oxygen tensions whereas the sensitivity is low at high oxygen tensions. The sensor is autoclavable, free of maintenance requirements, stable over long periods and gives reliable measurements of low oxygen tensions in dense microbial cultures.

Dissolved oxygen concentrations may also be determined by a tubing method. The probe consists of a coil of permeable teflon or propylene tubing within the fermenter through which is passed a stream of helium or nitrogen. The oxygen, which diffuses from the fermentation medium through the tubing wall into the inert gas stream, is then determined using a paramagnetic gas analyser. Times of 2 to 10 minutes are required before taking readings. The tubing will withstand repeated sterilisation and has been used continuously for up to 1000 hours at pilot scale. If it is necessary to increase the dissolved oxygen concentration in a medium this may be achieved by increasing the airflow rate or the rpm of the impeller or a combination of both processes. Another way is to increase the ratio of oxygen to nitrogen in the input gas using a variable proportionating valve while maintaining a constant gas flow rate. The cost of this technique would normally restrict its use to laboratories and pilot plants.

## Inlet and Exit-Gas Analysis

The measurement and recording of the inlet and/or exit gas composition is important in many fermentation studies. By observing the concentrations of carbon dioxide and oxygen in the entry and exit gases in the fermenter and knowing the gas flow rate it is possible to determine the oxygen uptake of the system, the carbon dioxide evolution rate and the respiration rate of the microbial culture.

The oxygen concentration can be determined by a paramagnetic gas analyser. Oxygen has a strong affinity for a magnetic field, a property which is shared with only nitrous and nitric oxides. The analysers may be of a deflection or thermal type.

In the deflection analyser, the magnetic force acts on a dumb-bell test body that is free to rotate about an axis (Fig. 17.8). The magnetic force which is created around the test body is proportional to the oxygen concentration. When the test body swings out of the magnetic field a corrective electrostatic force must be applied to return it to the original position. Electrostatic force readings can therefore be used as a measure of oxygen concentrations.

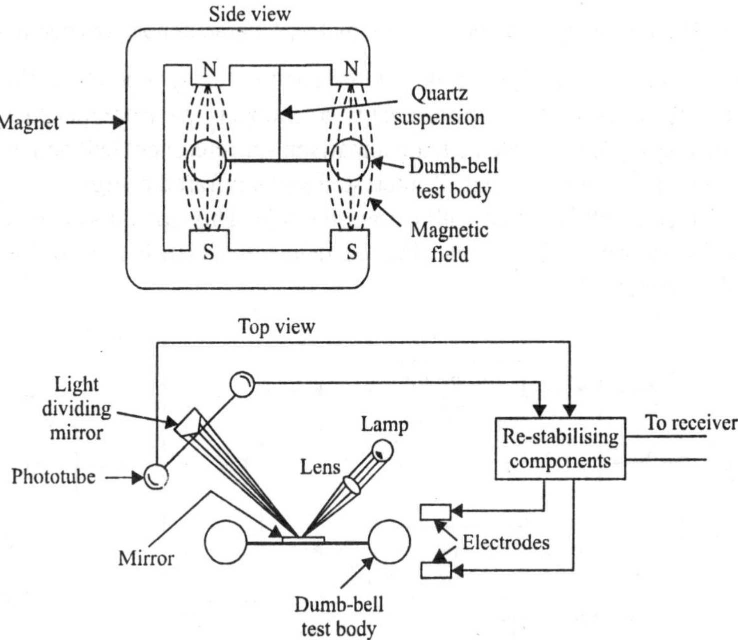


**Fig. 17.8.** A deflection-type paramagnetic oxygen analyser.

In a thermal analyser, a flow through 'ring' element is the detector component (Fig. 17.9). After entering the ring, the paramagnetic oxygen content of the sample is attracted by the magnetic field to the central glass tube where resistors heat the gases. The resistors are connected into a Wheatstone bridge circuit to detect variations in resistance due to flow-rate changes. The oxygen in the heated sample loses a high proportion of its paramagnetism. Cool oxygen in the incoming gas flow will now be attracted and displace the hot oxygen. This displacement action produces a convection current. The flow rate of the convection current is a function of the oxygen concentration and can be detected by resistors. The resulting gas flow cools the winding A and heats the winding B and it is the resulting temperature difference that imbalances the Wheatstone bridge.

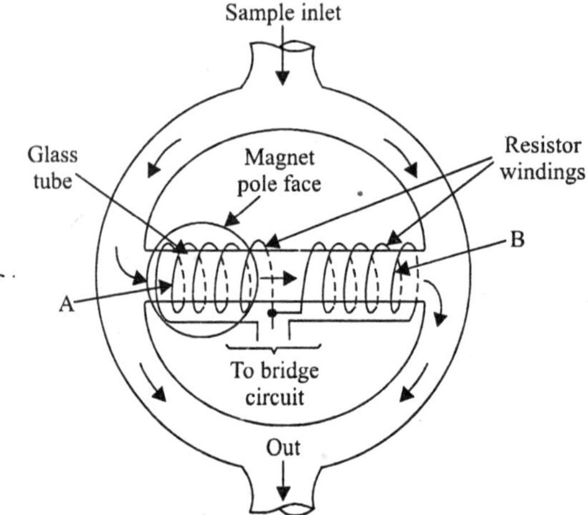


**Fig. 17.9.** The measuring element in a thermal-type paramagnetic oxygen analyser.

Carbon dioxide is commonly monitored by infrared analysis using a positive filtering method. The unit (Fig. 17.10) consists of a source of infrared energy, a 'chopper' to ensure that energy passes through each side of the optical system, a sample cell, a comparison (or reference) cell and an infrared detector sensitised at a wavelength at which the gas of interest absorbs infrared energy.

In this case the detector will be filled with carbon dioxide. This optical system senses the reduced radiation energy of the measuring beam reaching the detector, which is due to the absorption in the carbon dioxide in the sample cell.

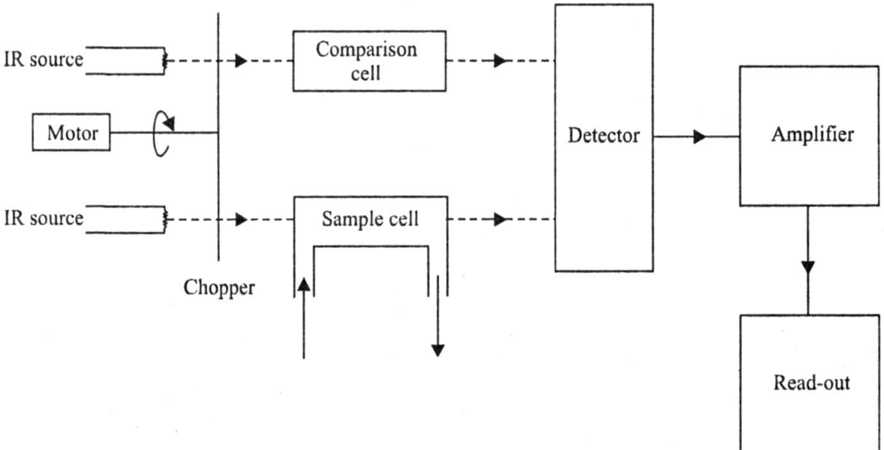


**Fig. 17.10.** Simple positive filtering infrared analyser.

It is expensive to have separate carbon dioxide and oxygen analysers for each separate fermenter. Therefore it may be possible to couple up a group of fermenters via a multiplexer to a single pair of gas analysers. Gas analysis readings can then be taken in rotation for each fermenter every 30 to 60 minutes. In many cases this will be adequate. Alternatively the gas analysers can be replaced by a mass spectrometer which can analyse a number of components as well as oxygen and carbon dioxide.

## pH Measurement and Control

In batch culture the pH of an actively growing culture will not remain constant for very long. In most processes there is a need for pH measurement and control during the fermentation if maximum yield of a product is to be obtained. Rapid changes in pH can often be reduced by the careful design of media, particularly in the choice of carbon and nitrogen sources and also in the incorporation of buffers or by batch feeding. The pH may be further controlled by the addition of appropriate quantities of ammonia or sodium hydroxide if too acidic or sulphuric acid if the change is to an alkaline condition. Normally the pH drift is only in one direction.

pH measurement is now routinely carried out using a combined glass reference electrode that will withstand repeated sterilisation at temperatures of 121°C and pressures of 138 kN $m^{-2}$. The electrodes may be silver/silver chloride with potassium chloride or special formulations (e.g., Friscolyt by Ingold) as an electrolyte. Occasionally calomel/mercury electrodes are used. The electrode is connected via leads to a pH meter/controller. If the electrode and its fermenter have to be sterilised in an autoclave then the associated leads and plugs to a pH meter must be able to withstand autoclaving and retain their electrical resistance. Repeated sterilisation may gradually change the performance of the electrode. The long culture times associated with animal cell culture or continuous culture of any cells makes a withdrawal option highly desirable to allow for servicing the electrode or troubleshooting without interrupting the fermentation. The housing for this option needs to be carefully designed to ensure that the fermentation does not become contaminated when the electrode is withdrawn, serviced, re-sterilised and inserted into the housing.

Ingold electrodes contain a ceramic housing in the reference half cell which has pore dimensions capable of preventing fungal or bacterial infections. However, it is often desirable in animal-cell culture to sterilise the reference electrolyte as well as the electrode surfaces and seals. Both liquid and gel filled electrodes are available. The liquid system gives a faster response, is most stable and accurate. When the electrode is pressurised above the operating pressure in a fermenter the liquid electrolyte will gradually flow out of the ceramic diaphragm and prevent fouling, particularly from proteins in the fermentation broth precipitating on the membrane after contact with the electrolyte.

Control units, to be discussed later in this chapter, may be simple ON/OFF or more complex. In the case of the ON/OFF controller, the controller is set to a pre-determined pH value. When a signal actuates a relay, a pinch valve is opened or a pump started and acid or alkali is pumped into the fermenter for a short time which is governed by a process timer (0 to 5 seconds). The addition cycle is followed by a mixing cycle which is governed by another process timer (0 to 60 seconds) during which time no further acid or alkali can be added. At the end of the mixing cycle another pH reading will indicate whether or not there has been adequate correction of the pH drift. In small volumes the likelihood of overshoot is minimal. A recording unit may be wired to the pH meter to monitor the pH pattern throughout a process cycle.

Shinskey has discussed pH control of batch processing using proportional and proportional plus derivative control (see later section) when overshoot of the set point is to be avoided. In the case of proportional action, the controller must be adjusted so that a valve on an acid feed-line is shut when the error is zero. However, overshoot is possible as there may be delay in closing the valve once the set point is achieved. In some cases the overshoot cannot be corrected because of lack of alkali nor may it be desirable. Therefore the preclude an overshoot, the valve must be closed before the controlled variable reaches the set point. This may be done using proportional plus derivative control. The derivative action will need careful adjustment. Too little derivative action will cause some overshoot while too much will lead to the premature closure of the valve. This premature closure may be only for a short time before the valve opens again to give a response pattern as shown in Fig. 17.11.

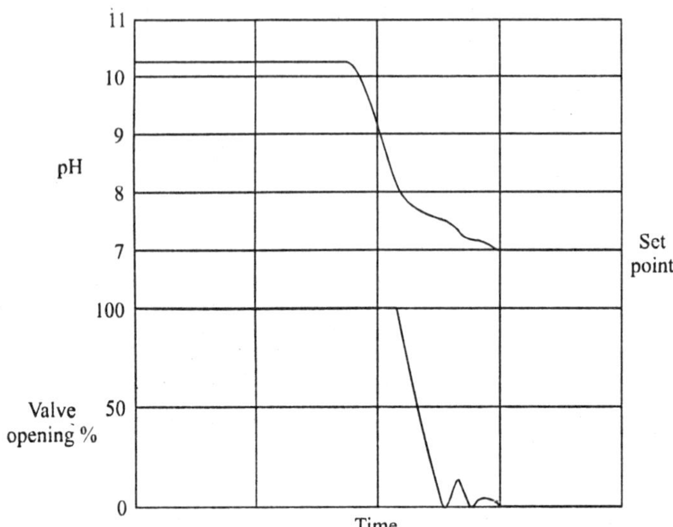


**Fig. 17.11.** Valve opening and pH changes with proportional plus derivative control.

## Redox

Redox is a measure of the oxidation-reduction potential of a biological system and can be determined as a voltage (mV), the value in any system depending on the equilibrium of:

$$\underset{\text{(negative value)}}{\text{Reduced form}} \rightleftharpoons \underset{\text{(positive value)}}{\text{Oxidised form}} + \text{electron(s)}$$

The measuring electrode consists of gold, platinum or iridium which is welded to a copper lead. The interpretation of results presents difficulties and is confusing. The culture is not at redox equilibrium until possibly the end of a growth cycle. During the cycle, although some redox half-reactions may be in equilibrium they cannot all be in a dynamic system. The micro-organisms can also be at a different redox potential from the broth. It has been speculated that the probe signal is indicating something about the relative concentrations of uncertain and probably varying chemical species. This is far from the ideal for a sensor which should only be measuring a specific factor. If the broth contains traces of oxygen this will probably dominate the signal. It is the ability to detect low concentrations of oxygen in media (1 ppm) where redox electrodes may have a good application for determining oxygen availability in anaerobic or micro-aerophilic processes operated on a small scale.

## Carbon Dioxide Electrodes

The measurement of dissolved carbon dioxide is possible with an electrode, since a pH or voltage change can be detected as the gas goes into solution. The first available electrode consisted of a combined pH electrode with a bicarbonate buffer (pH 5) surrounding the bulb and ceramic plug, with the solution being retained by a PTFE membrane held by an O-ring. Unfortunately, this electrode was not steam sterilisable. This basic design has been modified so that dissolved $CO_2$ from the sample permeates a stainless-steel reinforced silicone bi-layer membrane and dissolves in the internal bicarbonate electrolyte of the electrode. The subsequent pH shift is determined by an internal pH element. This system is extremely sensitive to shifts in the pH element as one pH unit change represents a ten-fold increase in $pCO_2$, but precautions have been taken in the design to ensure minimal drift of the sensor. It is possible to calibrate the electrode on-line using specially formulated buffer solutions. This new version can be steam sterilised.

## ON-LINE ANALYSIS OF OTHER CHEMICAL FACTORS

If good control of a fermentation is to be obtained, then all chemical factors which can influence growth and product formation ought to be continuously monitored. This ideal situation has not yet been achieved but a number of techniques are currently being developed.

### Ion-Specific Sensors

Ion-specific sensors have been developed by Orion research and radiometer to measure $NH_4^+$, $Ca^{2+}$, $K^+$, $Mg^{2+}$, $PO^{3-}$, $SO^{2-}$, etc. The response time of these electrodes varies from 10 seconds to several minutes depending on the concentration of the ion species, the composition of the sample, etc. However, none of these probes is steam sterilisable.

### Enzyme and Microbial Electrodes

Enzyme or microbial cell electrodes can be used in some analyses. A suitable enzyme or microbial cell which produces a change in pH or forms oxygen in the enzyme reaction is chosen and immobilised on a membrane held in close contact to a pH or oxygen electrode. Unfortunately, the oxygen demand of an enzyme may restrict the maximum substrate concentration which might be detected in a medium. Enfors overcame this problem for glucose determination by co-immobilising glucose oxidase and catalase. More recently, it has been possible to use a ferrocene derivative as an artificial redox carrier to shuttle electrons from glucose oxidase to a carbon electrode, thus making the device largely independent of oxygen concentrations. Enzyme electrodes are also commercially available to monitor cholesterol, triglycerides, lactate, acetate, oxalate, methanol, ethanol, creatine, ammonia, urea, amino acids, carbohydrates and penicillin.

Ideally, an electrode that can be inserted into a fermenter and steam sterilised is required, but none is yet available which can be used in this manner, even when using enzymes which are stable at high temperatures which can be obtained from thermophilic micro-organisms. Hewetson prepared sterilised penicillinase electrodes by assembling sterile components or by standing assembled components in chloroform before placing in a fermenter.

### Near Infrared Spectroscopy

Hammond and Brookes have described the development of near infrared spectroscopy (NIR; 460–1200 nm) for rapid, continuous and batch analysis of components of fermentation broths. In samples from an antibiotic fermentation, they used NIR absorbance bands to simultaneously estimate fat (in the medium), techoic acid (biomass) and antibiotic (the product). Fat analysis has been made possible with a fibre optic sensor placed *in situ* through a port in the fermenter wall. The assay time for an antibiotic has been reduced from 2 hours to 2 minutes. A method has also been developed to measure alkaline protease production in broths. Vaccari has used this technique to measure glucose, lactic acid and biomass in a lactic acid fermentation.

### Mass Spectrometers

The mass spectrometer can be used for on-line analysis since it is very versatile and has a response time of less than 5 seconds for full-scale response and taking about 12 seconds for a sample stream. It allows for monitoring of gas partial pressures ($O_2$, $CO_2$, $CH_4$, etc.), dissolved gases ($O_2$, $CO_2$, $CH_4$, etc.) and volatiles (methanol, ethanol, acetone, simple organic acids, etc.). Heinzle combined a data processor with a mass spectrometer equipped with a capillary gas inlet to measure gas partial pressures and a membrane inlet to detect dissolved gases and volatiles. It is therefore possible to multiplex this type of system to analyse several fermenters sequentially.

## CONTROL SYSTEMS

The process parameters which are measured using probes described in the previous sections may be controlled using control loops.

A control loop consists of four basic components:

1. A measuring element.
2. A controller.
3. A final control element.
4. The process to be controlled.

In the simplest type of control loop, known as feedback control (Fig. 17.12), the measuring element senses a process property such as flow, pressure, temperature, etc., and generates a corresponding output signal. The controller compares the measurement signal with a pre-determined desired value (set point) and produces an output signal to counteract any differences between the two. The final control element receives the control signal and adjusts the process by changing a valve opening or pump speed and causing the controlled process property to return to the set point.

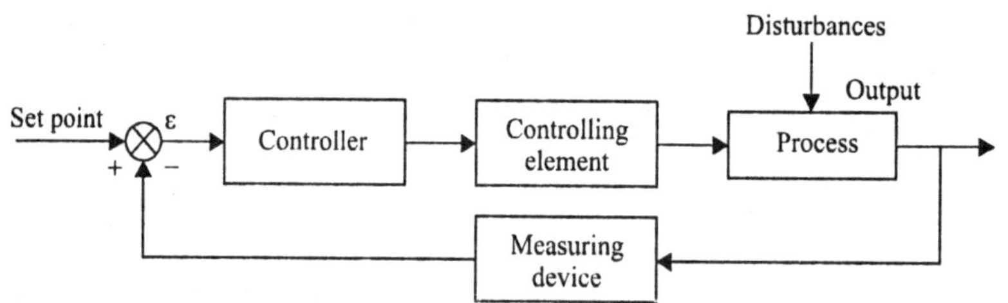


**Fig. 17.12.** A feedback control loop.

### Manual Control

A simple example of control is manual control of a steam valve to regulate the temperature of water flowing through a pipe (Fig. 17.13). Throughout the time of operation a plant operative is instructed to monitor the temperature in the pipe. Immediately the temperature changes from the set point on the thermometer, the operative will take appropriate action and adjust the steam valve to correct the temperature deviation. Should the temperature not return to the set point within a reasonable time, further action may be necessary. Much depends on the skill of individual operatives in knowing when and how much adjustment to make. This approach with manual control may be very costly in terms of labour and should always be kept to a strict minimum when automatic control could be used instead. A justifiable use of manual control may be in the adjustment of minor infrequent deviations.

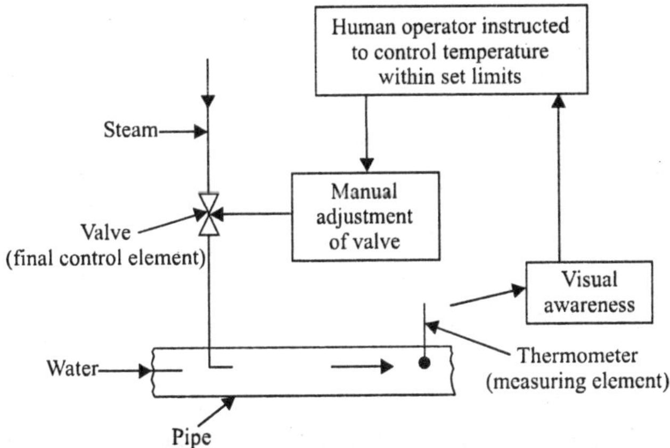


**Fig. 17.13.** Simple manual-control loop for temperature control.

## Automatic Control

When an automatic control loop is used, certain modifications are necessary. The measuring element must generate an output signal which can be monitored by an instrument. In the case of temperature control, the thermometer is replaced by a thermocouple, which is connected to a controller which in turn will produce a signal to operate the steam valve (Fig. 17.14).

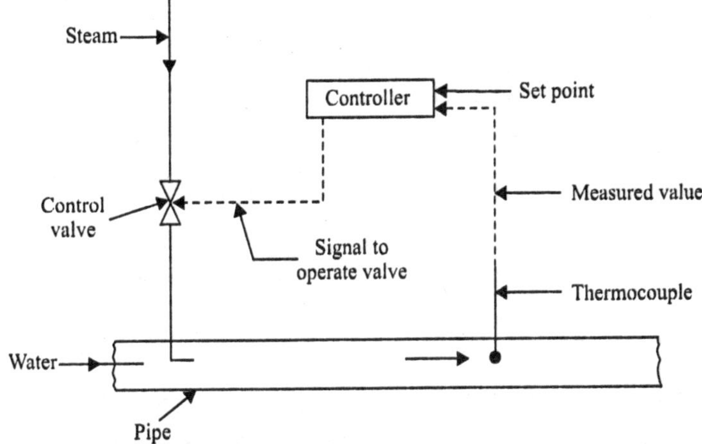


**Fig. 17.14.** Simple automatic control loop for temperature control.

Automatic control systems can be classified into four main types:

1. Two-position controllers (ON/OFF).
2. Proportional controllers.
3. Integral controllers.
4. Derivative controllers.

### *Two-position controllers (ON/OFF)*

The two-position controller, which is the simplest automatic controller, has a final control unit (valve, switch, etc.) which is either fully open (ON) or fully closed (OFF). The response pattern to such a change will be oscillatory. If there is instant response then the pattern will be as shown in Fig. 17.15.

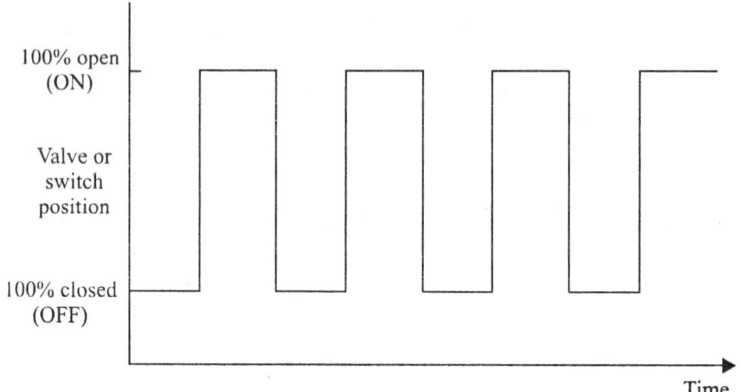


**Fig. 17.15.** Oscillatory pattern of a simple two-position valve or switch.

If one considers the example of the heating of a simple domestic water tank controlled by a thermostat operating with ON/OFF, then there will be a delay in response when the temperature reaches the set point and the temperature will continue rising above this point before the heating source is switched off. At the other extreme, the water will continue cooling after the heating source has been switched on. With this mode of operation an oscillatory pattern will be obtained with a repeating pattern of maximum and minimum temperature oscillating about the set point, provided that all the other process conditions are maintained at a steady level (Fig. 17.16).

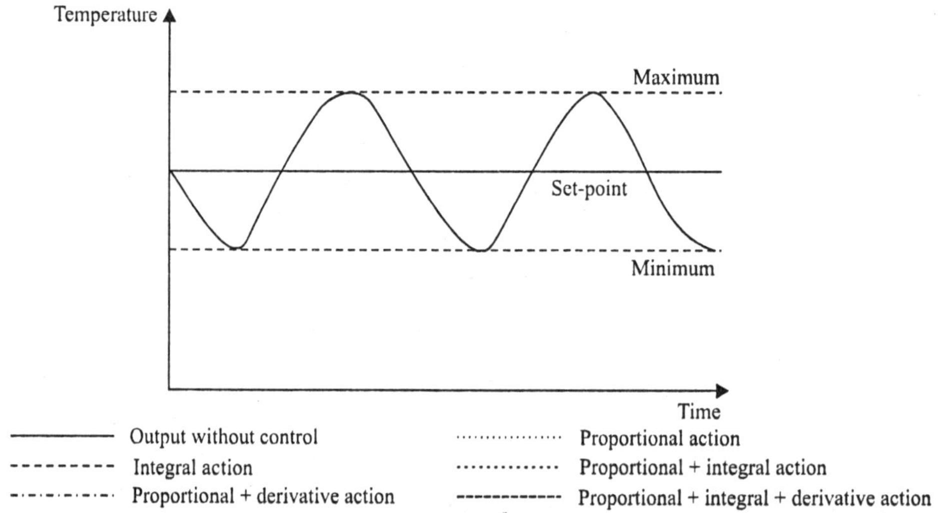


**Fig. 17.16.** Oscillatory pattern of the temperature of a domestic water tank (no water being drawn off) using ON/OFF control of the heating element.

If this type of controller is to be used in process control then it is important to establish that the maximum and minimum values are acceptable for the specific process and to ensure that the oscillation cycle time does not cause excessive use of valves or switches. ON/OFF control is not satisfactory for controlling any process parameter where there are likely to be large sudden changes from the equilibrium. In these cases alternative forms of automatic control must be used.

In more complex automatic control systems three different methods are commonly used in making error corrections. They are: proportional, integral and derivative. These control methods may be used singly or in combinations in applying automatic control to a process, depending upon the complexity of the process and the extent of control required. Since many of the controllers used in the chemical industries are pneumatic, the response to an error by the controller will be represented by a change in output pressure. Pneumatic controllers are still widely used because they are robust and reliable. In other cases, when the controller is electronic, the response to an error will be represented as change in output current or voltage.

### *Proportional control*

Proportional control can be explained as follows: the change in output of the controller is proportional to the input signal produced by the environmental change (commonly referred to as error) which has been detected by a sensor.

Mathematically it can be expressed by the following equation:

$$M = M_0 + K_c\Sigma$$

where, $M$ = output signal

$M_0$ = controller output signal when there is no error

$K_c$ = controller gain or sensitivity

$\Sigma$ = the error signal.

Hence, the greater the error (environmental change) the larger is the initial corrective action which will be applied. The response to proportional control is shown in Fig. 17.17, from which it may be observed that there is a time of oscillation which is reduced fairly quickly. It should also be noted that the controlled variable attains a new equilibrium value. The difference between the original and the new equilibrium value is termed the offset.

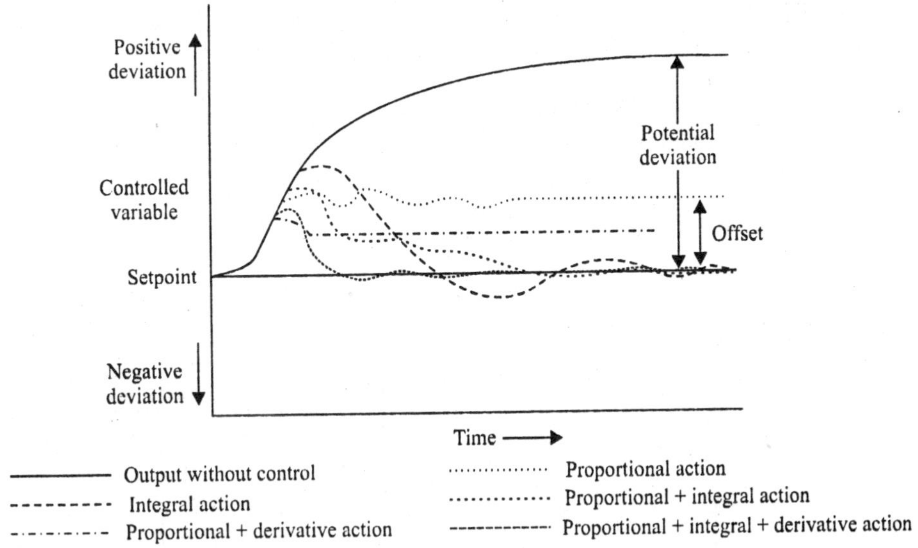


**Fig. 17.17.** Typical controlled plant responses.

The term $K_c$ (controller gain) is the multiplying factor (which may be dimensioned) which relates a change in input to the change in output.

$$\Delta I \rightarrow \boxed{\text{Controller}} \rightarrow \Delta O$$

(Change in input) (Change in output)

Then

$$\Delta I = K_c \Delta O$$

$K_c$ may contain conversion units if there is an electrical input and a pressure output or *vice versa*.

If the input to the controller is 1 unit of change, then:

1. With a controller gain of 1, the output will be 1 unit.
2. With a controller gain of 2, the output will be 2 units, etc.

On many controllers $K_c$ is graduated in terms of proportional band instead of controller gain.

Now
$$K_c \; \alpha \frac{1}{PB} \text{(proportional band)}$$

or
$$K_c = c' \frac{1}{PB} \text{ where } c' \text{ is a constant}$$

This quantity *PB* is defined as the error required to move the final control element over the whole of its range (e.g., from fully open to fully shut) and is expressed as a percentage of the total range of the measured variable (e.g., two extremes of temperature).

A fermenter with a heating jacket will be used as an example (Fig. 17.18). A thermocouple is connected to a temperature controller which has a span of 10°C covering the range 25°C to 35°C, with a set point at 30°C. The controller valve which is controlled by a pressure regulator, is fully open at 5 psig (46 kN $m^{-2}$) and fully closed at 15 psig (138 kN $m^{-2}$), while the set point of 30° corresponds to a control pressure on the valve of 10 psig (92 kN $m^{-2}$). When the controller gain is 1, a change of 10° will cause a pressure change of 10 psig when the valve will be fully open at 25°C and fully closed at 35°C. Thus, the proportional band is 100 per cent. If the controller gain is 2, a 5° change will cause the valve to go from fully open to fully closed, i.e., 27.5°C to 32.5°C. In this case the proportional band which will be:

$$\frac{\text{Actual band width}}{\text{Total band range}} = \frac{5°}{10°} \times 100 = 50 \text{ per cent.}$$

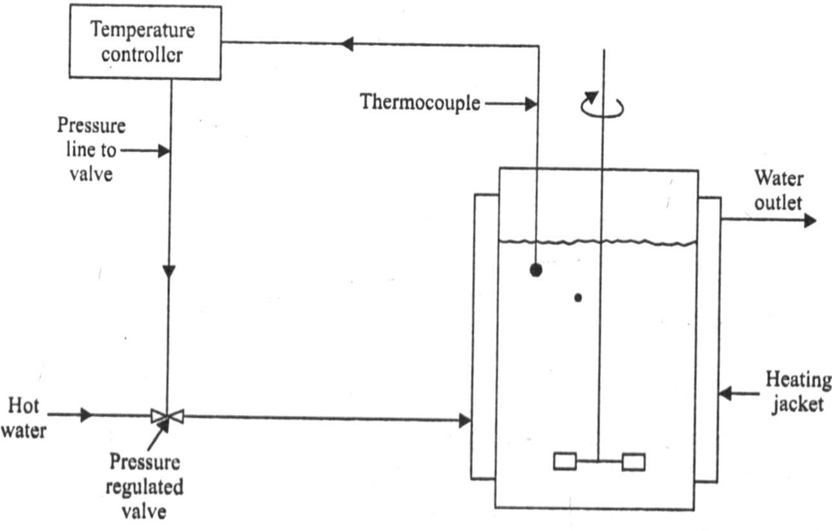


**Fig. 17.18.** A fermenter with a temperature-controlled heating jacket.

Either side of this band the pressure will be constant. In the case of a controller gain of 4, a 2.5°C change will cause a pressure change of 10 psig (92 kN m$^{-2}$), which will cause the control valve to go from fully open to fully closed (28.75° to 31.25°). In this case the proportional band will be

$$= \frac{2.5°}{10°} \times 100 = 25 \text{ per cent.}$$

These results are summarised in Table 17.2 for controller gains of 1, 2 and 4.

**Table 17.2.** The effect of controller gain ($K_c$) on band width of a proportional temperature controller.

| | *Measured temperature* | *Pressure output (psig) at $K_c$ = 1 psig/1°* | | *Pressure output (psig) at $K_c$ = 2 psig/1°* | | *Pressure output (psig) at $K_c$ = 4 psig/1°* | |
|---|---|---|---|---|---|---|---|
| | 35° | 15 | | 15 | | 15 | |
| | 34° | 14 | | 15 | | 15 | |
| | 33° | 13 | | 15 | | 15 | |
| | 32.5° | 12.5 | | 15 | | 15 | |
| | 32° | 12 | | 14 | | 15 | |
| | 31.25° | 11.25 | | 12.5 | | 15 | |
| | 31° | 11 | | 12 | | 14 | |
| Set → | 30° | 10 | Proportional band 100% | 10 | Proportional band 50% | 10 | Proportional band 25% |
| point | 29° | 9 | | 8 | | 6 | |
| | 28.75° | 8.75 | | 7.5 | | 5 | |
| | 28° | 8 | | 6 | | 5 | |
| | 27.5° | 7.5 | | 5 | | 5 | |
| | 27° | 7 | | 5 | | 5 | |
| | 26° | 6 | | 5 | | 5 | |
| | 25° | 5 | | 5 | | 5 | |

When the proportional band is very small (the controller gain is high) the control mode can be likened to simple ON/OFF, with a high degree of oscillation but no offset. As the proportional band is increased (low controller gain) the oscillations are reduced but the offset is increased. Settings for proportional band width are normally a compromise between degree of oscillation and offset. If the offset is not desirable it can be eliminated by the use of proportional control in association with integral control (Fig. 17.17 and later section).

### *Integral control*

The output signal of an integral controller is determined by the integral of the error input over the time of operation. Thus:

$$M = M_0 + \frac{1}{T_i} \int \Sigma \, dt$$

where $T_i$ = integral time.

It is important to remember that the controller output signal changes relatively slowly at first as time is required for the controller action to integrate the error. It is evident from Fig. 17.17 that the maximum deviation from the set point is significant when compared with the use of proportional control for control of the chosen parameter and the system takes longer to settle down. There is, however, no offset which is advantageous in many control processes.

### *Derivative control*

When derivative control is applied the controller senses the rate of change of the error signal and contributes a component of the output signal that is proportional to a derivative of the error signal. Thus:

$$M = M_0 + T_d \frac{d\Sigma}{dt}$$

where $T_d$ is a time rate constant.

It is important to remember that if the error is constant there is no corrective action with derivative control. In practice, derivative control is never used on its own. The response curve has, therefore, been deliberately omitted from Fig. 17.17.

Figure 17.19 demonstrates the response of derivative control to sinusoidal error inputs. The output is always in a direction to oppose changes in error, both away from and towards the set point, which in this example results in a 90° phase shift. This opposition to a change has a fast damping effect and this property is very useful in combination with other modes of control which will be discussed later.

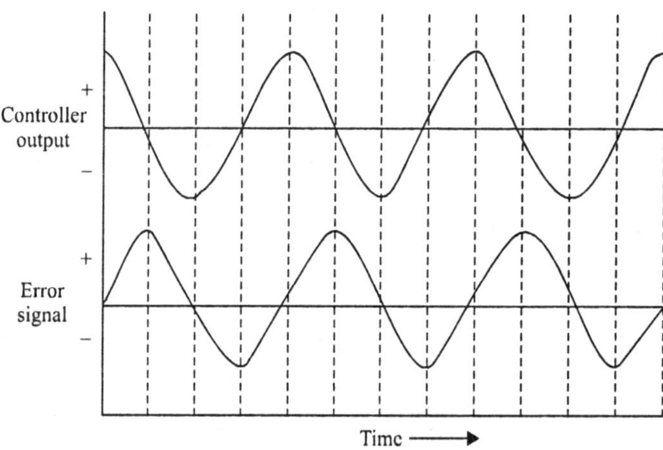


**Fig. 17.19.** Response of a derivative controller to sinusoidal error inputs.

## Combinations of Methods of Control

Three combinations of control systems are used in practice:

1. Proportional plus integral.
2. Proportional plus derivative.
3. Proportional plus integral plus derivative.

### *Proportional plus integral control*

When proportional plus integral control is used, the output response to an error gives rise to a slightly higher initial deviation in the output signal compared with that which would be obtained with proportional control on its own (Fig. 17.17). This is due to a contribution in the signal from integral control. However, the oscillations are soon reduced and there is finally no offset. This mode of control finds wide applications since the proportional component is ideal in a process where there are moderate changes, whereas the integral component will allow for large load changes and eliminate the offset that would have occurred.

### *Proportional plus derivative control*

If proportional plus derivative control is used, the output response to an error will lead to reduced deviations, faster stabilisation and a reduced offset (Fig. 17.17) compared with proportional control alone. Because the derivative component has a rapid stabilising influence, the controller can cope with rapid load changes.

### *Proportional plus integral plus derivative control*

The combination of proportional plus integral plus derivative normally provides the best control possibilities (Fig. 17.17). The advantages of each system are retained. The maximum deviation and settling time are similar to those for a proportional plus derivative controller whilst the integral action ensures that there is no offset. This method of control finds the widest applications because of its ability to cope with wide variations of patterns of changes which might be encountered in different processes.

## Controllers

The first primary automatic controllers were electronic control units which were adjusted manually to set up desired PID response patterns to a disturbance in a control loop. These controllers are relatively expensive and some knowledge of control engineering is necessary to make the correct adjustments to obtain the required control responses.

The availability of cheap computers and suitable computer programmes to mimic PID control and handle a number of control loops simultaneously has made it possible to use some very complex control techniques for process optimisation.

## More Complex Control Systems

In certain situations PID control is not adequate to control a disturbance. Control may be difficult when there is long time lag between a change in a manipulated variable and its effect on the measured variable. Consider the example of a heat exchanger where the water temperature is regulated by the flow rate of steam from a steam valve. By the time the effects of a change in steam flow influence the hot water temperature, a considerable energy change has occurred in the heat exchanger which will continue to drive the hot water temperature away from the set point after a correction has been made to the steam-flow valve. This lag will lead to cycling of the measured variable about the set point. Cascade control can solve this problem. When cascade control is used (Fig. 17.20), the output of one controller is the set point for another. Each controller has its own measured variable with only the primary controller having an individual set point and only the secondary controller providing an output to the process. In more complex cascade systems more loops may be included.

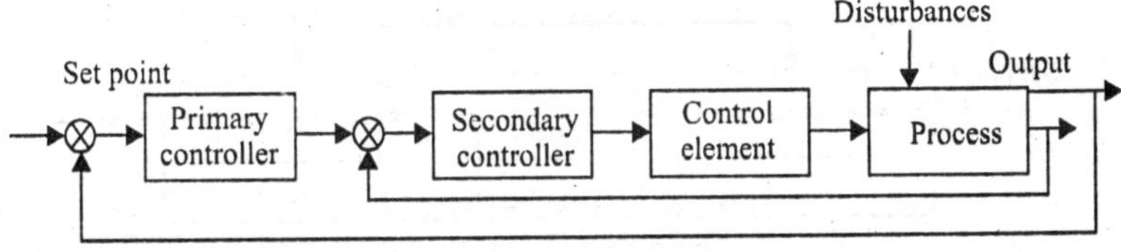


**Fig. 17.20.** A cascade feedback control loop.

When cascade control is used to control water temperature in a heat exchanger process (Fig. 17.21) there is the primary loop (slow or outer loop), consisting of the temperature sensor and the primary temperature controller of the process water temperature and the secondary loop (fast or inner loop) consisting of the steam flow sensor, the steam flow controller, its process variable (steam flow), the control valve and the process. The addition of the secondary controller, whose measured variable is steam flow, allows steam flow variations to be corrected immediately before they can affect the hot water temperature. This extra control loop should help to minimise temperature cycling.

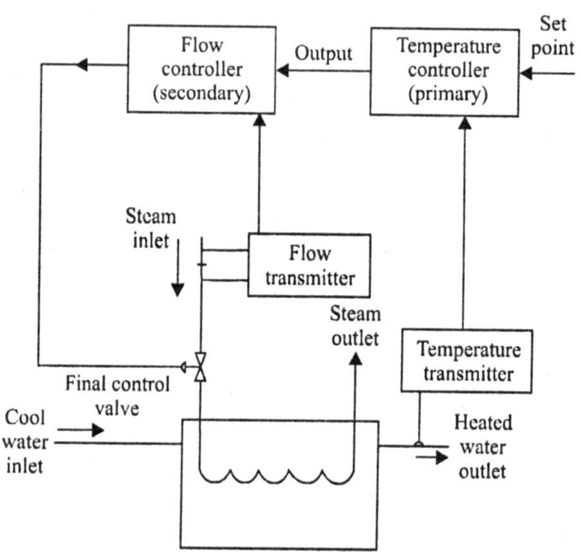


**Fig. 17.21.** Cascade control of a heat-exchange process.

Feed-forward control (Fig. 17.22) makes it possible to utilise other disturbances besides the measured values of the process that have to be controlled and enable fast control of a process. The cascade control system (Fig. 17.21) which has just been discussed would be adequate provided that the water inlet pressure and temperature do not fluctuate. When fluctuations occur, feed-forward control may be used, but this will require the measurement of both inlet water temperature and flow rate with both variables combined to control the steam flow rate (Fig. 17.23). In the example shown a computer is programmed to determine the heat energy required to be added to the cool inlet water to bring it to the appropriate temperature at the outlet and allow the correct amount of steam to enter the heating coils. Inputs for temperature and flow are made from the inlet pipe to the computer. Calculations can then be made to determine the required heat input to raise the water temperature in the heat exchanger. The output control signal to the valve A is computed to allow the correct amount of steam to enter the heating coils.

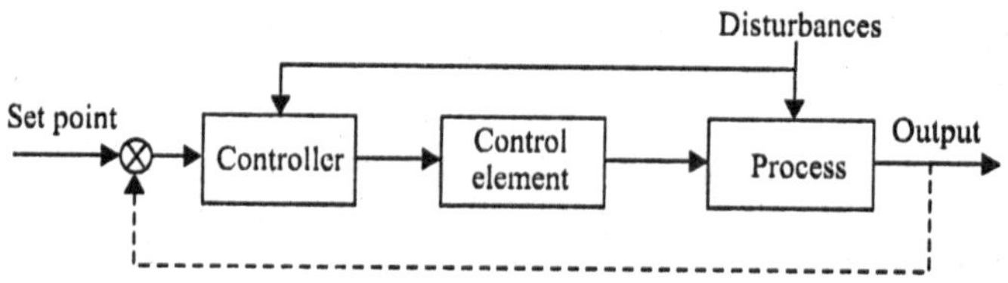


**Fig. 17.22.** A feed-forward control loop.

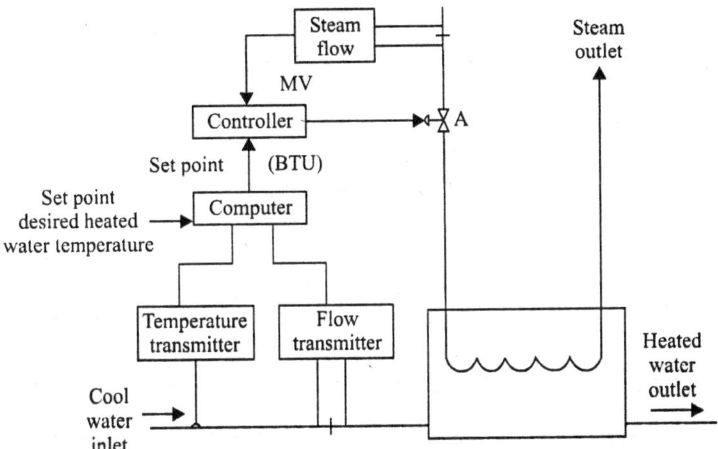


**Fig. 17.23.** Feed-forward control of a heat-exchange process.

When one or more of the process variables or characteristics is not known and cannot be measured directly, then adaptive control should be considered. The on-line identification of process characteristics and the subsequent use of this information to improve the process constitute adaptive control. The sequences and the interactions in the adaptive control loop are outlined in Fig. 17.24. Adaptive control is useful in circumstances where the process dynamics are not well defined or change with time. This may be most useful in controlling a batch fermentation where considerable and often complex changes may occur. Adaptive control strategies have been used in fed-batch yeast cultivation, amino acid production and penicillin production.

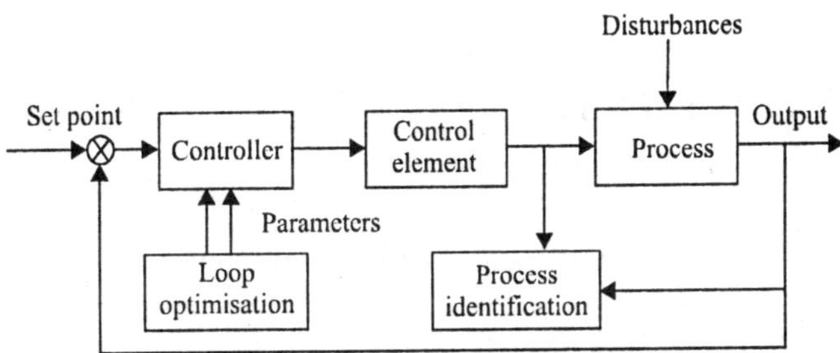


**Fig. 17.24.** Adaptive control.

It is also possible to prepare sequential programmes using a computer based control system. Consider control of the dissolved oxygen concentration in a fermentation broth. This may be changed by altering the agitation rate, the air or gas flow rate, the partial pressure of oxygen in the inlet gas or the total fermenter pressure. In practice, combinations of these variables may be used either sequentially or simultaneously using suitable computer programmes. Initially the agitation rate can be increased to respond to decreases in dissolved oxygen concentration. When a pre-determined maximum agitation rate is reached, the airflow can be steadily increased to a pre-set maximum, followed by the third and subsequent stages.

Chapter 18

# Computer Applications in Fermentation Technology

## INTRODUCTION

Since the initial use of computers in the 1960s for modelling fermentation process and in processes control for production of glutamic acid and penicillin, there have been numerous computer applications in fermentation technology. Initially, the use of large computers was restricted because of their cost but reductions in costs and the availability of cheaper small computers has widened interest in their possible applications.

The availability of efficient small computers has led to their use for pilot plants and laboratory systems since the financial investment for the on-line computer amounts to a relatively insignificant part of the whole system.

## DISTINCT AREAS OF COMPUTER FUNCTION

There are three distinct areas of computer function, they are:

1. Logging of process data: Data logging is performed by the data acquisition system which has both hardware and software components. There is an interface between the sensors and the computer. The software should include the computer program for sequential scanning of the sensor signals and the procedure of data storage.
2. Data analysis (reduction of logged data): Data reduction is performed by the data-analysis system, which is a computer program based on a series of selected mathematical equations. The analysed information may then be put on a print out, fed into a data bank or utilised for process control.
3. Process control: Process control is also performed using a computer program. Signals from the computer are fed to pumps, valves or switches via the interface. In addition the computer program may contain instructions to display devices or teletypes, to indicate alarms, etc.

At this point it is necessary to be aware that there are two distinct fundamental approaches to computer control of fermenters. The first is when the fermenter is under the direct control of the computer software. This is termed Direct Digital Control (DDC) and will be discussed in the next section. The second approach involves the use of independent controllers to manage all control functions of a fermenter and the computer communicates with the controller only to exchange information. This is termed Supervisory Set-Point Control (SSC) and will be discussed in more detail in the Process Control section.

It is possible to analyse data, compare it with model systems in a data store and use control programs which will lead to process optimisation. However, process optimisation by this method is not a widely used procedure in the fermentation industries at present. It is important to be aware of these different applications, since this will influence the size and type of computer system which will be appropriate for the precise role that it is intended to perform, whether in a laboratory, a pilot plant or manufacturing plant or a combination of these three.

## COMPONENTS OF A COMPUTER-LINKED SYSTEM

When a computer is linked to a fermenter to operate as a control and recording system, a number of factors must be considered to ensure that all the components interact and function satisfactorily for control and data logging. A DDC system will be used as an example to explain computer controlled addition of a liquid from a reservoir to a fermenter. A simple outline of the main components is given in Fig. 18.1. A sensor S in the fermenter produces a signal which may need to be amplified and conditioned in the correct analogue form. At this stage it is necessary to convert the signal to a digital form which can be subsequently transmitted to the computer. An interface is placed in the circuit at this point. This interface serves as the junction point for the inputs from the fermenter sensors to the computer and the output signals from the computer to the fermenter controls such as a pump T attached to an additive reservoir. Digital to analogue conversion is necessary between the interface and the pump T.

A sensor will generate a small voltage proportional to the parameter it is measuring. For example, a temperature probe might generate 1 V at 10°C and 5 V at 50°C. Unfortunately, the signal cannot be understood by the computer and must be converted by an analogue to digital converter (ADC) into a digital form.

The accuracy of an ADC will depend on the number of bits (the unit of binary information) it sends to the computer. An 8-bit converter will work in the range 0–255 and it is therefore able to divide a signal voltage into 256 steps. This will give a maximum accuracy of 100/256, which is approximately 0.4 per cent. However, a 10-bit converter can give 1024 steps with an accuracy of 100/1024, which is approximately 0.1 per cent. Therefore when a parameter is to be monitored very accurately a converter of the appropriate degree of accuracy will be required. The time taken for an ADC to convert voltage signals to a digital output will vary with accuracy, but improved accuracy leads to slower conversion and hence slower control responses. However, cycle times of about 1 second may be adequate in many fermentation systems. It is also important to ensure that the voltage ranges of the sensors are matched to the ADC input range. A digital to analogue converter (DAC) converts a digital signal from the computer into an electrical voltage which can be used to drive electrical equipment, e.g., a stirrer motor. Like the ADC, the accuracy of the DAC will be determined by whether it is 8 bit, 10 bit, 12 bit, etc., and will for example determine the size of steps in the control of rpm of a stirrer motor.

The small computer itself is dedicated solely to one or more fermenters. This computer is coupled to a real-time clock, which determines how frequently readings from the sensor(s) should be taken and possibly recorded. The other ancillary equipment linked directly to the computer might include a visual display unit, a data store, a tele-type, a graphic display unit, a print out, alarms and a barometer.

The small computer is often connected to a large main frame computer for random access, not on a real-time scale, but for long-term data storage and retrieval and for complex data analysis which will not be utilised subsequently in real-time control.

It is also possible to develop programs so that on-line instruments can be checked regularly and re-calibrated when necessary. Swartz and Cooney were able to routinely re-calibrate a paramagnetic oxygen analyser and an infrared carbon dioxide analyser every 12 hours utilising a program which connected a gas of known composition to the analyser and subsequently monitored the analyser outputs.

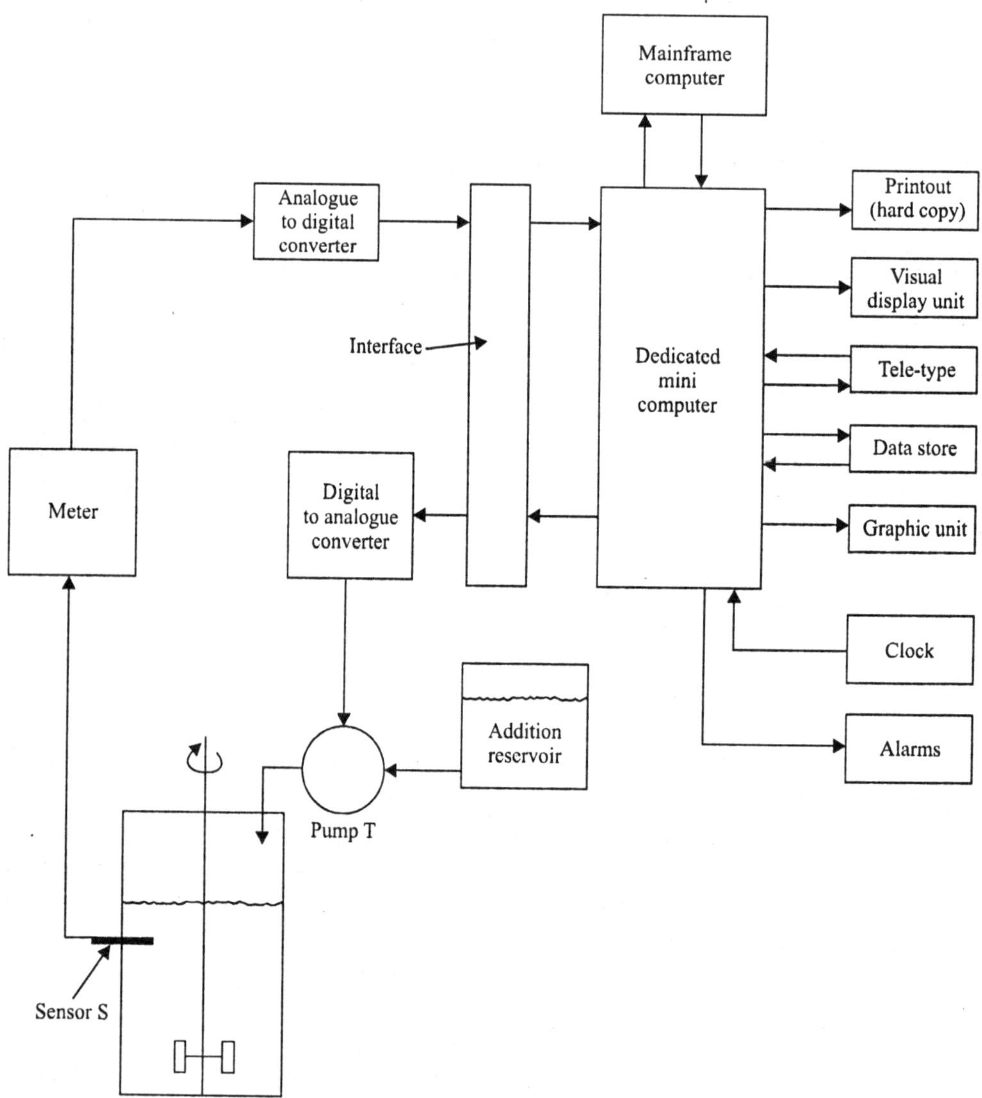


**Fig. 18.1.** Simplified layout of computer-controlled fermenter with only one control loop shown.

## DATA LOGGING

The simplest task for a computer is data logging. Parameters such as those already listed in Chapter 17 'Instrumentation and Control' (Table 17.1) can be measured by sensors which produce a signal which is compatible with the computer system.

Programs have been developed so that by reference to the real-time clock, the signals from the appropriate sensors will be scanned sequentially in a pre-determined pattern and logged in a data store. Typically, this may be 2 to 60 seconds intervals and the data is printed out on a visual display unit. In preliminary scanning cycles the values are compared with pre-defined limit values and deviations from these values result in an error print out or if more extreme then an alarm may be activated. In the final cycle of a sequence, say every 5 to 60 minutes, the program instructs that the sensor readings are permanently recorded on a print out or in data store.

At the same time as on-line data is being recorded from sensors, analytical data for broth viscosity, microbial growth, substrate and precursor utilisation and product formation, which have to be determined separately may be logged into the data store for specific known times.

Thus, it is now possible to record data continuously for a range of parameters from a number of fermenters simultaneously using minimal manpower, provided that the capital outlay is made for fermenters with suitable instrumentation coupled with adequate computer facilities.

## DATA ANALYSIS

Because a computer can undertake so many calculations very rapidly, it is possible to design programs to analyse fermentation data in a number of ways. A linked main-frame computer may be used for part of this analysis as well as the dedicated small computer.

A number of the monitoring systems are given in Table 18.1. Gateway sensors are so-called because the information they yield can be processed to give further information about the fermentation.

**Table 18.1.** Gateway sensors.

| *Sensor* | *Information that may be determined from the sensor signal* |
|---|---|
| pH | Acid product formation |
| Dissolved oxygen | Oxygen-transfer rate |
| Oxygen in exit gas | Oxygen-uptake rate |
| Gas-flow rate | Oxygen-uptake rate |
| Carbon dioxide in exit gas | Carbon dioxide evolution rate |
| Gas-flow rate | Carbon dioxide evolution rate |
| Oxygen-uptake rate | Respiratory quotient |
| Carbon dioxide evolution rate | Respiratory quotient |
| Sugar-level and feed rate | Yield and cell density |
| Carbon dioxide evolution rate | Yield and cell density |

The respiratory quotient of a culture may be calculated from the metered gas-flow rates and analyses for oxygen and carbon dioxide leaving a known volume of culture in the fermenter. This procedure was used to monitor growth of *Candida utilis* in a 250 $dm^3$ fermenter, to follow or forecast events during operation.

If one defines the fraction of substrate which is converted to product then it is possible to write mass balances for C, H, O and N with the measurement of only a few quantities ($O_2$, $CO_2$, $NH_3$, etc.). All the other quantities can be calculated, including biomass and yield, if the biomass elemental composition is known. This procedure was used for the analysis of a baker's yeast fermentation. Biomass production can be regarded as a stoichiometric relationship in which substrate is converted, in the presence of oxygen and ammonia to biomass, carbon dioxide and water:

$$\text{Carbon source-energy} + \text{oxygen} + \text{ammonium} \rightarrow \text{Cells} + \text{water} + \text{carbon dioxide}$$

Thus, the equation can be written in the form:

$$aC_xH_yO_z + bO_2 + cNH_3 \rightarrow dC_rH_sO_tN_u + H_2O + fCO_2$$

where a, b, c, d, e and f are moles of the respective reactants and products. $C_xH_yO_z$ is the molecular formula of the substrate where x, y and z are the specific carbon, hydrogen and oxygen atom numbers. Biomass is represented by $C_rH_sO_tN_u$ where r, s, t and u are the corresponding numbers of each element in the cell. This technique was developed to use with bakers' yeast fermentations.

## PROCESS CONTROL

Arminger and Moran recognised three levels of process control that might be incorporated into a system. Each higher level involves more complex programs and needs a greater overall understanding of the process. The first level of control, which is already routinely used in the chemical industries, involves sequencing operations, such as manipulating values or starting or stopping pumps, instrument recalibration, on-line maintenance and fail-safe shut-down procedures. In most of these operations the time base is at least in the order of minutes, so that high-speed manipulations are not vital. Two applications in fermentation processes are sterilisation cycles and medium batching.

The next level of computer control involves process control of temperature, pH, foam control, etc., where the sensors are directly interfaced to a computer [Direct Digital Control (DDC); Fig. 18.1]. When this is done separate controller units are not needed. The computer program determines the set point values and the control algorithms, such as PID, are part of the computer software package. Better control is possible as the control algorithms are mathematically stored functions rather than electrical functions. This procedure allows for greater flexibility and more precise representation of a process control policy. The system is not very expensive as separate electronic controllers are no longer needed, but computer failure can cause major problems unless there is some manual back-up facility.

The alternative approach is to use a computer in a purely supervisory role. All control functions are performed by an electronic controller using a system illustrated in Fig. 18.2 where the linked computer only logs data from sensors and sends signals to alter set points when instructed by a computer program or manually. This system is known as Supervisory Set Point Control (SSC) or Digital Set Point Control (DSC). When SSC is used, the modes of control are limited to proportional, integral and derivative because the direct control of the fermenter is by an electronic controller. However, in the event of computer failure the process controller can be operated independently.

The most advanced level of control is concerned with process optimisation. This will involve understanding a process, being able to monitor what is happening and being able to control it to achieve and maintain optimum conditions. Firstly, there is a need for suitable on-line sensors to monitor the process continuously. A number are now available for dissolved oxygen, dissolved carbon dioxide, pH, temperature, biomass (the bug meter, NADH fluorescence, near infrared spectroscopy) and some metabolites (mass spectroscopy and near infrared spectroscopy). All these sensors have been discussed earlier in this chapter. Secondly, it is important to develop a mathematical model that adequately describes the dynamic behaviour of a process. Shimizu has stressed the vital role which these models play in optimisation and reviewed the use of this approach in batch, fed-batch and continuous processes for biomass and metabolites. This approach with appropriate on-line sensors and suitable model programs has been used to optimise bakers' yeast production, an industrial antibiotic process and lactic acid production.

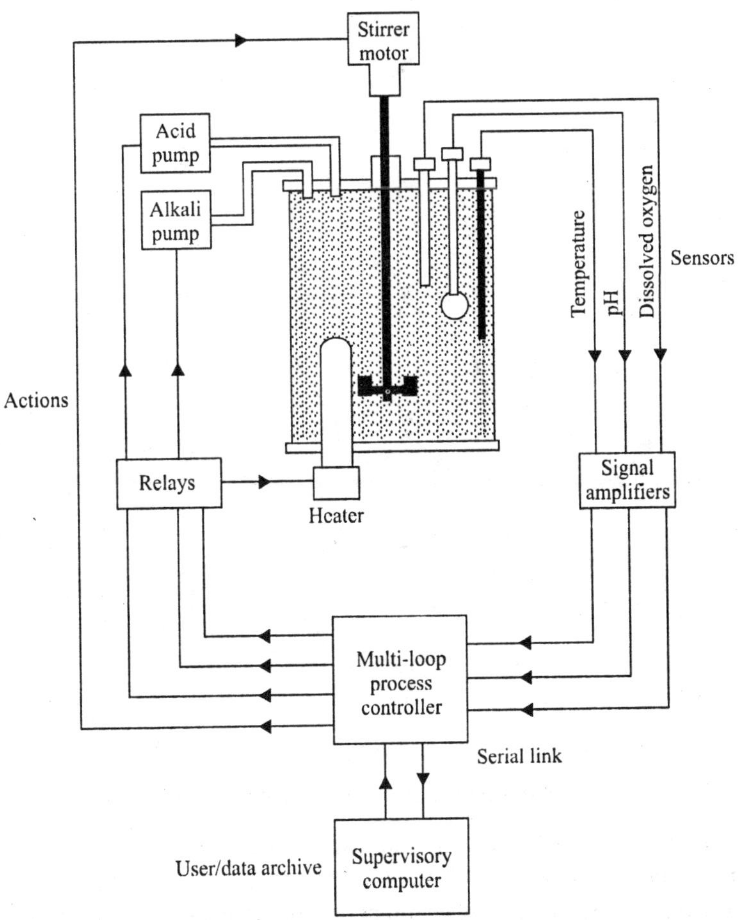


**Fig. 18.2.** Diagrammatic representation of supervisory set point control (SSC) system for fermenters. This example illustrates a system controlling temperature by means of heating only, dissolved oxygen tension by stirrer speed and pH by the addition of acid and alkali. All control functions are performed by the intelligent process controller and the computer only communicates with this in order to log data and send new set points when instructed to do so by the user.

Although much progress has been made in the ability to control a process, few sensors are yet available to monitor on-line for many metabolites or other parameters in a fermentation broth thus delaying or making a fast response difficult for on-line control action. Also, it is possible that not all the important parameters in a process have been identified and the mathematical model derived to describe a process may be inadequate. Because of these limitations, an artificial neural network may be used to achieve better control. These are highly interconnected networks of non-linear processing units arranged in layers with adjustable connecting strengths (weights). In simpler neural networks there is one input layer, one hidden layer and one output layer (Fig. 18.3). Unlike recognised knowledge-based systems, neural networks do not need information in the form of a series of rules, but learn from process examples from which they derive their own rules. This makes it possible to deal with non-linear systems and approximate or limited data.

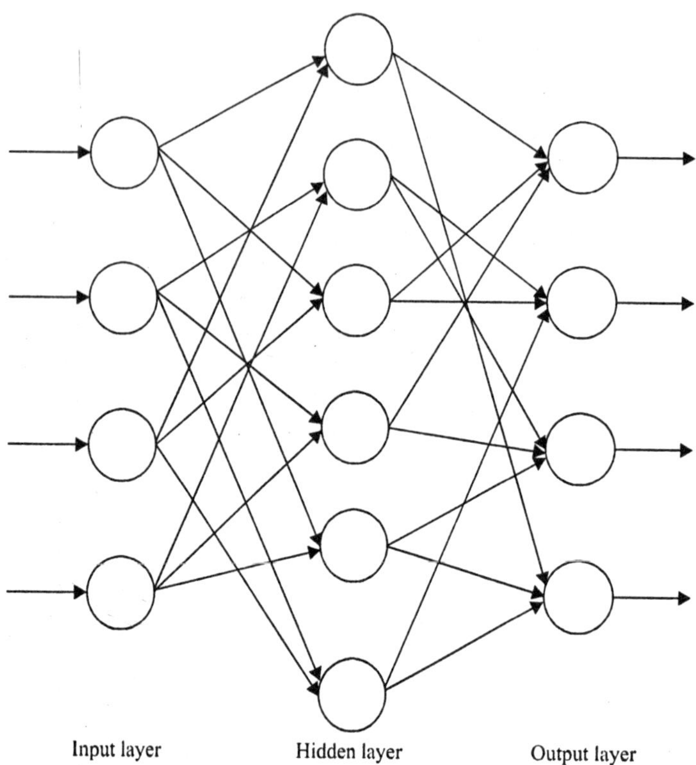


**Fig. 18.3.** Two-layer neural network (not all the possible interconnections are shown).

When training a neural network the aim is to adjust the strengths of the interconnections (neurons) so that a set of inputs produces a desired set of outputs. The inputs may be process variables such as temperature, pH, flow rates, pressure and other direct or indirect measurements which give information about the state of the process. The process outputs obtained (biomass, product, etc.) produce the teacher signal(s) which trains the network. The difference between the desired output and the value predicted by the network is the prediction error. Adjustments are made to minimise the total prediction error by modifying the interconnection strengths until no further decrease in error is achieved. Commercial computer packages are now available to help to determine which of the input variables to use for training and to determine the optimum number of interconnections and hidden layers.

This method of control is still at an early stage of development, but it has already been used in a case study on ethanol production by *Zymomonas mobilis*, in real-time variable estimation and control of a glucoamylase fermentation and recombinant *Escherichia coli* fermentations.

In industrial systems where a significant amount of on-line and off-line process data may be available, but there are tight time restraints imposed on process optimisation, the potential for developing a relatively accurate neural network model within short time scales become very attractive.

## COMPUTER-AIDED DESIGN OF INTEGRATED BIOCHEMICAL PROCESS

As a result of the advances in molecular biology and genetic engineering, the scientific community has come to realise the great potential for developing new products and systems through novel use of micro-organisms and enzymes. The challenge for the biochemical industry is now to scale-up and commercialise those products. This is a difficult task, especially for small corporations, because it is a complex problem that requires co-ordination of a large number of activities across many disciplines. Computer-aided process design tools have been successfully used in the chemical process industries for over three decades to scale-up and optimise integrated processes for the production of petrochemicals and other products. Similar benefits can be expected from the use of such tools in the biochemical industries. This section describes the architecture and key components of computer-aided bioprocess design tools with particular emphasis on bioprocess designer.

A case study on β-galactosidase production illustrates how such tools can be used in an industrial environment to analyse and optimise integral biochemical processes.

## Computer-Aided Process Design

Design is the creative process that leads from the identification of a need to an end product, system or process that satisfies that need. Process design consists of two main sub-activities, process synthesis and process analysis.

### *Process synthesis*

Process synthesis deals with the selection and assembly of a set of process steps capable of producing the desired product and satisfying a number of product quality, economic and environmental/safety constraints. The result of process synthesis is a number of feasible flowsheets.

### *Process analysis*

Process analysis deals with the analysis and evaluation of individual or integrated processes. The results of process analysis for a flowsheet commonly include the calculation of material and energy balances, estimation of size and cost of equipment, economic evaluation, process scheduling and environmental product/process safety evaluation.

The two sub-activities of process design are not independent. Process design is an evolutionary activity that requires a chain of synthetic and analytic tasks. Based on the results of process analysis, for instance, one may want to modify either the structure or the operating conditions of a flowsheet aiming at a better solution. This flowsheet modification is a synthetic activity.

The automation of process synthesis and process analysis using computers is the focus of computer-aided process design. Automated design tools have found limited applications in the bioprocess industry till now because: (i) the biotech industry is relatively new and primarily dominated by scientists that are not process oriented; (ii) most biological products, especially the high value ones, are complex and labile materials (e.g., proteins and polysaccharides), poorly characterised with respect to physical properties (e.g., diffusivity, solubility, viscosity, etc.) and these values are difficult to predict from thermodynamic and micro transport models; (iii) several important bioprocess unit operations are poorly understood and predictive models which could be used for design do not exist; and (iv) most bioprocesses operate in batch or semi-continuous mode introducing time as independent variable.

## Benefits of Computer-Aided Bioprocess Design

Computer-aided process design tools can play a key role at various stages of product and process development. The benefits from the use of such tools vary depending on the type of product, the size of the investment, the location of the facility and other parameters.

#### Product and project selection

During project selection, they can be used to rapidly evaluate a large number of project ideas and discard the non-promising ones. This can drastically reduce engineering time because experience has shown that less than 5 per cent of new project ideas are ever commercialised.

#### R & D planning

In research and development they can be used to plan and prioritise pilot plant work. More specifically, before serious process development is initiated, the engineer develops on the computer a conceptual model of the entire process that requires scale-up. Through the initialisation steps, the user identifies all the missing information (e.g., partition coefficients for extractors, rejection coefficients for membrane filters, etc.) that is needed to develop an accurate model of the entire process. Then, based on the missing information, the engineer designs the appropriate experiments to gather that data and complete the modelling work. Having a good model of the entire process, the engineer starts asking 'what–if' questions and carrying out sensitivity analysis and optimisation with respect to key design variables. If simulation yields any unexpected results, experiments are planned to test them. This systematic procedure of designing and running pilot plant experiments saves development time and yields more efficient and robust processes.

#### Environmental and safety issues

Environmental and process safety problems are best solved when such issues are considered during the early stages of process development and built into the initial process design. Computer-aided process design tools, when equipped with the appropriate features, enable engineers to consider the environmental impact of new products and processes at the early stages of process development when it is still easy to make process modifications. This is very important especially for therapeutic biological products because after a new product is approved as safe and effective by the Food and Drug Administration (FDA) it is very costly to make process modifications. Using such tools, for instance, the engineer can systematically select extraction solvents or chaotropic agents for solubilisation of inclusion bodies that meet all process requirements and at the same time are environmentally benign. The engineer also can readily evaluate process modifications (i.e., addition of recycle streams, replacement of chemicals, etc.) that minimise waste generation.

#### Improved communication

When such tools are used by all groups involved in process development and design, they introduce a common language of communication that drastically reduces duplicate work. In a sense, their use constitutes a first step towards concurrent engineering and integrated product development with reduced cost.

#### Retrofit of existing facilities

Computer-aided process design tools are not only useful for designing new processes but also for retrofitting existing ones. More specifically, they can be used to evaluate process modifications of existing manufacturing facilities in order to increase capacity, reduce production cost, reduce waste generation (and consequently cost of waste treatment and disposal), improve process safety, etc.

#### Process simulators

Process simulators are mainly used to analyse and evaluate complex processing systems, carry out sensitivity analysis and optimise given flowsheets. They have been successful in minimising production cost of existing plants and improving the design of new plants. These tools, however, lack synthetic capabilities. Recent approaches make use of Knowledge Based Expert Systems (KBES). KBES capture experimential design knowledge that is primarily used in process synthesis to constrain the solution space. However, KBES do not offer the best environment to handle algorithmic and 'number crunching' tasks. Conventional programming languages are more appropriate for those tasks. To combine the best of the two approaches, hybrid systems are often developed. In such systems, a KBES that handles synthetic tasks is interfaced to a number crunching program that handles the analytic tasks of design.

Substantial research and development efforts in computer-aided design of the integrated biochemical processes began in the mid 1980's following the development efforts and commercialisation of new high value bioproducts.

### *Bioprocess simulator (BPS)*

BPS is an extension of the Aspen Plus process simulator. It uses the infrastructure and the facilities provided by Aspen Plus. BPS, for a given flowsheet, carries out material and energy balances, estimates the size and cost of equipment and carries out an economic evaluation.

### *Bioseparation Designer*

Bioseparation designer is a knowledge based system that carries out synthesis of protein separation systems. It is written in Common Lisp and runs on the symbolics computer. Based on a given set of input data, bioseparation designer generates possible flowsheets and employs an evaluation technique to find the best one. Its synthetic knowledge base contains information mainly focused on the production of proteins by *E. coli*. It has facilities to estimate a number of protein properties from their amino acid sequence. It provides a database to store protein properties. Bioseparation designer also includes some analytic and economic evaluation capabilities.

### *Biodesigner*

Biodesigner is a design tool that combines synthetic with analytic capabilities and the emphasis is on interactive analysis. It carries out synthesis using a KBES based on the Nexpert object expert system shell. The knowledge utilised for process synthesis is based on the properties of the producing micro-organism, product, contaminants, etc. Biodesigner runs on personal computers and features an interactive and user friendly interface.

## Components of a Bioprocess Design Tool

Figure 18.4 shows the typical architecture of a computer-aided bioprocess design tool that combines synthetic with analytic capabilities. Such tools usually consist of three components, the user interface, the synthetic component and the analytic component.

### *User Interface*

Contemporary computer-aided process design tools use graphical user interfaces that facilitate the communication of the user with the computer. Graphical interfaces eliminate the need for development of Fortran-like input files and shorten the learning period. Figure 18.5, for instance, shows the main window of the analytic component of bioprocess designer. The user initialises a flowsheet interactively through appropriate dialog windows. Most contemporary computer-aided process design tools also feature on-line 'Help' facilities.

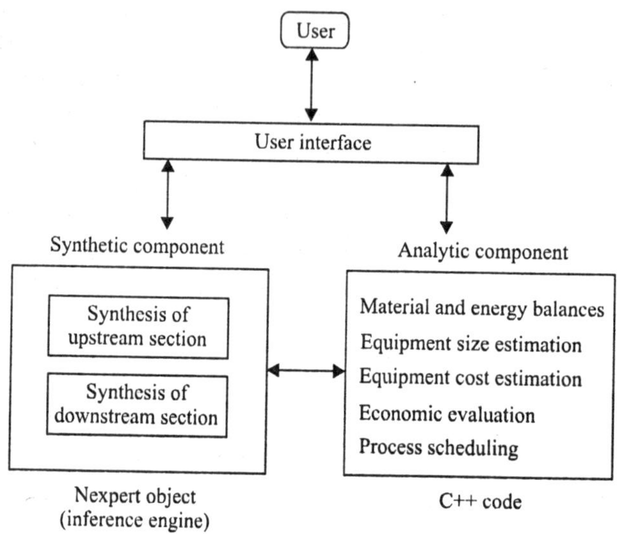


**Fig. 18.4.** Biodesigner's system architecture.

### *Synthetic component*

Process synthesis deals with the selection and assembly of integrated processes that are capable of producing, recovering and purifying the desired product(s) while meeting financial and environmental constraints. Synthesis of entire flowsheets is a considerably more difficult problem compared to process analysis. This explains why the development of process synthesis tools is still a challenging research problem and why no synthetic tools (for entire flowsheets) are currently commercially available. As mentioned previously, one of the most successful techniques for process synthesis makes use of knowledge based expert systems (KBES). KBES offer a natural way to capture and utilise experiential design knowledge that is primarily used in process synthesis to reduce the solution space. Knowledge in the form of heuristics is stored in KBES mainly in the form of production rules. An example of the typical syntax of a rule is 'if the type of micro-organism is an animal cell, then use membrane filtration for media sterilisation'. The synthetic component of biodesigner (Fig. 18.4) consists of two such knowledge bases. The first deals with synthesis of upstream sections (media preparation and bioreaction) and contains forty rules. The second knowledge base deals with synthesis of downstream sections (product recovery and purification) and contains ninety rules. Industrial practice, common sense and rules reported in the literature constitute the main sources of synthetic knowledge in biodesigner.

### *Analytic component*

The analytic (simulation) component of a computer-aided process design tool calculates the material and energy balances of an entire flowsheet, estimates the size and cost of equipment and does the economic evaluation. The heart of any analytic component is the unit operation models. Models for traditional as well as for unit operations specific to biochemical processes should be included in bioprocess design tools. The analytic component also generates a number of reports, such as the stream report, the economic evaluation report(s), the input/output data report, etc. The analytic component of the IBM version of bioprocess designer is written in Visual C++. The material and energy balances around unit operations are calculated in sequential-modular approach.

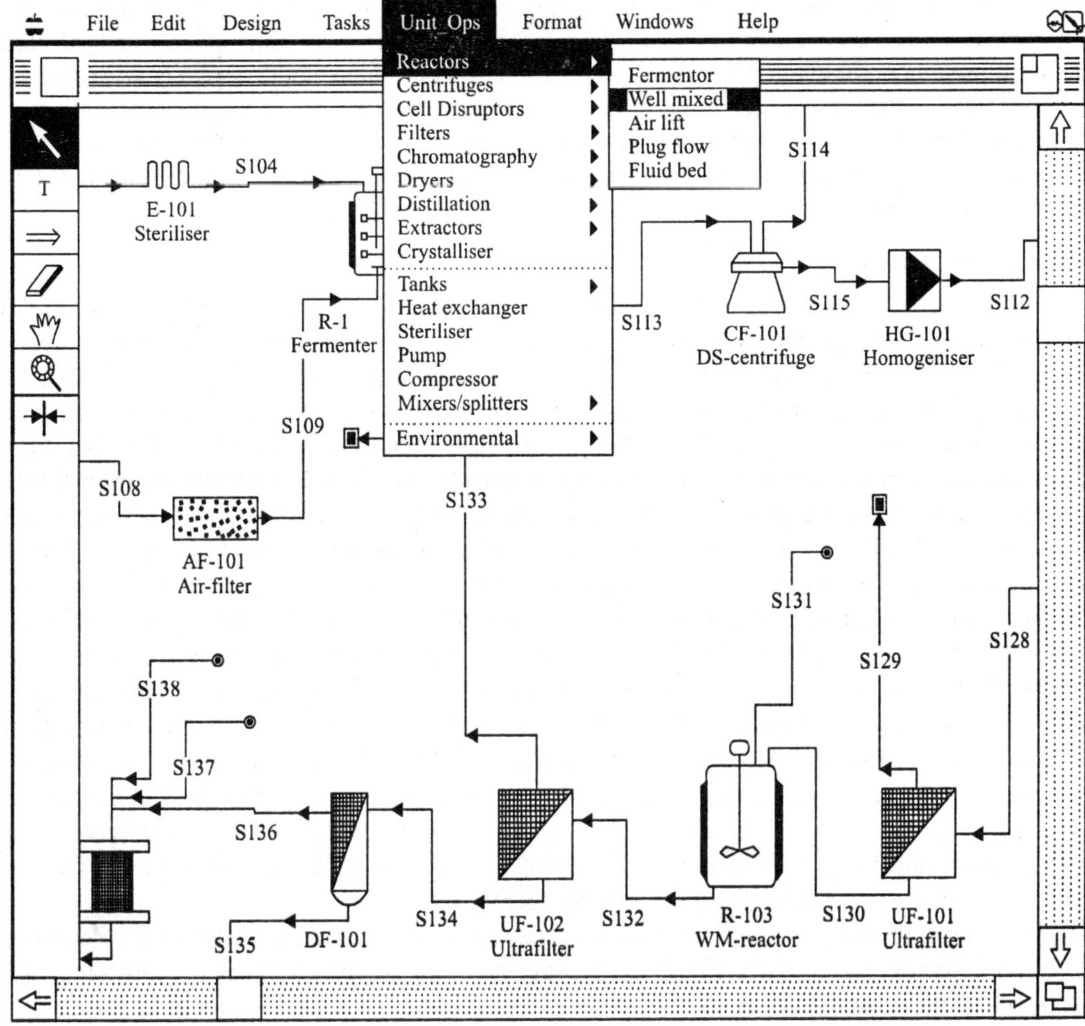


**Fig. 18.5.** Main dialog window of the analytic component of bioprocess designer.

## *Process scheduling*

A number of unit operations in the biochemical industries operate in batch or semi-continuous mode. Also, many biochemical plants are multiproduct facilities that utilise the same equipment-line for the production of multiple products. To accommodate this time-dependency of equipment activity, a bioprocess design tool must be equipped with process scheduling capabilities and dynamic models for several unit operations. Process scheduling information is also needed for the accurate sizing of utilities and the estimation of volatile organic compound (VOC) emissions from process equipment. An important issue in the development of process simulator is validation of the unit operation models.

## An Example—Simulation and Evaluation of β-Galactosidase Production

Bioprocess designer was used to analyse and evaluate the production of an intracellular enzyme, β-galactosidase, by *E. coli*. The analysis of the production of any other intracellular protein that does not form inclusion bodies would be very similar. β-Galactosidase has found limited industrial applications till now. It is mainly used in the utilisation of cheese whey to convert lactose into glucose and galactose. An annual production of about 5 tonnes of pure enzyme is considered as the design basis in this case study. The production rate has been selected to allow the size of the plant and the corresponding cost values to be high enough and representative of the values of a relatively large plant that produces an intracellular recombinant protein. It is assumed that the plant batch time is 72 hours and that 96 batches are processed per year.

### *Process description*

Fermentation media are prepared in the stainless steel tank and are continuously sterilised. The filling of the production fermenter with sterilised media takes 4 hours. The axial compressor CMP101 and the absolute filter AFL101 provide sterile air to the fermenters at an average rate of 0.4 VVM. The fermentation time in the production fermenter is about 18 hours, while the turnaround time is 12 hours. The final concentration of *E. coli* in the production fermenter is about 35 g/liter (dcw). The first step of the downstream section is cell harvesting and is carried out by a hollow-fibre membrane microfilter of 0.45 μm pore diameter. A high pressure homogeniser is used to break the cells and release the intracellular product. The broth undergoes two passes under a pressure drop of 800 bars. The removal of cell debris particles, created by homogenisation, is accomplished by a disk-stack centrifuge in a period of 8 hours. The dead-end polishing filter removes any remaining particulate material. The dilute protein solution is concentrated by a hollow-fibre membrane ultrafilter of 1,00,000 molecular weight cut-off to a total protein concentration of 5–6 per cent w/w. The ion exchanger is used for high resolution purification. Anion exchange is used because β-galactosidase is negatively charged in the range of pH that is stable. The type of resin is fast flow Mono Q (from Pharmacia). Four cycles are used each of 6 hours. The ultrafilter is used to further concentrate (after its dilution by the ion exchanger) the protein solution to a final concentration of total protein of approximately 5 per cent w/w. Finally, purification is completed by the gel filter unit. The type of resin is Sephadex G25 or Sepharose CL (from Pharmacia). The final product purity is approximately 99.5 per cent (Fig. 18.6).

For the plant of the design basis (5 tonnes of β-galactosidase per year) the fermenters and compressors are the most expensive pieces of equipment in the upstream section, while centrifuges and chromatography units dominate the downstream section.

## Sensitivity Analysis

After a model for the entire process is developed on the computer, tools like bioprocess designer can be used to ask and readily answer 'what if' questions and carry out sensitivity analysis with respect to key design variables. In this case study, the effect of the two parameters on the production cost was examined.

### *Effect of dilution after cell harvesting on process economics*

Dilution after cell harvesting and prior to cell disruption is done to mix the concentrated broth with a buffer solution that will ensure the stability of the product after its release by cell disruption. Dilution plays a negative role on process economics because the higher the dilution, the higher the equipment size and cost further downstream and consequently the higher the production cost. However, there is also a positive effect associated with dilution. More specifically, in the removal of cell debris particles by the centrifuge, the particles in the heavy phase of the centrifuge comprise only about 20–30 per cent of total volume. The rest of the volume is liquid, with product protein dissolved in it. Consequently, the

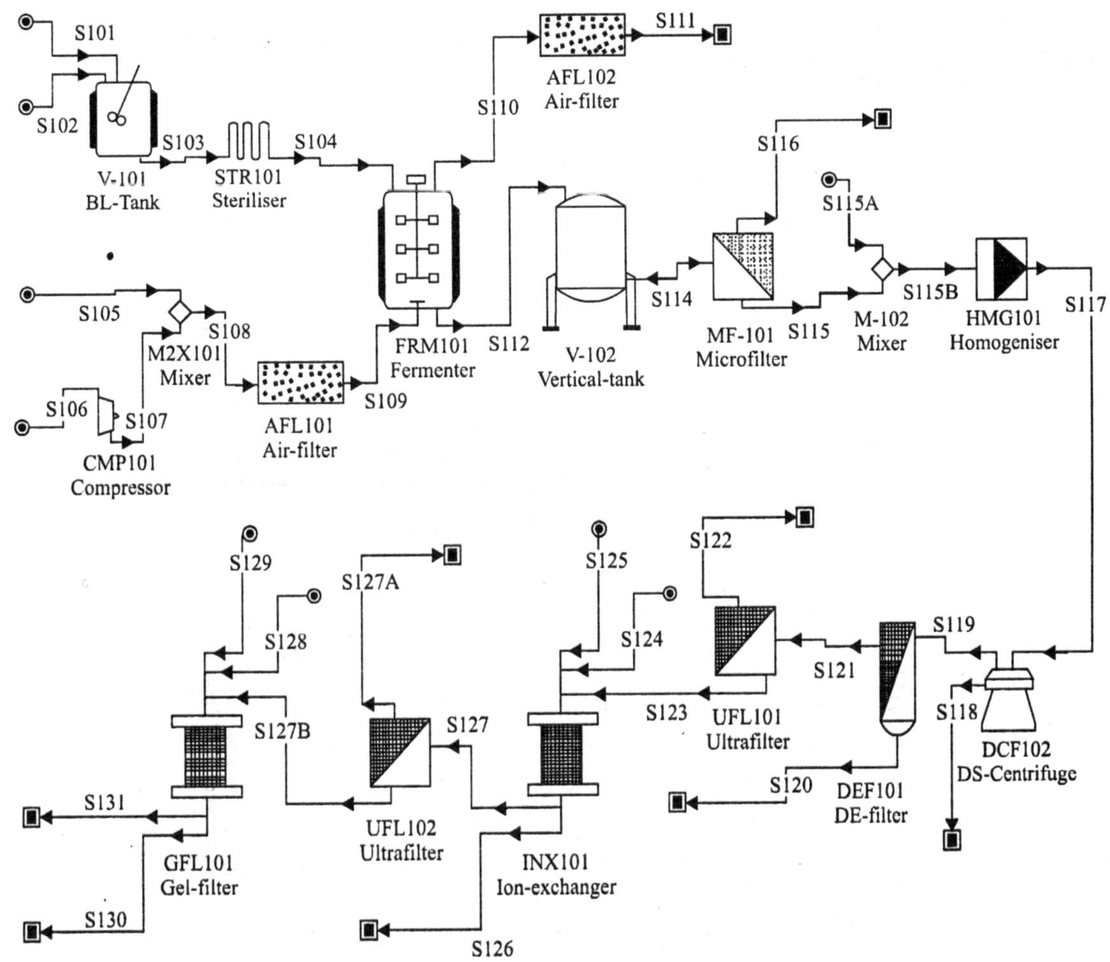


**Fig. 18.6.** Production of β-galactosidase-process flowsheet.

higher the extent of dilution, the lower the product concentration in stream (feed to centrifuge) and as a result the lower the product loss in the heavy phase of the centrifuge. The results of the effect of dilution on process economics are shown in Fig. 18.7(a). As can be seen, the combined effect of dilution initially is positive but after a certain level (20000 litres/batch) becomes negative.

### *Effect of protein solution concentration on process economics*

After cell debris removal, the dilute protein solution is concentrated by ultrafiltration to reduce the volume of material that needs to be processed by the expensive chromatography units. In this case study, the effect of protein solution concentration by the UFL102 unit on process economics was analysed. The higher the concentration factor, the lower the capital and operating cost of the expensive gel filtration unit (GFL 101). The results of the analysis are shown in Fig. 18.7(b). The y-axis on the left shows the product cost that is reduced drastically by increasing the concentration factor. The y-axis on the right shows the total protein solution concentration (as % w/w) in the feed stream to gel filtration. Chromatography manufacturers recommend that the total protein concentration in the feed stream to a chromatography unit should not exceed 5 per cent w/w. This information can be used to set an upper bound to the extent of concentration.

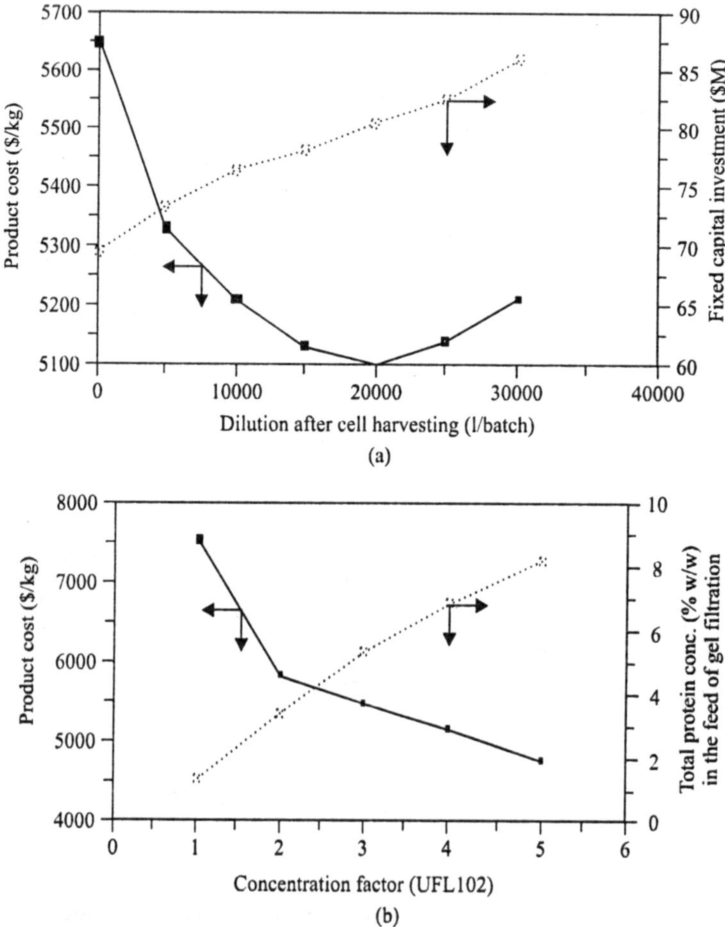


**Fig. 18.7.** (a) Effect of dilution after cell harvesting on process economics; and (b) effect of protein solution concentration by the UF102 ultrafilter on process economics.

Thus, computer-aided bioprocess design tools can play an important role in the development of biochemical processes and commercialisation of biological products. At the early stages of project selection, such tools can be used to screen the large number of projects ideas from a profitability point of view and help focus development efforts on the most promising projects. During process development, such tools can be used to analyse and evaluate alternative processing schemes, reduce the impact of the whole process on the environment, interpret experimental results and help design experimental protocols. During final design and plant construction, such tools can be used to optimise the entire process from a total-systems point of view. All these benefits substantially reduce cost and development time associated with a biological product.

## COMPUTER CONTROLLED ENZYMATIC REACTOR

This section deals with control and design of a medical reactor to be used in the extracorporeal treatment of laeukemia in children. In this work, an adaptive controller based on state space approach is developed and tested for the control of a medical reactor, using L-asparaginase attached on the internal open nylon tube. As a medical reactor, the design must be simple and safe to be operated. It was shown that the control algorithm was adequate to maintain the system in a specified operational condition when the manipulated variable was the electric current, even when significant disturbances took place as well as setpoint changes (servo control). This makes this system to be a promising method in the extracorporeal treatment of leukaemia. The enzyme L-asparaginase (L-asparagine amidohydrolase) catalyses the hydrolysis of L-asparagine to L-aspartic acid and ammonia. Interest in this enzyme arose a few decades ago when it was discovered its value in the treatment of acute lymphoblastic leukaemia in children. Unlike normal cells, the malignant cells can only synthetise L-asparagine slowly and are dependent on an exogenous supply. The antineoplastic activity results from depletion of circulating pools of the amino acid by the L-asparaginase. The way of use has been through intramuscular applications of the free enzyme solution which must be pure. However, in many cases high doses are necessary for an effective treatment and as the enzyme is a foreign protein, troublesome side effects can arise such as nausea, vomiting and anaphylatic effects consequently limiting its use. This enzyme can be obtained from several microbial sources and recently was demonstrated the ability of *Erwinia aroideae* to produce the enzyme growing in supplemented cheese way.

Alternatively, the beneficial effects of the L-asparaginase can be used in this therapeutic without the side effects problems, if the extracorporeal treatment is used. It has been shown that the L-asparaginase covalently attached to nylon tubing are useful in clinical works. This tube method shows great promise for medical use for three main reasons: (i) it allows enzyme reactions to be carried out on a living system without the need to introduce into the systemic circulation (for example, blood could be perfused through an enzyme-activated nylon coil); (ii) there is no loss of enzyme, so that a very limited quantity of enzyme can be used over extended periods of time; and (iii) it may readily be incorporated into clinical systems, many of which already make use of tubes. Furthermore, we can design a system which must be operationally simple and operated under conditions to provide high performance, stability and safe. The ability to control enzymatic reactors at their desired point of operation accurately and automatically is of considerable interest in the design of medical reactors, since it can provide high safety performance. For the control system design of bioreactors, the following characteristics must be taken into account: (i) the lack of accurate mathematical models which describe the system; (ii) the time-varying and non-linear nature of the process; and (iii) the lack of reliable sensors which can detect the state variables such as the concentration of reactants and metabolites. With such a complex problem, simple control schemes without automatic adjustments in the parameters means that the efficiency of the process is low, so the use of more sophisticated adaptive control schemes with a state and/or parameter estimator must be considered.

Distributed parameter control problems can be treated through non-linear stochastic techniques but the solution is very complex. On the other hand, it is possible to approximate the equations in such a way to enable linear techniques to be used, which can then use, for example, self-tuning regulators (STR). These form part of the general class of adaptive controllers dealing with uncertainties, non-linearities and time varying process parameters.

In the present study we consider the problem associated with the maintenance of adequate operational conditions in an open tubular reactor with L-asparaginase attached in the internal surface, which can be used as a medical reactor in the leukaemia treatment. Using an appropriate mathematical model to represent the system, the process is simulated operating under control. A servo STR type, based in the state space approach is developed and is shown that this control algorithm is adequate to maintain the system in a specified operational condition, even when significant disturbances take place as well as a change in the setpoint. Also, we have demonstrated the reconstruction of the states by the estimator, based on the Kalman filtering technique.

## Process Model

The L-asparaginase reactor consists of a nylon tube with the enzyme attached to the internal tube surface. The internal diameter is 1 mm and a variable electric resistance is adjusted along the center of the reactor. An input electric current is supplied at the terminals of the resistance to heat generation. The chemical reaction taking place in the system is the following:

$$\text{L-Asparagine} \xrightarrow{E} \text{L-Aspartic acid} + NH_3$$

where $E$ is the L-Asparaginase enzyme. This reaction is weakly exothermic but very sensitive to temperature variation. The mathematical equations which describe the dynamic behaviour of the system may be expressed as follow:

$$\frac{\partial C_i}{\partial t} = D_i \frac{\partial^2 C_i}{\partial z^2} - v\frac{\partial C_i}{\partial z} - r_i \qquad \text{... (18.1)}$$

$$\frac{\partial T}{\partial t} = \alpha_F \frac{\partial^2 T}{\partial z^2} - v\frac{\partial T}{\partial z} - \frac{4\alpha_F h_i}{k_F d_i}(T - T_T) + \frac{\alpha_F P_u}{k_F A_T dz} \qquad \text{... (18.2)}$$

with initial and boundary conditions:

$$t = 0, \forall z : C_i = 0 \quad T = T_0$$
$$z = 0, t > 0 : C_A = C_0 \quad C_B = 0 \quad T = T_0$$

The reaction rate is represented by the Michaelis-Menten model, which is valid for this system:

$$r_i = \frac{V_M(T,E)C_i}{K_M + C_i} \qquad \text{... (18.3)}$$

The maximal reaction rate ($V_M$) is a function of the enzyme concentration and the temperature (in the Arrhenius form). The enzyme is deactivated irreversible with the age time and temperature, expressed in the following manner:

$$E = E_0 \exp(-K_d t) \qquad \text{... (18.4)}$$

$$K_d = K_{d0} \exp\left(-\frac{E_d}{RT}\right) \qquad \text{... (18.5)}$$

The system of differential equation is conveniently solved by a numerical algorithm based on a finite difference method.

## Adaptive Control Algorithm

A general strategy to design an adaptive control system is to estimate on-line the model parameters and then adjust the controller parameters based on the model estimation. This technique is often referred too as self-tuning controller and is showed schematically in Fig. 18.8.

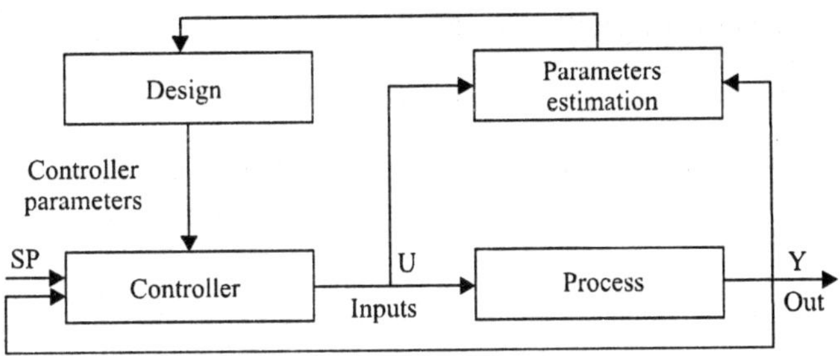


**Fig. 18.8.** Self-tuning control scheme.

The control law in the STC algorithm and the adaptive estimation can be performed as follows. A non-linear system given by:

$$x_{k+1} = \int (x_k, \Phi_k, u_k) \quad \text{... (18.6)}$$

can be modelled stochastically with varying parameters given by:

$$\Phi_{k+1} = \Phi_k + w_k \quad \text{... (18.7)}$$

so that the estimator stays permanently active if the process covariance is set to a positive number.

Linearisation of the system (Eq. 18.6) leads to:

$$x_{k+1} = \Phi_k x_k + \Gamma_k u_k + w_k \quad \text{... (18.8)}$$

and the state measurements ($y_k$) are obtained through the equation:

$$y_k = Hx_k + v_k \quad \text{... (18.9)}$$

where x are the states of the system and y are the observations.

The plant model (Eq. 18.8), together with the state observations (Eq. 18.9) can be used in a recursive Kalman filter and the following adaptive prediction is obtained:

$$\bar{x}_{k+1} = \bar{x}_k + G_{k+1}[y_{k+1} - H\bar{x}_k] \quad \text{... (18.10)}$$

with $\bar{x}_k$ given by the linear model:

$$\bar{x}_k = \Phi_k \bar{x}_{k-1} + \Gamma_k u_k \quad \text{... (18.11)}$$

and with $G_{k+1}$ being the Kalman filter gain.

For the Kalman filter the calculation of $G$ is done via the least squares method, minimising an objective function $J$ which is a weighted sum of the variances of the estimation errors:

$$J = \sum_{j=1}^{n} a_{jj} P_{jj} \quad \text{... (18.12)}$$

in which $a_{jj}$ are weighting coefficients.

Thus, the equation for G can be written as:

$$G_{k+1} = (\Phi P_k H^T + \psi N \Omega^T)\,(HP_k H^T + \Omega N \Omega^T + R)^{-1} \quad \text{... (18.13)}$$

with the actual value for the covariance P given by:

$$P_{k+1} = (\Phi - G_{k+1}H)P_k\Phi^T + (\psi - G_{k+1}\Omega)N\psi^T + \Gamma Q\Gamma^T \quad \text{... (18.14)}$$

The equation 18.15 indicates that the new variances are smaller than the prediction variance one step ahead. In the same way it is possible to estimate the controller parameters by their inclusion in the form of an extended state:

$$X_k = \begin{bmatrix} x \\ \hline x_{kc} \end{bmatrix}_k \quad \text{... (18.15)}$$

where $x_{kc}$ represents the elements of $K_c$ (controller parameters).

An improved self-tuning controller algorithm can be obtained if the error associated with the controlled variable and the desired value is considered as an input to the extended states. The controller gain $K_c$ can be calculated by an embedded procedure within the Kalman filter which gives a minimum variance estimate (Fig. 18.9).

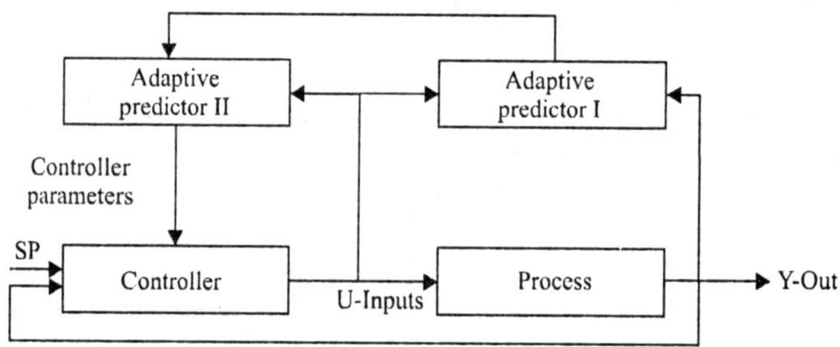


**Fig. 18.9.** Improved self-tuning controller.

For the implementation of the control strategy shown above to the system, it is necessary to find a simplified model of the form of Eq. 18.8.

Empirical models are not suitable because they are not representative of all the phenomena taking place in the system, but a discretised linear approximation can be developed which is satisfactory. Using a Taylor expansion series together with finite difference method makes it possible to generate a reduced model (resulting in a lumped parameter system), which is robust. Thus, the linear discrete time model then becomes:

$$\frac{dx_i}{dt} = Ax_i + Bu + w \quad \text{... (18.16)}$$

where w is a white noise term, assumed to be a zero mean gaussian process which takes into account modelling errors, uncertainties in parameters and neglected non-linearities.

## Controller Performance

The control objective is to maintain constant the temperature at the reactor exit, despite disturbances in the input variables. The manipulated variable must be very well defined so that the system can be controllable. Only one manipulated variable was used and only the temperature was used as measured output variable so that a SISO controller was established.

Figure 18.10 shows the reactor exit temperature when the system is in open loop as well as when it is under control. The system is heated by an imposed initial electric current, stabilising far from the desired value when it is without control. However, if the system is operated under control, using the electric current as a manipulated variable, the desired condition is achieved fastly. Also, when the inlet fluid temperature is disturbed, the controller is able to drive the system to the desired conditions.

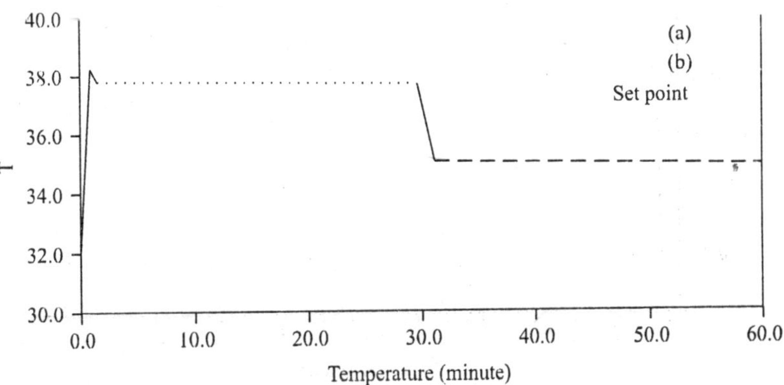


**Fig. 18.10.** Fluid temperature at the reactor exit without control (a) and with control (b) using the electric current as manipulated variable.

As a nonlinear and time varying system, its performance can be reduced during operation, mainly due to the enzyme deactivation. Thus, a new point of operation can be achieved and the controller must be able to track this new condition, acting as a servo controller. Figure 18.11 shows the performance of the controller acting as a servo controller, leading the system to the new condition fastly.

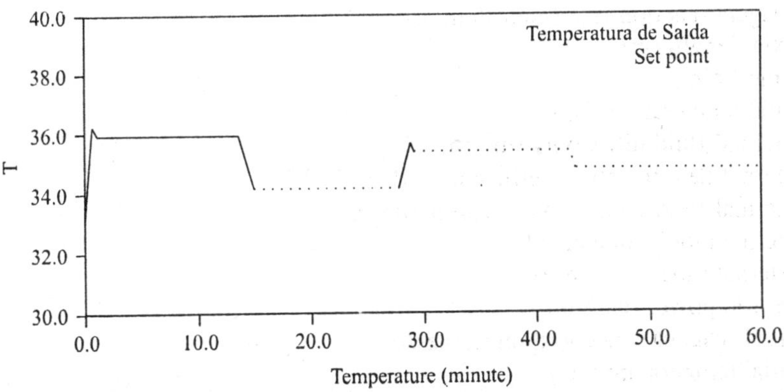


**Fig. 18.11.** Fluid temperature at the reactor exit operated under control using a programd set point change.

Figure 18.12 shows the process concentration of the reactant and estimated by the filter. Initially, while the process outlet concentration was zero, the estimate was 0.1 moles/m$^3$. After about 2 minutes the estimated concentration tracks the concentration at the process outlet. This is very important when the information of such state variables is desirable but the measurement device is time consuming or does not exists.

Thus, the problem of on-line control of a medical reactor is represented by a non-linear model with time varying distributed parameters. It has been shown that the self-tuning controller proposed in this work with the control strategy used here had a good performance, even under different operating conditions and using only one manipulated variable and only one measurement point (reactor exit temperature). The controller was also able to maintain the desired operational conditions when a programed set point was used. Also, good performance of the filter have been shown, making the system stable under control even under different operating conditions.

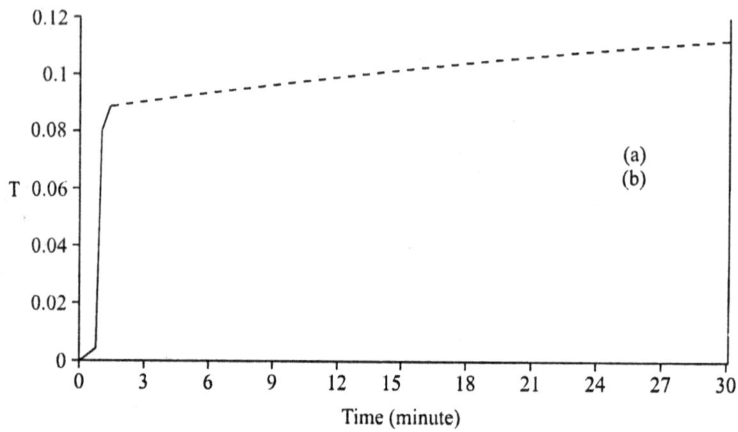


**Fig. 18.12.** Reactant concentration of the process (a) and estimated (b) at the reactor exit.

## Nomenclature

| | |
|---|---|
| $C_i$ | Reagent and product concentration (moles/l) |
| $D_A$ | Reagent diffusional coefficient ($m^2$/min.) |
| $D_B$ | Product diffusional coefficient ($m^2$/min.) |
| v | Reagent flow rate (m/min.) |
| $r_A$ | Reagent reaction rate (moles/l.min.) |
| z | Axial length (m) |
| t | Time (min.) |
| T | Fluid temperature (°C) |
| $\alpha_f$ | Thermal fluid diffusivity ($m^2$/min.) |
| $h_i$ | Internal heat transfer coefficient (cal/$m^2$min.°C) |
| $k_f$ | Thermal fluid conductivity (cal/m.min.°C) |
| $d_i$ | Internal tube diameter (m) |
| $d_e$ | External tube diameter (m) |
| $P_w$ | Electric power (cal/min.) |
| $C_0$ | Initial reagent concentration (moles/l) |
| $T_0$ | Initial temperature (°C) |
| $E_0$ | Initial enzyme concentration (UI/$m^2$) |
| $V_M$ | Maximum reaction rate (moles/l.min.) |
| $K_M$ | Michaelis-Menten constant (moles/l) |
| A | System coefficient matrix |
| B | System input matrix |
| x | States of the system |
| u | Inputs of the system |
| y | Output variable |
| Γ | Input matrix |
| ν | Measurement noise |
| Φ | Transition matrix. |

Chapter 19

# Principles of Effluent Treatment

## INTRODUCTION

Every fermentation plant utilises raw materials which are converted to a variety of products. Depending on the individual process, varying amounts of a range of waste materials are produced. Typical wastes might include unconsumed inorganic and organic media components, microbial cells and other suspended solids, filter aids, waste wash water from cleansing operations, cooling water, water containing traces of solvents, acids, alkalis, human sewage, etc.

The organic content of domestic, agricultural and industrial wastes causes the largest polluting problem as it is metabolised by organisms present in the receiving water, thereby causing a rapid depletion of oxygen and elimination of the natural aquatic flora and fauna. This organic content of waste may be determined by three tests. The chemical oxygen demand (COD) and total organic carbon (TOC) tests measure total organic carbon by chemical oxidation and pyrolysis, respectively, both using infrared $CO_2$ detection. The biological oxygen demand ($BOD_5$) test determines the fraction of the waste which is biologically oxidisable by a 5-day incubation using an adapted inoculum. Waste suspended solids measurements, using filtration and turbidity methods, are designed to assess the effects of pollution on light penetration (affecting algal growth) and on solids deposition (affecting river-bed life). While nitrogen is an essential nutrient in waste treatment processes, residual ammonia may be toxic to aquatic life. Oxidised products of ammonia can result in excessive algal growth and in cases where the waste-water is recycled for water supply, nitrate can be toxic to growing children.

With liquid wastes, it may be possible to dispose of untreated effluents to a municipal sewage treatment works (STW). Obviously, much will depend on the composition, strength and volumetric flow rate of the effluent. STWs are planned to operate with an effluent of a reasonably constant composition at a steady flow rate. Thus, if the discharge from an industrial process is large in volume and intermittently produced it may be necessary to install storage tanks on site to regulate the effluent flow. In some locations, municipal sewers are not available or the effluent may be of such a composition that the waste-water treatment company or regulatory authority requires some form of pre-treatment before discharge to its sewers. In these cases an effluent-treatment plant will have to be installed at the factory. Whatever the pollutant load of the liquid effluent, its discharge to a sewer will be a cost centred activity and will incur charges from the treatment company.

Normally, fermentation effluents do not contain toxic materials which directly affect the aquatic flora or fauna. Unfortunately, most of the effluents do contain high levels of organic matter which are readily oxidised by microbial attack and so drastically deplete the dissolved oxygen concentration in the receiving water unless there is a large dilution factor. This can be shown by the oxygen sag curve in Fig. 19.1. Different aquatic species have varying tolerances to depleted oxygen levels and as a consequence some species will die off in specific stretches of the receiving water and in other regions a different population capable of growth at lower oxygen levels will develop.

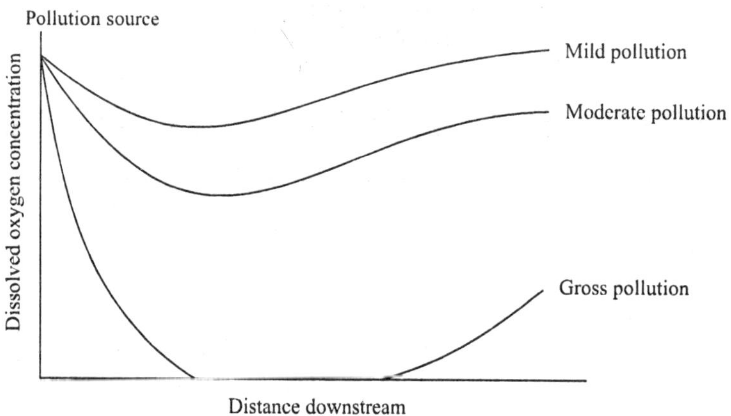


**Fig. 19.1.** The oxygen sag cure.

Effluents may be treated in a variety of ways, as will be outlined later in this section. In a number of processes it may be possible to recover waste organic material as a solid and sell it as a by-product which may be an animal feed supplement or a nutrient to use in fermentation media. The marketable by-product helps to offset the cost of the treatment process. It is now recognised that water is no longer a cheap raw material, hence there are considerable advantages in reducing the quantities used and in recycling whenever it is feasible. Obviously the introduction of good 'housekeeping' will lead to reduction in the volume of water used and the volume of effluent for treatment and final discharge. Recycling and re-use of materials, waste minimisation, waste reduction at source and integrated pollution control are now very important factors to consider in the design and operation of any manufacturing facility and may be the subject of new legislation in this field.

## DISSOLVED OXYGEN CONCENTRATION AS AN INDICATOR OF WATER QUALITY

Since oxygen is essential for the survival of most macro-organisms, it is important to ensure that there are adequate levels of dissolved oxygen in rivers, lakes, reservoirs, etc., if they are to be managed satisfactorily. Ideally, the oxygen concentration should be at least 90 per cent of the saturation concentration at the ambient temperature and salinity of the water. It is, therefore, important to know how effluents containing soluble and particulate organic matter can influence the dissolved oxygen concentration. One widely used method of assessment is the 'biochemical oxygen demand' (BOD), which is a measure of the quantity of oxygen required for the oxidation of organic matter in water, by micro-organisms present, in a given time interval at a given temperature. The oxygen concentration of the effluent, or a dilution of it, is determined before and after incubation in the dark at 20°F for 5 days. The oxygen decrease can then be determined titrimetrically and the results presented as mg of oxygen consumed per $dm^3$ of sample. Mineral nutrients and a suitable bacterial inoculum are usually added to the initial sample to ensure optimal growth conditions. This test is only an estimate of biodegradable material, hence recalcitrant or inhibitory compounds might be overlooked.

Because the BOD test takes 5 days it may be necessary to resort to the 'chemical oxygen demand' (COD), a chemical test which only takes a few hours to complete. The test is based on treating the sample with a known amount of boiling acidic potassium dichromate solution for 2.5 to 4 hours and then titrating the excess dichromate with ferrous sulphate or ferrous ammonium sulphate. The oxidised organic matter is taken as being proportional to the potassium dichromate utilised. Most compounds are oxidised virtually to completion in this test, including those which are not biodegradable. In circumstances where substances are toxic to micro-organisms, the COD test may be the only suitable method available for assessing the degree of treatment required. The BOD:COD ratios for sewage are normally between 0.2:1 and 0.5:1.

The ratio values for domestic sewage may be fairly steady. When industrial effluents of variable composition and loading are discharged, the ratio may fluctuate considerably. Very low BOD:COD ratios will indicate high concentrations of non-biodegradable organic matter and consequently biological effluent treatment processes may be ineffective. A number of alternative tests are available to indicate the 'oxygen demand' or a waste-water, including total organic carbon (TOC) and permanganate value.

## SITE SURVEYS

A complete survey of industrial operations is essential for any individual site before an economical waste treatment programme can be planned. It is desirable to divide the facility into as many units as possible, as knowledge of the various material streams may show unexpected losses of finished product, solvent wastage, excessive use of water or unnecessary contamination of water which might be recycled, recovered or re-used within the site. The factors and concentrations where appropriate, listed in Table 19.1 ought to be known at all production rates under which an individual unit may operate in a representative time period.

**Table 19.1.** Factors to investigate in a site survey.

| |
|---|
| Daily flow rate |
| Fluctuations in daily, weekly and seasonal flow |
| BOD/COD |
| Suspended solids |
| Turbidity |
| pH range |
| Temperature range |
| Odours and tastes |
| Colour |
| Hardness |
| Detergents |
| Radioactivity |
| Presence of specific toxins or inhibitors (e.g., heavy metals, phenolics, etc.) |

The survey may indicate a need for better control of water usage and should identify sources of uncontaminated and contaminated water that might be re-used in the factory. Concentrated waste streams should be kept separate if they contain materials that can be profitably recovered. It is also often more economical to treat a concentrate rather than a large volume of a dilute effluent because of the saving on pumps and settling tank capacities, provided that concentrations do not reach toxic or inhibitory levels in biological treatment processes.

The various wastes may be tested in a laboratory and on a pilot scale to assess the best potential methods of chemical and biological treatment. Once the pHs of the effluents are known samples may be mixed to see if a neutral pH is reached. A variety of tests may be used to establish methods for reducing salt concentrations, coagulating suspended particles and colloidal materials and for breaking emulsions.

The commonly used biological tests include respirometry, aeration-flask tests and continuous-culture experiments. Small flask respirometers and oxygen electrodes are used initially to establish the conditions to use in bio-oxidation of the effluent and to test for the presence of toxic materials. Large respirometers are useful for predicting effluent treatment rates and oxygen requirements. The residues in the flasks can be analysed to see if there are any recalcitrant materials. The use of laboratory continuous-culture vessels fitted with sludge-return pumps and settling tanks can provide detailed information. Proposed large-scale operating conditions for feed and aeration rates can be tested and their effectiveness assessed. The results from all these experiments may help in the design of a full-scale plant. If the survey is comprehensive it should be possible to plan an overall treatment programme for a site and to establish:

1. Water sources which can be combined or re-used.
2. Concentrated waste streams which contain valuable wastes to be recovered as food, animal feed, fertiliser or fuel.
3. Toxic effluents needing special treatment or acids or alkalis needing neutralisation.
4. The effluent loading expected under maximum production conditions.
5. The effluent(s) which might be discharged directly, without treatment, on to land or to a water course and not cause any pollution.
6. The effluent(s) which might be discharged into municipal sewers.

When all the relevant information has been obtained one can predict the size and type of effluent treatment plant required and thus its capital and operational costs. This can then be compared with water company charges to treat the waste at an STW with and without on-site treatment. It should be remembered that the water company may insist on on-site treatment before a waste is discharged to the sewer and will in most cases set consent limits for maximum flow rates and concentrations of specific analytes.

## STRENGTHS OF FERMENTATION EFFLUENTS

It is already evident from earlier sections of this chapter that the presence of high levels of particulate or soluble organic matter in water will result in potential high BODs. This is precisely what is being achieved in all large-scale fermentation processes. An initial medium rich in organic matter is converted to biomass and primary and secondary metabolites. Unfortunately, the product often represents a small proportion of the initial raw material, even in an efficiently operated fermentation. The spent wastes remaining after the distillation of whisky may account for 90 per cent of the initial raw organic materials, while in a antibiotic fermentation the effluent may represent in excess of 95 per cent.

Data for a variety of fermentation effluents are summarised in Table 19.2. The BODs of many of these samples are much higher than that of domestic sewage and some may be comparable with strong effluents such as sulphite paper mill liquor. It is evident from these data that fermentation effluents may present serious potential pollution problems and may be expensive to dispose of unless well planned processes are used. A number of steps may be taken to reduce BODs in a process. Some of these will be discussed in this section. Careful selection of raw materials may have a significant effect on the type and quantities of effluent being produced. The cheapest raw material which meets the nutrient requirements of the micro-organisms may not be ideal if product yield, recovery cost, effluent disposal cost and possible by-product value are considered together. The high BOD value of fungal mycelium (40000 to 70000 mg $dm^{-3}$) would indicate that any biomass should normally be kept separate from the remainder of an effluent and some of it may be sold as a by-product.

It may also be worthwhile to concentrate liquid fractions, for example, industrial alcohol and distillery stillagcs (10000 to 25000 mg $dm^{-3}$) will both produce dried soluble fractions which can be sold.

**Table 19.2.** BOD strength of effluents (mg $dm^{-3}$).

| *Effluent* | *BOD* |
|---|---|
| Domestic sewage | 350 |
| Sulphite liquor from paper mill | 20000–40000 |
| Beer: | |
| (a) Spent grain press | 15000 |
| (b) Hop-press liquor | 7430 |
| (c) Yeast wash water | 7400 |
| (d) Spoil beer | Up to 1,00,000 |
| (e) Bottle washings | 550 |
| Maltings: | |
| (a) Suspended solids | 1240 |
| (b) Wastes | 20–204 |
| (c) Grain washings | 1500 |
| Brewery effluent | 1400–1800 |
| Industrial alcohol stillage | 10000–25000 |
| Distillery stillage | 10000–25000 |
| Yeast production | 3000–14000 |
| Antibiotic waste | 5000–30000 |
| Penicillin: | |
| (a) Wet mycelium from filter | 40000–70000 |
| (b) Filtrate | 2150–10000 |
| (c) Wash water | 210–13800 |
| Streptomycin spent liquor | 2450–5900 |
| Aureomycin spent liquor | 4000–7000 |
| Solvents | Up to 20,00,000 |

Metabolites or components of some fermentation effluents may be extremely toxic and polluting and will require complete destruction, for example by chemical or thermal methods, before disposal. The need for such a treatment strategy will therefore make a significant contribution to the overall cost of the process. One such metabolite is avermectin produced by *Streptomyces avermitilis* fermentations. Here all effluent streams from the process are captured and any avermectin present chemically degraded.

## TREATMENT AND DISPOSAL OF EFFLUENTS

The effluent disposal procedure which is finally adopted by a particular manufacturer is obviously determined by a number of factors, of which the most important is the control exercised by the relevant authorities in many countries on the quantity and quality of the waste discharge and the way in which it might be done. The range of effluent-disposal methods which can be considered is:

1. The effluent is discharged to land, river or sea in an untreated state.
2. The effluent is removed and disposed of in a landfill site or is incinerated.
3. The effluent is partially treated on-site (e.g., by lagooning) prior to further treatment or disposal by one of the other routes indicated.
4. Part of the effluent is untreated and discharged as in 1 or 2, the remainder is treated at a sewage works or at the site before discharge.
5. All of the effluent is sent to the sewage works for treatment, although there might be reluctance by the sewage works to accept it, possibly resulting in some preliminary on-site treatment being required and discharge rates and effluent composition defined.
6. All the effluent is treated at the factory before discharge.

## DISPOSAL

### Sea and River

The simplest way of disposal will be on a sea coast or in a large estuary where the effluent is discharged through a pipeline (installed by the factory or local authorities) extending below the low-water mark. In such a case there may be little preliminary treatment and one relies solely on the degree of dilution in the sea water.

If effluents are to be discharged into a river they must meet the requirements of the local river or drainage authorities. In Britain there is a Royal Commission standard requiring a maximum BOD (5 days) of 20 mg $dm^{-3}$ and 30 mg $dm^{-3}$ of suspended solid matter (the 20:30 standard). Stricter standards are often applied, depending on the use of the receiving water, such as a 10:10 standard; in addition, levels of ammoniacal nitrogen may be stipulated. There are, as well, often stringent upper limits for toxic metals and chemicals which might kill the fauna (particularly fish) and flora, e.g., sulphites, cyanides, phenols, copper, zinc, cadmium, arsenic, etc. It is highly unlikely that one would be able to discharge an industrial waste today without some form of pre-treatment.

### Lagoons (Oxidation Ponds)

Lagoons, holding ponds, oxidation ponds, etc., may be used by a number of industries if land is available at a reasonable cost. It is a method often used in seasonal industries where capital investment in effluent plant is difficult to justify. The lagoon normally consists of a volume of shallow water enclosed by watertight embankments. Oxidation ponds are typically 1–2 meters deep. They can be designed to maintain aerobic conditions throughout, but more commonly decomposition at the surface is aerobic and that nearer the bottom is anaerobic and they are then known as facultative ponds. Oxygen for aerobic degradation is provided both from the surface of the pond and from algal photosynthesis. Deeper ponds (known as lagoons) are mechanically agitated to provide aeration. Lagoons are simple to build and operate, but are expensive in terms of land requirements. They may be used as the sole method of treatment, incorporating both physical (sedimentation) and biological processes, but the effluent produced may not reach locally acceptable standards. Alternatively they can provide an initial pre-treatment or can be used to 'polish' effluent from secondary treatment processes.

### Spray Irrigation

Liquid wastes can be applied directly to land as irrigation water and fertiliser when they are claimed to have a number of beneficial effects on the soil and plants. If this method of disposal is to be used, then it is necessary to have a large area of land near to a manufacturing plant in an area of low to medium rainfall. Pipeline costs will often restrict use of this technique. Colovos and Tinklenberg described the disposal of antibiotic and steroid wastes with BODs of 5000 to 20000 mg $dm^{-3}$. These wastes were initially chlorinated to lower the BOD and reduce unpleasant odours and then sprayed on the land until the equivalent of 38 mm of rainfall was reached. This process was repeated at monthly intervals and improved plant growth.

When appropriate, solid wastes may be spread onto land as a fertiliser and soil conditioner. This practice is common with sewage sludges and Ekwealor report the use of spent brewers grains to improve the productivity of fragile soils. Irrespective of whether the waste is liquid or solid, the concentration of heavy metals and certain organic components will require careful monitoring and control to safeguard the environment and public health.

### Well Disposal

Disused wells, boreholes or mine shafts may provide an ideal, cheap method for disposal when the volume of waste is limited, the underground strata are suitable and the chances of contamination of water supplies utilised by water authorities are negligible. Melcher has described the use of wells 500 meters deep for the daily disposal of:

| | |
|---|---|
| Acetic acid | 900 kg |
| Ammonium acetate | 900 kg |
| Sodium acetate | 760 kg |
| Sodium chloride | 450 kg |
| Sodium and ammonium bromide | 225 kg |
| Methanol, xylene, tars and organic compounds | |

This mixture had a pH of 4 to 5 and a COD of 40000 to 60000 mg $dm^{-3}$ rising to 1,00,000 mg $dm^{-3}$ and was pumped into the wells at 50 to 100 $dm^3$ $min^{-1}$.

Careful hydrogeological surveys will be needed to prove that waste disposal in wells will not cause pollution of aquifers and threaten groundwater supplies.

### Landfilling

Landfilling is a disposal method for municipal solid waste (MSW) and industrial waste. It utilises natural or manmade voids (e.g., disused clay pits) into which the waste is deposited. Both solid and liquid wastes can be deposited depending on restrictions imposed by the site licence. Strict controls exist on the amount of liquid and toxic materials which can be accepted because of the threat of groundwater pollution if leachate (a liquid having BOD levels up to 30000 mg $dm^{-3}$) escapes from the site. Leachate is generated from liquid deposited in the site, water entering the site naturally via precipitation or surface run-off and by anaerobic microbial action as organic matter in the landfill is degraded. Microbial action similar to that in anaerobic digesters leads to the production of landfill gas (LFG) which, being 50–60 per cent methane can, if collected efficiently, provide a useful source of energy.

### Incineration

A number of designs exist for the incineration of solid and/or liquid wastes either on site or at a commercial incinerator, including rotary kilns, fluidised beds and multiple hearth furnaces. Combustion temperatures need to be carefully controlled to destroy and prevent the formation of dioxins and furans, formation of which occurs at between 300°F and 800°F and total destruction is effected at temperature above 1000°F with a retention time of 1 second. Flue gases from the incinerator require cleaning to remove particulates, acid vapours, etc., using electrostatic precipitators, cyclones and wet scrubbers to comply with local environmental protection standards. Waste disposal by incineration is currently significantly more expensive than landfilling.

### Disposal of Effluents to Sewers

Municipal authorities and water treatment companies which accept trade effluents into their sewage systems will want to be sure that:

1. The sewage works has the capacity to cope with the estimated volume of effluent.
2. The effluent will not interfere with the treatment processes used at the sewage works.
3. There are no compounds present in the effluent which will pass through the sewage works unchanged and then cause problems when discharged into a watercourse.

It is common practice for local authorities to demand preliminary on-site pre-treatment before discharge into sewers to minimise the effects of industrial wastes. The actual pre-treatment required will depend on the precise nature of the waste and may range from simple sedimentation to complex chemical and biological processes.

### Treatment Processes

Fermentation wastes may be treated on-site or at an STW by any or all of the three following methods:

1. Physical treatment.
2. Chemical treatment.
3. Biological treatment.

The final choice of treatment and disposal processes used in each individual factory will depend on local circumstances.

Treatment processes may also be described in the following manner:

1. Primary treatment; physical and chemical methods, e.g., sedimentation, coagulation, etc.
2. Secondary treatment; biological methods (e.g., activated sludge) conducted after primary treatment.
3. Tertiary treatment; physical, chemical or biological methods (e.g., microstrainers, sand filters and grass plot irrigation) used to improve the quality of liquor from previous stages.
4. Sludge conditioning and disposal; physical, chemical and biological methods. Anaerobic digestion is often used to condition (make it more amenable to dewatering) the sludge produced in previous stages. Following dewatering (e.g., by centrifugation using a decanter centrifuge) the sludge can then be disposed of by incineration, landfilling, etc.

### Physical Treatment

The removal of suspended solids by physical methods before subsequent biological treatment will considerably reduce the BOD of the resulting effluent. In nearly all fermentation processes the cells are separated from the liquid fraction in recovery processes. Obviously, biomass processes need not be considered. Yeast cells from other processes may be a marketable product, but microbial cells may not always be marketable, particularly when contaminated with filter aid. In these instances, when the cells and filter aid are a waste, the recovered material may be dealt with in two basic ways:

1. The waste is disposed of without any further treatment.
2. The waste bulk is reduced by mechanical dewatering with a filter press, centrifuge, rotary vacuum filter of belt press. The compressed waste is then incinerated or disposed of in a landfill site.

Solid wastes are produced in some processes before inoculation. In breweries, where malted grain is still used, coarse screens or 'whirlpool' centrifuges may be used to remove spent grain from the wort after it is mashed. About 5 kg (wet weight) of grain are produced per barrel (180 $dm^3$) of beer. If hops are used, rather than hop extracts, they will also be recovered on screens in a 'hop back'. This residue may amount to 250 grams per barrel. Both the spent grain and hop waste may then be mechanically dewatered before being sold or dumped.

The stillage (after distillation) in whisky distilleries may be passed through screens (1 mm openings). These screenings are then removed, mechanically dewatered and dried in rotary driers to yield a potentially marketable residue known as Distillers' grains. According to a survey in Scotland, about half the whisky distilleries were evaporating the spent waste to a syrup containing 45 per cent solids, mixing with spent grain, drying and selling the final product, 'Distillers' Dark Grains', as a low-grade cattle food.

Physical processes installed for primary effluent treatment may include the following stages:

1. Screens, to remove larger suspended and floating matter.
2. Comminutors, to reduce particle size.
3. Constant velocity channels (~0.3 m $s^{-1}$) for grit removal, to prevent damage to plant in later processes.
4. Sedimentation tanks for the removal of finer suspended matter. These are generally circular or rectangular continuous flow tanks operating at retention times of 6–15 hours (and designed to have a minimum retention time of 2 hours), with facility for the continuous removal of settled sludge. Sedimentation tanks can remove 70 per cent of the incoming suspended solids and depending on the nature of the waste, up to 40 per cent of its BOD load. They can be operated with or without prior chemical coagulation/flocculation. Similar settlement processes are also conducted after secondary (biological) treatment.

Physical processes used in tertiary treatment to produce an effluent of better quality than the 30:20 standard include microstrainers, slow sand filters, upflow sand filters and rapid gravity sand filters. Throughputs vary between around 3 $m^3$ $m^{-2}$ $day^{-1}$ for slow sand filters and 700 $m^3$ $m^{-2}$ $day^{-1}$ for microstrainers. Suspended solids removal is generally 50–70 per cent and BOD removal around 30–50 per cent, depending on the technique used.

### Chemical Treatment

Fine suspended particles in an effluent may be removed by coagulation and/or flocculation. Coagulation is essentially instantaneous whereas flocculation requires some more time and gentle agitation to achieve 'aggregation' of the particles. Ferrous or ferric sulphate, aluminium sulphate (alum), calcium hydroxide (lime) and polyelectrolytes are often used as chemical coagulant. A solution of coagulant of the appropriate strength for effective treatment is added to the effluent in a vigorously mixed tank, a precipitate or floc forms almost immediately and carries down the suspended solids to form a sludge. This sludge may be drawn off, mechanically dewatered and subjected to further treatment. The flocs formed on coagulation may be small and will therefore require an extended period to settle, and as a consequence, for a given throughput of effluent a large sedimentation tank will be needed. Increasing the particle diameter by encouraging small flocs to coalesce (flocculation) increases the rate of sedimentation and thus, for a given throughput, a smaller vessel can be operated. Polyelectrolytes are commonly used as flocculants and following addition the effluent is gently mixed (turbulent mixing would break up the flocs) by passage through sinuous flocculation channels, hydrodynamic flocculators or mechanically mixed flocculators.

## Biological Treatment

Most organic-waste materials may be degraded biologically. This process may be achieved aerobically or anaerobically in a number of ways. The most widely used aerobic processes are trickling filters, rotating disc contactors, activated sludge processes and their modifications. The anaerobic processes (digestion, filtration and sludge blankets) are used both in the treatment of specific waste-water and in sludge conditioning.

### Aerobic Processes

#### *Trickling filters*

The term filter in this unit operation is a misnomer, as the action of a trickling filter is not one of filtration, but rather it is a fixed film bioreactor. Settled effluent to be treated is passed down through a packed bed counter-current to a flow of air. Micro-organisms adhering to the packing matrix adsorb oxygen from the upflowing air and organic matter from the downflowing effluent; the latter is then metabolised and the effluent stream's BOD reduced.

A conventional trickling filter (Fig. 19.2) usually consists of a cylindrical concrete tank 2 to 3 meters in depth and 8 to 16 meters in diameter. Some filters are rectangular in shape, but a rotary system allows more uniform hydraulic loading. The tank is packed with a bed of stone (usually granite) or special plastic packings, the bed being underlaid with drains. The packing material diameter should be 50 to 100 mm to give a specific surface of around 100 $m^2$ $m^{-3}$ and the material should be packed to give a voidage (per cent air space of total bed volume) of 45 to 55 per cent, which should minimise the risk of the spaces between the packing material becoming blocked by the microbial film.

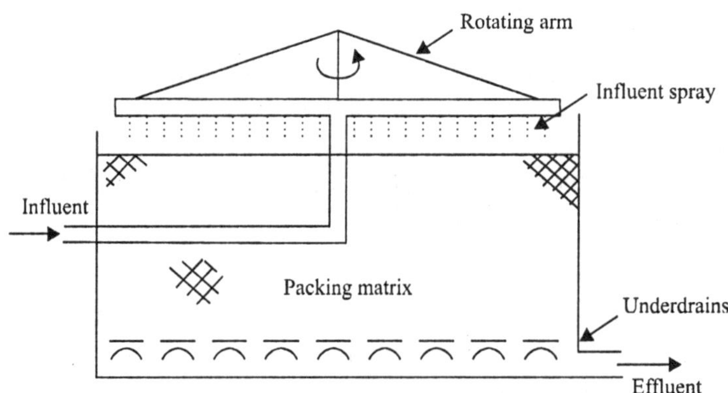


**Fig. 19.2.** Schematic diagram of a trickling filter.

Synthetic packing material, although more expensive, has a higher surface area and voidage, allowing higher treatment rates per unit volume of bed and reducing the likelihood of blockages. The trickling filter is always followed by a secondary sedimentation tank or humus tank to remove suspended matter (e.g., biofilm sloughed off the packing) from the treated effluent. In conventionally loaded or low-rate filters, the effluent from which the suspended solids have been removed, is fed on to the upper surface of the bed by spray nozzles or mechanical distributor arms. The effluent trickles gradually through the bed and a slime layer of biologically active material (bacteria, fungi, algae, protozoa and nematodes) forms on the surface of the support material. The large surface area created in the bed permits close contact between air flowing upwards through the bed, the descending effluent and the biologically active growth. The bacteria in the biological film remove the majority of the organic loading. Complex organic materials are broken down and utilised, nitrogenous matter and ammonia are oxidised to nitrates and sulphides and other compounds are similarly oxidised. The higher organisms (protozoa, etc.) control the accumulation of the biological film (prevents the filter from blocking) and improve the settling characteristics of the solids (humus) discharged with the filter effluent. In low-rate filters the scouring action of the hydraulic load normally has a minor role in removal of any loose microbial film. The active slime takes time to develop and can be poisoned by the addition of toxic chemicals. The simple filter is inefficient when operated at abnormally high organic loading rates. Initially, there is a very rapid build up of bacteria, fungi and algae at the top of the filter which cannot be controlled by the resident population of worms and larvae. The voids, therefore, block up, resulting in ponding (untreated effluent accumulating on the surface of the filter bed). Film growth can be limited by reducing the dosing frequency, which gives better liquid distribution deeper into the bed.

A trickling filter bed should remove 75 to 95 per cent of the BOD and 90 to 95 per cent of the suspended solids at organic loading rates of 0.06–0.12 kg BOD $m^{-3}$ $day^{-1}$ for conventional trickling filtration. When part of the treated effluent is being recirculated to dilute the feed and increase the hydraulic load placed on the unit the organic loading can be increased to 0.9–0.15 kg BOD $m^{-3}$ $day^{-1}$. The increase in hydraulic load thus applied causes greater hydraulic scouring of the bed (preventing blockage), but does not reduce treatment efficiency due to improved wetting of the packing surface and thus, more efficient use of the biofilm. To achieve the Royal Commission (20:30) Standard together with a high degree of nitrification, filters being supplied with domestic sewage should receive organic loading rates of 0.07–0.1 kg BOD $m^{-3}$ $day^{-1}$ and hydraulic loading rates of 0.12–0.6 $m^3$ $m^{-3}$ $day^{-1}$. It is possible to modify the trickling filter to increase the capacity for organic loading by the use of two sets of filters and settling tanks in series; this is known as alternating double filtration (ADF). Effluent is applied to the first filter at a high hydraulic and organic loading rate, it passes from this filter through the first settling tank and then on to a second filter and settler. After a period of one to two weeks the sequence of the filters is reversed and the second filter receives the higher loading. In this way heavy film growth is promoted in the first filter to receive the effluent, but when the filter sequence is reversed it becomes nutrient limited, encouraging excess film removal. Loading rates of 0.32–0.47 kg BOD $m^{-3}$ $day^{-1}$ have been claimed but recommended rates for design purposes are 0.15–0.26 kg BOD $m^{-3}$ $day^{-1}$.

Alternatively, enclosed deep beds of 3.5 to 5.5 meters depth may be used in which air is blown through the beds by fans. Loading rates up to 12 times that of the ordinary filter have been claimed.

Cook has stressed the need to consider the possible intermittency of a factory waste-water treatment process. It was shown that starving of a laboratory-scale trickling filter beyond 48 hours resulted in near failure of the filter. This indicated the need to supplement or artistically load the filter to maintain a viable biomass in the system. Stages which may be included in aerobic microbiological waste treatment are shown in Table 19.3.

**Table 19.3.** Stages which may be included in aerobic microbiological waste treatment.

| |
|---|
| Adsorption of substrate to biological surface |
| Degradation of adsorbed substrate by extracellular enzymes |
| Cellular absorption of dissolved materials |
| Excretion of breakdown products |
| Cell lysis and ingestion by a secondary cell population |

### Towers

Because trickling filters do not have both a high specific area and a high voidage, they are less suitable for the treatment of large volumes of strong industrial effluents (Table 19.2). Large areas of land would be required for the expensive and extensive traditional filter beds. Towers, 6–9 meters in height, packed with lightweight (40–80 kg $m^{-3}$) plastic multi-faced modules or small random packing units have provided to be a space saving and relatively inexpensive solution to the problem. These packings have a relatively open structure for oxygen transfer (specific surface of 100–300 $m^2$ $m^{-3}$) and high voidage (90–98 per cent), but are expensive compared with the conventional filter packings. They are capable of coping with high BOD lodgings. At a loading of 3.2 kg BOD $m^{-3}$ $day^{-1}$ a 50 per cent BOD removal may be achieved and at 1.5 kg $m^{-3}$ $day^{-1}$ 70 per cent removal is possible. The biological film is similar to that formed on the conventional packing and scouring is due to the hydraulic load applied rather than the predation of higher organisms.

### Biologically aerated filters (BAFs)

Biologically aerated filters are a relatively recent development based on the trickling filter. They consist of a packed bed which provides sites for microbial growth through which air is passed but, unlike trickling filters, the reactor volume is flooded with the effluent to be treated which is passed upwards or downwards through the reactor (i.e., co- or counter-currently to the air supply) depending on the design. The packing matrix may be natural (e.g., pumice) or synthetic (e.g., polyethylene) and may be either a fixed structure or randomly packed.

The combination of aeration and filtration allows high rates of BOD and ammonia removal together with solids capture, so that sedimentation tanks may not be required. However, regular backwashing is essential to remove filtered solids and excess biomass. Organic loading rates for 90 per cent BOD removal are significantly greater than those obtained for trickling filters, being in the range 0.7–2.8 kg BOD $m^{-3}$ $day^{-1}$. They are versatile treatment systems and of those currently in operation, design capacities vary between 600 and 70000 $m^3$ $day^{-1}$. In addition to their use as a secondary treatment process they can also be utilised for tertiary treatment or modified to allow denitrification in a manner similar to that of activated sludge system.

### Rotating biological contactors (rotating disc contactors)

In this treatment method (Fig. 19.3) a unit composed of closely spaced discs (2 to 3 m diameter with 1 to 2 cm spacing between discs), on a central drive shaft are rotated slowly (0.5 to 15 rpm) through the effluent so that 40 to 50 per cent of the disc surfaces are submerged. The discs, usually made from synthetic materials (e.g., polystyrene, PVC), are arranged in stages or groups separated by baffles to minimise short circuiting or surging and to enhance specific treatment requirements such as nitrification. The discs may be flat or corrugated to increase surface area. A microbial film forms on the discs; this is aerated during the exposed part of the cycle and absorbs nutrients during the submerged part. Shear forces produced as the discs rotate through the liquid control the thickness of the biofilm, with excess biofilm being sloughed from the discs. A sedimentation tank following the biological stage is therefore required to remove biological solids. Loading rates of 13 gram BOD $m^{-2}$ $day^{-1}$ for domestic sewage and partial treatment of loads of 400 grams BOD $m^{-2}$ $day^{-1}$ have been used. To achieve the 20:30 standard the loading rate should not exceed 6 grams BOD $m^{-2}$ $day^{-1}$. Rotating biological contactors are compact, easily covered for health and aesthetic reasons, available as packaged units, simpler to operate under varying loads than trickling filters (the biofilm being wetted at all times) and are easily added onto existing treatment processes. As such they can provide a cost effective method of on-site treatment.

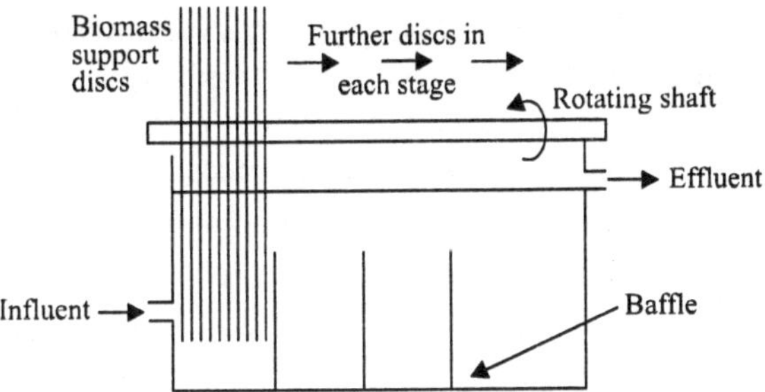


**Fig. 19.3.** Schematic diagram of a rotating biological contactor.

Ware and Pescod describe the use of full scale anaerobic/aerobic rotating biological contactors for treating brewery waste-water. Greater than 85 per cent COD removal was obtained in the aerobic stage, but difficulties were experienced in maintaining anaerobic populations.

### *Rotating drums*

Large rotating drums packed with random plastic packing materials or spheres have been manufactured as a development of rotating discs. Loading rates for the random packing were similar to those of rotating discs, while plastic spheres used in partial treatment could cope with loads of 6 kg BOD $m^{-2}$ $day^{-1}$.

### *Fluidised-bed systems*

Fluidised-bed reactors in waste-water treatment are relatively recent innovation. The support matrix (sand, anthracite, reticulated foam) has a large surface area on which the biofilm adheres and thus they are able to operate at high biomass concentration with high rates of treatment. This allows strong waste-water to be treated in small reactors. They are also useful for the treatment of industrial waste-water when variable loadings are encountered. The support matrix is fluidised by the upflow of effluent through the reactor and the degree of bed expansion is controlled by the flow rate of waste-water. The treated effluent can thus be decanted off without loss of the support matrix and with careful operation a secondary sedimentation tank may not be needed. The support matrix is regularly withdrawn to remove excess biomass. Fluidised-bed systems can be operated aerobically, anaerobically (see later section) or anoxically for denitrification.

### *Activated sludge processes*

The basic activated-sludge process (Fig. 19.4) consists of aerating and agitating the effluent in the presence of a flocculated suspension of micro-organisms on particulate organic matter—the activated sludge. This process is now the most widely used biological treatment process for both domestic and industrial waste-waters. The raw effluent enters a primary sedimentation tank where coarse solids are removed. The partially clarified effluent passes to a second vessel, which can be of a variety of designs, into which air or oxygen is injected by bubble diffusers, paddles, stirrers, surface aerators, etc. Vigorous agitation is used to ensure that the effluent and oxygen are in contact with the activated sludge. After a pre-determined residence time of several hours, the effluent passes to a second sedimentation tank to remove the flocculated solids. Part of the sludge from the settlement tank is recycled to the aeration tank to maintain the biological activity. The overflow obtained from the settlement tank should be of a 20:30 standard or better and be suitable for discharge to inland water. The excess sludge is dewatered and dried, to be sold as a fertiliser, incinerated or landfilled. In conventional activated sludge processes, organic loading rates are 0.5–1.5 kg BOD $m^{-3}$ $day^{-1}$ with hydraulic retention times of 5–14 hours depending on the nature of the waste-water, giving BOD reductions of 90–95 per cent. High-rate activated-sludge processes can be used as a partial treatment for strong wastes prior to further treatment or discharge to a sewer and are widely used in the food processing and dairy industries. The organic loading rate is 1.5–3.5 kg BOD $m^{-3}$ $day^{-1}$ and with hydraulic retention times of only 1–2 hours, BOD reductions of 60–70 per cent are possible.

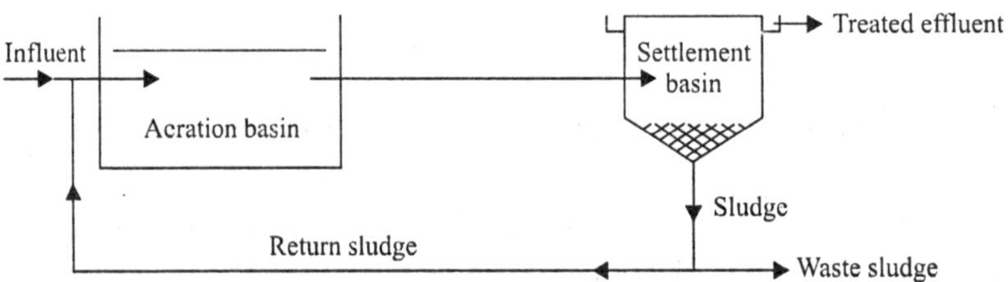


**Fig. 19.4.** Simplified cross-section of an activated sludge process.

A number of modifications of the basic process can be used to improve treatment efficiency, or for a more specific purpose such as denitrification. Tapered aeration and stepped feed aeration are used to balance oxygen demand (which is greatest at the point of waste-water entry to the aeration basin) with the amount of oxygen supplied. Contact stabilisation exploits biosorption processes and thereby allows considerable reduction in basin capacity (~50 per cent) for a given waste-water throughput. Denitrification (the biological reduction of nitrate to nitrite and on to nitrogen gas under anoxic conditions) can be accomplished in an activated-sludge plant when the first part of the basin is not aerated.

In advanced activated-sludge systems the amount of dissolved oxygen available for biological activity is increased to improve treatment rate. One vessel of this type is the 'deep shaft' which is quite distinctive from the other aeration tanks. The 'deep shaft' (Fig. 19.5) consists of a shaft 50 to 150 meters deep, separated into a down-flow section (down-comer) and an up-flow section (riser).

The shaft may be 0.5 to 10 meters diameter, depending on capacity. Fresh effluent is fed in at the top of the 'deep shaft' and air is injected into the down-flow section at a sufficient depth to make the liquid circulate at 1 to 2 m $s^{-1}$. The driving force for circulation is created by the difference in density (due to air bubble volume) between the riser and down-flow sections. For starting up, circulation of liquid is stimulated by injecting air at the same depth in the riser. Air injection is then gradually all transferred to the air injection point in the down-comer. Because of the pressure created in the down-comer, oxygen-transfer rates of 10 kg $O_2$ $m^{-3}$ $hr.^{-1}$ can be achieved and bubble contact times of 3 to 5 minutes are possible instead of 15 seconds in diffused air systems.

BOD removal rates of 90 per cent are achievable at organic loadings of 3.7–6.6 kg BOD $m^{-3}$ $day^{-1}$ at hydraulic retention times of 1.18–1.75 hours. Sludge production was found to be much less than that for conventional sewage-treatment processes.

Two types of pure oxygen systems have also been developed to increase the rate of oxygen transfer:

1. Closed systems which operate in oxygen-rich atmospheres.
2. Open systems employing fine bubble diffusers.

An example of the closed system is the UNOX Process developed by the Union Carbide Corporation in the USA (Fig. 19.6). The enclosed oxygenation tank is compartmentalised by baffles. Settled waste-water and returned sludge are fed into the first stage and oxygen pumped into the headspace. The oxygen and waste-water move sequentially through the compartments and the oxygen concentration in the gas phase decreases in each stage as it is consumed by the micro-organisms. As the nutrient concentration also falls stage by stage, oxygen supply and demand is balanced.

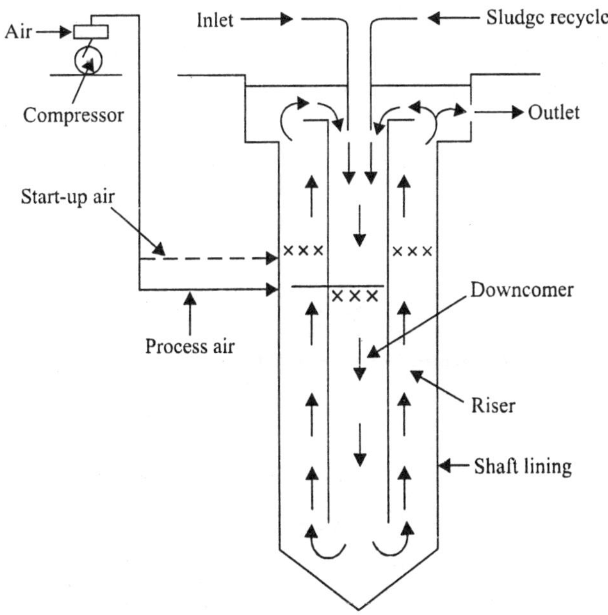


**Fig. 19.5.** Deep-shaft effluent treatment plant.

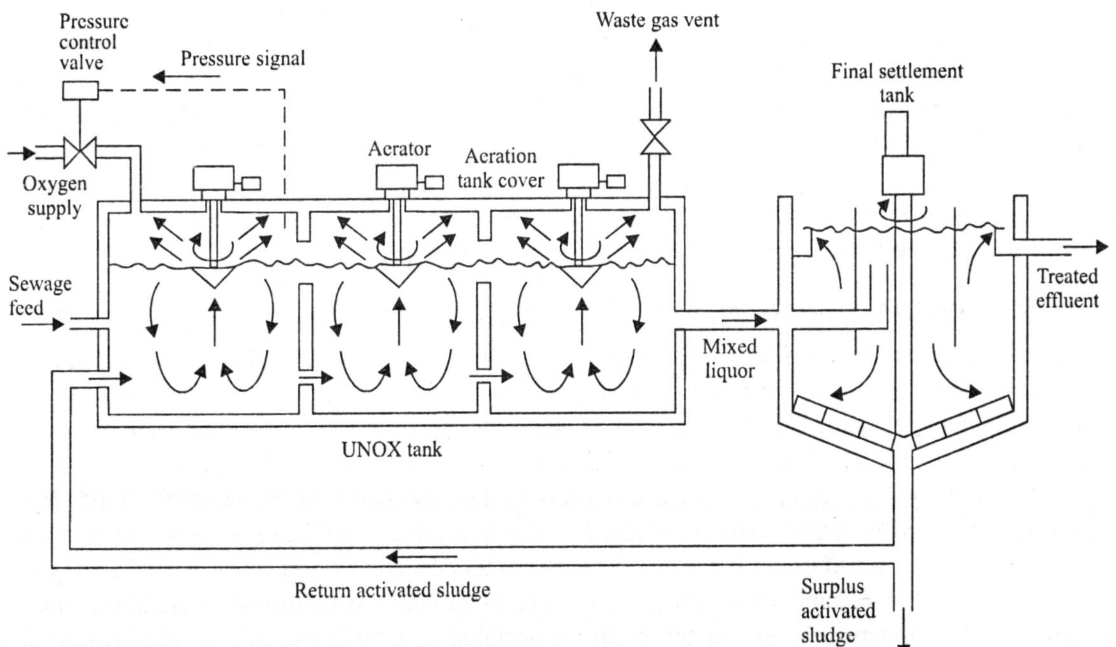


**Fig. 19.6.** Schematic diagram of a multistage UNOX system.

Organic loading rates (when treating municiple waste-water) are 3–4 times higher than those of aerated systems at 2.5–4.0 kg BOD $m^{-3}$ $day^{-1}$. Pure oxygen systems also have shorter residence times and produce less sludge with better settling qualities than conventional systems. The UNOX process has been successfully used for the treatment of brewery waste-water containing 2000 mg BOD $dm^{-3}$ at flow rates of 2269 $m^3$ $day^{-1}$ with a 6.6 hours retention time.

The VITOX aeration system (Fig. 19.7) developed by the British Oxygen Company is an example of an open tank oxygenation system. Its main advantage is that it can be used in existing aeration tanks either to replace or upgrade conventional aeration without the need to replace or modify existing units. Oxygenation is achieved by pumping settled waste-water at high pressure through a venturi where oxygen is injected.

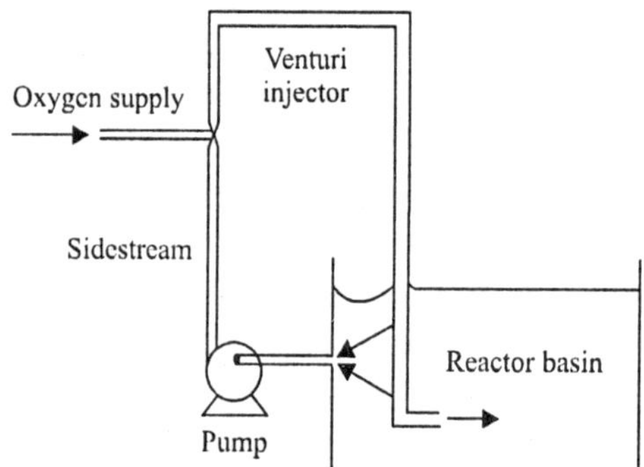


**Fig. 19.7.** Simplified VITOX sidestream aeration system.

Turbulence and high pressures thus created ensure high levels of oxygen dissolution. The flexibility of this system means that it is particularly useful for the treatment of high strength intermittently produced waste-water such as those generated by food processing industries.

Recently fluidised-bed aerobic waste treatment systems have been designed which in essence combined the operating principles of the percolating filter and activated sludge systems. Biological support materials such as plastic or sand are used and fluidisation is achieved by injection of air or oxygen at the bottom of the bed (Fig. 19.8). High biomass concentrations can be maintained in the system. These new systems exhibit great promise for the high-intensity removal of wastes.

## ANAEROBIC WASTE TREATMENT

Anaerobic digestion of wastes, which originated with use of septic tanks, now involves a range of high rate digesters. Some advantages which anaerobic systems have over aerobic systems include the lower power requirement per unit BOD treated, digester load capacity, production of combustible gas as an end-product and production of reduced biomass volume for disposal. Despite these advantages, it is not possible to obtain total degradation of waste using this method and polishing by aerobic methods may be required. The biochemical sequences involved during anaerobic degradation of waste to methane and carbon dioxide are illustrated in Fig. 19.9. These processes depend on the development of a complex interdependent bacterial community involving three microbial groups: the hydrolytic fermentative group; the hydrogenic acetogenic group and the hydrogenotrophic methanogenic group. The hydrolytic fermentative bacteria convert complex polymers to sugars, organic acids, alcohols and esters generating carbon dioxide and hydrogen. The hydrogenic, acetogenic bacteria transform the fermentation products of the first group into acetate, carbon dioxide and hydrogen. The methanogens then covert the acetate and hydrogen into methane and carbon dioxide.

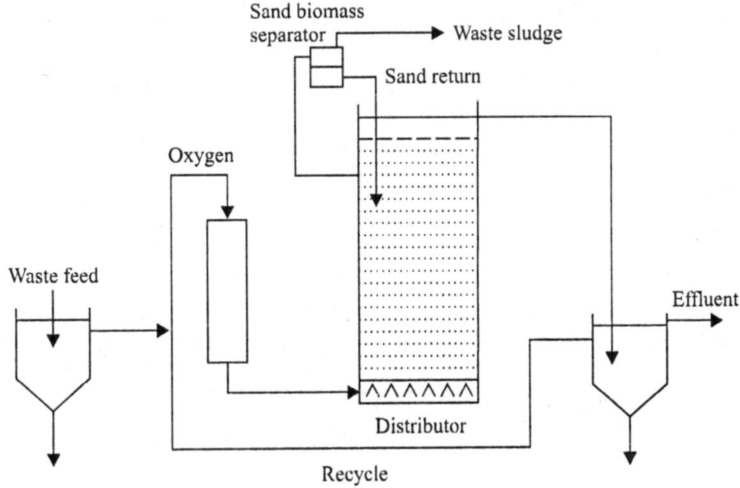


**Fig. 19.8.** Oxygen aerated fluidised-bed waste treatment plant using sand support particles.

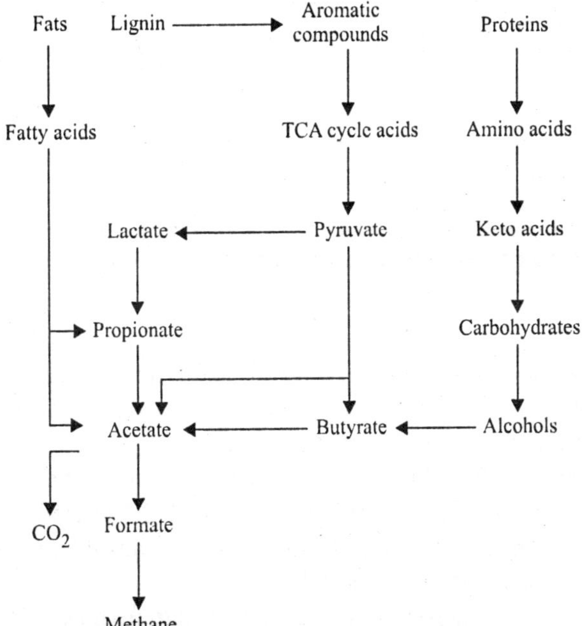


**Fig. 19.9.** Anaerobic metabolism of compounds to methane and carbon dioxide.

The slow growing methanogens, which are susceptible to stress from hydrogen accumulation and from the influent load, cause most of the operational problems associated with anaerobic digestion systems. It is only at hydrogen partial pressures of less than $10^{-3}$ atm, that carbohydrates can be degraded successfully under anaerobic conditions. The hydrogenotrophic methanogens which are responsible for scavenging hydrogen have an estimated doubling time of 8–10 hours whereas the acid-forming hydrolytic bacteria have a doubling time of about 30 minutes. A stressful situation leading to a decline in the methanogen population and a relative increase in hydrolytic bacteria will result in an accumulation of hydrogen. In the presence of an increased hydrogen concentration, the metabolism of the acid-forming bacteria is altered and propionic, butyric, caproic, valeric and lactic acids are produced instead of acetate and hydrogen. Consequently, since digestion overloading is indicated by a rise in concentration of hydrogen and certain acid, continuous monitoring of these metabolites may be used to control the feed rate.

### Anaerobic Reactor Design

Some basic anaerobic reactor designs are illustrated in Fig. 19.10. The conventional digester (a) which is a completely mixed system without biomass recycle has a low water treatment efficiency. Systems (b) to (e) are designed to retain high biomass concentrations within the reactor, (b) by recovery and recycling from the effluent stream, (c) by flocculation and (d) and (e) by microbial film attachment to support material.

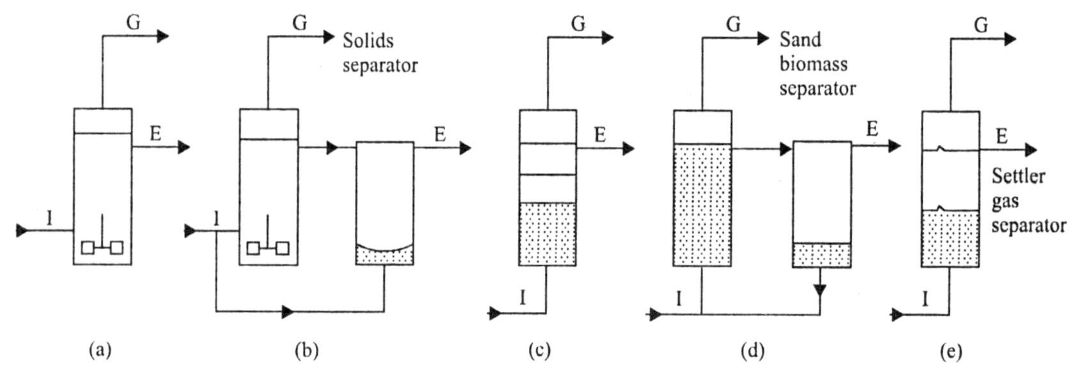


**Fig. 19.10.** Reactor design for anaerobic digestion. (a) Standard digester; (b) anaerobic contact process; (c) anaerobic filter; (d) fluidised bed and; (e) upflow anaerobic sludge type. I, Influent; G, gas; and E, effluent.

### Anaerobic Treatment

Anaerobic treatment of waste organic materials originated with the use of septic tanks and Imhoff tanks, which have now been replaced by a variety of high-rate digesters. Loehr has listed the following reasons for using anaerobic processes for waste treatment:

1. Higher loading rates can be achieved than are possible for aerobic treatment techniques.
2. Lower power requirement may be needed per unit of BOD treated.
3. Useful end-products such as digested sludge and/or combustible gases may be produced.
4. Organic matter is metabolised to a stable form.
5. There is an alteration of water-binding characteristics to permit rapid sludge dewatering.
6. The reduced amount of microbial biomass leads to easier handling of sludge.
7. Low levels of microbial growth will decrease the possible need for supplementary nutrients with nutritionally unbalanced wastes.

#### *Anaerobic digestion*

Large volumes of wet sludge which are produced in primary and secondary sedimentation tanks may have to be reduced in volume before disposal. This volume of sludge can be reduced by anaerobic digestion. In sludges containing 20000–60000 mg $dm^{-3}$ solid matter, 80 per cent of the degradable matter may be digested, which will reduce the solids content by 50 per cent. During anaerobic digestion acid fermenting bacteria degrade the waste to free volatile fatty acids, mainly acetic and propionic acid, which are then converted to methane (~60 per cent) and carbon dioxide (~40 per cent). The gas produced (biogas) is a very useful by-product and can be burnt as a heating fuel, fed to gas engines to generate electricity or used as a vehicle fuel. As well as being used in sludge digestion and conditioning, anaerobic digesters are also used directly in the treatment of many high strength waste-water, for example from the food and agricultural industries.

A number of anaerobic processes have been developed; completely mixed reactors (often these are simply described as anaerobic digesters) with or without sludge recycle, anaerobic filters, up-flow anaerobic sludge blankets (UASB) and anaerobic fluidised-beds are the most common. Two stage anaerobic treatment systems utilising a short retention time completely mixed acidification reactor followed by a UASB methanogenic reactor have been reported. Benefits claimed include improved reliability, higher BOD/COD reductions and greater methane productivity.

### *Anaerobic digesters*

The digester tanks used for this process may be up to 12000 $m^3$ volume and equipped with heating coils for accurate temperature control normally in the mesophilic range (25°F–38°F, usually 28°F–32°F in anaerobic digestion) to increase the rate of digestion, which is also improved by mechanical agitation. Some digesters are operated thermophilically (50°F–70°F, usually 55°F–63°F in anaerobic digestion) to increase the rate of degradation and biogas production, though much of the extra gas produced will be consumed in maintaining the digester temperature. The use of thermophilic anaerobic digestion has been investigated for the treatment of high strength (16–25 grams COD $dm^{-3}$), low pH (~3.8) and high temperature winery effluents. Wiegant reported the successful application of thermophilic anaerobic digestion of vinasse in completely mixed and UASB reactors, although methane production was found to decrease at high loading rates due to the presence of inhibitory compounds. Retention times are generally between 10 and 30 days, although with solids recycle retention times may be as low as 0.5 day. Most digesters operate on a semi-continuous basis with the appropriate volume of digested material being removed and replaced with 'fresh' material on a daily basis.

### *Anaerobic filters*

In anaerobic filters, as with the aerobic filters, there is a microbial film growing on an inert support. Anaerobic filters may be operated in an upflow or downflow mode and a wide variety of packings are used; both natural and synthetic. The first full-scale anaerobic filters brought into operation a 1,00,000 $m^3$, plant to treat distillery effluents. Many other effluents are found to be amenable to treatment using anaerobic filter systems; antibiotic fermentation wastes, citric acid fermentation wastes, yeast production waste-water and brewery and winery waste-water, molasses distillery slops and fermentation and pharmaceutical wastes.

### *Up-flow anaerobic sludge blankets (UASB)*

In the system (originally developed in the Netherland) high levels of active biomass are retained in the reactor by flocculation. No support media is added to the reactor, instead the flocculated sludge develops in the reactor and acts as a fluidised-bed. Feed is pumped through the bed (the sludge blanket), above which fine particles flocculate and settle back to the blanket as sludge, thus preventing washout of organisms. Anaerobic sludge blankets have been found to be effective in the treatment of many waste-water, including sugar-beet wastes, domestic sewage, slaughterhouse wastes, agricultural wastes, brewery wastes, winery wastes and distillery wastes.

## BY-PRODUCTS

In any fermentation recovery process there is a need to recognise whether there is a potential marketable waste and, if necessary, to develop a market. Obviously the marketability and cost of reclamation of by-products will be very important in deciding upon a policy of waste recovery. Under favourable market conditions it has been claimed that the profit on animal feed by-products from a completely integrated distillery may almost pay the cost of the grain. This claim would now be more difficult to substantiate, due to fuel cost and capital outlay on plant.

Some microbial processes yield residues which are difficult or impossible to sell. In these processes it may be possible and worthwhile, to change to low residue raw materials such as refined sugars or other pure compounds as sources of soluble nutrients.

### Distilleries

In grain-based distilleries it has been common practice to recover spent grain and stillage, the waste liquor after the alcohol has been distilled off. The stillage is first passed through screens with 1 mm openings. The screenings are then dewatered with mechanical filter presses and dried in rotary driers. This product is termed Distillers' Dried Grains (light grains). The screened stillage is concentrated in evaporators to give 25 to 35 per cent solids in a thick syrup. Stillages which have been prefiltered may be concentrated to 35 to 50 per cent solids. This syrup can then be mixed with the pressed screenings and dried to give Distillers' Dried Grains with Solubles (dark grains). Alternatively the evaporated stillage may be dried completely in drum driers to produce Distillers' Dried Solubles. Dark and light grains and dried solubles have all been used as animal-feed supplements. Flachowsky and others reported their use following chemical treatment as a replacement feed for sheep. Dried solubles have also been used as a medium adjunct in the preparation of antibiotics.

In distilleries using cane molasses as a feedstock, evaporated spent wash has been used as fuel for boilers. It has proved worthwhile to recover potassium salts from sugar-beet stillage. The market for the evaporated product must be considered within a range of 50 km of the evaporation plant.

In recent years the traditional distillers' by-products from stillage have become increasingly uneconomic. A number of processes to produce SCP from stillage using *Geotrichum candidum, Candida utilis* or *C. tropicalis* have been evaluated.

### Breweries

The three marketable wastes from breweries are spent grain, spent hops and yeast. The spent grain is recovered from the mash tun and is then sold as animal feed either after pressing in a wet state or after drying in rotary driers. Alternatively, the wet grain may be used in the preparation of silage for cattle. The possible markets for hops are restricted. Some are used as a fertiliser or as a low-grade fuel.

Yeast can be separated from beer by filtration or centrifuging. The yeast slurry is then dried in drum driers. Some of the yeast may be mixed with brewers' spent grains to produce a feed material with a slightly higher protein content than normal brewers' grains. The yeast may also be used directly as a source of vitamins. If it is to be used as a human food it must be debittered to remove the hop bitter substances absorbed on to the yeast cells. The cells are then washed in an alkaline solution, washed with water and drum dried. Although bakers' yeast was originally obtained as a brewery by-product, this market has diminished considerably. Most bakers' yeast is now produced directly by a distinct production process. Dewatered sludge from brewery waste-water treatment operations has been reported to increase agricultural yields when used as fertiliser. Lyons reports on the potential use of brewery effluents as a feedstock for fuel and industrial ethanol production.

### Amino Acid Wastes

The main wastes from glutamic acid or lysine fermentations are cells, a liquor with a high amino acid content which can be used as an animal-feed supplement and the salts removed from the liquor by crystallisation, which is a good fertiliser.

### Fuel Alcohol Wastes

The stillage from ethanol production and wastes from starchy fermentations are, following concentration, saleable as an animal-feed supplement. Wastes from sugar fermentations and distillation can be digested anaerobically and the methane generated used as an energy source. Faust also suggest the use of $CO_2$ rich off gases in the food and beverage industries.

### Microbial Inoculants and Enzymes for Waste Treatment

During the past twenty years a series of microbial inoculants and enzyme additives have been developed for waste treatment applications. Commercial microbial enzymes and the micro-organisms that produce them may be added to effluent treatment systems to accelerate breakdown of cellulose, hemicellulose, fat, proteins, etc. Microbial inoculants may also be used during start-up of a waste treatment system or to accelerate recovery of a fermentation which has been stressed by an influent load or by a toxic feed. In addition, new genetic techniques are enabling inoculant producers to develop more efficient strains which may aerobically or anaerobically digest waste materials at faster rates than their counterparts present in the natural waste treatment microbial population.

Apart from applications of inoculants in normal municipal and agricultural waste treatment, there is substantial interest in the use of biological system for degradation of hydrocarbons and recalcitrant chemical compounds. Microbial inoculants are available for degradation of crude oils and selected toxic materials such as phenol. Microbial enrichment culture isolation methods and genetic manipulation investigations are currently being directed toward the development of suitable microbial strains for degradation of compounds including azo dyes, stilbenes, chloroaromatics, chlorinated hydrocarbons and DDT.

## BIOPROCESS DEVELOPMENT FOR MERCURY DETOXIFICATION AND AZO DYE DECOLOURISATION

Biological treatment of environmental pollutants is often very complex and requires intensive control of operation parameters. Fermentation of micro-organisms able to degrade or destroy pollutants is a key technology to ensure appropriate cell concentration and populations are maintained in the treatment system. Optimal operation strategies, especially policy for substrates feeding and oxygen supply, need to be developed to enhance the rates of biodegradation or biotransformation for efficient cleanup of the target pollutants. In this section, two unique types of biological treatment processes, namely, mercury detoxification and azo dye decolourisation, are introduced as the example to show how fermentation technology and bioreactor approaches are used to develop bioprocesses for environmental applications. Detoxification of mercury is conducted aerobically using wildtype and recombinant strains as the biocatalyst. The performance and stability of fed-batch and chemostat bioreactors for mercury detoxification are greatly correlated to the dosage of the toxic mercury substrate. In contrast, azo dye is less toxic but decolourisation of azo dyes requires aerobic/anaerobic sequential environments. Thus, bioreactor design for bacterial decolourisation focused on strategies for the supply of oxygen and retention of high cell concentration within the reactors to achieve optimal combination of aerobic cell growth and anaerobic decolourisation. The effectiveness and feasibility of various bioreactor configurations and strategies are assessed.

Environmental Biotechnology has become a promising means to clean up pollutants from the contaminated environments in a more natural and cost-effective way. In biological treatment processes, micro-organisms capable of degrading or transforming organic or inorganic pollutants into simpler or less toxic forms play a major role. An efficient biological treatment involves effective growth of biodegraders and control of the bioprocess for an efficient biodegradation or biotransformation of the pollutants. Therefore, fermentation technology is obviously a crucial tool to enable an efficient biological waste treatment process. However, it is great challenge for fermentation technology when it is applied in waste treatment, since the substrates (pollutants) are more complex and often toxic to the micro-organisms as compared with those used in food or pharmaceutical industry. In addition, the performance of waste degradation is usually very sensitive to the fermentation conditions [pH, aeration, temperature, food to micro-organism (F/M) ratio, nutrients, etc.] and may require aerobic/anaerobic sequential environments for mineralisation of some complex pollutants. For instance, biodegradation of chlorinated aromatic compounds requires anaerobic dechlorination step followed by aerobic ring cleavage of the chlorine-free aromatic compounds and subsequent degradation of the resulting fatty acids via β-oxidation steps. The other typical example is biological nitrogen removal that converts ammonia to nitrates via an aerobic nitrification step and the nitrates are reduced anaerobically to $N_2$ with a denitrification step. As a result, an appropriate bioprocess design and process control seems to be the key to a successful biological treatment system. This section provides examples that demonstrate strategies to develop bioprocesses for mercury detoxification and for azo dye decolourisation. The former is an aerobic process that deals with a toxic substrate. The latter involves an aerobic/anaerobic sequential environment, in which cell growth and degradation of substrate have distinct oxygen requirement.

### Bioprocess Development for Microbial Mercury Detoxification

Mercury and mercurial compounds have been used in a variety of industries, causing severe mercury pollution in aquatic systems and soils. Conventional mercury-removal methods mainly via physical and chemical routes to either trap and collect mercury from contaminated sites or to remove mercury by chemical precipitation. The drawback of physical methods has been the requirement for additional treatments, while chemical methods often leave hazardous by-products or residual sludges. Thus, biological methods allowing more natural, efficient and economical clean-up of mercury waste have emerged.

Mercury-resistant micro-organisms can resist mercury due to their ability to volatilise soluble forms of mercury from the environment via a sequence of enzymatic reactions, which are recognised as mercury detoxification. The genetic basis of mercury resistance is encoded in *mer* operons located on either plasmids or transposable elements. Organomercurial lyase (*mer*B gene product) catalyses cleavage of C-Hg bonds in organomercurial compounds to release mercuric ions, which are enzymatically reduced to less toxic and more volatile metallic mercury ($Hg^o$) by *mer*A-product mercuric reductase. Cysteine-rich transport proteins originated from *mer*P and *mer*T genes locate on the periplasmic space and inner membrane, respectively. The *mer*P and *mer*T proteins are responsible for the specific delivery of ambient mercuric ions toward mercuric reductase in the cytoplasm for reduction of $Hg^{2+}$ to volatile $Hg^{o}$. The constitutive *mer* operon is induced by the sub-toxic level of mercuric ions. The mechanism of mercury detoxification is illustrated in Fig. 19.11.

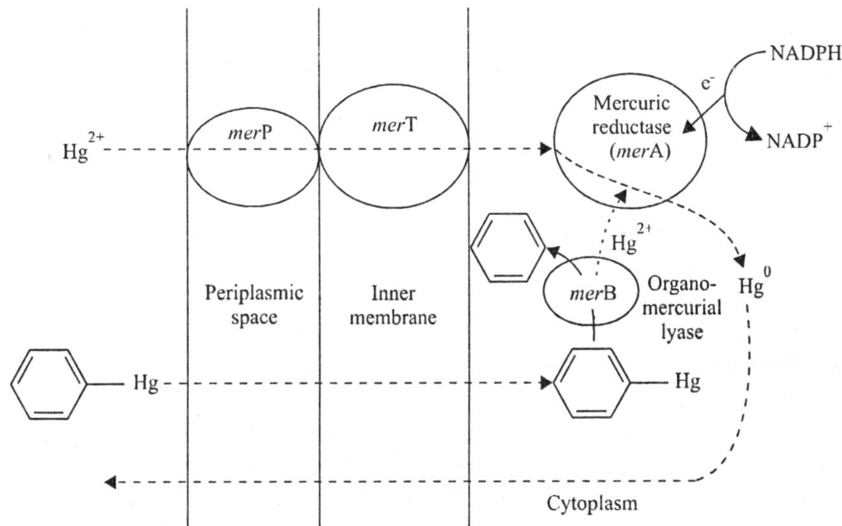


**Fig. 19.11.** Mechanism of bacterial mercury resistance.

In contrast to intensive studies and understanding of the molecular genetics of *mer* operon and the biochemistry of mercury detoxification, much less effort has been devoted to develop microbial mercury detoxification processes. Only recently, Saouter and others reported their preliminary investigation of using $Hg^{2+}$-reducing strains to decontaminate a polluted freshwater pond in Tennessee and demonstrated their studies on volatilisation of mercury using resting or immobilised-cell systems.

Recent attempts investigated the kinetic of mercury detoxification by a wild-type mercury-resistant strain (*P. aeruginosa* PU21) and a recombinant. *E. coli* strain (*E. coli* PWS1). The two mercury-resistant strains were used to develop batch, fed-batch and continuous bioreactors to demonstrate operation strategies for efficient and practical detoxification of mercury in waste-water. A mercury vapour recovery device was also designed to recover metallic mercury ($Hg^{o}$), the end product of mercury detoxification.

Mercury detoxification occurred when the wild-type or the recombinant strain was cultivated in mercury-containing media. Serial adaptation may be needed to prepare mercury-hyperresistant strain of *P. aeruginosa* PU21, while the mercury-reducing activity of the recombinant strain *E. coli* PWSl can be turned on by IPTG induction. Since mercury is toxic to the mercury-resistant cells when its concentration is higher than some threshold value, the loading of mercury into the bioreactor need to be well controlled for a stable mercury detoxification operation. In addition, a sufficient initial cell concentration to start up the bioreactor is also critical to enable total survival of the cells for efficient mercury detoxification.

It was found that *P. aeruginosa* PU21 (Rip64) is able to detoxify mercury rapidly during its transient growth. The reduction of mercury with growing cells exhibited a maximal rate (approximately $1.11 \times 10^{-6}$ μm Hg/cell/hr.) at the substrate concentration of 8 mg Hg/l. The mercury detoxification performance of *P. aeruginosa* PU21 was also correlated to the bacterial growth phase, with cells at lag phase and exponential phase having higher specific detoxification rates. The recombinant strain *E. coli*

PWSl exhibited better specific mercury detoxification activity than *P. aeruginosa* PU21 at batch operations, while the PWSl strain became unstable during long-term fed-batch operations primarily due to plasmid instability.

Two fed-batch strategies were used for mercury detoxification operations: step-wise fixed-interval (SWFI) feeding and repeated constant-rate (RCR) feeding. For SWFI feeding, mercury detoxification efficiency 2.9 mg/l/hr. and 3.3 mg/l/hr. for 2 and 5 mg/l feeding, respectively. In contrast, for RCR feeding strategy, the mercury detoxification of efficiency was closely related to the mercury feeding rate ($F_{Hg}$), initial inoculum size ($X_o$), as well as the initial culture volume ($V_o$) and the best mercury detoxification efficiency was 5.6 mg Hg/l/hr., which occurred when $F_{Hg}$ = 16.9 mg Hg/hr., $X_0 = 4.45 \times 10^9$ cells/ml and $V_o$ = 1200 ml.

Chemostat operations for mercury detoxification were also conducted but gave lower detoxification efficiencies than those obtained from fed-batch mode. At a dilution rate of 0.18 and 0.325 $hr.^{-1}$ and mercury feeding concentrations of 6.15 mg/l, the mercury detoxification efficiency was only 1.09 and 1.94 mg/l/hr., respectively. However, chemostat operations were able to continuously reduce mercury concentrations in the feed (1.0–6.15 mg/l) to nearly undetectable amounts and the steady-state operation can be maintained for a prolonged period (nearly 55 hour).

Mercuric reductase was purified from a recombinant strain *E. coli* PWSl able to overexpress *mer*A product. The purified mercuric reductase was covalently immobilised on a porous support Celite 545 and the immobilised enzyme was utilised for mercury detoxification with a fixed-bed process. Enzymatic reduction of $Hg^{2+}$ with immobilised mercuric reductase requires continuous feeding of cofactor NADPH, which is very costly. Therefore, compounds that have similar structure to NADPH were used as the artificial cofactor. Neutral Red (a dye) was found to be a satisfactory alternative of NADPH as it achieved nearly 40 per cent mercury reduction efficiency compared to using NADPH. The best mercury detoxification performance of the immobilised-enzyme column was 5.41 μmol/hr./l.

It is of importance to recover the end product of mercury detoxification process, metallic mercury. The end product is mainly in the gas stream and can be recovered by adsorption on metal (e.g., gold, silver, copper) or charcoal adsorbents or by absorption and stabilisation with acids. In recent research, mercury vapour was collected by a device consisted of three gas/liquid absorption columns in series using concentrated $HNO_3$ as the oxidant, followed by an activated carbon adsorption trap. With continuous feeding of up to 10 mg/l of $Hg^{2+}$, the recovery efficiency in acid columns was over 80 per cent and the mercury concentration at the exit of activated carbon trap was below 0.2 μg/l.

### Bioprocess Development for Bacterial Decolourisation of Azo Dyes

Azo dyes are one of the most commonly used synthetic dyes in textile, food, papermaking and cosmetic industries. Release of azo dyes into the environment has become a major concern in waste-water treatment since some azo dyes or their metabolites may be mutagens or carcinogens. Azo dyes are often considered xenobiotics and recalcitrant aerobically and conventional aerobic waste-water treatment processes usually cannot efficiently decolourise azo dye contaminated effluents. Therefore, in some cases traditional biological processes are combined with physical and chemical treatment, such as flocculation, chemical coagulation precipitation, to achieve a better decolourisation of the waste-water. However, as more efficient bacterial decolourisers have been isolated or constructed, biological decolourisation may become an effective method to transform, degrade or mineralise azo dyes.

As indicated in Fig. 19.12, bacterial decolourisation of azo dyes is typically initiated by azoreductase-catalysed anaerobic reduction or cleavage of azo bonds, followed by aerobic or anaerobic degradation of the resulting aromatic amines by a mixed bacterial community. Decolourisation by fungi (e.g., *Phanerochaete chrysosporium*) is achieved via the ligninolytic-degradation activity to degrade azo dyes aerobically with the aid of an extracellular enzyme, lignin peroxidase. Most biological decolourisation processes have been fungi systems, while the development of bacterial decolourisation processes has been relatively less studied. Hu recently isolated a *Pseudomonas luteola* strain able to efficiently decolourise a group of azo dyes. The strain was subject to detail investigations in terms of decolourisation kinetics and mechanism and decolourisation processes including a fed-batch process and an immobilised cell system. Saouter also obtained an *E. coli* mutant (*E. coli* $NO_3$) showing excellent decolourisation rate for a variety of azo dyes. In addition, a dye-decolourising recombinant strain *E. coli* CY1 harbouring azo dye decolourising genes of *Rhodococcus* sp., was also constructed in laboratory.

$$C_6H_5{-}N{=}N{-}C_6H_5 \xrightarrow[\text{NADH} \rightarrow \text{NAD}^+ \text{ (Anaerobic)}]{\text{Azo reductase}} C_6H_5{-}NH_2 + C_6H_5{-}NH_2 \xrightarrow{\text{Mixed bacterial consortia (aerobic or anaerobic)}} CO_2 + H_2O$$

**Fig. 19.12.** Mechanism of bacterial decolourisation of azo dyes.

### *General consideration of bacteria decolourisation processes*

Many aerobic (e.g., *P. luteola*) or facultative anaerobic bacteria (e.g., *E. coli*) show excellent decolourisation activity for azo dyes when the dissolved oxygen is limiting, since the presence of free oxygen seems to inhibit azoreductase activity due to competing with azo bonds for electron donor (e.g., NADH). Therefore, during the course of decolourisation, cell growth essentially stops or becomes very limited as a result of lack of oxygen supply. This appears to cause difficulty in bioreactor design, because maintaining sufficient cells concentration of the bacterial decolouriser in the reactor would be a problem if the cells do not grow at a satisfactory rate under the reaction conditions. As a result, for stable operation of a long-term continuous decolourisation bioprocess, it is necessary to develop strategies to provide oxygen for cell growth without hindering decolourisation activity or strategies to maintain an adequate amount of cells in the bioreactor when cell growth is insignificant.

On the other hand, pH value is also a critical parameter that affects the efficiency of bacterial decolourisation. For instance, a pH value of 7–9 favoured decolourisation of Reactive Red 22 by *P. luteola* and *E. coli*, while a low pH severely limited the decolourisation activity. It was also found that when glucose was used as the carbon source for *P. luteola* and *E. coli* $NO_3$, the pH dropped significantly, probably due to acid formation (e.g., acetic acid), to inhibit their decolourisation activity. The inhibitory effect was not observed when an alternative carbon source, such as glycerol, was used.

### *Decolourisation with fed-batch processes utilising high-cell-density fermentation*

Fed-batch bioprocesses using *P. luteola* as the decolouriser were shown to effectively decolourise an azo dye, Reactive Red 22. To avoid oxygen inhibition on the bacterial decolourisation, the bioreactor was operated with gentle agitation and without aeration, but the condition led to insignificant cell growth. Initial biomass loading, the feeding strategy and the yeast extract to dye ratio (Y/D) all affected the performance of the fed-batch bioreactor, with the dye loading rate the most critical one. Since cells grew poorly under oxygen-deficient environment required for decolourisation, a sufficient biomass loading is crucial to provide enough total decolourisation activity for colour removal. High cell-density fermentation strategies was applied to increase the initial biomass loading to up to 15 g/l and resulted in a two-fold increase in overall decolourisation efficiency. Within a certain limit, the decolourisation efficiency was enhanced significantly by increasing the dye loading rate, while overloading of the dye may cause substrate inhibition and may also result in higher residual dye concentration at the end of feeding, raising difficulty in the follow-up treatment. The Y/D ratio is also an important factor to ensure sufficient supply of yeast extract, which serves as the raw material for electron-donor (NADH) production. It was found that Y/D ratio should exceed 0.5 to maintain a stable decolourisation. Without using high-cell-density fermentation approaches (initial biomass loading = 2 g/l), the best specific decolourisation rate and overall decolourisation efficiency obtained for *P. luteola* were 113.7 mg dye/g cell/hr. and 86.3 mg dye/l/hr., respectively, with a constant dye loading rate of 200 mg/hr. and a Y/D ratio of 1.25. In contract, when initial biomass loading was increased to 15 g/l, the overall efficiency raised to 194 mg dye/g cell/hr., while best specific rate was comparable to that obtained without employing high-cell-density fermentation.

### *Continuous decolourisation with aerobic/anaerobic sequences*

Since cell growth is poor during conditions for efficient decolourisation, it is not feasible to apply a regular continuous culture for decolourisation of azo dyes. A modified process that includes an aerobic growth stage and an anaerobic decolourisation stage was examined for its feasibility for persistent and stable decolourisation operations. Description of the operation strategy is indicated in Fig. 19.13. The cells were first grown aerobically in the reactor at batch mode until cell concentration exceeded 2.5 g/l. The reactor was then switched to continuous mode without air supply but with gentle agitation. The continuous culture was operated at different dilution rates to examine the effect of substrate (medium and dye) loading on decolourisation. Because the continuous culture was carried out under oxygen-limiting condition, cell concentration in the bioreactor decreased progressively with operation time. When cell concentration was lower than 0.4 g/l, the feeding and effluent were terminated and the aeration was turned on for aerobic cell growth. Again when cell concentration reached the designated value, the reactor was returned to continuous mode. This procedure was repeated several times.

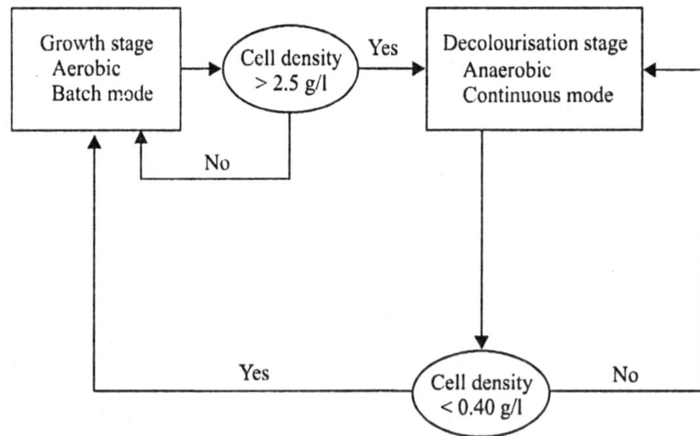


**Fig. 19.13.** Operation strategy for continuous decolourisation with aerobic/anaerobic sequences.

The intermittent anaerobic-aerobic process was used to decolourise mixed day with *E. coli* $NO_3$ strain at hydraulic retention time (HRT) of 32, 18 and 10 hr. and a constant dye feeding of 5000 ADMI. The colour-removal efficiency was around 70–80 per cent at all runs for nearly 120 hours of operation.

The total decolourisation efficiency was 93, 156 and 65 ADMI·L/hr. at HRT = 32, 18 and 10 hours, respectively. If combination of one aerobic stage and one anaerobic stage is defined as one cycle, the decolourisation rate of *E. coli* $NO_3$ appeared to become higher after the original cycle. This may be attributed to the acclimation effect since the *E. coli* $NO_3$ cells were repeatedly exposed to the azo dyes.

### Decolourisation with continuous cultures combined with microfiltration membrane

Since bacterial decolourisation normally proceeds under anaerobic conditions, which limit cell growth of the bacterial decolourisers. Therefore, retention of cells within the bioreactor becomes a critical issue in performing successful continuous decolourisation processes. One way to enhance cell retention is to completely recycle of the cells with a microfiltration membrane. Cross-flow microfiltration modules (e.g., hollow fibre membrane) have been frequently used due to its capability for a long-term operation. The membrane reactor causes no mass transfer limitations and also raises the cell concentration within the reactor (even though the growth rate is low); thus has the potential to enhance the operation efficiency.

In one laboratory, a membrane bioreactor was designed for continuous decolourisation of dye-containing water using. *P. luteola* as the decolouriser. Figure 19.14 shows organisation and set-up of the bioreactor. Cells were first grown in the reaçtor at a batch mode until cell density reached a desired level, then the reactor was switched into a continuous mode as the feeding (medium and dye) started and the effluent was introduced into a hollow-fibre membrane module, where cells were blocked and returned into the bioreactor, but the decolourised water penetrated through the membrane. The optimal operation strategy was determined by investigating the effects of operation parameters [e.g., temperature, hydraulic retention time (HRT), dye loading rate, medium loading rate and biomass concentration, etc.] on decolourisation efficiency. The results show that the best decolourisation rate occurred at a temperature of 40°C. The overall decolourisation efficiency increased when biomass concentration was increased. The optimal HRT was 13 hours.

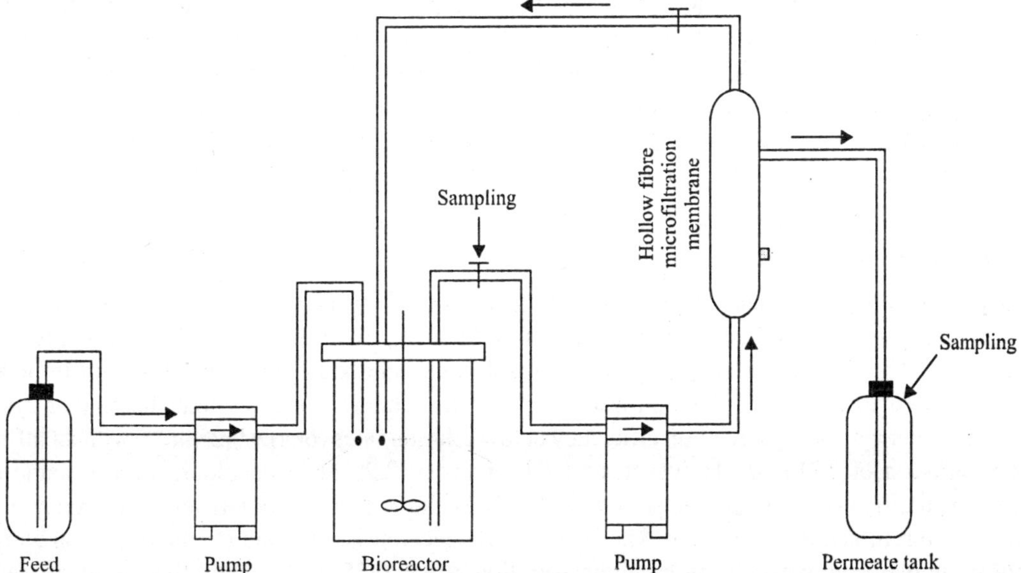


**Fig. 19.14.** Schematic description of membrane bioreactor for continuous decolourisation of azo dye by *P. luteola*.

The decolourisation rate increased rapidly with an increase in dye loading rate. The best specific decolourisation rate (90 mg dye/g cell/hr.) was obtained at a dye loading of 50 mg/hr. addition of extracellular metabolites of a dye-decolourising bacterium *E. coli* $NO_3$ appeared to enhance the decolourisation ability of *P. luteola* and the maximal decolourisation was elevated from 90 to 125 mg dye/g cell/hr.

### *Decolourisation with fixed-bed bioreactor containing immobilised cells*

Cell immobilisation is an alternative way to retain cells inside the bioreactor for continuous decolourisation. Cell immobilisation systems provide an easier solid/liquid separation, which is often encountered in bioprocess operations. Entrapment of cells in natural or synthetic gel matrices allows efficient cell retention and is usually cheaper than the membrane systems. However, the mechanical stability and the effect of the matrix materials or synthesis procedures on cell viability need to be concerned. Entrapment of cells may also raise mass transfer resistance between cells and substrates. Biofilm systems, in which cells attached on the surface of the carrier support, are also one of the most commonly applied cell immobilisation approaches, especially in biological treatment of environmental pollutants. Establishment of a biofilm system does not require extra physical or chemical treatment, making it very friendly to the environment and is also cost-effective. However, biofilm formation is usually time-consuming and the attached cells may come off easily due to shear stress caused by high volumetric flow rates.

#### *Gel entrapment systems*

Two natural gel matrices—calcium alginate (CA) and κ-carrageen (CN) and a synthetic polymer matrix (polyacrylamide; PAA) were used to entrap the cells of *P. luteola* and the resulting immobilised cells were packed into glass columns for fixed-bed decolourisation of an azo dye. The two natural gels are easy to prepare but the beads are less stable mechanically. In addition, the CA beads are sensitive to single-charged cations (e.g., $Na^+$, $K^+$), while elevated temperature causes damages to the structure of CN beads. The synthetic gel (PAA) has better mechanical strength but its monomer (arcylamide) is toxic to the cells. The bioprocess was conducted at different dye loading ($C_{dye}$), bed length (L), representing biomass loading and volumetric flow rate (F). The CA and PAA columns gave much better decolourisation performance than the CN column did. The best decolourisation efficiency for CA column at $L = 20$ cm was 15 mg/hr./l ($C_{dye} = 50$ mg/l, $F = 30$ ml/hr.) and for PAA column was 37 mg/hr./l ($C_{dye} = 200$ mg/l, $F = 15$ ml/hr.).

#### *Biofilm systems*

Decolourisation was also conducted with *P. luteola* cells forming biofilm on ceramic supports. The support material were packed into a column (5 cm in diameter, 30 cm in length), which was fed with inoculated growth medium for ca. 2 days to form sufficient amount of biofilms. The biofilm-loaded column was used for fixed-bed decolourisation operations. The results show that the reactor system maintained 95 per cent decolourisation efficiency at low dilution rates (below 0.25 $hr.^{-1}$) with an influent dye concentration of 200 mg/l. However, as HRT exceeded 0.25 $hr.^{-1}$, decolourisation became less efficient, but addition of the metabolites from *E. coli* $NO_3$ allowed the decolourisation efficiency of the biofilm system returned to 95 per cent. The best specific decolourisation rate for the operation without metabolite supplements was 2.2 mg/g cell/hr. (dilution rate = 0.25 $hr.^{-1}$), while the rate increased to 5.6 mg/g cell/hr. (dilution rate = 0.5 hour) when 40 per cent (v/v) of metabolites from *E. coli* $NO_3$ were added into the feed.

Thus, the bioreactor strategies for microbial detoxification of mercury and decolourisation of azo dyes is due to differences in the characteristics of the target substrate and the environment required for the biotransformation, the key concerns for bioreactor design in the two cases are different. For mercury detoxification, fed-batch bioreactors were shown to be favourable due to more efficient control of mercury concentration for optimal detoxification and also to avoid substrate inhibition arising from toxic effects of mercury. The best mercury detoxification efficiency was 5.6 mg Hg/hr./l, which occurred when the substrate was fed with a repeated-continuous-rate strategy. For azo dye decolourisation, since cell growth and decolourisation take place at distinct oxygenic states, the challenge becomes to determine the timing of oxygen supply for cell growth and to maintain high cell density during decolourisation stage, in which cells essentially do not grow. Among the bioreactor strategies examined, the immobilised-cell systems utilising hollow fibre membrane and biofilm on porous supports are better bioprocesses for continuous decolourisation. Although the membrane reactors are able to maintain high cell density for high decolourisation rates, the biofilm systems are more feasible in practical applications due to the advantages of being cost effective and easier to operate and scale up. Moreover, development of better decolourising biocatalysts by molecular genetic approaches would have the potential to enable further improvement in decolourisation efficiency.

Chapter 20

# Recovery and Purification of Fermentation Products

## INTRODUCTION

The extraction and purification of fermentation products may be difficult and costly. Ideally, one is trying to obtain a high-quality product as quickly as possible at an efficient recovery rate using minimum plant investment operated at minimal costs. Unfortunately, recovery costs of microbial products may vary from as low as 15 per cent to as high as 70 per cent of the total manufacturing costs. Obviously, the chosen process and therefore its relative cost, will depend on the specific product. Atkinson and Mavituna indicated percentage of total cost being 15 per cent for industrial ethanol, 20–30 per cent for bulk penicillin G and up to 70 per cent for enzymes. The high (and sometimes dominant) cost of downstream processing will affect the overall objective in some fermentations.

If a fermentation broth is analysed at the time of harvesting it will be discovered that the specific product may be present at a low concentration in an aqueous solution that contains intact micro-organisms, cell fragments, soluble and insoluble medium components and other metabolic products. The product may also be intracellular, heat labile and easily broken down by contaminating micro-organisms. All these factors tend to increase the difficulties of product recovery. To ensure good recovery or purification, speed of operation may be the overriding factor because of the labile nature of a product. The processing equipment must therefore be of the correct type and also the correct size to ensure that the harvested broth can be processed within a satisfactory time limit.

The choice of recovery process is based on the following criteria:

1. The intracellular or extracellular location of the product.
2. The concentration of the product in the fermentation broth.
3. The physical and chemical properties of the desired product (as an aid to selecting separation procedures).
4. The intended use of the product.
5. The minimal acceptable standard of purity.
6. The magnitude of bio-hazard of the product or broth.
7. The impurities in the fermenter broth.
8. The marketable price for the product.

The main objective of the first stage for the recovery of an extracellular product is the removal of large solid particles and microbial cells usually by centrifugation or filtration (Fig. 20.1). In the next stage, the broth is fractionated or extracted into major fractions using ultrafiltration, reverse osmosis, adsorption/ion-exchange/gel filtration or affinity chromatography, liquid-liquid extraction, two phase aqueous extraction or precipitation. Afterwards, the product-containing fraction is purified by fractional precipitation, further more precise chromatographic techniques and crystallisation to obtain a product which is highly concentrated and essentially free from impurities. Other products are isolated using modifications of this flow-stream.

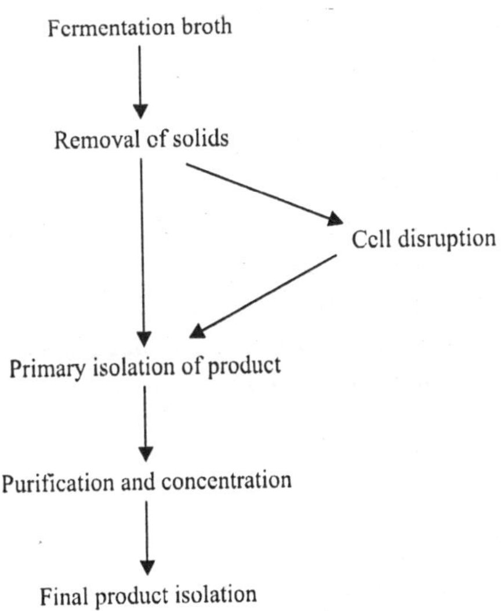


**Fig. 20.1.** Stages in the recovery of product from a harvested fermentation broth.

Attempts to simplify this outline extraction procedure for antibiotic recovery using 'whole broth' processing have met with limited success. The technique of 'whole broth' processing involves initial removal of large particles, which is then followed by passage of the broth (including cells) through, for example, well mixed ion-exchange columns or counter-current liquid-liquid extraction units to extract the product directly. This topic will be discussed in more detail in a later section of this chapter.

It may be possible to modify the handling characteristics of the broth so that it can be handled faster with simpler equipment making use of a number of techniques:

1. Selection of a micro-organism which does not produce pigments or undesirable metabolites.
2. Modification of the fermentation conditions to reduce the production of undesirable metabolites.
3. Precise timing of harvesting.
4. pH control after harvesting.
5. Temperature treatment after harvesting.
6. Addition of flocculating agents.
7. Use of enzymes to attack cell walls.

It must be remembered that the fermentation and product recovery are integral parts of an overall process. Because of the interactions between the two, neither stage should be developed independently, as this might result in problems and unnecessary expense. Darbyshire has considered this problem with reference to enzyme recovery. The parameters to consider included time of harvest, pigment production, ionic strength and culture medium constituents. Large volumes of supernatants containing extracellular enzymes need immediate processing while harvesting times and enzyme yields might not be predictable. This can make recovery programmes difficult to plan. Pigment production might make some recovery procedures difficult, when the pigment binds to the same resin as the enzyme. Changes in fermentation conditions may reduce pigment formation. Certain antifoams remain in the supernatant and affect ultrafiltration or ion-exchange resins used in recovery stages. Trials may be needed to find the most suitable antifoam. The ionic strength of the production medium may be too high, resulting in the harvested supernatant needing dilution with demineralised water before it can be processed. Such a negative procedure should be avoided if possible by unified research and development programmes. Media formulation is dominated by production requirements, but the protein content of complex media should be critically examined in view of subsequent enzyme recovery. This view is also shared by Khosrovi, and others when considering water recycle in biomass production. They stated that the interaction between the different unit operations in a recycle process made it imperative that commercial plant design and operation should be viewed in an integrated fashion.

Flow sheets for recovery of penicillin, cephamycin C, citric acid and micrococcal nuclease are given in Figs. 20.2, 20.3, 20.4 and 20.5, to illustrate the range of techniques used in microbiological recovery processes. In the selection of processes for the recovery of biological products it should always be understood that recovery and production are interlinked and that good recovery starts in the fermentation by the selection of, amongst other factors, the correct media and time of harvesting.

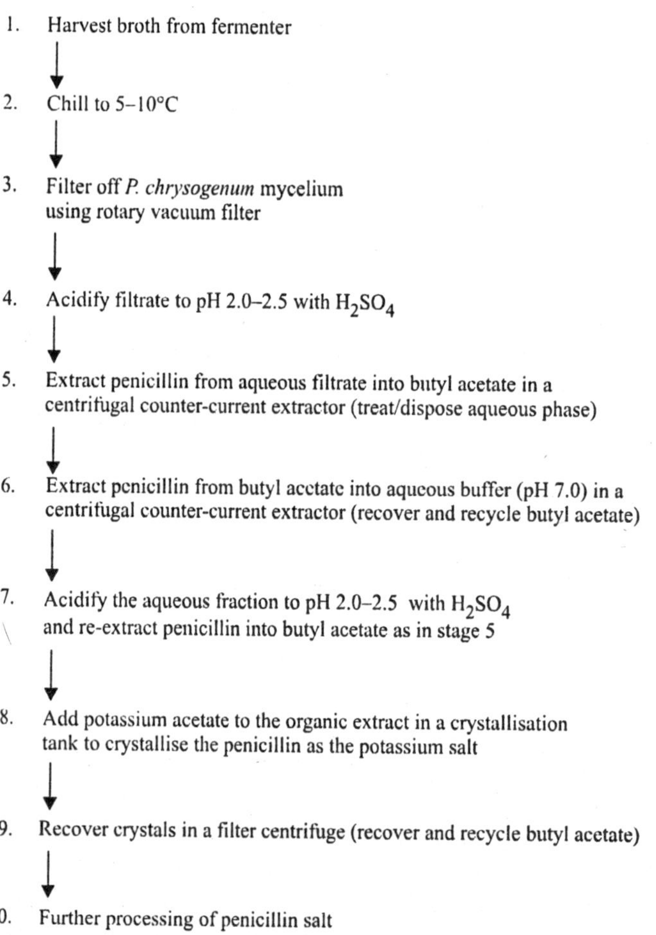


**Fig. 20.2.** Recovery and partial purification of penicillin G.

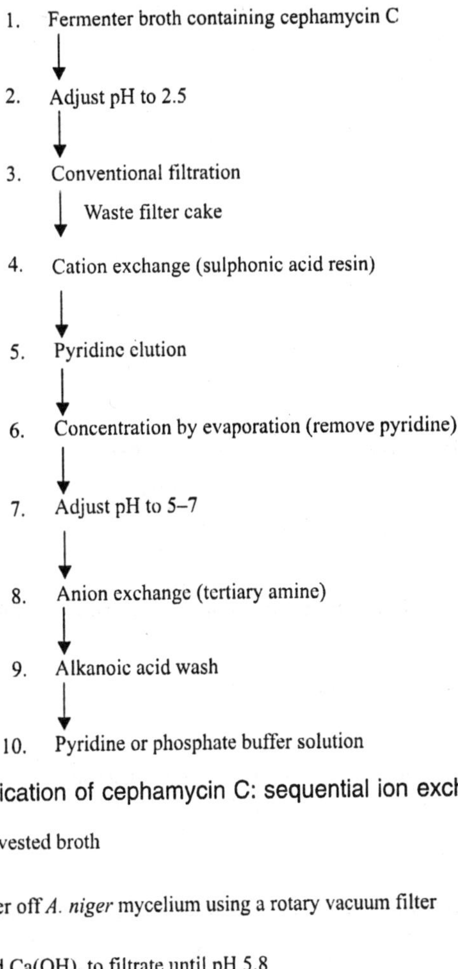


**Fig. 20.3.** Purification of cephamycin C: sequential ion exchange process.

1. Harvested broth
↓
2. Filter off *A. niger* mycelium using a rotary vacuum filter
↓
3. Add $Ca(OH)_2$ to filtrate until pH 5.8
↓
4. Calcium citrate
↓
5. Add $H_2SO_4$ while at 60°C
↓
6. Filter on rotary vacuum filter to recover $CaSO_4$
↓
7. Activated charcoal to decolourise
↓
8. Cation and anion exchange resins
↓
9. Evaporate to point of crystallisation at 36°C
↓
10. Crystals of citric monohydrate separated in continuous centrifuges
↓
11. Driers at 50–60°C

**Fig. 20.4.** Recovery and purification of citric acid.

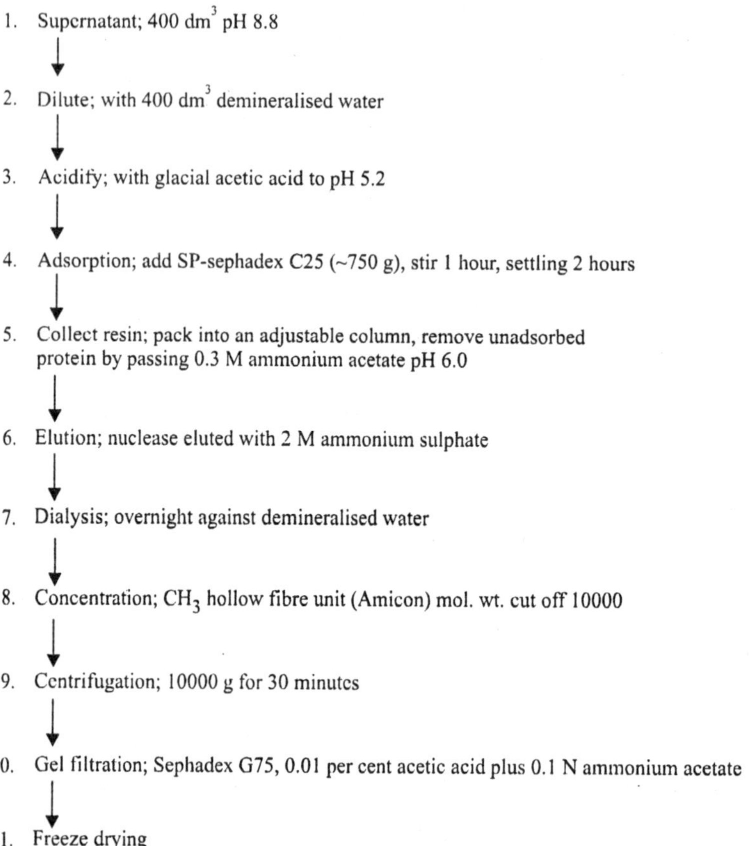


**Fig. 20.5.** Purification of micrococcal nuclease.

The recovery and purification of many compounds may be achieved by a number of alternative routes. The decision to follow a particular route involves comparing the following factors to determine the most appropriate under a given set of circumstances: (i) capital costs; (ii) processing costs; (iii) throughput requirements; (iv) yield potential; (v) product quality; (vi) technical expertise available; (vii) conformance to regulatory requirements; (viii) waste treatment needs; (ix) continuous or batch processing; (x) automation; and (xi) personnel health and safety.

The major problem currently faced in product recovery is the large-scale purification of biologically active molecules. For a process to be economically viable large-scale production is required and therefore, large-scale separation, recovery and purification. This then requires the transfer of small-scale preparative/analytical technologies (e.g., chromatographic techniques) to the production scale whilst maintaining efficiency of the process, bio-activity of the product and purity of the product so that it conforms with safety legislation and regulatory requirements.

## REMOVAL OF MICROBIAL CELLS AND OTHER SOLID MATTER

Microbial cells and other insoluble materials are normally separated from the harvested broth by filtration or centrifugation. Because of the small size of many microbial cells it will be necessary to consider the use of filter aids to improve filtration rates, while heat and flocculation treatments are employed as techniques for increasing sedimentation rates in centrifugation. The methods of cell and cell debris separation described in the following sections have been practised for many years. Bowden reviews some potential developments in cell recovery. These include the use of electrophoresis and dielectrophoresis to exploit the charged properties of microbial cells, ultrasonic treatment to improve flocculation characteristics and magnetic separations. All these techniques suffer from high cost and scale-up difficulties and currently are not appropriate technologies. Of more current interest is the use of two-phase liquid extraction. Though still most appropriately used for separation of selected soluble components, it is easy to scale-up and use conditions which are gentle on the product.

## FOAM SEPARATION

Foam separation depends on using methods which exploit differences in surface activity of materials. The material may be whole cells or molecular such as a protein or colloidal and is selectively adsorbed or attached to the surface of gas bubbles rising through a liquid, to be concentrated or separated and finally removed by skimming (Fig. 20.6). It may be possible to make some materials surface active by the application of surfactants such as long-chain fatty acids, amines and quaternary ammonium compounds. Materials made surface active and collected are termed colligends whereas the surfactants are termed collectors. When developing this method of separation, the important variables which may need experimental investigation are pH, air-flow rates, surfactants and colligend-collector ratios.

Rubin investigated foam separation of *E. coli* starting with an initial cell concentration of $7.2 \times 10^8$ cells $cm^{-3}$. Using lauric acid, stearyl amine or *t*-octyl amine as surfactants, it was shown that up to 90 per cent of the cells were removed in 1 minute and 99 per cent in 10 minutes. The technique also proved successful with *Chlorella* sp. and *Chlamydomonas* sp. In other work with *E. coli*, Grieves and Wang were able to achieve cell enrichment ratios of between 10 and $1 \times 10^6$ using ethyl-hexadecyl-dimethyl ammonium bromide.

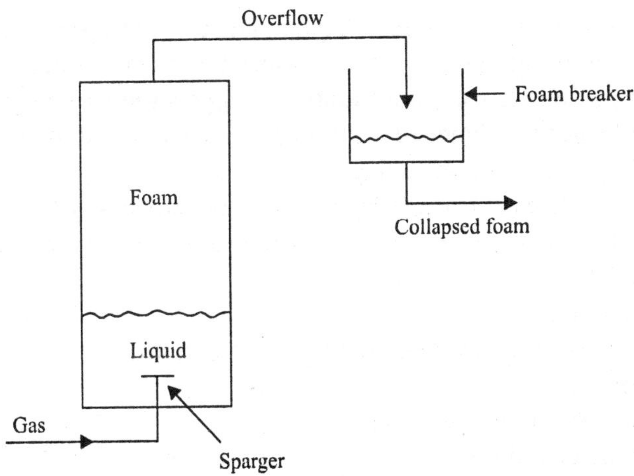


**Fig. 20.6.** Schematic flow diagram for foam fractionation.

## PRECIPITATION

Precipitation may be conducted at various stages of the product recovery process. It is a particularly useful process in that it allows enrichment and concentration in one step, thereby reducing the volume of material for further processing.

It is possible to obtain some products (or to remove certain impurities) directly from the broth by precipitation or to use the technique after a crude cell lysate has been obtained.

### Agents Used in Precipitation

Typical agents used in precipitation render the compound of interest insoluble and these include:

1. Acids and bases to change the pH of a solution until the isoelectric point of the compound is reached and pH equals pI, when there is then no overall charge on the molecule and its solubility is decreased.
2. Salts such as ammonium and sodium sulphate are used for the recovery and fractionation of proteins. The salt removes water from the surface of the protein revealing hydrophobic patches which come together causing the protein to precipitate. The most hydrophobic proteins will precipitate first, thus allowing fractionation to take place.
3. Organic solvents: Dextrans can be precipitated out of a broth by the addition of methanol. Chilled ethanol and acetone can be used in the precipitation of proteins mainly due to changes in the dielectric properties of the solution.
4. Non-ionic polymers such as polyethylene gylcol (PEG) can be used in the precipitation of proteins and are similar in behaviour to organic solvents.
5. Polyelectrolytes can be used in the precipitation of a range of compounds, in addition to their use in cell aggregation.
6. Protein binding dyes (triazine dyes) bind to and precipitate certain classes of protein.
7. Affinity precipitants are an area of much current interest in that they are able to bind to and precipitate, compounds selectively.

## FILTRATION

Filtration is one of the most common processes used at all scales of operation to separate suspended particles from a liquid or gas, using a porous medium which retains the particles but allows the liquid or gas to pass through. It is possible to carry out filtration under a variety of conditions, but a number of factors will obviously influence the choice of the most suitable type of equipment to meet the specified requirements at minimum overall cost, including:

1. The properties of the filtrate, particularly its viscosity and density.
2. The nature of the solid particles, particularly their size and shape, the size distribution and packing characteristics.
3. The solids: liquid ratio.
4. The need for recovery of the solid or liquid fraction or both.
5. The scale of operation.
6. The need for batch or continuous operation.
7. The need for aseptic conditions.
8. The need for pressure or vacuum suction to ensure an adequate flow rate of the liquid.

### Theory of Filtration

A simple filtration apparatus is illustrated in Fig. 20.7, which consists of a support covered with a porous filter cloth. A filter cake gradually builds up as filtrate passes through the filter cloth. As the filter cake increases in thickness, the resistance to flow will gradually increase. Thus, if the pressure applied to the surface of the slurry is kept constant, the rate of flow will gradually diminish. Alternatively, if the flow rate is to be kept constant the pressure will gradually have to be increased. The flow rate may also be reduced by blocking of holes in the filter cloth and closure of voids between particles, if the particles are soft and compressible. When particles are compressible it may not be feasible to apply increased pressure.

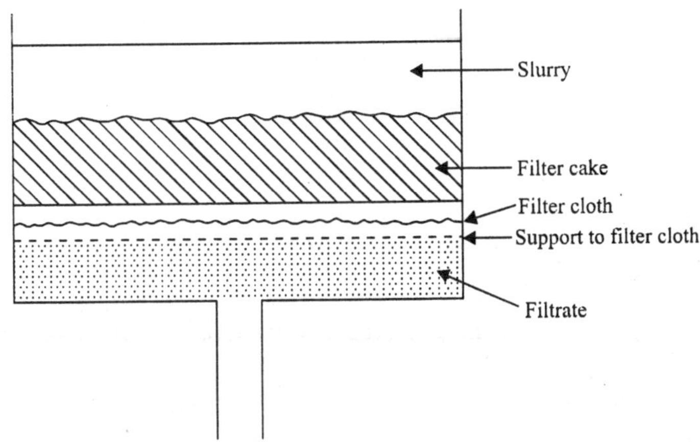


**Fig. 20.7.** Diagram of a simple filtration apparatus.

Flow through a uniform and constant depth porous bed can be represented by the Darcy equation:

$$\text{Rate of flow} = \frac{dV}{dt} = \frac{KA\Delta P}{\mu L} \qquad \text{... (20.1)}$$

where,

$\mu$ = liquid viscosity
$L$ = depth of the filter bed
$\Delta P$ = pressure differential across the filter-bed
$A$ = area of the filter exposed to the liquid
$K$ = constant for the system

$K$ itself is a term which depends on the specific surface area s (surface area/unit volume) of the particles making up the filter-bed and the voidage $\Sigma$ when they are packed together. The voidage is the amount of filter-bed area which is free for the filtrate to pass through. It is normally 0.3 to 0.6 of the cross-sectional area of the filter bed. Thus K (Kozeny's constant) can be expressed as:

$$K = \frac{\Sigma^2}{5(1-\Sigma)^2 s^2}$$

Unfortunately s and $\Sigma$ are not easily determined.

In most practical cases $L$ is not readily measured but can be defined in terms of:

$V$ = Volume of filtrate passed in time t
$v$ = volume of cake deposited per unit volume of filtrate

Then, $$L = \frac{vV}{A}$$

Substituting in equation (20.1):

$$\frac{dV}{dt} = \frac{KA^2 \Delta P}{\mu v V} \quad \text{... (20.2)}$$

This is a general equation relating rate of filtration to pressure drop, cross-sectional area of the filter and filtrate retained. Equation 20.2 can be integrated for filtration at constant pressure.

$$V dV = \frac{KA^2 \Delta P \, dt}{\mu v} \quad \text{... (20.3)}$$

Integrating equation (20.3):

$$V^2 = \frac{2KA^2 \Delta P t}{\mu v} \quad \text{... (20.4)}$$

Now in equation (20.4), $\Delta P$ is constant, $\mu$ is generally equal to 1, $v$ can be determined by laboratory investigation and $A^2$ remains approximately constant. Thus there is a linear relationship between $V^2$ and $t$. By carrying out small-scale filtration trials it is therefore possible to obtain a value for $K$. It is then possible to re-apply the equation for large-scale filtration calculations.

Although it is also possible to derive the equation for the pressure necessary to maintain a constant filtration rate, it has little practical application. The pressure is made up of two components. Firstly the pressure needed to pass the constant volume through the filter resistance and secondly, an increasing pressure component which is proportional to the resistance from the increasing cake depth. This filtration procedure would be complex to perform practically and other methods of filtration are used to achieve constant flow rates, e.g., vacuum drum filters.

## Use of Filter Aids

It is common practice to use filter aids when filtering bacteria or other fine or gelatinous suspensions which prove slow to filter or partially block a filter, Kieselguhr (diatomaceous earth) is the most widely used material. It has a voidage of approximately 0.85 and, when it is mixed with the initial cell suspension, improves the porosity of a resulting filter cake leading to a faster flow rate. Alternatively, it may be used as an initial bridging agent in the wider pores of a filter to prevent or reduce blinding. The term 'blinding' means the wedging of particles which are not quite large enough to pass through the pores, so that an appreciable fraction of the filter surface becomes inactive.

The minimum quantity of filter aid to be used in filtration of a broth should be established experimentally. Kieselguhr is not cheap and it will also absorb some of the filtrate, which will be lost when the filter cake is disposed. The main methods of using the filter aid are:

1. A thin layer of Kieselguhr is applied to the filter to form a pre-coat prior to broth filtration.
2. The appropriate quantity of filter aid is mixed with the harvested broth. Filtration is started, to build up a satisfactory filter-bed. The initial raffinate is returned to the remaining broth prior to starting the true filtration.
3. When vacuum drum filters are to be used which are fitted with advancing knife blades, a thick pre-coat filter is initially built up on the drum.

In some processes such as microbial biomass production, filter aids cannot be used and cell pre-treatment by flocculation or heating must be considered. In addition it is not normally practical to use filter aids when the product is intracellular and its removal would present a further stage of purification.

## Batch Filters

### *Plate and frame filters*

A plate and frame filter is a pressure filter in which the simplest form consists of plates and frames arranged alternately. The plates are covered with filter cloths (Fig. 20.8) or filter pads. The plates and frames are assembled on a horizontal framework and held together by means of a hand screw of hydraulic ram so that there is no leakage between the plates and frames which form a series of liquid-tight compartments. The slurry is fed to the filter frame through the continuous channel formed by the holes in the corners of the plates and frames. The filtrate passes through the filter cloth or pad, runs down grooves in the filter plates and is then discharged through outlet taps to a channel. Sometimes, if aseptic conditions are required, the outlets may lead directly into a pipe. The solids are retained within the frame and filtration is stopped when the frames are completely filled or when the flow of filtrate becomes uneconomically low.

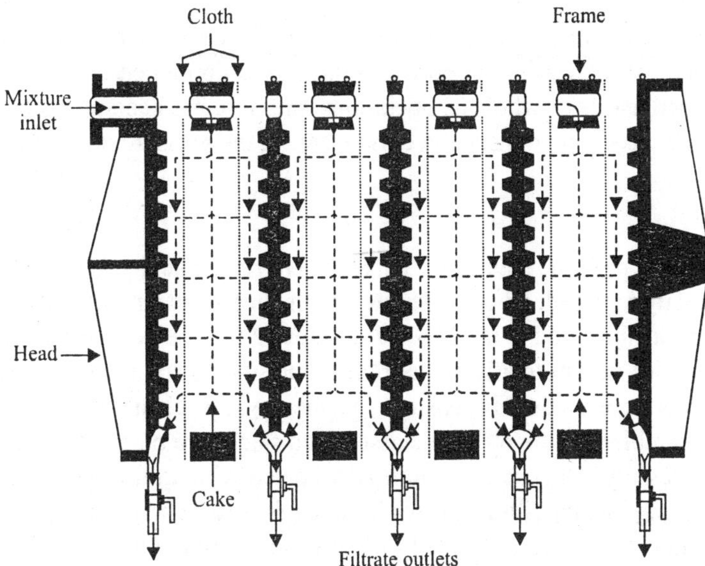


**Fig. 20.8.** Flush plate and frame filter assembly. The cloth is shown away from the plates to indicate flow of filtrate in the grooves between pyramids.

On an industrial scale the plate and frame filter is one of the cheapest filters per unit of filtering space and requires the least floor space, but it is intermittent in operation (a batch process) and there may be considerable wear of filter cloths as a result of frequent dismantling. This type of filter is most suitable for fermentation broths with a low solids content and low resistance to filtration. It is widely used as a 'polishing' device in breweries to filter out residual yeast cells following initial clarification by centrifugation or rotary vacuum filtration. It may also be used for collecting high value solids that would not justify the use of a continuous filter. Because of high labour costs and the time involved in dismantling, cleaning and re-assembly, these filters should not be used when removing large quantities of worthless solids from a broth.

### *Pressure leaf filters*

There are a number of intermittent batch filters usually called by their trade names. These filters incorporate a number of leaves, each consisting of a metal framework of grooved plates which is covered with a fine wire mesh or occasionally a filter cloth and often pre-coated with a layer of cellulose fibres. The process slurry is fed into the filter which is operated under pressure or by suction with a vacuum pump. Because the filters are totally enclosed it is possible to sterilise them with steam. This type of filter is particularly suitable for 'polishing' large volumes of liquids with low solids content or small batch filtrations of valuable solids.

*Vertical metal-leaf filter*

This filter consists of a number of vertical porous metal leaves mounted on a hollow shaft in a cylindrical pressure vessel. The solids from the slurry gradually build up on the surface of the leaves and the filtrate is removed from the plates via the horizontal hollow shaft. In some designs the hollow shaft can be slowly rotated during filtration. Solids are normally removed at the end of a cycle by blowing air through the shaft and into the filter leaves.

*Horizontal metal-leaf filter*

In this filter the metal leaves are mounted on a vertical hollow shaft within a pressure vessel. Often, only the upper surfaces of the leaves are porous. Filtration is continued until the cake fills the space between the disc-shaped leaves or when the operational pressure has become excessive. At the end of a process cycle, the solid cake can be discharged by releasing the pressure and spinning the shaft with a drive motor.

*Stacked-disc filter*

One kind of filter of this type is the metafilter. This is a very robust device and because there is no filter cloth and the bed is easily replaced, labour costs are low. It consists of a number of precision-made rings which are stacked on a fluted rod (Fig. 20.9). The rings (22 mm external diameter, 16 mm internal diameter and 0.8 mm thick) are normally made from stainless steel and precision stamped so that there are number of shoulders on one side. This ensures that there will be clearances of 0.025 mm to 0.25 mm when the rings are assembled on the rods. The assembled stacks are placed in a pressure vessel which can be sterilised if necessary. The packs are normally coated with a thin layer of Kieselguhr which is used as a filter aid. During use, the filtrate passes between the discs and is removed through the grooves of the fluted rods, while solids are deposited on the filter coating. Operation is continued until the resistance becomes too high and the solids are removed from the rings by applying back pressure via the fluted rods. Metafilters are primarily used for 'polishing' liquids such as beer.

### Continuous Filters

#### *Rotary vacuum filters*

Large rotary vacuum filters are commonly used by industries which produce large volumes of liquid which need continuous processing. The filter consists of a rotating, hollow, segmented drum covered with a fabric or metal filter which is partially immersed in a trough containing the broth to be filtered (Fig. 20.10). The slurry is fed on to the outside of the revolving drum and vacuum pressure is applied internally so that the filtrate is drawn through the filter, into the drum and finally to a collecting vessel. The interior of the drum is divided into a series of compartments, to which the vacuum pressure is normally applied for most of each revolution as the drum slowly revolves (~1 rpm). However, just before discharge of the filter cake, air pressure may be applied internally to help ease the filter cake off the drum. A number of spray jets may be carefully positioned so that water can be applied to rinse the cake. This washing is carefully controlled so that dilution of the filtrate is minimal.

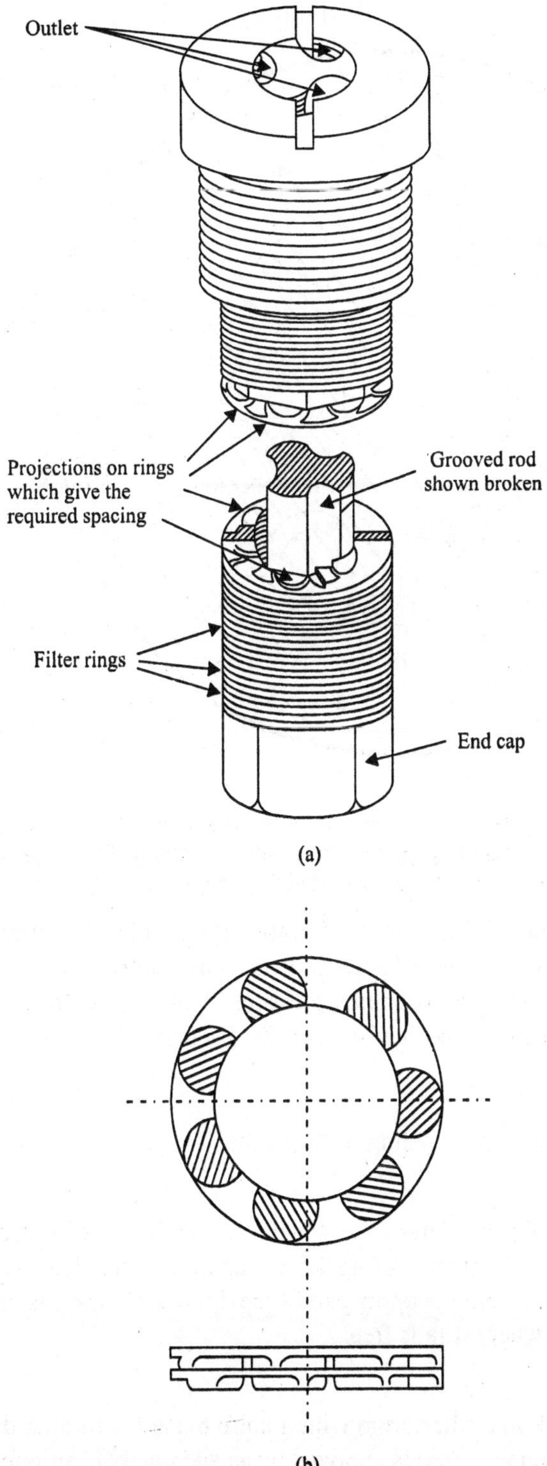


**Fig. 20.9.** (a) Metafilter pack; and (b) rings for metafilter.

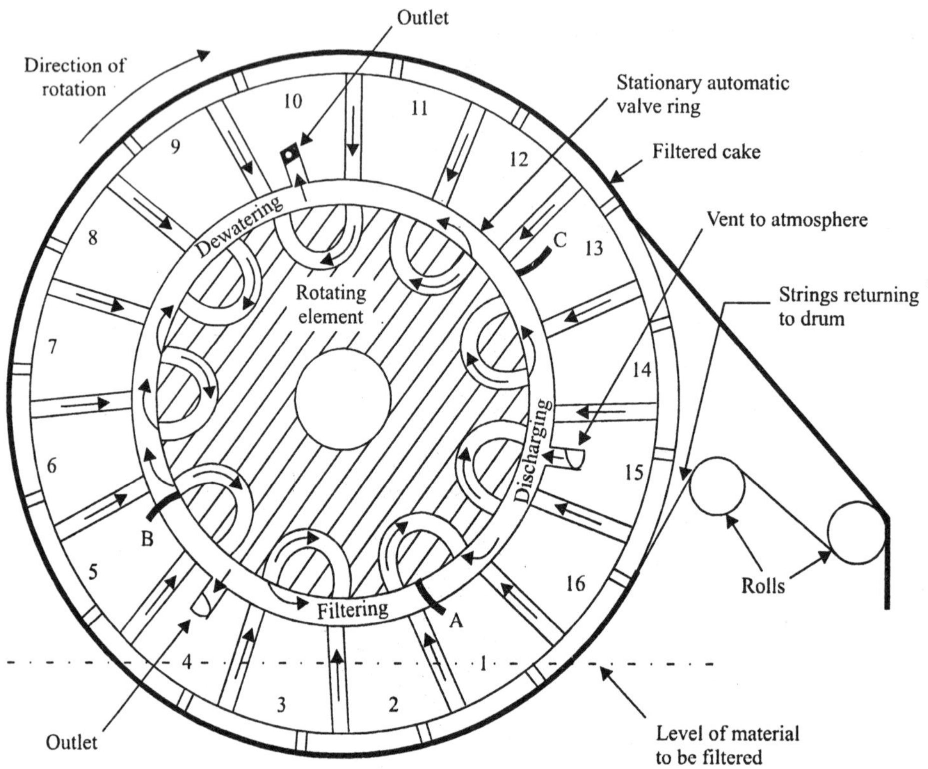


**Fig. 20.10.** Diagram of string-discharge filter operation. Sections 1 to 4 are filtering; sections 5 to 12 are dewatering; and section 13 is discharging the cake with the string discharge. Sections 14, 15 and 16 are ready to start a new cycle. A, B and C represent dividing members in the annular ring.

It should be noted that the driving force for filtration (pressure differential across the filter) is limited to one atmosphere (100 kN $m^{-2}$) and in practice it is significantly less than this. In contrast, pressure filters can be operated at many atmospheres pressure. A number of rotary vacuum drum filters are manufactured, which differ in the mechanism of cake discharge from the drum:

1. String discharge.
2. Scraper discharge.
3. Scraper discharge with pre-coating of the drum.

### *String discharge*

Fungal mycelia produce a fibrous filter cake which can easily be separated from the drum by string discharge (Fig. 20.11). Long lengths of string 1.5 cm apart are threaded over the drum and round two rollers. The cake is lifted free from the upper part of the drum when the vacuum pressure is released and carried to the small rollers where it falls free.

### *Scraper discharge*

Yeast cells can be collected on a filter drum with a knife blade for scraper discharge (Fig. 20.12). The filter cake which builds up on the drum is removed by an accurately positioned knife blade. Because the knife is close to the drum, there may be gradual wearing of the filter cloth on the drum.

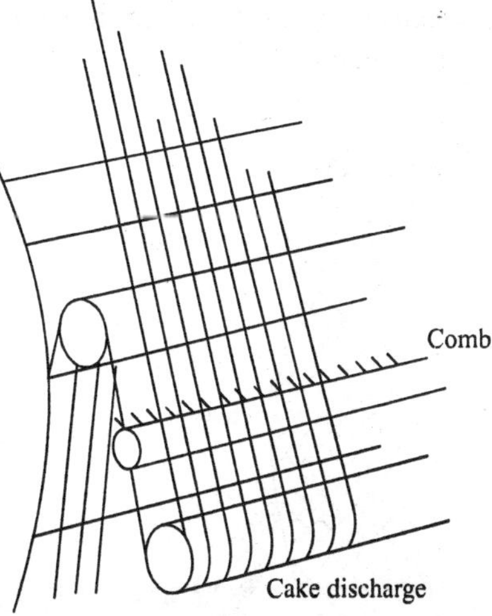


**Fig. 20.11.** Cake discharge on a drum filter using strings.

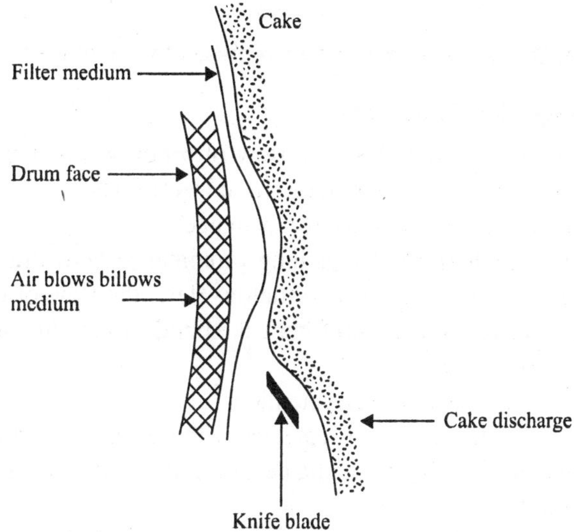


**Fig. 20.12.** Cake discharge on a drum using a scraper.

### *Scraper discharge with pre-coating of the drum*

The filter cloth on the drum can be blocked by bacterial cells or mycelia of actinomycetes. This problem is overcome by precoating the drum with a layer of filter-aid 2 to 10 cm thick. The cake which builds up on the drum during operation is cut away by the knife blade (Fig. 20.13) which mechanically advances towards the drum at a controlled slow rate. Alternatively, the blade may be operated manually when there is an indication of 'blinding' which may be apparent from a reduction in the filtration rate. In either case the cake is removed together with a very thin layer of pre-coat. The operating variables studied included drum speed, extent of drum submergence, knife advance speed and applied vacuum. The work indicated that optimisation for a new process might require prolonged trials. Although primarily used for the separation of micro-organisms from broth, studies have indicated that rotary vacuum filters can be effective in the processing of disrupted cells.

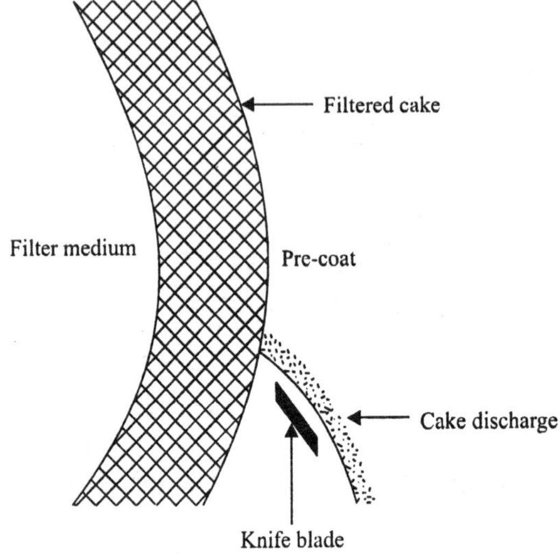


**Fig. 20.13.** Cake discharge on a pre-coated drum filter.

## Cross-flow Filtration (Tangential Filtration)

In the filtration processes previously described, the flow of broth was perpendicular to the filtration membrane. Consequently, blockage of the membrane led to lower rates of productivity and/or the need for filter aids to be added and these were serious disadvantages.

In contrast, an alternative which is rapidly gaining prominence both in the processing of whole fermentation broths and cell lysates is cross-flow filtration. Here, the flow of medium to be filtered is tangential to the membrane (Fig. 20.14a) and no filter cake builds up on the membrane.

The benefits of cross-flow filtration are:

1. Efficient separation, > 99.9 per cent cell retention.
2. Closed system; for the containment of organisms with no aerosol formation.
3. Separation is independent of cell and media densities, in contrast to centrifugation.
4. No addition of filter aid.

The major components of a cross-flow filtration system are a media storage tank (or the fermenter), a pump and a membrane pack (Fig. 20.14b). The membrane is usually in a cassette pack of hollow fibres or flat sheets in a plate and frame type stack or a spiral cartridge. In this way and by the introduction of a much convoluted surface, large filtration areas can be attained in compact devices. Two types of membrane may be used; microporous membranes with a specific pore size (0.45, 0.22 μm, etc.) or an ultrafiltration membrane with a specified molecular weight cut-off (MWCO). The type of membrane chosen is carefully matched to the product being harvested, with microporous and 1,00,000 MWCO membranes being used in cell separation.

The output from the pump is forced across the membrane surface; most of this flow sweeps the membrane, returning retained species back to the storage tank and generally less than 10 per cent of the flow passes through the membrane (permeate). As this process is continued the cells or other retained species are concentrated to between 5 and 10 per cent of their initial volume. More complex variants of the process can allow *in situ* washing of the retentate and enclosed systems for containment and sterilisation.

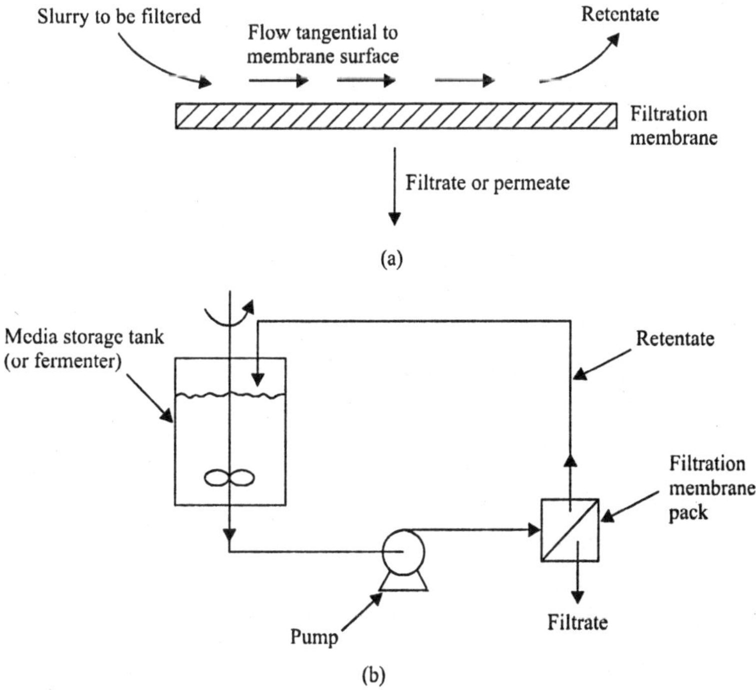


**Fig. 20.14.** (a) Schematic diagram of cross-flow filtration; and (b) major components of a cross-flow filtration system.

Many factors influence filtration rate. Increased pressure drop will, up to a point increase flow across the membrane, but it should be remembered that the system is based on a swept clean membrane. Therefore, if the pressure drop is too great the membrane may become blocked. The filtration rate is therefore, influenced by the rate of tangential flow across the membrane; by increasing the shear forces at the membrane's surface retained species are more effectively removed, thereby increasing filtration rate. Higher temperatures will increase filtration rate by lowering the viscosity of the media, though this is clearly of limited application in biological systems. Filtration rate is inversely proportional to concentration and media constituents can influence filtration rate in three ways. Low molecular weight compounds increase media viscosity and high molecular weight compounds decrease shear at the membrane surface, both leading to a reduction in filtration rate. Finally, broth constituents can 'foul' the membrane, primarily by adsorption onto the membrane's surface, causing a rapid loss in efficiency. This can be controlled by modification of the membrane or media formulation in particular by reducing the use of antifoaming agents. Lee has shown that the pulses of air injected into the flow to a cross-flow filter increase the shear rate at the membrane surface reducing the effects of membrane fouling.

## CENTRIFUGATION

Micro-organisms and other similar sized particles can be removed from a broth by using a centrifuge when filtration is not a satisfactory separation method.

Although a centrifuge may be expensive when compared with a filter it may be essential when:

1. Filtration is slow and difficult.
2. The cells or other suspended matter must be obtained free of filter aids.
3. Continuous separation to a high standard of hygiene is required.

Non-continuous centrifuges are of extremely limited capacity and therefore not suitable for large-scale separation. The centrifuges used in harvesting fermentation broths are all operated on a continuous or semi-continuous basis. Some centrifuges can be used for separating two immiscible liquids yielding a heavy phase and light phase liquid, as well as a solids fraction. They may also be used for the breaking of emulsion.

According to Stoke's law, the rate of sedimentation of spherical particles suspended in a fluid of Newtonian viscosity characteristics is proportional to the square of the diameter of the particles, thus the rate of sedimentation of a particle under gravitational force is:

$$V_g = \frac{d^2 g\,(\rho_P - \rho_L)}{18\,\mu} \qquad \text{... (20.5)}$$

where,

$V_g$ = rate of sedimentation (m s$^{-1}$)
$d$ = particle diameter (m)
$g$ = gravitational constant (m s$^{-2}$)
$\rho_P$ = particle density (kg m$^{-3}$)
$\rho_L$ = liquid density (kg m$^{-3}$)
$\mu$ = viscosity (kg m$^{-1}$ s$^{-1}$)

This equation can then be modified for sedimentation in a centrifuge:

$$V_c = \frac{d\omega^2 r\,(\rho_p - \rho_L)}{18\,\mu} \qquad \text{... (20.6)}$$

where,

$V_c$ = rate of sedimentation in the centrifuge (m s$^{-1}$)
$\omega$ = angular velocity of the rotor (s$^{-1}$)
$r$ = radial position of the particle (m)

Dividing equation (20.6) by equation (20.5) yields:

$$\frac{\omega^2 r}{g}$$

This is a measure of the separating power of a centrifuge compared with gravity settling. It is often referred to as the relative centrifugal force and given the symbol 'Z'.

It is evident from this formula that factors influencing the rate of sedimentation over which one has little or no control are the difference in density between the cells and the liquid (increased temperature would lower media density but is of little practical use with fermentation broths), the diameter of the cells (could be increased by coagulation/flocculation) and the viscosity of the liquid. Ideally, the cells should have a large diameter, there should be large density difference between cell and liquid and the liquid should have a low viscosity. In practice, the cells are usually very small, of low density and are often suspended in viscous media. Thus, it can be seen that the angular velocity and diameter of the centrifuge are the major factors to be considered when attempting to maximise the rate of sedimentation (and therefore throughput) of fermentation broths.

## Cell Aggregation and Flocculation

Following an industrial fermentation it is quite common to add flocculating agents to the broth to aid de-watering. The use of flocculating agents is widely practised in the effluent-treatment industries for the removal of microbial cells and suspended colloidal matter.

It is well known that aggregates of microbial cells, although they have the same density as the individual cells, will sediment faster because of the increased diameter of the particles (Stoke's law). The sedimentation process may be achieved naturally with selected strains of brewing yeasts, particularly if the wort is chilled at the end of fermentation and leads to a natural clearing of the beer.

Micro-organisms in solution are usually held as discrete units in three ways. Firstly, their surfaces are negatively charged and therefore repulse each other. Secondly, because of their generally hydrophilic cell walls a shell of bound water is associated with the cell which acts as a thermodynamic barrier to aggregation. Finally, due to the irregular shapes of cell walls (at the macromolecular level) steric hindrance will also play a part.

During flocculation one or more mechanisms besides temperature can induce cell flocculation:

1. Neutralisation of anionic charges, primarily carboxyl and phosphate groups, on the surfaces of the microbial cells, thus allowing the cells to aggregate. These include changes in the pH and the presence of a range of compounds which alter the ionic environment.
2. Reduction in surface hydrophilicity.
3. The use of high molecular weight polymer bridges. Anionic, non-ionic and cationic polymers can be used, though the former two also require the addition of a multivalent cation.

Flocculation usually involves the mixing of a process fluid with the flocculating agent under conditions of high shear in a stirred tank, although more compact and efficient devices have been proposed. This stage is known as coagulation and is usually followed by a period of gentle agitation when flocs developed initially are allowed to grow in size.

Nakamura described the use of various compounds for flocculating bacteria, yeasts and algae, including alum, calcium salts and ferric salts. Other agents which are now used include tannic acid, titanium tetrachloride and cationic agents such as quaternary ammonium compounds, alkyl amines and alkyl pyridinium salts. Gasner and Wang reported a many hundred-fold increase in the sedimentation rate of *Candida intermedia* when recoveries of over 99 per cent were readily obtained. They found that flocculation was very dependent on the choice of additive, dosage and conditions of floc formation, with the most effective agents being mineral colloids and polyelectrolytes. Nucleic acids, polysaccharides and proteins released from partly lysed cells may also bring about agglomeration. In SCP processes, phosphoric acid has been used as a flocculating agent since it can be used as a nutrient in medium recycle with considerable savings in water usage.

The majority of flocculating agents currently in use are polyelectrolytes, which act by charge neutralisation and hydrophobic interactions to link cells to each other. In processes where the addition of some toxic chemicals is to be avoided, alternative techniques have been adopted. One method is to coagulate microbial protein which has been released from the cells by heating for short periods. Kurane reported the use of bioflocculants obtained from *Rhodococcus erythropolis*. They are suggested as being safer alternatives to conventional flocculants. Warne and Bowden suggest the use of genetic manipulation to alter cell surface properties to aid aggregation. Flocculating agents such as cross-linked cationic polymers may also be used in the processing of cell lysates and extracts prior to further downstream processing. Bentham utilised borax as a flocculating agent for yeast cell debris prior to decanter centrifugation.

## Range of Centrifuges

A number of centrifuges will be described which vary in their manner of liquid and solid discharge, their unloading speed and their relative maximum capacities. When choosing a centrifuge for a specific process it is important to ensure that the centrifuge will be able to perform the separation at the planned production rate and operate reliably with minimum manpower. Large-scale tests may therefore be necessary with fermentation broths or other materials to check that the correct centrifuge is chosen.

### *Basket centrifuge (Perforated-bowl basket centrifuge)*

Basket centrifuges are useful for separating mould mycelia or crystalline compounds. The centrifuge is most commonly used with a perforated bowl lined with a filter bag of nylon, cotton, etc. (Fig. 20.15). A continuous feed is used and when the basket is filled with the filter cake it is possible to wash the cake before removing it. The bowl may suffer from blinding with soft biological materials so that high centrifugal forces cannot be used. These centrifuges are normally operated at speeds of up to 4000 rpm for feed rates of 50 to 300 $dm^3 min^{-1}$ and have a solids holding capacity of 30 to 500 $dm^3$. The basket centrifuge may be considered to be a centrifugal filter.

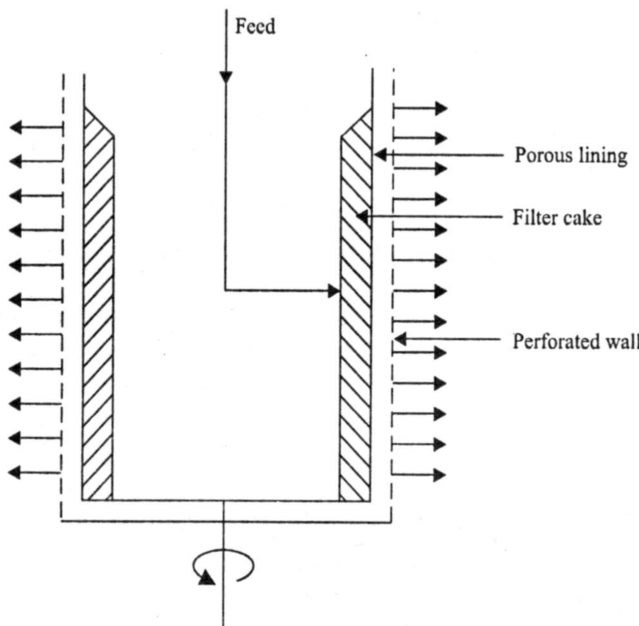


**Fig. 20.15.** Diagram of basket centrifuge.

### *Tubular-bowl centrifuge*

This is a centrifuge to consider using for particle size ranges of 0.1 to 200 μm and up to 10 per cent solids in the in-going slurry. Figure 20.16 shows an arrangement used in a Sharples super-centrifuge. The main component of the centrifuge is a cylindrical bowl (or rotor) (A in Fig. 20.16), which may be of a variable design depending on application, suspended by a flexible shaft (B), driven by an overhead motor or air turbine (C). The inlet to the bowl is via a nozzle attached to the bottom bearing (D). The feed which may consist of solids and light and heavy liquid phases is introduced by the nozzle (E). During operation solids sediment on the bowl wall while the liquids separate into the heavy phase in zone (G) and the light phase in the central zone (H). The two liquid phases are kept separate in their exit from the bowl by an adjustable ring, with the heavy phase flowing over the lip of the ring. Rings of various sizes may be fitted for the separation of liquids of various relative densities. Thus the centrifuge may be altered to use for:

1. Light-phase/heavy-phase liquid separation.
2. Solids/light-liquid phase/heavy-liquid phase separation.
3. Solids/liquid separation (using a different rotor).

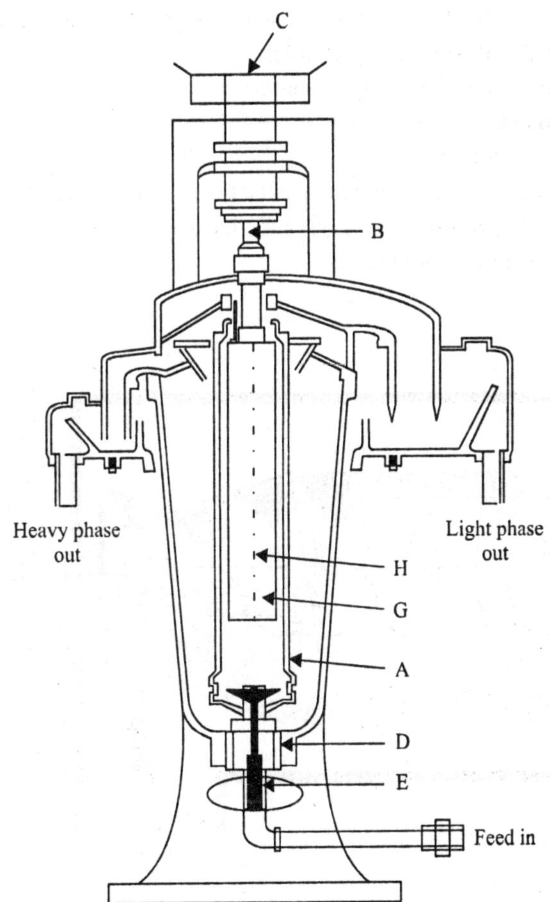


**Fig. 20.16.** Section of a Sharples super-centrifuge.

Advantages of this design of centrifuge are: (i) high centrifugal force; (ii) good dewatering; and (iii) ease of cleaning.

Disadvantages are limited solids capacity, difficulties in the recovery of collected solids, gradual loss in efficiency as the bowl fills, solids being dislodged from the walls as the bowl is slowing down and foaming. Plastic liners can be used in the bowls to help improve batch cycle time. Alternatively a spare bowl can be changed over in about 5 minutes.

### Solid-bowl scroll centrifuge (Decanter centrifuge)

This type of centrifuge is used for continuous handling of fermentation broths, cell lysates and coarse materials such as sewage sludge (Fig. 20.17). The slurry is fed through the spindle of an archimedean screw within the horizontal rotating solids bowl. Typically the speed differential between the bowl and the screw is in the range 0.5 to 100 rpm. The solids settling on the walls of the bowl are scraped to the conical end of the bowl. The slope of the cone helps to remove excess liquid from the solids before discharge. The liquid phase is discharged from the opposite end of the bowl. The speed of this type of centrifuge is limited to around 5000 rpm in larger models because of the lack of balance within the bowl, with smaller models having bowl speeds of up to 10000 rpm. Bowl diameters are normally between 0.2 and 1.5 meters, with the length being up to five times the diameter. Feed rates range from around 200 $dm^3$ $hr.^{-1}$ to 200 $m^3$ $hr.^{-1}$ depending on scale of operation and material being processed. A number of variants on the basic design are available:

1. Cake washing facilities (screen bowl decanters).
2. Vertical bowl decanters.
3. Facility for in-place cleaning.
4. Bio-hazard containment features; steam sterilisation *in situ*, two or three stage mechanical seals, control of aerosols, containment casings and the use of high pressure sterile gas in seals to prevent the release of micro-organisms.

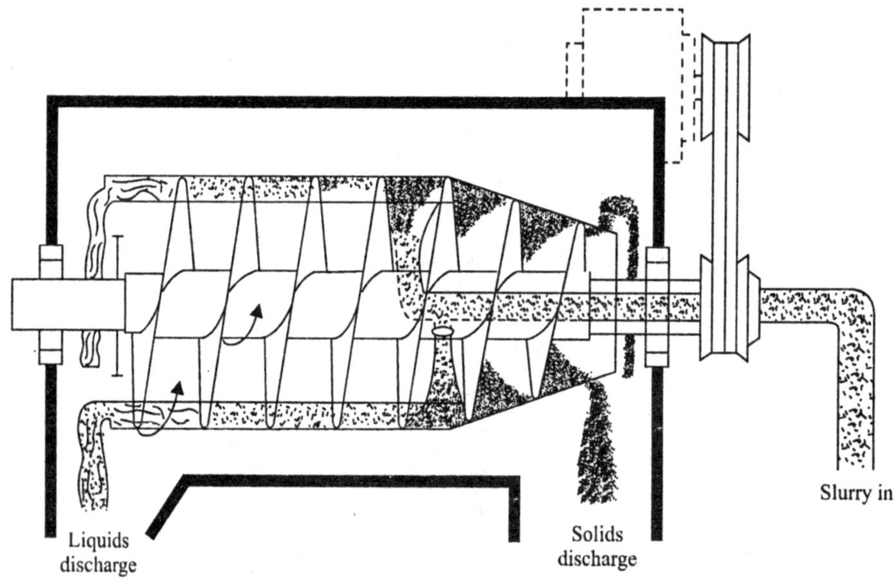


**Fig. 20.17.** Diagram of a solid-bowl scroll centrifuge.

### Multichamber centrifuge

Ideally, this is a centrifuge for a slurry of up to 5 per cent solids of particle size 0.1 to 200 μm diameter. In the multichamber centrifuge (Fig. 20.18), a series of concentric chambers are mounted within the rotor chamber. The broth enters via the central spindle and then takes a circuitous route through the chambers. Solids collect on the outer faces of each chamber. The smaller particles collect in the outer chambers where they are subjected to greater centrifugal forces (the greater the radial position of a particle, the greater the rate of sedimentation).

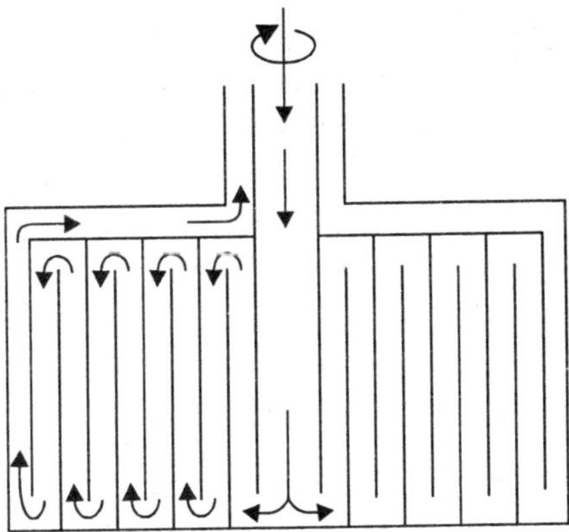

**Fig. 20.18.** LS of a multichamber centrifuge.

Although these vessels can have a greater solids capacity than tubular bowls and there is no loss of efficiency as the chamber fills with solids, their mechanical strength and design limits their speed to a maximum of 6500 rpm for a rotor 46 cm diameter with a holding capacity of up to 76 $dm^3$. Because of the time needed to dismantle and recover the solids fraction, the size and number of vessels must be of the correct volume for the solids of a batch run.

### *Disc-bowl centrifuge*

This centrifuge relies for its efficiency on the presence of discs in the rotor or bowl (Fig. 20.19). A central inlet pipe is surrounded by a stack of stainless-steel conical discs. Each disc has spacers so that a stack can be built up. The broth to be separated flows outwards from the central feed pipe, then upwards and inwards between the discs at an angle of 45° to the axis of rotation. The close packing of the discs assists rapid sedimentation and the solids then slide to the edge of the bowl, provided that there are no gums or fats in the slurry and eventually accumulate on the inner wall of the bowl. Ideally, the sediment should form a sludge which flows, rather than a hard particulate or lumpy sediment. The main advantages of these centrifuges are their small size compared with a bowl without discs for a given throughput. Some designs also have the facility for continuous solids removal through a series of nozzles in the circumference of the bowl or intermittent solids removal by automatic opening of the solids collection bowl. The arrangement of the discs makes this type of centrifuge laborious to clean. However, recent models are designed to allow for cleaning *in situ*. In addition this and similar plant have the facility for *in situ* steam sterilisation and total containment, incorporating double seals to comply with containment regulations. Feed rates range from 45 to 1800 $dm^3$ $min^{-1}$, with rotational speeds typically between 500 and 10000 rpm.

## CELL DISRUPTION

Micro-organisms are protected by extremely tough cell walls. In order to release their cellular contents a number of methods for cell disintegration have been developed. Any potential method of disruption must ensure that labile materials are not denatured by the process or hydrolysed by enzymes present in the cell. Huang report the use of a combination of different techniques to release products from specific locations within yeast cells. In this way the desired product can be obtained with minimum contamination. Although many techniques are available which are satisfactory at laboratory scale, only a limited number have been proved to be suitable for large-scale applications, particularly for intracellular enzyme extraction. Containment of cells can be difficult or costly to achieve in many of the methods described below and thus containment requirements will strongly influence process choice. Methods available fall into two major categories:

Physico-mechanical methods:

1. Liquid shear.
2. Solid shear.
3. Agitation with abrasives.
4. Freeze-thawing.
5. Ultrasonication.

Chemical methods:

1. Detergents.
2. Osmotic shock.
3. Alkali treatment.
4. Enzyme treatment.

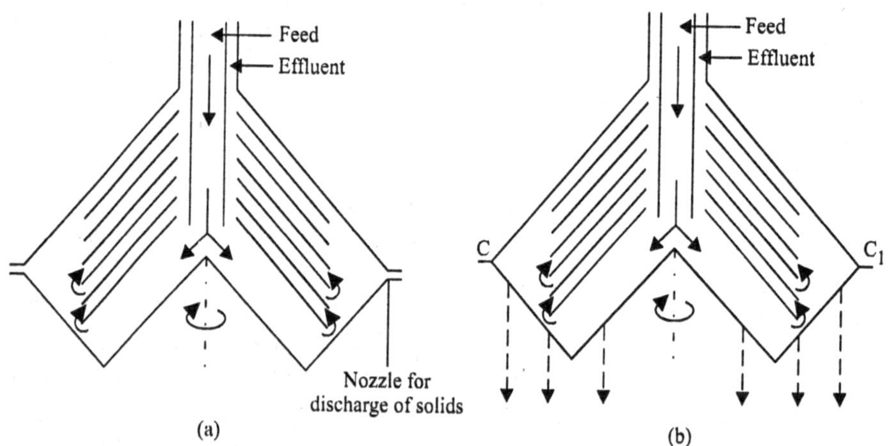


**Fig. 20.19.** (a) LS of disc-bowl centrifuge with nozzle discharge; and (b) LS of disc-bowl centrifuge with intermittent discharge. (Solids discharged when rotor opens intermittently along the section $C–C_1$.)

## Physical-Mechanical Methods

### *Liquid shear*

Liquid shear is the method which has been most widely used in large scale enzyme purification procedures. High-pressure homogenisers used in the processing of milk and other products in the food industry have proved to be very effective for microbial cell disruption. A high-pressure positive displacement pump, incorporates an adjustable valve with a restricted orifice (Fig. 20.20). The smallest model has one plunger, while there are several in larger models. During use, the microbial slurry passes through a non-return valve and impinges against to operative valve set at the selected operating pressure. The cells then pass through a narrow channel between the valve and an impact ring followed by a sudden pressure drop at the exit to the narrow orifice. The large pressure drop across the valve is believed to cause cavitation in the slurry and the shock waves so produced disrupt the cells. Brookman considered the size of the pressure drop to be very important in achieving effective disruption and as with all

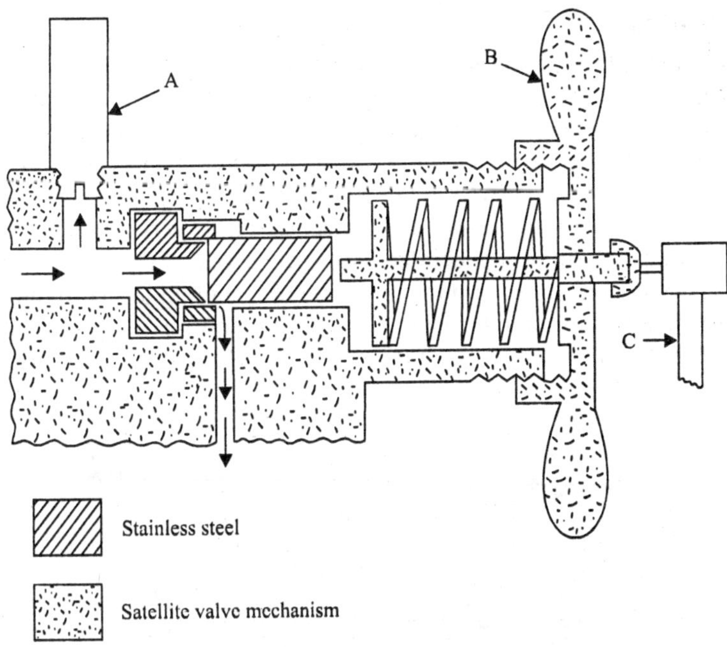


**Fig. 20.20.** Details of homogeniser valve assembly (A) 0–50,000 psi pressure transducer; (B) pressure-control handwheel; (C) linear variable displacement transformer; ( → ) direction of flow.

mechanical methods, cell size and shape influence ease of disruption. The working pressure are extremely high. Hetherington used a pressure of 550 kg $cm^{-2}$ for a 60 per cent yeast suspension. A throughput of 6.4 kg soluble protein $hr.^{-1}$ with 90 per cent disruption could be achieved with a small industrial machine. In larger models, flow rates of up to 600 $dm^3$ $hr.^{-1}$ are now possible and operating pressures of 1200 bar are utilised in some processes. Darbyshire has stressed the need for cooling the slurry to between 0 and 4°C to minimise loss in enzyme activity because of heat generation during the process. The increase in slurry temperature is approximately proportional to the pressure drop across the valve. Because of problems caused by heat generation and because cell suspensions can be surprisingly abrasive, it is common practice to operate such homogenisers in a multi-pass mode but at a lower pressure.

The degree of disruption and consequently the amount of protein released will influence the ease of subsequent separation of the product from the cell debris in high-pressure homogenisers and bead mills. A careful balance must therefore be made between percentage release of product and the difficulty and cost of further product purification.

### *Solid shear*

Pressure extrusion of frozen micro-organisms at around –25°C through a small orifice is a well established technique at a laboratory scale using a Hughes press or an X-press to obtain small samples of enzymes or microbial cell walls. Disruption is due to a combination of liquid shear through a narrow orifice and the presence of ice crystals. Magnusson and Edebo developed a semi-continuous X-press operating with a sample temperature of –35°C and an X-press temperature of –20°C. It was possible to obtain 90 per cent disruption with a single passage of *S. cerevisiae* using a throughput of 10 kg yeast cell paste $hr^{-1}$. This technique might be ideal for microbial products which are very temperature labile.

### *Agitation with abrasives*

Mechanical cell disruption can also be achieved in a disintegrator containing a series of rotating discs and a charge of small beads. The beads are made of mechanically resistant materials such as glass, alumina ceramics and some titanium compounds (Fig. 20.21). In a small disintegrator, the Dyno-Muhle KD5, using a flow rate of 180 $dm^3$ $hr.^{-1}$, 85 per cent disintegration of an 11 per cent w/v suspension of *S. cerevisiae* was achieved with a single pass. Although temperatures of up to 35°C were recorded in the disintegrator, the specific enzyme activities were not considered to be very different from values obtained by other techniques. Dissipation of heat generated in the mill is one of the major problems in scale-up, though this can generally be overcome with the provision of a cooling jacket. In another disintegrator, the agitator blades were alternately mounted vertically and obliquely on the horizontal shaft (Fig. 20.22). A flow rate of up to 400 $dm^3$ $hr.^{-1}$ was claimed for a vessel with a nominal capacity of 20 $dm^3$.

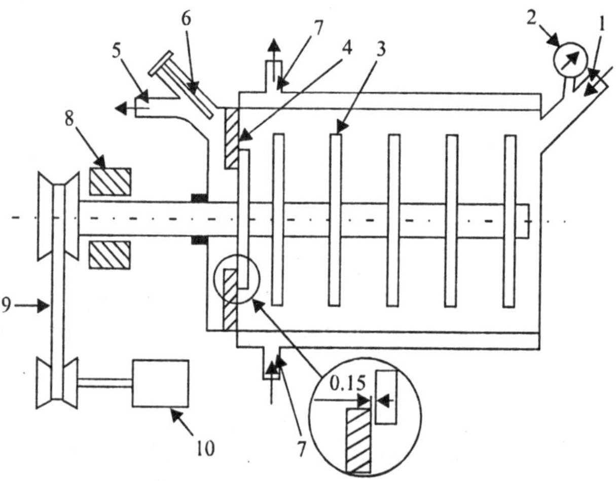


**Fig. 20.21.** Simplified drawing of the Dyno-Muhle KD5 (1) Inlet of suspension; (2) manometer; (3) rotating disc; (4) slit for separation of glass beads from the suspension; (5) outlet of suspension; (6) thermometer; (7) cooling water, inlet and outlet; (8) bearings; (9) variable V-belt drive; and (10) drive motor. Cylinder dimensions: inside length 33 cm; inside diameter 14 cm.

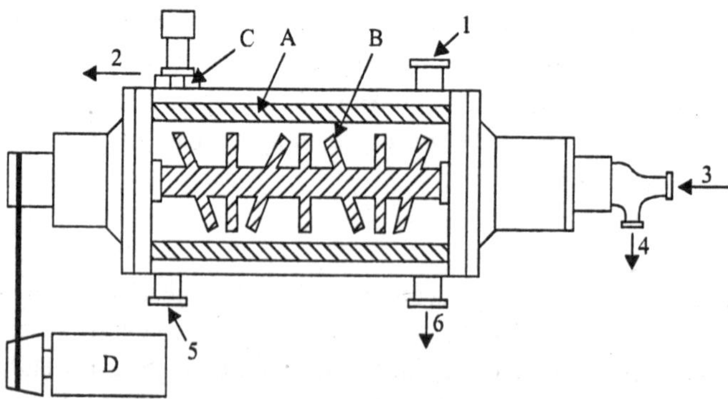


**Fig. 20.22.** Simplified drawing of the Netzsch model LM-20 mill: A, cylidrical grinding vessel with cooling jacket; B, agitator with cooled shaft and discs; C, annular vibrating slot operator; D, variable-speed drive motor; 1 and 2, product inlet and outlet; 3 and 4, agitator cooling inlet and outlet; 5 and 6, vessel-cooling inlet and outlet.

### *Freezing-thawing*

Freezing and thawing of a microbial cell paste will inevitably cause ice crystals to form and their expansion followed by thawing will lead to some subsequent disruption of cells. It is slow, with limited release of cellular materials and has not often been used as a technique on its own, although it is often used in combination with other techniques. β-Glucosidase has been obtained from *S. cerevisiae* by this method. A sample of 360 grams of frozen yeast paste was thawed at 5°F for 10 hours. This cycle was repeated twice before further processing.

### *Ultrasonication*

High frequency vibration (~ 20 kHz) at the tip of an ultrasonication probe leads to cavitation and shock waves thus produced cause cell disruption. The method can be very effective on a small scale, but a number of serious drawbacks make it unsuitable for large-scale operations. Power requirements are high, there is a large heating effect so cooling is needed, the probes have a short working life and are only effective over a short range. Continuous laboratory sonicators with hold-up volumes of around 10 $cm^3$ have been shown to be effective.

## Chemical Methods

### *Detergents*

A number of detergents will damage the lipoproteins of the microbial cell membrane and lead to release of intracellular components. The compounds which can be used for this purpose include quaternary ammonium compounds, sodium lauryl sulphate, sodium dodecyl sulphate (SDS) and Triton X-100. Unfortunately, the detergents may cause some protein denaturation and may need to be removed before further purification stages can be undertaken. Pullulanase is an enzyme which is bound to the other membrane of *Klebsiella pneumoniae*. The cells were suspended in pH 7.8 buffer and 1 per cent sodium cholate was added. The mixture was stirred for 1 hour to solubilise most of the enzyme. The use of Triton X-100 in combination with guanidine-HCl is widely and effectively used for the release of cellular protein. Hettwer and Wang obtaining greater than 75 per cent protein release in less than one hour from *Escherichia coli* under fermentation conditions.

### *Osmotic shock*

Osmotic shock caused by a sudden change in salt concentration will cause disruption of a number of cell types. However, the effect on microbial cells is normally minimal. It has proved to be a successful technique for the extraction of luciferase from *Photobacterium fischeri*. A batch of 120 $dm^3$ of broth was harvested and the cells collected as a cell paste in a Sharples centrifuge. Enzyme extraction was achieved by osmotic lysis using a ratio of 1 gram of cell paste to 4 $cm^3$ of cold distilled water with stirring for 15 to 30 minutes. A second extraction gave a small additional yield of enzyme. Only low levels of soluble protein were released using this technique.

### *Alkali treatment*

Alkali treatment might be used for hydrolysis of microbial cell wall material provided that the desired enzyme will tolerate a pH of 11.5 to 12.5 for 20 to 30 minutes. Darbyshire has reported the use of this technique in the extraction of L-asparaginase.

### *Enzyme treatment*

There are a number of enzymes which hydrolyse specific bonds in cell walls of a limited number of micro-organisms. Enzymes shown to have this activity include lysozyme and enzyme extracts from leucocytes, *Streptomyces* spp., *Micromonospora* spp., *Penicillium* spp., *Trichoderma* spp. and snails. Although this is probably one of the most gentle methods available unfortunately it is relatively expensive and the presence of the enzyme (s) may complicate further downstream purification processes. The use of immobilised lysozyme has been investigated by a number of workers and may provide the solution to such problems.

Chemical and enzymic methods for the release of intracellular products have not been used widely on a large scale, with the exception of lysozyme. However, their potential for the selective release of product and that they often yield a cleaner lysate mean that they are potentially invaluable tools in the recovery of fermentation products.

Enzymes may also be used as a pre-treatment to partially hydrolyse cell walls prior to cell disruption by mechanical methods.

## LIQUID-LIQUID EXTRACTION

The separation of a component from a liquid mixture by treatment with a solvent in which the desired component is preferentially soluble is known as liquid-liquid extraction. The specific requirement is that a high percentage extraction of product must be obtained but concentrated in a smaller volume of solvent.

Prior to starting a large-scale extraction, it is important to find out on a small scale the solubility characteristics of the product using a wide range of solvents. A simple rule to remember is that 'like dissolves like'. The important 'likeness' as far as solubility relations are concerned is in the polarities of molecules. Polar liquids mix with each other and dissolve salts and other polar solids. The solvents for non-polar compounds are liquids of low or nil polarity.

The dielectric constant is a measure of the degree of molar polarisation of a compound. If this value is known it is then possible to predict whether a compound will be polar or non-polar, with a high value indicating a highly polar compound. The dielectric constant $D$ of a substance can be measured by determining the electrostatic capacity $C$ of a condenser containing the substance between the plates. If $C_0$ is the value for the same condenser when completely evacuated then:

$$D = \frac{C}{C_0}$$

Experimentally, dielectric constants are obtained by comparing the capacity of the condenser when filled with a given liquid with the capacity of the same condenser containing a standard liquid whose dielectric constant is known very accurately. If $D_1$ and $D_2$ are the dielectric constants of the experimental and standard liquids and $C_1$ and $C_2$ are the electrostatic capacities of a condenser when filled with each of the liquids, then:

$$\frac{D_1}{D_2} = \frac{C_1}{C_2}$$

The value of $D_1$ can be calculated since $C_1$ and $C_2$ can be measured and $D_2$ is known. The dielectric constants for a number of solvents are given in Table 20.1.

**Table 20.1.** Dielectric constants of solvents at 25°C (arranged in order of increasing polarity).

| *Solvent* | *Dielectric constant* |
|---|---|
| Hexane | 1.90 (least polar) |
| Cyclohexane | 2.02 |
| Carbon tetrachloride | 2.24 |
| Benzene | 2.28 |
| Diethyl ether | 4.34 |
| Chloroform | 4.87 |
| Ethyl acetate | 6.02 |
| Butan-2-ol | 15.8 |
| Butan-1-ol | 17.8 |
| Propan-1-ol | 20.1 |
| Acetone | 20.7 |
| Ethanol | 24.3 |
| Methanol | 32.6 |
| Water | 78.5 (Most polar) |

The final choice of solvent will be influenced by the distribution or partition coefficient $K$ where:

$$K = \frac{\text{Concentration of solute in extract}}{\text{Concentration of solute in raffinate}}$$

The value of $K$ defines the ease of extraction. When there is a relatively high $K$ value, good stability of product and good separation of the aqueous and solvent phases, then it may be possible to use a single-stage extraction system (Fig. 20.23). A value of 50 indicates that the extraction should be straightforward whereas a value of 0.1 shows that the extraction will be difficult and that a multistage process will be necessary.

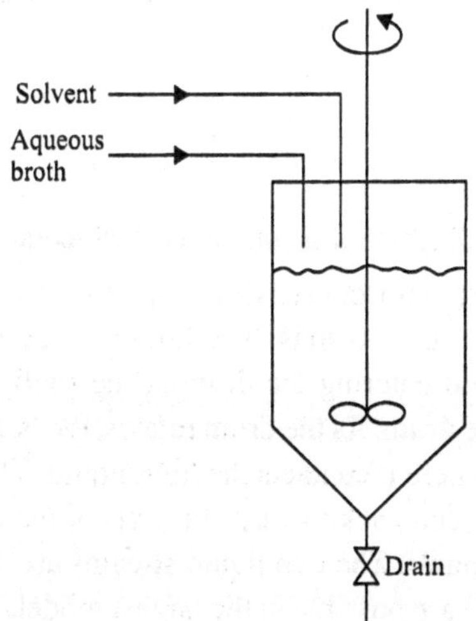


**Fig. 20.23.** Diagram of a single-stage extraction unit.

Unfortunately, in a number of systems the value of K is low and co-current or counter-current multistage systems have to be utilised. The co-current system is illustrated in Fig. 20.24. There are n mixer/separator vessels in line and the raffinate goes from vessel 1 to vessel n. Fresh solvent is added to each stage, the feed and extracting solvent pass through the cascade in the same direction. Extract is recovered from each stage. Although a relatively large amount of solvent is used, a high degree of extraction is achieved.

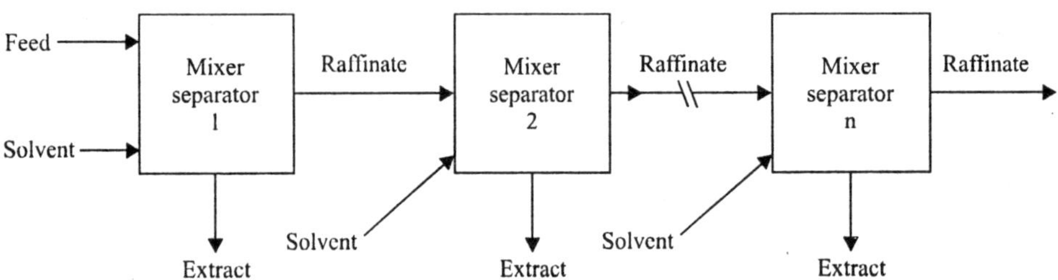


**Fig. 20.24.** Diagram of a co-current flow extraction system.

A counter-current system is illustrated in Fig. 20.25. There are a number of mixers/separators connected in series. The extracted raffinate passes from vessel 1 to vessel n while the product-enriched solvent is flowing from vessel n to vessel 1. The feed and extracting solvent pass through the cascade in opposite directions. The most efficient system for solvent utilisation is counter-current operation, showing a considerable advantage over batch and co-current systems. Unless there are special reasons the counter-current system should be used.

In practice, the series of counter-current extractions are conducted in a single continuous extractor using centrifugal forces to separate the two liquid phases. The two liquid streams are forced to flow counter-current to each other through a long spiral of channels within the rotor.

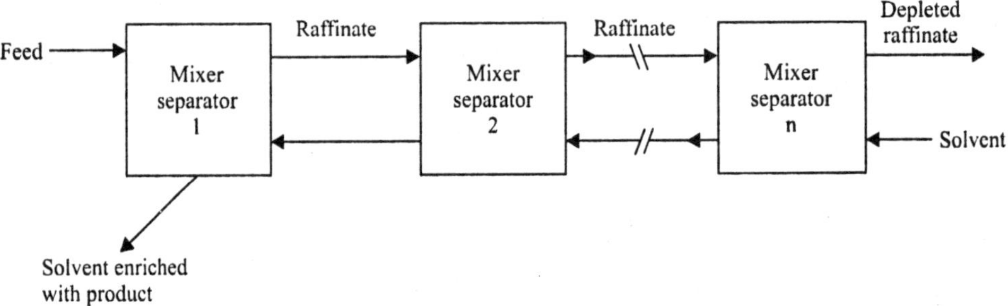


**Fig. 20.25.** Diagram of a counter-current extraction system.

The Podbielniak centrifugal extractor consists of a horizontal cylindrical drum revolving at up to 5000 rpm about a shaft passing through its axis. The liquids to be run counter-current are introduced into the shaft, with the heavy liquid entering the drum at the shaft while the light liquid is led by an internal route to the periphery of the drum. As the drum rotates, the heavy liquid is forced to the periphery of the drum by centrifugal action where it contacts the light liquid. The solute is transferred between the liquids and the light liquid is displaced back towards the axis of the drum. The heavy liquid is returned to the drum's axis via internal channels. The two liquid streams are then discharged via the shaft. Flow rates in excess of 1,00,000 $dm^3$ $hr.^{-1}$ are possible in the largest models. Probably the most useful property of this type of extractor is the low hold-up volume of liquid in the machine compared with the throughput.

Penicillin G is an antibiotic which is recovered from fermentation broths by centrifugal counter-current solvent extraction. At neutral pHs in water penicillin is ionised:

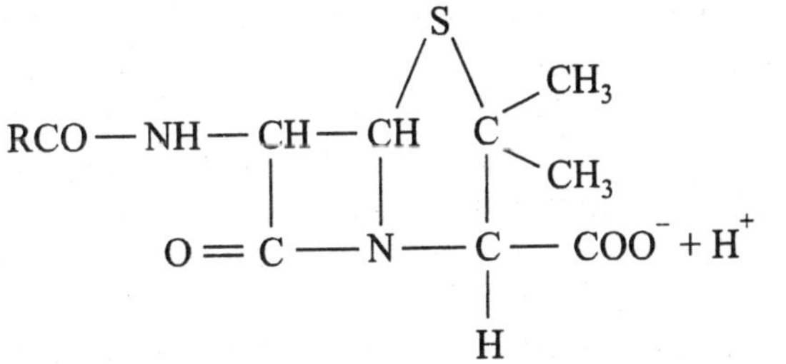

$$
\begin{array}{ccccccccc}
 & & & & & \text{S} & & & \\
 & & & & \diagup & & \diagdown & & \text{CH}_3 \\
\text{RCO}-\text{NH}- & \text{CH} & - & \text{CH} & & & & \text{C} & \\
 & | & & | & & & & | & \text{CH}_3 \\
\text{O}= & \text{C} & - & \text{N} & - & & & \text{C} & -\ \text{COO}^{-} + \text{H}^{+} \\
 & & & & & & & | & \\
 & & & & & & & \text{H} &
\end{array}
$$

In acid conditions this ionisation is suppressed and the penicillin is more soluble in organic solvent. At pH 2 to 3 the distribution ratio of total acid will be:

$$K = \frac{\text{(RCOOH)org}}{\text{(RCOOH)aq} + \text{(RCOO )aq}}$$

For penicillin this value may be as high as 40 in a suitable solvent. The penicillin extraction process may involve the four following stages:

1. Extraction of the penicillin G from the filtered broth into an organic solvent (amyl or butyl acetate of methyl iso-butyl ketone).
2. Extraction from the organic solvent into an aqueous buffer.
3. Extraction from aqueous buffer into organic solvent.
4. Extraction of the solvent to obtain the penicillin salt.

At each extraction stage progressively smaller volumes of extractant are used to achieve concentration of the penicillin (see also Fig. 20.2). Unfortunately, penicillin G has a half-life of 15 minutes at pH 2.0 at 20°F. The harvested broth is therefore initially cooled to 0°F to 3°F. The cooled broth is then acidified to pH 2 to 3 with sulphuric or phosphoric acid immediately before extraction. This acidified broth is quickly passed through a Podbielniak centrifugal counter-current extractor using about 20 per cent by volume of the solvent in the counter flow. Ideally, the hold-up time should be about 60 to 90 seconds. The penicillin-rich solvent then passes through a second Podbielniak extractor counter-current to an aqueous NaOH or KOH solution (again about 20 per cent by volume) so that the penicillin is removed to the aqueous phase (pH 7.0 to 8.0) as the salt.

$$\text{RCOOH (org)} + \text{NaOH(aq)} \longrightarrow \text{RCOO}^{-}\text{Na}^{+} + \text{H}_2\text{O}$$

These two stages may be sufficient to concentrate the penicillin adequately from a broth with a high titre. Penicillin will crystallise out of aqueous solution at a concentration of approximately $1.5 \times 10^6$ units $cm^{-3}$. If the broth harvested initially contains 60,000 units $cm^{-3}$ and two five-fold concentrations are achieved in the two extraction stages, then the penicillin liquor should crystallise. If the initial broth titre is lower than 60000 units $cm^{-3}$ or the extractions are not so effective, the solvent and buffer extractions will have to be repeated.

At each stage the spent liquids should be checked for residual penicillin and solvent usage carefully monitored. Since the solvents are expensive and their disposal is environmentally sensitive they are recovered for recirculation through the extraction process. The success of a process may depend on efficient solvent recovery and re-use.

## SOLVENT RECOVERY

A major item of equipment in an extraction process is the solvent-recovery plant which is usually a distillation unit. It is not normally essential to remove all the raffinate from the solvent as this will be recycled through the system. In some processes the more difficult problem will be to remove all the solvent from the reaffinate because of the value of the solvent and problems which might arise from contamination of the product. Distillation may be achieved in three stages:

1. Evaporation, the removal of solvent as a vapour from a solution.
2. Vapour-liquid separation in a column, to separate the lower boiling more volatile component from other less volatile components.
3. Condensation of the vapour, to recover the more volatile solvent fraction.

Evaporation is the removal of solvent from a solution by the application of heat to the solution. A wide range of evaporators is available. Some are operated on a batch basis and others continuously. Most industrial evaporators employ tubular heating surfaces. Circulation of the liquid past the heating surfaces may be induced by boiling or by mechanical agitation. In batch distillation (Fig. 20.26) the vapour from the boiler passes up to column and is condensed. Part of the condensate will be returned as the reflux for counter-current contact with the rising vapour in the column. The distillation is continued until a satisfactory recovery of the lower-boiling (more volatile) component(s) has been accomplished. The ratio of condensate returned to the column as reflux to that withdrawn as product is, along with the number of plates or stages in the column, the major method of controlling the product purity. A continuous distillation (Fig. 20.27) is initially begun in a similar way as with a batch distillation, but no condensate is withdrawn initially. There is total reflux of the condensate until ideal operating conditions have been established throughout the column. At this stage the liquid feed is fed into the column at an intermediate level. The more volatile components move upwards as vapour and are condensed, followed by partial reflux of the condensate. Meanwhile, the less volatile fractions move down the column to the evaporator (re-boiler). At this stage part of the bottoms fraction is continuously withdrawn and part is reboiled and returned to the column. Counter-current contacting of the vapour and liquid streams is achieved by causing:

1. Vapour to be dispersed in the liquid phase (plate or tray column).
2. Liquid to be dispersed in a continuous vapour phase (packed column).

The plate or tray column consists of a number of distinct chambers separated by perforated plates or trays. The rising vapour bubbles through the liquid which is flowing across each plate and is dispersed into the liquid from perforations (sieve plates) or bubble caps. The liquid flows across the plates and reaches the re-boiler by a series of overflow wiers and down pipes.

A packed tower is filled with a randomly packed material such as rings, saddles, helices, spheres or beads. Their dimensions are approximately one-tenth to one-fiftieth of the diameter of the column and are designed to provide a large surface area for liquid-vapour contacting and high voidage to allow high throughput of liquid and vapour.

The heat input to a distillation column can be considerable. The simplest ways of conserving heat are to pre-heat the initial feed by a heat exchanger using heat from:

1. The hot vapours at the top of the column.
2. Heat from the bottoms fraction when it is being removed in a continuous process.
3. A combination of both.

## TWO-PHASE AQUEOUS EXTRACTION

Liquid-liquid extraction is a well established technology in chemical processing and in certain sectors of biochemical processing. However, the use of organic solvents has limited application in the processing of sensitive biologicals. Aqueous two-phase systems, on the other hand, have a high water content and low interfacial surface tension and are regarded as being biocompatible. Two-phase aqueous systems have been known since the late nineteenth century and a large variety of natural and synthetic hydrophilic polymers are used today to create two (or more) aqueous phases. Phase separation occurs when hydrophilic polymers are added to an aqueous solution and when the concentrations exceed a certain value two immiscible aqueous phases are formed. Settling time for the two phases can be prolonged, depending on the components used and vessel geometry. Phase separation can be improved by using centrifugal separators or novel techniques such as magnetic separators. Many systems are available:

1. Non-ionic polymer/non-ionic polymer/water, e.g., polyethylene glycol/dextran.
2. Polyelectrolyte/non-ionic polymer/water, e.g., sodium carboxymethyl cellulose/polyethylene glycol.
3. Polyelectrolyte/polyelectrolyte/water, e.g., sodium dextran sulphate/sodium carboxymethyl cellulose.
4. Polymer/low molecular weight component/water, e.g., dextran/propyl alcohol.

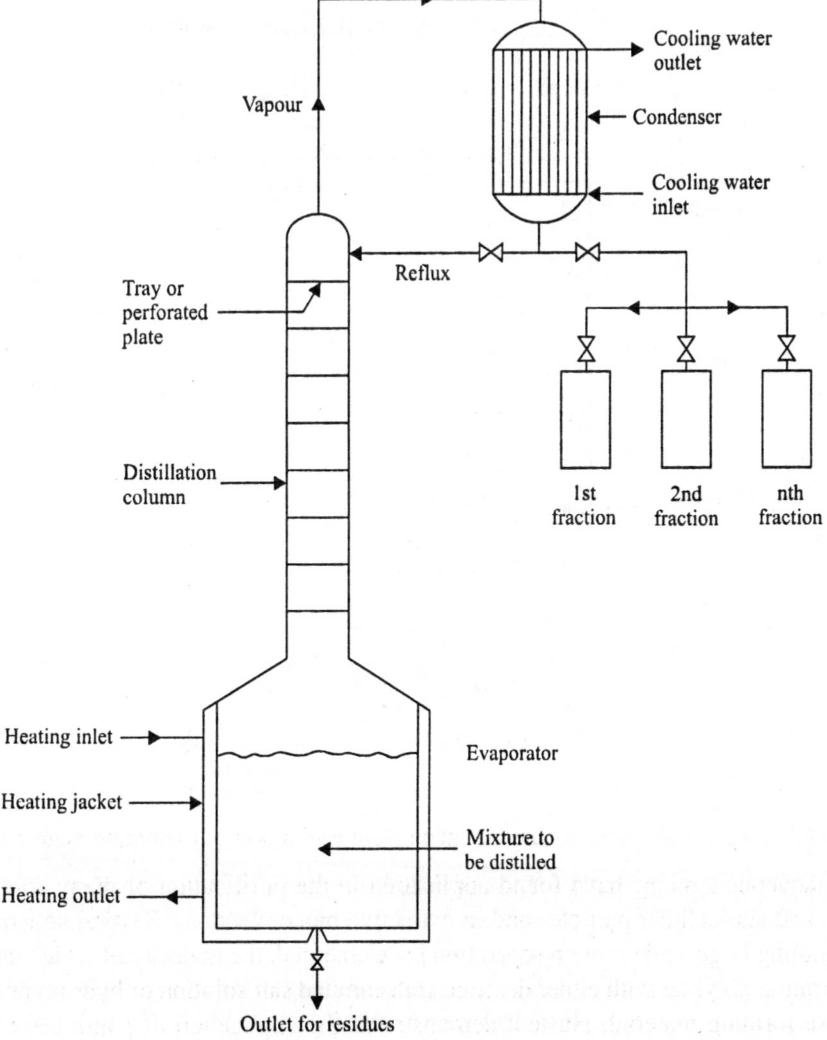


**Fig. 20.26.** Diagram of a batch distillation plant with a tray or perforated-plate column.

The distribution of a solute species between the phases is characterised by the partition coefficient and is influenced by a number of factors such as temperature, polymer (type and molecular weight), salt concentration, ionic strength, pH and properties (e.g., molecular weight) of the solute. As the goal of any extraction process is to selectively recover and concentrate a solute, affinity techniques such as those applied in chromatographic processes can be used to improve selectivity. Examples include the use of PEG-NADH derivatives in the extraction of dehydrogenases, *p*-aminobenzamidine in the extraction of trypsin and cibacron blue in the extraction of phosphofructokinase. It is possible to use different ligands in the two phases leading to an increase in selectivity or the simultaneous recovery and separation of several species.

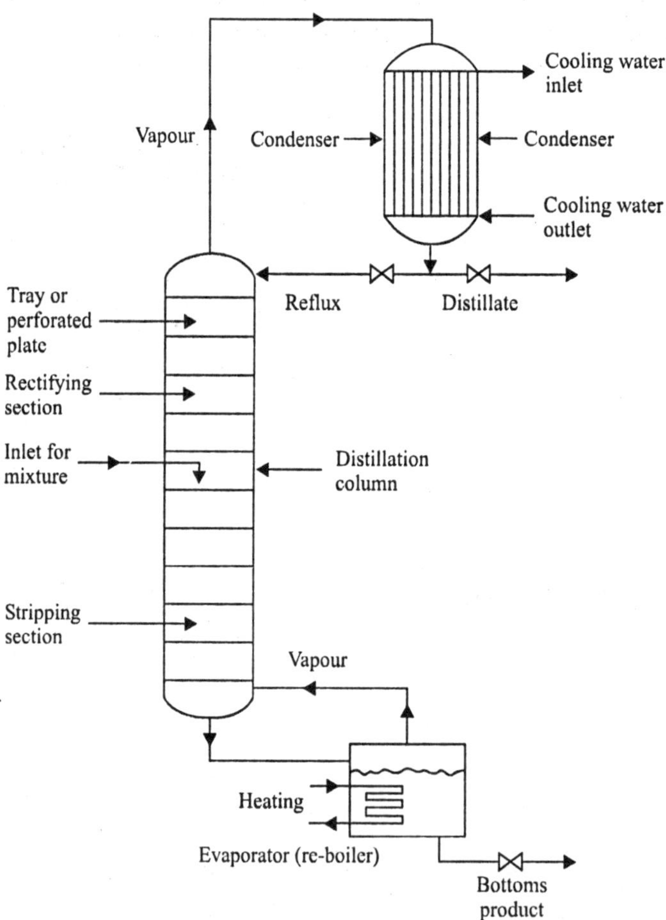


**Fig. 20.27.** Diagram of a continuous distillation plant with a tray or perforated-plate column.

Two phase aqueous systems have found application in the purification of many solutes; proteins, enzymes, cells and sub-cellular particles and in extractive bioconversions. Several aqueous two-phase systems for handling large-scale protein separation have emerged, the majority of which use PEG as the upper phase forming polymer with either dextran, concentrated salt solution or hydroxypropyl starch as the lower phase forming material. Hustedt demonstrated the application of continuous cross-current extraction of enzymes (fumarase and penicillin acylase) by aqueous two-phase systems at production scale.

## SUPERCRITICAL FLUID EXTRACTION

The technique of supercritical fluid extraction utilises the dissolution power of supercritical fluids, i.e., fluids above their critical temperature and pressure. Its advantages include the use of moderate temperatures and that several cheap and non-toxic fluids are available.

Supercritical fluids are used in the extraction of hop oils, caffeine, vanilla, vegetable oils and β-carotene. It has also been shown experimentally that the extraction of certain steroids and chemotherapeutic drugs can be achieved using supercritical fluids. Other current and potential uses include the removal of undesirable substances such as pesticide residues, removal of bacteriostatic agents from fermentation broths, the recovery of organic solvents from aqueous solutions, cell disintegration, destruction and treatment of industrial wastes and liposome preparation. There are, however, a number of significant disadvantages in the utilisation of this technology:

1. Phase equilibria of the solvent/solute system is complex, making design of extraction conditions difficult.
2. The most popular solvent (carbon dioxide) is non-polar and is therefore most useful in the extraction of non-polar solutes. Though co-solvents can be added for the extraction of polar compounds, they will complicate further downstream processing.
3. The use of high pressures leads to capital costs for plant and operating costs may also be high.

Thus, the number of commercial processes utilising supercritical fluid extraction is relatively small, due mainly to the existence of more economical processes. However, its use is likely to increase in some sectors, for example the recovery of high value biologicals, when conventional extractions are inappropriate and in the treatment of toxic wastes.

## CHROMATOGRAPHY

In many fermentation processes, chromatographic techniques are used to isolate and purify relatively low concentrations of metabolic products. In this context, chromatography will be concerned with the passage and separation of different solutes as liquid is passed through a column, i.e. liquid chromatography. Depending on the mechanism by which the solutes may be differentially held in a column, the techniques can be grouped as follows:

1. Adsorption chromatography.
2. Ion-exchange chromatography.
3. Gel permeation chromatography.
4. Affinity chromatography.
5. Reverse phase chromatography.
6. High performance liquid chromatography.

Chromatographic techniques are also used in the final stages of purification of a number of products. The scale-up of chromatographic processes can prove difficult and there is much current interest in the use of mathematical models and computer programmes to translate data obtained from small-scale processes into operating conditions for larger scale applications.

### Adsorption Chromatography

Adsorption chromatography involves binding of the solute to the solid phase primarily by weak van der Waals forces. The materials used for this purpose to pack columns include inorganic adsorbants (active carbon, aluminium oxide, aluminium hydroxide, magnesium oxide, silica gel) and organic macro-porous resins. Adsorption and affinity chromatography are mechanistically identical, but are strategically different. In affinity systems selectivity is designed rationally whilst in adsorption selectivity must be determined empirically.

Dihydro-streptomycin can be extracted from filtrates using activated charcoal columns. It is then eluted with methanolic hydrochloric acid and purified in further stages. Active carbon may be used to remove pigments to clarify broths. Penicillin-containing solvents may be treated with 0.25 to 0.5 per cent active carbon to remove pigments and other impurities.

Macro-porous adsorbants have also been tested. The first synthetic organic macro-porous adsorbants, the Amberlite XAD resins, were produced by Rohm and Haas. These resins have surface polarities which vary from non-polar to highly polar and do not possess any ionic functional groups. Voser considers his most interesting application to be in the isolation of hydrophilic fermentation products. He stated that these resins would be used at Ciba-Geigy in recovery of cephalosporin C (acidic amino acid), cefotiam (basic amino acid), desferrioxamine B (basic hydroxamic acid) and paramethasone (neutral steroid).

## Ion Exchange

Ion exchange can be defined as the reversible exchange of ions between a liquid phase and a solid phase (ion-exchange resin) which is not accompanied by any radical change in the solid structure. Cationic ion-exchange resins normally contain a sulphonic acid, carboxylic acid or phosphonic acid active group. Carboxy-methyl cellulose is a common cation exchange resin. Positively charged solutes (e.g., certain proteins) will bind to the resin, the strength of attachment depending on the net charge of the solute at the pH of the column feed. After deposition solutes are sequentially washed off by the passage of buffers of increasing ionic strength or pH. Anionic ion-exchange resins normally contain a secondary amine, quaternary amine or quaternary ammonium active group. A common anion exchange resin, DEAE (diethylaminoethyl) cellulose is used in a similar manner to that described above for the separation of negatively charged solutes. Other functional groups may also be attached to the resin skeleton to provide more selective behaviour similar to that of affinity chromatography. The appropriate resin for a particular purpose will depend on various factors such as bead size, pore size, diffusion rate, resin capacity, range of reactive groups and the life of the resin before replacement is necessary. Weak-acid cation ion-exchange resins can be used in the isolation and purification of streptomycin, neomycin and similar antibiotics. In the recovery of streptomycin, the harvested filtrate is fed on to a column of a weak-acid cationic resin such as Amberlite IRC 50 which is in the sodium form. The streptomycin is adsorbed on to the column and the sodium ions are displaced.

$$\mathrm{RCOO^-Na^+_{(resin)}} + \text{streptomycin} \longrightarrow \mathrm{RCOO^-}\ \text{streptomycin}^+_{\mathrm{(resin)}} + \mathrm{NaOH}$$

Flow rates of between 10 and 30 bed volumes per hour have been used. The resin bed is now rinsed with water and eluted with dilute hydrochloric acid to release the bound streptomycin.

$$\mathrm{RCOO^-}\ \text{streptomycin}^+_{\mathrm{(resin)}} + \mathrm{HCl} \longrightarrow \mathrm{RCOOH_{(resin)}} + \text{streptomycin}^+\ \mathrm{Cl^-}$$

A slow flow is used to ensure the highest recovery of streptomycin using the smallest volume of eluant. In one step the antibiotic has been both purified and concentrated, may be more than 100-fold. The resin column is regenerated to the sodium form by passing an adequate volume of NaOH slowly through the column and rinsing with distilled water to remove excess sodium ions.

$$\mathrm{RCOOH_{(resin)}} + \mathrm{NaOH} \longrightarrow \mathrm{RCOO^-Na^+_{(resin)}} + \mathrm{H_2O}$$

The resin can have a capacity of 1 gram of streptomycin $g^{-1}$ resin. Commercially, it is not economic to regenerate the resin completely, therefore the capacity will be reduced. In practice, the filtered broth is taken through two columns in series while a third is being eluted and regenerated. When the first column is saturated, it is isolated for elution and regeneration while the third column is brought into operation. Ion-exchange chromatography may be combined with HPLC in, for example, the purification of somatotropin using DEAE cellulose columns and β-urogastrone in multi-gram quantities using a cation exchange column.

### Gel Permeation

This technique is also known as gel exclusion and gel filtration. Gel permeation separates molecules on the basis of their size. The smaller molecules diffuse into the gel more rapidly than the larger ones and penetrate the pores of the gel to a greater degree. This means that once elution is started, the larger molecules which are still in the voids in the gel will be eluted first. A wide range of gels are available, including cross-linked dextrans (Sephadex and Sephacryl) and cross-linked agarose (Sepharose) with various pore sizes depending on the fractionation range required.

One early industrial application, although on a relatively small scale, was the purification of vaccines. Tetanus and diphtheria broths for batches of up to 100,000 human doses are passed through a 13 $dm^3$ column of G 100 followed by a 13 $dm^3$ column of G 200. This technique yields a fairly pure fraction which is then concentrated ten-fold by pressure dialysis to remove the eluant buffer ($Na_2HPO_4$).

### Affinity Chromatography

Affinity chromatography is a separation technique with many applications since it is possible to use it for separation and purification of most biological molecules on the basis of their function or chemical structure. This technique depends on the highly specific interactions between pairs of biological materials such as enzyme-substrate, enzyme-inhibitor, antigen-antibody, etc. The molecule to be purified is specifically adsorbed from, for example, a cell lysate applied to the affinity column by a binding substance (ligand) which is immobilised on an insoluble support (matrix). Eluent is then passed through the column to release the highly purified and concentrated molecule. The ligand is attached to the matrix by physical absorption or chemically by a covalent bond. The pore size and ligand location must be carefully matched to the size of the product for effective separation. The latter method is preferred whenever possible.

Coupling procedures have been developed using cyanogen bromide, bisoxiranes, disaziridines and periodates, for matrixes of gels and beads. Four polymers which are often used for matrix materials are agarose, cellulose, dextrose and polyacrylamide. Agarose activated with cyanogen bromide is one of the most commonly used supports for the coupling of amino ligands. Silica based solid phases have been shown to be an effective alternative to gel supports in affinity chromatography.

Purification may be several thousand-fold with good recovery of active material. The method can, however, be quite costly and time consuming and alternative affinity methods such as affinity cross-flow filtration, affinity precipitation and affinity partitioning may offer some advantages. Affinity chromatography was used initially in protein isolation and purification, particularly enzymes. Since then many other large-scale applications have been developed for enzyme inhibitors, antibodies, interferon and recombinant proteins and on a smaller scale for nucleic acids, cell organelles and whole cells. In the scale-up of affinity chromatographic processes bed height limits the superficial velocity of the liquid, thus scale-up requires an increase in bed diameter or adsorption capacity.

### Reverse Phase Chromatography (RPC)

This chromatographic method utilises a solid phase (e.g., silica) which is modified so as to replace hydrophilic groups with hydrophobic alkyl chains. This allows the separation of proteins according to their hydrophobicity. More-hydrophobic proteins bind most strongly to the stationary phase and are therefore eluted later than less-hydrophobic proteins. The alkyl groupings are normally eight or eighteen carbons in length ($C_8$ and $C_{18}$). RPC can also be combined with affinity techniques in the separation of, for example, proteins and peptides.

### High Performance Liquid Chromatography (HPLC)

HPLC is a high resolution column chromatographic technique. Improvements in the nature of column packing materials for a range of chromatographic techniques (e.g., gel permeation and ion-exchange) yield smaller, more rigid and more uniform beads. This allows packing in columns with minimum spaces between the beads, thus minimising peak broadening of eluted species. It was originally known as high pressure liquid chromatography because of the high pressures required to drive solvent through silica based packed beds. Improvements in performance led to the name change and its widespread use in the separation and purification of a wide range of solute species, including bio-molecules. HPLC is distinguished from liquid chromatography by the use of improved media (in terms of their selectivity and physical properties) for the solid (stationary) phase through which the mobile (fluid) phase passes.

The stationary phase must have high surface area/unit volume, even size and shape and be resistant to mechanical and chemical damage. However, it is factors such as these which lead to high pressure requirements and cost. This may be acceptable for analytical work, but not for preparative separations. Thus, in preparative HPLC some resolution is often sacrificed (by the use of larger stationary-phase particles) to reduce operating and capital costs. For very high value products large-scale HPLC columns containing analytical media have been used.

Affinity techniques can be merged with HPLC to combine the selectivity of the former with the speed and resolving power of the latter.

### Continuous Chromatography

Although the concept of continuous enzyme isolation is well established the stage of least development is continuous chromatography. Fox developed a continuous-fed column for this purpose (Fig. 20.28). It consisted of two concentric cylindrical sections clamped to a base plate. The space (1 cm wide) between the two sections was packed with the appropriate resin or gel giving a total column capacity of 2.58 $dm^3$. A series of orifices in the circumference of the base plate below the column space led to collecting vessels. The column assembly was rotated in a slow-moving turntable (0.4–2.0 rpm). The mixture for separation was fed to the apparatus by an applicator rotating at the same speed as the column, thus allowing application at a fixed point, while the eluent was fed evenly to the whole circumference of the column. The components of a mixture separated as a series of helical pathways, which varied with the retention properties of the constituent components.

This method gave a satisfactory separation and recovery but the consumption of eluent and the unreliable throughput rate were not considered to be satisfactory for a large-scale method. However, the development of such continuous separation equipment suitable for large scale extraction would considerably simplify the use of chromatographic separation.

## MEMBRANE PROCESSES

### Ultrafiltration and Reverse Osmosis

Both processes utilise semi-permeable membranes to separate molecules of different sizes and therefore act in a similar manner to conventional filters.

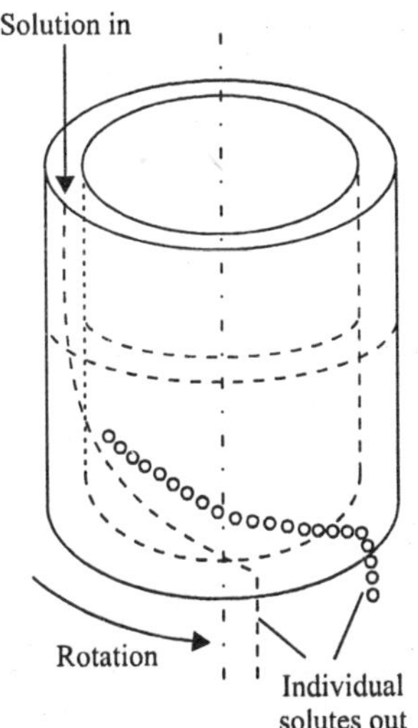


**Fig. 20.28.** The principle of continuous-partition chromatography. – – , faster-moving component; and o o, slower-moving component.

## Ultrafiltration

Ultrafiltration can be described as a process in which solutes of high molecular weight are retained when the solvent and low molecular weight solutes are forced under hydraulic pressure (around 7 atmospheres) through a membrane of a very fine pore size. It is therefore used for product concentration and purification. A range of membranes made from a variety of polymeric materials, with different molecular weight cut-offs (500 to 500,000), are available which makes possible the separation of macro-molecules such as proteins, enzymes, hormones and viruses. It is practical only to separate molecules whose molecular weights are a factor of ten different due to variability in pore size. Because the flux through such a membrane is inversely proportional to its thickness, asymmetric membranes are used where the membrane (~0.3 μm thick) is supported by a mesh around 0.3 mm thick.

When considering the feasibility of ultrafiltration it is important to remember that factors other than the molecular weight of the solute affect the passage of molecules through the membranes. There may be concentration polarisation caused by accumulation of solute at the membrane surface which can be reduced by increasing the shear forces at the membrane surface either by conventional agitation or by the use of a cross-flow system. Secondly, a slurry of protein may accumulate on the membrane surface forming a gel layer which is not easily removed by agitation. Formation of the gel layer may be partially controlled by careful choice of conditions such as pH. Finally, equipment and energy costs may be considerable because of the high pressures necessary; this also limits the life of ultrafiltration membranes.

There are numerous examples of the use of ultrafiltration for the recovery of bio-molecules: viruses, antibiotics. Affinity ultrafiltration is a novel separation process developed to circumvent difficulties in affinity chromatography. It offers high selectivity, yield and concentration, but it is an expensive batch process and scale-up is difficult.

## Reverse Osmosis

Reverse osmosis is a separation process where the solvent molecules are forced by an applied pressure to flow through a semi-permeable membrane in the opposite direction to that dictated by osmotic forces and hence is termed reverse osmosis. It is used for the concentration of smaller molecules than is possible by ultrafiltration. Concentration polarisation is again a problem and must be controlled by increased turbulence at the membrane surface.

### Liquid Membranes

Liquid membranes are insoluble liquids (e.g., an organic solvent) which are selective for a given solute and separate two other liquid phases. Extraction takes place by the transport of solute from one liquid to the other. They are of great interest in the extraction and purification of biologicals for the following reasons:

1. Large area for extraction.
2. Separation and concentration are achieved in one step.
3. Scale-up is relatively easy.

Their use has been reported in the extraction of lactic acid and citric acid using a supported liquid membrane. The utilisation of selective carriers to transport specific components across the liquid membrane at relatively high rates has increased interest in recent years. Liquid membranes may also be used in cell and enzyme immobilisation and thus provide the opportunity for combined production and isolation/extraction in a single unit. The potential use of liquid membranes has also been described for the production of alcohol reduced beer as having little effect on flavour or the physico-chemical properties of the product.

## DRYING

The drying of any product (including biological products) is often the last stage of a manufacturing process. It involves the final removal of water from a heat-sensitive material ensuring that there is minimum loss in viability, activity or nutritional value. Drying is undertaken because:

1. The cost of transport can be reduced.
2. The material is easier to handle and package.
3. The material can be stored more conveniently in the dry state.

It is important that as much water as possible is removed initially by centrifugation or in a filter press to minimise heating costs in the drying process. Driers can be classified by the method of heat transfer to the product and the degree of agitation of the product. In contact driers the product is contacted with a heated surface. An example of this type is the drum drier (Fig. 20.29), which may be used for more temperature stable bio-products. A slurry is run onto a slowly rotating steam heated drum, evaporation takes place and the dry product is removed by a scraper blade in a similar manner as for rotary vacuum filtration. The solid is in contact with the heating surface for 6–15 seconds and heat transfer coefficients are generally between 1 and 2 kW $m^{-2}$ $K^{-1}$. Vacuum drum driers can be used to lower the temperature of drying.

A spray drier (Fig. 20.30) is most widely used for drying of biological materials when the starting material is in the form of a liquid or paste. The material to be dried does not come into contact with the heating surfaces, instead, it is atomised into small droplets through for example a nozzle or by contact with a rotating disc. The droplets then fall into a spiral stream of hot gas at 150°F to 250°F. The high surface area: volume ratio of the droplets results in a rapid rate of evaporation and complete drying in a few seconds, with drying rate and product size being directly related to droplet size produced by the atomiser. The evaporative cooling effect prevents the material from becoming overheated and damaged. The gas-flow rate must be carefully regulated so that the gas has the capacity to contain the required moisture content at the cool-air exhaust temperature (75°F to 100°F). In most processes the recovery of

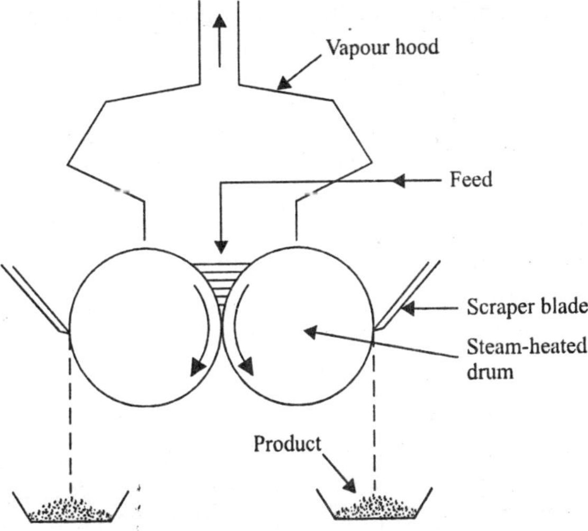


**Fig. 20.29.** Cross-section of a drum drier.

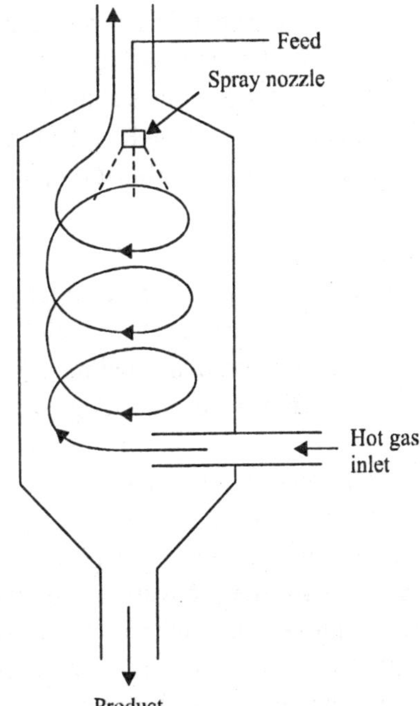


**Fig. 20.30.** Counter-current spray drier.

very small particles from the exit gas must be conducted using cyclones or filters. This is especially important for containment of biologically active compounds.

The jet spray drier is particularly suited to handling heat sensitive materials. Operating at a temperature of around 350°F, residence times are approximately 0.01 second because of the very fine droplets produced in the atomising nozzle.

Spray driers are the most economical available for handling large volumes and it is only at feed rates below 6 kg $min^{-1}$ that drum driers become more economic.

Freeze drying is an important operation in the production of many biologicals and pharmaceuticals. The material is first frozen and then dried by sublimation in a high vacuum. The great benefit of this technique is that it does not harm heat sensitive materials. The process is often termed lyophilisation when the solvent being evaporated is water.

Fluidised bed driers are used increasingly in the pharmaceutical industry. Heated air is fed into a chamber of fluidised solids, to which wet materials is continuously added and dry material continuously removed. Very high mass-transfer rates are achieved, giving rapid evaporation and allowing the whole bed to be maintained in a dry condition.

## CRYSTALLISATION

Crystallisation is an established method used in the initial recovery of organic acids and amino acids and more widely used for final purification of a diverse range of compounds.

In citric acid production, the filtered broth is treated with $Ca(OH)_2$ so that the relatively insoluble calcium citrate crystals will be precipitated from solution. Checks are made to ensure that the $Ca(OH)_2$ has a low magnesium content, since magnesium citrate is more soluble and would remain in solution. The calcium citrate is filtered off and treated with sulphuric acid to precipitate the calcium as the insoluble sulphate and release the citric acid. After clarification with active carbon, the aqueous citric acid is evaporated to the point of crystallisation. Crystallisation is also used in the recovery of amino acids. Samejima has reviewed methods for glutamic acid, lysine and other amino acids. The recovery of cephalosporin C as its sodium or potassium salt by crystallisation has been described by Wildfeuer.

## WHOLE BROTH PROCESSING

The concept of recovering a metabolite directly from an unfiltered fermentation broth is of considerable interest because of its simplicity, the reduction in process stages and the potential cost savings. It may also be possible to remove the desired fermentation product continuously from a broth during fermentation so that inhibitory effects due to product formation and product degradation can be minimised throughout the production phase.

Bartels developed a process for adsorption of streptomycin on to a series of cationic ion-exchange resin columns directly from the fermentation broth, which had only been screened to remove large particles so that the columns would not become blocked. This procedure could only be used as a batch process. Belter developed a similar process for the recovery of novobiocin. The harvested broth was first filtered through a vibrating screen to remove large particles. The broth was then fed into a continuous series of well-mixed resin columns fitted with screens to retain the resin particles, plus the absorbed novobiocin, but allow the streptomycete filaments plus other small particulate matter to pass through. The first resin column was removed from the extraction line after a pre-determined time and eluted with methanolic ammonium chloride to recover the novobiocin.

Karr developed a reciprocating plate extraction column (Fig. 20.31) to use for whole broth processing of a broth containing 1.4 g $dm^{-3}$ of a slightly soluble organic compound and 4 per cent undissolved solids provided that chloroform or methylene chloride were used for extraction. Methyl-iso-butyl ketone, diethyl ketone and iso-propyl acetate were shown to be more efficient solvents than chloroform for extracting the active compound, but they presented problems since they also extracted impurities from the mycelia, making it necessary to filter the broth before beginning the solvent extraction. Considerable economies were claimed in a comparison with a process using a Podbielniak extractor, in investment, maintenance costs, solvent usage and power costs but there was no significant difference in operating labour costs.

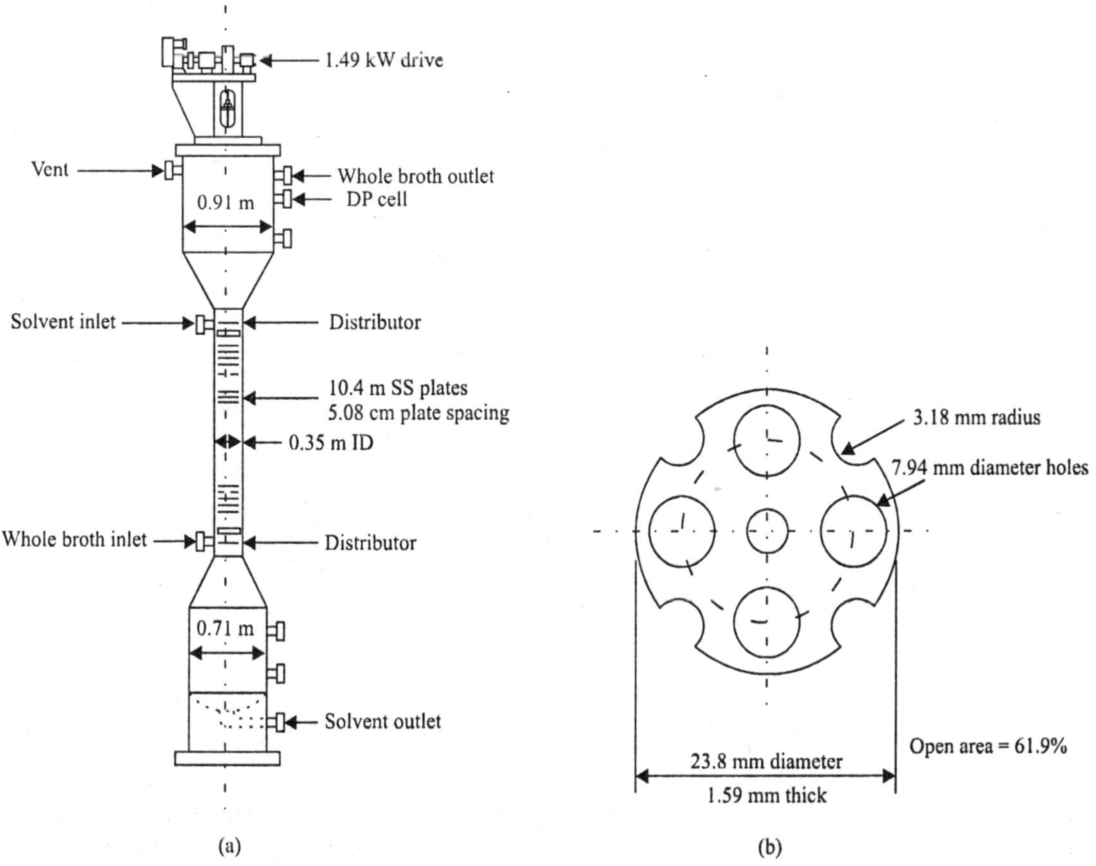


**Fig. 20.31.** (a) Diagram of a 0.35 m internal diameter reciprocating plate column; and (b) plan of a 23.8 m stainless-steel plate for a 25 mm diameter reciprocating plate test column.

An alternative approach is to remove the metabolite continuously from the broth during the fermentation. Cycloheximide production by *Streptomyces griseus* has been shown to be affected by its own feed-back regulation. Wang have tested two techniques at laboratory scale for improving production of cycloheximide. In a dialysis method (Fig. 20.32), methylene chloride was circulated in a dialysis tubing loop which passed through a 10 $dm^{-3}$ fermenter. Cycloheximide in the fermentation broth was extracted into the methylene chloride. It was shown that the product yield could be almost doubled by this dialysis-solvent extraction method to over 1200 μg $cm^{-3}$ as compared with a control yield of approximately 700 μg $cm^{-3}$. In a resin method, sterile beads of XAD-7, an acrylic resin, as dispersed beads or beads wrapped in an ultrafiltration membrane, were put in fermenters 48 hours after inoculation. Some of the cycloheximide formed in the broth is absorbed by the resin. Recovery of the antibiotic from the resin is achieved by solvents or by changing the temperature or pH.

When assayed after harvesting, the control (without resin) had a bioactivity of 750 μg $cm^{-3}$. Readings of total bioactivity (from beads and broth) for the bead treatment and the membrane-wrapped bead treatments were 1420 μg $cm^{-3}$ and 1790 μg $cm^{-3}$ respectively. Roffler reviewed the use of a number of techniques for the *in situ* recovery of fermentation products:

1. Vacuum and flash fermentations for the direct recovery of ethanol from fermentation broths.
2. Extractive fermentations (liquid-liquid and two-phase aqueous) or the recovery of ethanol, organic acids and toxin produced by *Clostridium tetani*.
3. Adsorption for the recovery of ethanol and cycloheximide.
4. Ion-exchange in the extraction of salicylic acid and antibiotics.
5. Dialysis fermentation in the selective recovery of lactic acid, salicylic acid and cycloheximide.

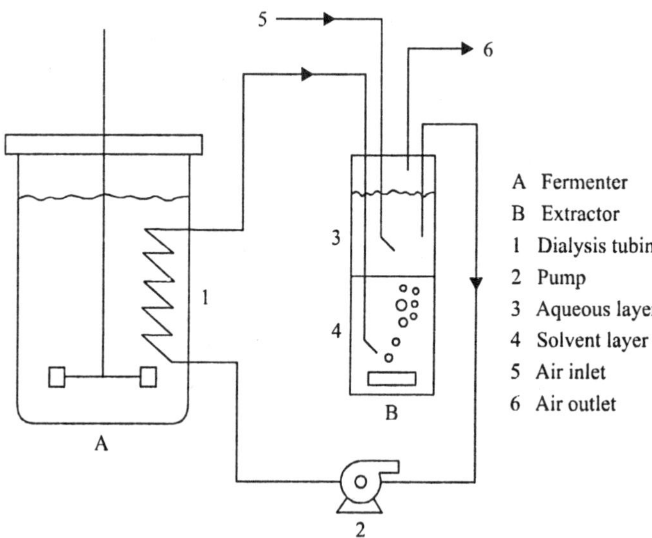


**Fig. 20.32.** Dialysis-extraction fermentation system.

Hansson has used an expanded adsorption bed for the recovery of a recombinant protein produced by *E. coli* directly from the fermentation broth. The protein was produced in high yields (550 mg $dm^{-3}$) and > 90 per cent recovery together with concentration (volume reduction) and removal of cells was achieved on the expanded bed. Affinity chromatography was used for further purification and again an overall yield of > 90 per cent obtained.

Chapter 21

# Fermentation Economics

## INTRODUCTION

If a fermentation process is to yield a product at a competitive price, the chosen micro-organism or animal cell culture should give the desired end-product in predictable and economically adequate, quantities. A number of basic objectives are commonly used in developing a successful process which will be economically viable.

1. The capital investment in the fermenter and ancillary equipment should be confined to a minimum, provided that the equipment is reliable and may be used in a range of fermentation processes.
2. Raw materials should be as cheap as possible and utilised efficiently. A search for possible alternative materials might be made, even when a process is operational.
3. The highest-yielding strain of micro-organisms or animal cell culture should be used.
4. There should be a saving in labour whenever possible and automation should be used where it is feasible.
5. When a batch process is operated, the growth cycle should be as short as possible to obtain the highest yield of product and allow for maximum utilisation of equipment. To achieve this objective it may be possible to use fed-batch culture.
6. Recovery and purification procedures should be as simple and rapid as possible.
7. The effluent discharge should be kept to a minimum.
8. Heat and power should be used efficiently.
9. Space requirements should be kept to a minimum, but there should be some allowance for potential expansion in production capacity.
10. All the above must comply with safety guidelines and regulations.

The consideration of so many criteria means that there may have to be a compromise for the particular set of circumstances relating to an individual process. Winkler identified three key economic objectives in the fermentation stage: maximum product yield, process productivity and substrate utilisation. However, these criteria may be overridden by the demands of subsequent purification stages for high product concentration and high product purity which may in turn be overridden by safety considerations.

In any process it is important to know the cost breakdown, so that it may be seen where the biggest potential savings may be achieved. In a review of a number of processes, Nyiri and Charles concluded that four basic components contributed to the process cost in the following decreasing order: raw materials, fixed costs, utilities, labour. When raw materials are a major part of the total cost it is obvious that media and microbial strain-improvement research should form a major part of a development programme. Development work on components contributing little to the cost of the product could not be justified.

Caution is necessary when examining some examples of potential fermentation processes, since the capital and operating costs normally have been estimated using chemical engineering costing principles.

Government aid or taxation can determine the viability of many fermentation processes. Support in construction of plant, development and production programmes of acetone-butanol and penicillin during wartime led to production of compounds at an earlier stage than might normally be regarded as economically feasible.

## ISOLATION OF MICRO-ORGANISMS OF POTENTIAL INDUSTRIAL INTEREST

The most appropriate micro-organism for a potential process is usually found by isolation from a variety of sources, most commonly soil. The classical method of screening to obtain a suitable organism tended to be very time consuming, expensive and often without a very clear objective. Recently Berdy has speculated that the screening of 100,000 soil micro-organisms may lead to the isolation of 5 to 50 new compounds, but there is no guarantee after evaluation that a useful new drug or other product will be found. If a desired characteristic, which gives the organism a selective advantage, has already been recognised, a screen might be designed incorporating this characteristic as a selective factor. The isolation may begin with pre-treatment of samples which favour the survival of the preferred organism. This is followed by growth on selective or non-selective media and often associated with batch or continuous enrichment. Important factors which will be of economic significance which might be selected in a well planned screen could include:

1. Growth on a simple cheap medium.
2. Growth at a higher temperature (to reduce cooling costs).
3. Better resistance to contamination.

However, the synthesis of other microbial products (e.g., antibiotics) does not give the producing organism any selective advantage which might be used in an isolation procedure. Therefore a collection of these organisms must be made before testing for the desired characteristic. Because many isolation procedures will lead to the re-discovery of known organisms with known activities, it is important to use well planned, efficient isolation procedures which can prove very productive.

A number of approaches are currently being used to improve isolation procedures. Numerical taxonomic databases are being exploited to design selective media for certain microbial taxa. Using these databases it has been possible to design media to encourage the growth of uncommon streptomycetes or discourage the growth of common species. Knowledge of antibiotic sensitivity gained from taxonomic databases has led to the design of other selective media which will select for resistant groups of organisms. The selection of antibiotic producing soil isolates has also been achieved using media designed by a stepwise discrimination analysis technique.

These 'designed' isolation media are now being used extensively for the isolation of novel and rare micro-organisms. For example, within the actinomycetes, *Streptomyces* spp. have been extensively screened since the 1940s for antibiotic production with subsequent notable commercial success. More interest has now been shown in isolation of strains of the less common genera such as *Actinomadura, Actinoplanes, Kitosatosporia, Streptoalloteichus*, etc., which are producing other novel bioactive compounds. It has become a common practice to obtain isolates from unusual habitats, which may include extreme environment, to ensure that the greatest microbial diversity is being examined.

The screening tests which have been developed to detect new useful compounds of potential industrial interest have become much more selective and sensitive. Better knowledge of cell biochemistry has enabled the design of screens which are much more precisely targeted to detect the desired activity using specific detector strains. Many large companies which undertake screening programmes have introduced some automation to enable high throughput rates which are cost effective and less labour intensive.

## STRAIN IMPROVEMENT

Strain improvement using a mutation/selection programme for improving an organism being used in an established process or a potential process can be very cost effective. Historically, mutation/selection programmes to improve strains of *Penicillium chrysogenum* were time consuming, labour intensive and very random because of the lack of knowledge about penicillin biosynthesis. These mutation programmes did, however, contribute significantly to increases in penicillin yields from less than 100 units $cm^{-3}$ in the 1940s to over 51,000 units $cm^{-3}$ by 1976 and a four-fold increase in yields between 1970 and 1985 at Gist Brocades.

It is always very important to decide if a strain improvement programme can be justified on financial grounds to improve the overall economy of a process.

A better understanding of cell metabolism and its regulation has enabled the development of more logical targeted methods to be introduced to select for mutants with desirable 'blue prints' where there may be a need to block undesirable enzyme activities or eliminate negatively acting control mechanisms. This approach is much more efficient and economic in terms of resources and time. It was first employed extensively in the preparation of mutant strains used in amino acid fermentations.

Although the main targets in strain improvement are normally to increase the product yield or specific production rates, it is also important to consider strain stability, resistance to phage infection, response to dissolved oxygen, tolerance to medium components, production of foam and the morphological form of the organism. These are very important in helping to achieve targets in a research and development programme as they can have a significant impact on the process and/or product (Table 21.1). A study of the range of targets given in this table, which might be aided by strain improvement, has economic effects on all aspects of a fermentation process.

**Table 21.1.** Criteria for strain improvement.

| *Target* | *Impact on process or product* |
|---|---|
| Improvement of titre and/or specific production rate | General decrease of production costs, improved exploitation of reactor capacity, lower investment costs, increased efficiency in downstream processing steps |
| Improvement of yield | Lower costs for substrates, decreased production of heat and $CO_2$, lower cooling costs, less waste and pollution |
| Change in catabolic capabilities | Use of more favourable substrates (less expensive, better availability, etc.), omission of pre-treatment steps (e.g., enzymatic hydrolysis of polysaccharides) |
| Improvement of technological features of micro-organisms (e.g., flocculation behaviour, structure of mycelium, sporulation, foaming, strain stability, etc.) | Less energy costs for mixing and oxygen transfer, improved separation characteristics, fewer problems in inoculum preparation or scale-up of the process |

*(Cont'd ...)*

| *Target* | *Impact on process or product* |
|---|---|
| Improvement of product quality | Decreased production of specific by-products (fewer impurities), prevention of product degradation (e.g., pectinases) |
| Modification of products | Improvement of solubility in extraction solvents (e.g., addition of specific side chains), increased thermic stability of altered enzymatic properties of proteins |
| Changing of the locus of product accumulation (e.g., intra-cellular to extra-cellular) | Improved product recovery (e.g., omission of cell disruption), correct products (e.g., fully processed proteins) |

## MARKET POTENTIAL

The fermentation technologist should be aware of the problem of assessing market potential, although he/she may not be primarily involved in collating or assessing the necessary data.

Four categories of microbial product can be recognised economically and it is important to consider to which category of compound belongs:

1. Low price bulk chemicals, e.g., solvents, biomass, high fructose syrups.
2. Mid price chemicals, e.g., organic acids, amino acids, bipolymers.
3. High price microbial and animal-cell products, e.g., enzymes, vitamins, antibiotics, corticosteroids, vaccines, etc.
4. Very high-price animal-cell products, e.g., monoclonals, tissue plasminogen activator, etc.

The third and fourth groups can normally be produced only by a microbial or animal-cell based process and therefore do not have to compete with an alternative chemical process which is usually much cheaper. This includes compounds which have complicated structures, are chemically or thermally unstable or for which a multi-stage chemical synthesis would be expensive. Many microbial products are not exploited because cheaper synthetic processes are available.

Hepner has examined the factors that determine the feasibility of large-scale ethanol production by fermentation. He considered that ethanol produced by fermentation would only be competitive with synthetic ethanol from crude oil if the fermentation plant was in an area where cheap supplies of carbohydrate were available. It is necessary to estimate the size of the present and potential market and the increase in demand for a compound. This type of exercise was undertaken for single-cell protein.

## EFFECTS OF LEGISLATION ON PRODUCTION OF ANTIBIOTICS AND RECOMBINANT PROTEINS IN THE USA

There are a number of differences in regulatory requirements for production facilities for antibiotics and recombinant proteins. In both cases, the products must be licensed by the FDA before they can be marketed.

The recombinant protein is produced by a precisely specified process using high quality substrates and processes and purified in an aseptic pharmaceutical facility that has been inspected and licensed by the CBER (Centre of Biological Evaluation and Research) of the FDA. Steps must also be taken to ensure that there is no cross-contamination if more than one product is produced in the same facility. Any changes in a process or facilities must be approved by the CBER before implementation. It may take 7 years to obtain approval by the CBER for a production plant. In contrast, the requirements for an antibiotic production plant are less stringent and it may be operational in 4 years. The delay in start up to produce the recombinant protein will result in much higher costs.

## PLANT AND EQUIPMENT

It is most logical to build equipment as large as possible because of the economy of scale. There is an empirical relationship between cost and size of an item of equipment. According to this relationship, as facility size increases, its cost increases thus:

$$\frac{\text{cost}_1}{\text{cost}_2} = \left(\frac{\text{size}_1}{\text{size}_2}\right)^n$$

where n is an exponent or scale factor. Scale factors have been estimated to be 0.6 for brewing, 0.7 to 0.8 for a single cell protein plant, 0.6 for antibody production, 0.75 for fermentation processes and 0.6 for waste-water treatment. In brewing, when cylindrico-conical vessels are used, there would appear to be no economic advantage in scaling up above 1,08,000 $dm^3$ capacity, although vessels of 360,000 $dm^3$ have been installed. In some breweries a number of smaller vessels have been installed to allow for fluctuating demand of different beers. The operational vessel volume is a critical factor when considering high volume-low cost products. In processes carried out at volumes greater than 100 $m^3$, the use of an air-lift fermenter is more economical as the relative investment costs per unit of output decreased more rapidly than for a stirred type of fermenter (Table 21.2).

**Table 21.2.** Estimate of relative costs of a fermentation process as a function of scale (100 $m^3$ = 1 unit of volume/unit of time).

| *Scale ($m^3$)* | *Cost scale factor* |
|---|---|
| 1000 | 0.3 |
| 200 | 0.6 |
| 100 | 1.0 |
| 10 | 10.0 |
| 1 | 30.0 |

However, a number of restraints have to be considered before deciding on the scale of operation. Such restrictions include cooling and aeration requirements and the method of fermentation vessel construction. The need for cooling provisions in many fermentations has already been described. At this stage it is important to remember that the volume of a fermenter is proportional to $r^3$ (where r is the fermenter radius), whereas the increase in surface area is proportional to $r^2$. Therefore the scaling up of a vessel will lead to a decrease in the surface area to volume ratio and therefore a decrease in the effectiveness of a cooling jacket. There may be a fermenter volume above which it will not be possible to remove enough heat to maintain constant temperature, unless the cooling capacity is increased by incorporating internal cooling coils or by using external heat exchangers. These modifications may prove costly or else interfere with mixing in the vessels.

There is possibly the alternative of using micro-organisms with higher optimum temperatures. The oxygen requirements of a process may limit the size of vessel which can be operated successfully. In acetic acid (vinegar) production, where very efficient aeration is critical, because *Acetobacter* spp. are very sensitive to oxygen depletion in the broth, the maximum acetator capacity which could be used was about 50,000 $dm^3$.

Because of the capital investment and operational costs there is now a trend to consider unconventional fermenter designs of simple construction with very efficient oxygen-transfer to be used for specific purposes, particularly single-cell protein. In this context, Schugerl and others have considered the case of a single-cell protein plant for 100,000 tonnes annum$^{-1}$ using methanol as the main substrate. The plant investment was only 20 per cent of the production cost. In this investment cost only 20 to 25 per cent was due to the fermenter. Of this, the vessel accounted for 5 to 10 per cent, the stirrer and aerator 5 to 10 per cent and cooling 10 to 15 per cent.

Energy, water, aeration and auxiliaries were estimated to be a further 10 per cent of the production costs. Since the fermenter investment costs are approximately 4 to 5 per cent and the operating costs are 10 per cent, it was concluded that the type of vessel could influence only 15 per cent of the total production costs. If a tower fermenter were used, the main advantage would appear to be lower operating costs, particularly reduced energy and cooling-water costs, as the removal of mechanical stirring would diminish the cooling water requirements.

A useful guide has been prepared by the Institution of Chemical Engineers which outlined, in reasonably quantitative terms, many factors making up a check list which must be incorporated into the final capital costing of a chemical-plant project.

It is possible from a knowledge of a proposed plant location, a sketch of the chemical plant process flow sheet, the size of the major items of equipment and the service requirements, to estimate capital and operating costs to ±15 per cent. Unfortunately, comparable literature for fermentation plants is not available. It is essential to remember the most of the data are estimates for proposed processes. Some equipment-cost breakdowns are quoted in Table 21.3.

**Table 21.3.** Capital cost breakdown for fermentation plant.

| *Item* | *% of total* |
|---|---|
| (a) *Penicillium plant*, estimated for five 225,000 $dm^3$ fermenters with ancillary equipment | |
| Process equipment | 23.6 |
| Installation | 5.2 |
| Insulation | 1.9 |
| Instruments | 2.7 |
| Piping | 11.8 |
| Electrical | 15.8 |
| Building | 11.3 |
| Utilities | 21.3 |
| Site | 2.4 |
| Laboratory equipment | 3.8 |
| Spare parts | 0.5 |
| (b) *Norprotein plant* | |
| Raw materials storage | 10 |
| Medium preparation and utilities | 17 |
| Fermentation | 41 |
| Cell recovery and drying | 22 |
| Product storage | 10 |

*(Cont'd ...)*

| *Item* | *% of total* |
|---|---|
| (c) *ICI plc. single-cell protein plant* | |
| Raw materials | 3 |
| Storage and packing | 12 |
| Off-site services | 16 |
| On-site services | 11 |
| Fermentation | 14 |
| Compression | 9 |
| Dewatering | 19 |
| Drying | 12 |
| Effluent treatment | 4 |

If a vessel is to be used to produce genetically engineered products, the costs will increase significantly because of the extra containment provisions which must be incorporated during construction. It has been estimated that the basic fermenter costs will increase by 10–30 per cent for each increase in Containment level. At level 2 or B3 the provision of containment of an instrumented 20 $dm^3$ fermenter cost may increase three times. The cheaper alternative might be to enclose an ordinary fermenter in a suitable containment cabinet. Hambleton and others have reported the use of a Class 3 microbiological cabinet for fermenters of up to 50 $dm^3$. Buckland has stated that it now costs as much to build a 3 $m^3$ facility for a recombinant protein which will need extensive containment provisions as for a 2000 $m^3$ scale facility for an antibiotic. In a containment facility only 30 per cent may be usable space. The remaining 70 per cent of the building is needed to accommodate and service the facility.

Reisman has stressed the need for the correct level of containment. More stringent requirements may exist in a laboratory than are strictly needed. The same high level of containment in a production facility may be onerous, costly and unnecessary. Containment needs should be determined jointly by a group of appropriately trained staff.

The designing of vessels which can be converted for multi-use may have economic advantages, but descriptions of such vessels except at pilot scale are not common.

The life of a proposed plant has to be predicted. MacLennan has estimated the life of a single-cell protein plant to be 10 years. This is probably the lower limit. Hamer claimed that a planned SCP plant life of 15 years is typical. Acetone and butanol have been produced in 25-year-old vessels, while in long-established breweries in Great Britain, equipment of 50 to 100 years old is still used.

Allowance must be made for the service provisions for a fermentation process. In a moderate sized antibiotic plant, there may be the need for the provision of 5 tonnes $hr.^{-1}$ of steam, 5000 kW $hr.^{-1}$ of electricity, 57,000 $m^3$ $hr.^{-1}$ of compressed air and 200,000 $dm^3$ $hr.^{-1}$ of water. When producing 1 tonne of acetic acid, 480 $m^3$ of cooling water, 10 $m^3$ of process water, 12 tonnes of steam and 570 kW of electricity would be utilised.

## MEDIA

The cost of the various components of a production medium can have a profound effect on the overall cost of a fermentation process, since these account for 38 to 73 per cent of the total production cost. The organic-carbon source in microbial processes is usually the most expensive component contributing to the cost of the process. Ratledge has made a detailed analysis of annual price and availability of major carbon substrates. The price of a natural material may fluctuate due to other competing demands and the annual variation in the quantity harvested. Big capital investment may be tied up in natural materials if they are seasonal and require storage. Hastings and Jackson stated that up to $23 \times 10^6$ $dm^3$ of molasses may have to be stored for an industrial alcohol-production process. A particular material may be selected because it is cheap locally, rather than the best substrate.

In a cost study analysis for tissue plasminogen activator by a mammalian-cell process, fermentation materials accounted for 75 per cent of the total raw materials cost. Of this cost, calf serum was estimated to contribute 73 per cent or 55 per cent of the total cost. However, it has been possible to develop a serum-free medium for economic production of antibodies from hybridoma cells.

In the search for suitable nutrient components for media, six criteria were considered which ought to be satisfied whenever possible. Ratledge has stressed the need to note the amount of carbon in a carbon substrate when costing. Higher yields might be achieved by changing from a carbohydrate to an alternative carbon source, but increased aeration and/or agitation rates may be necessary with the change in substrate. The cost of this extra provision must therefore be less than the savings from the change of substrate if the process is to be feasible. When cheapness is added as a further restraint the number of potential major nutrients which could be used on an industrial scale is limited. Carbohydrates from beet, sugar cane or grain are the major carbon and energy sources in most media and comply with the requirements of economy.

During the 1970s there was considerable interest in using petrochemicals as substrates for SCP production as protein animal feed by a number of major chemical and petroleum companies. A number of processes were developed using methane or methanol as the main carbon substrate. None are now being operated. Major factors which contributed to making these processes uncompetitive included the increased cost of the substrate and the availability of cheaper alternative animal feeds.

A variety of waste materials would seem to be potential cheap carbon sources. Unfortunately, it has been shown that their use is very restricted because they cannot compete economically with conventional substrates. This may be due to a number of possible reasons including variability of the material, impurities which make downstream processing more difficult, high water content making transport costly, geographical location, quantities produced and limited seasonal availability. The possible use of waste substrates may depend on the cost of alternative methods of disposal or on the availability of government grants to diminish pollution.

Mineral components normally constitute a smaller part of the cost of media, e.g., they account for 4 to 14 per cent of the manufacturing cost of single-cell protein. Although feed grades of phosphates are more expensive than fertiliser grades, they do not contain impurities such as iron, arsenic and fluoride. This is an important consideration in the production of foods and drugs. The hydroxides and sulphates of potassium, magnesium, manganese, zinc and iron are preferred to the chlorides to minimise corrosion of stainless steel. The source of basic materials can cause considerable variation in product yield. Corbett compared six samples of calcium carbonate and found that five of them reduced the titre of penicillin G in a production medium. Problems concerned with the storage, handling and mixing of media should not be neglected. Powders must be kept in dry conditions because of the possibility of substances becoming rock-like or glutinous. Some bulk liquids with a high solids content need to be kept warm to prevent them solidifying, e.g., glucose and corn-steep liquor. If storage temperatures are too high there could be degradation. It is also vital for workers to follow instructions for media preparation very carefully to prevent 'lumpy' media, etc.

## AIR STERILISATION

The problems associated with producing large volumes of sterile air for aerobic fermentations are unique. Although sterilisation by heating is technically possible, it has generally been regarded as too costly for full-scale operation, although it might be used in the treatment of exhaust gases.

Absolute fixed-pore membrane systems using pleated membranes of PTFE are now widely used in the fermentation industry and have proved to be very reliable. This is very important when considering the costs associated with loss of fermentation batches due to contamination and production downtime due to filter failure. Banks has reported that a contamination probability of 1 in 1000 is economically acceptable for microbial batch fermentations, while in large scale animal-cell culture processes contamination rates as low as 2 per cent are now achievable.

Operating costs will be based on the estimated life of the filters. Factors to consider include the cost of replacement filters or filter materials, servicing and labour. Even if the filters could be cleaned there must be an allowance for depreciation due to normal wear and tear. Savings may also be made by introducing series filtration whereby the major part of the foreign matter from the air stream is taken out by varying degrees of coarse filtration, thus reducing renewals of the more expensive high-efficiency filter media such as membrane filters.

The treatment of fermenter exhaust gases to satisfy containment requirements is also important. Treatment is normally by filtration with 0.2 μm hydrofilters, but in-line incinerators may be an alternative approach. Filtration is usually cheaper but it may be necessary to supplement filtration with incineration depending on the process and scale.

## HEATING AND COOLING

Ideally there should be no heating or cooling at any stage in a fermentation process, but because this is virtually impossible, heat should be conserved and cooling minimised by careful process design. A fermentation may include the following heating or cooling stages:

1. Sterilisation or boiling of the medium to 100°C or above followed by cooling to 35°C or below.
2. Heating the fermenter and ancillary equipment to sterilise it, followed by cooling.
3. Heat may be generated during the fermentation. This heat output has to be removed by cooling to maintain the growth temperature of the micro-organisms within prescribed limits.
4. After harvesting, heat may be required to remove water from the product.

Cooling requirements will be influenced by the size and type of an individual process. British Petroleum Ltd. estimated that the cooling requirements for a 100,000 tonnes annum$^{-1}$ single-cell protein plant to be $110 \times 10^6$ kcal hr.$^{-1}$ using *n*-alkanes or gas-oil as the primary substrate. To reduce cooling requirements, the specific energy input may be minimised through the use of air-lift fermenters. Cooling equipment has been estimated at 10 to 15 per cent of the investment cost for single-cell protein. Another way to minimise cooling costs is to use micro-organisms with higher optimum growth temperatures, if it is feasible. The selection and use of thermophiles and thermotolerant organisms would have obvious advantages to reduce cooling demands.

## AERATION AND AGITATION

Nearly all fermentations require some form of mixing to maintain a constant environment and many also need aerating. Fermentations may be broadly classified into:

1. Fermentations which are anaerobic where oxygen is undesirable, e.g., acetone-butanol.

2. Fermentations which have a minimal oxygen demand, e.g., ethanol.
3. Fermentations which have a high oxygen demand, e.g., antibiotics, acetic acid, single-cell protein.

In categories 1 and 2, aeration is not generally regarded as a major economic consideration. During an acetone-butanol fermentation carbon dioxide and hydrogen are evolved. Once this gas production starts it will help to maintain anaerobic conditions and stir the mass of broth without the need for mechanical agitation. Anaerobic conditions are achieved initially in a production fermenter by maintaining a positive pressure of filtered carbon dioxide and hydrogen obtained from another established fermentation.

For ethanol production, the yeast inoculum in the vessel is initially dispersed in the medium by compressed air or by mechanical stirring. Aeration or agitation is stopped once the biomass concentration reaches a pre-determined level. A vigorous anaerobic fermentation commences and the evolution of carbon dioxide bubbles stirs the contents of the vessel and disperses the cells in the medium so that mechanical agitation is unnecessary. In this process aeration and agitation are considered to be a minor component of the total production costs.

Fermentations having a high oxygen demand must be agitated with sufficient power to maintain a uniform environment and to disperse the stream of air introduced by aeration. In an early reference it was stated that the cost of energy necessary to compress air for yeast production proved that a considerable amount (10 to 20 per cent) of the total production expenses was due to aeration. Swartz has reported that the mixing costs in a penicillin fermentation are >15 per cent of the total production costs. Energy consumption for a stirred aerobic fermentation to provide agitation, air compression and chilled water is approximately 8.2 kW $m^{-3}$.

In a single-cell protein processes, the carbon substrate yield coefficient is the most critical physiological factor. It is also well documented that much higher carbon-substrate yield coefficients are obtained with methane or *n*-alkanes instead of carbohydrates. Unfortunately, cells grown on hydrocarbons have greater oxygen requirements. The oxygen requirements of hydrocarbon yeast fermentation is almost triple that of a yeast fermentation grown on a carbohydrate substrate and producing an equal quantity of cells. Therefore, if there is to be effective utilisation of a hydrocarbon substrate, which can account for over 50 per cent of total production costs, the production fermenter must have a high oxygen-transfer capacity. The demands on fermenter design are further complicated by the hydrocarbon fermentation being highly exothermic, which necessitates the provision of good cooling facilities if a constant temperature is to be maintained in the fermenter.

A few companies developed SCP processes using mechanically stirred fermenters with sparged air. In the Swedish Norprotein process it has been estimated that the total utilities costs, which included aeration and agitation for 100,000 tonnes $year^{-1}$ of SCP would only be 16 per cent of total production costs.

A number of companies, decided to develop fermenters based on the air-lift principle. The main advantages of these fermenters are simpler design and reduced energy and cooling water costs. Since the energy supplied to an air-lift fermenter is only supplied with the air, it is crucial to obtain a fermenter design which minimises the energy requirement for biomass production yet creates high oxygen-transfer facilities to ensure efficient substrate utilisation.

## BATCH-PROCESS CYCLE TIMES

In a batch process productivity must be determined for the complete process cycle. Here productivity is defined as grams of product $dm^{-1}$ $hr.^{-1}$. This productivity is based on a combination of the time for the actual fermentation and the time to prepare the fermenter ready for the next run. Heijnen estimated this to take 15–25 hours. Thus, the total time for a fermentation may be calculated as:

$$t = \frac{1}{\mu_m} \cdot \ln \frac{X_f}{X_0} + t_T + t_L + t_D$$

where, $\mu_m$ = maximum specific growth rate
$X_0$ = initial cell concentration
$X_f$ = final cell concentration
$t_T$ = turn-around time (washing, sterilising, filling with media)
$t_D$ = delay time until inoculation
$t_L$ = lag time after inoculation

The overall productivity $P$ is given by:

$$P = \frac{X_f}{\frac{1}{\mu_m} \ln \frac{X_f}{X_0} + t_T + t_D + t_L}.$$

It will be possible from this equation to determine the effects of process changes on the overall productivity. A larger initial inoculum would increase $X_0$ and shorten the process time. Actively growing inocula would reduce the lag time $t_L$. It is also worthwhile to isolate faster-growing organisms and/or higher yielding strains. In many processes the growth phase and/or the production phase has been extended by the use of fed-batch or continuous feed with improved productivity. In fermentations with short growth cycles such as bakers' yeast (14 to 24 hours), the turnaround time will be as important as the time between inoculation and harvesting. When the production cycle is long, as with penicillin (6 to 7 days), a few extra hours for turnaround will have little influence on overall productivity.

## CONTINUOUS-CULTURE

At the present time, very few large-scale continuous-culture processes are being operated. These are primarily for the production of microbial biomass, glucose isomerase, buttermilk souring and yoghurt.

It is appropriate at this stage to compare batch-culture productivity and continuous-culture productivity. Wang derived an equation to quantify this relationship:

$$\frac{\text{Continuous-culture productivity}}{\text{Batch-culture productivity}} = \frac{\ln (X_m/X_0) + \mu_m t_L}{(X_m - X_0)/X_m} D_c Y$$

where $\mu_m$ = maximum specific-growth rate
$X_0$ = initial cell density
$X_m$ = maximum cell density
$t_L$ = turn-around time
$D_c$ = critical dilution rate
$Y$ = cellular yield coefficient for the limiting nutrient.

In an example Wang used an inoculum size of 5 per cent ($X_0/X_m = 0.05$), a process turn-around time of 10 hours, a cellular yield of 0.5 gram cells per gram of substrate and a final cell concentration of 30 grams $dm^{-3}$. Productivity was then calculated for a series of maximum specific growth rates (Table 21.4). It is clear from these data that the faster the growth of the organism, the more favourable is a continuous process over a batch process.

**Table 21.4.** Comparison of productivity in batch and continuous culture.

| *Maximum specific growth rate in batch culture ($hr.^{-1}$)* | *Continuous productivity batch productivity* |
|---|---|
| 0.05 | 0.09 |
| 0.10 | 0.21 |
| 0.20 | 0.53 |
| 0.40 | 1.5 |
| 0.80 | 4.6 |
| 1.0 | 6.8 |
| 1.2 | 9.5 |

When assessing feasibility of a continuous process for product formation it is necessary to know: volumetric productivity, conversion yield of product from the most expensive substrate in the medium and the product concentration. In processes for cell biomass, some alcohols and organic acids where the major production cost is at the fermentation stage, volumetric productivity and conversion yield will be most important. A continuous process may be more economic than a batch process if higher productivities at higher efficiencies can be achieved. Unfortunately, in some processes, the final product concentration in the effluent broth from a continuous-culture will be less than that obtained in a batch process, which will create a need for greater concentration at the recovery stage.

Heijnen argue that low dilution rates are favoured if high product concentrations are to be obtained as recovery cost may be 50–80 per cent of total costs. However, the concentration preference is not valid in processes which can start with a simple concentration step (SCP or intracellular glucose isomerase) or with products that need no concentrating (beer, yoghurt and buttermilk).

In a production plant, continuous culture offers the advantages of constant flow, product quality and simple automation and control. Disadvantages are due to specific production facilities that cannot be used for other purposes, lack of continuous recovery techniques and lack of constant market demand.

The continued increase in efficiency of fed-batch culture processes for antibiotics and other non-growth-associated products makes manufacturers reluctant to make radical alterations to established processes such as the introduction of continuous culture.

## RECOVERY COSTS

The costs of product recovery and purification are rarely quoted, though in some processes they are obviously considerable. It is now accepted that cost analyses of 'older' fermentation processes split approximately 1:1 between fermentation costs and isolation/purification costs. However, the change in the last 20 years to use recombinant organisms to produce exotic compounds at extremely low titres has meant that the fermentation/isolation-purification split may be 1:8 or 1:10. This means that the fermentation may only be 10 per cent of the costs, while the recovery accounts for 90 per cent. In this case the correct choice of the recovery-purification procedure can be crucial to the success of a process and early evaluation of alternative techniques may be very important. Stowell and Bateson identified a number of factors contributing to these costs which are given below:

1. Yield losses, even if only modest, are certain to occur at each stage of the recovery process.
2. High energy and maintenance costs associated with running filtration and centrifugation equipment.
3. High costs of solvents and other raw materials used in recovery and refining of products.

Atkinson and Mavituna reported losses of 8 per cent for citric acid and 4 per cent for penicillin G in the recovery and purification stages before conversion to the potassium salt in production processes. They also stressed the importance of trying to reduce the number of downstream stages as much as possible to reduce capital and operating costs.

It is thought that depreciation, return on capital and maintenance can account for over 80 per cent of the overall cost for a large-scale rotary filtration or centrifugation plant. However, it is considered that removing cells by filtration is less energy consuming than by centrifuging. If filter aids are to be used, in the most economical way, this will still increase the cost for a product at a 10 per cent concentration in the broth. One of the main factors affecting centrifuge economics is the size of the particle to be separated. Filtration costs are less dependent on particle size. At a particle size greater than 1 to 2 µm, centrifugation is more economical. Below this size ultrafiltration becomes more economical.

Maiorella has compared extraction processes in 11 alternative ethanol fermentation processes. Selective ethanol removal by flash distillation was thought to be the most economic technique.

When a product may be made by a microbial process or obtained from an alternative source there are cost limits on product recovery to be considered. When high-value end-products have been produced it has been acceptable to use relatively large weights of filter aid to achieve initial clarification to remove small amounts of solids.

The solvents used in subsequent extraction have then been recovered in a high energy-consuming distillation plant. In this case, the manufacturer has only to make his product as economically as those of other fermentation companies.

Before certain products may be marketed the extraction/purification procedure will have to be validated (approved) by the FDA or similar regulatory bodies. Any changes in extraction procedures will have to be checked for re-validation, which will incur further costs. Therefore, if validation is to become an issue in processing it may be worthwhile having alternative procedures validated at an early stage in development.

## WATER USAGE AND RECYCLING

Many fermentations have a high daily water usage (Table 21.5). As charges for water increase, many of these processes will become vulnerable to cost escalation because of the relatively large volumes of water required per unit volume of product.

**Table 21.5.** Daily water usage in fermentation processes.

| *Industry* | *$m^3$ of water used $day^{-1}$* |
|---|---|
| Maltings | 230 |
| Brewing | 10000 |
| Distilling | 320 |
| Antibiotics | 5200 |
| Acetic acid | 700 |
| Single cell protein (methanol substrate) | 4000–12000 |
| Yeast (alkane substrate) | 18200 |
| Bacteria (methanol substrate) | 45500 |
| Bacteria (methane substrate) | 18200 |

There is now a widespread interest in reducing overall consumption. In a bacitracin plant, the water from the mash cooler was collected and re-used to charge the mashing vessels and wash the fermenters. Water from the cooler coils was used to wash down the discharge cake from the filter presses. Bernstein designed a 'closed-loop' system for fermenting cheese whey in which effluents were completely recycled.

Recycling of water was an integral stage of large scale SCP processes developed during the 1980s to minimise water consumption, reduce effluent treatment costs and reduce media costs by recycling of spent media. When ICI plc's SCP Pruteen plant was operating, it was designed to recycle most of the fermenter medium water. Under optimum conditions Bernstein claimed that the water loss could be reduced to 3 per cent of the flow through the fermenter using water recovery systems.

## EFFLUENT TREATMENT

Before deciding on the most economic form of treatment for wastes it is important to make a factory survey. The information should include the water volume, the organic and solids loading, range of pH variation, nutrient level, temperature fluctuation and the presence of any toxic compounds. It will also be necessary to consider company finance policies, the site location and government legislation for waste disposal.

In the majority of fermentation processes it is impossible to dispose of effluents at zero cost. Whether the waste is incinerated, dumped on waste land or discharged to sewers, rivers or tidal water, some expenditure will be necessary for treatment that ensures that minimal harm is done to the environment.

The various alternative disposal procedures may be compared using economic considerations. Pape claimed that the cheapest treatment method was controlled dumping, followed by waste incineration or dumping in salt mines. The most expensive method was biological degradation in a waste-water-treatment plant. The last method has often to be used because the effluents usually contain only a few per cent of organic matter which would be costly to separate, concentrate and incinerate.

The possibility of direct disposal of pharmaceutical waste into the sea is now very restricted, especially if the waste is untreated.

Avermectin is an antihelmintic compound produced by *Streptomyces avermitilis*. During development of this compound it was recognised that it was very potent and could have a potential impact on aquatic fauna. All possible exit streams from the process at the factory, both fermentation and downstream purification, are therefore captured and chemically degraded. In this case, environmentally safe waste treatment is a major component in production costs. An alternative method for disposing of wastes of this type would be to absorb all waste streams in a suitable material and incinerate this and solid wastes at an appropriate temperature.

# References

Allen, K.M., *Bioreactors and Biotransformation*, Butterworths, London.

Arceivala, K.J., *Media for Industrial Fermentations*, Marcel Dekker Inc., New York.

Benaim Pinto, C., *Microbial Biomass and Economic Microbiology*, Prentice-Hall, London.

Budyko, M.I., *Bioactive Microbial Products,* D. Van Nostrand, New York.

Cambell, K.E. and Lemer, H.A., *Industrial Fermentation*, Academic Press, London.

Chang, J.C., *Biochemical Engineering and Biotechnology Handbook*, John Wiley & Sons, New York.

Coolingwood, R.W., *Fermentation Technology and its Industrial Applications*, John Wiley & Sons, New York.

Downe, S.A., *Industrial Microbiology and Biotechnology*, John Wiley & Sons, New York.

Dugan, P.R., *Animal Cell Biotechnology*, Plenum Publishing Corporation, London.

Goldman, M., *Economic Analysis of Fermentation Process*, Gordon and Breach, Science Publishers, New York.

Gould, G.W., *Textbook of Microbiology*, D. Van Nostrand, New York.

Harding, G., *Fundamentals of Biochemistry*, Prentice-Hall, London.

Hidy, G.M., *Energetics and Kinetics in Biotechnology*, London.

Jarvis, B., *Biological Methods for Waste Treatment*, John Wiley & Sons, New York.

Jencks, W.P., *Biological Treatment of Aqueous Wastes*, John Wiley & Sons, New York.

Kim, C.K., *Bioseparation Techniques*, Marcel Dekker, New York.

Lewis, R.L., *Industrial Processing with Membranes*, Chilton Book Co., USA.

McCaull, J. and Crossland, J., *Molecular Biology and Biotechnology*, Harcourt Brace Jovanovich, New York.

Mitchell, R., *Industrial Aspects of Biochemistry and Microbiology,* McGraw-Hill, New York.

Nyer, E.K., *Cell Biology*, Van Nostrand Reinhold, New York.

Odum, S.K., *Down Stream Processing in Biotechnology*, W.B. Saunders and Co., New York.

Phillips, D.J.H., *Effluent Treatments in Biochemical Industries*, Applied Science Publishers, London.

Reid, G.K., *Modelling and Control of Fermentation Processes*, Reinhold Publishing Corporation, New York.

Samson, R.A., *Instruments for Monitoring and Controlling Reactors*, Pergamon Press, Oxford, London.

Teal, J.M., *Computers in Fermentation Technology*, Pergamon Press, New York.

Vollenweider, R.A., *Scale-up Methodology for Chemical Principles*, Blackwell Scientific Publications, New York.

Wyatt, G.M., *Fermentation and Enzyme Technology*, Reston Publishing Co., Reston, Virginia.

# Index

**K**



**L**



**M**



**N**